AF736567

Integration of Advanced Systems in Environmental Science and Water Desalination

Integration of Advanced Systems in Environmental Science and Water Desalination explores cutting-edge technologies revolutionizing environmental monitoring and desalination processes. The book bridges the gap between research and real-world applications, offering insights into machine learning, IoT-based sensor networks, remote sensing, and advanced materials that enhance efficiency and sustainability. Readers will discover how modern innovations—such as satellite imaging, smart sensors, and renewable energy-powered desalination—are tackling critical water and environmental challenges. Designed for professionals and researchers, this comprehensive resource provides practical solutions for improving water resource management and fostering a more resilient, technology-driven future.

- Provides a well-rounded perspective for professionals and researchers, combining theoretical advancements with real-world case studies in environmental science and desalination.
- Highlights energy-efficient desalination techniques, novel filtration materials, and renewable energy applications to improve water resource management.
- Covers the use of smart sensors, AI-driven analytics, and remote sensing for more accurate and efficient environmental monitoring and water desalination.

Integration of Advanced Systems in Environmental Science and Water Desalination

Edited by Jamal Mabrouki

CRC Press
Taylor & Francis Group
Boca Raton London New York

CRC Press is an imprint of the
Taylor & Francis Group, an informa business

Designed cover image: Shutterstock

First edition published 2026
by CRC Press
2385 NW Executive Center Drive, Suite 320, Boca Raton FL 33431

and by CRC Press
4 Park Square, Milton Park, Abingdon, Oxon, OX14 4RN

CRC Press is an imprint of Taylor & Francis Group, LLC

© 2026 selection and editorial matter, Jamal Mabrouki, individual chapters, the contributors

Reasonable efforts have been made to publish reliable data and information, but the author and publisher cannot assume responsibility for the validity of all materials or the consequences of their use. The authors and publishers have attempted to trace the copyright holders of all material reproduced in this publication and apologize to copyright holders if permission to publish in this form has not been obtained. If any copyright material has not been acknowledged please write and let us know so we may rectify in any future reprint.

Except as permitted under U.S. Copyright Law, no part of this book may be reprinted, reproduced, transmitted, or utilized in any form by any electronic, mechanical, or other means, now known or hereafter invented, including photocopying, microfilming, and recording, or in any information storage or retrieval system, without written permission from the publishers.

For permission to photocopy or use material electronically from this work, access www.copyright.com or contact the Copyright Clearance Center, Inc. (CCC), 222 Rosewood Drive, Danvers, MA 01923, 978-750-8400. For works that are not available on CCC please contact mpkbookspermissions@tandf.co.uk.

For Product Safety Concerns and Information please contact our EU representative GPSR@taylorandfrancis.com. Taylor & Francis Verlag GmbH, Kaufingerstraße 24, 80331 München, Germany.

Trademark notice: Product or corporate names may be trademarks or registered trademarks and are used only for identification and explanation without intent to infringe.

Library of Congress Cataloging-in-Publication Data
Names: Mabrouki, Jamal editor
Title: Integration of advanced systems in environmental science and water desalination / edited by Jamal Mabrouki.
Description: First edition. | Boca Raton FL : CRC Press, 2026. | Includes bibliographical references. | Summary: "Integration of Advanced Systems in Environmental Science and Water Desalination explores cutting-edge technologies revolutionizing environmental monitoring and desalination processes. The book bridges the gap between research and real-world applications, offering insights into machine learning, IoT-based sensor networks, remote sensing, and advanced materials that enhance efficiency and sustainability"-- Provided by publisher.
Identifiers: LCCN 2025021409 | ISBN 9781041088592 hardback | ISBN 9781041088585 paperback | ISBN 9781003647263 ebook
Subjects: LCSH: Environmental monitoring--Technological innovations | Saline water conversion--Technological innovations | Artificial intelligence--Industrial applications
Classification: LCC QH541.15.M64 I55 2026
LC record available at https://lccn.loc.gov/2025021409

ISBN: 9781041088592 (hbk)
ISBN: 9781041088585 (pbk)
ISBN: 9781003647263 (ebk)

DOI: 10.1201/9781003647263

Typeset in Times
by KnowledgeWorks Global Ltd.

Contents

Preface

There is great potential to improve sustainability and resilience through the integration of advanced systems in environmental science and water desalination. This initiative aims to provide comprehensive answers to some of the most pressing environmental and water resource issues by using state-of-the-art technologies: improved speed and accuracy in detecting threats and changes in the environment, leading to more informed decisions. Modern materials and technologies in the meticulous engineering design of desalination and monitoring systems.

The goal of the book *Advanced Systems in Environmental Science and Water Desalination* is to give readers a thorough understanding of the most recent developments in technology and how they are being applied to environmental monitoring and water desalination. It aims to close the knowledge gap between state-of-the-art research and useful application by providing insights into creative fixes for urgent environmental problems and water scarcity situations. The book's scope includes a thorough examination of machine learning applications in environmental science, data analytics, IoT-based sensor networks, and remote sensing technologies.

This book takes you on a tour of the newest innovations that are significantly advancing the disciplines of water desalination and environmental science. It describes how sophisticated instruments and methods are employed to keep an eye on our surroundings. For example, satellite photos are used to monitor changes in cities, forests, and water quality from space. In general, the book aids readers in comprehending how these cutting-edge systems collaborate to address some of the most urgent environmental and water-related issues that we face today. It is an excellent resource for anyone curious about how technology is being used to build a more sustainable future because it is written in an easy-to-understand manner.

Professionals with expertise in desalination, water management, and related sectors seek new approaches to improve sustainability and efficiency. Professionals are interested in the latest technical developments and useful applications working in environmental monitoring, conservation, and sustainability.

List of Contributors

Aa. Aangri
Faculty of Sciences
University Ibn Tofail
Kenitra, Morocco

Aa. Aarab
FSJES-Agdal, Mohammed V University
Morocco

F. Z. Abbadi
Laboratory of Applied Chemistry and Environment, Faculty of Sciences and Technology
Hassan 1st University
Settat, Morocco

Z. Abdelali
Hassan II University
Casablanca, Morocco

Y. Abrouki
Laboratory of Spectroscopy, Molecular Modelling, Materials, Nanomaterials, Water and Environment, CERN2D, Faculty of Science
Mohammed V University in Rabat
Rabat, Morocco

R. Aftiss
Laboratory of Engineering Sciences and Biosciences LSIB, Faculty of Science
Technology Hassan II University
Casablanca, Morocco

M. Aharchi
Intelligent Automation & Bio Med Genomics Laboratory, STSM Doctoral Center
Abdelmalek Essaadi University
Morocco

O. Ait Layachi
Laboratory of Physical Chemistry and Biotechnology of Biomolecules and Materials, Faculty of Sciences and Technology
Hassan II University of Casablanca
Mohammedia, Morocco

A. O. Alaoui
L-IMIA, T-IDMS
FST Errachidia/Moulay Ismail University
Meknes, Morocco

M. Algheriani
Department of Geography and Geographic Information Systems, Faculty of Arts and Languages
University of Tripoli
Libya

Aa. Alitane
Geoengineering and Environment Laboratory, Research Group "Water Sciences and Environment Engineering", Geology Department, Faculty of Sciences
Moulay Ismail University
Meknes, Morocco
Hydrology and Hydraulic Engineering Department
Vrije Universiteit Brussels VUB
Brussels, Belgium

Y. Al Masmoudi
Faculty of Science
Choaïb Doukkali University
El-Jadida, Morocco

Aa. Alraiani
Libyan Authority for Scientific Research
Tripoli, Libya

M. Amraouy
National Institute of Posts and Telecommunications
Rabat, Morocco

H. Anas
TIM Team, ENSA Marrakech
Cadi Ayyad University
Marrakech, Morocco

Anass Ariss
Department of Computer Science, Intelligent Processing Systems & Security Team, Faculty of Sciences
Mohammed V University in Rabat
Rabat, Morocco

D. Azdem
Laboratory of Spectroscopy, Molecular Modeling, Materials, Nanomaterial's, Water and Environment, CERNE2D, Faculty of Science
Mohammed V University in Rabat
Agdal, Rabat, Morocco

K. Azoulay
Laboratory of Applied Chemistry and Environment, Faculty of Sciences and Technology
Hassan 1st University
Settat, Morocco

M. Azrour
IDMS Team, Faculty of Sciences and Techniques
Moulay Ismail University of Meknes
Errachidia, Morocco

S. Belkacemi
Faculty of Science
Ibn Tofail University
Kénitra, Morocco

S. Benkirane
SISAR Team, LaRTiD Laboratory, Technology Higher School Essaouira
Cadi Ayyad University
Marrakech, Morocco

ELM. Benmalik
Mohammed V University of Rabat
Morocco

M. Bortali
National School of Applied Science
Chouaib Doukkali University
El Jadida, Morocco

Aa. Boudouma
Laboratory of Physical Chemistry and Biotechnology of Biomolecules and Materials, Faculty of Sciences and Technology
Hassan II University of Casablanca
Mohammedia, Morocco

S. Boudraham
Laboratory of Spectroscopy, Molecular Modeling, Materials, Nanomaterials, Water and Environment, CERNE2D, Faculty of Science
Mohammed V University in Rabat
Rabat, Morocco

D. Bougmoum
FSBM – UH2C, Condensed Matter Physics Laboratory
Casablanca, Morocco

M. A. Bourkha
TIM Team, ENSA Marrakech
Cadi Ayyad University
Marrakech, Morocco

M. Boutaib
Bouregreg-Chaouia Water Agency
Benslimane, Morocco

J. J. Alarcón Cabañero
Department of Irrigation, Centro de Edafología y Biología Aplicada del Segura CEBAS, Campus de Espinardo
Murcia, Spain

M. Chahid
Laboratory of Applied and Marine Geosciences, Geotechnics and Geohazards LR3G, Faculty of Sciences
Abdelmalek Essaadi University
Tetouan, Morocco

K. Chawqi
School of Science and Engineering
Al Akhawayn University
Ifrane, Morocco

T. Chtioui
Faculty of Sciences
University Ibn Tofail
Kenitra, Morocco

Y. Chtouki
School of Science and Engineering
Al Akhawayn University
Ifrane, Morocco

Wafaa Dachry
Laboratory of Advanced Systems Engineering & Innovation, Faculty of Science and Technology
Hassan 1st University
Settat, Morocco
Foundation for Research Development and Innovation in Science and Engineering
Casablanca, Morocco
Chouaib Doukkali University, National School of Applied Science
El Jadida, Morocco

O. Doumkou
Department of Irrigation, Centro de Edafología y Biología Aplicada del Segura CEBAS, Campus de Espinardo
Murcia, Spain

H. M. Driss
Faculty of Letters and Human Sciences
Mohammed First University
Oujda, Morocco

M. Ech-Chebaby
Department of Computer Sciences, Faculty of Sciences and Techniques
Moulay Slimane University
Beni-Mellal, Morocco

Ahmed Ekhmaj
Department of Soil and Water Science, Faculty of Agriculture
University of Tripoli
Libya

K. El Azzouzi
School of Science and Engineering
Al Akhawayn University
Ifrane, Morocco

Aa. El Bouari
FSBM – UH2C Applied Materials Physicochemistry Laboratory
Casablanca, Morocco

Aa. El Brouzi
Laboratory of Physical Chemistry and Biotechnology of Biomolecules and Materials, Faculty of Sciences and Technology
Hassan II University of Casablanca
Mohammedia, Morocco

E. El Hachimi Hassan II
Faculty of Sciences Ain Chock, Laboratory of Geosciences Applied to the Engineering of Amenagement
University of Casablanca
Casablanca, Morocco

S. El Hajjaji
Laboratory of Spectroscopy, Molecular Modeling, Materials, Nanomaterial's, Water and Environment, CERNE2D, Faculty of Science
Mohammed V University in Rabat
Rabat, Morocco

Souad El Hajjaji
Laboratory of Spectroscopy, Molecular Modeling, Materials, Nanomaterial, Water and Environment, CERNE2D, Faculty of Science
Mohammed V University in Rabat
Rabat, Morocco

Y. El Hassouani
New Energies and Materials Engineering, Faculty of Science and Technology
Moulay Ismail University of Meknes
Errachidia, Morocco

M. El Marrakchi
Laboratory of Physical Chemistry and Biotechnology of Biomolecules and Materials, Faculty of Sciences and Technology
Hassan II University of Casablanca
Mohammedia, Morocco

M. El Mouedden
Laboratory of Spectroscopy, Molecular Modeling, Materials, Nanomaterials, Water and Environment, CERNE2D, Faculty of Science
Mohammed V University in Rabat
Agdal, Rabat, Morocco

Aa. El Ouali
Geoengineering and Environment Laboratory, Research Group "Water Sciences and Environment Engineering", Geology Department, Faculty of Sciences
Moulay Ismail University, Presidency
Meknes, Morocco

N. El Yanboiy
Optoelectronics and Applied Energy Techniques, Faculty of Science and Technology
Moulay Ismail University of Meknes
Errachidia, Morocco

I. Elabbassi
Optoelectronics and Applied Energy Techniques, Faculty of Science and Technology
Moulay Ismail University of Meknes
Errachidia, Morocco

B. Z. Elabidine
Mohammed VI Polytechnic University
Benguerir, Morocco

S. Elbouamri
UH2 EST Casablanca, Laboratoire LAREMO
Morocco

Z. Elhadari
Department of Computer Sciences, Faculty of Sciences and Techniques
Moulay Slimane University
Beni-Mellal, Morocco

M. Y. Elhafidi
FSBM – UH2C, Condensed Matter Physics Laboratory
Casablanca, Morocco

N. Elhajrat
Optoelectronics and Applied Energy Techniques, Faculty of Science and Technology
Moulay Ismail University of Meknes
Errachidia, Morocco

F. Elkaoumi
Ibn Zohr University
Agadir, Morroco

M. Elkouche
Faculty of Letters and Human Sciences
Mohammed First University
Oujda, Morocco

A. Elmajidi
TIM Team, ENSA Marrakech
Cadi Ayyad University
Marrakech, Morocco

H. Elmoudnia
Laboratoire des Matériaux Innovants, Énergie et Développement Durable IMED-Lab, Faculty of Science and Technology
Cadi Ayyad University
Marrakech, Morocco

K. El-Moustaqim
Improvement and Valuation of Plant Resources, Faculty of Sciences
Ibn Tofaïl University—KENITRA-University Campus
Kenitra, Morocco

O. Eloutassi
Materials and Modelling Laboratory, Department of Physics, Faculty of Science
Moulay Ismail University of Meknes
Meknes, Morocco

O. Enassiri
ENSA, Ibn Tofail University
Kenitra, Morocco

F. Z. Ennasri
Energy, Materials and Sustainable Development EMDD Laboratory—Higher School of Technology – SALE, Center of Water, Natural Resources, Environment and Sustainable Development CERN2D
University Mohammed V in Rabat
Rabat, Morocco

Imane Ennejjai
Department of Computer Science, Intelligent Processing Systems & Security Team, Faculty of Sciences
Mohammed V University in Rabat
Rabat, Morocco

A. Essahlaoui
Geoengineering and Environment Laboratory, Research Group "Water Sciences and Environment Engineering", Geology Department, Faculty of Sciences
Moulay Ismail University, Presidency
Meknes, Morocco

M. Faraji
FSAC – UH2C, Laboratory for Renewable Energy and Dynamic Systems
Casablanca, Morocco

P. Faria
CERIS, Department of Civil Engineering, NOVA School of Science and Technology
NOVA University Lisbon
Caparica, Portugal

Abdelilah Fatni
Laboratory of Analysis, Modeling, Engineering, Natural Substances and Environment, Polydisciplinary Faculty of Taroudant
University Ibn ZohrThe Institution Will Open in a New Tab
Morocco

M'barek Feddaoui
Laboratory of Energy Engineering, Materials and Systems, ENSA Agadir
Ibn Zohr University
Morocco

Ayoub Ghailane
Laboratory of Spectroscopy, Molecular Modeling, Materials, Nanomaterial, Water and Environment, CERNE2D, Faculty of Science
Mohammed V University in Rabat
Rabat, Morocco

Aa. Ghanam
Materials Science, Energy, and nano-Engineering MSN | Mohammed VI Polytechnic University UM6P.

M. J. González-Ortega
Department of Aerospace Engineering and Fluid Mechanics
University of Seville
Seville, Spain

A. Guezzaz
SISAR Team, LaRTiD Laboratory, Technology Higher School Essaouira
Cadi Ayyad University
Marrakech, Morocco

H. Gziri
Laboratory of Advanced Systems Engineering & Innovation, Faculty of Science and Technology
Hassan 1st University
Settat, Morocco
Foundation for Research Development and Innovation in Science and Engineering
Casablanca, Morocco

H. Haddaoui
Scientific Institute
Mohammed V University in Rabat
Rabat, Morocco

M. Hadine
Laboratory of Applied Geophysics, Geotechnics, Geology of the Engineer and the Environment, Mohammadia School of Engineers EMI
Rabat, Morocco

Aa. Hajjaji
National School of Applied Science
Chouaib Doukkali University
El Jadida, Morocco

M. Hakkou
Scientific Institute
Mohammed V University in Rabat
Rabat, Morocco

M. Halimi
Optoelectronics and Applied Energy Techniques, Faculty of science and technology
Moulay Ismail University of Meknes
Errachidia, Morocco

A. Hantem
Laboratory MADILOG
Ibn Zohr University
Morocco

J. Hicham

I. Hilal
6 National Center for Energy Sciences and Nuclear Techniques CNESTEN
Rabat, Morocco

M. Hissouf
GEMS Laboratory
ENSA Ibn Zohr University
Agadir, Morocco

Mohamed Hissouf
Laboratory of Energy Engineering, Materials and Systems, ENSA Agadir
Ibn Zohr University
Morocco

D. Hmouni
Improvement and Valuation of Plant Resources, Faculty of Sciences
Ibn Tofaïl University—KENITRA-University Campus
Kenitra, Morocco

A. Ait Hou
Research Team: Biodiversity, Environment and Plant Protection, Department of Chemistry, Faculty of Sciences and Technology
Moulay Ismail University
Errachidia, Morocco

H. Hrir
Laboratory of Physical Chemistry and Biotechnology of Biomolecules and Materials, Faculty of Sciences and Technology
Hassan II University of Casablanca
Mohammedia, Morocco

A. Hsain
School of Science and Engineering
Al Akhawayn University
Ifrane, Morocco

M. Idbenali
Laboratory of Sciences and Didactics of Sciences, Center for Education & Training Profession
Souss-Massa, Morocco
Laboratory of Energy Engineering, Materials and Systems, National School of Applied Sciences of Agadir
Ibn Zohr University
Agadir, Morocco

N. Idboufker
Department of Telecommunications and Computer Sciences, National School of Applied Sciences
Cady Ayyad University
Marrakech, Morocco

R. Jalal
Laboratoire de Recherche en Développement Durable et Santé LRDDS, Faculty of Science and Technology
Cadi Ayyad University
Marrakech, Morocco
Centre d'Agrobiotechnologie et Bioingénierie, Unité de Recherche Labellisée CNRST Centre AgroBiotech
Cadi Ayyad University
Marrakech, Morocco

O. Jbaihi
Foundation for Research Development and Innovation in Science and Engineering, SUPTECH
Casablanca, Morocco

S. Jemjami
Laboratory of Applied Chemistry and Environment, Faculty of Sciences and Technology
Hassan 1st University
Settat, Morocco

A. Jouamaa
National School of Applied Sciences
Mohammed First University
Oujda, Morocco

I. I. Jouicha
Laboratory of Advanced Systems Engineering & Innovation, Faculty of Science and Technology
Hassan 1st University
Settat, Morocco
Foundation for Research Development and Innovation in Science and Engineering
Casablanca, Grand Casablanca, Morocco

H. Jounaid
Laboratory of Geosciences Applied to the Engineering of Amenagement, Faculty of Sciences Ain Chock
Hassan II University of Casablanca
Casablanca, Morocco

A. Kabbori
SISAR Team, LaRTiD Laboratory, Technology Higher School Essaouira
Cadi Ayyad University
Marrakech, Morocco

O. Kaibi
Geoengineering and Environment Laboratory, Research Group "Water Sciences and Environment Engineering", Geology Department, Faculty of Sciences
Moulay Ismail University, Presidency
Meknes, Morocco

H. Kably
Faculty of Letters and Human Sciences
Ibn Tofail University
Kenitra, Morocco

A. Karbal
FSAC – UH2C, Laboratory for Renewable Energy and Dynamic Systems
Casablanca, Morocco

M. Ait Kbir
Intelligent Automation & Bio Med Genomics Laboratory, STSM Doctoral Center
Abdelmalek Essaadi University
Morocco

M. Khala
Optoelectronics and Applied Energy Techniques, Faculty of Science and Technology
Moulay Ismail University of Meknes
Errachidia, Morocco

Fatima Zahra Khaled
Laboratory of Spectroscopy, Molecular Modeling, Materials, Nanomaterial, Water and Environment, CERNE2D, Faculty of Science
Mohammed V University in Rabat
Rabat, Morocco

K. Khallaki
LIPIM, Process Engineering, Computer Science and Mathematics Laboratory, National School of Applied Sciences of Khouribga
University Sultan Moulay Slimane
Morocco

E. Khoumri
Laboratory of Physical Chemistry and Biotechnology of Biomolecules and Materials, Faculty of Sciences and Technology
Hassan II University of Casablanca
Mohammedia, Morocco

K. Lalam
Mohammed VI Polytechnic University
Benguerir, Morocco

C. Lamiri
Laboratory of Spectroscopy, Molecular Modelling, Materials, Nanomaterials, Water and Environment, CERN2D, Faculty of Science
Mohammed V University in Rabat
Rabat, Morocco

Ahmed Loudini
Cadi Ayyad University
Marrakech, Morocco

Jamal Mabrouki
Laboratory of Spectroscopy, Molecular Modelling, Materials, Nanomaterial, Water and Environment, CERNE2D, Faculty of Science
Mohammed V University in Rabat
Rabat, Morocco

M. Machkour

F. Z. A. Mahrez
National School of Commerce and Management
Mohammed First University
Oujda, Morocco

M. R. A. Mansour Loukkos
Hydraulic Basin Agency ABHL
Tetouan, Morocco

H. Medromi
Foundation for Research Development and Innovation in Science and Engineering
Casablanca, Morocco
Engineering Research Laboratory LRI, System Architecture Team EAS, National High School of Electricity and Mechanic ENSEM
Hassan II University
Casablanca, Morocco

Hicham Meftah
Laboratory of Energy Engineering, Materials and Systems, ENSA Agadir
Ibn Zohr University
Morocco

N. Mejjad
National Center for Energy Sciences and Nuclear Techniques
Rabat, Morocco

A. Mellouki
Ibn Zohr University
Morocco

C. Messaoudi
Optoelectronics and Applied Energy Techniques, Faculty of Science and Technology
Moulay Ismail University of Meknes
Errachidia, Morocco

J. Eddine Stitou El Messari
Laboratory of Applied and Marine Geosciences, Geotechnics and Geohazards LR3G, Faculty of Sciences
Abdelmalek Essaadi University
Tetouan, Morocco

R. Mghaiouini
FSBM – UH2C Applied Materials Physicochemistry Laboratory
Casablanca, Morocco
ENSA, Laboratory of Advanced Systems Engineering
IbnTofail University
Kénitra, Morocco

A. Moufti
Physical Science Modeling Team, Regional Center for Education and Training Professions
Casablanca-Settat, Morocco

K. Mouzdahir
University Hassan 1st
Settat, Morocco

A. Nait
Alla Laboratory of Energy Engineering, Materials and Systems, National School of Applied Sciences of Agadir
Ibn Zohr University
Agadir, Morocco

M. Najim
GEMS Laboratory
ENSA Ibn Zohr University
Agadir, Morocco

K. Ibno Namr
Faculty of Science
Choaïb Doukkali University
El-Jadida, Morocco

S. Nasrdine
Energy, Materials and Sustainable Development EMDD Laboratory—Higher School of Technology—SALE, Center of Water, Natural Resources, Environment and Sustainable Development CERN2D
University Mohammed V in Rabat
Rabat, Morocco

M. Niranjanamurthy
Department of Artificial Intelligence and Machine Learning, BMS Institute of Technology & Management
Visvesvaraya Technological University Belagavi
Bengaluru, India

I. Nori
LIPIM, Process Engineering, Computer Science and Mathematics Laboratory, National School of Applied Sciences of Khouribga
University Sultan Moulay Slimane
Morocco

S. Nouh
Hassan II University
Casablanca, Morocco

H. Noureddine
Laboratory of Geosciences Applied to the Engineering of Amenagement, Faculty of Sciences Ain Chock
Hassan II University of Casablanca
Casablanca, Morocco

L. Oubaouzine
UH2 EST Casablanca, Laboratoire LAREMO
Morocco

S. Oubdil
Civil Engineering and Environment Laboratory LGCE, Mohammadia, School of Engineers EMI
Mohammed V University in Rabat
Rabat, Morocco

I. Ait Oujallal
Laboratory of Spectroscopy, Molecular Modelling, Materials, Nanomaterials, Water and Environment, CERN2D, Faculty of Science
Mohammed V University in Rabat
Rabat, Morocco

A. Oumou
Geoengineering and Environment Laboratory, Research Group "Water Sciences and Environment Engineering", Geology Department, Faculty of Sciences
Moulay Ismail University, Presidency
Meknes, Morocco

K. Outaaza
Laboratory MADILOG
Ibn Zohr University
Morocco

M. Rabouli
National School of Applied Science
Ibn Tofail University
Kenitra, Morocco

T. Rachiq
Laboratory of Spectroscopy, Molecular Modeling, Materials, Nanomaterial, Water and Environment, CERNE2D, Faculty of Science
Mohammed V University in Rabat
Rabat, Morocco

J. E. Rafiq
Hassan II University
Casablanca, Morocco

A. V. Rompaey
Geography and Tourism Research Group, Department of Earth and Environmental Sciences, KU Leuven
Heverlee, Belgium

L. Saâdi
Laboratoire des Matériaux Innovants, Énergie et Développement Durable IMED-Lab, Faculty of Science and Technology
Cadi Ayyad University
Marrakech, Morocco

E. Sagar
Department of Soil and Water Science, Faculty of Agriculture
University of Tripoli
Libya

F. P. Salcedo
Department of Irrigation, Centro de Edafología y Biología Aplicada del Segura CEBAS, Campus de Espinardo
Murcia, Spain

A. Saydtahiri
TIM Team, ENSA Marrakech
Cadi Ayyad University
Marrakech, Morocco

A. Semmar
Foundation for Research Development and Innovation in Science and Engineering, SUPTECH
Casablanca, Morocco

A. S. Sherif
Department of Soil and Water Science, Faculty of Agriculture
University of Tripoli
Libya

M. Slaoui
Energy, Materials and Sustainable Development EMDD Laboratory—Higher School of Technology—SALE, Center of Water, Natural Resources, Environment and Sustainable Development CERN2D
University Mohammed V in Rabat
Rabat, Morocco

S. Souiri
Scientific Institute
Mohammed V University in Rabat
Rabat, Morocco

M. S. Sumathi
Department of Electronics and Telecommunication Engineering, BMS Institute of Technology & Management
Visvesvaraya Technological University Belagavi
Bengaluru, India

H. Taghlaoui
Laboratory of Territories, Environment, and Development Kenitra
Ibn Tofail University
Morocco

A. Tallou
IRTA Institute of Agrifood Research and Technology, Carretera de Cabrils
Cabrils, Spain

M. Taoufik
Physical Science Modeling Team, Regional Center for Education and Training Professions
Casablanca-Settat, Morocco

T. Tbatou
Laboratory of engineering Sciences and Biosciences LSIB, Faculty of Science and Technology
Hassan II University
Casablanca, Morocco

S. Touil
LIPIM, Process Engineering, Computer Science and Mathematics Laboratory, National School of Applied Sciences of Khouribga
University Sultan Moulay Slimane
Morocco

B. R. Vinod
Department of Civil Engineering, BMS Institute of Technology & Management
Visvesvaraya Technological University Belagavi
Bengaluru, India

G. A. Vivaldi
Department of Soil, Plant and Food Science
University of Bari Aldo Moro
Bari, Italy

A. Wahbi
FSJES-Soussi
Mohammed V University
Morocco

M. Waqif
Laboratoire des Matériaux Innovants, Énergie et Développement Durable IMED-Lab, Faculty of Science and Technology
Cadi Ayyad University
Marrakech, Morocco

M. Yessari
National School of Applied Science
Chouaib Doukkali University
El Jadida, Morocco

N. Youssoufi
Laboratory MADILOG
Ibn Zohr University
Morocco

K. Zabour
Laboratory of Energy Engineering, Materials and Systems, ENSA Agadir
Ibn Zohr University
Morocco

Soumia Ziti
Department of Computer Science, Intelligent Processing Systems & Security Team, Faculty of Sciences
Mohammed V University in Rabat
Rabat, Morocco

H. Zougagh
Department of computer sciences, Faculty of Sciences and Techniques
Sultan Moulay Slimane University
Beni-Mellal, Morocco

A. Zouhri
University Hassan 1st
Settat, Morocco

O. Zytoune
1ENSA, Ibn Tofail University
Kenitra, Morocco

1 Multidisciplinary Approach on Reclaimed Water Use Projects in Agriculture

F. P. Salcedo, J. J. Alarcón Cabañero, M. J. González-Ortega, O. Doumkou, A. Tallou, and G. A. Vivaldi

1.1 INTRODUCTION: BACKGROUND AND DRIVING FORCES

In the next decades, a large percentage of the world population will live in areas with water scarcity and poor water quality due to climate change and increasing population pressure; thus, agriculture will face the challenge of supplying an increasing demand for food resulting from the forecasted population growth rates (Tallou et al., 2022). One strategy for maintaining or increasing productivity under resource scarcity is to make greater use of marginal-quality lands and waters. Irrigated agriculture worldwide (2.7 times more productive than rainfed agriculture) will play a crucial role in the food production system as well as in the use of non-conventional water resources such as saline groundwater and saline reclaimed water (RW) (Vivaldi et al., 2019, 2024). In implementing such a strategy, key factors for sustainability are soil salinity and water quality.

In the Mediterranean Region, irrigated agriculture contributes 75% to the final production. The continued use of these water resources for irrigation will probably put the agro-systems and the environment at risk from salinization, soil compaction, and undesirable ions toxicity. Soil secondary salinization affects an estimated 1–3 million hectares in the enlarged European Union (EU), mainly in the Mediterranean countries. It is regarded as a major cause of desertification. A new strategy of water management is a key method for desertification prevention. Actually, there is a need for technologies that increase water use efficiency and make additional (non-conventional) water sources available for fertigation, thereby decreasing water scarcity and the discharge of water and nutrients to the environment (Harmanny & Malek, 2019).

Moreover, countries more exposed to water scarcity could play a central role in the adoption and success of effective land and water management policies. In this respect, they require new technological capacity and innovative solutions to increase water sources available for agricultural activities in order to provide more food and avoid health crises. The Region of Murcia, located in southeastern Spain, is an example in Europe in the use of RW for agricultural irrigation, illustrating the important role that treated wastewater can play in areas with scarce water resources, improving agricultural sustainability indices (Maestre-Valero et al., 2019a).

Wastewater refers to polluted and degraded water in several forms. It can come from industries, houses, agriculture, and other resources. RW refers to wastewater coming from industries at pretreated or untreated levels. Blackwater and greywater refer to wastewater coming from houses. Wastewater coming from industries, houses, and stormwater represent municipal wastewater, which can be used for irrigation after proper treatment (Khan et al., 2022).

An innovative solution to the problem of high-salinity water is the combination of agriculture and aquaculture, which is attracting more and more interest nowadays because of the multiple advantages of the system. Those techniques have been successfully used in various countries such as Brazil (Sánchez et al., 2015), Egypt, and Jordan (Crespi & Lovatelli, 2011), where desalination units

DOI: 10.1201/9781003647263-1

have been installed and the desalinated water used for irrigation. The rejected brine was utilized to grow marine species such as fish and algae (Lyra et al., 2019). Furthermore, using RW helps reduce the demand for freshwater resources and provides essential nutrients to crops, potentially reducing the need for chemical fertilizers and lowering agricultural costs (Maestre-Valero et al., 2019b).

Although RW water is commonly and successfully used in many countries (e.g., USA, Australia, etc.), in the EU, water reuse faces numerous barriers. Among them, safety risks, economic concerns, and social acceptance can be currently defined as the main barriers, considering that 1) safety risks have been traditionally linked to the use of improperly treated wastewater, 2) cost is probably the first driven force in fresh produce production, and 3) public acceptance of RW by the public varies according to its potential use.

1.2 OBJECTIVES

The main aim of this paper is to show a successful multidisciplinary approach to RW use projects in agriculture, laying the foundations of a novel and more efficient crop production management by enabling RW use for irrigation. To achieve this main aim, the following methodology was implemented: evaluate the horticultural crops' performance, determine the water use efficiency (WUE), establish new agronomic thresholds, use plant uptake models to assess the short- and long-term effects of reclaimed irrigation water, and verify this under Mediterranean field conditions. Finally, there is the integration of new tools as Geographic Information System (GIS), environmental value, and public perception into the water resource use studies. All this works at different scales.

1.3 RESEARCH PILOT PLANTS AT PLOT LEVEL

In the region of Murcia, a network of experimental plots is used in different Mediterranean locations and different Mediterranean types of crops, with the effect of using RW from a nearby wastewater treatment plant in combination with deficit irrigation strategies on the physiology, performance, quality, and safety of crops. The evaluation of the long- and medium-term effects on soil salt accumulation, as well as the influence of the use of this water on drainage fractions and groundwater pollution, is being studied. In these experimental platforms, the effects of treated wastewater are compared with results obtained with conventional water usually used by farmers. The last goal is to optimize the use of such RW, showing that such use may be a strategy aimed at optimizing water use, especially in areas with limited conventional water resources. Studies have demonstrated the effectiveness of RW in reducing fertilizer use and enhancing crop performance under various growing conditions (Maestre-Valero et al., 2019b).

1.3.1 Long-Term Research Experimental Farm (Citrus) on Salty RW

The first investigation approximation into RW was on lemon trees (Pedrero & Alarcón, 2009). This work aimed to study the effects of RW on lemon tree performance. In particular, this research was to compare two sources of treated wastewater, one obtained with a secondary treatment (Cartagena) and the other with a tertiary treatment blending with well water (Campotejar), and to study their effects on soil chemical properties, on the leaf mineral status, crop production, and fruit quality and safety. The main results obtained show that the possibility of blending RW with well water is a good solution to avoid the problems associated with wastewater use in agriculture. The high salinity and Cl and B concentration were the main restrictions associated with treated wastewater used in the experiment. The fruit-quality indexes (weight, firmness, soluble solids, pH, total acidity and maturity index, etc.) were significantly better in the plots irrigated with secondary treated wastewater (Pedrero et al., 2012a). The soil microbiological analysis revealed an absence of fecal coliforms, *Escherichia coli*, and helminth eggs in the experimental plots irrigated with tertiary treated wastewater, but with secondary treated wastewater, the soil accumulation of fecal coliforms exceeded

FIGURE 1.1 Research citrus experimental farm in Campotejar (Molina de Segura, Murcia).

health standards (Pedrero et al., 2012a). In both cases, the use of RW as irrigation water for lemon trees did not represent a microbial risk for lemon fruit.

The second investigation approximation was on an experimental plot conducted at a commercial orchard located in Campotejar (Murcia), Spain (Figure 1.1). In this case, the study was more complex because it was conducted with three different irrigation water qualities and two different irrigation doses were applied. The experimental plot of 0.5 ha was cultivated with 7-year-old "Star Ruby" grapefruit trees (Citrus Paradisi Macf) grafted on Macrophylla rootstock [Citrus Macrophylla] and 0.5 ha with 11-year-old "Orogrande" mandarin trees grafted on Carrizo rootstock. The irrigation head was equipped and supplied with three irrigation water sources; the first (TW) was pumped from the Tajo-Segura water transfer (EC $\approx$ 1 dS m^{-1}), the second water source (RW) was pumped from the "North of Molina de Segura" wastewater treatment plant (WWTP) (EC $\approx$ 3-4 dS m^{-1}), and the third one was delivered by the irrigators' association of Campotejar (IW) (1 $\approx$ EC $\approx$ 4 dS m^{-1}). The time-variable quality water depended on the proportional mix of the available water resources needed to cover the actual irrigation demand (underground water, transfer water, and RW). The treatment in the WWTP was a conventional activated sludge with ultraviolet tertiary treatment. Two irrigation treatments were carried out for each water source. The first was a control treatment (C) irrigated throughout the growing season to recover 100% of the soil water lost by crop evapotranspiration (ETc). The second was a regulated deficit irrigation treatment (RDI) irrigated similarly to C, except during the second stage of fruit growth when it received 50% of the water applied to the control. The stage selected to apply the RDI was identified as being unsusceptible to moderate water stress (Gonzalez-Altozano & Castel, 2003). In our case, this stage extended over a time period of 55–65 days between late June and late August. No leaching fraction was added to the irrigation doses.

Little is known about the interaction between deficit irrigation and saline water. For this reason, the main objectives of the second study, during the first four years, were to evaluate the combined effects of different irrigation water qualities (Tajo Segura water transfer canal, irrigators association, and RW) and the regulated deficit irrigation strategies on Mandarin tree crop performance under Mediterranean climate conditions and their effects on plant physiology and leaf mineral status, soil chemical properties and water content, and yield and fruit quality. Yield reductions were not significant between treatments, although a tendency to reduce a number of fruits was detected in the RW treatments (Pedrero et al., 2012b). This reduction was more pronounced under combined conditions of reduced irrigation quality water and regulated deficit irrigation (Mounzer et al., 2013). The combined effects of RDI strategies and the use of RW can increase some fruit quality parameters, such as vitamin C, on mandarin trees. Plant water relations and vegetative growth were not significantly affected by the use of RW. However, water stress in the RDI treatments induced some reductions in the stem water potential that immediately recovered when the deficit period disappeared (Pedrero et al., 2012b, 2012c).

Although no leaf-toxic visual symptoms were seen during the experiment, a tendency was identified in terms of salt accumulation in the soil during the last season of RW-RDI treatment. In this sense, it is important to remark that in arid and semi-arid areas, the combination of RDI strategies and RW use can be affected in the long term because of salt and boron accumulation (Pedrero et al., 2014a, 2015b).

During the next five years (2010–2015), research studies on the same pilot farm continued to evaluate different indicators on salt stress, establishing phytotoxicity thresholds for Na, Cl, and B and evaluating salt accumulation at the leaf level. After assessing the validity of three indicators (gas exchange parameters, stem water potential, and leaf chlorophyll content), the results showed the need for seasonal measurements of leaf chlorophyll content as an important diagnostic indicator of salt stress on field crops of grapefruit (Romero-Trigueros et al., 2014a). Additionally, it was observed salt accumulation at the leaf level could eventually lead to possible risks, establishing thresholds at which a reduction in yields occurs at Na: 0.1 g/100 g, Cl: 0.6 g/100 g, and B: 100 ppm. In fact, recent research showed that irrigation water sources with high levels of boron could affect citrus production with similar thresholds (Grattan et al., 2014).

The quality of irrigation water on mandarins was assessed based on a fuzzy model in the same orchard in Campotejar (Murcia) with a measurement of two years. There was application of three irrigation treatments: TW from Tajo-Segura channel (EC = 1 dS m^{-1}), irrigators association water of Campotejar (EC = 1–3 dS m^{-1}), and RW pumped from the "North of Molina de Segura" wastewater treatment plant (EC = 3 dS m^{-1}). The best results were found in irrigation with the irrigators association water (Kalavrouziotis et al., 2016).

The long-term effects have been studied in mandarin and grapefruit trees for six years. The two irrigation treatments applied were transfer water (TW) and RW (RW). The water productivity and soil water content were measured in both treatments, as well as fruit yield, tree water status, and phytotoxic elements in leaves. Only the B in leaf concentrations was significant according to the results. A yield reduction in mandarin production and water productivity was observed after the third season. The effects on soil chemical properties in mandarin trees were salinity and increased after RW irrigation (Leogrande et al., 2022). In comparison, the yield was not affected by RW in grapefruit production. On top of that, the water productivity between treatments was similar and the amount of water used for irrigation was the same (Nicolás et al., 2017, 2018; Romero-Trigueros et al., 2020).

In 2019, the water status of grapefruit in a commercial orchard in Campotejar was estimated using thermal indicators and infrared thermometry (IRT). The irrigation treatment applied was RW with RDI. The objective of this research was to use canopy and air temperatures to determine the water status of the trees. As shown by the results, the canopy temperature was higher than the air temperature, which results in positive differences between canopy temperature (Tc)-Ta for a wide range of VPD during the middle hours of the day (Romero-Trigueros et al., 2019).

Moreover, intense studies have been done on grapefruit because trees received saline RW from which they were planted. These studies used nutritional and structural traits measurements at the leaf level and isotopic measurements to assess sustainability in the medium to long term. From this study, it was hypothesized that (i) the positive correlation between leaf δ15N content and leaf salt content suggested that causal links exist between δ15N and salt stress; (ii) excess nitrates provided by the reclaimed irrigation water were lost in the ecosystem through leaching, denitrification, etc., enriching the medium with δ15N and increasing the δ15N value in plants. Therefore, the usefulness of isotopic discrimination measures to predict crop sustainability in the medium to long term when using water sources of different qualities combined with deficit irrigation strategies have been demonstrated (Romero-Trigueros et al., 2014b).

It is interesting to know that the lowest relative yields occurred in the RW-RDI and TW-RDI treatments, where ECe values were highest. In addition, the threshold at which yield is reduced in this research (2 dS m^{-1}) compared to the TW-C treatment coincides with the values proposed by Ayers and Westcot (1989) for citrus (Ece = 1.7 dS m^{-1}), 25 years before. Non-microbiological risks were observed by the use of RW for citrus tree production during the experiment (Pedrero et al., 2012a, 2012b) as was reported by different authors (Pereira et al., 2011; Reboll et al., 2000; Zekri & & Koo, 1993, 1994).

Actually, two new treatments have been added to the experimental design in Campotejar. 1) The dual-water treatment will be irrigated to restore 100% ETc using the following criteria: Fresh water from the Tajo-Segura canal (EC ≤ 1 dS m^{-1}) will be used whenever the actual ETc is higher than 3 mm/day. Under our experimental conditions in Murcia, this period may extend from mid-May to mid-August. Reclaimed saline water (EC > 4 dS m^{-1}) will be utilized during the remaining period (i.e., when the daily ETc is lower than 3 mm) and 2) regulated saline irrigation will be irrigated similarly to the control treatment (100% ETc) with fresh water, except during the second stage of fruit growth when saline-treated water will be used for irrigation. The objective is to test new management practices to substitute the deficit stress by osmotic stress during the non-critical fruit growth stage.

More recently, an experiment that concerned the calculation of total inorganic carbon (TIC) took place in the orchard of Campotejar. Since the experiment collected 117 soil samples from three sites: two of them being in south Italy and the other one located in Campotejar (Murcia). This experiment aimed to compare the results of calcimetric (reference method) and dry combustion methods to estimate the TIC. The dry combustion showed slightly greater values of TIC. It is proven that this method is reliable and can provide accurate estimates of TIC in soils (Leogrande et al., 2021).

In conclusion, irrigation with RW can increase soil salt accumulation, so blending RW with low EC groundwater is a good solution. Also, the combination of deficit irrigation techniques (RDI) and RW can affect production with lower relative yields. The use of secondary RW presents no microbiological risk in the citrus and there are no leaf-toxic visual symptoms. It can also improve some fruit quality indexes.

1.3.2 Agricultural and Water Research Platform at Roldán-Balsicas WWTP

The Agricultural and Water Research Platform from CEBAS-CSIC situated inside of the wastewater treatment plant Roldán-Balsicas (37°47´48.88´´N,0°57´33.63´´O), is based on a 700 m^2 plastic greenhouse with a screen and heating system, irrigation, and different automatisms.

1.3.2.1 Horticultural Crops Experiment under the Greenhouse

The use of RW in controlled greenhouse conditions has led to significant savings in mineral fertilizers and has demonstrated microbiological safety, ensuring the viability of RW for horticultural crops (Maestre-Valero et al., 2019b). The study was carried out on a plastic greenhouse with a screen and heating system, irrigation, and different automatisms in the WWTP of Roldan-Balsicas, Murcia, Spain (Figure 1.2). Plants were grown under greenhouse conditions to determine the effects

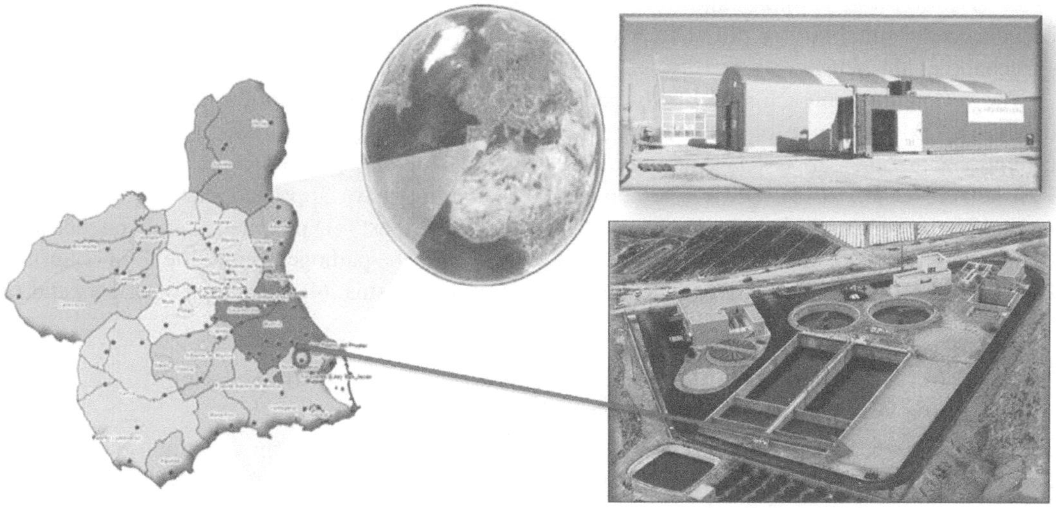

FIGURE 1.2 Agricultural and water research platform at Roldán-Balsicas WWTP.

of different irrigation water sources and growing media on growth, yield, and fruit quality. Crops (tomato, pepper, and zucchini) were grown in two commercial culture systems, rock wool and coconut fiber, and two irrigation water qualities, irrigation community (CW) and RW from the Roldan wastewater treatment plant (RW). The water was supplied by drip irrigation with three compensated pressure drips per bag, each with a flow rate of 3 Lh^{-1} and spaced 0.3 m apart. A total of 1080 plants were distributed into the 680 m^2 greenhouse surface. During the three years of experiments, one of biggest benefits was that the use of RW for irrigation was a significant savings in mineral fertilizers (N, P_2O_5, K_2O, and CaO) (6, 5, 8, and 40%, respectively) for both substrates (Pedrero et al., 2014b). After the analysis was performed on the plants and fruit, it was concluded that the use of RW for hydroponic tomato and pepper crop irrigation was suitable under conventional agricultural practices (hydroponics, drip irrigation, polyethylene greenhouse, etc.) and, most importantly, offers a microbiological food safety.

1.3.2.2 Agricultural System Microbiological Safety

Health and environmental aspects are susceptible issues and essential prerequisites, since wastewater effluent will not be used and/or be accepted to replace conventional or possibly other nonconventional water sources for irrigation, unless it is adequately treated and safely applied (Gerba & Rose, 2003). Systematic monitoring of indicator microorganisms, such as *E. coli*, can help identify potential risk factors for fecal contamination in the agricultural environment.

On horticultural crops under greenhouse conditions, profound microbiological studies were developed. In a pepper study in Roldan greenhouses, levels of *E. coli* in irrigation water (both reclaimed and surface water) and water sprayed in humidifiers to regulate ambient humidity and pepper fruits were assessed. Additionally, the fertilizer solutions' role as a potential contamination vector was investigated. One exciting result showed that regarding the fertilizer solutions, *E. coli* was detected more frequently and in higher levels in the fertilizer solution richer in micronutrients (Gil et al., 2014). These results suggested a low risk of contamination in this agricultural system despite the combination of risky irrigation water sources (reclaimed and surface water) and the hydroponic-grown system. Nevertheless, special care should be taken regarding the microbiological quality of the agricultural solutions that directly contact the edible parts of the crop (Allende et al., 2016; Gil et al., 2014).

Finally, the election of the quantification technique for microbiological indicator has been demonstrated on a zucchini experiment under greenhouse conditions. The suitability of two *E. coli* quantification techniques (plate count and qPCR) was examined for irrigation water and fresh produce. *E. coli* levels using the qPCR assay were significantly higher than that obtained by plate count in all irrigation water samples and fresh produce. However, differences in the water characteristics influenced the suitability of qPCR as a tool to predict potential contamination in irrigation water (Truchado et al., 2016).

The following approach will be studied based on the quantitative microbial risk assessment (QMRA). QMRA is one of the preferred tools for determining those risks. QMRAs overcome limitations of other approaches (e.g., epidemiological studies), such as costs and logistics (De Keuckelaere et al., 2015). QMRA is considered a data-intensive approach that requires several inputs. These are categorized by De Keuckelaere et al. (2015) as: "(1) prevalence and concentration of microorganisms of concern; (2) transmission routes (how the pathogens enter the food chain); (3) growth, removal, survival, and/or inactivation of microorganisms; (4) consumer behavior; and (5) dose-response relationship."

1.3.2.3 Accumulation of Contaminants of Emerging Concern (CECs) in a Wastewater Treatment Plant (WWTP) Using RW

Two different experiments took place at the CEBAS-CSIC research platform, greenhouse Roldán, Lo Ferro, and Balsicas WWTP (Murcia, Spain) concerning the accumulation of CECs in lettuce plants irrigated with wastewater from the communities of Lo Ferro, Roldán and Balsicas (Murcia, Spain). The first experiment was carried out during the summer of 2016 in three varieties of lettuce: Iceberg var. Gitana (IL); Mini Romaine var. Jabera (MRL); and Oak Leaf var. Kiprien (OLL). The aim was to study the accumulation of two widely used drugs, the anticonvulsant carbamazepine (CBZ) and the anti-inflammatory diclofenac (DCF), and to investigate whether the risk of the transfer into the food chain via edible tissues of lettuces exists. The experiment was divided into 5 experimental plots, a blank 1 with drug-free irrigation and the other 4 were irrigated at a different drug concentration, with 8 liters of water per day. According to the results, CBZ was found at the highest levels in the leaves; instead, the highest levels of DCF were in the roots. As a conclusion, the concentration that was found in the edible part of the plants did not imply any risk to human health (González García et al., 2018).

Similar results of the following experiment were carried out from November 2019 until March 2021. This time there was only one variety of lettuce, the red oak baby lettuce (*Lactuca sativa L.*), but the accumulation of four CECs was determined: CBZ, DCF, ketoprofen (KTP), and naproxen (NPX). Three irrigation water qualities were applied with different compositions and concentrations of the pharmaceutical compounds. The experiment was carried out in soil and trays, with drip and sprinkler irrigation, respectively. The concentration was evaluated after each crop (three in total), and also from the regrowth of the first harvest. The compound with the highest accumulation in lettuce leaves was CBZ. But, unexpectedly, after the regrowth practices, DCF had the highest accumulation. No risk to human health was detected, except CBZ for those on vegetarian diets in the tray scenario. Last but not least, the experiment proved that agricultural practices such as soil cropping and drip irrigation have an impact on pharmaceutical uptake and can limit the presence of the compounds (Ponce-Robles et al., 2022).

In this line of investigation, to reduce the risks related to the presence of CECs and pathogenic microorganisms, a self-sustainable pilot prototype based on three combined modules was installed as an alternative tertiary treatment in a conventional WWTP. The three modules are disc-filtration, granular activated carbon (GAC) bed adsorption, and UV disinfection. Seven CECs were selected for the study (acetaminophen [ACT], CBZ, DCF, erythromycin [ERY], KTP, NPX, and sulfamethoxazole [SMX]). The efficiency of the proposed prototype for CECs' removal has been confirmed; the optimal results were at a workflow of 500 L/h. Additionally, a high removal efficiency for different microorganisms was obtained, and those are *E. coli*, F-specific coliphages, somatic coliphages, total coliphages, and *Clostridium perfringens*. Finally, it is a sustainable and economically viable methodology for implementing conventional WWTPs (Ponce-Robles et al., 2020).

1.3.3 Pilot Plant Irrigation at the District Level. Miraflores Irrigators Community and Irrigation Community of Campo De Cartagena (CRCC)

Such pilot plants could be used as demonstration platforms where the continuous assessment of the quality of irrigation water was used, and will be assessed, based on trials carried out on the plot, in the use of this RW in the different crops of the study area. The last goal is to optimize this new water resource at a larger production scale, assessing the effects of RW on plants and soil, and estimating how the distribution's reservoirs, water pipes systems, and all the associated infrastructures could affect the effluent quality. The project, "Feasibility study of using RW from the WWTP of Jumilla in the Miraflores irrigators community," was developed in 2012 (Nicolás et al., 2012), and assessed the irrigation community of Miraflores (Jumilla, Murcia) on the best management practices to use the RW in their different crops and irrigation infrastructure. During the project's second phase, the main objective was the management and control of RW from the Jumilla WWTP in Miraflores. For that purpose, continuous measurements with probes (pH, electrical conductivity, turbidity, nitrates, and phosphates) were done to analyze and interpret water quality from the water reuse law in an agricultural point of view. The information was collected by a head programmer to close the valve, when water conditions were deficient and did not comply with the requirements of RD 1620/2007 quality 2.1. level. Also, training courses on the agricultural use of RW for irrigators have been developed.

The Irrigation Community of Campo de Cartagena (CRCC) was founded in 1952 to resolve problems including the overexploitation of the deep aquifers and the rising groundwater level in the shallow aquifer which generating drainage problems, and also the high salinity of the groundwater. The CRCC integrates the management of five different water resources because it is the largest irrigation community in Europe. The results of this project are primarily the distribution and storage infrastructure with a 64-km irrigation channel longitude with a capacity of 300.000 m^3 and a reservoir capacity of 2.090.863 m^3, including the modernization of irrigation, as 98% use drip irrigation. Lastly, this project used four different water resources, including surface water external to the system (the Tajo-Segura Water Transfer), partially desalinated groundwater, RW, and desalinated seawater (Drechsel et al., 2023).

One of the problems of storing different irrigation water qualities with nutrients as RW is the algae growth. In 2014, Maestre and Pedrero developed a project, "Assessment of the effect of ultrasounds on irrigation ponds storing RW from tertiary treatment," for assessing the irrigation community of Campo de Cartagena about the use of ultrasounds for algae growth control. In arid and semi-arid regions, where water scarcity and availability vary seasonally, irrigation ponds are a common facility to guarantee water supply to crops. A particular case was the Segura basin, a semi-arid region in southeastern Spain characterized by an extraordinary abundance of irrigation ponds, more than 15,000 (Valero, 2012). The large amount of nutrients contained in RW stored in ponds, together with the appropriate climatic conditions (high temperature and light), leads to the growth and development of algae of both macroscopic and microscopic size that can cause technical and water quality problems (Brainwood et al., 2004; Kotopoulis et al., 2009; von Sperling et al., 2008). Ultrasonic treatment systems for algae growth control are highly specific underwater sound systems that emit ultrasonic sound waves to destroy algae. They offer the possibility of eliminating a wide variety of common algae and cyanobacteria that cause problems in the water, emit in different frequencies, and are inaudible. In the project, they were used in three irrigation ponds of RW and one with water from the Tajo-Segura water transfer. The samples were collected fortnightly from the end of May to September 2014. The project proved that implementing ultrasound equipment in irrigation ponds controls the growth of microalgae and macroalgae, improving the quality of stored water. The results showed no effect on the concentration of macronutrients, micronutrients and anions, pH, and electrical conductivity.

1.3.4 Simplified Treatments, Multi-Barrier Approach, and Alternative Agriculture for Agricultural Water Reuse

On the agricultural and water research platform at Roldán-Balsicas WWTP, several European research projects have been developed in the last five years, both with the agricultural testing of different prototypes of wastewater treatment ("Intelligent Reclaim Irrigation System" [IRIS] and "Low-cost water desalination and sensor technology compact module" [DESERT]) (Figure 1.3), and the agronomic development of irrigation programming tools with water of different qualities and database in soil, plants, and environment ("Professional online irrigation system of expert programming" [OPIRIS], KBBE.2013.1.4-07 EU). The objective of the IRIS project was to convert wastewater compounds into renewable sources for agriculture. The technology converts raw domestic wastewater into irrigation water containing fertilizers, CO_2, and biogas, which will be used in greenhouse production. The main aim of the DESERT project is more ambitious because it introduces a new approach and vision of agriculture, creating an innovative concept as a smart system combining sustainable technology on water treatment and water quality sensors tuned to the local situation.

According to Intriago et al. (2018), the DESERT project combines water treatment technologies based on filtration and solar-based renewable energy. During this experiment, six treatments were applied. The irrigation water was RW through the DESERT-prototype (PW) from a secondary effluent of a wastewater treatment plant, as well as conventional irrigation water (CW) and the secondary effluent (SW) itself using both drip and sprinkler irrigation methods. The experiment took place in the greenhouse located in the Roldán, Lo Ferro y Balsicas municipal WWTP facilities in

FIGURE 1.3 Simplified treatments for agricultural water reuse at Roldán-Balsicas WWTP.

Murcia, where baby romaine lettuces (*Lactuca sativa* var. romana) were grown between November 2016 and January 2017 (60 days). Evaluations of the physicochemical and microbiological quality of irrigation water, including the agronomic and microbiological quality of crops from all treatments, demonstrated that PW-related results complied in all cases with the relevant standards and guidelines. In conclusion, due to those assessments, the DESERT prototype proved to be an appropriate technology for safe water reclamation with the aim of agricultural use (Intriago et al., 2018).

1.3.5 Sustainable Agriculture: Eco-friendly Biofertilizers and RW Use

Biofertilizers are known to offer an eco-friendly, cost-effective, and sustainable alternative to chemical fertilizers in agriculture, providing numerous benefits. They can increase soil fertility and support plant growth through their important biological properties and nutrient content (Garcia-gonzalez and Sommerfeld 2016; Zaffar et al. 2020). On the other hand, chemical fertilizers contribute to environmental deterioration and pollution such as eutrophication, soil contamination, and health hazards. Unlike chemical fertilizers, biofertilizers can align nutrient release with plant needs by gradually mineralizing organic matter. They advocate leveraging the naturally existing biological system for nutrient mobilization.

In a study conducted in Murcia (Tallou et al., 2022), tomatoes (*Solanum lycopersicum*) plants were grown in a greenhouse, irrigated with RW, and provided with biofertilizers issued from anaerobic digestion of olive mill wastewater without and with 1% and 5% of phosphate residues in mesophilic conditions for 25 days. The different percentages (1% of raw substrates [OMW raw; OMW+1% PR raw; olive mill wastewater+5% phosphate residues raw]; and phosphate residues) and digestates (olive mill wastewater digestate, olive mill wastewater+1% phosphate residues digestate and olive mill wastewater+5% phosphate residues digestate) were provided every two weeks. The source of RW was from Roldán, Lo Ferro y Balsicas municipal WWTP facilities in Murcia and it was used for automatic irrigation (Figure 1.3). The control RW contained a low level of chemical fertilizers to compare tomato plant growth, leaf analysis, steam water potential, production yield, and fruit quality parameters to plants fed with biofertilizers. Generally, parameters progressively increased during the growing and harvesting stage, which refer to the essential elements that cover the plants' needs. Plants fed with biofertilizers showed the most extended plant height (olive mill wastewater+5% phosphate residues raw), and the best accumulation of essential elements in leaves (olive mill wastewater+1% phosphate residues digestate and olive mill wastewater+5% phosphate residues digestate). The maximum average fruit weight per treatment (35.5 g) was recorded for plants fed with olive mill wastewater raw and olive mill wastewater+5% phosphate residues. The maximum yield production per plant was obtained when applying phosphate residues. Biofertilizers (digestates) showed good performance, high fruit quality, and perfect tomato yield production compared to the control plants. The results obtained in this study were promising and confirm the existing literature on the benefits of biofertilizers that can be combined or used instead of chemicals as a strategy for sustainable agriculture.

These simplified treatments for agricultural water reuse can contribute to mitigating the negative effects of intensive surface and groundwater abstraction, improving water quality and increasing farmers' income by saving energy costs and reducing their water and fertilizer needs.

More specifically, saline agriculture is also an alternative strategy to use in high saline conditions or to combine with traditional agriculture. For that, as wild plants, halophytes can cope with a wide range of environmental conditions, including salinity, drought, and extreme temperatures, and can adopt operating methods and mechanisms regulated by their genetic code. Halophytes are able to absorb large quantities of salts and regulate them in various plant organs. These halophytes are also good candidates to remediate polluted soils containing heavy metals and petroleum hydrocarbons. Most halophytes have a great economic value: some can be used as food or for oil and medicine production and others can be used as vegetables and as a raw material for food processing.

The halophytes experiment also took place in this greenhouse, which had the following characteristics: average temperature 15°C, surface area 680 m^2, relative air humidity 67%, and transpiration 0.5 Lm^{-2}. The irrigation water was RW obtained from the Balsicas, Roldan and Lo Ferro municipal WWTP. The RW was used in this study as a model water with poor microbiological quality. Halophytes were selected because they tolerate extremely high salt concentrations (Li et al., 2020). It has been proven that irrigation with RW increases salinity. In addition, some aquifers in Campo de Cartagena, in the region of Murcia, have been abandoned due to salinization and seawater intrusion caused by excessive groundwater pumping.

The halophyte plants used were *Salicornia* species, with six plants per m^2 in April 2021 and harvested in July of the same year. The experiment was divided into four plots (DS = non-saline soil-saline water; DD = non-saline soil-surface water; SS = saline soil-saline water; SD = saline soil-surface water). During each period, the initial, mid, and harvesting season, three soil samples were collected from measurements. Measurements included the EC, pH, alkalinity, cations and anions, organic matter, total organic carbon (TOC), total nitrogen (TN), and C/N ratio.

In addition, relative water content (RWC), chlorophyll, polyphenols, and steam water potential were assessed. In DD, treatment found the highest amount of polyphenol content (mg/g), followed in descending order in the treatments DS, SS, and SD (with the lowest amount of polyphenol content). The harvesting season of *Salicornia* plants starts usually after two to three months of the growing season, when the plants reach a height of 30 cm. Total production and production per plant for each treatment were weighted. The results were positive where the water content for *Salicornia* plants was in the optimal range for all treatments. Also, the production recorded some differences but they were not significant, meaning that *Salicornia* can tolerate a saline environment, which increases the water and energy savings.

1.3.6 Tools for Planning and Management of RW Projects

1.3.6.1 Public Acceptance and Education of RW Use

Successful implementation of a water reuse project depends not only on its economic and environmental feasibility, but mainly on the general public's support, who, ultimately, pays for and might be affected by the reuse project. All environmental benefits – including non-market benefits – should be included in this economic assessment to ensure an efficient and sustainable water resource allocation. A specific study to estimate the non-market benefits that society attaches to the use of RW for agricultural purposes, which is expected to play a growing role in areas where conventional water resources are scarce, was developed in Murcia. It was shown that the use of RW for irrigation has significant non-market environmental benefits (mean willingness to pay of €5.13 per month per household, which adds up to a total annual value of €23.3 million) (Alcón et al., 2012). This study, the first of its kind in Spain and among the first in Europe, is believed to help policymakers implement water management measures to achieve the water framework directive environmental objectives in an efficient, sustainable, and equitable manner.

Farmers' acceptance of well-treated (reclaimed) wastewater is impacted by many factors, such as location, culture, religion, and socio-economic conditions, including the water costs, the structure of irrigation networks, and cropping patterns (Ganoulis, 2012). More recently, according to the FAO, three scenarios have been implemented, with a view of adopting the use of marginal quality water (MQW) – like saline water or wastewater for irrigation. The first scenario aims to normalize the use of reclaimed (safely treated) wastewater where freshwater is becoming scarce by overcoming barriers. For instance, it can offer irrigation water, cover fertilizer needs at a lower cost, and offer RW with a high concentration of nutrients but low salinity. Thus, farmers' trust in water treatment prevails. The second scenario is about risk mitigation measures, where polluted water is already used in the absence of any alternative. Lastly, the third scenario concerns the acceptance of salinity mitigation measures. Some examples that could lead to the solution are the demand of the market for salt-tolerant crops, but also by the selection of soils suitable for drainage and alternative crops (Drechsel et al., 2023).

1.3.6.2 Geographic Information Systems (GIS), Remote Sensing, and Machine Learning

GISs based on multi-criteria analysis can be used as a tool in the planning and management of water reuse projects, helping in the identification of potential sites for RW use. The first work was developed in the Beira Interior Region (Portugal) with the aim of identifying potential sites for RW use, using a GIS-based multi-criteria analysis (Pedrero et al., 2011). A constructed wetland (CW) system located in the northwest part of the Beira Interior region was used as the source of RW (Figure 1.4). The same methodology improved and was developed in the northwest part of the region of Murcia

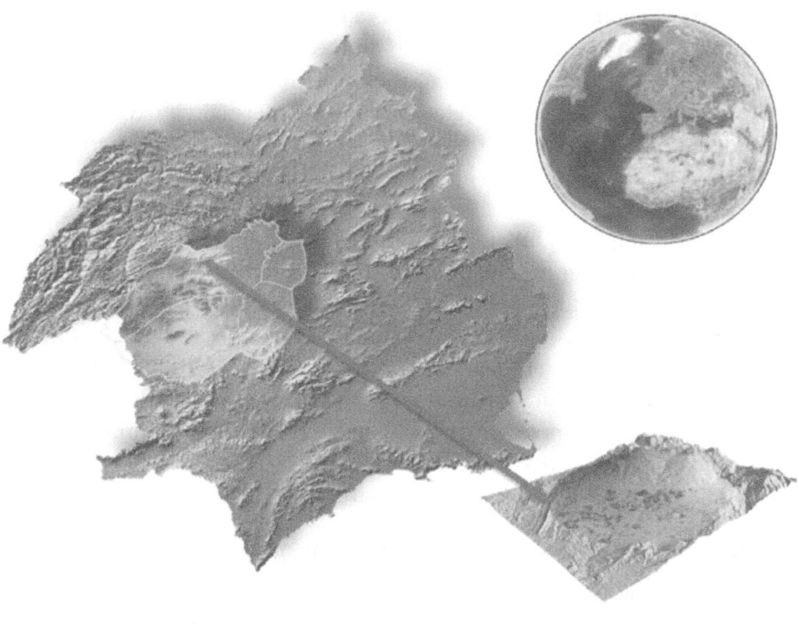

FIGURE 1.4 Multi-criteria analysis with GIS. The optimal area for aquifer recharge (purple area).

(Spain) to optimize the use of RW through integrating the water resources management. After analyzing all the criteria established, of the total area studied (237,960 ha), only 2.7% (6442 ha) was considered optimal for aquifer recharge (Pedrero et al., 2013) (Figure 1.4). So, after two different works on different locations, the application of multi-criteria analysis resulted in a final, optimal map showing areas appropriate for the infiltration of RW.

New technologies in agriculture, such as the use of hyperspectral sensing techniques, were also studied in RW studies to evaluate the effects on spectral properties and leaf traits of citrus orchards. After a long period of applying saline RW on citrus trees in Campotejar, RW did not affect either NIR–NDVI or anatomy traits, but these techniques are suitable for detecting chlorophyll dynamics (Contreras et al., 2013).

In 2022, an experiment was implemented using remote sensing to evaluate the effects of environmental factors on soil salinity. It was carried out in Campo de Cartagena (Murcia), a semi-arid area where citrus trees, a sensitive crop to salinity, cover 30% of the region. The study was focused on nine representative citrus farms. Soil samples were analyzed and different remote-sensing indices and environmental data sets were applied. The results showed that applied salinity spectral indices were valid for detecting soil salinity in citrus trees, despite the heterogeneity of the data. However, to be able to apply soil salinity predictive models, the incorporation of other variables related to agricultural management practices is crucial (Pedrero Salcedo et al., 2022).

During the same year, the prediction of soil salinity using remotely piloted aircraft systems in lettuce under irrigation with non-conventional water sources was also investigated. The experiment was carried out in a greenhouse and in an open-air commercial field. The greenhouse was located on a research platform from CEBAS-CSIC in the Balsicas, Roldan, and Lo Ferro municipal WWTP in the Murcia. There, 121 m^2 of mini-Romaine lettuce (*Lactuca sativa* var. romanica) were transplanted in October 2019 and were harvested in December 2019. The irrigation applied was RW produced in the DESERT prototype, with a secondary effluent of the WWTP. The control plot was irrigated with non-saline water (EC = 0.7–0.8 dS/m) and the saline plot with saline water with an EC of 3–4 dS/m. As concerns the commercial plot, the planting occurred in October 2019, occupying 1.17 ha of lettuce of the variety *Mestiza*. The harvest occurred in January 2020. Drip irrigation was applied with a mix of different conventional and non-conventional sources: Tajo-Segura water transfer (88.7%), the Segura River basin (3.0%), RW from WWTPs (6.7%), and the Mojon desalination plant (1.6%). In conclusion, saline irrigation has a negative impact on lettuce growth by decreasing the fresh head weight. Also, negative impacts in the leaf mineral content were observed, such as mineral deficiency or toxicity; for instance, the higher concentrations of sodium and chloride. The results showed a moderate linear correlation with the thermal camera. The second vegetative parameter, normalized difference vegetation index (NDVI), had minimal correlations. In contrast, soil salinity indicators had higher correlations, with the SI the best of all the measures (Pedrero Salcedo et al., 2022).

More recently, high-resolution planet satellite images and machine learning techniques to predict the olive water status in order to improve olive orchard irrigation management were used. This study was conducted in an irrigated olive orchard in southern Italy during the years 2021–2022. The Ψ_{stem} was measured in the field using reflectance data and measured Ψ_{stem} data from 24 olive trees and derived a Ψ_{stem} prediction model for an in-season estimation of Ψ_{stem} at the olive orchard. The model with the more accurate prediction was the Random Forest model with $R^2 = 0.78$. This was the first study which combined remote sensing and machine learning techniques to predict the water status of an olive orchard and thus paved the way for deeper studies incorporating these models with more information and more complex algorithms (Garofalo et al., 2023).

1.3.6.3 Models on Medium to Long-Term Saline RW Application

The use of new model predictions helps to determine the impact of using saline RW for citrus irrigation. The model for simultaneous salt and water uptake by plants can be coupled with groundwater models in order to predict the salt budget of ecosystems and to manage agriculture in arid zones of the Mediterranean region. During the collaboration with Denmark Technical University (DTU), it has been demonstrated to be a useful tool to measure the water uptake of growing plants and salt concentrations in a solution simultaneously, and in parallel with toxic effects on the plants. The model used is a coupled model based on a coupling of tipping buckets of water and a solute transport model for a soil water balance model at the surface and plant uptake model (based on transpiration) substituted with an uptake factor (F) for exclusion of salt, which is based on measured data for a willow tree in the laboratory (Trapp et al., 2008). The comparison of measured and simulated data has value for both inconsistencies in the data seen, but also model shortcomings. A calibration and validation of the model was done, and the verification of the model under Mediterranean field conditions with other compounds will be done in the near future.

On the other hand, fuzzy logic is a very flexible model since it is capable of handling both qualitative and quantitative, correlated or completely uncorrelated variables, used widely (Aroba et al., 2007; De Gruijter et al., 2011; D'Ercole et al., 2000; Geographic Information Systems for Irrigation Management in Southwest Europe, 2012; Shahmoradi & Isalou, 2013; Skarlatos et al., 2004; Skarlatos et al., 2013). It seems that the rule-based fuzzy modeling is a promising approach because it can be applied to many topics of agriculture as the selection of a site for a wastewater treatment plant (Shahmoradi & Isalou, 2013) or studied the effect of wastewater evapotranspiration on citrus cultivation (Skarlatos et al., 2014). In collaboration with the colleges from Hellenic Open University (Patras, Greece), a study to show the long-term effects of different irrigation water qualities under Mediterranean conditions through a Mamdani fuzzy modeling scheme was done.

The application of fuzzy modeling showed that the fuzzy model can describe the effect of irrigation in mandarin cultivation well. This is because the models take into account many parameters, whatever the degree of correlation is between them. The division of the fuzzy model into submodels enables us to compare the effect of irrigation on individual characteristics of the system, such as the quality of the leaves and soil. The model showed that the overall effect of watering with irrigation community water gives the best results, although as the water quality from this treatment is not always the same because of the different water sources dotation, intensive soil status monitoring is needed to avoid the salt accumulation and a reduction in the physical soil properties. Sensitive indicators of soil status are, therefore, critical to allow monitoring of the effect of the application of these water sources. Among them, labile forms of soil organic carbon and variations in biological soil properties can be more efficient in highlighting slight changes in soil quality status (Zhang et al., 2015).

The environmental performance of nectarine orchards irrigated with treated municipal wastewater were quantified using a database based on field experimental data. The nectarine orchards located in southern Italy perform better when irrigated by RW than irrigated by surface water. The only downside caused by the substitution of surface water with treated RW was the generation of eutrophication and acidification burdens. The irrigation with RW increased water consumption, but it also made a positive contribution to water stress. The results showed a better understanding of the environmental consequences of RW reuse in agriculture, and therefore, can also increase public acceptance (Moretti et al., 2019).

A study on an index-based approach described the various aspects which affect the water and the nutrient recovery after specific combinations between the treatment facility and the irrigation area. Data sets were collected for 11 pilot reuse schemes in southern Italy and tested with the index-based approach. The objective was to evaluate the potential for water and nutrient recovery from irrigation with RW. As stated by the results, indicators allowed to quantify the environmental benefits of RW irrigation leading to the diversion of less freshwater from conventional sources and the disposal of less RW into natural water receptors. Additionally, irrigation with RW provides the advantage of partially supplying the needs of irrigation and nutrition requirements. In summary, the proposed set of indicators can offer useful information for effectively implementing water reuse policies for irrigated agriculture (Vivaldi et al., 2022).

1.4 CONCLUSION

The developed work shows the success of a multidisciplinary approach to RW use projects in agriculture through an integrated methodology. This long-term interdisciplinary work is the crossover of environmental sciences, biology, chemistry, economics, agriculture, and engineering and has generated novel strategies to describe and quantify the interactions of the use of saline RW for horticultural crop irrigation in agriculture. This approach will result in better understanding of the short- and long-term effects of using saline RW in horticultural crop irrigation and will be a key challenge in promoting the expansion of water recycling for agricultural purposes, ensuring field sustainability and public safety. It will have a strong impact in the agriculture, but also in other fields where water management is of relevance. Therefore, for the preservation of profitable intensive agriculture that respects the environment, innovative agricultural projects are needed for incorporating state-of-the-art technologies in water reuse with multidisciplinary studies on salinity and toxic effects on crops at the field and laboratory scale and long-term prediction is necessary to establish new reclaimed irrigation water strategies for the conservation of the natural environment.

REFERENCES

Alcón, F., Pedrero, F., Martín-Ortega, J., Arcas, N., Alarcón, J. J., & de Miguel, M. D. (2012). The non-market value of reclaimed waste water for use in agriculture: A contingent valuation approach. *Spanish Journal of Agricultural Research*, *8*, 187–196.

Allende, A., Lopez-Galvez, F., Gil, M., Pedrero, F., & Alarcon, J. J. (2016). Monitoring generic *E. coli* in reclaimed and surface water used in hydroponically cultivated greenhouse peppers and the influence of fertilizer solutions. *Food Control*, *67*, 90–95.

Aroba, J., Grande, J. A., Andujar, J. M., de la Torre, M. L., & Riquelme, J. C. (2007). Application of fuzzy logic and data mining techniques as tools for qualitative interpretation of acid mine drainage processes. *Environmental Geology*, *53*(1), 135–145.

Ayers, R. S., & Westcot, D. W. (1989). *Water Quality for Agriculture*. FAO Irrigation and Drainage, Rome, Italy.

Brainwood, M. A., Burgin, S., & Maheshwari, B. (2004). Temporal variations in water quality of farm dams: Impacts of land use and water sources. *Agricultural Water Management*, *70*(2), 151–175. https://doi.org/10.1016/j.agwat.2004.03.006

Contreras, S., Cutillas, P., Santoni, C. S., Romero-Trigueros, C., Pedrero, F., & Alarcón, J. J. (2013). Detection of salinity stress on citrus orchards: A field experiment using reclaimed waters, hyperspatial remote sensing and leaf-anatomy and physiology measurements. *Water Environment Research*, *86*(11), 2242–2250.

Crespi, V., & Lovatelli, A. (2011). *Aquaculture in Desert and Arid Lands: Development Constraints and Opportunities*. Food and Agriculture Organization of the United Nations.

D'Ercole, C., Groves, D. I., & Knox-Robinson, C. M. (2000). Using fuzzy logic in a geographic information system environment to enhance conceptually based prospectivity. *Functional Plant Biology*, *37*, 613–620.

De Gruijter, J. J., Walvoort, D. J. J., & Bragato, G. (2011). Application of fuzzy logic to Boolean models for digital soil assessment. *Geoderma*, *166*, 15–33.

De Keuckelaere, A., Jacxsens, L., Amoah, P., Medema, G., McClure, P., Jaykus, L.-A., & Uyttendaele, M. (2015). Zero risk does not exist: Lessons learned from microbial risk assessment related to use of water and safety of fresh produce. *Comprehensive Reviews in Food Science and Food Safety*, *14*(4), 387–410. http://doi.org/10.1111/1541-4337.12140

Drechsel, P., Marjani Zadeh, S., & Pedrero, F. (2023). *Water Quality in Agriculture: Risks and Risk Mitigation*. FAO & IWMI. https://doi.org/10.4060/cc7340en

Ganoulis, J. (2012). Risk analysis of wastewater reuse in agriculture. *International Journal of Recycling of Organic Waste in Agriculture*, *1*(1), 3. https://doi.org/10.1186/2251-7715-1-3

Garcia-Gonzalez, J., & Sommerfeld, M. (2016). Biofertilizer and biostimulant properties of the microalga *Acutodesmus dimorphus*. *Journal of Applied Phycology*. https://doi.org/10.1007/s10811-015-0625-2

Garofalo, S. P., Giannico, V., Costanza, L., Alhajj Ali, S., Camposeo, S., Lopriore, G., Pedrero Salcedo, F., & Vivaldi, G. A. (2023). Prediction of stem water potential in olive orchards using high-resolution planet satellite images and machine learning techniques. *Agronomy*, *14*(1), 1. https://doi.org/10.3390/agronomy14010001

Geographic Information Systems for Irrigation Management in Southwest Europe. (2012). *Option Mediterraneennes*, 223–231. https://www.researchgate.net/publication/256482369_The_use_of_remote_sensing_and_geographic_information_systems_for_irrigation_management_in_Southwest_Europe

Gerba, C. P., & Rose, J. B. (2003). International guidelines for water recycling microbiological considerations. *Water Science and Technology: Water Supply*, *3*(4), 311–316.

Gil, M., Lopez-Galvez, F., Allende, A., Pedrero, F., & Alarcon, J. J. (2014). Safety assessment of greenhouse hydroponic tomatoes irrigated with reclaimed and surface water. *International Journal of Food Microbiology*, *17*, 191-97-102.

González-Altozano, P., & Castel, J. R. (2003). Riego deficitario controlado en 'Clementina de Nules' I. Efectos sobre la producción y la calidad de la fruta. *Spanish Journal of Agricultural Research*, *1*(2), 93–101.

González García, M., Fernández-López, C., Pedrero-Salcedo, F., & Alarcón, J. J. (2018). Absorption of carbamazepine and diclofenac in hydroponically cultivated lettuces and human health risk assessment. *Agricultural Water Management*, *206*, 42–47. https://doi.org/10.1016/j.agwat.2018.04.018

Grattan, S. R., Díaz, F. J., Pedrero, F., & Vivaldi, G. A. (2014). Assessing the suitability of saline wastewaters for irrigation of citrus: Emphasis on boron and specific-ion interactions. *Agricultural Water Management*, *157*, 48–58.

Harmanny, K. S., Malek, Ž. (2019). Adaptations in irrigated agriculture in the mediterranean region: An overview and spatial analysis of implemented strategies. *Regional Environmental Change*, *19*, 1401–1416. https://doi.org/10.1007/s10113-019-01494-8

Intriago, J. C., López-Gálvez, F., Allende, A., Vivaldi, G. A., Camposeo, S., Nicolás Nicolás, E., Alarcón, J. J., & Pedrero Salcedo, F. (2018). Agricultural reuse of municipal wastewater through an integral water reclamation management. *Journal of Environmental Management*, *213*, 135–141. https://doi.org/10.1016/j.jenvman.2018.02.011

Kalavrouziotis, I. K., Pedrero, F., & Skarlatos, D. (2016). Water and wastewater quality assessment based on fuzzy modeling for the irrigation of mandarin. *Desalination and Water Treatment*, *57*(43), 20159–20168. https://doi.org/10.1080/19443994.2015.1110050

Khan, M. M., Siddiqi, S. A., Farooque, A. A., Iqbal, Q., Shahid, S. A., Akram, M. T., Rahman, S., Al-Busaidi, W., & Khan, I. (2022). Towards sustainable application of wastewater in agriculture: A review on reusability and risk assessment. *Agronomy*, *12*(6), Article 6. https://doi.org/10.3390/agronomy12061397

Kotopoulis, S., Schommartz, A., & Postema, M. (2009). Sonic cracking of blue-green algae. *Applied Acoustics*, *70*(10), 1306–1312. https://doi.org/10.1016/j.apacoust.2009.02.003

Leogrande, R., Pedrero, F., Nicolas, E., Vitti, C., Lacolla, G., & Stellacci, A. M. (2022). Reclaimed water use in agriculture: Effects on soil chemical and biological properties in a long-term irrigated citrus farm. *Agronomy*, *12*(6), 1317. https://doi.org/10.3390/agronomy12061317

Leogrande, R., Vitti, C., Castellini, M., Mastrangelo, M., Pedrero, F., Vivaldi, G. A., & Stellacci, A. M. (2021). Comparison of two methods for total inorganic carbon estimation in three soil types in Mediterranean area. *Land*, *10*(4), 409. https://doi.org/10.3390/land10040409

Li, L., Zhao, Y., Han, G., Guo, J., Meng, Z., & Chen, M. (2020). Progress in the study and use of seawater vegetables. *Journal of Agricultural and Food Chemistry*, *68*(22), 5998–6006. https://doi.org/10.1021/acs.jafc.0c00346

Lyra, D. A., Robertson, S., Mateo-Sagasta, J., & Shoaib, I. (2019). Financial analysis of halophytes cultivation in a desert environment using different saline water resources for irrigation. (p. 347). https://link.springer.com/chapter/10.1007/978-981-13-3762-8_17

Maestre-Valero, J. F., Gonzalez-Ortega, M. J., Martinez-Alvarez, V., Gallego-Elvira, B., Conesa-Jodar, F. J., & Martin-Gorriz, B. (2019b). Revaluing the nutrition potential of reclaimed water for irrigation in southeastern Spain. *Agricultural Water Management*, *218*, 174–181.

Maestre-Valero, J. F., González-Ortega, M. J., Martínez-Álvarez, V., & Martin-Gorriz, B. (2019a). The role of reclaimed water for crop irrigation in southeast Spain. *Water Supply*, *19*(5), 1555–1562.

Moretti, M., Van Passel, S., Camposeo, S., Pedrero, F., Dogot, T., Lebailly, P., & Vivaldi, G. A. (2019). Modelling environmental impacts of treated municipal wastewater reuse for tree crops irrigation in the mediterranean coastal region. *Science of the Total Environment*, *660*, 1513–1521. https://doi.org/10.1016/j.scitotenv.2019.01.043

Mounzer, O., Pedrero, F., Bayona, J. M., Nicolás, E., & Alarcón, J. J. (2013). Transient soil salinity under the combined effect of reclaimed water and regulated deficit drip irrigation of mandarin trees. *Agricultural Water Management*, *120*, 23–29.

Nicolás, E., Maestre-Valero, J. F., Pedrero, F., Mounzer, O., Nortes, P. A., Romero-Trigueros, C., Bayona, J. M., Alcobendas, R., & Alarcón, J. J. (2017). Long-term effect of irrigation with saline reclaimed water on adult mandarin trees. *Acta Horticulturae*, *1150*, 407–412. https://doi.org/10.17660/ActaHortic.2017.1150.57

Nicolás, E., Pedrero, F., Alarcón, J. J., Mounzer, O., Martinez, V., Nortes, P. A., Alcón, F., Egea, G., & De Miguel, M. D. (2012). Estudio de viabilidad de uso de las aguas regeneradas procedentes de la EDAR de Jumilla en la Comunidad de Regantes Miraflores. *Universidad Politécnica de Cartagena Servicio de Documentación.*

Nicolás, E., Romero Trigueros, C., Nortes Tortosa, P. A., Pedrero Salcedo, F., Bayona Gambín, J. M., Maestre Valero, J. F., & Alarcón Cabañero, J. J. (2018). Long-term physiological and agronomic responses of citrus irrigated with saline reclaimed water. In *Water Scarcity and Sustainable Agriculture in Semiarid Environment* (pp. 131–147). Elsevier. https://doi.org/10.1016/B978-0-12-813164-0.00007-7

Pedrero, F., & Alarcón, J. J. (2009). Effects of treated wastewater irrigation on lemon trees. *Desalination, 246*, 631–639.

Pedrero, F., Alarcón, J. J., Abellán, M., & Perez, P. (2013). Reclaimed water use optimization through groundwater recharge using GIS. *Desalination and Water Treatment.* https://doi.org/10.1080/19443994.2014.1003979

Pedrero, F., Alarcón, J. J., García, A. J., Nortes, P. A., & Abellán, M. (2014b). From agricultural wastewater use to intelligent reclaim irrigation system (IRIS) in the region of Murcia. In IWA Regional Symposium on Water, Wastewater and Environment: Traditions and Culture. Patras, Greece.

Pedrero, F., Albuquerque, A., Marecos Do Monte, H., Cavaleiro, V., & Alarcón, J. J. (2011). Potencial use of reclaimed wastewater with GIS-based multi-criteria analysis in Portugal. *Resources, Conservation and Recycling, 56*, 105–116.

Pedrero, F., Allende, A., Gil, M. I., & Alarcón, J. J. (2012a). Effects of variations in quality of treated wastewater used in irrigation on soil chemical properties, leaf mineral status and crop production of lemon trees. *Agricultural Water Management, 109*, 54–60.

Pedrero, F., Maestre-Valero, J. F., Mounzer, O., Alarcón, J. J., & Nicolás, E. (2014a). Physiological and agronomic mandarin trees performance under saline reclaimed water combined with regulated deficit irrigation. *Agricultural Water Management.* https://doi.org/10.1016/j.agwat.2014.08.013

Pedrero, F., Maestre-Valero, J. F., Romero-Trigueros, C., Nortes, P., Mounzer, O., Bayona, J. M., Alarcón, J. J., & Nicolás, E. (2015a). Response of young 'Star Ruby' grapefruit trees to regulated deficit irrigation with saline reclaimed water. *Agricultural Water Management, 158*, 51–60.

Pedrero, F., Maestre-Valero, J. F., Romero-Trigueros, C., Nortes, P., Mounzer, O., Bayona, J. M., Alarcón, J. J., & Nicolás, E. (2015b). Long-term physiological and agronomic responses of mandarin trees to irrigation with saline reclaimed water. *Agricultural Water Management, 166*, 1–8.

Pedrero, F., Mounzer, O., Alarcón, J. J., Bayona, J. M., & Nicolás, E. (2012b). The viability of irrigating mandarin trees with saline reclaimed water in a semi-arid mediterranean region: A preliminary assessment. *Irrigation Science, 31*(4), 759–768.

Pedrero, F., Mounzer, O., Nicolás, E., & Alarcón, J. J. (2012c). Influence of irrigation with saline reclaimed water on young grapefruits. *Desalination and Water Treatment, 51*(10–12), 2488–2496.

Pedrero Salcedo, F., Pérez Cutillas, P., Aziz, F., Llobet Escabias, M., Boesveld, H., Bartholomeus, H., & Tallou, A. (2022). Soil salinity prediction using remotely piloted aircraft systems under semi-arid environments irrigated with salty non-conventional water resources. *Agronomy, 12*(9), 2022. https://doi.org/10.3390/agronomy12092022

Pereira, B. F. F., He, Z. L., Stoffella, P. J., & Melfi, A. J. (2011). Reclaimed wastewater: Effects on citrus nutrition. *Agricultural Water Management, 98*, 1828–1833.

Ponce-Robles, L., Benelhadj, L., García-García, A. J., Pedrero-Salcedo, F., Nortes-Tortosa, P. A., Albacete, J., & Alarcón, J. J. (2022). Risk assessment for uptake and accumulation of pharmaceuticals by baby leaf lettuce irrigated with reclaimed water under commercial agricultural activities. *Journal of Environmental Management, 324*, 116321. https://doi.org/10.1016/j.jenvman.2022.116321

Ponce-Robles, L., Masdemont-Hernández, B., Munuera-Pérez, T., Pagán-Muñoz, A., Lara-Guillén, A. J., García-García, A. J., Pedrero-Salcedo, F., Nortes-Tortosa, P. A., & Alarcón-Cabañero, J. J. (2020). WWTP effluent quality improvement for agricultural reuse using an autonomous prototype. *Water, 12*(8), 2240. https://doi.org/10.3390/w12082240

Reboll, V., Cerezo, M., Roig, A., Flors, V., Lapeña, V., & García-Agustín, P. (2000). Influence of wastewater vs groundwater on young citrus trees. *Journal of the Science of Food and Agriculture, 80*, 1441–1446.

Romero-Trigueros, C., Bayona Gambín, J. M., Nortes Tortosa, P. A., Alarcón Cabañero, J. J., & Nicolás Nicolás, E. (2019). Determination of crop water stress index by infrared thermometry in grapefruit trees irrigated with saline reclaimed water combined with deficit irrigation. *Remote Sensing, 11*(7), 757. https://doi.org/10.3390/rs11070757

Romero-Trigueros, C., Cabañero, J. J., Tortosa, P. A., Gambín, J. M., Maestre-Valero, J. F., & Nicolás, E. N. (2020). Medium-long term effects of saline reclaimed water and regulated deficit irrigation on fruit quality of citrus. *Journal of the Science of Food and Agriculture, 100*(3), 1350–1357. https://doi.org/10.1002/jsfa.10091

Romero-Trigueros, C., Nortes, P., Pedrero, F., Mounzer, O., Alarcón, J. J., Bayona, J. M., & Nicolás, E. (2014a). Assessment of the viability of using saline reclaimed water in grapefruit in medium to long term. *Spanish Journal of Agricultural Research*, *12*(4).

Romero-Trigueros, C., Nortes, P. A., Alarcón, J. J., & Nicolás, E. (2014b). Determination of 15n stable isotope natural abundances for assessing the use of saline reclaimed water in grapefruit. *Environmental Engineering and Management Journal*, *13*(10), 2525–2530.

Salcedo, F. P., Cutillas, P. P., Cabañero, J. J. A., & Vivaldi, A. G. (2022). Use of remote sensing to evaluate the effects of environmental factors on soil salinity in a semi-arid area. *Science of the Total Environment*, *815*, 152524. https://doi.org/10.1016/j.scitotenv.2021.152524

Sánchez, A. S., Nogueira, I. B. R., & Kalid, R. A. (2015). Uses of the reject brine from inland desalination for fish farming, Spirulina cultivation, and irrigation of forage shrub and crops. *Desalination*, *364*, 96–107. https://doi.org/10.1016/j.desal.2015.01.034

Shahmoradi, B., & Isalou, A. (2013). Site selection for wastewater treatment plant using integrated fuzzy logic and multicriteria decision model: A case study in Kahak, Iran. *Journal of Advances in Environmental Health Research*, *1*(1), 51–61.

Skarlatos, A., Kalavrouziotis, I., Skarlatos, D., & Papadopoulos, A. (2013). A fuzzy model for the determination of the suitability of a given area for Vitis vinifera L. cultivation. *Fresenius Environmental Bulletin*, *22*(5), 1472–1481.

Skarlatos, D., Kalavrouziotis, I. K., Montes, C. R., Melfi, A. J., & Pereira, B. F. F. (2014). Wastewater reuse in citrus: A fuzzy logic model for optimum evapotranspiration. *Desalination and Water Treatment*. https://doi.org/10.1080/19443994.2014.915387

Skarlatos, D., Karakasis, K., & Trochidis, A. (2004). Railway fault diagnosis using a fuzzy logic method. *Applied Acoustics*, *65*, 951–966.

Tallou, A., Aziz, F., Garcia, A. J., Salcedo, F. P., El Minaoui, F. E., & Amir, S. (2022). Bio-fertilizers issued from anaerobic digestion for growing tomatoes under irrigation by treated wastewater: Targeting circular economy concept. *International Journal of Environmental Science and Technology*, *19*(4), 2379–2388. https://doi.org/10.1007/s13762-021-03265-7

Trapp, S., Feificova, D., Feldbech Rasmussen, N., & Bauer-Gottwein, P. (2008). Plant uptake of NaCl in relation to enzyme kinetics and toxic effects. *Environmental and Experimental Botany*, *64*, 1–7.

Truchado, P., Lopez-Galvez, F., Gil, M. I., Pedrero-Salcedo, F., Alarcon, J. J., & Allende, A. (2016). Suitability of different Escherichia coli enumeration techniques to assess the microbial quality of different irrigation water sources. *Food Microbiology*, *58*, 29–35.

Valero, J. F. M. (2012). *Efectos de la aplicación de coberturas de sombreo suspendidas sobre balsas de riego* [Tesis doctoral, Universidad Politécnica de Cartagena]. https://dialnet.unirioja.es/servlet/tesis?codigo=50455

Vivaldi, G. A., Camposeo, S., Lopriore, G., Romero-Trigueros, C., & Salcedo, F. P. (2019). Using saline reclaimed water on almond grown in Mediterranean conditions: Deficit irrigation strategies and salinity effects. *Water Supply*, *19*(5), 1413–1421. https://doi.org/10.2166/ws.2019.008

Vivaldi, G. A., Salcedo, F. P., & Tallou, A. (2024). Application of biological effluent. In *Microirrigation for Crop Production* (pp. 393–410). Elsevier. https://doi.org/10.1016/B978-0-323-99719-5.00012-5

Vivaldi, G. A., Zaccaria, D., Camposeo, S., Pasanisi, F., Salcedo, F. P., & Portoghese, I. (2022). Appraising water and nutrient recovery for perennial crops irrigated with reclaimed water in Mediterranean areas through an index-based approach. *Science of the Total Environment*, *820*, 152890. https://doi.org/10.1016/j.scitotenv.2021.152890

von Sperling, E., da Silva Ferreira, A. C., & Nunes Ludolf Gomes, L. (2008). Comparative eutrophication development in two Brazilian water supply reservoirs with respect to nutrient concentrations and bacteria growth. *Desalination*, *226*(1), 169–174. https://doi.org/10.1016/j.desal.2007.02.105

Zaffar, N., et al. (2020). Concentration of selected plant nutrients and target elements for phytoremediation and phytomining in biogas digestate. *Carpathian Journal of Earth and Environmental Sciences*, *15*(1), 43–48.

Zekri, M., & Koo, R. C. J. (1993). A reclaimed water citrus irrigation project. *Proceedings of the Florida State Horticultural Society*, *106*, 30–35.

Zekri, M., & Koo, R. C. J. (1994). Treated municipal wastewater for citrus irrigation. *Journal of Plant Physiology*, *17*, 693–708.

Zhang, M., He, Z., Zhao, A., Zhang, H., Endale, D. M., & Schomberg, H. H. (2015). Water-extractable soil organic carbon and nitrogen affected by tillage and manure application. *Soil Science*, *180*(1), 37–44. https://doi.org/10.1097/SS.0b013e31821d6d63

2 Thermodynamic Calculation on the Thermal Decomposition of Different Types of Calcium Carbonates Using the Computational Thermochemistry Tool

HSC Chemistry

F. Z. Abbadi, K. Azoulay, A. Moufti, A. Ait Hou, and S. Jemjami

2.1 INTRODUCTION

The study of thermal processes in inorganic materials is of crucial importance in various fields of sciences and engineering. Among these processes, the thermal decomposition of calcium carbonate is one of the processes that have significant implications in a distinct area such as the cement industry, regarding its significance in the production and use of heat in cement clinker [1]. In environmental science, the sequestration processes help reduce carbon dioxide emissions [2]. And the generation at diverse temperatures has different types of crystals including calcite, vaterite, and aragonite [3].

Calcium carbonate, represented chemically by $CaCO_3$, is a chemical substance composed of three primary elements such as carbon, oxygen, and calcium by 12%, 47.96%, and 40.04%, respectively [4]. This occurs in the most stable polymorphs, including aragonite and calcite [5]. Furthermore, calcium carbonate ($CaCO_3$) is widely recognized as a major constituent of sedimentary rocks like limestone with more than 90%, and dolomite ($CaCO_3.MgCO_3$) with 54.35%. This relevance emphasized the significance of this substance in geological processes [6–8].

Through the years, there has been extensive research into the thermal decomposition of calcium carbonate [9], which leads to the formation of calcium oxide (CaO) and carbon dioxide gas (CO_2) at elevated temperatures [10, 11]. Generally, carbonate's thermal decomposition is an endothermic reaction [12], influenced by such factors as the increase in heat and conductivity of the atmosphere, carbon dioxide pressure, and the highest concentration of carbon dioxide in argon (Ar) and nitrogen (N) [13–15].

Thermodynamic analysis plays a pivotal role in understanding and predicting the behavior of chemical reactions including the thermal decomposition of calcium carbonate because it offers a comprehensive overview of the mechanisms involved in chemical reactions [16]. Besides, it can determine the probability of the occurrence of a reaction, as well as the direction in which the stasis will move under specific circumstances [17]. By studying the thermodynamic properties of substances such as the fundamentals of decreasing Gibbs free energy [18], the enthalpy of production

DOI: 10.1201/9781003647263-2

and entropy [19], research can predict reaction spontaneity, equilibrium conditions, and feasibility of reaction.

Understanding the thermodynamic properties of calcium carbonate advances our knowledge of the various polymorphous phases such as calcite, aragonite, and vaterite. In the change of phase with different enthalpy and standard entropy [20], this comprehensive is essential, given that it aims to minimize the amount of greenhouse gas in the atmosphere [21].

This study aims to address a detailed analysis of the thermal decomposition reactions of several types of calcium carbonates, highlighting the use of the advanced thermochemistry tool: HSC Chemistry which allows precise thermodynamic calculations to be carried out, thus providing an in-depth understanding of the reaction mechanisms involved in these processes. It also explores the fundamental aspects of the thermochemistry of calcium carbonates and emphasizes the significant results obtained through the application of this approach.

2.2 METHODOLOGY

It is widely known that calculating the hypothetical thermodynamic properties of a system can be achieved through its equilibration, which can be employed to address varied concerns in chemistry and chemical manufacturing [22]. Thermochemical calculations based on the principles of thermodynamics provide a systematic approach to figuring out the behavior of chemical systems in various circumstances [23]. Modeling those thermochemical calculations can be carried out by using HSC Chemistry software, which supplies a wide range of thermochemical databases and software for computing a balance thermodynamic, in complicated chemical equilibria and diagrammed phases calculations [24].

The utilization of HSC Chemistry evolved by Outokumpo Oy aims to generate quick and easy conventional thermodynamic calculations. In addition, it handles all the theoretical thermodynamic issues [25, 26].

Consequently, HSC Chemistry enables the simulation of the equilibration of chemical systems with precision and accuracy, providing valuable insights into the thermodynamic properties and behavior of the studied substance. From this perspective, the objective of this study is to investigate the thermal decomposition of different types of calcium carbonate as an indication of temperature using the HSC Chemistry chemical reactions and equilibrium software.

2.3 RESULTS AND DISCUSSION

2.3.1 First Study: Monitoring the Thermal Decomposition of $CaCO_3$

In the first study, we tracked the thermodynamic behavior of calcium carbonate ($CaCO_3$) thermal decomposition reactions. The experiment was carried out using HSC Chemistry to calculate firstly the thermodynamic data of this reaction, generate a $CaCO_3$ predominance diagram, and determine the decomposition of $CaCO_3$ by varying the temperature from 900°C to 1400°C. In addition, we examined the decomposition of $MgCO_3$ to comprehend its behavior compared to $CaCO_3$. The results obtained are presented in Table 2.1 and Figures 2.1, 2.2, 2.3, and 2.4.

In the context of thermodynamic calculations to monitor the thermal decomposition of different types of calcium carbonate, the table provided is instrumental in understanding the fundamental properties of its thermal decomposition.

These thermodynamic data show that as the temperature rises, the standard variance in Gibbs free energy (ΔG) goes progressively negative, particularly above 900°C, suggesting that the decomposition reaction becomes thermodynamically favorable at elevated temperatures.

In addition, the data demonstrate positive enthalpy and entropy variations with increasing temperature, indicating that the decomposition reaction of calcium carbonate is endothermic. This leads to the formation of calcium oxide (CaO) and carbon dioxide at the highest temperature, which involves the use of heat energy and results in an enhanced disorder of the system.

TABLE 2.1
Thermodynamic Data Calculated by HSC Chemistry for the Reaction $CaCO_3 = CaO + CO_2(g)$

T C	deltaH kcal	deltaS cal/K	deltaG kcal	K	Log(K)
200.000	42.295	37.158	24.714	3.832E–012	–11.417
300.000	42.025	36.640	21.025	9.602E–009	–8.018
400.000	41.741	36.184	17.384	2.268E–006	–5.644
500.000	41.437	35.763	13.787	1.266E–004	–3.897
600.000	41.103	35.358	10.231	2.748E–003	–2.561
700.000	40.736	34.960	6.715	3.103E–002	–1.508
800.000	40.332	34.565	3.239	2.190E–001	–0.660
900.000	39.889	34.171	–0.198	1.089E+000	0.037
1000.000	39.405	33.775	–3.595	4.142E+000	0.617
1100.000	38.879	33.377	–6.953	1.279E+001	1.107
1200.000	38.310	32.977	–10.271	3.341E+001	1.524
1300.000	37.697	32.575	–13.548	7.627E+001	1.882
1400.000	37.040	32.170	–16.786	1.559E+002	2.193
1500.000	36.338	31.763	–19.982	2.905E+002	2.463

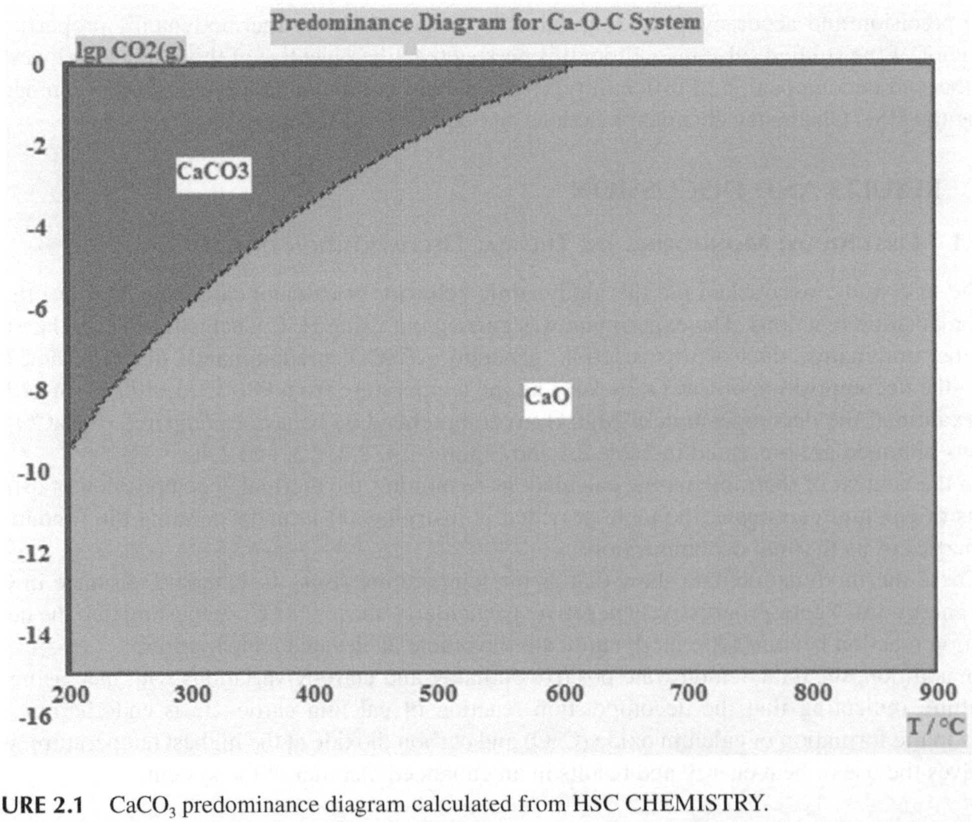

FIGURE 2.1 $CaCO_3$ predominance diagram calculated from HSC CHEMISTRY.

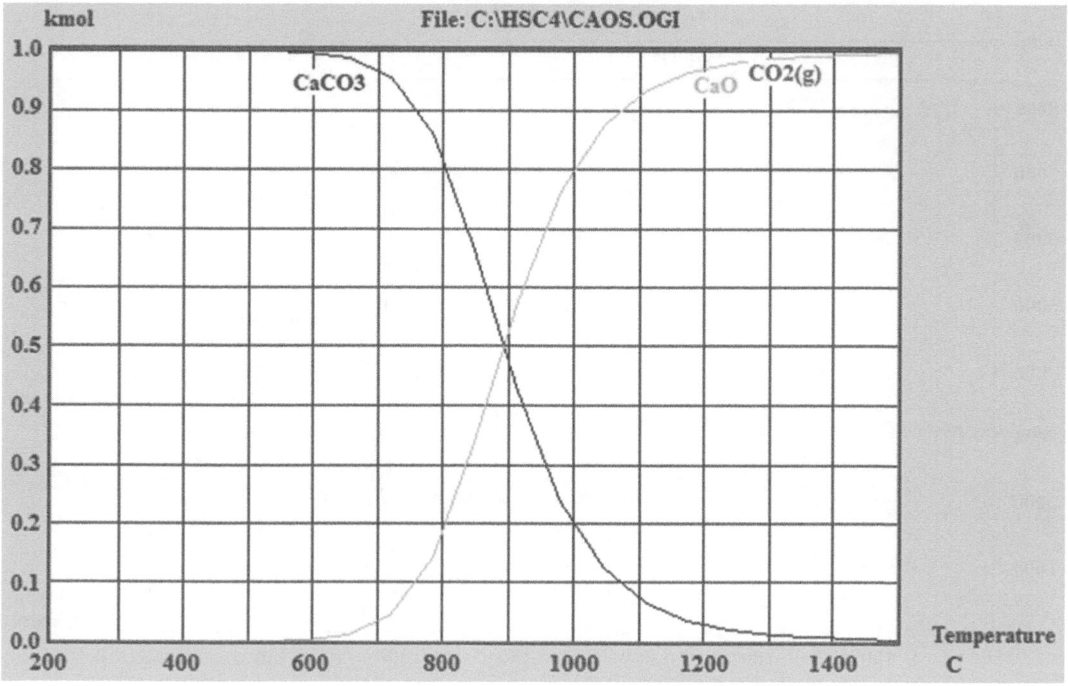

FIGURE 2.2 $CaCO_3$ decomposition as a function of temperature.

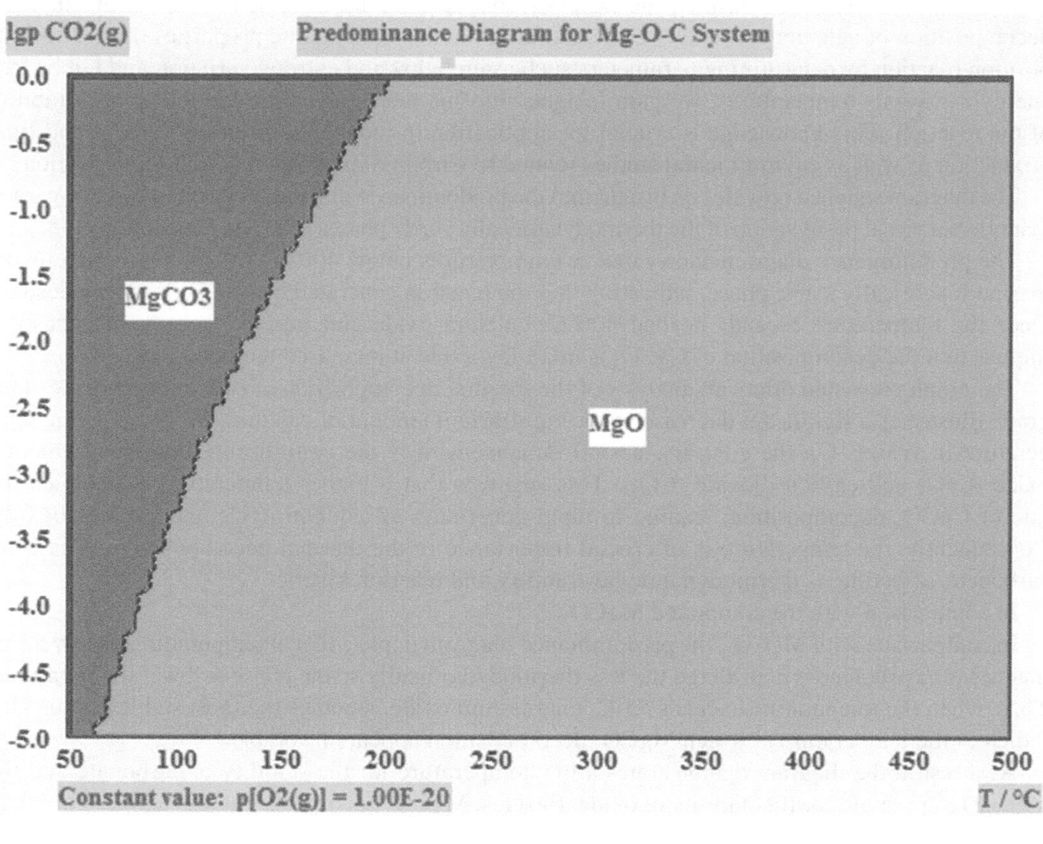

FIGURE 2.3 $MgCO_3$ predominance diagram calculated from HSC CHEMISTRY.

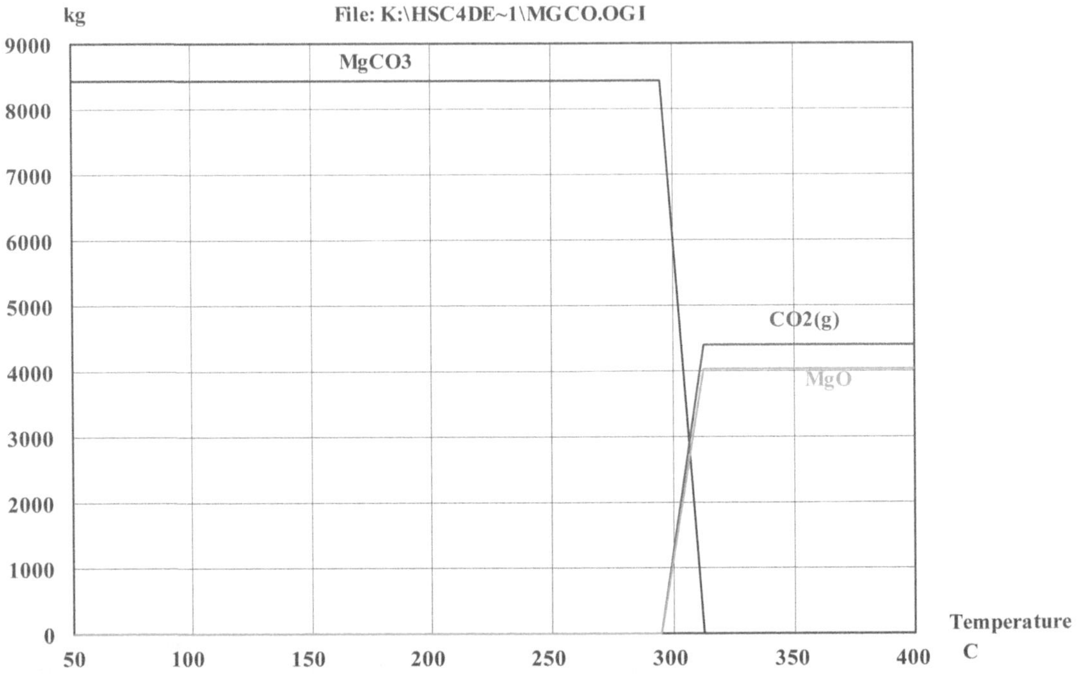

FIGURE 2.4 $MgCO_3$ decomposition as a function of temperature.

These observations emphasized the necessity of thermodynamic calculations in the thermal decomposition of calcium carbonate. Understanding the thermodynamic properties of this decomposition reaction by determining parameters such as enthalpy and entropy variation and Gibbs free energy at various temperatures, we gain insights into the thermodynamic feasibility and stability of the reaction. This knowledge is crucial for application in industries including cement and lime production as well as environmental studies related to carbon dioxide capture and sequestration.

The diagram supplied provides an insight into the predominance diagram of calcium-oxygen-carbon that offers a visual presentation of the thermodynamically stable phases at varied temperatures.

The predominance diagram shows that at temperatures below 400°C, calcium carbonate is the thermodynamically stable phase, indicating that the reaction generated by $CaCO_3$ is highly desired. Once the temperature ascends beyond 400°C, calcium oxide emerges the most stable phase. It implies that the decomposition of $CaCO_3$ is more favorable at increased temperatures.

The graph presented offers an analysis of the thermal decomposition of calcium carbonate. This graph illustrates a significant decrease in the equilibrium amount of calcium carbonate as the temperature increases. On the contrary, a vital enhancement is the equilibrium amount of calcium oxide (CaO) and carbon dioxide (CO_2). This suggests that a higher temperature accelerates the rate of $CaCO_3$ decomposition, leading to rapid generation of calcium oxide and carbon dioxide. Consequently, the temperature is of crucial importance in the thermal decomposition of calcium carbonate, affecting its thermodynamic favorability and reaction kinetic.

In comparison with the compound $MgCO_3$:

In comparison with $MgCO_3$, the predominance diagram depicts that at temperatures below 55°C, magnesium carbonate is considered the less thermodynamically stable phase at lower temperatures. Thus, when the temperature exceeds 55°C, magnesium oxide becomes the most stable phase. This indicates the temperature at which $MgCO_3$ decomposition appears favorable.

As a result, the diagram demonstrates a low temperature for the stability of carbonate and the highest temperature for the stability of oxide. Besides, $MgCO_3$ decomposes at less temperature than $CaCO_3$ due to its lower thermal stability.

Those thermodynamic diagrams of $CaCO_3$ and $MgCO_3$ are of crucial importance because it helps in predicting the phase behavior of the system at varied temperatures.

The graph provided shows an analysis of the thermal decomposition of magnesium carbonate. The graph depicts an enhanced temperature with a decrease in the equilibrium amount of magnesium carbonate. Conversely, the equilibrium amount of magnesium oxide and carbon dioxide increases. This indicates that at lower temperatures, the decomposition of magnesium carbonate is conducted to produce magnesium oxide (MgO) and carbon dioxide (CO_2).

Correspondingly, the decomposition of calcium carbonate starts at a lower temperature than calcium carbonate, revealing its high thermal reactivity. Moreover, the equilibrium amount of undecomposed $CaCO_3$ is bigger than $MgCO_3$, demonstrating $CaCO_3$'s greater stability.

Conclusion of the first study: It appears from the interpretation of the above curves that:

- The thermodynamically advantageous decomposition reaction of $CaCO_3$ occurs beyond 900°C.
- Calcium carbonate ($CaCO_3$) is stable at temperatures below 400°C, as well as calcium oxide, which remains most stable above 400°C, indicating its decomposition.
- By comparison with the compound $MgCO_3$, which is less stable than $CaCO_3$, it decomposes at a lower temperature.
- The decomposition of $CaCO_3$ leads to CaO and CO_2 at higher temperatures, compared to the $MgCO_3$ decomposition into MgO and CO_2 at lower temperatures.

2.3.2 Second Study: Monitoring the Thermal Decomposition of CaMg $(CO_3)_2$

In the second study, we monitored the thermal decomposition of CaMg $(CO_3)_2$. The experiment was performed using HSC Chemistry at temperatures between 100°C and 1000°C. The obtained results are displayed in Figure 2.5.

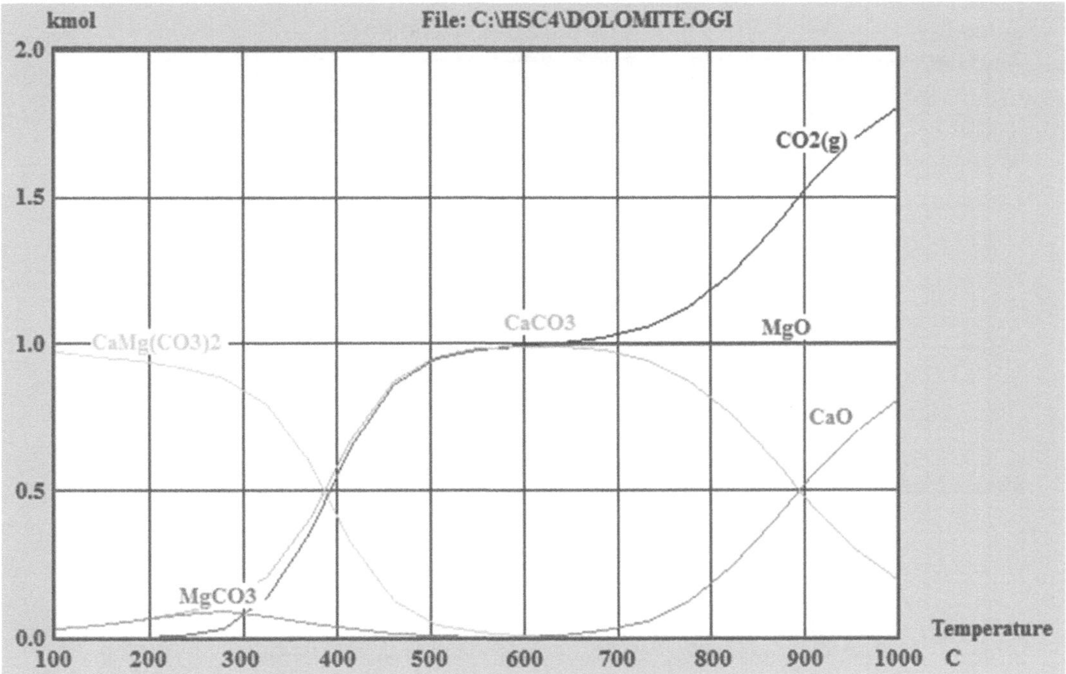

FIGURE 2.5 $CaMg(CO_3)_2$ decomposition as a function of temperature.

The revised graph analyzes the thermal decomposition properties of dolomite CaMg $(CO_3)_2$ and the equilibrium quantities resulting from decomposition by-products over a temperature range from 100°C to 1000°C. Primarily, at approximately 400°C, dolomite $CaMg(CO_3)_2$ decomposes into calcium carbonate ($CaCO_3$) and magnesium oxide (MgO), releasing carbon dioxide (CO_2). Nonetheless, beyond 700°C, the decomposition of $CaCO_3$ into calcium oxide could become significant, resulting in a decline in $CaCO_3$'s equilibrium amount and an overall increase in MgO. These observations provide valuable insights into the method by which dolomite decomposes based on temperature, aiding in the prediction of the most favorable conditions for obtaining the desired products CaO and MgO. Besides, these findings emphasized the importance of figuring out the temperature-dependent behavior of different types of calcium carbonate in industrial processes.

Conclusion of the second study: It appears from the interpretation of the above curve that dolomite $CaMg(CO_3)_2$ decomposes primarily at around 400°C into calcium carbonate ($CaCO_3$) and magnesium oxide (MgO) while emitting carbon dioxide (CO_2).

2.3.3 Third Study: Monitoring the Thermal Decomposition of $CaCO_3$ in the Presence of SiO_2

In the third study, we investigated the thermal decomposition of $CaCO_3$ in the presence of SiO_2. The experiment was conducted using HSC Chemistry and the temperature varied from 100°C to 1500°C. The results are shown in Figure 2.6.

The graph depicted provides a crucial perspective on tracking the thermal decomposition of calcium carbonate ($CaCO_3$) in the presence of silicon dioxide (SiO_2). A significant decrease in $CaCO_3$ observed indicates its decomposition, while the equilibrium amount of CaO and CO_2 exhibit an increase as temperatures rise. Besides, silicon dioxide appears to maintain an almost stable equilibrium within the temperature range shown. Nevertheless, SiO_2's presence is anticipated to alter an

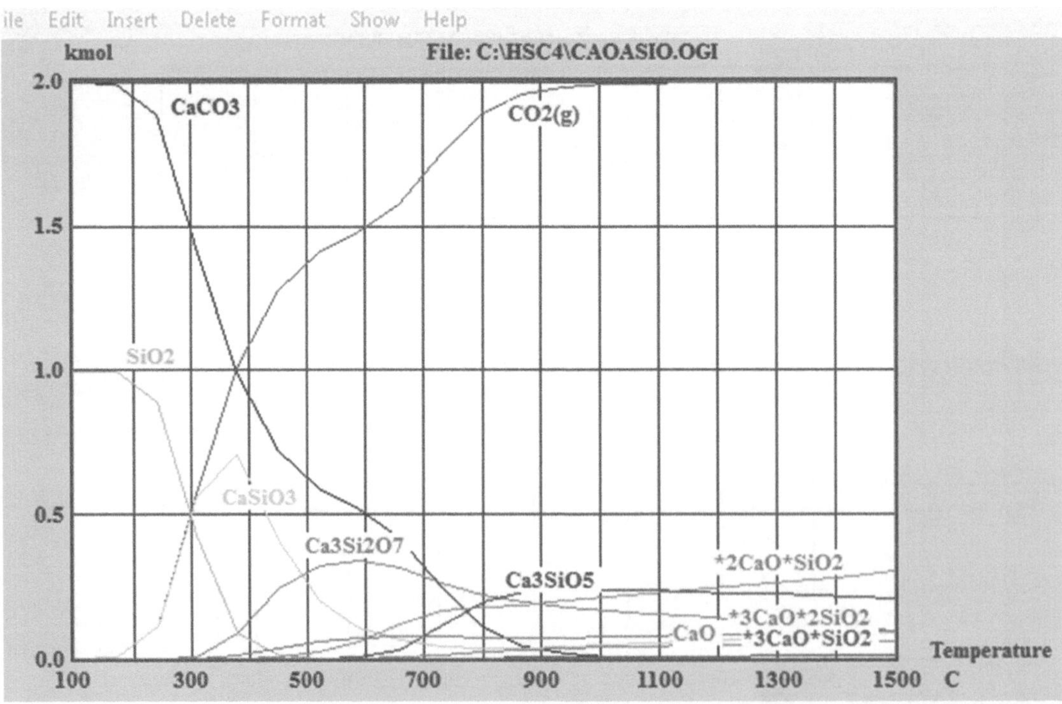

FIGURE 2.6 $CaCO_3$ decomposition in the presence of SiO_2 as a function of temperature.

impact at the highest temperature, which could contribute to the formation of calcium carbonate, including $CaSiO_3$ (calcium orthosilicate), Ca_3SiO_7 (dicalcium silicate), and Ca_3SiO_5 (tricalcium silicate). This suggests a crucial decrease in $CaCO_3$'s equilibrium amount compared to a system without SiO_2, as well as a substantial rise in the equilibrium amount of calcium silicate. Understanding the dynamic of the decomposition reaction of $CaCO_3$ in the presence of SiO_2 leads to optimizing industrial processes, reducing environmental impact, and predicting product formation.

Conclusion of the third study: It appears from the interpretation of the above curve that the presence of SiO_2 at the high-temperature decomposition of $CaCO_3$ promotes the formation of calcium silicates ($CaSiO_3$, Ca_3SiO_7, Ca_3SiO_5), consequently minimizing the equilibrium amount of $CaCO_3$.

2.3.4 Fourth Study: Monitoring the Thermal Decomposition of $CaCO_3$ in the Presence of Al_2O_3 and Fe_2O_3

In the fourth study, we tracked the thermal decomposition of $CaCO_3$ in the presence of Al_2O_3 and Fe_2O_3. The experiment was done by using HSC Chemistry at varied temperatures from 100°C to 1500°C. The acquired results are illustrated in Figure 2.7.

The graph presented offers an analysis of the thermal decomposition of calcium carbonate ($CaCO_3$) in the presence of aluminum oxide (Al_2O_3) and iron (III) oxide (Fe_2O_3). As the temperature rises, the graph exhibits a declining equilibrium amount of $CaCO_3$ implying its decomposition into carbon dioxide and calcium carbonate. Thus, the equilibrium amount of Al_2O_3 and Fe_2O_3 remains constant and then decreases beyond 500°C, which implies that their presence seems to promote the decomposition of $CaCO_3$, possibly through the formation of new calcium-aluminium and calcium-iron oxide. Those new compounds including $CaO*Al_2O_3$, $CaO*_2Al_2O_3$, $3CaO*Al_2O_3$, and $2CaO*Fe_2O_3$ showed an increasing equilibrium amount within elevated temperatures, probably involving the decomposed $CaCO_3$ and Al_2O_3 or Fe_2O_3. These findings' data reveal a potential influence of Al_2O_3 and Fe_2O_3, reducing the activation energy needed for $CaCO_3$ decomposition.

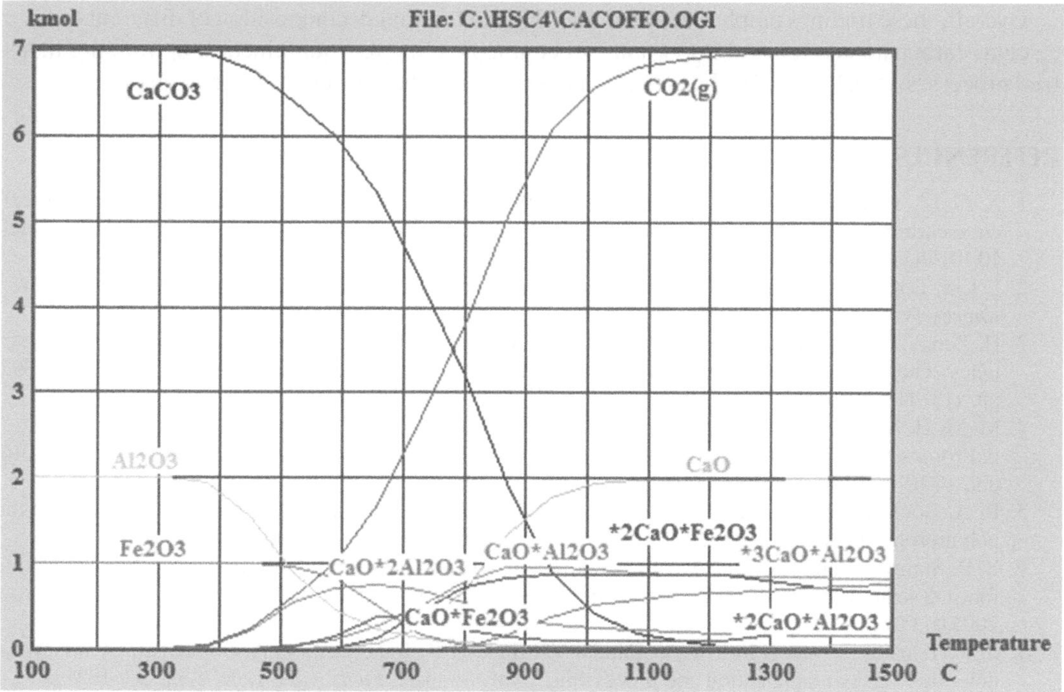

FIGURE 2.7 $CaCO_3$ decomposition in the presence of Al_2O_3 and Fe_2O_3 as a function of temperature.

Analyzing the thermal decomposition of $CaCO_3$ with Al_2O_3 and Fe_2O_3 gives important insights into material behavior, reaction mechanisms, and process optimization in a variety of scientific and engineering fields.

Conclusion of the fourth study: It appears from the interpretation of the above curve that the presence of aluminum oxide (Al_2O_3) and iron (III) oxide (Fe_2O_3) enhances the thermal decomposition of calcium carbonate ($CaCO_3$), potentially facilitated by the formation of novel calcium-aluminum and calcium-iron oxide compounds at the highest temperature.

2.4 CONCLUSION

The utilization of HSC Chemistry software for thermodynamic calculations enabled the monitoring of the thermal decomposition processes of various forms of calcium carbonate. The primary results are as follows:

- At temperatures above 900°C, calcium carbonate ($CaCO_3$) is thermodynamically decomposed into calcium oxide (CaO) and carbon dioxide (CO_2). In comparison magnesium carbonate ($MgCO_3$), decays at lower temperatures to yield magnesium oxide (MgO) and carbon dioxide (CO_2).
- At a temperature that occurs around 400°C, dolomite $CaMg(CO_3)_2$ decomposes to generate calcium carbonate ($CaCO_3$) and magnesium oxide (MgO) while releasing carbon dioxide (CO_2).
- The presence of silicon oxide SiO_2 in the decomposition of $CaCO_3$ at elevated temperatures enhances the production of calcium silicate, such as $CaSiO_3$, Ca_3SiO_7, and Ca_3SiO_5, running mitigates of $CaCO_3$'s equilibrium amount.
- The decomposition of $CaCO_3$ at higher temperatures is improved with the existence of aluminum oxide (Al_2O_3) and iron (III) oxide (Fe_2O_3) by generating the new components of calcium-aluminum and calcium-iron oxide.

Overall, these findings emphasized that tracking the thermal decomposition of different types of calcium carbonate is of crucial importance in comprehending reaction kinetics, optimizing industrial processes, developing novel materials, and evaluating the impact on the environment.

REFERENCES

1. X. G. Li, Y. Lv, B. G. Ma, W. Q. Wang, and S. W. Jian, "Decomposition kinetic characteristics of calcium carbonate containing organic acids by TGA," *Arab. J. Chem.*, vol. 10, pp. S2534–S2538, 2017. doi: 10.1016/j.arabjc.2013.09.026.
2. S. Lin, T. Kiga, Y. Wang, and K. Nakayama, "Energy analysis of CaCO3 calcination with CO2 capture," *Energy Procedia*, vol. 4, pp. 356–361, 2011. doi: 10.1016/j.egypro.2011.01.062.
3. H. Zeng, Z. Yan, M. Jiao, D. Xu, and J. Jiang, "A novel method for preparing calcium carbonate particles: Thermal decomposition from calcium hydrogen carbonate solution," *Key Eng. Mater.*, vol. 697, pp. 113–118, 2016.
4. M. M. H. Al Omari, I. S. Rashid, N. A. Qinna, A. M. Jaber, and A. A. Badwan, "Calcium Carbonate," in Profiles of Drug Substances, Excipients and Related Methodology, 1st ed., vol. 41. Elsevier Inc., 2016. doi: 10.1016/bs.podrm.2015.11.003.
5. R. A. Boulos, F. Zhang, E. S. Tjandra, A. D. Martin, D. Spagnoli, and C. L. Raston, "Spinning up the polymorphs of calcium carbonate," *Sci. Rep.*, vol. 4, pp. 1–6, 2014. doi: 10.1038/srep03616.
6. B. R. Stanmore, and P. Gilot, "Review-calcination and carbonation of limestone during thermal cycling for CO2 sequestration," *Fuel Process. Technol.*, vol. 86, no. 16, pp. 1707–1743, 2005. doi: 10.1016/j.fuproc.2005.01.023.
7. S. K. Tripathy, Y. Rama Murthy, L. Panda, V. Singh, P. K. Banerjee, and A. Bhatnagar, "A review on dolomite: Genesis, application and processing," *J. Mines, Met. Fuels*, vol. 61, no. 4, pp. 93–102, 2013.
8. J. W. Morse, R. S. Arvidson, and A. Lüttge, "Calcium carbonate formation and dissolution," *Chem. Rev.*, vol. 107, no. 2, pp. 342–381, 2007. doi: 10.1021/cr050358j.

9. C. B. A. G. S. J. Khinast, and G. F. Krammer, "Decomposition of limestone: The influence of CO2 and particle size on the reaction rate," *Science (80-.)*, vol. 51, no. 4, pp. 623–634, 1996.
10. I. Galan, F. P. Glasser, and C. Andrade, "Calcium carbonate decomposition," *J. Therm. Anal. Calorim.*, vol. 111, no. 2, pp. 1197–1202, 2013. doi: 10.1007/s10973-012-2290-x.
11. X. K. Tian, S. C. Lin, J. Yan, and C. Y. Zhao, "Sintering mechanism of calcium oxide/calcium carbonate during thermochemical heat storage process," *Chem. Eng. J.*, vol. 428, p. 131229, 2022. doi: 10.1016/j.cej.2021.131229.
12. J. Akwilapo, L.D. Kyobe, J. M. P., and Buchweishaija, "The decomposition kinetics of solids, part 2: Thermodynamic evidence for the enthalpy driven cooling of reaction environment," *Tanzania J. Eng. Technol.*, vol. 1, no. 2, 2006.
13. D. W. Johnson, "Effect of thermal transport mechanisms on the thermal decomposition of CaCO3," *Thermochim. Acta*, vol. 18, no. 1, pp. 15–19, 1977, doi: 10.1016/0040-6031(77)80031-2.
14. J. M. Criado, M. González, J. Málek, and A. Ortega, "The effect of the CO2 pressure on the thermal decomposition kinetics of calcium carbonate," *Thermochim. Acta*, vol. 254, pp. 121–127, 1995. doi: 10.1016/0040-6031(94)01998-V.
15. H. Wei, and Y. Luo, "A study on the kinetics of thermal decomposition of CaCO3," *J. Therm. Anal.*, vol. 45, no. 1–2, pp. 303–310, 1995. doi: 10.1007/BF02548694.
16. F. C. Meunier, J. Scalbert, and F. Thibault-Starzyk, "Unraveling the mechanism of chemical reactions through thermodynamic analyses: A short review," *Appl. Catal. A Gen.*, vol. 504, pp. 220–227, 2015. doi: 10.1016/j.apcata.2014.12.028.
17. J. S. Laskowski, "Thermodynamic and kinetic flotation criteria," *Miner. Process. Extr. Metall. Rev.*, vol. 5, no. 1–4, pp. 25–41, 1989. doi: 10.1080/08827508908952643.
18. Q. Liu, C. Proust, F. Gomez, D. Luart, and C. Len, "The prediction multi-phase, multi reactant equilibria by minimizing the Gibbs energy of the system: Review of available techniques and proposal of a new method based on a Monte Carlo technique," *Chem. Eng. Sci.*, vol. 216, p. 115433, 2020. doi: 10.1016/j.ces.2019.115433.
19. W. Bains, J. J. Petkowski, Z. Zhan, and S. Seager, "A data resource for prediction of gas-phase thermodynamic properties of small molecules," *Data*, vol. 7, no. 3, 2022. doi: 10.3390/data7030033.
20. G. Wolf, and C. Günther, "Thermophysical investigations of the polymorphous phases of calcium carbonate," *J. Therm. Anal. Calorim.*, vol. 65, no. 3, pp. 687–698, 2001. doi: 10.1023/A:1011991124181.
21. L. Fedunik-Hofman, A. Bayon, and S. W. Donne, "Comparative kinetic analysis of CaCO3/CaO reaction system for energy storage and carbon capture," *Appl. Sci.*, vol. 9, no. 21, pp. 5–9, 2019. doi: 10.3390/app9214601.
22. S. Jemjami, and M. Taoufik, "The Thermochemical Calculations to Follow up Water Treatment by the Calcium Carbonate (Calcium Hydroxide and Calcium Oxide)," *ACM Int. Conf. Proceeding Ser.*, 2020, doi: 10.1145/3399205.3399233.
23. M. C. Galetz, B. Rammer, and M. Schütze, "Refractory metals and nickel in high temperature chlorine-containing environments - thermodynamic prediction of volatile corrosion products and surface reaction mechanisms: A review," *Mater. Corros.*, vol. 66, no. 11, pp. 1206–1214, 2015. doi: 10.1002/maco.201408130.
24. C. W. Bale, and G. Eriksson, "Metallurgical thermochemical databases-a review," *Can. Metall. Q*, vol. 29, no. 2, pp. 105–132, 1990. doi: 10.1179/cmq.1990.29.2.105.
25. A. Roine, *Hsc - Software Ver. 3.0 for Thermodynamic Calculations*. The Canadian Institute of Mining and Metallurgy, 1989. doi: 10.1016/b978-0-08-036087-4.50007-0.
26. "HSC Chemistry." [Online]. Available: https://www.hsc-chemistry.com/hscchemistry. Consulted June 2024.

3 Comparative Study of ML Models for Predicting Solar Radiation Components in Representative Moroccan Climates

M. Khala, N. El Yanboiy, I. Elabbassi, O. Eloutassi, M. Halimi, Y. El Hassouani, and C. Messaoudi

3.1 INTRODUCTION

The demand for renewable power generation is on the rise, prompting many countries to commit to reducing carbon emissions in the coming years [1]. To reduce environmental pollution and address growing concerns about climate change, there is an extreme need for a global transition to a green energy. This transition focuses on reducing dependence on fossil fuels and cutting greenhouse gas emissions. Consequently, technologies that produce electricity while preserving ecological values must be prioritized and developed [2]. Solar energy emerges as a crucial solution due to its abundant and nature-friendly characteristics [3]. Morocco is a pioneer in the adoption of solar energy in Africa, with a sustainable policy to generate 20% solar electricity by 2030 [4]. This progress is supported by its advantageous location, presenting substantial potential for photovoltaic (PV) solar generation, detailed in Table 3.1 with long-term annual averages of photovoltaic and climate data for five representative cities highlighting the country's varied climates. GTI represents the global tilted irradiation, TEMP is the air temperature, PVOUT is the photovoltaic electricity production, and PR is the performance ratio [5].

Given the intermittent nature of climate patterns, accurate solar energy prediction becomes imperative. Solar radiation is composed of essential components, including the direct normal irradiance (DNI), diffuse horizontal irradiation (DHI), and GHI, while the GHI is the summation of DNI and DHI [6]. The significance of solar radiation forecasting is crucial in the realm of solar energy prediction. This process encompasses diverse steps, integrating weather conditions and energy consumption considerations. Various data-driven approaches, including neural networks, regression models, and other machine learning (ML) models, have played a key role in this field [7, 8]. Along the same line of research, numerous environmental parameters, including humidity, temperature, wind speed, wind direction, pressure, and cloudiness, affect SR forecasts to varying degrees. However, these parameters are subject to continuous variations, presenting challenging difficulties that impact the accuracy of multi-scale SR and PV energy output forecasting [6].

In this context, various methods have been employed for SR prediction, including convolutional methods, ML and deep learning (DL) methods, hybrid models, and numerical weather prediction (NWP). Many studies have been conducted to predict SR prediction with different scales using classical statistical models, such as Gaussian process regression (GPR) [9], and exponential smoothing [10]. Other studies focus on auto-regressive (AR), auto-regressive moving average (ARMA), and auto-regressive integrated moving average (ARIMA) models [11–13]. ML models are frequently used. To predict daily global SR in Fez, Morocco, ML models with RF feature selection and

 DOI: 10.1201/9781003647263-3

TABLE 3.1
Long-Term Annual Averages of Photovoltaic and Climate Data for the Locations

Location	GHI (kW/m^2)	DNI (kW/m^2)	GTI (kW/m^2)	TEMP(°C)	PVOUT (kWh/kWp)	PR (%)
ERRACHIDIA	2112	2287	2443	20.4	1906	78
TANGIER	1838	1921	2083	18.3	1654	79.4
CASABLANCA	1854	1775	2075	18.9	1653	79.7
TIZNIT	2057	2093	2316	18.8	1840	79.5
DAKHLA	2159	1973	2320	19.6	1858	80.1

Bayesian optimization (BO) techniques were used [14]. Other researchers have used support vector machine (SVM) and extreme gradient boosting (XGB) [15]. Voyant et al. reviewed several ML techniques for SR forecasting, including genetic algorithms (GAs), ANNs, support vector regressor (SVR), hidden Markov model (HMM), linear regression (LR), Naive bayes (NB), k-nearest neighbors (k-NN) algorithm, and decision trees (DTs).

Recently, DL models have been applied to address the complex nonlinear relationships in SR data and meteorological parameters, especially with the availability of large databases. Various models, including long short-term memory (LSTM) and bidirectional LSTM, have been explored [8, 16–18]. DL, as a subset of artificial intelligence (AI), proves to be efficient for estimating nonlinear functions compared to traditional ML methods. However, this type of learning requires more computing time and more complexity.

Hybrid networks, as an alternative learning strategy for neural networks, are particularly effective in combining the functionalities of two or more ML/DL methods [14, 19]. For instance, an approach employing LSTM and genetic programming (GP) is introduced for short-term global SR prediction by utilizing weather data. Ultimately, NWP employs mathematical models to simulate physical processes in atmospheric transformation and predict weather based on current conditions. While NWP models have succeeded in meteorology, their complexity poses challenges when applied to SR for everyday users due to challenges arising from their complexity. Issues include uncertain initial conditions, unresolved atmospheric phenomena, intricate local orography, and the propagation of numerical errors and approximations. The main contribution of this research is systematically comparing the performance of three multi-output ML algorithms—LR, XGB, and RF—in predicting GHI and DNI. This work also provides valuable insights into the selection of appropriate models and features for SR components forecasting. Additionally, the research reveals the significance of meteorological variables through feature importance rankings and hyperparameter tuning through RandomizedSearchCV optimization, highlighting the critical role of predictors such as DNI, in enhancing the accuracy of SR predictions. The practical implications of this research are far-reaching, offering significant contributions to the field of solar radiation forecasting. The study's insights carry direct applicability to sustainable energy planning, enabling more informed decisions in harnessing solar power.

The rest of the chapter is organized as follows: Section 3.2 introduces the materials and methods, presenting the overall framework of the study. Section 3.3 presents and discusses the results and their interpretation. Finally, the chapter is concluded in Section 3.4.

3.2 MATERIALS AND METHODS

3.2.1 Data Analysis

The data set used for forecasting the two components of SR, GHI, and DNI includes weather feature values obtained from the National Solar Radiation Data Base (NSRDB), a public database. The data set spans three years, from 2017 to 2019, with a temporal resolution of one hour.

TABLE 3.2
Geographical Coordinates of the Five Studied Locations

Location	Errachidia	Tangier	Casablanca	Tiznit	Dakhla
Latitude	31.877558°	35.770000°	33.610000°	29.690000°	23.690000°
Longitude	–4.449463°	–5.820000°	–7.620000°	–9.740000°	–15.940000°

It comprises seven parameters: GHI, DNI, temperature, cloud type (CT), relative humidity (RH), DHI, pressure (P), wind direction (WD), and wind speed (WS). Moreover, the test study areas are five locations distributed over Morocco. The choice of these five locations is due to the fact that Morocco is categorized into 6 climate zones based on a 10-year analysis of climate data from 37 weather stations, as reported by the Moroccan Energy Efficiency Agency (AMEE). The locations reflect the variety of the Moroccan climate, from arid deserts to coastal Mediterranean influences. Table 3.2 shows the geographical coordinates of the five studied locations. The resulting data set comprising 26,280 data records for each location has been split into training and testing sets. The training set is 70%; the test set is 30%. The training and testing subsets are then concatenated. The code was written in Python 3.8 language and executed on a computer with a 2.53 GHz processor and 8 GB RAM.

3.2.2 Proposed Method

In essence, an SR forecast model considers predictor meteorological parameters to offer an informed estimation of the key components of solar irradiance. However, the selection of predictors plays a crucial role in improving the accuracy of ML prediction models, as demonstrated in several research studies and areas such as Errachidia, Morocco. In this study, based on the correlation heatmap examined in Figure 3.1, five features were selected to forecast GHI and DNI components. For forecasting purposes, LR, RF, and XGB models are proposed.

LR is categorized into three approaches, including univariate simple, multiple, and multivariate linear regression, where the multivariate linear regression analysis aims to decipher potential linear relationships among several response and predictor variables in multivariate linear regression.

The RF algorithm exemplifies ensemble learning by constructing multiple DTs through distinct data subsets, aggregating their judgment results for a final output. Introducing a new method involving random sampling of samples and features enhances model accuracy and stability compared to traditional landslide evaluation methods, surpassing the performance of individual DTs. Numerous studies indicate that RF exhibits high fault tolerance, effectively handling algorithmic challenges, outliers, and noise.

The XGB algorithm is a sophisticated supervised ML approach rooted in the DT method. Differing from traditional gradient boosting, XGB incorporates regularization in the objective function's loss function, promoting smoother learning weights and mitigating the risk of overfitting. This DT-based algorithm employs a tree ensemble model for predictive output.

In the realm of ML, improving modeling accuracy involves two crucial tasks. The first task focuses on fine-tuning the selected approaches, within this study, the RandomizedSearchCV method which explores the optimal hyperparameter combination by sampling values from a predefined distribution, is employed for this purpose. The second task involves the evaluation and validation of models using commonly used metrics for time series forecasting, such as RMSE and R^2 [15], which are applied in this case study. Increased R^2 values, close to 1, suggest a better performance of the predictive model, which fits the regression line well. Conversely, lower values of RMSE suggest better model performance (Figures 3.1 and 3.2).

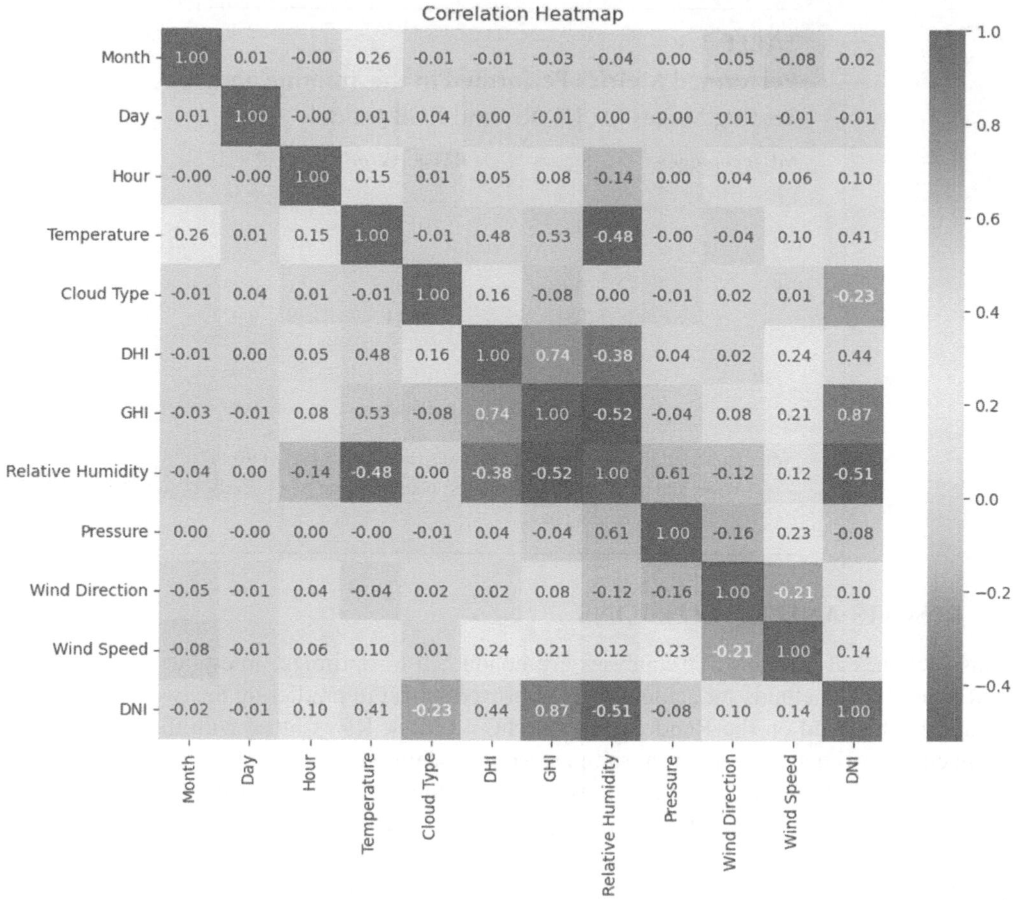

FIGURE 3.1 Correlation heatmap between variables.

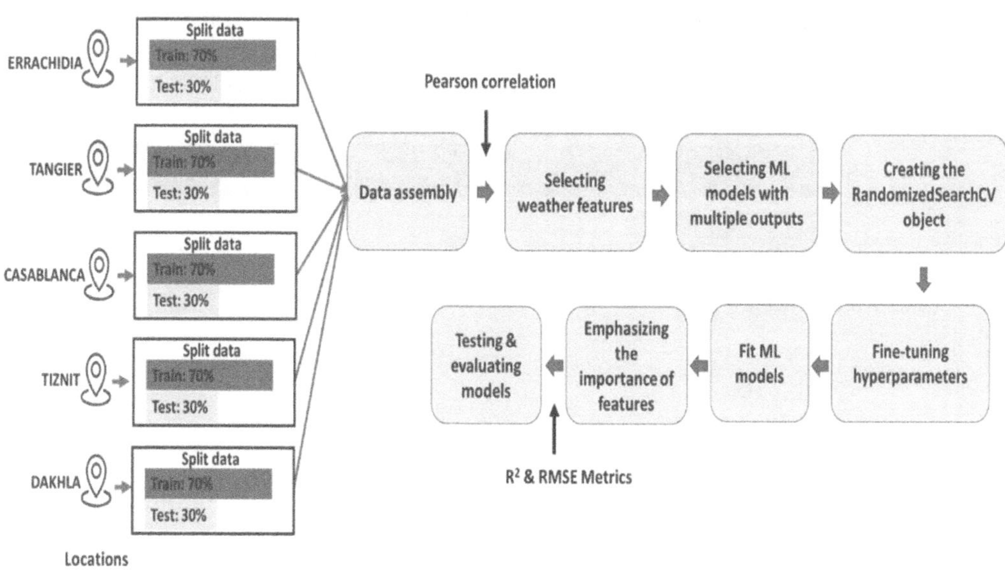

FIGURE 3.2 Overall process of SR forecasting and optimization.

TABLE 3.3
Performed Metrics Performed in the Training and Testing Stages for LR, RF, and XGB Models

ML Techniques	Stages	RMSE (W/m^2)	R^2 (%)
LR	Train	209.18	0.56
	Test	210.10	0.56
RF	Train	25.39	0.99
	Test	**90.11**	**0.92**
XGB	Train	101.45	0.90
	Test	103.85	0.89

Note: Bold demonstrates superior accuracy with an RMSE of 90.11 W/m^2 and an R^2 of 92% in the test phase, and an R^2 of nearly 99% in the training phase.

3.3 RESULTS AND DISCUSSIONS

In this section, calculations and analyses are conducted to optimize, fine-tune, and evaluate the LR, RF, and XGB methods for achieving an improved model in predicting the two SR components, GHI and DNI. Based on the RandomizedSearchCV object, RF is fitted with the best parameters fine-tuned: n_estimators=300, min_samples_split=2, min_samples_leaf=1, max_features='sqrt', and max_depth=20. XGB is adjusted using the following best hyper-parameters: colsample_bytree=0.656, learning_rate=0.057, max_depth=5, n_estimators=137, and subsample=0.834.

The test results obtained for the three algorithms are presented in Table 3.3, with a comparison based on the R^2 and RMSE values calculated in both the test and learning phases. Table 3.3 reveals that the RF model outperforms LR and XGB, demonstrating superior accuracy with an RMSE of 90.11 W/m^2 and an R^2 of 92% in the test phase, and an R^2 of nearly 99% in the training phase. However, LR exhibited the lowest predictive capability, achieving R^2 = 56% and RMSE = 210.10 W/m^2 in the training section, and maintaining similar performance in the test section. On the other hand, XGB yielded an R^2 of 85% and an RMSE of 35 W/m^2 in the test part, slightly lower than the values obtained in the training part. Both XGB and RF are ensemble learning techniques that provide feature importance as

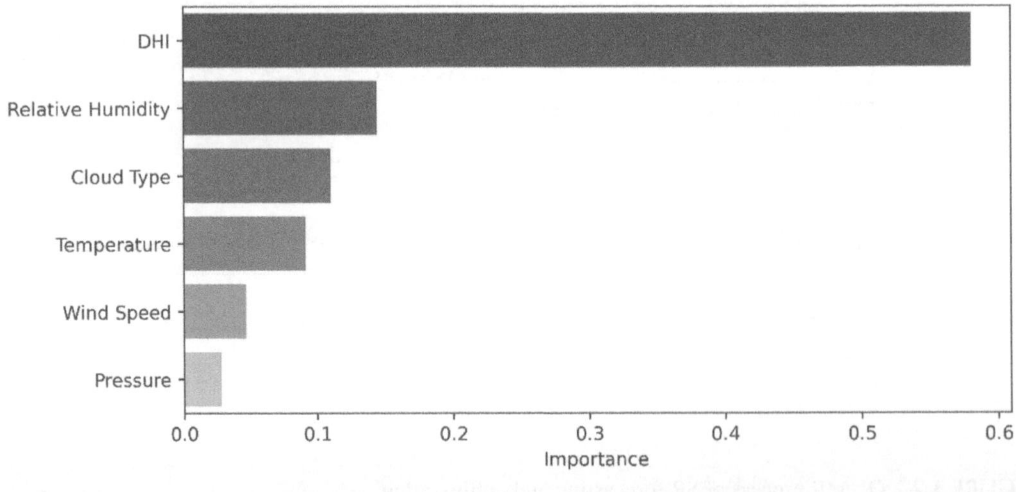

FIGURE 3.3 Importance features for RF.

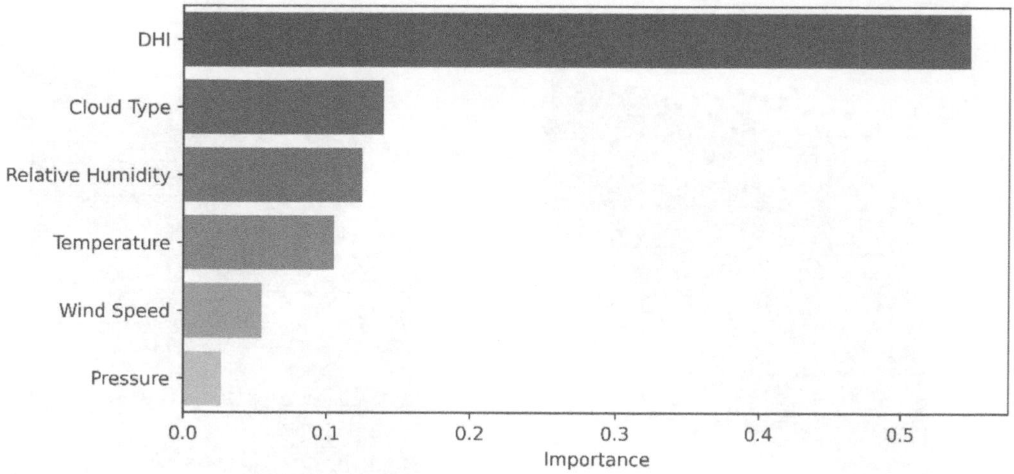

FIGURE 3.4 Importance features for XGB.

a valuable output. These methods are widely used in ML for predictive modeling, and understanding feature importance helps in interpreting model results and making informed decisions.

Feature importance scores are relative within each model and provide insights into the internal workings of the algorithm. Figures 3.3 and 3.4 display the feature importance for XGB and RF, respectively. As the feature importance scores are relative within each model, they provide an insight into the inner workings of the algorithms. Both models similarly prioritized the importance of characteristics, with one notable difference: RF assigned higher priority to RH than CT, while XGB reversed their order. In both models, DHI was consistently ranked as the most crucial feature, while P held the position of the least important feature.

Typically, analysis of predicted DNI and GHI involves visualizing the relationship between the actual values and the values predicted by the LR, XBG, and RF models. This is done using a scatter plot in Figures 3.5–3.7. In the context of solar radiation forecasting, such scatter plots help assess

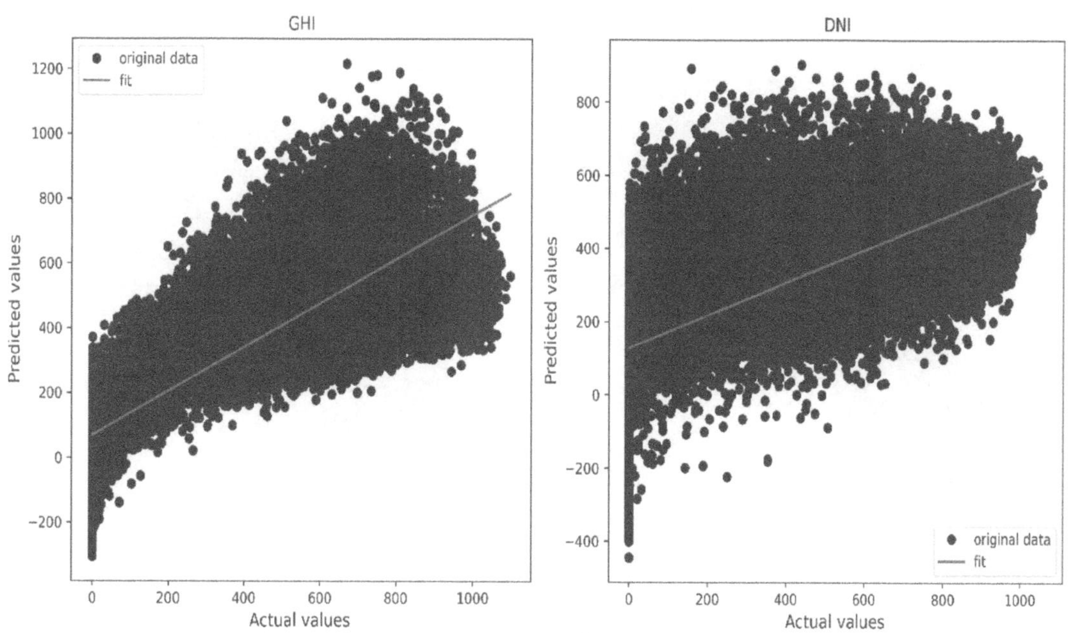

FIGURE 3.5 Scatter plot and fitted line between predicted and actual values of SR components for LR.

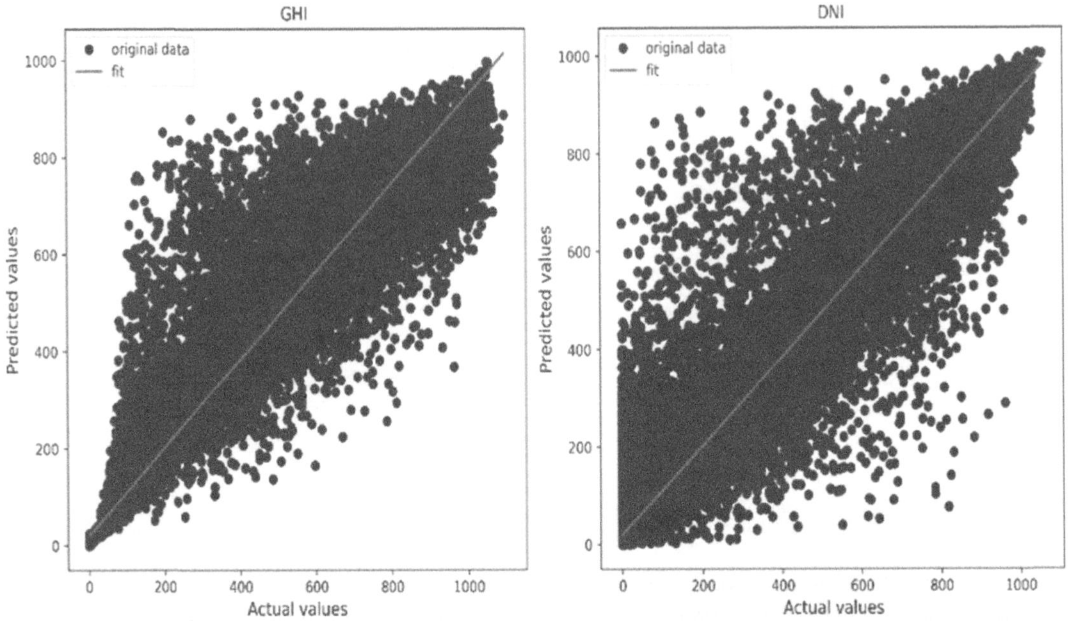

FIGURE 3.6 Scatter plot and fitted line between predicted and actual values of SR components for XGB.

how well the model predictions align with the actual observed values. A well-fitted model would result in points clustering along a diagonal line, indicating a strong correlation between predictions and actual measurements. As a result, predicted and actual DNI and GHI values are well fitted by the RF model, as illustrated in Figure 3.7.

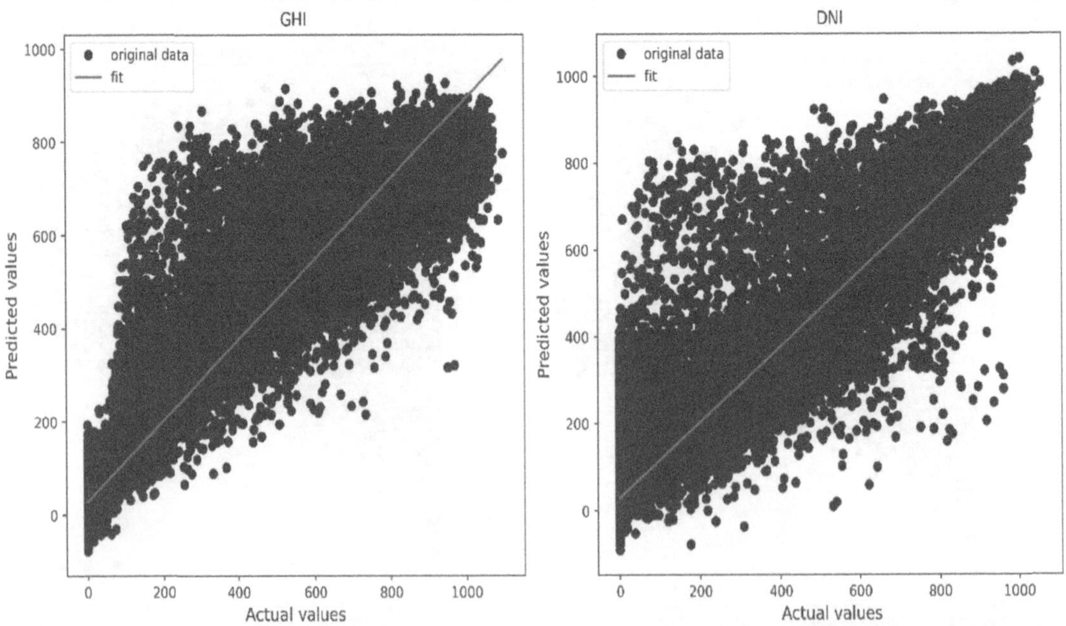

FIGURE 3.7 Scatter plot and fitted line between predicted and actual values of SR components for RF.

3.4 CONCLUSION

This study presents a comprehensive analysis of ML models for predicting GHI and DNI solar features in the context of solar energy improvement. Leveraging a data set spanning three years and featuring correlated meteorological parameters, including SR and climate data, the study explores the capabilities of RF, LR, and XGB algorithms. Feature importance rankings underline the significance of meteorological variables, emphasizing the critical role of predictors like DHI. Scatter plots visually depict the alignment between predicted and actual values, providing insights into model performance and potential areas for improvement. The proposed methodology, incorporating RandomizedSearchCV for hyper-parameter tuning, ensures robust model optimization. As a consequence, the findings demonstrate the superior accuracy of the RF model over LR and XGB, in both training and test phases, for predicting solar radiation components. RF exhibits a precision of R^2 of 92% and RMSE of 90.11 W/m^2 in the testing phase, providing valuable insights for selecting appropriate models in solar energy forecasting. This work contributes to advancing solar radiation forecasting methodologies, vital for optimizing renewable energy systems.

REFERENCES

1. Tian, J., Yu, L., Xue, R., Zhuang, S. and Shan, Y. Global low-carbon energy transition in the post-COVID-19 era. Applied Energy 307, 118205 (2022).
2. Pegels, A. and Altenburg, T. Latecomer development in a "greening" world: Introduction to the Special Issue. World Development 135, 105084 (2020).
3. Agbo, E. P., Edet, C. O., Magu, T. O., Njok, A. O., Ekpo, C. M. and Louis, H. Solar energy: A panacea for the electricity generation crisis in Nigeria. Heliyon 7(5), e07016 (2021).
4. IRES. https://www.ires.ma/en/publications/general-reports. (last access: February 12, 2024).
5. MASEN. "https://solaratlas.masen.ma/map?c=27.059126:-9.074707:5&s=31.625815:-7.989137". (last access: February 12, 2024).
6. Tian, J., Ooka, R. and Lee, D. Multi-scale solar radiation and photovoltaic power forecasting with machine learning algorithms in urban environment: A state-of-the-art review. Journal of Cleaner Production 426, 139040 (2023).
7. Gupta, N. and Tomar, P. Multi-model approach applied to meteorological data for solar radiation forecasting using data-driven approaches. Optik 286, 170957 (2023).
8. Tian, P., Zhang, C., Zhou, J. and Nazir, M. S. An integrated framework of Bi-directional long- short term memory (BiLSTM) based on sine cosine algorithm for hourly solar radiation forecasting. Energy 221, 119887 (2021).
9. Rohani, A., Taki, M. and Abdollahpour, M. A novel soft computing model (Gaussian process regression with K-fold cross validation) for daily and monthly solar radiation forecasting (Part: I). Renewable Energy 115, 411–422 (2018).
10. Zhang, D., Yang, D., Reindl, T. and Walsh, W. M. Short-term solar irradiance forecasting using exponential smoothing state space model. Energy 55, 1104–1113 (2013).
11. Carpentieri, A., Folini, D., Nerini, D., Pulkkinen, S., Wild, M. and Meyer, A. Intraday probabilistic forecasts of surface solar radiation with cloud scale-dependent autoregressive advection. Applied Energy 351, 121775 (2023).
12. Belmahdi, B., Louzazni, M. and Bouardi, A. E. One month-ahead forecasting of mean daily global solar radiation using time series models. Optik 219, 165207 (2020).
13. Yang, D., Jirutitijaroen, P. and Walsh, W. M. Hourly solar irradiance time series forecasting using cloud cover index. Solar Energy 86, 3531–3543 (2012).
14. Chaibi, M., Benghoulam, E. M., Tarik, L., Berrada, M. and Hmaidi, A. E. Machines learning models based on feature selection and Bayesian optimization for predicting daily global solar radiation. Social Science Research Network (2021).
15. Fan, J., Wang, X., Wu, L., Zhou, H., Zhang, F., Yu, X., Lu, X. and Xiang, Y. Comparison of Support Vector Machine and Extreme Gradient Boosting for predicting daily global solar radiation using temperature and precipitation in humid subtropical climates: A case study in China. Energy Conversion and Management 164, 102–111 (2018).
16. Khala, M., Abouzid, H., Teidj, S., Eloutassi, O. and Messaoudi, C. LSTM deep learning method for radiation short and long-term prediction. Lecture notes in networks and systems 6, 676–696 (2023).

17. Neshat, M., Nezhad, M. M., Mirjalili, S., Garcia, D. A., Dahlquist, E. and Gandomi, A. H. Short-term solar radiation forecasting using hybrid deep residual learning and gated LSTM recurrent network with differential covariance matrix adaptation evolution strategy. Energy 278, 127701 (2023).
18. Gallo, R., Castangia, M., Macii, A., Macii, E., Patti, E. and Aliberti, A. Solar radiation forecasting with deep learning techniques integrating geostationary satellite images. Engineering Applications of Artificial Intelligence 116, 105493 (2022).
19. Huang, X., Li, Q., Tai, Y., Chen, Z., Zhang, J., Shi, J., Gao, B. and Liu, W. Hybrid deep neural model for hourly solar irradiance forecasting. Renewable Energy 171, 1041–1060 (2021).

4 Innovative Waste Management Strategies for Advancing Sustainable Development

A Comparative Analysis of El Jadida and Trieste

Y. Al Masmoudi, A. Tallou, N. Mejjad, S. Belkacemi, A. Jouamaa, S. Bel-Lahbib, and K. Ibno Namr

4.1 INTRODUCTION

Coastal ecosystems are dynamic and fragile environments due to their natural characteristics and can become easily vulnerable to different types of pollution released by different human activities. Several economic activities rely on marine resources. Therefore, protecting and preserving the marine environment is a priority. Since the 1970s, there has been an increasing trend toward protecting the environment by forbidding waste dumping in the ocean (the U.S. Environmental Protection Agency [EPA]). Indeed, many studies have been carried out since the 70s, mainly after the convention of London was signed (U.S. EPA, 2015). These studies aimed at assessing and monitoring the coastal environment quality while preventing the dumping of more waste in the ocean (e.g., Tanjung and Hamuna, 2019; Yusuf and Abiye, 2019; Tian et al., 2020; Vasiliu et al., 2020; Yoon et al., 2020; Mejjad, 2023). It has been reported that marine pollution causes heavy damage to the environment, and biodiversity, which may consequently impact the socio-economic structures worldwide (Mejjad and Rovere, 2021). In this sense, retaining under control the different land-based and sea-based activities and their associated waste is an important step toward protecting and conserving the marine environment and biodiversity. Therefore, waste management is a major issue that has preoccupied stakeholders worldwide for decades, where waste type and composition remain the main challenge as each type of waste requires a specific analysis and treatment, allowing efficient management. Indeed, the coastal environment receives a combination of biological, chemicals, radioactive, and trash from different sources, most of which derive from land-based sources or different activities at sea, which make the management processes challenging and complicated.

The present work then sought to assess, analyze and compare the management process of waste released in two coastal regions, El Jadida (Morocco) and Trieste (Italy). Those two cities play interesting socio-economic roles in both countries and host different marine ecosystems of ecological and economic interest. The main objectives of the present review are the following:

- Identification of the different types of waste in the two coastal regions.
- Examination of the importance of waste management in preserving the environment and sustaining the use of ocean resources.
- Exploration of the role of waste disposal in improving socio-economic conditions (e.g., creating job opportunities).
- Suggestions and recommendations of innovative and suitable solutions that can help the two coastal regions overcome their challenges and mutually benefit each other.

DOI: 10.1201/9781003647263-4

4.2 MATERIALS AND METHODS

4.2.1 The Study Area and Regional Setting

4.2.1.1 El Jadida City-Morocco

El Jadida is located in the region of Abda-Doukkala that enjoys a strategic position with a 350 km long seafront. The region has around 2 million inhabitants (RGPH, 2004), with more than half living in rural areas. The climate in the Doukkala Abda region is semi-arid with a mild temperate winter with a generally hot and dry summer and oceanic influences on the coast. The temperatures are moderate in areas near the ocean and more contrasted in the interior. The rainy season covers, on average, the period from October to May. The rains regularly fall in autumn and winter. Doukkala region is one of the most authentic regions in Morocco. It is one of the few lowlands in Morocco that has always been heavily populated. It can be argued that the Doukkala region has often experienced demographic pressure, given the nature of the natural resource potential it offers compared to the number of populations it hosts. As shown in Figure 4.1, the population of El Jadida is estimated at 786,716. Around 80% of economic activities are concentrated along the Atlantic coast (Casablanca-El Jadida and Safi). Mainly in the large cities of this area, there are around 80% of jobs and more than 75% of firms. There are three types of the most dominant industries. There are leather and textiles, metallurgical and electrochemical, and chemical industries. Furthermore, while tourism embodies 53% of the Moroccan Atlantic seaboard, maritime transport represents 92% of the Atlantic coast and commercial transport fishing activities.

4.2.1.2 Trieste-Italy

The province of Trieste is situated in Italy, in the Friuli-Venezia Giulia region. It has 236,000 inhabitants and Trieste is the capital. The province of Trieste is the smallest of the Italian provinces in terms of surface area. It is also the one with the fewest municipalities (Horvat et al., 1999). The

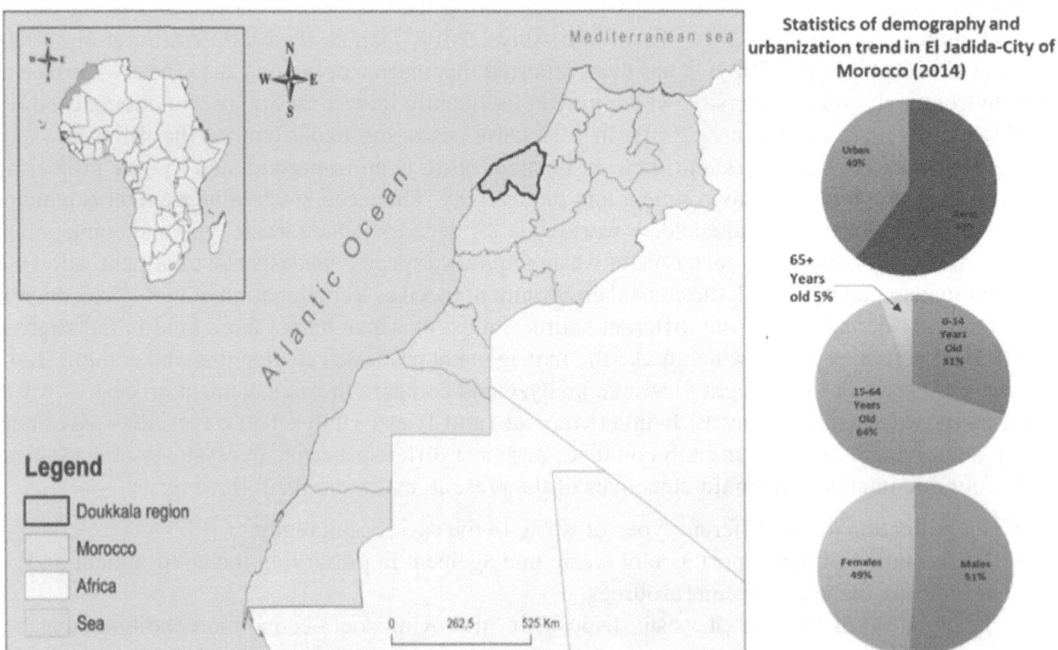

FIGURE 4.1 Geographic location, statistics of demography and urbanization trend in El Jadida-City of Morocco (2014).

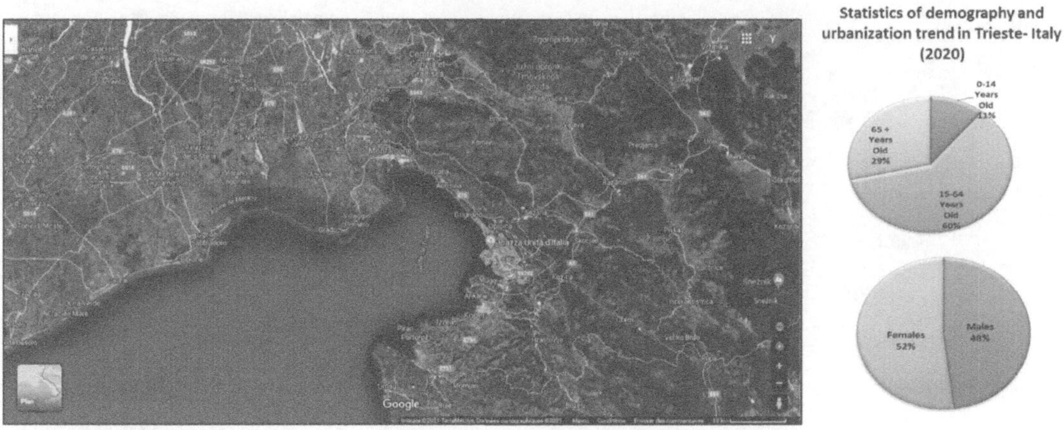

FIGURE 4.2 Geographic location and statistics of demography trend in Trieste, Italy (2020).

climate of Trieste can be described as transitional Mediterranean (Gatto et al., 2021). It has a relatively mild climate on the coast and outside the Po Valley. Still, it is exposed to the Bora, the channeled wind passing through the corridor among the mountains and rushing directly into the Gulf of Trieste with gusts greater than 100 kilometers per hour and possible frosts from December to early March. The average temperature of the coldest month (January) is 5.8°C, and the hottest months (July and August) is 28°C. Trieste is a fairly rainy city: the average annual rainfall is 1000 mm and thunderstorms can occur in late spring and summer. As presented in Figure 4.2, in 2013, the population of Trieste reached 204,849, of whom 53.3% were females, and 46.7% were males.

4.2.2 Data Collection

This review was based on investigating scientific papers, proceeding papers, chapters, thesis reports, and technical reports. In this order, over 100 documents were consulted to perform the present research work. The consulted databases are Google Scholar, Researchgate, Scopus, and Web of Sciences. Keywords used to search the internet engines were as follows: "waste management", "waste management processes", "waste management concept", "waste management policies", "waste management & sustainability", "the waste economy", "circular economy", "waste recycling", "waste in El Jadida", "waste in Trieste", "waste management in Morocco", and "waste management in Morocco". All search queries were performed in English and French.

4.2.3 Analysis Results

4.2.3.1 Waste Management: An Overview

Waste management is a combination of processes and actions allowing safe and correct waste disposal while considering the ecological and environmental components. It can also be defined as strategies that seek to minimize environmental impacts, conserve resources, and ensure sustainable waste management practices (Demirbas, 2011). Waste management includes handling waste from collection, transportation, recycling, or disposal, as indicated in Figure 4.3. It has been noticed then that the waste management (WM) processes impact the three components of sustainability (social, economic, and environmental), where the good planning and adoption of advanced strategies in the WM field remain indispensable to meet and ensure the sustainable growth of future needs.

From the social dimension, the statistics revealed that waste produced per capita increased from 2.68 pounds in 1960 to 4.48 pounds per day in 2015. Noting that people's attitudes toward waste disposal and management vary from place to place, we cannot overgeneralize because of today's

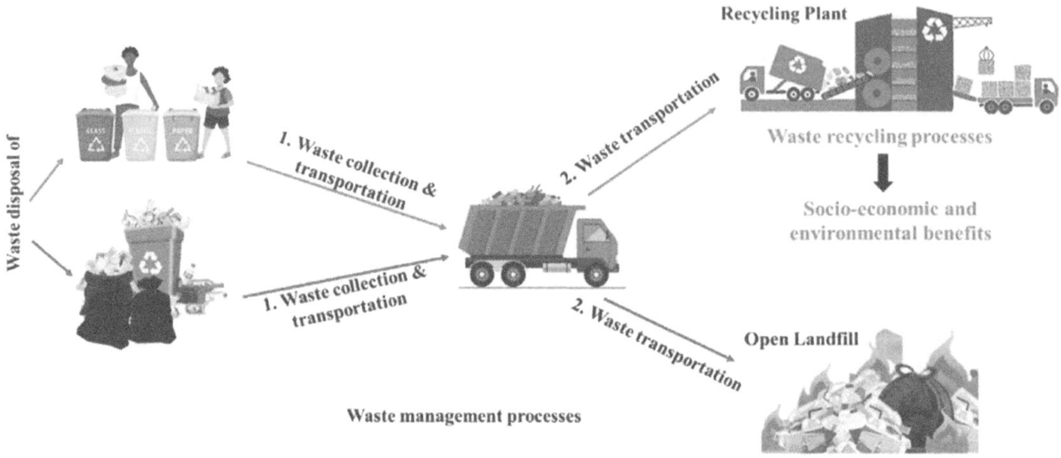

FIGURE 4.3 The main waste management steps.

globalized network of values. However, society is the main cause of environmental quality deterioration, which means efficient waste management relies on human behavior and attitudes. In this order, increasing awareness toward avoiding inadequate waste disposal would help solve the WM issues. Many developed countries are attempting to form a closed system by transforming consumers into users to achieve a non-generation of waste (Vlachos, 1975). In this context, in their article "Resource Recovery from Waste: Restoring the Balance between Resource Scarcity and Waste Overload", Anne P. M. Velenturf and Phil Purnell have coined the concept of ecosystem stewardship. It states that people are responsible for the depletion of resources and waste overload because the economy is part and parcel of society and depends on the environment (Velenturf and Purnell, 2017). That is to say, people's uncontrolled activities can disrupt the circular flows of the economy and environment. Still, it cannot be denied that stakeholders are the initiators and the driving force toward environmental awareness and behavioral change (Gayialis et al. 2019). They only have to design methods that fit into a particular context of a specific society to capture people's attention and increase their interest. This can be done by using rewards and reinforcements wisely (Asmawati et al., 2010).

At the economic level, according to the World Bank's report on waste management worldwide, while 93% of waste in developing countries is mismanaged, wealthy countries, which form only 16% of the global population, produce 34% of the world's waste and only 2% of mismanaged waste (Ellis, 2018). This reflects the costly dimension of waste management and shows the urge for a gap closure for the sake of global benefits. In 1994, uncollected wastes blocked the drains, causing flooding in India, and people lost their jobs and died because of the plague outbreak. The economic losses cost 2 billion USD. Hence, not addressing waste management is more expensive than applying the waste-to-resource strategy (Wilson et al., 2015). Thus, the term "circular economy" was invented to create business and job opportunities for society while preserving the environment. The sustainable circular economy approach emanates from the perception of wastes as unexploited resources, which aims to extract the maximum value from resources. It can reduce waste at source, help fulfill the Sustainable Development Goals (SDGs) and create healthy life cycles. Therefore, the world cannot fulfill economic growth through waste management without perceiving waste as a valuable resource. From the environmental dimension, studies differentiate between two kinds of air pollution: indoor (the uncontrolled burning of waste within poor households) and outdoor air pollution (Roser, 2013), which threaten oceans' biodiversity, human beings, and local economies. In 2019, the earth's surface temperature reached 0.95 Celsius. The increasing degrees of temperature are considered to be the leading cause of rising sea water levels, shrinking glaciers, floods, droughts, and other environmental catastrophes (Ian, 2020a, 2020b).

In 2018, the global carbon dioxide increased to 36.6 billion metric tons (Ian, 2019). These increasing numbers endanger the planet, mainly coastal and sea areas. Thus, today's world is in desperate need of effective waste management. Human activities and the tradition of a linear economy are the main reasons behind the current environmental conditions. However, the unprecedented year 2020, due to the COVID-19 pandemic, has known many socio-economic and environmental shifts. Quarantined spaces have witnessed a decrease in emissions of smoke, waste, and nitrogen dioxide air pollution because of reduced oil consumption, controlled human activities, and travel restrictions (Cherif et al., 2020; El Zowalaty et al., 2020; Sbai et al., 2021; Mejjad et al., 2021a). Still, these positive changes do not have a lasting effect, but they are an excellent lesson to learn from. In other words, in the waste management context, the COVID-19 pandemic can give new insights into managing waste better and shifting from reducing its quantities to achieving the zero-waste objective.

4.2.4 Main Goals and Challenges Related to Waste Management

4.2.4.1 Sustainable Use of Sea Resources and Reduction of Sea Pollution

Studies (Hansel and Diaz, 2021; Norrie et al., 2001) have revealed that marine plants contribute 70% of the oxygen we breathe, while oceans absorb around 30% of CO_2 emissions and receive trillions of plastic waste. Apart from the 70% of debris that sinks into the oceans' ecosystem, plastic waste is considered the most dangerous because its degradation takes 500 to 1000 years, and marine creatures mistake microplastics for food. This intoxicates marine life and can cause severe health issues to humans by consuming seafood and water supplies (MOPS, 2020). Also, the dumpsites on land alongside the sea can pollute the surface and groundwater (Mejjad et al., 2021b) and influence tourism (Wilson et al., 2015). Hence, sea pollution and the depletion of sea resources not only impact marine life alone, but they also go beyond the environmental dimension to cause devastating social and economic impacts. Thus, ensuring the sustainability of sea resources and reducing sea pollution is part and parcel of the SDGs. This is embodied in the 14th goal of sustainable development. It aims to prevent marine pollution by 2050, achieve sustainable management of coastal ecosystems, eliminate IUU fishing, enhance healthy marine life, and increase economic benefits (UNDP, 2016). In addition, wastewater management is vital in preserving sea resources and reducing sea pollution by decreasing all types of toxic particles, mainly marine microplastics and/or nanoplastics. However, studies have found that wastewater treatment plants (WWTPs) are ineffective in removing microplastic particles less than 100 μm. This makes the aim to sustain healthy marine life in low-income countries one of the biggest challenges because they cannot afford the costs of innovative solutions.

4.2.4.2 Reduction of the Waste Quantities from the Source

Reducing the number of sources of waste is one of the most effective waste management techniques and the first step towards trespassing the issue of waste mismanagement and all of its devastating effects on the environment, society, and economy. According to the Global Waste Management Outlook, waste reduction techniques can be applied at any stage of a product's life cycle, as observed in Figure 4.4. According to Wilson et al. (2015), these techniques can be stated as follows:

- Production stage: extending the product's life span by designing them to be durable, repairable, and reusable. Also, using less materials during the production processes.
- Transport/purchase stage: refilling and reducing packaging.
- User/Consumption stage: in this stage, the consumer is the stakeholder who can repair and reuse food waste to achieve a zero-food waste.
- End-of-use stage: the consumer/community has to enhance and maintain the culture of donation, exchangeability, reusing products, and buying from flea markets or online websites.

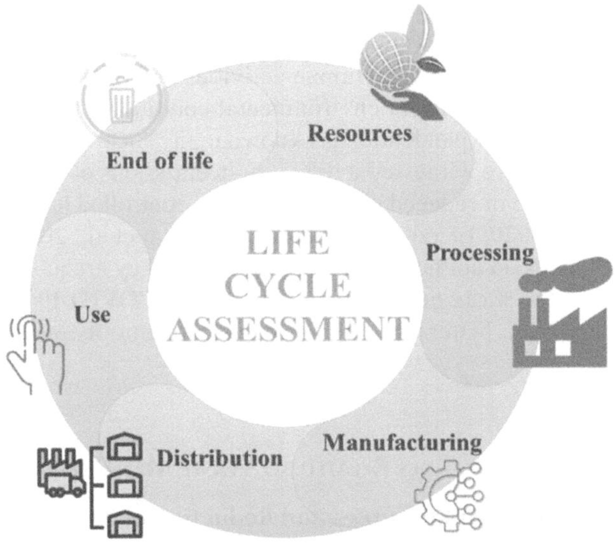

FIGURE 4.4 The life cycle of products.

- Collection stage: to remanufacture the product, the local authorities and waste companies have to organize events and campaigns to raise "reuse awareness" and increase people's interest in swapping objects and buying from secondhand shops.

Still, waste reduction cannot be addressed without shedding light on the reusing and recycling processes. The 3Rs emanate from the perception of wastes as unexploited resources and embody the transformation of products from linear flows to circular ones.

4.2.4.3 Recycling and Reutilization of Produced Wastes

Reusing and recycling waste is essential to sustainable waste management and environmental preservation. Reusing produced wastes is embodied in swapping or donating already used items and trading in secondhand goods. These concepts pass through different steps, including sorting the items at source, checking, cleaning, renovating, and repairing them (Kenneth and Twumasi, 2015). Consequently, through the reusing process, the world will not need to make new products (Bartl, 2014). However, today's economic systems are not highly interested in repairing already used items and abandoning the need to produce new ones because it is unprofitable and does not serve their interests. While the reuse of products does not include any kind of treatment, recycling reprocesses wastes into new products that can be used again. It starts with collecting unwanted products to be processed and sold. In the third step, the recyclables get manufactured and turned into consumer goods. At the end of the process, the consumer purchases new products to start new life cycles. Still, without reducing and reutilization of waste, recycling remains insufficient to achieve effective waste management.

4.2.5 Waste Management and Ocean Quality

4.2.5.1 El Jadida-Morocco

The rapid development of the Moroccan coast has always been compatible with the socio-economic interests of the country. Population growth, urbanization, and the development of the industrial sector have worsened the situation. The coastal waters have become more exposed to contaminant discharges that consist of many dangerous chemicals. Thus, many Moroccan researchers aim to track these changes by evaluating ocean water's physicochemical quality and the state of the local environment. The Atlantic coast of the El Jadida-Safi area is subject to continuous upwelling phenomena; it is rich in nutrients and trace elements, including rare earth elements (Mejjad et al., 2016). This area is important in Morocco because of its large population and intensive industrial development. According to Benchekroun and Bouchama (1991), 55% of urban discharges come from Moroccan coastal towns. Untreated wastewater is discharged into the sea, causing a huge load of contaminants. Kaimoussi (1996) reported that the amount of wastewater evacuated is expected to reach around 109 m^3 within 30 years. It is worth noting that El Jadida is the second industrial pole in Morocco. It is located on the Atlantic coast between Casablanca (90 km southwest of Casablanca) and Jorf Lasfar (one of the largest ports in Africa), and it covers an area of 2480 hectares. It is limited to the north by the Atlantic Ocean and the east by the rural municipality of Haouzia. El Jadida is a city of industries and pollution. Since years ago, many studies have been done on water pollution in El Jadida and the danger of drinking tap water (Chafik et al., 2001; Kaimoussi et al., 2004; Maanan, 2007; Salama et al., 2013; Hanif et al., 2014; Mountadar et al., 2018; Belahbib et al., 2020). However, very few studies have been conducted on marine pollution. In contrast, several studies have been carried out in the coastal environment of Doukkala-Ab El Jadida in the Oualidia lagoon (Maanan et al., 2014; Mejjad et al., 2016, 2018) and in the Sidi Moussa lagoon (Maanan et al., 2004; Benmhammed et al., 2021). El Jadida has witnessed a huge expansion in terms of the growing population, urbanization, and establishment of an industrial zone on the highway of Marrakech. This transformation is indeed a cause of the toxins and pollutants that have damaged humans and animals and contributed to environmental degradation, especially in the level of marine and maritime sectors (Wurl and Phillip Obbard, 2004; Khan et al., 2013; Verma et al., 2016). The inflows of chemicals, solid waste from industrial and agricultural effluents, human-made sedimentation, and oil spills are the major pollutants and sources of marine environment degradation in the city (Kaimoussi et al., 2002; Lamghari Moubarrad and Assobhei, 2007; Mouradi et al., 2014), causing challenges in terms of waste management.

4.2.5.2 Trieste-Italy

The port of Trieste is situated within the site of national interest (SIN). This site is considered a polluted and contaminated area that needs surface water and groundwater remediation (Solis-Weiss et al., 2004; Cibic et al., 2008; Kristan et al., 2014; Cibic et al., 2017). The SIN covers around 12,000,000 m^2 of marine coastal areas. The fact that the site consists of five areas (port, shipbuilding, iron foundry, petroleum, and a residential area) makes it highly exposed and vulnerable to environmental damage due to many different human activities. Still, the local authorities have restricted many activities, such as fishing, aquaculture, swimming, and urban development. Continuous environmental assessments and an effective management scheme can enhance the area's economic development by creating productive activities and protecting the environment. The managing body of the SIN adopted the integrated ecosystem approach as part of a well-planned strategy to implement the SSI's new Environmental Characterization Plan. Thus, a continuous assessment of the ecosystem functioning in the harbor area is needed to verify the possibility of excluding dome areas from the SIN. The effluent discharges in coastal environments characterized by large industrial settlements and urban areas have a tremendous impact on the environment. These discharges cause an accumulation of contaminants like heavy metals and organic pollutants (Ridgway and Shimmield, 2002) in the ocean and the environment. While human activities are considered the leading factor of organic pollutants, natural factors such as volcanic activity, forest fires, soil erosion, and so on are considered the source of heavy metals (Zwolsman et al., 1993). Also, when organic pollutants are degradable, heavy metals cannot be degraded either on the biological level or the chemical one; they are rather persistent and can be accumulated or transported. Here, the organic matter contents, the redox conditions, and the association with minerals' surface are the main factors that influence their behavior (Nissenbaum and Swaine, 1976; Guo et al., 1997; Quesmerais et al., 1998; Dong et al., 2000).

4.3 PROPOSALS AND DISCUSSIONS

4.3.1 Waste between Effective Management and Mismanagement—El Jadida

Waste management is dictated by health and environmental imperatives. This is the most important issue that should be dealt with, mainly because the amount of waste generated continues to increase and diversify. This heterogeneous nature of the waste represents the main difficulty in generalizing a management and treatment channel for these refusals. Indeed, each type of waste has its own chemical and physical properties, making managing heterogeneous waste more challenging and calling for adopting and applying the concept of waste sorting as a priority. The ultimate objective of waste management is to reduce the volume of materials intended for the final landfill to minimize the risks of pollution they can cause for health and the environment (polluting potential, biogas emission, leachate, pathogens, etc.). Management strategies must apply simple principles, making achieving specific objectives possible (FCQ, 1999; Wicker, 2000). The rapid urbanization of southern cities of the Doukkala-Abda region has resulted in rising waste generation rates that city managers were unprepared to deal with. For decades, these officials were more concerned with collecting and disposing of refuse without trying to approach waste as a resource that may have economic value. The gradual deterioration of living conditions, youth unemployment, and poverty are all factors that have favored the emergence of an informal sector, which has the main tasks in waste management, namely the recovery and recycling of waste. This informal sector plays the greatest role in reducing flows and contributes significantly to the local and national economy (El Mjiri et al., 2020; Saadani et al., 2020). Here, the issue of industrial waste management will be dealt with by focusing on the Somadir Company and the phosphate treatment company of Jorf Lasfr because these two companies are possible contributors to ocean water pollution and the deterioration of the whole ecosystem in the region (Chofqi et al., 2004; Essedaoui et al., 2017; Belahbib et al., 2021)

4.3.1.1 Jorf Lasfar Company (JFC)

The JFC (17 km southwest of El Jadida) is a production unit for fertilizers, phosphoric acid, and sulfuric acid. Like any other industry, this activity generates waste, which must be managed. Consequently, this management necessarily follows the decrees of the regulatory review linked to waste management and its elimination. The Quality, Safety, and Environment department, where this work was carried out, has to ensure the safety of the group's staff, those in charge, and the various machines used.

4.3.1.1.1 Integration of a New Management Mode

The current state of solid waste management at JFC requires developing a new management plan to comply with environmental regulations and standards at the national and international levels. Table 4.1 shows that the most suitable waste management model is the 5S Model, as it is efficient in workplaces where high work quality demands a clean, safe, and rigorous environment.

4.3.1.1.2 The Advantages of 5S Are Numerous

1. Less material losses
2. Fewer accidents
3. More pleasant working environment
4. Openness to more elaborated methods.

4.3.1.2 SOMADIR Company

The factory (SOMADIR) located in the industrial zone of El Jadida represents a great source of marine pollution. The SOMADIR Company alone discharges a high percentage of wastewater into the sea. From among the 1870 cubic meters that all industrial units throw daily into the sea, the percentage of the yeast factory alone reaches 860 cubic meters. According to the aforementioned report,

TABLE 4.1
A Descriptive of the 5S Model

Stage	The Objective	Why This S	How to Do
Stage 1: Seiri, Sort	Sort to only keep what is strictly related to the performance of the work	A cluttered work station presents a potential risk of accidents and does not promote quality and performance	Sorting the waste necessary for its reuse in place of storage bins. Indeed, waste sorting aims to avoid unsuitable mixtures to better reuse or recycle and protect the environment and the ocean. Each entity generating one of the wastes must imperatively sort and store the waste by category at the appropriate stage in the scrap yard, for more and more categories of waste (packaging, electrical and electronic waste, used oil, scrap metal, etc.).
Step 2: Seiton, Put away	Determine the places and store the waste or waste bins in these places in the workshops	Clear identification and a reasonable arrangement of things facilitate research and avoid unit	It is necessary to define the rules of arrangement in order to arrange to visualize and easily locate the waste; each waste is identified explicitly and has a bucket determined and known in the most functional way possible after unloading each bucket must return to its place.
Seiso, Clean	Maintaining the job in terms of cleanliness to create good operational conditions	Every anomaly can be detected easily and more quickly in a clean environment	Cleaning can be considered under three successive approaches: the initial major cleaning, regular cleaning, and avoiding soiling.
Step 4: Seiketsu, Standardizer	Standardize and respect the previous 3S	The first three Ss are actions to be carried out on the fourth S.	The rules must be displayed and visible, understandable by everyone, relevant, and easy to implement.
Step 5: Sitsuke, Follow and evolve	The sustainability and improvement of 5S	Stabilize and maintain the first 4 to keep them alive	It requires rigor in the application and regular monitoring of rules and practice in memory to correct the drifts and make them evolve according to the progress made.

the wastewater discharged by the yeast factory contains organic materials such as molasses, which are considered raw materials for manufacturing soft yeast—noting that molasses harm marine life and act as poison for the biodiversity and the whole marine ecosystem because of their liquid sugar properties (Nikodinovic-Runic et al., 2013), the liquid sinks to the bottom of the ocean floor and suffocates fish by absorbing the oxygen, forcing the fish to shallow waters. After manufacture, it excretes liquid materials that contain significant concentrations of dangerous bacterial and organic materials. In addition to using chemical substances that cause an imbalance in the ecological systems, especially marine ones, the leakage of toxic nitrogen is transformed into harmful "nitrates". The yeast produced by the SOMADIR factory is included in the agricultural food industries that use raw materials such as water, molasses, and other materials, including nitrogen and phosphor, which release liquid materials. These materials are loaded with nitrogen, phosphorus, and other salts, increasing the water's acidity and temperature. This process negatively affects the reception field, which is the sea which contributes to the environmental imbalance and leads to water contamination and the death of fish, algae, and other plants. According to several studies, cadmium levels are greater along the El Jadida shoreline in the mussel Mytilusgalloprovincialis (Essedaoui et al., 2001) as well as in saltwater and sediment (Kaimoussi et al., 2001; Ferssiwi et al., 2004). This metallic pollution can degrade the quality of the environment and jeopardize its ecological balance, and it is a human health risk. The worst part is that the company's absence of a treatment plant aggravates the situation because the wastewater passes through the same municipal canals to the beach of El Jadida. Apart from the negative impact of industrial waste, household waste influences marine biodiversity, the quality of the oceans, the environment, and tourism. Thus, the coastal area of El Jadida is suffering from the adverse effects of household waste mismanagement. Accordingly, there is a need to first increase awareness among Eljadida inhabitants toward managing their household waste; secondly, the El Jadida municipality is called to provide the city with needed equipment for adequate household waste management while considering adopting selective recycling bins to allow more efficient management.

4.3.1.3 Household Waste in El Jadida City

The choice of El Jadida's landfill site has impacted human health (diseases), ocean water quality (leaching), the environment, and the economy (tourism) (Figure 4.5). After carrying out several questionnaires with the city inhabitants and the company responsible for managing household waste in the city, it was concluded that the leading cause is the lack of selective sorting bins, making it possible to carry on automatic sorting. Unlike several European countries, people in Morocco are not used to sorting their houses, possibly due to the lack of awareness and equipment related to waste sorting. Thus, preserving the environment, including the marine environment and ocean resources, must be a priority through increasing awareness about the impact of the mismanagement of waste on socio-economic sectors and human well-being.

People are used to disposing of their waste and random garbage in traditional bins. This makes collecting waste very harmful and makes it difficult to properly sort between organic, plastic, metal, and other types of waste for further recycling purposes. For this reason, the Laboratory of Geosciences and Techniques of the Environment of the Faculty of Sciences (El Jadida) in collaboration with the Moroccan association of environment, proposed an experimental protocol (prototype) of a smart bin for selective sorting as demonstrated in the Figures 4.6–4.8. This bin will allow for automatic sorting of the produced waste and separate the metal and the leaching with the rest of the organic matter so that it will be possible to produce a well-degraded mixture of organic matter, which will then be used as compost that will be beneficial in the agronomic sector and green spaces. It should be highlighted that the Moroccan kingdom works hard to develop its economy by adopting renewable energy and developing green sectors that depend on the sustainable use and exploitation of natural resources and the conservation of the environment. Similarly, the blue economy is

FIGURE 4.5 Waste production in El Jadida City (2020).

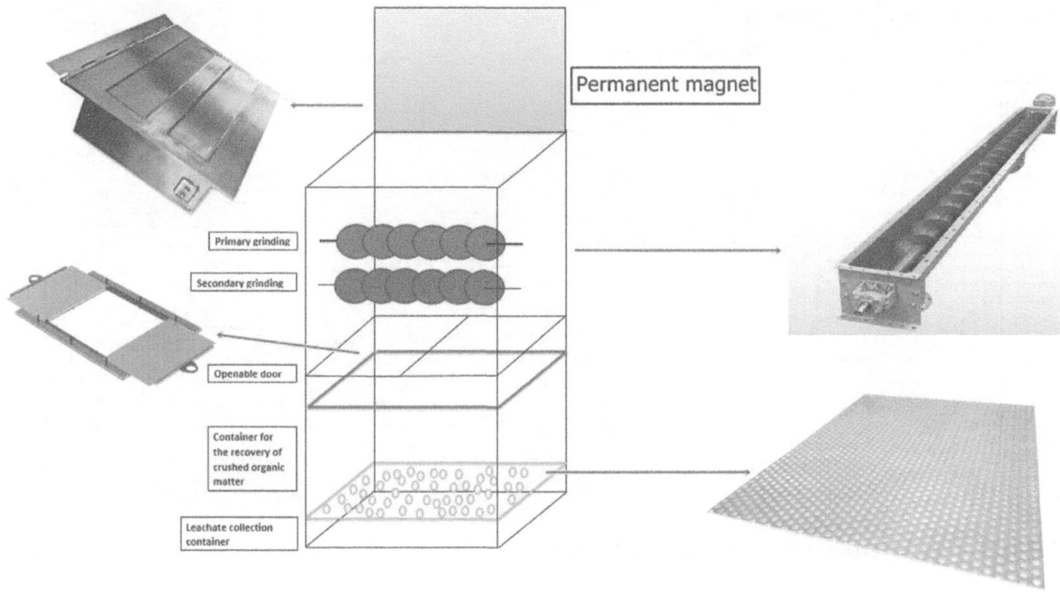

FIGURE 4.6 Scheme of the smart sorting bin.

a vital sector that relies upon the quality of the marine environment. In this sense, protecting and conserving the marine environment has become a priority to avoid more impacts on the economic sectors. Indeed, in terms of Sustainable Development and Environmental Protection, Morocco undertook a series of strategic actions in order to reform the sector of waste management, including the implementation of programs related to solid waste management, strengthening of the legal framework, development of a national master plan for hazardous waste management, and reform of local taxation. The actions undertaken included the implementation of landfills in different cities of the kingdom, including El Jadida, and the national capacity building through organizing a series of workshops for the benefit of local authorities' staff of 16 regions, including El Jadida. These initiatives and actions show the importance of the environmental component as a motor engine for a sustainable economy.

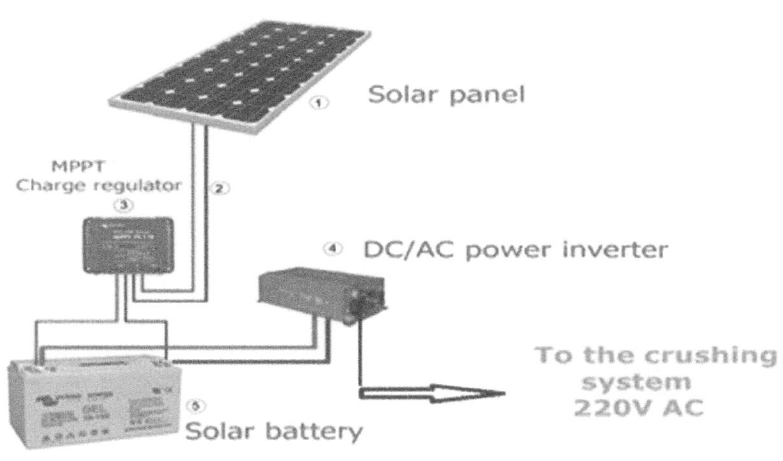

FIGURE 4.7 Photovoltaic system for supplying the smart sorting bin.

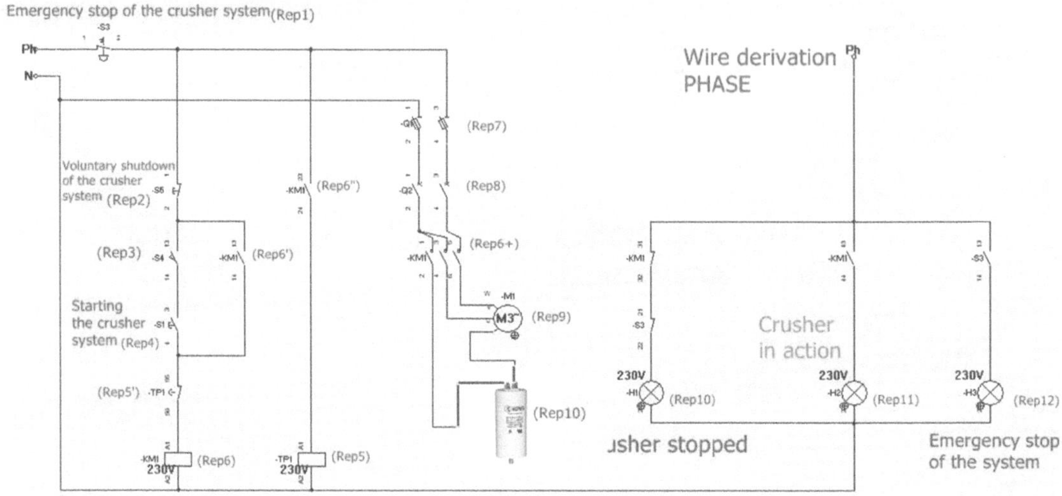

FIGURE 4.8 Electrical scheme of the smart sorting bin.

4.3.2 Trieste-Italy as a Case Study

Establishing regulatory and legal tools in certain countries, particularly the north, has been essential in rational and healthy waste management. In general, waste management in these countries is the responsibility of local communities. In Europe, as provided by law, municipalities must organize the disposal of municipal waste (ADEME, 2000-b). In several countries, the intervention of private companies approved by public services has proved to be of capital importance in managing certain components of waste. This mainly concerns sales packaging. These companies are responsible for organizing, supervising, and financing the selective collection, sorting, and recycling the waste; for example, the Duales System Deutschland (DSD GmbH) in Germany, Altstoff Recycling Austria, and Fost Plus in Belgium. Figure 4.9 illustrates waste management in northern countries.

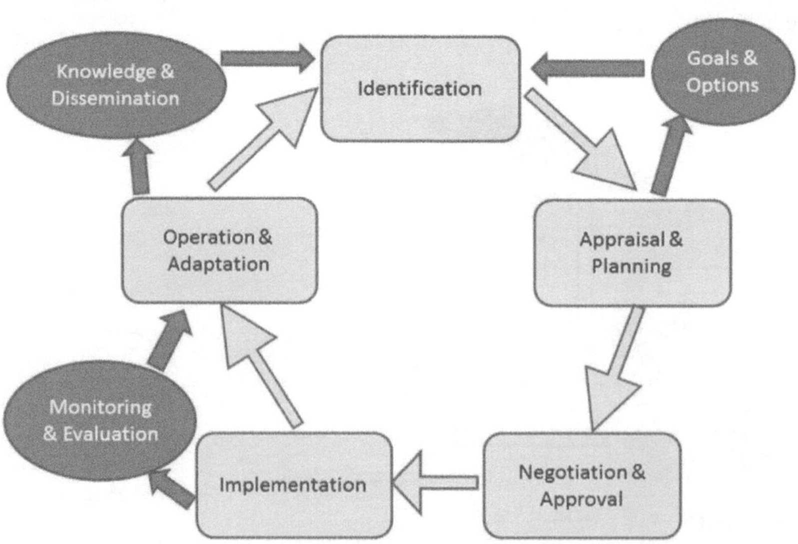

FIGURE 4.9 Typical diagram of waste management in northern countries.

4.3.2.1 Establishment of Interregional Cooperation

The excellent relationship and the spirit of cooperation between Morocco and Italy go back to the 1990s. Indeed, in 1991, Morocco and Italy signed the Agreement of Friendship and Cooperation, which bettered the political and economic relations between the two countries. Lately, in 2019, through the visit of Moroccan Minister Nasser Bourita to Roma, Morocco and Italy have signed a multi-dimensional strategic partnership agreement that aims to deepen the cooperation between the two countries by including new areas of investments like in the sectors of renewable energy, health, food industry, technologies, and ecosystems. They are two countries make from the exchangeability of experiences and knowledge as well as a mutual benefit as the basis of this partnership. Thus, the two countries have confirmed the continuation of their political relations in the post-COVID 19. Therefore, since Morocco is the fifth commercial partner at the level of Italian foreign trade and they are maintaining the spirit of cooperation and win-win multi-dimensional strategies, thinking about institutional cooperation between the two countries in the context of waste management seems possible and attainable. In fact, the relationship between the two countries will be much more fruitful through adopting long-term visions and waste-to-resource initiatives and cooperation. Waste management is the first issue to be addressed and highlighted because it is inclusive. It includes the three vehicles that maintain the planet's life: environment, society, and economy. Its long-term benefits are not limited to the local but have wide global benefits. The unprecedented act of initiating waste-to-resource cooperation between a developed country like Italy and a developing one like Morocco is a win-win multi-dimensional strategy that can lead to a gap closure and become a global icon to encourage similar initiatives. Here, while many researchers interpret the act of importing waste to a developing country as a new form of colonialism, the world needs to shift its vision and pave the way for mutual benefits and waste-to-resource interregional initiatives.

4.4 CONCLUSION

The excessive disposal of different types of waste into the ocean for decades has heavily impacted the marine environment and biodiversity. Trieste and El Jadida are both coastal cities whose economies depend on marine resources and environmental quality levels. Those cities are tourist destinations that receive many national and international visitors yearly. In this order, waste management is a priority for sustainable growth of the socio-economic sectors of those cities. Our review provided an overview of the two cities and waste management in both cities. The analysis highlights the need for developing new strategies for waste management in both cities, considering each city's historical, cultural, social-economic, and cultural specificities. Adopting circular economy concepts is indispensable for ensuring the sustainable use of marine resources, mainly in El Jadida city, while changing population behavior toward waste management. Obviously, protecting our oceans requires good preliminary management of industrial and household waste and the development of several technologies as management assistance tools for a healthy environment and the sustainable development of our countries. Above all, achieving the zero-waste objective should be the world's ultimate goal.

REFERENCES

ADEME (2000-b) Municipal waste: Key figures, 2nd editions, 12 pages.

Asmawati D, Abd Kadir NB and Yusooff F (2010) A study on the knowledge, attitudes, awareness status and behaviour concerning solid waste management. Procedia Social and Behavioral Sciences 18, 643–648. Elsevier.

Bartl A (2014) Moving from recycling to waste prevention. A review of barriers and enables. Waste Management & Research 32, 3–18.

Belahbib L, Arhouni F, Boukhair A, Essadaoui A, Ouakkas S, Hakkar M, Maged ASA, Benjelloun M, Bitar A and Nourreddine A (2020) Impact of phosphate industry on natural radioactivity in sediment, seawater, and coastal marine fauna of El Jadida Province, Morocco. Journal of Hazardous, Toxic, and Radioactive Waste 25(1). ASCE.

Belahbib L, Arhouni FE, Boukhair A, Essadaoui A, Ouakkas S, Hakkar M, Abdo MAS, Benjelloun M, Bitar A and Nourreddine A (2021) Impact of phosphate industry on natural radioactivity in sediment, seawater, and coastal marine fauna of El Jadida province, Morocco. Journal of Hazardous, Toxic, and Radioactive Waste 25, 04020064.

Benchekroun T and Bouchama M (1991) Reuse of wastewater in agriculture in Morocco. In: Proceedings of the World Water Resources Congress, pp. 13–18.

Benmhammed A, Laissaoui A, Mejjad N, Ziad N, Chakir E, Benkdad A, AitBouh H and El Yahyaouia A (2021) Recent pollution records in Sidi Moussa coastal lagoon (Western Morocco) inferred from sediment radiometric dating. Journal of Environmental Radioactivity 227, 106464.

Chafik A, Cheggour M, Cossa D, Benbrahim S and Sifeddine M (2001) Quality of Moroccan Atlantic coastal waters: water monitoring and mussel watching. Aquatic Living Resources 14, 239e249.

Cherif EK, Vodopivec M, Mejjad N, Esteves da Silva JCG, Simonovič S and Boulaassal H (2020) COVID-19 pandemic consequences on coastal water quality using WST sentinel-3 data: Case of Tangier, Morocco. Water 12(9), 2638. https://doi.org/10.3390/w12092638

Chofqi A, Younsi A, Lhadi E, Mania J, Mudry J and Veron A (2004) Environmental impact of an urban landfill on a coastal aquifer (El Jadida, Morocco). Journal of African Earth Sciences 39(2004), 509–516.

Cibic T, Acquavita A, Aleffi F, Bettoso N, Blasutto O, De Vittor C, Falconi C, Falomo J, Faresi L, Predonzani S, Tamberlich F and Fonda Umani S (2008) Integrated approach to sediment pollution: A case study in the Gulf of Trieste. Marine Pollution Bulletin 56(2008), 1650–1667.

Cibic T, Franzo A, Nasi F, Auriemma R and Del Negro P (2017) The port of Trieste (Northern Adriatic Sea)—A case study of the "Ecosystem Approach to Management. Frontiers in Marine Science, Sec. Marine Pollution 4, 336. https://doi.org/10.3389/fmars.2017.00336

Demirbas A (2011) Waste management, waste resource facilities and waste conversion processes. Energy Conversion and Management 52, 1280–1287.

Dong D, Nelson YM, Lion LW, Shuler ML and Ghiorse WC (2000) Adsorption of Pb and Cd onto metal oxides and organic material in natural surface coatings as determined by selective extractions: New evidence for the importance of Mn and Fe oxides. Water Research 34, 427–436.

Ellis C (2018) World Bank: Global waste generation could increase 70% by 2050 PUBLISHED Sept. 23.

El Mjiri I, Rahimi A and Bouasria A (2020) Urban sprawl evolution and soil artificialization assessment by using satellite data from 1985 to 2019: Case of El Jadida metropolitan in Morocco. In: Second International Sustainability and Resilience Conference: Technology and Innovation in Building Designs.

El Zowalaty M, Sean GY and Jarhult JD (2020) Environmental impact of the COVID-19 pandemic – A lesson for the future. Infection Ecology & Epidemiology 10(1). https://doi.org/10.1080/20008686.2020.1768023

Essedaoui A, Kerambrun P, Alliot E and Sif J (2001) Evaluation de l'état de santé de la lagune de Sidi Moussa (Maroc). Marine Life, 11, 1–2: 21–31. https://www.amstes.org/wp-content/uploads/2020/09/Essedaoui-Aafaf-Poster-CIIES-20.pdf

Essedaoui A, Massar R, Ferssiwi A and Bitar A (2017) Impact of industrial pollution on the population dynamics of *Mytilusgalloprovincialis* in the coastal region of El Jadida (Atlantic coast, Morocco). JMES 8(11), 4007–4014.

FCQ (1999) Front Commun Québécois. Project work by the Front commun québécois pour une management ecologique de waste on the public consultation on water management in Quebec, 33 pages.

Ferssiwi A, Sif J, El Hamri H, Rouhi A and Amiard JC (2004) J. Rech. Océanogr. 29 (3–4): 59–64

Gatto E, Buccolieri R, Perronace L and Santiago JL (2021) The challenge in the management of historic trees in urban environments during climate change: The case of Corso Trieste (Rome, Italy). Atmosphere 12(4), 500.

Gayialis SP, Konstantakopoulos GD and Tatsiopoulos IP (2019) Vehicle Routing Problem for Urban Freight Transportation: A Review of the Recent Literature. In: Sifaleras A and Petridis K (eds) Operational Research in the Digital Era – ICT Challenges. Cham: Springer, 89–104.

Guo T, DeLaune RD and Patrick WH (1997) The influence of sediment redox chemistry on chemically active forms of arsenic, cadmium, chromium, and zinc in estuarine sediment. Environment International 23, 305–316.

Hanif N, Chair M and Chbani Idrissi M (2014) Study of *Gelidium sesquipedale* overexploitation effects on Sargassum muticum proliferation of in El Jadida area, Morocco. International Journal of Innovation and Scientific Research 8(2), 283–290.

Hansel CM and Diaz JM (2021) Production of extracellular reactive oxygen species by marine biota. Annual Review of Marine Science 13, 177–200.

Horvat M, Covelli S, Faganeli J, Logar M, Mandića V, Rjar R, Širc A and Žagar D (1999) Mercury in contaminated coastal environments; a case study: the Gulf of Trieste. Science of the Total Environment 237–238, 43–56.

Ian T (2019) Global historical CO2 emissions 1758–2018. https://www.iea.org/articles/global-co2-emissions-in-2019

Ian T (2020a) Global climate change - Statistics & facts. https://www.statista.com/topics/1148/global-climate-change/#topicOverview

Ian T (2020b) Global land and ocean temperature anomalies 1880–2019.

Kaimoussi A (1996) Étude de la variabilité de l'accumulation des métaux lourds dans les différents compartiments (sédiments, mollusques et algues) du littoral de la région d'El Jadida, thèse 3e cycle, université Chouaib-Doukkali, faculté des Sciences, El Jadida, Maroc, 1996, 147 p.

Kaimoussi A, Chafiq A, Mouzdahir A and Bakkas S (2001) Comptes rendus de l'académie des sciences séries. Earth and Planetary Science 333(6), 337–341.

Kaimoussi A, Chafik A, Mouzdahira A and Bakkas S (2002) Diagnosis on the state of healthiness, quality of the coast and biological resources 'case of the Moroccan Atlantic coast' (City of El Jadida). Éditions scientifiques et médicales Elsevier SAS, C. R. Biologies 325, 253–260.

Kaimoussi A, Mouzdahir A and Saih A (2004) Variations saisonnières des teneurs en métaux (Cd, Cu, Fe, Mn et Zn) chez l'algue Ulva lactuca prélevée au niveau du littoral de la ville d'El Jadida (Maroc). C. R. Biologies 327, 361–369. Elsevier.

Kenneth P and Twumasi AS (2015). Assessing '3Rs' Model in Relation to Municipal Solid Waste Management in Wa, Ghana. p,114.

Khan MGM, Cong-Qiang L, Davide V, Kunshan G and Hiroshi O (2013) Sources, factors, mechanisms and possible solutions to pollutants in marine ecosystems. Environmental Pollution. Elsevier. http://dx.doi.org/10.1016/j.envpol.2013.08.005

Kristan U, Kanduč T, Osterc A, Šlejkovec Z, Ramšak A and Stibilj V (2014) Assessment of pollution level using *Mytilus galloprovincialis* as a bioindicator species: The case of the Gulf of Trieste. Marine Pollution Bulletin 89(1–2), 455–463. Elsevier.

Lamghari Moubarrad FZ and Assobhei O (2007) Health risks of raw sewage with particular reference to *Ascaris* in the discharge zone of El Jadida (Morocco). Desalination 215(1–3), 120–126.

Maanan M (2007) Biomonitoring of heavy metals using the intertidal mussel *Mytilus galloprovincialis* in Safi coastal waters, Morocco. Environmental Toxicology 22, 525e531.

Maanan M, Ruiz-Fernandez AC, Maanan M, Fattal P, Zourarah B and Sahabi M (2014) A long-term record of land use change impacts on sediments in Oualidia lagoon, Morocco. International Journal of Sediment Research 29, 1–10.

Maanan M, Zourarah B, Carruesco C, Aajjane A and Naud J (2004) Distribution of heavy metals in Sidi Moussa lagoon sediments (Atlantic Moroccan Coast). Journal of African Earth Sciences 39, 473e484

Mejjad N (2023). Bibliometric Analysis of the Literature on Coastal Sediment Pollution. In: Jayaraju, N., Sreenivasulu, G., Madakka, M., Manjulatha, M. (eds) Coasts, Estuaries and Lakes. Springer, Cham. https://doi.org/10.1007/978-3-031-21644-2_1

Mejjad N, Cherif EK, Rodero A, Krawczyk DA, El Kharraz J, Moumen A, Laqbaqbi M and Fekri A (2021a) Disposal Behavior of Used Masks during the COVID-19 Pandemic in the Moroccan Community: Potential Environmental Impact. International Journal of Environmental Research and Public Health 18(8), 4382. https://doi.org/10.3390/ijerph18084382

Mejjad N, Fekri A, El Hammoumi O, El Aouidi S, El Kharraz J, Kaya S and Moumen A (2021b) Application of DPSIR framework to analyze the groundwater pollution threats of municipal solid waste: Case study Médiouna Landfill, Morocco. E3S Web of Conferences 314, 06004. EDP Sciences.

Mejjad N, Laissaoui A, El-Hammoumi O, Benmansour M, Benbrahim S, Bounouira H, Benkdad A, Bouthir FZ, Fekri A and Bounakhla M (2016) Sediment geochronology and geochemical behavior of major and rare earth elements in the Oualidia Lagoon in the western Morocco. Journal of Radioanalytical and Nuclear Chemistry 309, 1133–1143.

Mejjad N, Laissaoui A, El-Hammoumi O, Fekri A, Amsil H, El-Yahyaoui A and Benkdad A (2018) Geochemical, radiometric, and environmental approaches for the assessment of the intensity and chronology of metal contamination in the sediment cores from Oualidia lagoon (Morocco). Environmental Science and Pollution Research 25, 22872–22888.

Mejjad N and Rovere M (2021) Understanding the impacts of blue economy growth on deep-sea ecosystem services. Sustainability 13(22), 12478. https://doi.org/10.3390/su132212478

MOPS (2020) Marine and ocean pollution statistics & facts 2020.

Mountadar S, Younsi A, Hayani A, Siniti M and Tahiri S (2018) Groundwater salinization process in the coastal aquifer Sidi Abed-Ouled Ghanem (Province of El Jadida, Morocco). Journal of African Earth Sciences147, 169–177.

Mouradi A, Bennasser L, Gloaguen V, Mouradi A, Zidane H and Givernaud T (2014) Accumulation of heavy metals by macroalgae along the Atlantic coast of Morocco between El Jadida and Essaouira. World Journal of Biological Research 6, 1.

Nikodinovic-Runic J, Guzik M, Kenny ST, Babu R, Werker A and Connor KE (2013) Carbon-rich wastes as feedstocks for biodegradable polymer (polyhydroxyalkanoate) production using bacteria. Advances in Applied Microbiology 84, 139–200.

Nissenbaum A and Swaine DJ (1976) Organic–metal interactions in recent sediments: The role of humic substances. Geochimica et Cosmochimica Acta 40, 809–816.

Norrie J, Branson T and Keathley PE (2001) Marine plant extracts impact on grape yield and quality. International Symposium on Foliar Nutrition of Perennial Fruit Plants 594, 315–319.

Quesmerais B, Cossa D, Rondeau B, Pham TT and Fortin B (1998) Mercury distribution in relation to iron and manganese in the water of the St. Laurent River. The Science of the Total Environment 213, 193–201.

RGPH (2004) Recensement général de la population et de l'habitat haut-commissariat au plan.

Ridgway J and Shimmield G (2002b) Estuaries as repositories of historical contamination and their impact on shelf areas. Estuarine Coastal and Shelf Sciences 55, 903–928.

Roser M (2013) Hannah Ritchie and Esteban Ortiz-Ospina: 2013. World Population Growth, Our World in Data.

Saadani S, Laajaj R and Maanan M (2020) Simulating spatial–temporal urban growth of a Moroccan metropolitan using CA–Markov model. Spatial Information Research 28, 609–621. https://doi.org/10.1007/s41324-020-00322-0

Salama Y, Chennaoui M, Mountadar M, Rihan M and Assobhei O (2013) The physicochemical and bacteriological quality and environmental risks of raw sewage rejected in the coast of the city of El Jadida (MOROCCO). Carpathian Journal of Earth and Environmental Sciences 8(2), 39–48.

Sbai SE, Mejjad N and Norelyaqine A (2021) Air quality change during the COVID-19 pandemic lockdown over the Auvergne-Rhône-Alpes region, France. Air Qual Atmos Health 14, 617–628. https://doi.org/10.1007/s11869-020-00965-w

Solis-Weiss V, Aleffi F, Bettoso N, Rossin P, Orel G and Fonda-Umani S (2004) Effects of industrial and urban pollution on the benthic macrofauna in the Bay of Muggia (industrial port of Trieste, Italy). Science of the Total Environment 328, 247–263.

Tanjung RHR and Hamuna B (2019). Assessment of water quality and pollution index in coastal waters of Mimika, Indonesia. Journal of Ecological Engineering, 20(2), 87–94.

Tian K, Wu Q, Liu P, Hu W, Huang B, Shi B and Wang T (2020) Ecological risk assessment of heavy metals in sediments and water from the coastal areas of the Bohai Sea and the Yellow Sea. Environment international 136, 105512.

U.S. EPA (2015) Report of Environmental Protection Agency (EPA) International Decontamination Research and Development Conference. U.S. Environmental Protection Agency, Washington, DC, EPA/600/R-15/283, 2015.

UNDP (2016) United Nations Development Programme Conserve and sustainably use the oceans, seas and marine resources for sustainable development, p.2 Bangkok Regional Hub 3rd Floor United Nations Service Building Rajdamnern Nok Avenue Bangkok 10200, Thailand

Vasiliu D, Bucse A, Lupascu N, Ispas B, Gheablau C and Stanescu I (2020) Assessment of the metal pollution in surface sediments of coastal Tasaul Lake (Romania). Environmental Monitoring and Assessment 192(12), 1–16.

Velenturf APM and Purnell P (2017) Resource recovery from waste: Restoring the balance between resource scarcity and waste overload, P2.

Verma R, Vinoda KS, Papireddy M and Gowda ANS (2016) Toxic pollutants from plastic waste- A review. International Conference on Solid Waste Management, 5IconSWM 2015, Procedia Environmental Sciences 35, 701–708.

Vlachos E (1975) Social Impact Assessment: An Overview. Army Corps of Engineers, Institute of Water Resources, Fort Belvoir, VA.

Wicker A (2000, November 27–28) Chapter 22: Waste management in "Statistics for the environmental policy," Munich; 12 pages.

Wilson DC, Rodic L and Modak P (2015) Global waste management outlook report UNEP.

Wurl O and Phillip Obbard J (2004) A review of pollutants in the sea-surface microlayer (SML): A unique habitat for marine organisms. Marine Pollution Bulletin 48, 1016–1030.

Yoon SJ, Hong S, Kim S, Lee J, Kim T, Kim B and Khim JS (2020) Large-scale monitoring and ecological risk assessment of persistent toxic substances in riverine, estuarine, and coastal sediments of the Yellow and Bohai seas. Environment international 137, 105517.

Yusuf MA and Abiye TA (2019) Risks of groundwater pollution in the coastal areas of Lagos, southwestern Nigeria. Groundwater for Sustainable Development 9, 100222.

Zwolsman IGJ, Berger WG and Vaneck MTG (1993) Sediment accumulation rates, historical input, post depositional mobility and retention of major elements and trace metals in salt marsh sediments of the Scheldt estuary, SW Netherlands. Marine Chemistry 44, 73–94.

5 Green Hydrogen in the Face of Climate Challenges

A Technology for Decarbonization

A. Ghailane, J. Mabrouki, K. El-Moustaqim, D. Azdem, and Y. Abrouki

5.1 INTRODUCTION

Climate change is one of the major challenges of the XXI century, with increasingly worrying environmental, economic and social consequences [1]. Global warming, rising sea levels and the intensification of extreme weather phenomena are all alarming signals that call for a rapid transformation of our energy model [2]. Moreover, the global increase in energy demand, largely driven by technological advances, population growth and economic development, has reinforced dependence on fossil fuels, the main source of greenhouse gas emissions, which are expected to continue to dominate the energy sector until at least 2050 [3]. Moving toward an environmentally friendly future requires us to find technical alternatives to fossil fuel consumption and greenhouse gas emissions. In this context, green hydrogen appears to be a promising alternative to support the energy transition. Although it is not a direct source of energy, it is always present in the form of compounds, such as water or hydrocarbons such as methane. Additionally, as the most abundant gas, hydrogen is an environmentally friendly fuel, as its conversion to energy produces only heat and water, thus reducing greenhouse gas emissions [4–6]. This makes it a key technology for decarbonizing several strategic sectors, including heavy industry, transport and energy storage [7–10]. Today, a national policy in favor of renewable energy production and zero net emissions is essential to ensure sustainability. Moreover, hydrogen's contribution to total energy production could reach 18% by 2050 [11]. However, despite its great potential, the development of green hydrogen faces a number of technical and economic challenges [12, 13]. Production costs are still high compared with other forms of hydrogen, and suitable infrastructures need to be put in place for large-scale transport and storage [9, 10]. What's more, integrating this technology into existing energy systems requires coordination among industrial, political and scientific players [14, 15].

This study aims to take an in-depth look at the role of green hydrogen in the fight against climate change and its potential as a decarbonization solution, highlighting its potential impact on reducing CO_2 emissions, the challenges it faces, possible solutions to accelerate its adoption and how green hydrogen could become a central pillar of a sustainable, decarbonized energy future.

5.2 CLIMATE CHANGE AND THE NEED FOR CLEAN ENERGY

Climate change is currently one of the most pressing environmental issues affecting ecosystems and the global economy [16, 17]. Rising average temperatures, ocean acidification, melting glaciers and an increase in extreme weather events (hurricanes, droughts, floods) are direct manifestations of global warming caused by rising greenhouse gas emissions [18, 19]. Among these gases, carbon dioxide (CO_2), from the combustion of fossil fuels (coal, oil, natural gas), is the main contributor to the anthropogenic greenhouse effect [20]. To limit these impacts and preserve future generations,

 DOI: 10.1201/9781003647263-5

the transition to clean, sustainable energy is an absolute necessity to reduce our carbon footprint and limit the impact of climate change. The use of renewable energy sources, such as solar, wind, hydro and biomass, offers a viable alternative to fossil fuels [21]. However, these energies face challenges linked to their intermittent nature and their storage, requiring the development of additional technologies to guarantee stable and continuous energy production. In this context, green hydrogen offers a promising solution [22, 23]. Produced by electrolysis of water from renewable energy sources, it not only reduces dependence on fossil fuels, but also facilitates the large-scale storage and transport of clean energy [24–26]. Consequently, the need for an effective energy transition relies on the integration of innovative and sustainable solutions to meet growing energy needs while minimizing environmental impact.

5.3 GREEN HYDROGEN PRODUCTION TECHNOLOGIES

5.3.1 Electrolysis

Water electrolysis is one of the most promising technologies for producing green hydrogen, as seen in Figure 5.1. This process uses electricity from renewable sources to break down water into hydrogen (H_2) and oxygen (O_2) with no carbon dioxide emissions (CO_2) [18–20]. The procedure is based on an electrochemical reaction that takes place inside an electrolyzer consisting of two electrodes (anode and cathode) immersed in an electrolyte [27]. Oxygen is recovered through an oxidation reaction at the anode, while hydrogen is recovered through a reduction reaction at the cathode. When the electric current is applied, water is split into its constituent elements according to the following reaction:

$$2H_2O \rightarrow 2H_2 + O_2 \tag{5.1}$$

There are three main electrolysis technologies, each with its advantages and challenges:

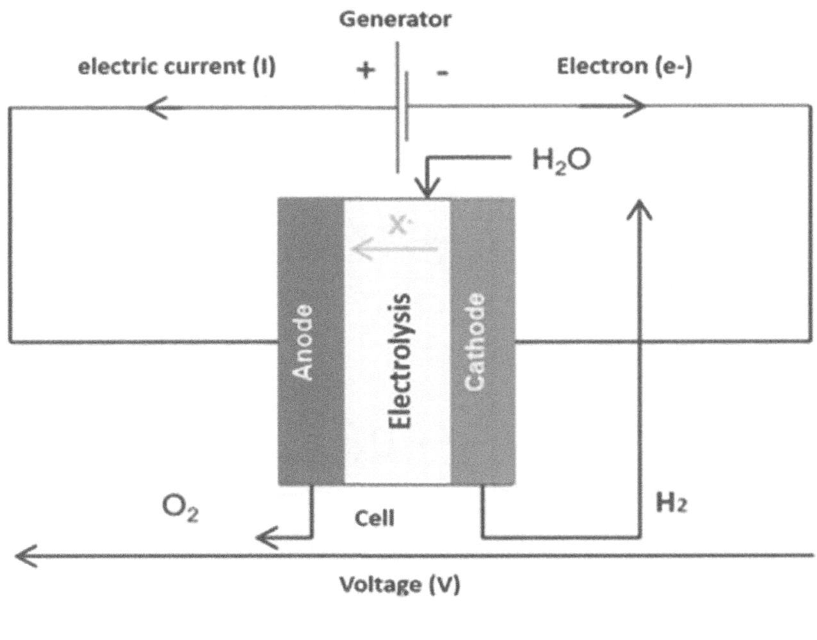

FIGURE 5.1 Schematic diagram of an electrolysis cell.

5.3.1.1 Alkaline Electrolysis

Alkaline electrolysis is the oldest and most widespread technology for hydrogen production [28]. It uses a solution of potassium hydroxide (KOH) or sodium hydroxide (NaOH) as an electrolyte, allowing hydroxide ions (OH^-) to move from the cathode to the anode, where oxygen is released, while hydrogen is generated at the cathode [29]. This technology is characterized by its robustness, long service life and relatively low cost. However, its energy efficiency remains lower than that of other more recent processes, and it operates optimally in steady-state conditions, which limits its adaptation to fluctuations in intermittent renewable energies.

5.3.1.2 PEM (Proton Exchange Membrane) Electrolysis

PEM technology uses a proton-conducting polymer membrane as an electrolyte [30–32]. This process stands out for its ability to adapt quickly to variations in the power output of renewable energies, its high current density and its superior energy yield. However, the main drawback is the high cost of materials, particularly the iridium and platinum used as catalysts, which hampers large-scale deployment [33].

5.3.1.3 Solid Oxide Electrolysis Cell (SOEC)

Solid oxide electrolysis (SOEC) is an advanced hydrogen production technology that operates at high temperatures (700–900°C), harnessing heat to improve energy efficiency and reduce electricity consumption [34–36]. It uses an oxygen-ion-conducting ceramic electrolyte, enabling the dissociation of water vapor into hydrogen (H_2) and oxygen (O_2). It proves particularly suitable for industrial environments with surplus heat. However, its deployment remains limited due to the complexity of the materials used and the challenges associated with long-term cell durability.

5.3.2 Photoelectrolysis

Photoelectrolysis of water is an innovative technology that enables the production of green hydrogen through the direct use of solar energy (Figure 5.2) [37]. This process is based on the use of semiconductor materials capable of absorbing light and generating an electric current without the need

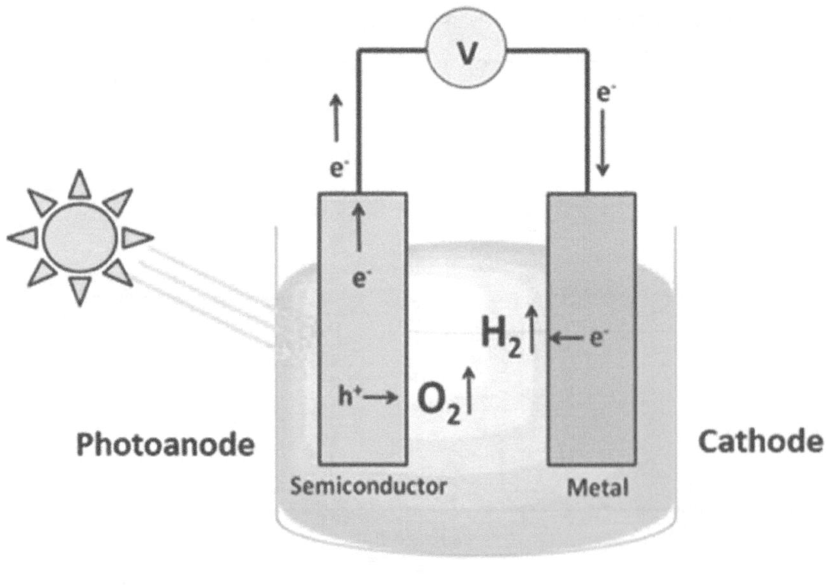

FIGURE 5.2 Schematic diagram of photoelectrolysis of water.

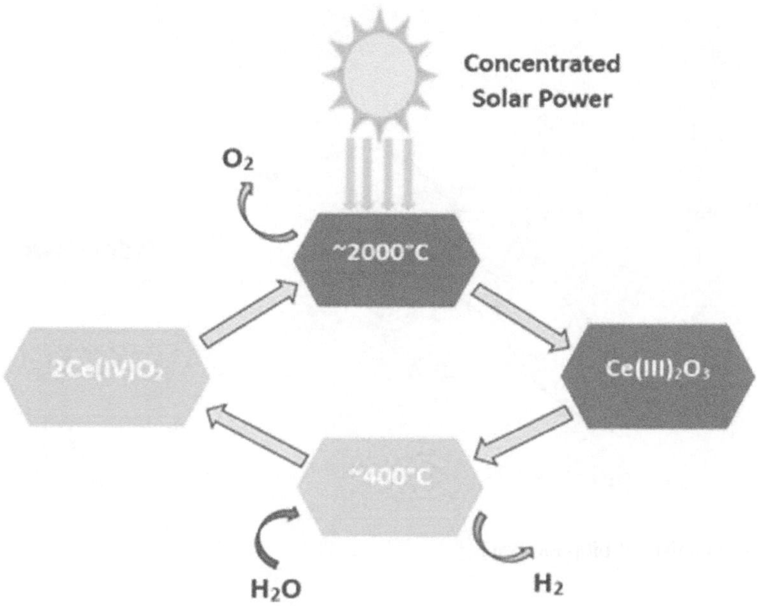

FIGURE 5.3 Cerium oxide cycle coupled with a concentrated solar thermal system.

for external fossil fuels [38]. The basic principle is based on a photoelectrochemical cell (PEC), consisting of a photoanode and a cathode, immersed in an aqueous solution. When light strikes the photoanode, it excites the electrons in the semiconductor material, creating an electron-hole pair [39–41]. Electrons travel to the cathode through an external circuit, where they convert protons into hydrogen (H_2), while water is oxidized to oxygen (O_2) by the gaps in the photoanode. The efficiency of photoelectrolysis is highly dependent on the electrodes' ability to capture solar energy, promote charge separation and minimize recombination. The most commonly studied materials include metal oxides and nitride- or phosphide-based semiconductors.

5.3.3 Thermochemical Dissociation of Water

Thermochemical water splitting is a process that exploits high temperatures, often supplied by renewable energy sources such as concentrated solar power [32, 33]. To induce chemical transformations, relying primarily on heat and cyclic chemical reactions to break down water (H_2O) into hydrogen (H_2) and oxygen (O_2). The process consists of a series of cyclic chemical reactions using intermediate compounds such as metal oxides, which are reconstituted with each cycle, as seen in Figure 5.3 [42]. Typically, these reactions take place in several phases, facilitating the separation of hydrogen and oxygen at temperatures lower than those required for the direct decomposition of water, which would require more than 2,500°C [38]. The major advantage of this technology lies in its potential to produce hydrogen efficiently and sustainably, without greenhouse gas emissions.

5.3.4 Biophotolysis

Biophotolysis (Figure 5.4), is a biological method of green hydrogen production that uses photosynthetic microorganisms, such as cyanobacteria or microalgae, to break down water into hydrogen and oxygen under the effect of sunlight [7, 43–46]. This process is based on specific enzymes, such as hydrogenases or nitrogenases, which catalyze the reduction of protons (H^+) to molecular hydrogen (H_2). Biophotolysis generally takes place in two main stages. First, photosynthesis

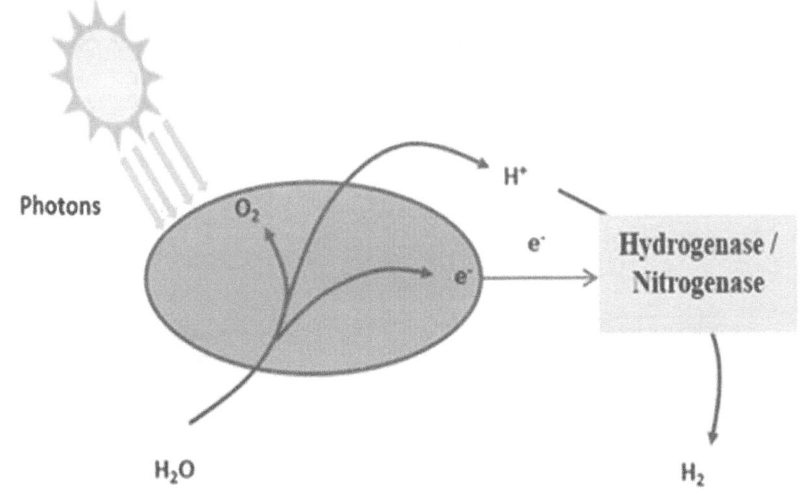

FIGURE 5.4 Mechanism of biophotolysis.

generates energy and electrons from light, then these electrons are transferred to reduce water to hydrogen [47, 48]. This technology has the advantage of being renewable and sustainable, since it uses only water, sunlight and biological entities, with no impact on climate.

5.4 INTEGRATING GREEN HYDROGEN IN THE FIGHT AGAINST CLIMATE CHANGE

Green hydrogen, considered an essential solution for sustainable development and the energy transition, offers a revolutionary solution to today's environmental and energy challenges [49, 50]. Its production from water and renewable energy sources not only provides a clean, sustainable alternative to conventional fuels, but also makes a significant contribution to combating climate change and reducing greenhouse gas emissions [51–53]. Its integration into various industrial and energy sectors represents a strategic advantage for the move towards a low-carbon economy. Green hydrogen has the potential to replace fossil fuels in various high-carbon emission sectors, such as industry, transport and power generation [54–56]. It also plays a key role in the storage and stabilization of renewable energies, enabling excess electricity to be converted into hydrogen, which can later be reused to produce clean energy via hydrogen fuel cells. In terms of mobility, its use as a fuel in hydrogen-powered vehicles contributes to the reduction of CO_2 emissions and atmospheric pollutants [57]. What is more, green hydrogen promotes a circular economy, optimizing the use of renewable resources and reducing energy waste, making it a key solution for accelerating the transition to a sustainable economy [58]. In this way, green hydrogen could become an essential pillar in achieving carbon neutrality and stepping up the fight against global warming.

5.5 CHALLENGES AND OBSTACLES TO THE PRODUCTION AND USE OF GREEN HYDROGEN

Green hydrogen, while a promising solution for the energy transition, faces a number of technical, economic and logistical challenges that hamper its large-scale deployment. Firstly, the high cost of production represents a considerable brake [59, 60]. The production of green hydrogen, mainly via water electrolysis, requires a considerable amount of electricity from renewable sources [27]. This leads to a higher cost compared with grey or blue hydrogen, produced from fossil fuels. What is more, the energy yields of electrolyzers remain relatively low, limiting the overall efficiency of the process. Another major challenge concerns transport and storage infrastructures [12]. Hydrogen is a light gas, extremely flammable and difficult to store in gaseous or liquid form due to the extreme pressures and temperatures required [61]. The development of suitable pipelines and safe storage solutions represents a significant additional cost. Technologically speaking, incorporating green hydrogen into energy and industrial systems remains a complex challenge. Existing infrastructures, designed for hydrocarbons, are not adapted to the massive use of hydrogen, requiring considerable investment to modify or replace them. In addition, regulatory and political challenges are slowing down the take-up of green hydrogen. The absence of a harmonized global standards framework, still-limited financial incentives and dependence on public policies are holding back investment and the implementation of large-scale projects. To overcome these obstacles, greater collaboration among governments, industries and researchers is essential to reduce costs, improve technologies and develop an infrastructure suited to the hydrogen economy [62, 63].

5.6 FUTURE PERSPECTIVES AND INNOVATIONS

Future developments and advances in green hydrogen open up promising horizons for accelerating the global energy transition. Technological advances, particularly in high-efficiency electrolyzers and innovative storage materials, could considerably reduce production and distribution costs [64]. In addition, the integration of artificial intelligence and optimized energy management systems would improve the efficiency of production and storage processes, while facilitating the use of intermittent renewable sources [65, 66]. Pilot projects and international collaborations are paving the way for large-scale hydrogen ecosystems, linking producers, distributors and consumers. In addition, innovations in fuel cells and industrial applications are expanding the outlets for green hydrogen. However, to fully realize this potential, it will be crucial to step up public and private investment, harmonize regulations and raise awareness of the benefits of this clean energy among economic and political players. In this way, green hydrogen could become a central pillar of a decarbonized, sustainable economy in the decades to come.

5.7 CONCLUSION

Green hydrogen represents an essential solution for meeting the challenges of energy transition and combating climate change. Thanks to its CO_2 emission-free production and its potential to decarbonize key sectors such as industry, transport and energy storage, it embodies a sustainable alternative to fossil fuels. However, its large-scale deployment requires overcoming major obstacles. Technological innovation, massive investment and international collaboration will be essential to accelerate its adoption and make it a pillar of the circular economy and carbon neutrality. In the future, green hydrogen could play a central role in building a cleaner, more resilient and sustainable energy system, paving the way for a society that respects the environment and future generations.

REFERENCES

1. J.-P. Maréchal, "Le changement climatique, un enjeu international majeur du XXIe siècle," *Géoéconomie*, vol. 50, no. 3, pp. 93–100, 2009. doi: 10.3917/geoec.050.0093.
2. C. Tebaldi *et al*, "Extreme sea levels at different global warming levels," *Nat. Clim. Change*, vol. 11, no. 9, pp. 746–751, 2021. doi: 10.1038/s41558-021-01127-1.
3. P. E. Brockway, A. Owen, L. I. Brand-Correa, and L. Hardt, "Estimation of global final-stage energy-return-on-investment for fossil fuels with comparison to renewable energy sources," *Nat. Energy*, vol. 4, no. 7, pp. 612–621, 2019. doi: 10.1038/s41560-019-0425-z.

4. S. Sharma, S. Agarwal, and A. Jain, "Significance of hydrogen as economic and environmentally friendly fuel," *Energies*, vol. 14, no. 21, 2021. doi: 10.3390/en14217389.
5. C. Ruocco, V. Palma, and A. Ricca, "Kinetics of oxidative steam reforming of ethanol over bimetallic catalysts supported on CeO2–SiO2: A comparative study," *Top. Catal.*, vol. 62, no. 5, pp. 467–478, 2019. doi: 10.1007/s11244-019-01173-2.
6. M. Benchrifa *et al.*, "Hydrogen Production Through Gasification of Organic Matter," in *Advanced Technology for Smart Environment and Energy*, J. Mabrouki, and A. Mourade, Eds., Cham: Springer Nature Switzerland, 2024, pp. 197–207. doi: 10.1007/978-3-031-50871-4_12.
7. K. El-Moustaqim, T. Rachiq, J. Mabrouki, M. Slaoui, and D. Hmouni, "Application of Microalgae in Green Hydrogen Production: A Diverse Biomass for Multiple Applications," in *Advanced Technology for Smart Environment and Energy*, J. Mabrouki, and A. Mourade, Éds., Environmental Science and Engineering, Cham: Springer Nature Switzerland, 2024, pp. 231–240. doi: 10.1007/978-3-031-50871-4_15.
8. K. El-Moustaqim, B. Mohammed, J. Mabrouki, and D. Hmouni, "Application of IoT-Based Intelligent Systems in Microalgae Biofuel Production," in *Advanced Technology for Smart Environment and Energy*, J. Mabrouki, and A. Mourade, Éds., Environmental Science and Engineering, Cham: Springer Nature Switzerland, 2024, pp. 293–305. doi: 10.1007/978-3-031-50871-4_20.
9. K. El-Moustaqim, J. Mabrouki, and D. Hmouni, "Applied Sciences Green Microalgae for Future Biomass Development," in *Advanced Systems for Environmental Monitoring, IoT and the application of Artificial Intelligence. Studies in Big Data*, J. Mabrouki and M. Azrour, Eds., vol 143. Cham: Springer, 2024. doi: 10.1007/978-3-031-50860-8_12..
10. K. El-Moustaqim, J. Mabrouki, M. Azrour, and D. Hmouni, "How Can Green Energy Be Improved by Integrating IoT into the Microalgae Process?," in *Sustainable and Green Technologies for Water and Environmental Management*, 2024, pp. 239–252. doi: 10.1007/978-3-031-52419-6_17.
11. B. Koirala, S. Hers, G. Morales-España, Ö Özdemir, J. Sijm, and M. Weeda, "Integrated electricity, hydrogen and methane system modelling framework: Application to the Dutch infrastructure outlook 2050," *Appl. Energy*, vol. 289, p. 116713, 2021. doi: 10.1016/j.apenergy.2021.116713.
12. N. Ma, W. Zhao, W. Wang, X. Li, and H. Zhou, "Large scale of green hydrogen storage: Opportunities and challenges," *Int. J. Hydrog. Energy*, vol. 50, pp. 379–396, 2024. doi: 10.1016/j.ijhydene.2023.09.021.
13. M. Younas, S. Shafique, A. Hafeez, F. Javed, and F. Rehman, "An overview of hydrogen production: Current status, potential, and challenges," *Fuel*, vol. 316, p. 123317, 2022. doi: 10.1016/j.fuel.2022.123317.
14. P. Muthukumar *et al.*, "Review on large-scale hydrogen storage systems for better sustainability," *Int. J. Hydrog. Energy*, vol. 48, no. 85, pp. 33223–33259, 2023. doi: 10.1016/j.ijhydene.2023.04.304.
15. K. El-Moustaqim, J. Mabrouki, M. Azrour, M. Hadine, and D. Hmouni, " Enabling Smart Agriculture Through Integrating the Internet of Things in Microalgae Farming for Sustainability," in *Smart Internet of Things for Environment and Healthcare*, M. Azrour, J. Mabrouki, A. Alabdulatif, A. Guezzaz, and F. Amounas, Eds., Cham: Springer Nature Switzerland, 2024, pp. 209–222. doi: 10.1007/978-3-031-70102-3_15.
16. M. Benchrifa, M. Elouardi, G. Fattah, J. Mabrouki, and R. Tadili, "Identification, Simulation and Modeling of the Main Power Losses of a Photovoltaic Installation and Use of the Internet of Things to Minimize System Losses," in *Advanced Technology for Smart Environment and Energy*, Cham: Springer International Publishing, 2023, pp. 49–60.
17. M. Yang *et al.*, "Circular economy strategies for combating climate change and other environmental issues," *Environ. Chem. Lett.*, vol. 21, no. 1, pp. 55–80, 2023. doi: 10.1007/s10311-022-01499-6.
18. M. El Alouani, B. Aouan, Y. Rachdi, S. Alehyen, E. H. El Herradi, H. Saufi, and N. Barka, "Porous geopolymers as innovative adsorbents for the removal of organic and inorganic hazardous substances: A mini-review," *Int. J. Environ. Anal. Chem.*, vol. 104, no. 16, pp. 4784–4796, 2024.
19. T. Rachiq, Y. Abrouki, J. Mabrouki, N. Samghouli, C. Fersi, S. Rahal, and S. El Hajjaji, "Evaluation of the efficiency of different materials to remove specific pollutants from landfill leachate," *Desalination Water Treat.*, vol. 238, pp. 240–250, 2021.
20. J. Min *et al.*, "The effect of carbon dioxide emissions on the building energy efficiency," *Fuel*, vol. 326, p. 124842, 2022. doi: 10.1016/j.fuel.2022.124842.
21. L. S. Paraschiv, and S. Paraschiv, "Contribution of renewable energy (hydro, wind, solar and biomass) to decarbonization and transformation of the electricity generation sector for sustainable development," *Energy Rep.*, vol. 9, pp. 535–544, 2023. doi: 10.1016/j.egyr.2023.07.024.
22. K. El-Moustaqim, S. E. Sbai, J. Mabrouki, and D. Hmouni, "Critical review of green microalgal biomass development through CO2 sequestration," *Euro-Mediterr. J. Environ. Integr.*, mai 2024, doi: 10.1007/s41207-024-00548-6.

23. K. El-Moustaqim, S. E. Sbai, Y. El Yousfi, J. Mabrouki, and D. Hmouni, "New strategy for the advancement of modern agriculture through the use of microalgae as biofertilizers," *Euro-Mediterr. J. Environ. Integr.*, Oct. 2024, doi: 10.1007/s41207-024-00678-x.
24. N. S. Hassan *et al.*, "Recent review and evaluation of green hydrogen production via water electrolysis for a sustainable and clean energy society," *Int. J. Hydrog. Energy*, vol. 52, pp. 420–441, 2024. doi: 10.1016/j.ijhydene.2023.09.068.
25. N. Al-Jadabi, M. Laaouan, S. El Hajjaji, J. Mabrouki, M. Benbouzid, and D. Dhiba, "The dual performance of moringa oleifera seeds as eco-Friendly natural coagulant and as an antimicrobial for wastewater treatment: A review," *Sustainability*, vol. 15, no. 5, p. 4280, 2023.
26. B. Amini Horri, and H. Ozcan, "Green hydrogen production by water electrolysis: Current status and challenges," *Curr. Opin. Green Sustain. Chem.*, vol. 47, p. 100932, 2024. doi: 10.1016/j.cogsc.2024.100932.
27. Q. Xu *et al.*, "Anion exchange membrane water electrolyzer: Electrode design, lab-scaled testing system and performance evaluation," *EnergyChem*, vol. 4, no. 5, p. 100087, 2022. doi: 10.1016/j.enchem.2022.100087.
28. K. Zeng, and D. Zhang, "Recent progress in alkaline water electrolysis for hydrogen production and applications," *Prog. Energy Combust. Sci.*, vol. 36, no. 3, pp. 307–326, 2010. doi: 10.1016/j.pecs.2009.11.002.
29. J. Brauns, and T. Turek, "Alkaline water electrolysis powered by renewable energy: A review," *Processes*, vol. 8, no. 2, 2020. doi: 10.3390/pr8020248.
30. H. K Loukili, J. Mabrouki, A. Anouzla, Y. Kouzi, S. A. Younssi, K. Digua, and Y. Abrouki, "Pre-treated Moroccan natural clays: Application to the wastewater treatment of textile industry," *Desalin. Water Treat.*, vol. 240, pp. 124–136, 2021.
31. Y. Nagao, "Proton-Conducting Polymers: Key to Next-Generation Fuel Cells, Electrolyzers, Batteries, Actuators, and Sensors," *ChemElectroChem*, vol. 11, no. 12. doi: 10.1002/celc.202300846.
32. H. R. Corti, "Polymer electrolytes for low and high temperature PEM electrolyzers," *Curr. Opin. Electrochem.*, vol. 36, p. 101109, 2022. doi: 10.1016/j.coelec.2022.101109.
33. K. Zhang *et al.*, "Status and perspectives of key materials for PEM electrolyzer," *Nano Res. Energy*, vol. 1, p. e9120032, 2022. doi: 10.26599/NRE.2022.9120032.
34. S. Mgalaa, J. Mabrouki, M. Elouardi, L. El Azzouzi, A. Moufti, S. El Hajjaji, and M. A. El Belghiti, "Study and evaluation of the degradation of procion blue dye by the ozonation method: Parametric and isothermal study," *Nanotechnol. Environ. Eng.*, vol. 7, no. 3, pp. 691–697, 2022.
35. M. Fallah Vostakola, H. Ozcan, R. S. El-Emam, and B. Amini Horri, "Recent advances in high-temperature steam electrolysis with solid oxide electrolysers for green hydrogen production," *Energies*, vol. 16, no. 8, 2023. doi: 10.3390/en16083327.
36. J. Mabrouki, M. Azrour, Y. Farhaoui, and S. El Hajjaji, "Intelligent System for Monitoring and Detecting Water Quality," in *Big Data and Networks Technologies*. Springer International Publishing, 2020, vol. 3, pp. 172–182.
37. X. M. C. Ta, R. Daiyan, T. K. A. Nguyen, R. Amal, T. Tran-Phu, and A. Tricoli, "Alternatives to water photooxidation for photoelectrochemical solar energy conversion and green H2 production," *Adv. Energy Mater.*, vol. 12, no. 42, p. 2201358, 2022. doi: 10.1002/aenm.202201358.
38. M. Gopinath, and R. Marimuthu, "A review on solar energy-based indirect water-splitting methods for hydrogen generation," *Int. J. Hydrog. Energy*, vol. 47, no. 89, pp. 37742–37759, 2022. doi: 10.1016/j.ijhydene.2022.08.297.
39. I. Bencheikh, K. Azoulay, J. Mabrouki, S. El Hajjaji, A. Moufti, and N. Labjar, "The use and the performance of chemically treated artichoke leaves for textile industrial effluents treatment," *Chemical Data Collections*, vol. 31, p. 100597, 2021.
40. R. B. Wexler, E. B. Stechel, and E. A. Carter, "Materials Design Directions for Solar Thermochemical Water Splitting," in *Solar Fuels*, John Wiley & Sons, Ltd, 2023, pp. 1–63. doi: 10.1002/9781119752097.ch1.
41. J. Mabrouki, M. Benbouzid, D. Dhiba, and S. El Hajjaji, "Simulation of wastewater treatment processes with Bioreactor Membrane Reactor (MBR) treatment versus conventional the adsorbent layer-based filtration system (LAFS)," *Int. J. Environ. Anal. Chem.*, vol. 102, no. 19, pp. 7458–7468, 2022.
42. P. Cavaliere, "Thermochemical Water Splitting Cycles", in *Water Electrolysis for Hydrogen Production*, P. Cavaliere, Ed., Cham: Springer International Publishing, 2023, pp. 105–157. doi: 10.1007/978-3-031-37780-8_3.
43. F. Ghizlane, J. Mabrouki, F. Ghrissi, and M. Azrour, "Proposal for a high-resolution particulate matter (PM10 and PM2. 5) capture system, comparable with hybrid system-based internet of things: Case of quarries in the western Rif," *Morocco. Pollution*, vol. 8, no. 1, pp. 169–180, 2022.

44. C. S. Madankar, P. Borde, and P. D. Meshram, "Studies on Extraction Techniques of Bio-Hydrogen - Biofuel Extraction Techniques - Wiley Online Library." Consulté le: 16 mars 2025. [En ligne]. Disponible sur: https://onlinelibrary.wiley.com/doi/abs/10.1002/9781119829522.ch10
45. A. Rather, and D.-A. K. Srivastav, " A study on biohydrogen production based on biophotolysis from cyanobacteria," *Ann. Romanian Soc. Cell Biol.*, vol. 25, pp. 12500–12509, 2021.
46. K. El-Moustaqim, J. Mabrouki, and D. Hmouni, "From Microalgal Biomass to Biohydrogen: A Novel Approach to Environmentally Sustainable Technology," in *Advancements in Climate and Smart Environment Technology*, IGI Global Scientific Publishing, 2024, pp. 29–38. doi: 10.4018/979-8-3693-3807-0.ch003.
47. M. Benchrifa, J. Mabrouki, M. Elouardi, M. Azrour, and R. Tadili, "Detailed study of dimensioning and simulating a grid-connected PV power station and analysis of its environmental and economic effect, case study," *Model. Earth Syst. Environ.*, vol. 9, no. 1, pp. 53–61, 2023.
48. K. El-Moustaqim, S. E. Sbai, J. Mabrouki, and D. Hmouni, "Critical review of green microalgal biomass development through CO2 sequestration," *Euro-Mediterr. J. Environ. Integr.*, vol. 10, no. 1, pp. 1–13, 2025. doi: 10.1007/s41207-024-00548-6.
49. A. Kovač, M. Paranos, and D. Marciuš, "Hydrogen in energy transition: A review," *Int. J. Hydrog. Energy*, vol. 46, no. 16, pp. 10016–10035, 2021. doi: 10.1016/j.ijhydene.2020.11.256.
50. A. G. Olabi *et al.*, "Green hydrogen: Pathways, roadmap, and role in achieving sustainable development goals," *Process Saf. Environ. Prot.*, vol. 177, pp. 664–687, 2023. doi: 10.1016/j.psep.2023.06.069.
51. H. Shin, D. Jang, S. Lee, H.-S. Cho, K.-H. Kim, and S. Kang, "Techno-economic evaluation of green hydrogen production with low-temperature water electrolysis technologies directly coupled with renewable power sources," *Energy Convers. Manag*, vol. 286, p. 117083, 2023. doi: 10.1016/j.enconman.2023.117083.
52. I. Bencheikh, K. Azoulay, J. Mabrouki, S. El Hajjaji, A. Dahchour, A. Moufti, and D. Dhiba, "The adsorptive removal of MB using chemically treated artichoke leaves: Parametric, kinetic, isotherm and thermodynamic study," *Sci. Afr.*, vol. 9, p. e00509, 2020.
53. I. Marouani *et al.*, "Integration of renewable-energy-based green hydrogen into the energy future," *Processes*, vol. 11, no. 9, 2023. doi: 10.3390/pr11092685.
54. A. Ahmed, A. Q. Al-Amin, A. F. Ambrose, and R. Saidur, "Hydrogen fuel and transport system: A sustainable and environmental future," *Int. J. Hydrog. Energy*, vol. 41, no. 3, pp. 1369–1380, 2016. doi: 10.1016/j.ijhydene.2015.11.084.
55. M. Alsunousi, and E. Kayabasi, "The role of hydrogen in synthetic fuel production strategies," *Int. J. Hydrog. Energy*, vol. 54, pp. 1169–1178, 2024. doi: 10.1016/j.ijhydene.2023.11.359.
56. M. Benchrifa, and J. Mabrouki, "Simulation, sizing, economic evaluation and environmental impact assessment of a photovoltaic power plant for the electrification of an establishment," *Adv. Build. Energy Res.*, vol. 16, no. 6, pp. 736–753, 2022.
57. S. K. Dash, S. Chakraborty, M. Roccotelli, and U. K. Sahu, "Hydrogen fuel for future mobility: Challenges and future aspects," *Sustainability*, vol. 14, no. 14, 2022. doi: 10.3390/su14148285.
58. Y. Ajiboye *et al.*, "The Future of Green Hydrogen in the Global Economy: Opportunities, Challenges and Implications," in *2024 International Conference on Science, Engineering and Business for Driving Sustainable Development Goals (SEB4SDG)*, avr. 2024, p. 1–28. doi: 10.1109/SEB4SDG60871.2024.10630372.
59. J. Mabrouki, A. Moufti, I. Bencheikh, K. Azoulay, Y. El Hamdouni, and S. El Hajjaji "Optimization of the Coagulant Flocculation Process for Treatment of Leachate of the Controlled Discharge of the City Mohammedia (Morocco)," in *Advanced Intelligent Systems for Sustainable Development (AI2SD'2019) Volume 7-Advanced Intelligent Systems for Sustainable Development Applied in Energy and Electrical Engineering*, Springer International Publishing, 2020, pp. 200–212.
60. F. Qureshi *et al.*, "Contemporary avenues of the Hydrogen industry: Opportunities and challenges in the eco-friendly approach," *Environ. Res.*, vol. 229, p. 115963, 2023. doi: 10.1016/j.envres.2023.115963.
61. O. Faye, J. Szpunar, and U. Eduok, "A critical review on the current technologies for the generation, storage, and transportation of hydrogen," *Int. J. Hydrog. Energy*, vol. 47, no. 29, pp. 13771–13802, 2022. doi: 10.1016/j.ijhydene.2022.02.112.
62. S. O. Bade, O. S. Tomomewo, A. Meenakshisundaram, P. Ferron, and B. A. Oni, "Economic, social, and regulatory challenges of green hydrogen production and utilization in the US: A review," *Int. J. Hydrog. Energy*, vol. 49, pp. 314–335, 2024. doi: 10.1016/j.ijhydene.2023.08.157.
63. A. Islam *et al.*, "Accelerating the green hydrogen revolution: A comprehensive analysis of technological advancements and policy interventions," *Int. J. Hydrog. Energy*, vol. 67, pp. 458–486, 2024. doi: 10.1016/j.ijhydene.2024.04.142.

64. J. Mabrouki, M. Benchrifa, M. Ennouhi, K. Azoulay, I. Bencheikh, T. Rachiq, and S. El Hajjaji, "Geographic information system for the study of water resources in Chaâba El Hamra, Mohammedia (Morocco)," in The International Conference on Artificial Intelligence and Smart Environment (pp. 469–474). Cham: Springer International Publishing, 2022, November.
65. A. Sai Ramesh *et al.*, "Artificial intelligence driven hydrogen and battery technologies – A review," *Fuel*, vol. 337, p. 126862, 2023. doi: 10.1016/j.fuel.2022.126862.
66. R. Ahmed, S. A. Shehab, O. M. Elzeki, A. Darwish, and A. E. Hassanein, "An explainable AI for green hydrogen production: A deep learning regression model," *Int. J. Hydrog. Energy*, vol. 83, pp. 1226–1242, 2024. doi: 10.1016/j.ijhydene.2024.08.064.

6 Technical Evaluation of Seawater Desalination Technologies

D. Azdem, A. Fatni, J. Mabrouki, M. El Mouedden, A. Ghailane, and S. El Hajjaji

6.1 INTRODUCTION

Desalination, often known as desalting is the process of removing salts dissolved from seawater by processes like membrane-based filtration or thermal distillation [1, 2]. Although it has potential, desalination is capital and energy intensive which limits its use in many desert and economically stressed regions [3, 4]. Based on the World Health Organization (WHO), the threshold for acceptable salinity for drinking water is around 500 parts of million (ppm) but in certain situations the salinity can go levels of up to 9,000 parts per million [5, 6]. Seawater generally has salinity levels that range between 35,010 and 45,010 ppm, which is far higher than recommended levels of consumption [7, 8]. In the world, water use is distributed unevenly: around 70% of it is used by the agriculture sector, 20% is utilized by industrial operations and 10% of it is devoted for domestic use [9, 10]. Morocco is among the most economically developed countries located in North Africa, serves as an example of the benefits and challenges that come with desalination [11]. Despite its vast coastline, Morocco has historically relied heavily on water reservoirs in the underground because of the saltiness of seawater [12]. This made it unsuitable for drinking as well as domestic consumption. In recent times, Morocco has made significant investments in reverse osmosis and thermal distillation techniques, which has resulted in an increased supply of potable water, as well as reducing the previous worries about water scarcity [13]. In evaluating desalination technology in general, it is important to take into account not just the cost and energy aspects but also aspects such as the endurance of semi-permeable membranes, as well as the sustainable and environmentally friendly removal of brine concentrated [14]. These issues are currently being investigated and their resolution may dramatically improve our understanding of the Human Development Index (HDI) especially in areas with constant water shortages. The principal goal of this research is to investigate the mechanism of reverse osmosis and distillation, analyze their strengths and weaknesses and examine how advances in these technologies can aid the development of socio-economics and raise the public's awareness in areas that are prone to drought.

6.2 HISTORICAL AND TECHNICAL BACKGROUND

Desalination of seawater has emerged as an important area of study in the field of scientific and technological research due to its ability to meet the increasing world demand for freshwater fueled by increasing population and the growing water scarcity [15]. Over time, many desalination techniques have been developed and improved, each based on different principles and methods [16]. Of these, thermal distillation and reverse osmosis (RO) have witnessed significant advances in terms of efficacy and practicality. Research is currently analyzing and evaluating these methods through looking at their capabilities in terms of technology, economic effectiveness and viability in light of new developments like multi-effect distillation at low temperatures. These efforts are aimed at optimizing desalination techniques to ensure sustainable and cost-effective production of freshwater

DOI: 10.1201/9781003647263-6

particularly in regions with acute water shortages. Desalination on a large scale started in the 1960s when the construction began of the very first large facility [17]. Since then, ongoing efforts by engineers and scientists have been focused on improving process efficiency and reducing costs for operations specifically due to the increasing energy costs. Desalination as a concept however, has earlier origins, and was first introduced in maritime operations, where the need to limit the dependence on freshwater storage was the catalyst for the development of the first desalination system [18]. One of the first known devices was invented by 1885 in Glasgow, Scotland, initially designed for use in military [19]. As time passed, it was developed and was able to be used in civilian settings. Although thermal distillation predated reverse osmosis's development however, its significantly lower energy consumption has been a major factor in its popularity across many industries, which has contributed to its rapid and widespread use.

At the time of its invention, desalination mostly was powered by steam, mostly used to ensure the supply of freshwater during naval operations [17]. In the 20th century, the desalination techniques were advancing rapidly and started to address the needs of the civilian population and address water shortages in coastal and arid communities [20]. In the midst of these advancements reverse osmosis was able to emerge as the most popular method of modern desalination because of its low energy usage and increased efficiency in comparison to previous thermal processes [21]. In spite of the fact around 97% of Earth's water is saline and only 1% of it is appropriate for consumption by humans, the problem of water scarcity is not caused by the earth's volume of water however [22]; it is due to an unbalanced access and distribution. The water cycle, which is natural which includes condensation, evaporation and precipitation, cleans water in general. The availability of clean drinking water is different widely across various regions [23]. Certain regions benefit from abundant water resources; however, others face constant water shortages. The coastal regions, particularly, are often afflicted with a dilemma because they are located in an area which is not suitable for human consumption because of its salinity levels. Prior to the introduction of desalination as well as advanced techniques for water purification and the inaccessibility to water that was potable in these areas posed major challenges to satisfying the demands of their population.

6.3 THERMAL DESALINATION PROCESSES

6.3.1 Definition and Principles

Thermal desalination, often called distillation, is among the earliest techniques that were developed to convert brackish and saltwater to potable water [15]. The entire process is founded on the concepts of boiling and evaporation and condensation. In the process the water is heated until its evaporation point, which allows water vapor to be separated from the salts that are dissolved. The vapor then evaporates into clean water leaving the salts in the form of a residue. The energy required to power this process is typically provided by modern steam generators wastewater heat boilers and by using back-pressure steam generated by turbines in power plants. Multiple effect distillation (MED) system are a crucial method for thermal desalination, can be set up in a variety of ways according to the design of the heat exchanger as well as the flow direction of vapor and brine [24]. To improve thermal efficiency and maximize heat recuperation, the systems can be set up in a line (linear arrangement) or as parallel lines that operate under different pressure conditions. Following the top brine temperature (TBT), MED systems are classified as low temperature (operating at temperatures below 90°C) and high temperature (operating over 90°C) (Figure 6.1).

6.3.2 Advantages and Limitations of the Thermal Distillation Process

A major benefit for thermal distillation lies in its insensitive towards the water quality. In contrast to reverse the process of osmosis (RO) which requires feedwater that has minimal suspended solids as well as the control of salinity in order to avoid the membrane from fouling up, thermal distillation

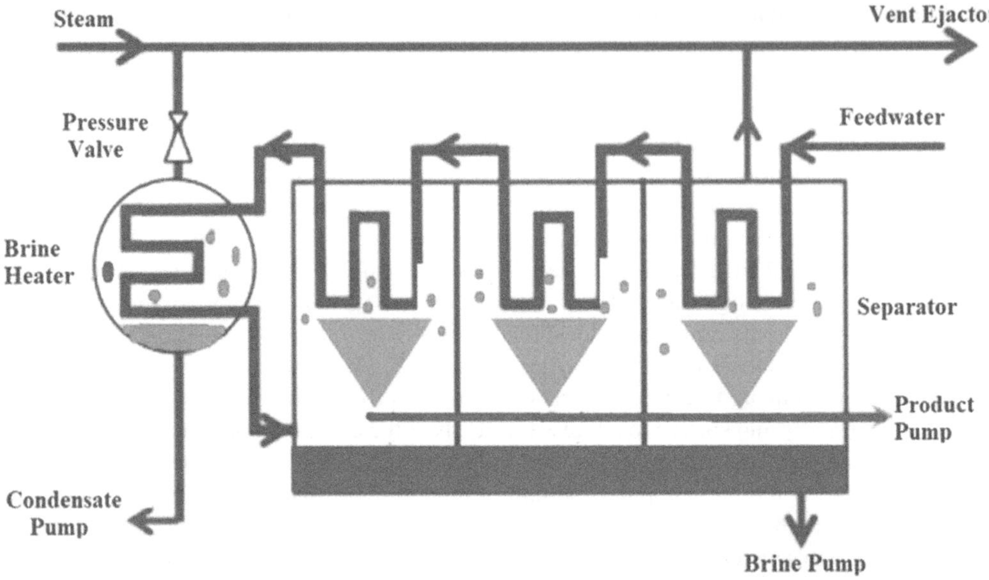

FIGURE 6.1 Thermal distillation process.

is able to effectively treat water with more impurities [25]. This means that the requirements for pre-treatment are not as stringent, which results in a reduction in the operating and maintenance costs that come with the conditioning of water before desalination. Furthermore, when compared to multi-stage flash (MSF) distillation systems, thermal distillation specifically as a result that of multiple-effect distillation (MED) demonstrates higher energy efficiency. MED systems use less thermal energy due to the reuse of heat through multiple factors, which improves the overall performance of desalination [24, 26]. Furthermore, thermal distillation units, in particular MED units, show higher efficiency in performance compared to MSF plants. The higher efficiency comes from better heat transfer mechanisms as well as optimized design layouts, which results in less specific energy usage and lower cost in terms of the amount produced by freshwater. This means that thermal distillation through MED provides a more economical and efficient alternatives in comparison to MSF systems, particularly in situations where there is waste heat recuperation or lower-quality heat sources are in place (Figure 6.2). It is crucial to keep in mind that although thermal distillation offers these benefits it has its own limitations, like reliance on the availability of thermal energy as well as corrosion and scaling problems in heat exchangers and, generally, higher capital cost when

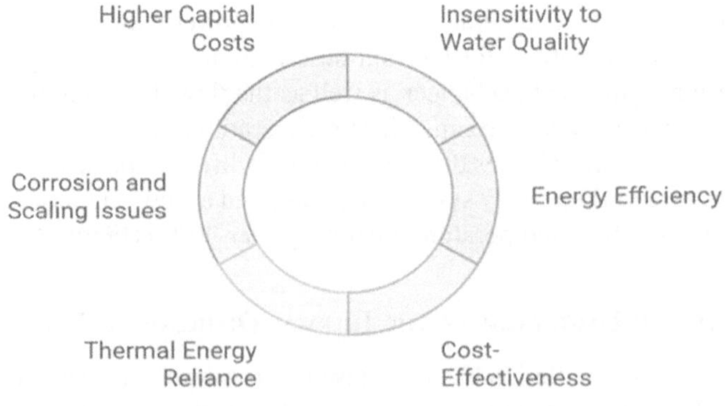

FIGURE 6.2 Thermal advantages and limitations.

compared with membrane-based systems. These elements need to be taken into consideration in selecting the best desalination technique based on the local resources for energy as well as water quality and economic limitations.

6.4 REVERSE-OSMOSIS FILTERING (RO)

Reverse osmosis (RO) can be described as a technique that uses the pressure of the membrane, which allows water molecules to move through and create pure water [27]. It requires only 25 percent more energy compared to the thermodynamic limits [28]. This technology was first developed in the late 1980s, and requires installing a water turbine (Pelton) that transfers energy generated by HPP's main pump to the fluid flow that is brine [29]. A battery-powered motor is used to satisfy the requirements of energy for the pump. The primary HPP boosts the pressure across the entire feedwater saline flow (Figure 6.3). RO is equipped with pipes that carry water from oceans but the bulk of the water supply will come through a pretreatment plant which will be described in the following paragraph. Seawater, pushed by high pressure is forced through an impermeable membrane. The pure water is pumped through a different pipe is sent to the pro-treatment plant to add minerals. The RO membrane of the plant generally consumes less energy than distillation processes [30]. In summary, the water source passes through the phases that include intake, preparation and final treatment. Seawater sources make up 63% of total capacity of desalination across the world [31]. Thermal processes create 61% of the water. The majority of the coastal countries' thermal filtration is used to desalinate seawater. The RO system is a main method alternative for many emerging and developed nations who are starting to consider saltwater as an option for many countries that are now beginning to think about desalination, even though it is the second most significant seawater treatment process in the world. Pro-treatment and pretreatment processes are crucial in RO filtration, since we must eliminate harmful substances to safeguard the RO membranes from harmful substances. Seawater comes with a broad variety of constituents that could harm the membranes, in particular the ones made from cellulose acetate [32] specifically designed to work at temperatures below 35% Celsius and possess a very strict tolerance to PH. Even without an initial treatment process, the expected life span of the membranes is between one to two years prior to replacement is necessary [33].

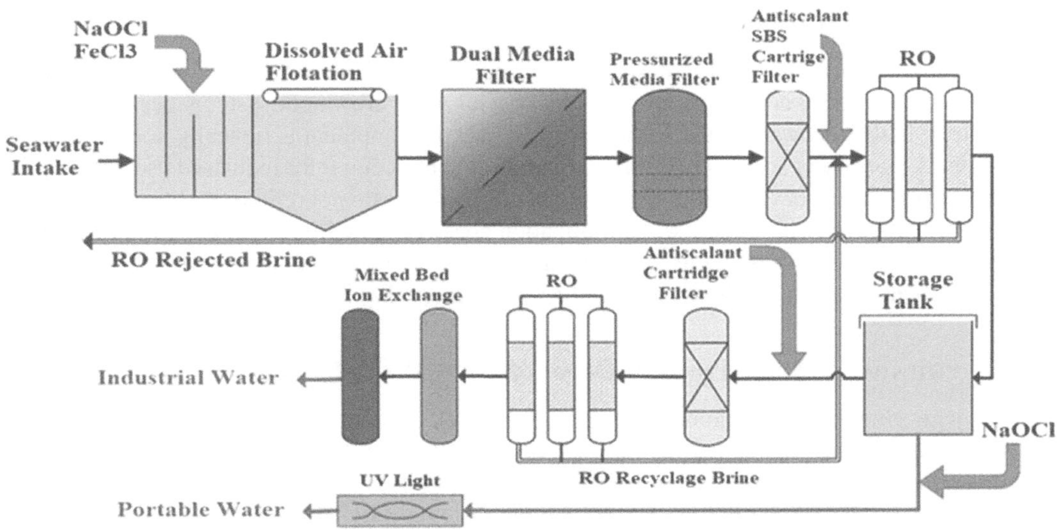

FIGURE 6.3 Schematic diagram of the seawater reverse osmosis (SWRO) process.

6.5 COMPARATIVE ANALYSIS OF OPERATIONAL, ECONOMIC, AND ENVIRONMENTAL ASPECTS OF DESALINATION TECHNOLOGIES

6.5.1 Energy Consumption in Desalination Processes

The process of osmosis shows significantly less energy consumption over single-purpose thermal distillation system [34]. But, when compared with dual purpose distillation systems those that integrate power generation the efficiency gain of RO is not as evident. One of the major drawbacks of RO is the limited life span of membranes, requiring regular replacement [35]. The replacements do not only come at expensive costs, but also add to the footprint of energy consumed by the system. However, the energy use of the thermal distillation process is extremely variable and largely dependent on the local environment, which makes it essential to conduct regional-specific economic and technical analyses.

6.5.2 Economic Evaluation of Large-Scale Seawater Desalination

Desalination using membranes, especially RO, has its own problems for large-scale applications because of the membrane's limited durability and the technical limitations that come with scaling up membrane systems [36]. While RO has the advantage of less energy consumption, this benefit is offset by the high maintenance costs and losses in efficiency when operating at higher capacities. It is well known that for desalination plants that produce more than 6,000 tons of freshwater every day dual-purpose distillation systems are an economical alternative RO, due to their greater capacity and economics for high-output levels.

6.5.3 Pretreatment Requirements for Thermal and Membrane-Based Desalination

Thermal distillation processes, seawater generally requires minimal pretreatment, usually just basic filtration to eliminate large particles [37]. However, RO systems demand extensive pretreatment, which includes clarification, flocculation and chlorination, in order to safeguard membrane integrity and guarantee maximum performance [38]. Due to the low rate of water recovery of RO system, massive and costly pretreatment facilities are needed, which take up a significant amount of physical space.

6.5.4 Infrastructure Needs and Operational Considerations

The thermal desalination system requires steam to begin operations, which calls for the addition of starter boilers that have adequate capacity, and special water treatment systems designed specifically to treat boiler feedwater [39]. Additionally, as backup equipment is typically not accessible in such systems, freshwater sources other than steam should be secured for industrial use during times of system maintenance or downtime. For instance, RO systems do not require steam to start. Lower temperatures of water feedwater at the initialization phase have no impact on the output of the system and no additional infrastructure for startup is needed. But, every desalination system should have a sufficient reserve capacity to sustain the water supply throughout maintenance intervals.

6.5.5 Potential of Solar-Powered Desalination in Water-Scarce Regions

In areas with a high concentration of solar energy, such as the areas of the Middle East and Africa, socio-economic development is often hindered because of the lack of access to potable water [40]. Desalination with sunlight in such regions is an exciting way to combat water scarcity while reducing carbon emissions. Solar-powered desalination systems provide an environmentally friendly approach to water production [41], which is compatible with global efforts to encourage renewable energy and ecological conservation.

6.6 CONCLUSION

As the need for fresh water around the globe is growing due to increasing population and economic growth seawater desalination has been gaining momentum as an essential solution to address water shortages, especially in desert and coastal regions. This study offers a comprehensive study of two well-known techniques for desalination, which include thermal distillation as well as reverse osmosis (RO). Both methods offer distinct advantages; however, they are also subject to limitations regarding the energy use and infrastructure requirements and viability for economics and environmental impact. RO is recognized as being energy efficient in comparison to distillation techniques that are solely for use. However, its advantages are diminished when compared to dual-purpose methods that utilize energy more effectively through integration into the power generation. Additionally, the less operating time of RO membranes, along with the large demands for pretreatment, results in an increase in maintenance costs and reduces the potential of this method in large-scale desalination plants.

Thermal distillation especially as a result of MED shows durability in the treatment of feedwater with various levels of impurity and requires only minimal pretreatment and gains from improved efficiency in thermal energy when the process is used to recover heat from waste. However, it is dependent to thermal power sources as well as higher capital expenses can be limiting particularly in areas that do not have access to cheap energy. The study of solar-powered desalination is an opportunity to sustainably increase water production in areas rich in solar energy regions like the Middle East and Africa. Through the use of renewable energy, these systems will not only reduce water shortages but also contribute to decrease in greenhouse gas emissions. The final decision to select the right desalination technique must be based on the context, taking into account the availability of energy, the economics as well as environmental concerns and the requirements for water quality. Research and development on membrane materials and energy recovery systems and brine management techniques will be crucial to improve the effectiveness, affordability and sustainability of desalination technology soon.

REFERENCES

1. D. Azdem, J. Mabrouki, A. Moufti, S. E. Hajjaji, and A. Fatni, "Present Situation and Environmental Impact of Seawater Desalination Plants in Morocco," in *Technical and Technological Solutions Towards a Sustainable Society and Circular Economy*, J. Mabrouki, and A. Mourade, Eds., Cham: Springer Nature Switzerland, 2024, pp. 581–592. doi: 10.1007/978-3-031-56292-1_46.
2. D. Azdem, J. Mabrouki, and S. El Hajjaji, "Methods and Technologies for the Treatment and Disposal of Brine in Seawater Desalination Plants," in *Advances in Environmental Engineering and Green Technologies*, J. Mabrouki, and M. Azrour, Eds., IGI Global, 2024, pp. 79–87. doi: 10.4018/979-8-3693-3807-0.ch007.
3. D. Azdem, J. Mabrouki, A. Moufti, S. El hajjaji, and A. Fatni, "Assessment of heavy metal contamination in seawater in Agadir coastline, Morocco," *Desalination and Water Treatment*, vol. 317, p. 100129, 2024. doi: 10.1016/j.dwt.2024.100129.
4. A. Fatni *et al.*, "Characterization and mitigation strategies for inorganic scaling in reverse osmosis system treating brackish groundwater," *Sustainable Chemistry One World*, vol. 5, p. 100035, 2025. doi: 10.1016/j.scowo.2024.100035.
5. H. A. Balogun *et al.*, "Cost and energy requirements of electrochemical membrane systems: A critical review and data analysis," *Journal of Environmental Chemical Engineering*, p. 113733, 2024.
6. C. Anastasi, *Desalination and Water Security*. Taylor & Francis, 2024. Consulté le: 19 mars 2025. [En ligne].
7. G. Fahdi, J. Mabrouki, A. Lamchaimech, D. Azdem, M. Benrhanem, and S. E. Hajjaji, "Impacts of climate change on groundwater quality and recharge in the Tensift basin, Morocco," *Water Conservation and Management*, vol. 8, no. 4, pp. 466–474, 2024. doi: 10.26480/wcm.04.2024.466.474.
8. G. Fahdi, D. Azdem, A. Lamchaimech, M. Benrhanem, J. Mabrouki, and S. El Hajjaji, "Analysis of the impact of climate change on water resources: Case of the Tensift basin (Morocco)," *Water Conservation and Management*, vol. 8, no. 4, pp. 461–465, 2024. doi: 10.26480/wcm.04.2024.461.465.

9. D. Azdem, A. Fatni, J. Mabrouki, and S. El Hajjaji, "The desalination strategy and economic integration of Morocco: Water scarcity, prospects, and challenges," *Circular Economy Applications in Energy Policy*, pp. 257–268, 2025.
10. A. Fatni, A. El Hammadi, R. Bouaddi, A. A. Taleb, N. El Baraka, and A. Laknifli, "Experimental design applied to improve the efficiency and the performance of the reverse osmosis process," *Desalination and Water Treatment*, vol. 236, pp. 45–54, 2021.
11. L. Migliore, "Addressing water scarcity in rural areas of the Souss-Massa region of Morocco: Comparative analysis of desalination, fog harvesting, and wastewater treatment methods with considerations of technical viability and human rights," 2024.
12. P. S. Minhas, and M. Qadir, *Irrigation Sustainability With Saline and Alkali Waters: Extent, Impacts and Management Guidelines*. Singapore: Springer Nature Singapore, 2024. doi: 10.1007/978-981-97-4102-1.
13. N. Zouhri, F. Z. Addar, M. Tahaikt, M. Elamrani, A. ELmidaoui, and M. Taky, "Techno-economic study and optimization of the performance of nanofiltration and reverse osmosis membranes in reducing the salinity of M′ rirt water city (Morocco)," *Desalination and Water Treatment*, vol. 317, p. 100042, 2024.
14. A. Politano *et al.*, " 2024 roadmap on membrane desalination technology at the water-energy nexus," *Journal of Physics: Energy*, vol. 6, no. 2, p. 021502, 2024.
15. S. M. J. S. Sabour, and B. Ghorashi, "A comprehensive review of major water desalination techniques and mineral extraction from saline water," *Separation and Purification Technology*, vol. 349, p. 127913, 2024.
16. M. Khashehchi *et al.*, "Solar Desalination Techniques: Challenges and Opportunities," in *Highly Efficient Thermal Renewable Energy Systems*, 2024, pp. 305–329. https://www.taylorfrancis.com/chapters/edit/10.1201/9781003472629-19/solar-desalination-techniques-morteza-khashehchi-sivasakthivel-thangavel-pooyan-rahmanivahid-milad-heidari-taleb-moazzeni-vikas-verma-ashwani-kumar
17. J. Birkett, "The Origins of Today's Desalination Technologies," in *A Multidisciplinary Introduction to Desalination*, River Publishers, 2022, pp. 265–293.
18. P. Kumar, A. Date, N. Mahmood, R. K. Das, and B. Shabani, "Freshwater supply for hydrogen production: An underestimated challenge," *International Journal of Hydrogen Energy*, vol. 78, pp. 202–217, 2024.
19. D. Hammadou, "Brackish water demineralization and brine fate. Case study: Ouled Djellal and El-Oued plants," 2024.
20. F. Ghizlane, J. Mabrouki, F. Ghrissi, and M. Azrour, "Proposal for a high-resolution particulate matter (PM10 and PM2. 5) capture system, comparable with hybrid system-based internet of things: Case of quarries in the western Rif, Morocco," *Pollution*, vol. 8, no. 1, pp. 169–180, 2022.
21. M. A. Elazab, H. A. Dahab, A. T. Elgohr, and M. S. Elhadidy, "A comprehensive review on hybridization in sustainable desalination systems," *Membranes*, vol. 20, p. 21, 2025.
22. S. Bhutto, A. Ahmed, and M. Hussain, "Hydro politics in Pakistan: A challenge to the National Unity," *International Research Journal of Management and Social Sciences*, vol. 5, no. 3, pp. 301–315, 2024.
23. N. Al-Jadabi, M. Laaouan, S. El Hajjaji, J. Mabrouki, M. Benbouzid, and D. Dhiba, "The dual performance of Moringa oleifera seeds as eco-friendly natural coagulant and as an antimicrobial for wastewater treatment: A review," *Sustainability*, vol. 15, no. 5, p. 4280, 2023.
24. H. A. H. Al-Hamzawi, and A. S. A. Al Sailawi, "Optimized multi-generation system for sustainable desalination and power production using solar-driven Multi-Effect Distillation (MED) in Basra, Iraq," *Energy for Sustainable Development | Journal*, vol. 85, p. 101684, 2025.
25. S. Mgalaa, J. Mabrouki, M. Elouardi, L. El Azzouzi, A. Moufti, S. El Hajjaji, and M. A. El Belghiti, "Study and evaluation of the degradation of procion blue dye by the ozonation method: Parametric and isothermal study," *Nanotechnology for Environmental Engineering*, vol. 7, no. 3, pp. 691–697, 2022.
26. Z. Rahimi-Ahar, and M. S. Hatamipour, "Exergy analysis of thermal desalination processes: A review," *Clean Technologies and Environmental Policy*, vol. 26, no. 5, pp. 1335–1362, 2024.
27. U. H. Nuhu *et al.*, "Rapid synthesis of charged covalent organic framework for sustainable reverse osmosis membranes," *Desalination*, vol. 607, p. 118797, 2025.
28. C. Liu, K. Rainwater, and L. Song, "Energy analysis and efficiency assessment of reverse osmosis desalination process," *Desalination*, vol. 276, no. 1–3, pp. 352–358, 2011.
29. T. K. Løken, *Design of a Portable Reverse Osmosis System*, NTNU, 2016.
30. R. R. Chandwankar, and J. Nowak, "Thermal Processes for Seawater Desalination: Multi-Effect Distillation, Thermal Vapor Compression, Mechanical Vapor Compression, and Multistage Flash," in *Handbook of Water and Used Water Purification*, Springer, 2024, pp. 465–502.

31. V. G. Gude, "Desalination and sustainability – An appraisal and current perspective," *Water Research*, vol. 89, pp. 87–106, 2016.
32. A. Etale, A. J. Onyianta, S. R. Turner, and S. J. Eichhorn, "Cellulose: A review of water interactions, applications in composites, and water treatment," *Chemical Reviews*, vol. 123, no. 5, pp. 2016–2048, 2023.
33. A. Alkhouzaam, and M. Khraisheh, "Towards sustainable management of end-of-life membranes: Novel transformation of end of life (EoL) reverse osmosis membranes for efficient dye/salt separation," *Desalination*, vol. 571, p. 117104, 2024.
34. M. Salajeghe, and M. Ameri, "Evaluation of the energy consumption of hybrid desalination RO-MED-FD to reduce rejected brine," *Environmental Progress & Sustainable Energy*, vol. 43, no. 2, p. e14312, 2024.
35. S. Wu *et al.*, "Next-generation desalination membranes empowered by novel materials: Where are we now?" *Nano-Micro Letters*, vol. 17, no. 1, pp. 1–24, 2025.
36. P. S. Goh, T. Matsuura, A. F. Ismail, and N. Hilal, "Recent trends in membranes and membrane processes for desalination," *Desalination*, vol. 391, pp. 43–60, 2016.
37. B. Van der Bruggen, and C. Vandecasteele, "Distillation vs. membrane filtration: overview of process evolutions in seawater desalination," *Desalination*, vol. 143, no. 3, pp. 207–218, 2002.
38. S. F. Anis, R. Hashaikeh, and N. Hilal, "Reverse osmosis pretreatment technologies and future trends: A comprehensive review," *Desalination*, vol. 452, pp. 159–195, 2019.
39. M. Al-Sahali, and H. Ettouney, "Developments in thermal desalination processes: Design, energy, and costing aspects," *Desalination*, vol. 214, no. 1–3, pp. 227–240, 2007.
40. A. B. Stambouli *et al.*, "Trends and challenges of sustainable energy and water research in North Africa: Sahara solar breeder concerns at the intersection of energy/water," *Renewable and Sustainable Energy Reviews*, vol. 30, pp. 912–922, 2014.
41. S. Kumar, M. Kumar, S. Chowdhury, B. S. Rajpurohit, and J. K. Randhawa, "Environmental concerns and long-term solutions for solar-powered water desalination," *Journal of Cleaner Production*, vol. 345, p. 131180, 2022.

7 Green Hydrogen Production through Water Desalination

C. Lamiri, K. El-Moustaqim, A. Ghailane, J. Mabrouki, and Y. Abrouki

7.1 INTRODUCTION

Hydrogen production, a topic attracting considerable attention worldwide, offers notable environmental benefits and energy flexibility. Indeed, hydrogen (H_2) has a gravimetric energy density 2.75 times greater than that of hydrocarbons, making it a relevant alternative to fossil fuels (Fallah Vostakola, Salamatinia, and Amini Horri 2022; Nasser et al. 2022). Overall, hydrogen production is categorized into 10 color codes, depending on the technology used, the energy source, and the greenhouse gas (GHG) emissions generated (Incer-Valverde et al. 2023; Panić, Cuculić, and Ćelić 2022).

Green hydrogen is a clean energy carrier, i.e., derived from renewable sources and generating no GHG emissions. It is considered an energy carrier of the future, with the promise of a net zero balance in terms of GHGs (El-Emam and Özcan 2019; Hassan et al. 2023). However, as hydrogen is a secondary energy source, the development of sustainable production technologies is of crucial importance in the transition to a hydrogen economy (El-Moustaqim, Mabrouki, and Hmouni 2024a, 2024c; El-Moustaqim et al. 2024b, 2024d). According to the International Energy Agency's (IEA) 2023 report, and a number of concordant studies, the cost of hydrogen production fluctuates according to the energy source and technology employed (Oni et al. 2022; Younas et al. 2022; Zainal et al. 2024).

In green H_2 production, the main components involved are renewable energy sources, electrolyzers, water feedstock, and storage systems (El-Moustaqim et al. 2024b, 2024d). These four components use a range of tools and technologies, allowing multiple production system configurations to be set up.

Renewable energies – solar, wind, geothermal, hydroelectric, biomass, and solar thermal – are used to generate electrical or thermal energy (El-Emam and Özcan 2019), while we look at the different types of electrolyzers, which mainly include alkaline electrolyzers (AK), proton exchange membrane electrolyzers (PEM), and solid oxide electrolyzer cells (SOEC) (Ishaq, Dincer, and Crawford 2022). With regard to raw water, ultrapure water is obtained by direct supply or by desalination of seawater (Khan et al. 2021). Ultimately, the method used to store hydrogen depends on its use, whether physical or chemical (Vidas and Castro 2021).

Basic stoichiometric calculations for water electrolysis show that 9 kg of ultrapure water are needed to produce 1 kg of hydrogen. However, it is important to note that the actual water requirements of electrolyzers are higher and can vary between 10.0 and 22.4 kg of ultrapure water for 1 kg of hydrogen produced (Simoes et al. 2021). In the current context, the use of salt water, which contains sodium chloride (NaCl), is proving unsuitable for electrolysis. This is due to a number of factors, including the oxidation reaction of chlorine at the anode, changes in the local hydrogen potential (pH) of the electrode, and the deleterious impact of bacteria and other impurities on the electrolyzer (Liu et al. 2021). Although a great deal of research has been devoted to the direct electrolysis of seawater, i.e., without prior desalination, no direct electrolyzer is currently available on the market. The first large-scale projects are currently in the development phase (Liu et al. 2024); that is why seawater currently undergoes a desalination and purification process before being subjected to electrolysis (Sztekler et al. 2024).

Hydrogen-based products are currently used as raw materials in industrial activities. However, if its potential as a complete energy carrier is to be fully exploited, it is likely to play a leading role in many other sectors. Of the 50 million metric tons produced worldwide each year, its main use is as a

DOI: 10.1201/9781003647263-7

raw material in the production of ammonia, with a 35% share in petroleum refining (Hanley, Deane, and Gallachóir 2018). Major applications can be envisaged in freight and passenger transport, i.e., internal combustion engines and fuel cell vehicles; storage (gaseous or liquid hydrogen); and thermal applications (natural gas mixtures and solid oxide fuel cells) (Hanley, Deane, and Gallachóir 2018; Yue et al. 2021). For electricity generation, two main methods can be distinguished: electricity generation on the one hand and gas-fired generation on the other (Bakenne, Nuttall, and Kazantzis 2016).

The aim of this project is to produce green hydrogen. This approach is based on water desalination to extract fresh water, followed by electrolysis powered by renewable energy sources; furthermore, the role of green hydrogen in future energy production and the decarbonization of the world.

7.2 WATER DESALINATION TECHNOLOGIES

The oceans and seas contain considerable reserves of water, making them ideal sources for electrolytic hydrogen production, but water salinity poses a major challenge to the use of seawater in this process (Ghizlane et al. 2022). To overcome this difficulty, water desalination and purification techniques are used to achieve the specifications required for hydrogen production. Water desalination technologies can be classified into two main categories according to their operating principle: thermal technologies and membrane technologies (Figure 7.1).

Thermal technologies require the use of heat or cold to operate, and desalination is generally achieved by phase change, such as evaporation or crystallization. Membrane technologies, on the other hand, are based on the use of semi-permeable membranes which, through a filtration mechanism, allow the passage of water molecules while limiting that of other components (Curto, Franzitta, and Guercio 2021). Thermal technologies encompass a range of methods and approaches, of which the following are representative, without claiming to be exhaustive:

- Multiple-effect distillation (MED)
- Multi-stage flash (MSF)
- Thermal vapor compression (MVC)
- Mechanical vapor compression (MVC)

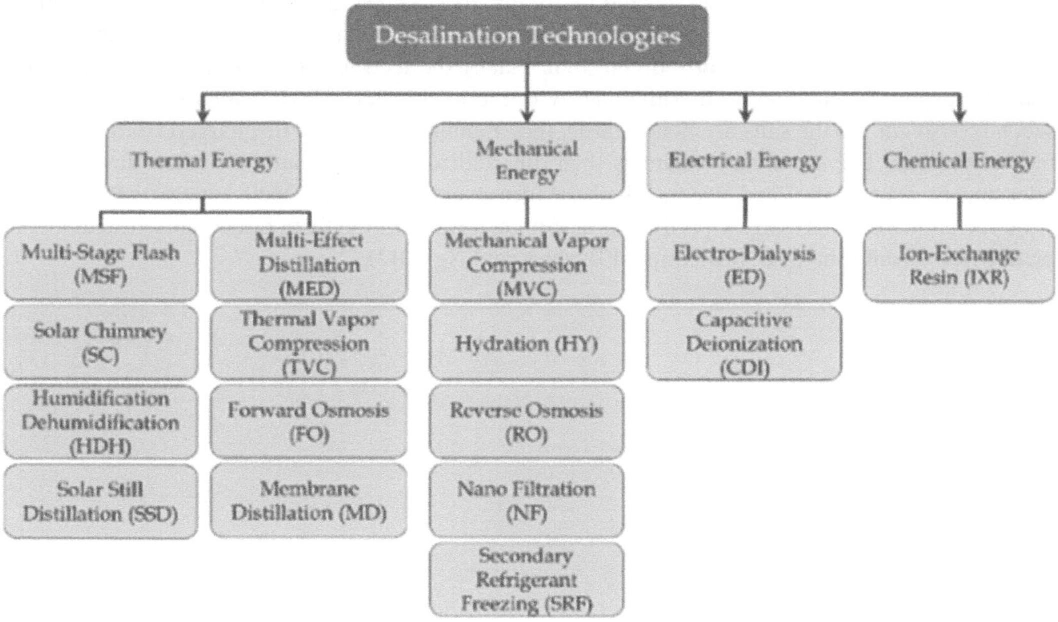

FIGURE 7.1 Water desalination technologies (Alkaisi, Mossad, and Sharifian-Barforoush 2017).

- Adsorption desalination (AD)
- Solar distillation (SD)
- Humidification-dehumidification (HDH)
- Freeze crystallization (FC)

The second group of desalination technologies are membrane technologies, such as:

- Reverse osmosis (RO)
- Electrodialysis (ED)
- Nanofiltration (NF)
- Direct osmosis (DO)

In today's context, RO, MSF, and MED technologies dominate the water desalination market, producing a cumulative 94% of all desalinated water (Jones et al. 2019). All three technologies are proven, widely adopted, and well documented. Consequently, the detailed review will focus exclusively on RO, MSF, and MED technologies.

7.2.1 Multi-effects Distillation (MED)

MED units are modular in design, enabling them to be configured in a variety of ways, depending on the shape of the heat exchangers or the orientation of the brine flow relative to the steam direction. In addition, these units can be grouped into one or two parallel lines, operating at different pressures, to optimize heat recovery (Blagin, Shimanov, and Gorshkalev 2019). MED units are classified according to their maximum brine temperature (MTB), which can be either below or above 90°C. MED units can be divided into two categories: low-temperature units, with a TBT below 90°C, and high-temperature units, with a TBT above 90°C (Blagin, Shimanov, and Gorshkalev 2019).

The maximum brine temperature is limited to 120°C due to calcium sulfate scaling. In addition, the minimum temperature is linked to the temperature available in the wave (Shatat, Riffat, and Ghabayen 2012). A conceptual diagram of the MED unit is shown in Figure 7.2. In this desalination process, the feed water, taken from the sea or ocean, undergoes a pretreatment process detailed in the following subsection. It is then heated to around 70°C and sprayed into evaporation chambers under reduced pressure. The first chamber is supplied with steam from a boiler or other heated fluid, whose temperature is slightly higher than that of the feed water. This steam condenses in a heat exchanger, transferring its thermal energy to the feed water, which evaporates and generates secondary steam. As the salts are non-volatile, they remain in the brine discharged from the evaporation chamber. The secondary steam is then directed to the next chamber, where it condenses, releasing its heat to the feed water, which in turn evaporates. This cycle repeats itself through several chambers, progressively producing desalinated water in the form of condensed steam, completely free of salt (Shahzad, Muhammad Burhan, and Ng 2017).

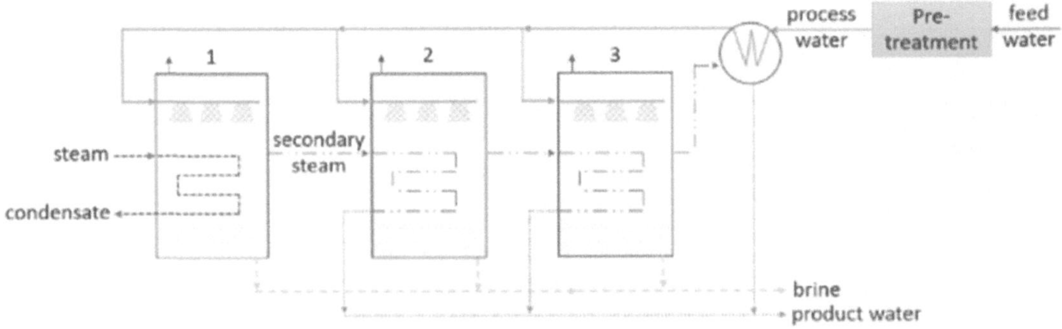

FIGURE 7.2 Diagram of a multiple-effect distillation unit.

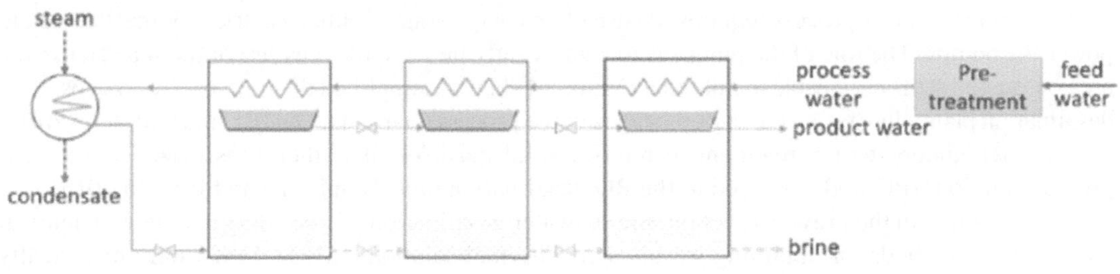

FIGURE 7.3 Diagram of an MSF desalination plant.

MED freshwater production units have a daily production capacity ranging from 2,000 to 20,000 m³/day. These ingenious systems efficiently meet the water needs of local populations. Moreover, they can be coupled with thermal or mechanical steam compression devices, an innovation that enhances their energy efficiency (Sharon and Reddy 2015).

7.2.2 Multi-stage Falsh (MSF)

MSF is a seawater-to-freshwater desalination process commonly used on ships and along coastlines in many parts of the world, including the United States, the Middle East, and Korea (Tenno and Nguyen 2016).

As shown in Figure 7.3, the MSF unit conceptual design includes a preliminary pretreatment step for the feed water, which is either taken from the sea or ocean or an alternative source. Subsequently, the process water is heated to a higher brine temperature of around 90 to 110°C, and then flows into the first evaporation chamber via the expansion valve (Rao et al. 2016; Shahzad, Muhammad Burhan, and Ng 2017). As a direct consequence of this process, the process water undergoes boiling, which leads to vaporization. The resulting salt-free steam condenses on the heat exchanger at the top of the evaporation chamber, transferring heat to the cold process water. At the same time, the brine flows from the first chamber to the second via the pressure reducer, undergoes boiling, and the whole process is repeated in the subsequent chambers in the successive evaporation chambers, the pressure decreases, allowing the water to boil at lower and lower temperatures.

7.2.3 Reverse Osmosis

Reverse osmosis (RO) is the most widely used desalination technology, as it has a relatively low energy consumption compared to other methods (Figure 7.4). Unlike thermal desalination processes, reverse osmosis does not require the use of heat, only electricity, to operate. The energy demand of reverse osmosis depends on various parameters, including the salinity of the feed water, the type of membranes employed, and the quality of the desired permeate. It is commonly estimated at between 0.3 and 7.5 kWh/m³ of permeate produced (Rao et al. 2016). In addition, the recovery rate (RR) of single-stage reverse osmosis units is generally between 0.75 and 0.80, while for three-stage units it is between 0.85 and 0.90.

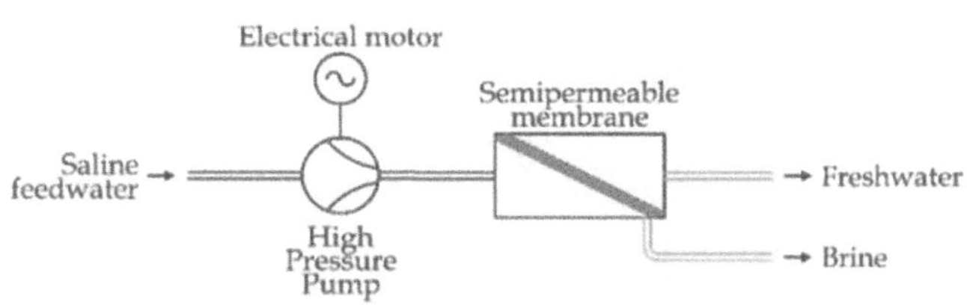

FIGURE 7.4 Diagram of a simple reverse osmosis (RO) desalination unit.

The reverse osmosis process requires the use of an energy source, either electrical or mechanical, to power the pumps. The role of the pumps is to significantly increase the pressure of the seawater in the area before the semi-permeable membrane. Figure 7.4 shows a simplified diagram of a reverse osmosis desalination plant, illustrating the various stages in the process (Garofalo, Giudici, and Guastella 2020).

Seawater undergoes pretreatment to remove solid particles. It is then pressurized by a high-pressure pump (HPP) and conveyed to the RO desalination unit (Henthorne and Boysen 2015).

As mentioned in the previous section, the seawater desalination sector offers a range of chemical and physical methods for obtaining fresh water. Thermal solutions (MED, MSF) were historically the first techniques adopted, but today membrane technologies (mainly RO) are spreading rapidly worldwide (Kasaeian, Rajaee, and Yan 2019).

7.3 GREEN HYDROGEN PRODUCTION THROUGH ELECTROLYSIS

An electrolyzer plays a crucial role in hydrogen production. This device is designed to convert water electrochemically, thereby separating the constituent elements of hydrogen and oxygen (Mohammadi and Mehrpooya 2018). At constant temperature (25°C) and pressure (1 bar), the change in Gibbs free energy is 237.2 kJ/mol. This value corresponds to the minimum electrical energy required to initiate the reaction. A further 48.6 kJ/mol represents the thermal energy required to cover the total enthalpy change (Lamy and Millet 2020).

There are several electrolytic systems available for water electrolysis, including anion exchange membrane electrolysis (AEM), alkaline water electrolysis (AWE), solid oxide water electrolysis (SOE), and proton exchange membrane electrolysis (PEM). Although their operating principles are similar, these systems operate under different conditions and with different materials (Chi and Yu 2018).

7.3.1 Electrolysis of Alkaline Water (AWE)

The method uses two non-platinum metals, nickel (Ni) and iron (Fe) electrodes, an electrolyte containing 30–40% potassium hydroxide (KOH), and a diaphragm membrane. Alkaline electrolysis is considered a reliable technology for hydrogen production up to the megawatt scale (Li and Baek 2021).

A typical AWE operates at a current density of between 300 and 400 mA cm^2, with a cell voltage of between 1.85 and 2.2 V, a moderate temperature of 70 to 90°C, and conversion efficiencies of between 60 and 80%. This technology has the significant advantage of not requiring the use of noble metal catalysts for hydrogen production. What is more, its relatively low temperature makes it easy to handle (Vincent and Bessarabov 2018).

7.3.2 Solid Oxide Water Electrolysis (SOWE)

In the study of solid oxide water electrolyzers (SOWEs), it is observed that water molecules undergo a reduction to hydrogen (H_2) and oxide ion (O_2^-) upon the addition of two electrons (e^-) at the cathode. Subsequently, the oxide ion propagates to the anode through the solid oxide electrolyte. This reaction leads to the formation of pure oxygen (O_2) at the anode (Shiva Kumar and Lim 2022).

7.3.3 Proton Exchange Membrane Electrolysis (PEM)

PEM is an essential component in fuel cells, where it plays a crucial role in electricity generation, as well as in electrolyzers, enabling hydrogen production. In addition, the PEM assumes a separating function between the cathode and anode, a crucial aspect for system stability and efficiency. The most widely used models are Nafion™ and Nafion™-based membranes, due to their remarkable physical and chemical properties. Indeed, they stand out for their high thermostability, high ionic conductivity, chemical stability, mechanical strength, and robustness at low temperatures and high relative humidity conditions (Ahmad Kamaroddin et al. 2021). The use of Nafion™ is subject to significant constraints, particularly due to its low proton conductivity when temperatures are high in a low-humidity environment, and the need for a longer synthesis time (Ahmad Kamaroddin et al. 2021; Kim et al. 2016).

In the membrane electrolysis reaction (PEM), water is introduced at the anode, where separation into proton and oxygen takes place. The proton then moves across the membrane to the cathode, where it recombines to form hydrogen (Holladay et al. 2009). for the separation of chemical elements, it is essential to understand the reaction process of oxygen gas with water. It has been found that the ionic resistance of PEM electrolyzers is low; as a result, a high current, more than 1,600 mA cm^{-2}, can be achieved while maintaining high yields of 55–70% (Holladay et al. 2009).

In this study, we will look at the simplified construction and operating principle of AWE, PEM, and SOEW electrolyzers. These different types of electrolyzers are shown in Figure 7.5.

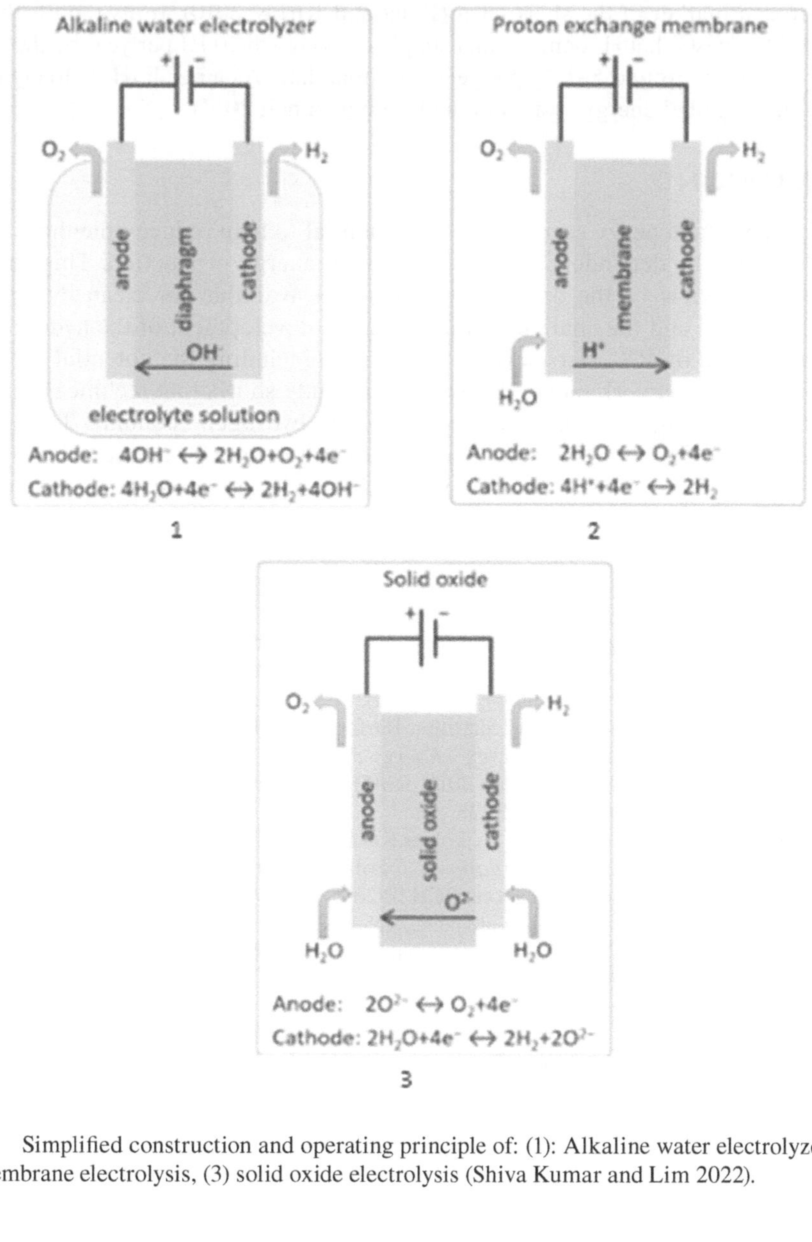

FIGURE 7.5 Simplified construction and operating principle of: (1): Alkaline water electrolyzer, (2) proton exchange membrane electrolysis, (3) solid oxide electrolysis (Shiva Kumar and Lim 2022).

7.4 HYDROGEN'S ROLE IN FUTURE ENERGY PRODUCTION AND GLOBAL DECARBONIZATION

Green hydrogen is a hot topic that is attracting growing interest from the scientific community. Considered a promising energy carrier for sustainable development and the energy transition, it is produced from water and renewable energy sources by electrolysis, with no greenhouse gas emissions. This characteristic makes green hydrogen a key player in the fight against climate change and the quest for carbon neutrality on a global scale. In addition, global demand for green hydrogen and its applications is expected to grow exponentially over the next decade, and international initiatives are underway to produce green hydrogen from renewable energy sources such as wind and solar power (Shiva Kumar and Lim 2022).

In the context of power grids with a high proportion of variable renewable electricity, hydrogen production is likely to help reduce production constraints. Water electrolysis, a well-established industrial technology, is a process that requires the use of grid electricity and is also likely to influence the hydrogen market in the short term (Balat and Kırtay 2010). In its report, the Hydrogen Council issued forecasts that H_2 demand and supply could reach 10 EJ per year by the end of 2050, with further growth of around 5–10% per year after that date. As a result, H_2 is likely to become a major player in the global energy system in the future (Council 2017).

7.5 CONCLUSION

Hydrogen is seen as an energy carrier with the potential to help reduce greenhouse gas (GHG) emissions generated by dependence on fossil fuels for energy production. This study presents a comprehensive review of the different technologies available for clean hydrogen production, their limitations and the challenges hindering the development of the hydrogen economy. Hydrogen production by RS offers several advantages, including the potential for a decentralized hydrogen supply network model. However, this study shows that technical challenges and other factors are holding back the development of the hydrogen economy. These include the lack of a value chain for clean hydrogen, difficulties in storing and transporting hydrogen, the high cost of production, the absence of international standards, investment risks, and the flammability of hydrogen.

REFERENCES

Ahmad Kamaroddin, Mohd Fadhzir, Nordin Sabli, Tuan Amran Tuan Abdullah, Shamsul Izhar Siajam, Luqman Chuah Abdullah, Aishah Abdul Jalil, and Arshad Ahmad. 2021. "Membrane-based electrolysis for hydrogen production: A review". *Membranes* 11 (11): 810. https://doi.org/10.3390/membranes11110810.

Alkaisi, Ahmed, Ruth Mossad, and Ahmad Sharifian-Barforoush. 2017. "A review of the water desalination systems integrated with renewable energy". *Energy Procedia*, 1st International Conference on Energy and Power, ICEP2016, 14-16 December 2016, RMIT University, Melbourne, Australia, 110: 268–74. https://doi.org/10.1016/j.egypro.2017.03.138.

Bakenne, Adetokunboh, William Nuttall, and Nikolaos Kazantzis. 2016. "Sankey-diagram-based insights into the hydrogen economy of today". *International Journal of Hydrogen Energy*, Special Issue on Progress in Hydrogen Production and Applications (ICH2P-2015), 3-6 May 2015, Oshawa, Ontario, Canada, 41 (19): 7744–53. https://doi.org/10.1016/j.ijhydene.2015.12.216.

Balat, Havva, and Elif Kırtay. 2010. "Hydrogen from biomass – Present scenario and future prospects". *International Journal of Hydrogen Energy* 35 (14): 7416–26. https://doi.org/10.1016/j.ijhydene.2010.04.137.

Blagin, E.V., A.A. Shimanov, and A.A. Gorshkalev. 2019. "Determination of the criteria for comparative analysis of desalination plant". *IOP Conference Series: Earth and Environmental Science* 264 (1): 012024. https://doi.org/10.1088/1755-1315/264/1/012024.

Chi, Jun, and Hongmei Yu. 2018. "Water electrolysis based on renewable energy for hydrogen production". *Chinese Journal of Catalysis* 39 (3): 390–94. https://doi.org/10.1016/S1872-2067(17)62949-8.

Council, Hydrogen. 2017. "Hydrogen scaling up: A sustainable pathway for the global energy transition". https://www.h2knowledgecentre.com/content/policypaper1201.

Curto, Domenico, Vincenzo Franzitta, and Andrea Guercio. 2021. "A review of the water desalination technologies". *Applied Sciences* 11 (2): 670. https://doi.org/10.3390/app11020670.

El-Emam, Rami S., and Hasan Özcan. 2019. "Comprehensive review on the techno-economics of sustainable large-scale clean hydrogen production". *Journal of Cleaner Production* 220: 593–609. https://doi.org/10.1016/j.jclepro.2019.01.309.

El-Moustaqim, Khadija, Jamal Mabrouki, and Driss Hmouni. 2024a. "Applied sciences green microalgae for future biomass development". In *Advanced Systems for Environmental Monitoring, IoT and the application of Artificial Intelligence*, 203–11. https://doi.org/10.1007/978-3-031-50860-8_12.

El-Moustaqim, Khadija, Jamal Mabrouki, Mourade Azrour, and Driss Hmouni. 2024b. "How can green energy be improved by integrating IoT into the microalgae process?" In *Sustainable and Green Technologies for Water and Environmental Management*. Springer, 239–52. https://doi.org/10.1007/978-3-031-52419-6_17.

El-Moustaqim, Khadija, Jamal Mabrouki, and Driss Hmouni 2024c. "From microalgal biomass to biohydrogen: A novel approach to environmentally sustainable technology". *Advancements in Climate and Smart Environment Technology*, 29–38. https://www.igi-global.com/chapter/from-microalgal-biomass-to-biohydrogen/346691.

El-Moustaqim, Khadija, Salah Eddine Sbai, Jamal Mabrouki, and Driss Hmouni. 2024d. "Critical review of green microalgal biomass development through CO2 sequestration". *Euro-Mediterranean Journal for Environmental Integration*. https://doi.org/10.1007/s41207-024-00548-6.

Fallah Vostakola, Mohsen, Babak Salamatinia, and Bahman Amini Horri. 2022. "A review on recent progress in the integrated green hydrogen production processes". *Energies* 15 (3): 1209. https://doi.org/10.3390/en15031209.

Garofalo, Elisabetta, Federico Giudici, and Salvatore Guastella. 2020. "A methodological framework for fostering renewable investments on photovoltaic in small Italian islands". *Frontiers in Energy Research* 8. https://doi.org/10.3389/fenrg.2020.560251.

Ghizlane, F., J. Mabrouki, F. Ghrissi, and M. Azrour. 2022. "Proposal for a high-resolution particulate matter (PM10 and PM2. 5) capture system, comparable with hybrid system-based internet of things: Case of quarries in the Western Rif, Morocco". *Pollution*, 8 (1): 169–80.

Hanley, Emma S., JP Deane, and BP Ó Gallachóir. 2018. "The role of hydrogen in low carbon energy futures–A review of existing perspectives". *Renewable and Sustainable Energy Reviews* 82: 3027–45. https://doi.org/10.1016/j.rser.2017.10.034.

Hassan, Qusay, Ammar M. Abdulateef, Saadoon Abdul Hafedh, Ahmed Al-samari, Jasim Abdulateef, Aws Zuhair Sameen, Hayder M. Salman, Ali Khudhair Al-Jiboory, Szymon Wieteska, and Marek Jaszczur. 2023. "Renewable energy-to-green hydrogen: A review of main resources routes, processes and evaluation". *International Journal of Hydrogen Energy* 48 (46): 17383–408. https://doi.org/10.1016/j.ijhydene.2023.01.175.

Henthorne, Lisa, and Buddy Boysen. 2015. "State-of-the-art of reverse osmosis desalination pretreatment". *Desalination*, State-of-the-Art Reviews in Desalination, 356: 129–39. https://doi.org/10.1016/j.desal.2014.10.039.

Holladay, J. D., J. Hu, D. L. King, and Y. Wang. 2009. "An overview of hydrogen production technologies". *Catalysis Today*, Hydrogen Production - Selected papers from the Hydrogen Production Symposium at the American Chemical Society 234th National Meeting & Exposition, August 19-23, 2007, Boston, MA, USA, 139 (4): 244–60. https://doi.org/10.1016/j.cattod.2008.08.039.

Incer-Valverde, Jimena, Amira Korayem, George Tsatsaronis, and Tatiana Morosuk. 2023. "Colors" of hydrogen: Definitions and carbon intensity". *Energy Conversion and Management* 291: 117294. https://doi.org/10.1016/j.enconman.2023.117294.

Ishaq, Haris, Ibrahim Dincer, and Curran Crawford. 2022. "A review on hydrogen production and utilization: Challenges and opportunities". *International Journal of Hydrogen Energy*, SI: Progress in Hydrogen Production, Storage and Distribution (Ahmadi), 47 (62): 26238–64. https://doi.org/10.1016/j.ijhydene.2021.11.149.

Jones, Edward, Manzoor Qadir, Michelle T. H. van Vliet, Vladimir Smakhtin, and Seong-mu Kang. 2019. "The state of desalination and brine production: A global outlook". *Science of The Total Environment* 657: 1343–56. https://doi.org/10.1016/j.scitotenv.2018.12.076.

Kasaeian, Alibakhsh, Fatemeh Rajaee, and Wei-Mon Yan. 2019. "Osmotic desalination by solar energy: A critical review". *Renewable Energy* 134: 1473–90. https://doi.org/10.1016/j.renene.2018.09.038.

Khan, M. A., Tareq Al-Attas, Soumyabrata Roy, Muhammad M. Rahman, Noreddine Ghaffour, Venkataraman Thangadurai, Stephen Larter, Jinguang Hu, Pulickel M. Ajayan, and Md Golam Kibria. 2021. "Seawater electrolysis for hydrogen production: A solution looking for a problem? *Energy & Environmental Science* 14 (9): 4831–39. https://doi.org/10.1039/D1EE00870F.

Kim, Deuk Ju, Dong Hee Choi, Chi Hoon Park, and Sang Yong Nam. 2016. "Characterization of the sulfonated PEEK/sulfonated nanoparticles composite membrane for the fuel cell application". *International Journal of Hydrogen Energy* 41 (13): 5793–802. https://doi.org/10.1016/j.ijhydene.2016.02.056.

Lamy, Claude, and Pierre Millet. 2020. "A critical review on the definitions used to calculate the energy efficiency coefficients of water electrolysis cells working under near ambient temperature conditions". *Journal of Power Sources* 447: 227350. https://doi.org/10.1016/j.jpowsour.2019.227350.

Li, Changqing, and Jong-Beom Baek. 2021. "The promise of hydrogen production from alkaline anion exchange membrane electrolyzers". *Nano Energy* 87: 106162. https://doi.org/10.1016/j.nanoen.2021.106162.

Liu, Tao, Zhiyu Zhao, Wenbin Tang, Yi Chen, Cheng Lan, Liangyu Zhu, and Wenchuan Jiang et al. 2024. "In-situ direct seawater electrolysis using floating platform in ocean with uncontrollable wave motion". *Nature Communications* 15 (1): 5305. https://doi.org/10.1038/s41467-024-49639-6.

Liu, Zhao, Beibei Han, Zhiyi Lu, Wanbing Guan, Yuanyuan Li, Changjiang Song, Liang Chen, and Subhash C. Singhal. 2021. "Efficiency and stability of hydrogen production from seawater using solid oxide electrolysis cells". *Applied Energy* 300: 117439. https://doi.org/10.1016/j.apenergy.2021.117439.

Mohammadi, Amin, and Mehdi Mehrpooya. 2018. "A comprehensive review on coupling different types of electrolyzer to renewable energy sources". *Energy* 158: 632–55. https://doi.org/10.1016/j.energy.2018.06.073.

Nasser, Mohamed, Tamer F. Megahed, Shinichi Ookawara, and Hamdy Hassan. 2022. "A review of water electrolysis–based systems for hydrogen production using hybrid/solar/wind energy systems". *Environmental Science and Pollution Research* 29 (58): 86994–18. https://doi.org/10.1007/s11356-022-23323-y.

Oni, A. O., K. Anaya, T. Giwa, G. Di Lullo, and A. Kumar. 2022. "Comparative assessment of blue hydrogen from steam methane reforming, autothermal reforming, and natural gas decomposition technologies for natural gas-producing regions". *Energy Conversion and Management* 254: 115245. https://doi.org/10.1016/j.enconman.2022.115245.

Panić, Ivan, Aleksandar Cuculić, and Jasmin Ćelić. 2022. "Color-coded hydrogen: Production and storage in maritime sector". *Journal of Marine Science and Engineering* 10 (12): 1995. https://doi.org/10.3390/jmse10121995.

Rao, Prakash, Arian Aghajanzadeh, Paul Sheaffer, William R. Morrow, Sabine Brueske, Caroline Dollinger, Kevin Price, Prateeti Sarker, Nicholas Ward, and Joe Cresko. 2016. "Volume 1: survey of available information in support of the energy-water bandwidth study of desalination systems". LBNL–1006424, 1342538. https://doi.org/10.2172/1342538.

Shahzad, Muhammad Wakil, Li Ang Muhammad Burhan, and Kim Choon Ng. 2017. "Energy-water-environment nexus underpinning future desalination sustainability". *Desalination* 413: 52–64. https://doi.org/10.1016/j.desal.2017.03.009.

Sharon, H., and K. S. Reddy. 2015. "A review of solar energy driven desalination technologies". *Renewable and Sustainable Energy Reviews* 41: 1080–118. https://doi.org/10.1016/j.rser.2014.09.002.

Shatat, Mahmoud, Saffa Riffat, and Said Ghabayen. 2012. "State of art water desalination technologies using conventional and sustainable energy sources". https://d1wqtxts1xzle7.cloudfront.net/76085641/2153-libre.pdf?1639167580=&response-content-disposition=inline%3B+filename%3DState_of_Art_Water_Desalination_Technolo.pdf&Expires=1757942399&Signature=eLuoVfZsYX2qZl-ICN~T1lKoXx2gVPC65JGE1nImoc~Q0ddWIEwH1aM2SdlB9HITI4MQB2JoDO6YRyzHHkXza9mvAj2gcwAYfwAH-76vnZ7Vbi5Gka8rrftOj4pr9lfbmJxb~FEznJ~6WRVonFy3uKSIQCepXlGb0kT2p-KLFxsV84X9YHsylnzv2MtqHV~dZ9kvbIMNztyXU07ITgeFxMGKhWoWLQ1G1Xbvs-bSssxjfhN3H1LtM7pkvxv8Ci1-091hrXhh-NIdJ3TF-ygEVUievd3X0KxczBY3Z00X5QvUEYdibzpbre25H~Wqy~75Mv8czZas4S8bDQv8QvNO60g__&Key-Pair-Id=APKAJLOHF5GGSLRBV4ZA.

Shiva Kumar, S., and Hankwon Lim. 2022. "An overview of water electrolysis technologies for green hydrogen production". *Energy Reports* 8: 13793–813. https://doi.org/10.1016/j.egyr.2022.10.127.

Simoes, Sofia G., Justina Catarino, Ana Picado, Tiago F. Lopes, Santino di Berardino, Filipa Amorim, Francisco Gírio, C. M. Rangel, and Teresa Ponce de Leão. 2021. "Water availability and water usage solutions for electrolysis in hydrogen production". *Journal of Cleaner Production* 315: 128124. https://doi.org/10.1016/j.jclepro.2021.128124.

Sztekler, Karol, Wojciech Kalawa, Tomasz Bujok, Piotr Boruta, Ewelina Radomska, Łukasz Mika, and Agata Mlonka-Mędrala et al. 2024. "Hybrid desalination system for baltic sea water: A preliminary study". *Desalination* 574: 117269. https://doi.org/10.1016/j.desal.2023.117269.

Tenno, R., and P. Nguyen. 2016. "Multistage flash evaporator control in PDE representation". *IFAC-PapersOnLine*, 2th IFAC Workshop on Thermodynamic Foundations for a Mathematical Systems Theory TFMST 2016, 49 (24): 70–5. https://doi.org/10.1016/j.ifacol.2016.10.757.

Vidas, Leonardo, and Rui Castro. 2021. "Recent developments on hydrogen production technologies: State-of-the-art review with a focus on green-electrolysis". *Applied Sciences* 11 (23): 11363. https://doi.org/10.3390/app112311363.

Vincent, Immanuel, and Dmitri Bessarabov. 2018. "Low cost hydrogen production by anion exchange membrane electrolysis: A review". *Renewable and Sustainable Energy Reviews* 81: 1690–704. https://doi.org/10.1016/j.rser.2017.05.258.

Younas, Muhammad, Sumeer Shafique, Ainy Hafeez, Fahad Javed, and Fahad Rehman. 2022. "An overview of hydrogen production: Current status, potential, and challenges". *Fuel* 316: 123317. https://doi.org/10.1016/j.fuel.2022.123317.

Yue, Meiling, Hugo Lambert, Elodie Pahon, Robin Roche, Samir Jemei, and Daniel Hissel. 2021. "Hydrogen energy systems: A critical review of technologies, applications, trends and challenges". *Renewable and Sustainable Energy Reviews* 146: 111180. https://doi.org/10.1016/j.rser.2021.111180.

Zainal, Bidattul Syirat, Pin Jern Ker, Hassan Mohamed, Hwai Chyuan Ong, I. M. R. Fattah, S. M. Ashrafur Rahman, Long D. Nghiem, and T M Indra Mahlia. 2024. "Recent advancement and assessment of green hydrogen production technologies". *Renewable and Sustainable Energy Reviews* 189: 113941. https://doi.org/10.1016/j.rser.2023.113941.

8 Artificial Intelligence for Optimizing Renewable Energy-Powered Desalination

A Smart Approach to Sustainable Water Solutions

I. Ait Oujallal, J. Mabrouki, S. Boudraham, and Y. Abrouki

8.1 INTRODUCTION

Access to clean and safe water is one of the most pressing challenges of the 21st century. Despite the fact that over 70% of the earth's surface is covered with water, only 2.5% of it is freshwater, and less than 1% is accessible for human use ("Renewable Energy Statistics 2022" s. d.). Rapid population growth, climate change, and increasing water demand for industrial, agricultural, and domestic purposes have led to severe water shortages in many regions worldwide. According to the United Nations World Water Development Report, global water demand is expected to rise by 55% by 2050, with more than 40% of the world's population projected to live in water-stressed areas. In response to these challenges, seawater desalination has emerged as a viable and essential solution for ensuring a sustainable water supply, particularly in arid and semi-arid regions (He et al. 2022).

Desalination, the process of converting saline water into freshwater, has gained significant attention in recent years due to its potential to alleviate water scarcity. Various desalination technologies, such as reverse osmosis (RO), multi-stage flash (MSF), and multi-effect distillation (MED), have been developed and widely implemented across the world. Among these, RO technology dominates the desalination market, accounting for nearly 70% of global desalination capacity, due to its high efficiency, cost-effectiveness, and modular design. However, despite technological advancements, conventional desalination remains highly energy-intensive. It is estimated that desalination plants consume between 2 and 5 kWh per cubic meter of water produced, leading to high operational costs and significant carbon emissions when powered by fossil fuels. This dependency on conventional energy sources has raised concerns regarding sustainability, environmental impact, and long-term economic feasibility.

To mitigate the environmental and economic drawbacks of fossil fuel–powered desalination, the integration of renewable energy resources (RERs) has emerged as a promising alternative. Solar, wind, geothermal, and ocean energy offer clean and sustainable solutions to power desalination processes, significantly reducing greenhouse gas emissions and dependence on non-renewable energy sources. However, the intermittency and variability of renewable energy present major challenges in ensuring a continuous and stable water supply.

Several hybrid renewable energy-desalination systems have been proposed to address these challenges. For instance, solar photovoltaic (PV) and wind energy combinations with battery storage, fuel cells, and electrolyzers have shown promising results in providing a stable energy supply for desalination plants. Additionally, geothermal and ocean thermal energy conversion (OTEC) have been explored as potential base-load energy sources for large-scale desalination operations. Despite these advancements, determining the optimal configuration of renewable energy systems

DOI: 10.1201/9781003647263-8

for desalination remains a complex problem due to the fluctuating energy availability, site-specific conditions, and economic constraints (Sayed et al. 2023).

To overcome these challenges, artificial intelligence (AI) has emerged as a powerful tool for optimizing the design, operation, and management of renewable energy-powered desalination systems. AI techniques, including machine learning (ML), artificial neural networks (ANNs), genetic algorithms (GAs), fuzzy logic (FL), and particle swarm optimization (PSO), have demonstrated significant potential in improving the efficiency, adaptability, and cost-effectiveness of desalination systems. These AI-driven approaches provide solutions for key challenges in desalination.

AI integration in desalination can be categorized into four main areas: decision-making and system design optimization, energy demand prediction and load balancing, process optimization and cost reduction, and real-time monitoring and adaptive control. In decision making and system design optimization, AI assists in selecting the optimal location for desalination plants based on environmental, economic, and logistical factors. Additionally, AI-based MCDM frameworks help identify the best renewable energy-desalination technology combinations. For energy demand prediction and load balancing, AI-driven forecasting models predict solar radiation, wind speed, and energy demand, improving system stability, while ML algorithms optimize battery energy storage systems, fuel cells, and hybrid energy sources to ensure uninterrupted operation. Regarding process optimization and cost reduction, AI techniques optimize desalination operating conditions (e.g., pressure, temperature, and flow rates) to maximize efficiency. AI-driven control systems further reduce energy consumption and operational costs, improving sustainability. Lastly, in real-time monitoring and adaptive control, AI enables automated fault detection and predictive maintenance, reducing downtime and improving system reliability. Moreover, AI-powered sensors and smart automation systems allow desalination plants to adjust parameters in real time for optimal performance.

Despite the numerous benefits of AI in desalination, several challenges remain. One of the primary barriers to large-scale AI integration in desalination is data availability and accuracy. AI models require large amounts of high-quality data to train and improve performance, but desalination plants often lack comprehensive data sets. Additionally, algorithmic complexity and computational demands can be a limitation, particularly for real-time applications in large-scale desalination plants.

Another concern is the economic feasibility of AI-powered desalination systems. While AI can significantly reduce operational costs in the long run, the initial investment in AI infrastructure, sensors, and data acquisition systems can be high. Furthermore, ethical and environmental considerations must be addressed, including concerns about automation replacing human jobs, energy consumption of AI models, and potential environmental impacts of desalination by-products.

This paper provides a comprehensive review of AI applications in optimizing renewable energy-powered desalination systems. It explores the key desalination technologies, discusses the role of AI in decision making, prediction, optimization, and control, and compares different AI techniques to evaluate their effectiveness. Furthermore, this study identifies current challenges and future research directions to improve AI-driven desalination processes.

By leveraging AI, the future of sustainable, efficient, and cost-effective desalination systems can be realized, addressing global water security concerns while minimizing environmental impacts. This paper aims to contribute to the growing body of research on AI-driven sustainable water solutions and provide practical insights for researchers, engineers, and policymakers working in the field of desalination and renewable energy integration.

8.2 RENEWABLE ENERGY–POWERED DESALINATION TECHNOLOGIES

8.2.1 Overview of Desalination Technologies

Desalination technologies can be categorized in terms of their energy sources into three primary types: electrical, mechanical, and thermal (Figure 8.1). Electrical desalination techniques, exemplified by capacitive deionization (CDI), electrodialysis (ED), and electrodialysis reversal

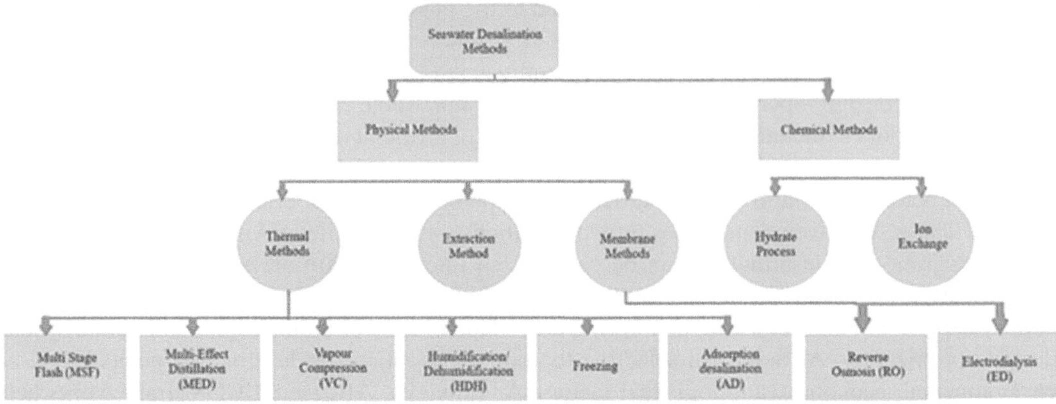

FIGURE 8.1 Schematic of desalination methods.

(EDR), use an electric field for the separation of salt ions from seawater. Mechanical desalination processes, exemplified by RO and mechanical vapor compression (MVC), use pressure or vacuum to drive the desalination process. Conversely, methods categorized under thermal desalination such as multi-stage flash distillation (MSF), multi-effect distillation (MED), and thermal vapor compression (TVC), employ thermal energy to promote evaporation of water, which is subsequently condensed to remove salts. Among this set of methodologies, RO, MSF, and MED dominate the global desalination market, together accounting for approximately 95% of the market share.

RO, in particular, has been widely used, representing 85% of the world's desalination plants, or 22,750 units, and contributing 70% of the total capacity, which was 107 million cubic meters per day in 2022. MSF and MED account for 18% and 7% of the water produced by desalination, respectively, of which MSF is 2% of the desalination plants and MED is 5%. Though an effective response to water scarcity, desalination has considerable environmental consequences. The process is energy-intensive and produces brine waste, which can impose severe ecological effects upon discharge into marine ecosystems. Enhancing the efficiency of desalination and reducing its environmental impact are thus key priorities for future research and technological development.

8.2.1.1 Reverse Osmosis (RO)

RO is a widely used membrane technology for desalination, which has been significantly improved since its development began in the 1970s. Today, it is the leading method in the global desalination industry. The process of RO works through a semi-permeable membrane that allows water molecules to pass through while, at the same time, blocking the flow of dissolved salt ions. By using hydraulic pressure greater than the osmotic pressure of the saline feed water, the system forces water to pass through the membrane, thus holding back the concentrated salt, which in turn creates a brine waste stream.

Based on the International Desalination Association (IDA), RO is the world's fastest-growing desalination technology today. The reasons for the fast uptake of this technology include the tremendous improvement in membrane technology, lowered energy consumption, and the fact that energy recovery devices (ERDs) are available. These improvements render RO applicable to a broad spectrum of feedwater types with different salinity levels, as well as to different system capacities. This versatility has allowed RO to be successfully applied in a variety of applications, from small-scale systems to large industrial plants (Wang et al. 2024).

Two major categories of RO desalination are: seawater reverse osmosis (SWRO) and brackish water reverse osmosis (BWRO). SWRO desalination plants usually work in a range of 55 –70 bars pressure with 40% –55% recovery rates.

For comparison, BWRO systems work at far lower pressures of 20 –25 bars but with higher recovery rates of 70% –85%. On the energy consumption front, SWRO systems tend to need 2 –5 kWh of energy for every cubic meter of water, with more energy being consumed for seawater desalination than for brackish water desalination (Alsarayreh et al. 2020).

These performance differences between SWRO and BWRO reflect how energy usage is directly proportional to feedwater salinity. SWRO systems, as they have to overcome seawater's higher osmotic pressure, need more energy compared to BWRO systems, which deal with water having lower salinity levels. Throughout the years, technological improvements, specifically in the form of the creation of energy recovery devices (ERDs), have led to huge energy savings in SWRO desalination plants. ERDs function by recovering and recycling energy from the brine stream that would otherwise be lost, thereby enhancing the energy efficiency of the desalination process and lowering operational expenditures. In addition, the modularity of RO systems has also enabled scalability, and they can now be used in diverse applications, ranging from small desalination plants for remote or rural communities to big plants for large urban cities. The ability to support different capacities and salinity has made RO a ready solution in addressing water scarcity problems worldwide.

8.2.1.2 Multi-Stage Flash (MSF)

MSF distillation is a more established thermal desalination method, developed in the 1950s, and remains a prevalent desalination method applied to seawater. Thermal desalination operates by inducing a phase change using heat, in this instance, evaporating water from brackish feedwater. The water vapor is collected as vapor and condensed into freshwater ("Desalination in the GCC Countries - A Review – ScienceDirect" s. d.).

In the MSF process, saltwater is first heated and then subjected to a sudden decrease in pressure "flashing" causing some of the water to evaporate. This water vapor is then condensed into freshwater. Simultaneously, the remainder of the feedwater is preheated using the latent heat of the vapor, and the operation is repeated in a number of stages; hence the term "multistage." The evaporation in MSF occurs in the bulk of the liquid, rather than directly at the heat exchange surface, as is the case in other thermal desalination methods like MED. This process minimizes the risk of scaling, which is a common problem linked to thermal desalination ("An Overview of Sustainable Desalination with Freezing Crystallization: Current Development, Future Challenges, and Prospects" s. d.).

Furthermore, MSF works under low-pressure conditions, which leads to decreased temperatures during evaporation. The process, therefore, enhances energy efficiency since it entails smaller temperature differences between the various stages. Contemporary MSF desalination plants consist of 10–30 stages with a temperature drop of around 2°C for each stage. Plants usually work within a temperature range of 90–120°C and have freshwater production capacities that vary from 4,000–30,000 cubic meters per day (Hendijanifard, HajAli, and Farhadi 2024).

Despite its efficacy and longevity, MSF, like all thermal desalination methods, is beset with scaling, corrosion, and power consumption issues. Scaling occurs when the salts settle out of the water and can plug the system, reducing its efficiency. Corrosion also occurs due to the high temperatures and saltiness of the water. To avoid corrosion, MSF plants also use corrosion-resistant materials and rely on regular maintenance to achieve optimum performance (Alamri and Obot 2019; Obot et al. 2020). On the other hand, scaling can be managed through appropriate pretreatment of feedwater, use of antiscalants, and operating parameter adjustment ("Advances in desalination for Water and Wastewater Treatment - Materials Research Forum" s. d.; Ait-Oujallal, Mabrouki, and Abrouki 2024; Ait-Oujallal et al. 2024). Nonetheless, the MSF process is characterized by its substantial energy requirements. The energy consumption specific to MSF desalination is considerable, necessitating around 4 kWh of electrical energy and 333 MJ of thermal energy per cubic meter of produced water ("An Overview of Sustainable Desalination with Freezing Crystallization: Current Development, Future Challenges, and Prospects" s. d.). This elevated energy requirement continues to be a significant constraint of the MSF methodology, despite the introduction of improvements in energy recovery systems and optimization strategies aimed at diminishing its total energy usage.

8.2.1.3 Multi-Effect Distillation (MED)

MED came into existence in the early 19th century, before the introduction of MSF desalination, and was mainly driven by the widespread use of steam power. MED is a thermal desalination method that employs a simple evaporator unit to produce freshwater. The underlying principle of MED is the multi-effect principle, where the vapor from one evaporator is used to supply heat to the later stages. This cascading process significantly enhances energy efficiency through the use of latent heat generated from the vapor created in earlier stages.

One of the key advantages of MED over other thermal desalination processes like MSF is its operation at low temperature. This can result in lower energy consumption per cubic meter of freshwater production and thus lower specific energy consumption. As far as energy efficiency is concerned, the gain output ratio (GOR) which measures the amount of freshwater produced per unit of steam used tends to range between 8 and 15 for MED systems. This is much better than the GOR of MSF systems, which tends to range between 8 and 10 ("Multi Effect Desalination and Adsorption Desalination (MEDAD): A Hybrid Desalination Method – ScienceDirect" s. d.). The higher GOR in MED systems means that they are more energy efficient, producing more freshwater per unit of energy consumed.

The lower operating temperatures in MED also reduce the tendency for scaling and corrosion, which are common problems with thermal desalination. All these render MED efficient and attractive for large-scale desalination applications, particularly when energy expenses are a significant consideration.

8.2.1.4 Electrodialysis

ED is a membrane desalination process utilizing an electric field for the elimination of dissolved salts from water solutions. ED originated in the mid-20th century and since then has been extensively applied in brackish water treatment, industrial wastewater cleaning, and selective ion extraction processes. Relative to other thermal desalination processes like MSF and MED based on phase change, ED is an electrochemical separation process that is not based on high temperatures and hence is more energy-efficient under some circumstances.

The fundamental mechanism of ED is the movement of salt ions through ion-exchange membranes under the influence of an electric field. The design of the system comprises a series of alternating cation-exchange membranes (CEMs) and anion-exchange membranes (AEMs) between two electrodes, i.e., an anode and a cathode. On applying an electric voltage, positively charged cations (e.g., Na^+ and Ca^{2+}) move toward the negatively charged cathode, passing through the cation-exchange membranes but encountering resistance from the anion-exchange membranes. On the other hand, negatively charged ions (e.g., Cl^-, SO_4^{2-}) migrate to the positively charged anode, traveling through the anion-exchange membranes but being rejected by the cation-exchange membranes. This process of separation produces alternating freshwater and concentrated brine compartments, enabling efficient desalination ("Electrodialysis Desalination for Water and Wastewater: A Review – ScienceDirect" s. d.).

One of the most important advantages of ED is its lower energy consumption compared to thermal desalination methods. In contrast to RO, in which energy needs rise with osmotic pressure, ED's energy needs are largely a function of feedwater salt concentration. Consequently, ED is most appropriately applied to brackish rather than seawater treatment. ED systems are also modular and scale up easily and so are suitable for a variety of applications, including decentralized treatment in remote locations. A further advantage of ED is that it can reduce scaling and fouling since the process is conducted without phase change. A modification of ED, called EDR, reverses the electrode polarity periodically to avoid the buildup of scaling and enhance the life of the membrane.

Despite its advantages, ED has some limitations that affect its widespread use. A major drawback of this process is its limited effectiveness when used in high-salinity water, such as seawater, due to the high energy required for the removal of large amounts of dissolved salts. While ED shows high efficiency for desalination of brackish water with salinity of up to 10,000 mg/L, its efficiency declines at higher salinity levels. Furthermore, membrane fouling is an issue; organic matter, biofilms, or precipitated salts can deposit on the membranes, which decreases overall efficiency over time. To prevent this issue, proper pretreatment of feed water and regular membrane cleaning are necessary. Additionally, while ED systems generally have lower operating energy costs, their initial capital costs can be greater than those of RO systems (Nikonenko et al. 2014).

In terms of energy usage, ED generally needs an energy consumption of 0.5–3 kWh per cubic meter of water treated, with the lower end for brackish water and the higher end for saltier sources. It is less energy-consuming than RO in some circumstances, particularly for low salinity feedwater. In contrast, BWRO consumes approximately 2–5 kWh/m^3, thus ED would be the better option for some applications. The efficiency of ED may be assessed based on GOR, an indicator of how much freshwater may be generated compared to every unit of electric power used. Due to its effectiveness and versatility, ED has found applications in numerous industries. It is widely used for desalination of brackish water in inland areas where desalination of seawater is not necessary. In industries, ED is used in wastewater treatment and in the recovery of valuable salts and chemicals. It is also used in the food and beverage industry for processes such as whey demineralization and beverage purification. Additionally, ED finds application in the desalination of agricultural water, where it reduces salinity in irrigation water to prevent soil degradation and improve crop production.

8.2.1.5 Other Desalination Technologies

In addition to well-established processes such as RO, ED, MSF, and MED, a number of alternative or emerging desalination techniques exist, each characterized by unique operational principles and performance profiles as shown in Figure 8.1. Vapor compression (VC), for instance, relies on the mechanical or thermal compression of vapor, followed by its condensation and subsequent recovery as freshwater. This approach is often suited to smaller-scale applications or scenarios where the capital investment required for MSF or MED is not economically feasible (Al-Mutaz and Wazeer 2014). Another thermally driven method, humidification–dehumidification (HDH), entails the evaporation of saline feedwater into an air stream (humidification) and the subsequent condensation of the resulting vapor (dehumidification). Owing to its relatively low operating temperatures, HDH is frequently integrated with solar collectors or other low-grade heat sources (Parekh et al. 2004; "Portable and Integrated Solar-Driven Desalination System using Membrane Distillation for Arid Remote Areas in Saudi Arabia – ScienceDirect" s. d.).

Freezing desalination exploits the principle that ice crystals formed from saline water contain minimal salt, allowing fresh water to be harvested upon careful management of freezing and melting stages. While theoretically capable of lower energy consumption, this method faces practical challenges in handling ice and ensuring energy efficiency, limiting its large-scale deployment (Rahimi, Modin, and Mijakovic 2020). Adsorption desalination (AD) employs materials such as silica gel or zeolite to adsorb water vapor selectively, followed by a regeneration step wherein heat drives off the adsorbed water, enabling its subsequent condensation. AD typically operates under low pressure and can utilize low-temperature heat sources, making it an attractive option in settings where conventional membranes or high-temperature processes are unsuitable (Ghaffour et al. 2014).

From a chemical perspective, ion exchange processes, sometimes grouped under broader chemical or hydrate-based desalination methods, rely on specialized resins or chemical agents to remove ions from saline water. Although valuable for the targeted removal of particular ions and for polishing low-salinity streams, ion exchange remains less prevalent in large-scale desalination than membrane- or thermal-based alternatives (Zhu et al. 2023). These technologies, while diverse in operating principles, share a common aim of providing reliable freshwater supplies under a variety of resource and environmental constraints. Ongoing research continues to refine their energy efficiency and cost-effectiveness, to extend their viability to more widespread applications.

8.2.2 Renewable Energy Resources and Their Role in Water Desalination

In the past several decades, RERs have become a vital alternative to conventional fossil fuels, as cleaner, more sustainable, and more widely available methods of providing energy (El-Moustaqim et al. 2024a, 2024b, 2024c). In recent decades, RERs have become more technology-advanced and the cost has consistently decreased to the point where RERs can be as competitive as conventional energy sources (AlMallahi et al. 2023; Zakaria et al. 2021). Also, the fluctuating prices of fossil fuels and geopolitical ramifications of energy dependence have hastened the switch to renewables (Ruhang et al. 2018). Over the past two decades, global generation from renewables has grown remarkably, and by 2020 was around 3 PWh, with the levelized cost of energy (LCOE) near \$0.039/kWh. The COVID-19 pandemic caused a brief slowing in the growth of new renewable capacity from 2020 to 2022, due to industrial and commercial slowdowns (Al-Othman et al. 2022). By 2020, of RERs other than hydropower, onshore wind was 49%, solar photovoltaic (PV) at 28%, and bioenergy at 19%.

The increasing global challenge of freshwater scarcity has prompted the exploration of renewable energy-powered desalination technologies as means to address the increasing need for water. Desalination is energy intensive, and RERs could have a significant impact in reducing the environmental impact and cost of operation of desalination. Examples of desalination technology that have been successfully powered by RER include RO, MED, MSF, and ED.

8.2.2.1 Wind Energy in Desalination

In the past 20 years, wind energy has undergone extraordinary growth, with its total global capacity in 2000 of 33 TWh rising to approximately 1,600 TWh in 2020 – equal to a nearly 50-fold increase (He et al. 2022). Wind energy is typically characterized as either onshore and offshore wind farms, with onshore wind farms accounting for approximately 94% of the urban wind energy capacity. Over the last decade, wind energy costs have decreased significantly, with the LCOE for onshore wind decreasing from \$0.089 to \$0.039/kWh and offshore from \$0.162 to \$0.093/kWh (Sayed et al. 2023). Wind turbines operate by converting wind's kinetic energy or moving energy into mechanical energy using rotating blades, which is then converted to electricity through a generator. A number of environmental and geographical factors contribute to the efficiency of wind energy generation including wind speed, terrain, and air density. Wind turbines are designed in two potential formats, vertical axis wind turbines (VAWTs) versus horizontal axis wind turbines (HAWT) (Li et al. 2023; Zwarteveen et al. 2021). For desalination, wind power can be used to operate reverse osmosis desalination systems, or provide electricity to hybrid renewable desalination plants that reduce the reliance of desalination on fossil fuel–produced energy (Cabrera, Lund, and Carta 2018).

8.2.2.2 Solar Energy in Desalination

Solar energy, which converts solar radiation into electricity, can be classified into two broad technologies: solar PV and solar concentrated power (CSP) (Benchrifa and Mabrouki 2022; Mabrouki et al. 2019, 2022). Solar PV consists of semiconductor materials that can directly convert sunshine into electricity through a process known as the photovoltaic effect. CSP approaches harness solar thermal energy and utilize it to generate steam to drive the power turbine. Solar power technology has grown at an impressive rate, increasing from 1.3 TWh in 2000 to approximately 845 TWh in 2020, with solar PV accounting for 98.5% of overall solar capacity. Associated with this capacity overall, the price for solar PV has markedly fallen (from a baseline of \$0.381/kWh in 2010 to \$0.057/kWh in 2020) and primarily CSP technology was also reduced (from \$0.34/kWh to \$0.108/kWh) during this time span. Among the various promising sustainable water treatment technologies, solar-powered desalination represents a cost-effective option, particularly for remote and out-of-grid areas characterized by limited freshwater resources. Reverse osmosis desalination technologies powered entirely by photovoltaic energy (PV-RO) and thermal solar desalination technologies making use of CSP technical advantages (e.g., MED and MSF) represent two compelling technologies with recent advances in practice (Sayed et al. 2023).

8.2.2.3 Bioenergy and Desalination

Bioenergy is energy obtained from biomass, which itself is defined based on organic material (derived from photosynthesis). The biomass can be processed to produce several biobased fuels, such as biogas, bioethanol, and biodiesel, which are produced via different means of conversion (pyrolysis, fermentation, transesterification) (Lee et al. 2020). The traditional first-generation biofuel, biodiesel, is derived from food crops (sugarcane, corn), while second-generation biofuels avoid the food and fuel debate by using agricultural residues and waste (Abdullah et al. 2024). More recently, there are now third-generation biofuels, which are based on algae, that have been introduced for more sustainable energy options as well. Bioenergy could possibly be integrated with desalination technology by providing energy to 'thermal' desalination plants. Biomass-fired steam plants are a way to provide the thermal heat energy necessary to perform the MED and MSF desalination processes, and they are a more sustainable, low-carbon alternative to traditional fossil fuel-based desalination plants (Herath et al. 2011).

8.2.2.3.1 Ocean Energy and Desalination

Ocean energy includes many types of energy derived from marine processes, including wave energy, tidal energy, OTEC, and salinity gradient energy. Ocean energy harnesses kinetic, potential, and chemical energy from seawater. Salinity gradient energy, also known as seawater chemical energy, is described as the energy due to salinity differences in bodies of water, and can be captured using different technologies like reverse electrodialysis (RED) and pressure-retarded osmosis (PRO) ("Alternate Storage of Opposite Charges in Multisites for High-Energy-Density Al–MOF Batteries - Guo - 2022 - Advanced Materials - Wiley Online Library" s. d.). Tidal and wave energy can power desalination plants, particularly in coastal areas with high tidal flows and wave energy. OTEC can utilize the warm surface water and cold deep water temperature gradient in thermal desalination processes (Khanmohammadi et al. 2020). Marine energy generation capacity grew from 547 GWh in 2000 to nearly 1 TWh in 2020 ("Renewable Energy Statistics 2022" s. d.).

8.2.2.4 Geothermal Energy and Desalination

Geothermal energy is defined as energy derived from heat stored below the surface of the Earth as a result of natural geological processes and radioactive decay. In geothermal power plants, deep underground reservoirs dissipate heat to working fluids to create steam to drive turbines and produce electricity. As of 2020, worldwide geothermal power capacity increased from approximately 52 TWh in 2000 to approximately 90 TWh as of 2020 with an average levelized cost of electricity (LCOE) of $0.05–$0.07/kWh ("Renewable Energy Statistics 2022" s. d.). Geothermal energy can also be used for desalination to provide heat to thermal desalination systems, where the most applicable and recognized systems include MED and MSF. This method to produce fresh water from geothermal energy provides an unlimited and steady source of energy which makes it particularly valuable for large desalination facilities in areas with geothermal activity.

8.2.3 Challenges in Renewable Energy–Powered Desalination

Despite the numerous benefits of integrating renewable energy with desalination, several challenges must be addressed to ensure efficiency and feasibility. One major challenge is the intermittency of renewable energy sources such as solar and wind, which can cause fluctuations in power supply and affect desalination plant operation. Energy storage solutions, such as batteries and thermal storage, are required to mitigate these fluctuations but add to the overall system cost (Ait-oujallal et al. 2024).

Another challenge is the high capital investment required for renewable energy–powered desalination infrastructure. Although operating costs are lower in the long term, the initial setup costs for solar, wind, and geothermal-powered desalination systems can be a barrier to widespread adoption, particularly in developing regions (Charcosset, Falconet, and Combe 2009). Additionally, technological limitations and efficiency losses in energy conversion and water treatment processes can impact overall system performance.

The geographical limitations of renewable energy sources also pose challenges. For example, solar desalination is highly dependent on sunlight availability, making it less effective in regions with limited solar exposure. Similarly, wind-powered desalination is restricted to locations with consistent wind patterns. Hybrid systems that combine multiple renewable sources, such as wind-solar desalination plants, can help overcome these geographical constraints (Calise et al. 2019).

Furthermore, desalination technologies require continuous maintenance and operational expertise. Renewable energy–powered desalination plants must be optimized to handle variations in energy input while maintaining water quality and production rates. Research and development efforts continue to focus on improving system efficiency, integrating energy storage, and developing cost-effective solutions for large-scale implementation ("Handbook of Solar Energy: Theory, Analysis and Applications | SpringerLink" s. d.).

8.3 ARTIFICIAL INTELLIGENCE IN DESALINATION SYSTEMS

Desalination systems involve complex interactions among numerous variables including feedwater quality, operating conditions, and the variability of renewable energy sources that often defy simple modeling approaches. Artificial intelligence (AI) addresses this complexity by offering tools capable of handling nonlinear dynamics, optimizing multiple objectives simultaneously, and providing predictive insights into system behavior.

One of the core reasons to employ AI in desalination is its ability to process large datasets and identify patterns that traditional techniques may miss. These capabilities facilitate multi-objective optimization, enabling designers to balance water production, energy consumption, and cost efficiency. AI also brings robust predictive capabilities that allow for better planning and proactive maintenance, ensuring more stable operation under variable conditions.

Several key AI techniques have emerged as especially useful in desalination applications.

8.3.1 Artificial Neural Networks (ANNs)

ANNs simulate the human brain's processing abilities, making them ideal for capturing complex relationships between input parameters such as solar irradiance, ambient temperature, and feedwater properties and outputs like permeate flow rate or thermal efficiency. Their ability to learn from historical data enables highly accurate modeling and forecasting. For example, ANN models in solar still applications have achieved prediction errors within 10–20% and high R^2 values (greater than 0.93) ("Cork Industry Wastewater Characterization: Assessment of the Biodegradability, Reuse and of the Relationship between BOD, COD and Tannins with TOC | Journal of Water Reuse and Desalination | IWA Publishing" s. d.; Mashaly et al. 2015).

8.3.2 Genetic Algorithms (GAs)

Inspired by natural evolution, GAs explore a wide range of solutions by iteratively "evolving" a population of candidate solutions. In desalination systems, GAs are particularly useful for optimizing system configurations such as the sizing of renewable energy components and the setting of operating parameters to reduce energy consumption and minimize costs.

8.3.3 Fuzzy Logic (FL) Systems

Fuzzy logic (FL) provides a framework for handling uncertainty and imprecision. In desalination, where operational variables can be highly variable and sometimes subjective (e.g., quality of feedwater or performance conditions), FL systems enable controllers to operate smoothly even in the face of ambiguous inputs. This makes them well suited for real-time control tasks ("Fuzzy Logic Systems for Engineering: A Tutorial | IEEE Journals & Magazine | IEEE Xplore" s. d.).

8.3.4 Particle Swarm Optimization (PSO)

PSO algorithms, modeled on the collective behavior of bird flocks or fish schools, allow the system to explore the solution space by sharing information among candidate solutions. PSO is effective in fine-tuning parameters in systems like RO, where it can optimize variables such as pressure and flow rate to achieve energy-efficient operation (Hesamamiri and Bourouni 2016).

8.4 AI APPLICATIONS IN RENEWABLE ENERGY DESALINATION

AI is applied throughout the life cycle of renewable energy–powered desalination systems, spanning from system design to real-time control. The applications can be broadly categorized into four areas: decision making and system design optimization, energy demand prediction and load balancing, process optimization and cost reduction, and real-time monitoring with adaptive control.

8.4.1 Decision Making and System Design Optimization

When designing desalination plants, AI facilitates complex decision-making processes that involve multiple criteria. For example, AI-driven frameworks are used to select optimal sites for desalination plants by evaluating environmental, technical, and economic factors. Tools such as Multi-Criteria Decision-Making (MCDM) algorithms including methods based on the Analytical Hierarchical Process (AHP) and fuzzy logic can rank various candidate sites and suggest the most suitable combinations of renewable energy resources and desalination technologies. This helps decision-makers choose an optimal RER-desalination technology combination (Ghassemi and Danesh 2013).

8.4.2 Energy Demand Prediction and Load Balancing

The intermittent nature of renewable energy sources necessitates accurate forecasting to ensure reliable desalination operations. AI-based forecasting models, particularly those using deep learning architectures such as Long Short-Term Memory (LSTM) networks and Gated Recurrent Units (GRU), predict fluctuations in solar irradiance, wind speed, and ambient temperature. These predictions enable operators to balance the load effectively by adjusting desalination processes in anticipation of energy shortfalls or surpluses. By aligning energy supply with operational demand, these models ensure a steady water production rate and reduce reliance on backup energy systems (Zhang et al. 2021).

8.4.3 Process Optimization and Cost Reduction

Optimizing process parameters is key to reducing both energy consumption and operational costs. AI-driven optimization techniques, including metaheuristic algorithms such as Genetic Algorithms (GA) and Particle Swarm Optimization (PSO), are employed to identify the best operational setpoints and system configurations. For instance, hybrid approaches that combine GA and PSO have been used to determine the optimal sizing of components like PV panels and wind turbines in hybrid systems, while also fine-tuning operating pressures and flow rates in RO systems. This leads to significant reductions in total annual cost and greenhouse gas emissions. Table 8.1 summarizes some of the key optimization techniques used in renewable energy desalination systems:

TABLE 8.1
Summary of AI Optimization Techniques for Renewable Energy–Energy-powered Desalination

Renewable Energy Source	Desalination Technology	AI Optimization Technique	Main Objective	Reference
Solar PV	Solar Stills	GA, ANN with HHO	Maximize water production and minimize operational cost	"Desalination Powered by Renewables: A Challenge and an AI Opportunity I Water Resources Management" (s. d.)
Wind	Reverse Osmosis (RO)	GA, PSO	Optimize operating pressures and flow rates	Bourouni, Ben M'Barek, and Al Taee (2011)
Hybrid PV-Wind	RO	Hybrid PSO–GWO	Minimize total cost and lifecycle emissions	Abdelshafy, Hassan, and Jurasz (2018)
Integrated Systems	Multiple Technologies	Simulated Annealing–Chaotic Search Hybrid	Optimize overall system configuration and component sizing	Ghanbari, Maleki, and Rezaei Ochbelagh (2024)

8.4.4 Real-Time Monitoring and Adaptive Control

Real-time control is essential to address the dynamic fluctuations in renewable energy supply and process conditions. AI-powered controllers use continuous sensor data from various system components to monitor parameters like feedwater quality, membrane performance, and energy input. By applying techniques such as fuzzy logic and reinforcement learning, these controllers can adjust operational variables in real-time. For example, in a wind-powered RO system, the controller might dynamically modulate pump speeds and pressure settings based on real-time wind data and energy forecasts. This ensures that the desalination process remains stable and efficient despite external fluctuations. Such adaptive control strategies also facilitate early fault detection and predictive maintenance, thereby reducing downtime and maintenance costs. Table 8.2 summarizes some of AI-Based Control Strategies in Desalination Systems.

TABLE 8.2
Summary of AI-Based Control Strategies in Desalination Systems

Control Strategy	AI Technique	Key Control Variables	Benefits	Reference
Real-Time Adaptive Control	Fuzzy Logic, RL	Flow rate, pressure, temperature	Enhances system stability and efficiency	"Advancements in Water Desalination Through Artificial Intelligence: A Comprehensive Review of AI-Based Methods for Reverse Osmosis Membrane Processes I Water Conservation Science and Engineering" (s. d.)
Hybrid Control	AI-PID Hybrid	Energy distribution, pump speed	Combines rapid response with adaptability	Mohd Ali et al. (2015)
Fault-Tolerant Control	ANN-Based Predictive Control	Membrane fouling, system alerts	Enables early fault detection and reduces downtime	Maldonado-Canca et al. (2024)

A typical implementation involves integrating these AI controllers with an array of sensors distributed throughout the plant. The sensor data are processed by the AI module, which then issues control signals to adjust operational parameters. Such systems have demonstrated marked improvements in maintaining optimal operating conditions and reducing energy waste.

8.5 COMPARATIVE ANALYSIS OF AI ALGORITHMS IN DESALINATION

A comparative evaluation of the AI techniques used in desalination reveals distinct advantages and limitations. Artificial Neural Networks (ANNs) offer exceptional modeling capabilities, capturing nonlinear dynamics with high accuracy. However, they require extensive training data and can be prone to overfitting. Genetic Algorithms (GAs) and Particle Swarm Optimization (PSO) excel in global optimization tasks but may suffer from slow convergence if not appropriately tuned. Fuzzy Logic systems provide robust control in uncertain environments but may lack the precision required for detailed parameter optimization.

Case studies in the literature illustrate that hybrid approaches, which combine the strengths of multiple algorithms, often outperform single-method strategies. For example, a hybrid PSO–GWO algorithm was applied to a hybrid PV-wind-RO system, resulting in significant cost reductions and enhanced energy efficiency compared to standalone methods ("Recent Advances in Detection and Control Strategies for Foodborne Bacteria in Raw and Ready-to-Eat Fruits and Vegetables - Abdelshafy - 2025 - Food Frontiers - Wiley Online Library" s. d.). These findings suggest that while no single AI technique is universally superior, a tailored combination of methods is likely to yield the best results in complex desalination applications.

8.6 CHALLENGES AND FUTURE RESEARCH DIRECTIONS

Despite significant progress, several challenges hinder the full integration of AI into renewable energy powered desalination systems. One major challenge is the integration of AI with existing desalination infrastructure. Reliable and high-quality sensor data are critical for AI models, yet many plants lack the comprehensive IoT networks needed to supply such data. Moreover, data accuracy and availability can vary widely between regions, limiting the effectiveness of AI-driven models.

There are also challenges related to the ethical and environmental implications of deploying AI at scale. Issues such as transparency, accountability, and the energy consumption of AI systems themselves must be carefully considered. Future research should focus on developing hybrid AI models that seamlessly integrate with IoT technologies, ensuring robust real-time performance. Additionally, advancements in data analytics and ML techniques will be essential to improve model accuracy and operational reliability. Researchers are also encouraged to incorporate sustainability metrics such as lifecycle cost analysis and environmental impact assessments into AI optimization frameworks to ensure that these systems not only perform efficiently but also contribute to broader sustainability goals (He et al. 2022).

8.7 CONCLUSION

In summation, the chapter demonstrates that a combination of renewable energy resources and desalination technologies, significantly enabled by advanced AI, is a credible method to tackle the growing water crisis our world is experiencing. The coupling of solar, wind, geothermal, and other renewable energy resources with established desalination methodologies, such as RO, MSF, and MED, reduces the greenhouse footprint of the source water production whilst also being a reliable method in greater efficiency and sustainability. This synthesis of innovative technologies has been studied in the chapter. For example, on one hand, dead technologies are being redesigned to utilize variable renewable energy sources more efficiently, while other AI technologies – ANNs, GAs, FL, and particle swarm optimization, for instance – are changing the process of systems design, system operation, and system maintenance. AI technologies have advanced and excel with complex interactions with multi-objective functions to consider trade-offs with water production, energy consumption, and operational costs. AI predictions of renewable energy variations will enable adaptive control parameters in operations as a real-time method for maximization and waste minimization at a source water production facility.

In addition, the discussion underscored the potential for hybrid systems, consisting of a number of renewable energy sources and distinct approaches to AI, with the potential to overcome limitations of all individual technologies. These hybrid systems addressed intermittency while also addressing initial capital expenditures and trustworthiness of sensor data. The combination of a robust IoT network and smart AI algorithms allows the desalination to be continuously monitored and faults repaired quickly to maintain resiliency under incalculable environmental conditions. The chapter also addressed the larger implications of these technological advances to society. The use of renewable energy to desalinate seawater, reduce dependence on fossil fuels, and lower greenhouse gas emissions, addressed the overall environmental impact that a water treatment process has on the environment. This transition to a sustainable model is central to long-term water security, especially in situations where water scarcity is compounded by other economic and environmental challenges.

Looking ahead there are numerous challenges that still remain. Future research needs to find a pathway to integrate AI into existing desalination infrastructure in a manner that is seamless, improves data quality, and reduces energy consumption by improving algorithm efficiency. Moreover, ethical and environmental concerns also need to be factored into the development of these technologies, ensuring that improved operational performance through AI and renewable energy integration is also sustainable. In sum, this chapter highlights the significant potential formed from integrating sustainable energy processes with intelligent and data-driven optimization. The progress made to date lays the groundwork for future development that will deliver less costly, more resilient, and more environmentally sustainable desalination – an important step towards meeting the world's future clean water demand.

REFERENCES

Abdelshafy, Alaaeldin M., Hamdy Hassan, and Jakub Jurasz. 2018. "Optimal design of a grid-connected desalination plant powered by renewable energy resources using a hybrid PSO–GWO approach". *Energy Conversion and Management* 173:331–47. https://doi.org/10.1016/j.enconman.2018.07.083.

Abdullah, Norulsamani, Nurul Atiqah Izzati Md Ishak, K. H. Tan, M. A. Zaed, R. Saidur, and A. K. Pandey. 2024. "Investigating the impact of various etching agents on Ti3C2Tx MXene synthesis for electrochemical energy conversion". *FlatChem* 47:100730. https://doi.org/10.1016/j.flatc.2024.100730.

"Advancements in Water Desalination Through Artificial Intelligence: A Comprehensive Review of AI-Based Methods for Reverse Osmosis Membrane Processes | Water Conservation Science and Engineering". s. d. Consulté le 25 mars 2025. https://link.springer.com/article/10.1007/s41101-023-00227-7.

"Advances in Desalination for Water and Wastewater Treatment - Materials Research Forum". s. d. Consulté le 21 mars 2025. https://mrforum.com/product/9781945291357-9/.

Ait-Oujallal, Ilham, Jamal Mabrouki, and Younes Abrouki. 2024. "The Effect of Renewable Energy Power Fluctuations on Water Electrolysis for Green Hydrogen Production". In *Technical and Technological Solutions Towards a Sustainable Society and Circular Economy*, édité par Jamal Mabrouki et Azrour Mourade, 413–25. Cham: Springer Nature Switzerland. https://doi.org/10.1007/978-3-031-56292-1_33.

Ait-Oujallal, Ilham, Jamal Mabrouki, Younes Abrouki, and Ahmed Moufti. 2024. "Evaluation of Seawater Desalination Technologies for Treatment: Energy and Economic Study". In *Smart Internet of Things for Environment and Healthcare*, édité par Mourade Azrour, Jamal Mabrouki, Abdulatif Alabdulatif, Azidine Guezzaz, et Fatima Amounas, 223–39. Cham: Springer Nature Switzerland. https://doi.org/10.1007/978-3-031-70102-3_16.

Alamri, Aeshah H., and I. B. Obot. 2019. "Highly efficient corrosion inhibitor for C1020 carbon steel during acid cleaning in multistage flash (MSF) desalination plant". *Desalination* 470:114100. https://doi.org/10.1016/j.desal.2019.114100.

AlMallahi, Maryam Nooman, Sara Maen Asaad, Abrar Inayat, K. Harby, and Mahmoud Elgendi. 2023. "Analysis of solar-powered adsorption desalination systems: Current research trends, developments, and future perspectives". *International Journal of Thermofluids* 20:100457. https://doi.org/10.1016/j.ijft.2023.100457.

Al-Mutaz, Ibrahim S., and Irfan Wazeer. 2014. "Comparative performance evaluation of conventional multi-effect evaporation desalination processes". *Applied Thermal Engineering* 73 (1):1194–203. https://doi.org/10.1016/j.applthermaleng.2014.09.025.

Al-Othman, Amani, Muhammad Tawalbeh, Remston Martis, Salam Dhou, Mehmet Orhan, Muhammad Qasim, and Abdul Ghani Olabi. 2022. "Artificial intelligence and numerical models in hybrid renewable energy systems with fuel cells: Advances and prospects". *Energy Conversion and Management* 253:115154. https://doi.org/10.1016/j.enconman.2021.115154.

Alsarayreh, Alanood A., M. A. Al-Obaidi, A. M. Al-Hroub, R. Patel, and I. M. Mujtaba. 2020. "Performance evaluation of reverse osmosis brackish water desalination plant with different recycled ratios of retentate". *Computers & Chemical Engineering* 135:106729. https://doi.org/10.1016/j.compchemeng.2020.106729.

"Alternate Storage of Opposite Charges in Multisites for High-Energy-Density Al–MOF Batteries - Guo - 2022 - Advanced Materials - Wiley Online Library". s. d. Consulté le 25 mars 2025. https://advanced.onlinelibrary.wiley.com/doi/abs/10.1002/adma.202110109.

"An Overview of Sustainable Desalination with Freezing Crystallization: Current Development, Future Challenges, and Prospects". s. d. Consulté le 21 mars 2025. https://www.mdpi.com/2071-1050/16/22/10138.

Benchrifa, Mohammed, and Jamal Mabrouki. 2022. "Simulation, sizing, economic evaluation and environmental impact assessment of a photovoltaic power plant for the electrification of an establishment". *Advances in Building Energy Research* 16 (6):736–53. https://doi.org/10.1080/17512549.2022.2096693.

Bourouni, K., T. Ben M'Barek, and A. Al Taee. 2011. "Design and optimization of desalination reverse osmosis plants driven by renewable energies using genetic algorithms". *Renewable Energy* 36 (3):936–50. https://doi.org/10.1016/j.renene.2010.08.039.

Cabrera, Pedro, Henrik Lund, and José A. Carta. 2018. "Smart renewable energy penetration strategies on islands: The case of Gran Canaria". *Energy* 162:421–43. https://doi.org/10.1016/j.energy.2018.08.020.

Calise, Francesco, Francesco Liberato Cappiello, Raffaele Vanoli, and Maria Vicidomini. 2019. "Economic assessment of renewable energy systems integrating photovoltaic panels, seawater desalination and water storage". *Applied Energy* 253:113575. https://doi.org/10.1016/j.apenergy.2019.113575.

Charcosset, C., C. Falconet, and M. Combe. 2009. "Hydrostatic pressure plants for desalination via reverse osmosis". *Renewable Energy* 34 (12):2878–82. https://doi.org/10.1016/j.renene.2009.02.026.

"Cork Industry Wastewater Characterization: Assessment of the Biodegradability, Reuse and of the Relationship between BOD, COD and Tannins with TOC | Journal of Water Reuse and Desalination | IWA Publishing". s. d. Consulté le 25 mars 2025. https://iwaponline.com/jwrd/article/2/1/33/28609/Cork-industry-wastewater-characterization.

"Desalination in the GCC Countries - A Review – ScienceDirect". s. d. Consulté le 21 mars 2025. https://www.sciencedirect.com/science/article/pii/S0959652622013300.

"Desalination Powered by Renewables: A Challenge and an AI Opportunity | Water Resources Management". s. d. Consulté le 25 mars 2025. https://link.springer.com/article/10.1007/s11269-024-03935-4.

"Electrodialysis Desalination for Water and Wastewater: A Review – ScienceDirect". s. d. Consulté le 25 mars 2025. https://www.sciencedirect.com/science/article/pii/S1385894719316250.

El-Moustaqim, Khadija, Benchrifa Mohammed, Jamal Mabrouki, and Driss Hmouni. 2024a. "Application of IoT-Based Intelligent Systems in Microalgae Biofuel Production". In *Advanced Technology for Smart Environment and Energy*, édité par Jamal Mabrouki and Azrour Mourade, 293–305. Environmental Science and Engineering. Cham: Springer Nature Switzerland. https://doi.org/10.1007/978-3-031-50871-4_20.

El-Moustaqim, Khadija, Jamal Mabrouki, Mourade Azrour, Mouhsine Hadine, and Driss Hmouni. 2024b. "Enabling Smart Agriculture Through Integrating the Internet of Things in Microalgae Farming for Sustainability". In *Smart Internet of Things for Environment and Healthcare*, édité par Mourade Azrour, Jamal Mabrouki, Abdulatif Alabdulatif, Azidine Guezzaz, and Fatima Amounas, 209–22. Cham: Springer Nature Switzerland. https://doi.org/10.1007/978-3-031-70102-3_15.

El-Moustaqim, Khadija, Jamal Mabrouki, Mourade Azrour, and Driss Hmouni. 2024c. "How Can Green Energy Be Improved by Integrating IoT into the Microalgae Process?" In Sustainable and Green Technologies for Water and Environmental Management, 239–52. https://doi.org/10.1007/978-3-031-52419-6_17.

"Fuzzy Logic Systems for Engineering: A Tutorial | IEEE Journals & Magazine | IEEE Xplore". s. d. Consulté le 25 mars 2025. https://ieeexplore.ieee.org/abstract/document/364485/.

Ghaffour, Noreddine, Sabine Lattemann, Thomas Missimer, Kim Choon Ng, Shahnawaz Sinha, and Gary Amy. 2014. "Renewable energy-driven innovative energy-efficient desalination technologies". *Applied Energy* 136:1155–65. https://doi.org/10.1016/j.apenergy.2014.03.033.

Ghanbari, Kamyar, Akbar Maleki, and Dariush Rezaei Ochbelagh. 2024. "Optimal design of solar/wind/energy storage system-powered RO desalination unit: Single and multi-objective optimization". *Energy Conversion and Management* 315:118768. https://doi.org/10.1016/j.enconman.2024.118768.

Ghassemi, Seyed Ali, and Shahnaz Danesh. 2013. "A hybrid fuzzy multi-criteria decision making approach for desalination process selection". *Desalination* 313:44–50. https://doi.org/10.1016/j.desal.2012.12.008.

"Handbook of Solar Energy: Theory, Analysis and Applications | SpringerLink". s. d. Consulté le 25 mars 2025. https://link.springer.com/book/10.1007/978-981-10-0807-8.

He, Qian, Hongfei Zheng, Xinglong Ma, Lu Wang, Hui Kong, and Ziye Zhu. 2022. "Artificial intelligence application in a renewable energy-driven desalination system: A critical review". *Energy and AI* 7:100123. https://doi.org/10.1016/j.egyai.2021.100123.

Hendijanifard, Mohammad, Amir HajAli, and Shahrokh Farhadi. 2024. "Comparing energy and exergy of multiple effect freeze desalination to MEE MSF RO". *Npj Clean Water* 7 (1):1–14. https://doi.org/10.1038/s41545-024-00395-6.

Herath, Indika, Markus Deurer, David Horne, Ranvir Singh, and Brent Clothier. 2011. "The water footprint of hydroelectricity: A methodological comparison from a case study in New Zealand". *Journal of Cleaner Production* 19 (14):1582–89. https://doi.org/10.1016/j.jclepro.2011.05.007.

Hesamamiri, Roozbeh, and Atieh Bourouni. 2016. "Customer support optimization using system dynamics: A multi-parameter approach". *Kybernetes* 45 (6):900–14. https://doi.org/10.1108/K-10-2015-0257.

Khanmohammadi, Shoaib, Mohammad Mehdi Baseri, Pouria Ahmadi, Abdullah A. A. A. Al-Rashed, and Masoud Afrand. 2020. "Proposal of a novel integrated ocean thermal energy conversion system with flat plate solar collectors and thermoelectric generators: Energy, exergy and environmental analyses". *Journal of Cleaner Production* 256:120600. https://doi.org/10.1016/j.jclepro.2020.120600.

Lee, Mengshan, Yi-Li Lin, Pei-Te Chiueh, and Walter Den. 2020. "Environmental and energy assessment of biomass residues to biochar as fuel: A brief review with recommendations for future bioenergy systems". *Journal of Cleaner Production* 251:119714. https://doi.org/10.1016/j.jclepro.2019.119714.

Li, Cruz Y., Zengshun Chen, Asiri Umenga Weerasuriya, Xuelin Zhang, Xisheng Lin, Lei Zhou, Yunfei Fu, and Tim K. T. Tse. 2023. "Best practice guidelines for the dynamic mode decomposition from a wind engineering perspective". *Journal of Wind Engineering and Industrial Aerodynamics* 241:105506. https://doi.org/10.1016/j.jweia.2023.105506.

Mabrouki, J., A. El Yadini, I. Bencheikh, K. Azoulay, A. Moufti, and S. El Hajjaji. 2019. "Hydrogeological and Hydrochemical Study of Underground Waters of the Tablecloth in the Vicinity of the Controlled City Dump Mohammedia (Morocco)". In *Advanced Intelligent Systems for Sustainable Development (AI2SD'2018)*, édité par Mostafa Ezziyyani, 913, 22–33. Advances in Intelligent Systems and Computing. Cham: Springer International Publishing. https://doi.org/10.1007/978-3-030-11881-5_3.

Mabrouki, Jamal, Ghizlane Fattah, Siham Kherraf, Younes Abrouki, Mourade Azrour, and Souad El Hajjaji. 2022. "Artificial Intelligence System for Intelligent Monitoring and Management of Water Treatment Plants". In *Emerging Real-World Applications of Internet of Things*, 69–87. CRC Press. https://www.taylorfrancis.com/chapters/edit/10.1201/9781003304203-4/artificial-intelligence-system-intelligent-monitoring-management-water-treatment-plants-jamal-mabrouki-ghizlane-fattah-siham-kherraf-younes-abrouki-mourade-azrour-souad-el-hajjaji.

Maldonado-Canca, Luis-Alfonso, Ana-María Casado-Molina, Juan-Pedro Cabrera-Sánchez, and Guillermo Bermúdez-González. 2024. "Beyond the post: An SLR of Enterprise artificial intelligence in social media". *Social Network Analysis and Mining* 14 (1):219. https://doi.org/10.1007/s13278-024-01382-y.

Mashaly, Ahmed F., A. A. Alazba, A. M. Al-Awaadh, and Mohamed A. Mattar. 2015. "Area determination of solar desalination system for irrigating crops in greenhouses using different quality feed water". *Agricultural Water Management* 154:1–10. https://doi.org/10.1016/j.agwat.2015.02.009.

Mohd Ali, Jarinah, M. A. Hussain, Moses O. Tade, and Jie Zhang. 2015. "Artificial intelligence techniques applied as estimator in chemical process systems – A literature survey". *Expert Systems With Applications* 42 (14):5915–31. https://doi.org/10.1016/j.eswa.2015.03.023.

"Multi Effect Desalination and Adsorption Desalination (MEDAD): A Hybrid desalination method – ScienceDirect". s. d. Consulté le 22 mars 2025. https://www.sciencedirect.com/science/article/pii/S1359431114002439

Nikonenko, Victor V., Anna V. Kovalenko, Mahamet K. Urtenov, Natalia D. Pismenskaya, Jongyoon Han, Philippe Sistat, and Gérald Pourcelly. 2014. "Desalination at overlimiting currents: State-of-the-art and perspectives". *Desalination*, Special Issue: Electromembrane Processes for Desalination, 342:85–106. https://doi.org/10.1016/j.desal.2014.01.008.

Obot, I. B., M. M. Solomon, I. B. Onyeachu, S. A. Umoren, Abdelkader Meroufel, A. Alenazi, and A. A. Sorour. 2020. "Development of a green corrosion inhibitor for use in acid cleaning of MSF desalination plant". *Desalination* 495:114675. https://doi.org/10.1016/j.desal.2020.114675.

Parekh, Sandeep, M. M. Farid, J. R. Selman, and Said Al-Hallaj. 2004. "Solar desalination with A humidification-dehumidification technique — A comprehensive technical review". *Desalination* 160 (2):167–86. https://doi.org/10.1016/S0011-9164(04)90007-0.

"Portable and Integrated Solar-Driven Desalination System Using Membrane Distillation for Arid Remote Areas in Saudi Arabia – ScienceDirect". s. d. Consulté le 25 mars 2025. https://www.sciencedirect.com/science/article/pii/S001191641400232X.

Rahimi, Shadi, Oskar Modin, and Ivan Mijakovic. 2020. "Technologies for biological removal and recovery of nitrogen from wastewater". *Biotechnology Advances* 43:107570. https://doi.org/10.1016/j.biotechadv.2020.107570.

"Recent Advances in Detection and Control Strategies for Foodborne Bacteria in Raw and Ready-to-Eat Fruits and Vegetables - Abdelshafy - 2025 - Food Frontiers - Wiley Online Library". s. d. Consulté le 25 mars 2025. https://iadns.onlinelibrary.wiley.com/doi/full/10.1002/fft2.541.

"Renewable Energy Statistics 2022". s. d. Consulté le 25 mars 2025. https://www.irena.org/publications/2022/Jul/Renewable-Energy-Statistics-2022.

Ruhang, Xu, Song Zixin, Tang Qingfeng, and Yu Zhuangzhuang. 2018. "The cost and marketability of renewable energy after power market reform in China: A review". *Journal of Cleaner Production* 204:409–24. https://doi.org/10.1016/j.jclepro.2018.09.018.

Sayed, Enas Taha, A. G. Olabi, Khaled Elsaid, Muaz Al Radi, Concetta Semeraro, Mohammad Hossein Doranehgard, Mohamed Elrayah Eltayeb, and Mohammad Ali Abdelkareem. 2023. "Application of artificial intelligence techniques for modeling, optimizing, and controlling desalination systems powered by renewable energy resources". *Journal of Cleaner Production* 413:137486. https://doi.org/10.1016/j.jclepro.2023.137486.

Wang, Liu, Xin Sun, Feng Gao, Yu Yang, and Riquan Song. 2024. "Solar membrane distillation: An emerging technology for reverse osmosis concentrated brine treatment". *Desalination* 592:118124. https://doi.org/10.1016/j.desal.2024.118124.

Zakaria, Muhammad, Abbas M. Sharaky, Al-Sayed Al-Sherbini, Mohamed Bassyouni, Mashallah Rezakazemi, and Yasser Elhenawy. 2021. "Water desalination using solar thermal collectors enhanced by nanofluids". *Chemical Engineering & Technolog* 45 (1):15–25. https://doi.org/10.1002/ceat.202100339.

Zhang, Chaobo, Junyang Li, Yang Zhao, Tingting Li, Qi Chen, Xuejun Zhang, and Weikang Qiu. 2021. "Problem of data imbalance in building energy load prediction: Concept, influence, and solution". *Applied Energy* 297:117139. https://doi.org/10.1016/j.apenergy.2021.117139.

Zhu, Shufang, Xiaoyu Jia, Yanpeng Ni, Bing Pan, Yunze Long, Dagang Miao, and Xu Yan. 2023. "A 3D solar-driven evaporator based on electrospun recycled PET film for efficient seawater desalination". *Journal of Cleaner Production* 408:137113. https://doi.org/10.1016/j.jclepro.2023.137113.

Zwarteveen, Margreet, Marcel Kuper, Cristian Olmos-Herrera, Muna Dajani, Jeltsje Kemerink-Seyoum, Cleaver Frances, Linnea Beckett, et al. 2021. "Transformations to groundwater sustainability: From individuals and pumps to communities and aquifers". *Current Opinion in Environmental Sustainability* 49:88–97. https://doi.org/10.1016/j.cosust.2021.03.004.

9 IoT-Based Monitoring and Faults Detection for Solar Panels Using Machine Learning Algorithms

N. El Yanboiy, M. Khala, I. Elabbassi, O. Eloutassi, Y. El Hassouani, and C. Messaoudi

9.1 INTRODUCTION

The global transition to sustainable energy sources has highlighted the essential contribution of solar power to meeting energy demand while minimizing the environmental effects [1]. Solar panels, as the main technology for harnessing solar energy, are an integral part of this evolution. However, ensuring high performance and durability of solar panels is crucial to boosting energy efficiency and supporting sustainable development [2]. Undetected faults in solar panels can result in decreased efficiency, lower energy output, and increased maintenance costs [3]. Therefore, effective monitoring and timely detection of faults are imperative. Recent advancements in technology, particularly in the fields of IoT and artificial intelligence (AI), have revolutionized the monitoring and fault detection in solar panels. The integration of IoT sensors and devices in photovoltaic (PV) systems enables real-time monitoring of various parameters, such as temperature, voltage, current, and irradiance levels. This continuous data collection allows for proactive maintenance and early detection of potential faults or anomalies in the system [4]. Furthermore, AI-driven algorithms, including ML and deep learning (DL) techniques, play a crucial role in analyzing the vast amount of data generated by IoT sensors. These algorithms can identify patterns, and deviations from normal operating conditions, thereby enabling intelligent fault detection and diagnosis [5]. By leveraging historical data and predictive analytics. The combination of IoT and AI technologies offers significant advantages in the detection and management of faults in photovoltaic systems. By providing real-time insights into system performance and system status, IoT-enabled monitoring systems enable operators to optimize energy production, minimize downtime, and reduce maintenance costs [6]. Numerous studies in the literature have made significant strides in the realm of intelligent fault detection in solar panels. In [7], a groundbreaking approach for anomaly detection in solar power plants is presented, integrating a hybrid support vector machine (SVM) with the grey wolf optimization (GWO) algorithm. The study showcases the superiority of SVM classification model optimized with GWO in accurately categorizing inverter states into normal and fault categories, achieving remarkable precision and performance. In [8], researchers introduced the extreme learning machine (ELM) technique integrated with IoT sensors. Classification models such as SVM and ELM techniques are utilized to effectively analyze sensor data and compare them for fault identification. In [8], authors present an intelligent solar panel monitoring system employing artificial neural network (ANN) technology for shading detection and fault identification in PV panels, supervised by IoT platform, this system offers real-time data analysis and alerts, accurately detecting shading to optimize energy generation efficiency and reduce maintenance costs. Muhammad A. et al. introduced an innovative method for fault detection in PV modules [9], leveraging the adaptive neuro-fuzzy inference system (ANFIS) methodology. This study focuses on effectively training

 DOI: 10.1201/9781003647263-9

the ANFIS model using grid partition (GP) and subtractive clustering (SC) strategies with research data, achieving high accuracy in predicting and classifying various faults in PV systems. The study performed in [10] utilizes a single voltage sensor for the entire PV array, enabling the detection of line-to-line (L-L) and line-to-ground (L-G) faults, as well as estimating the approximate number of faulty modules. Moreover, the circuit employs one voltage sensor for fault classification and another for precise identification of the faulty string. Additionally, the proposed algorithm is capable of discerning low and high impedance faults. Overall, these studies contribute valuable insights into enhancing fault detection in solar panels through innovative ML techniques and comprehensive analysis of influencing factors. However, persistent challenges remain in many existing approaches, particularly concerning data complexity, model scalability, and real-time performance. This paper aims to explore the effectiveness of an approach by presenting a comprehensive study on the integration of IoT sensors and ML techniques for monitoring and fault detection in solar panels. The study evaluates the performance of different ML models, including DTR, XGBoost, LightGBM, and SVR, in smart fault detection. The contribution of this study not only demonstrate the efficacy of the proposed methodology but also highlight its potential implications for advancing sustainable energy solutions in the context of solar energy. The subsequent sections of the paper are outlined as follows: Section 9.2 presents an overview of faults affecting PV systems. In Section 9.3, the IoT monitoring system and the ML approach for defect detection in solar panels are presented. Section 3.4 elaborates on the experimental results. Finally, Section 9.5 concludes this article.

9.2 PV SYSTEM FAULTS: OVERVIEW

Various types of faults can impact PV systems, encompassing issues such as shading, soiling, component degradation, electrical faults, and environmental factors [11], as illustrated in Figure 9.1. Shading occurs when objects block sunlight from reaching certain solar panels, resulting in decreased energy output. Soiling, caused by the accumulation of dirt or debris on panel surfaces, can reduce solar absorption and efficiency. Component degradation, such as aging or damage to solar panels or inverters, may also diminish system performance over time [12]. Electrical faults, including short circuits, open circuits, loose connections, or faulty wiring, pose risks to system integrity and safety [13]. Hot spots, arising from localized heating due to shading, soiling, or mismatched cells, can further decrease efficiency and potentially damage panels. Additionally, environmental factors such as temperature fluctuations, humidity, and weather events contribute to potential faults in

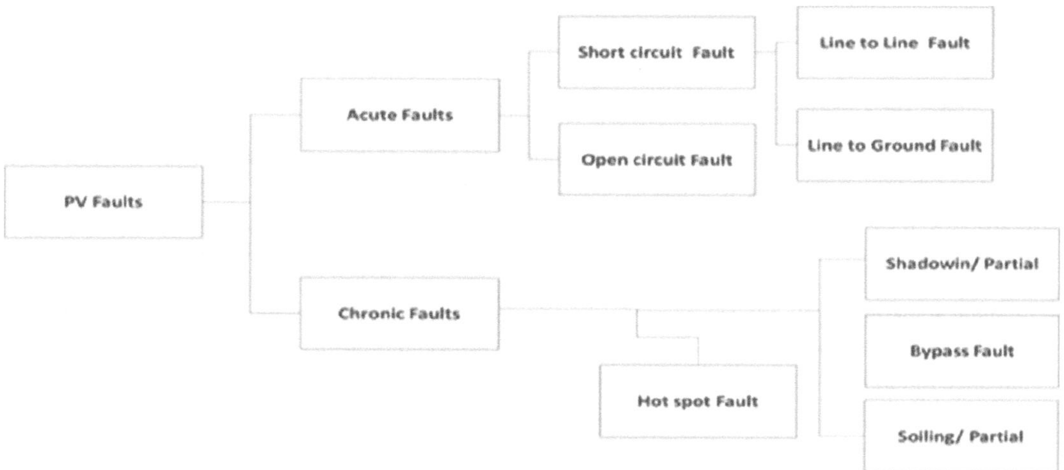

FIGURE 9.1 PV system faults: Overview.

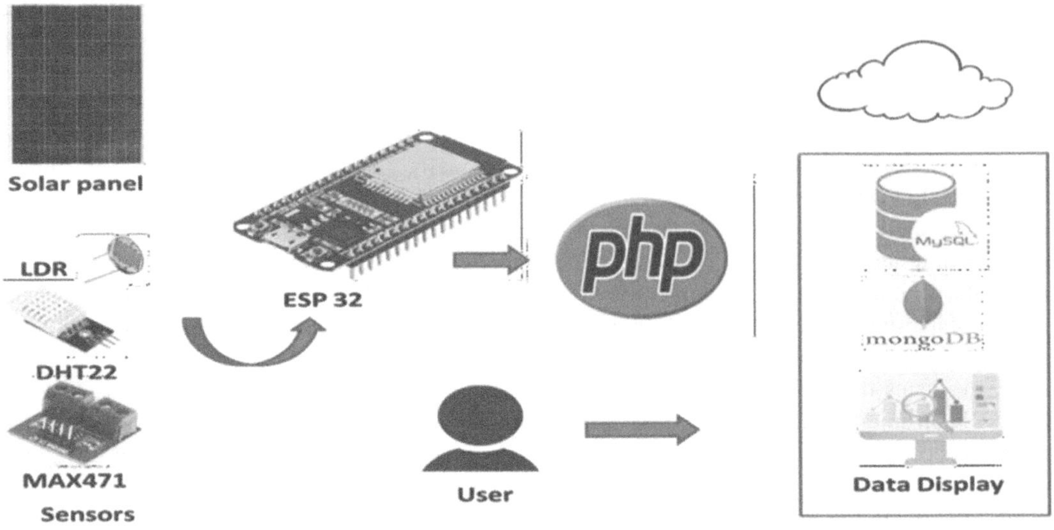

FIGURE 9.2 Proposed IoT architecture for solar PV systems.

PV systems [14]. Understanding these various fault types is essential for effectively managing and maintaining solar installations.

9.2.1 Solar PV Monitoring Systems

The solar PV monitoring system's structure consists of three separate tiers, as depicted in Figure 9.2: the data collection tier, the data analysis tier, and the data presentation and storage tier. At the data acquisition level, diverse sensor data including voltage, current, temperature, humidity, and irradiance are gathered. This system employs ESP32, a microcontroller with Wi-Fi capabilities, to execute the requisite calculations. The analog output of the sensors is routed to the ADC channels of the ESP32 module. Subsequently, the processed data is transmitted to the storage system, utilizing platforms such as MongoDB [15] and PhpMyAdmin [16], before being directed to the final stage for visualization of outcomes. Furthermore, the collected data is leveraged for fault detection within the PV system through the application of various ML models, facilitating the identification of faults such as soiling, shadowing, partial soiling, partial shadowing, and normal system operation.

9.2.2 Machine Learning Approaches for Detection Faults

In this study, a proposed method for fault detection in photovoltaic systems utilizes a combination of ML algorithms, each tailored to address specific aspects of fault detection, as shown in Figure 9.3. Initially, the DTR is employed to model the relationships between various input features and the occurrence of faults within the PV system. Hyperparameters such as maximum depth and minimum sample split are tuned to optimize the model's performance. Subsequently, the LightGBM model, a gradient-boosting framework known for its efficiency and high performance in large-scale data sets [17], is incorporated. Hyperparameters like learning rate and maximum depth are fine-tuned to enhance the model's fault detection capabilities. Following this, gradient boosting with XGBoost is employed, which iteratively improves the performance of weak learners. Parameters such as learning rate and maximum tree depth are adjusted to achieve optimal fault detection results. SVR is integrated to capture the complex relationships between input variables and fault occurrences. Parameters like kernel type and regularization parameter are optimized to enhance the fault detection accuracy of the SVR model. By leveraging these diverse approaches, the proposed method aims

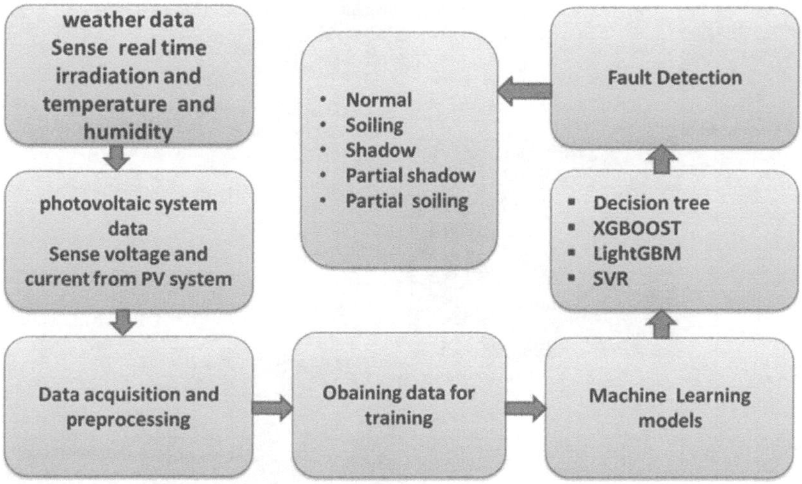

FIGURE 9.3 Comprehensive approach to general object detection framework.

to provide comprehensive fault detection capabilities for photovoltaic systems, thereby ensuring efficient maintenance and improved performance. In evaluating the effectiveness of the proposed fault detection method for PV systems, a set of key evaluation metrics are employed, including accuracy, precision, recall, and F1 score. Accuracy measures the proportion of correctly identified faults among all instances, providing an overall assessment of the model's correctness. Precision evaluates the ratio of accurately detected faults to all instances classified as faulty, demonstrating the model's capacity to minimize false positives. Recall, alternatively referred to as sensitivity, assesses the ratio of true faults correctly identified by the model among all existing faults, showcasing the model's effectiveness in detecting faults extensively. F1 score, which is the harmonic mean of precision and recall, offers a balanced assessment of the model's performance, particularly in scenarios where precision and recall have conflicting objectives. By collectively analyzing these evaluation metrics, insights are gained into the effectiveness and reliability of the proposed fault detection approaches. This enables optimization and fine-tuning of the models for enhanced performance in detecting faults within PV systems [18].

9.3 RESULTS AND DISCUSSION

9.3.1 Data Set

The data set, collected from an IoT monitoring system, encompasses measurements and environmental data related to a photovoltaic system. It includes readings from various sensors such as the light dependent resistor (LDR) sensor for light intensity, DHT22 for humidity and temperature, and MAX471 for voltage and current. Parameters within the data set consist of voltage (volt), current (A), power (W), temperature, humidity, and light intensity, as well as categorical variables indicating the state of the sky and the panel. These measurements provide valuable insights into both the electrical performance of the system and the surrounding environmental conditions influencing its operation. Voltage, current, and power readings offer crucial information about the system's electrical output, while temperature, humidity, and light intensity measurements provide context regarding prevailing environmental conditions. The categorical variables indicating the state of the sky and the panel contribute to understanding the system's operational status and its response to varying weather conditions. Figure 9.4 summarizes the correlation between these measurements, providing a visual representation of their relationships and interactions within the PV system.

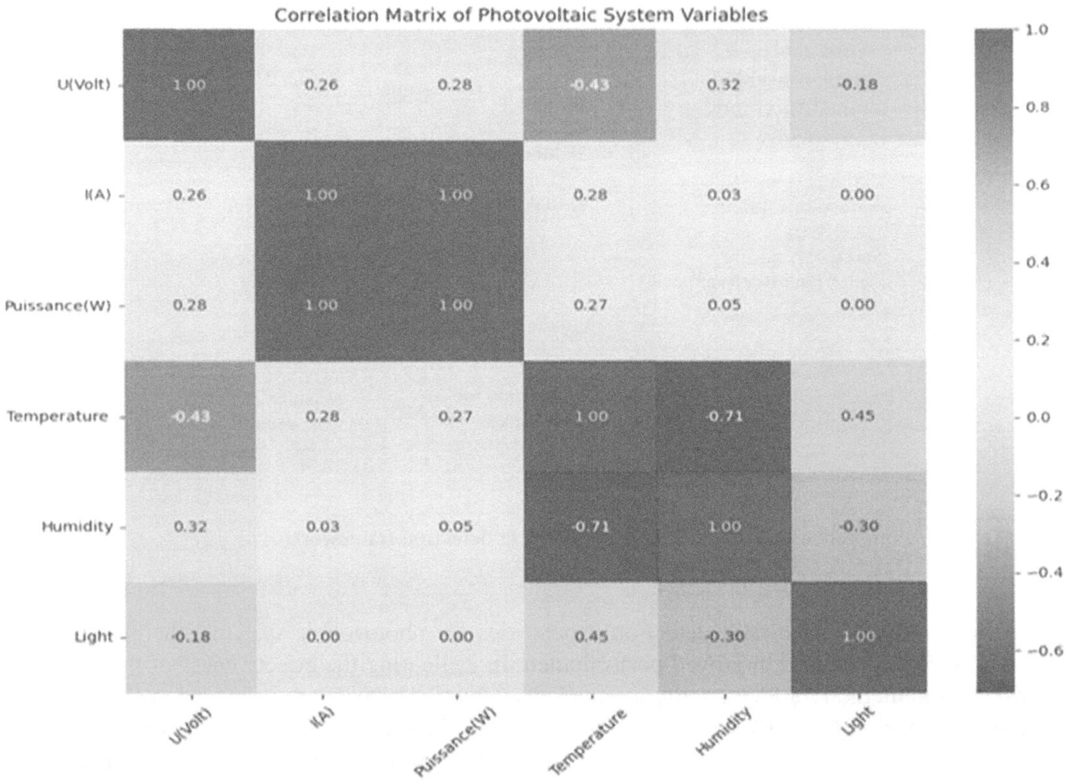

FIGURE 9.4 Correlation matrix of PV system variables.

The study evaluated different approaches for fault detection in PV systems, as shown in Table 9.1: DTR, XGBoost, LightGBM, and SVR. Among these, the DTR exhibited the highest performance, demonstrating impressive precision, recall, F1 score, and accuracy. It accurately identified faults in 93.95% of cases and captured 92.98% of true faults among all actual faults. Additionally, XGBoost performed commendably, displaying high accuracy and achieving a good balance between precision and recall. LightGBM showcased reasonable accuracy but exhibited lower precision, recall, and F1 score in comparison to the top-performing models. Conversely, SVR demonstrated the lowest performance, displaying lower precision, recall, and F1 score, indicative of its limitations in fault detection. Precision Analysis: Overall, ensemble methods like DTs and XGBoost demonstrated superior effectiveness for fault detection in PV systems compared to single models like SVR, as illustrated in Figure 9.5.

The confusion matrix, as depicted in Figure 9.6, provides a concise summary of the performance of the fault detection models. It presents the enumerations of true positive (TP), true negative (TN),

TABLE 9.1
Performance Assessment of Proposed Methods: Evaluation Metric Findings

Models	Precision	Recall	F1score	Accuracy
DTR	**0.94**	**0.92**	**0.93**	**0.92**
XGBoost	0.89	0.90	0.89	0.89
LightGBM	0.85	0.75	0.78	0.75
SVR	0.61	0.37	0.61	0.46

Note: DTR exhibited the highest performance, demonstrating impressive precision (in bold).

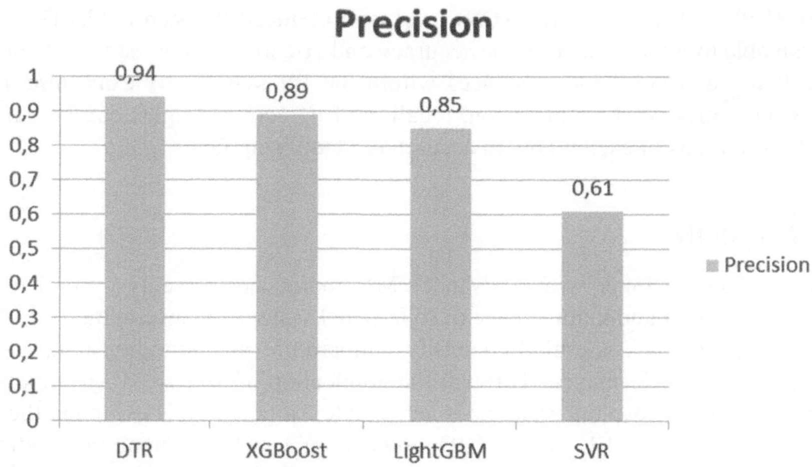

FIGURE 9.5 Precision for the proposed models.

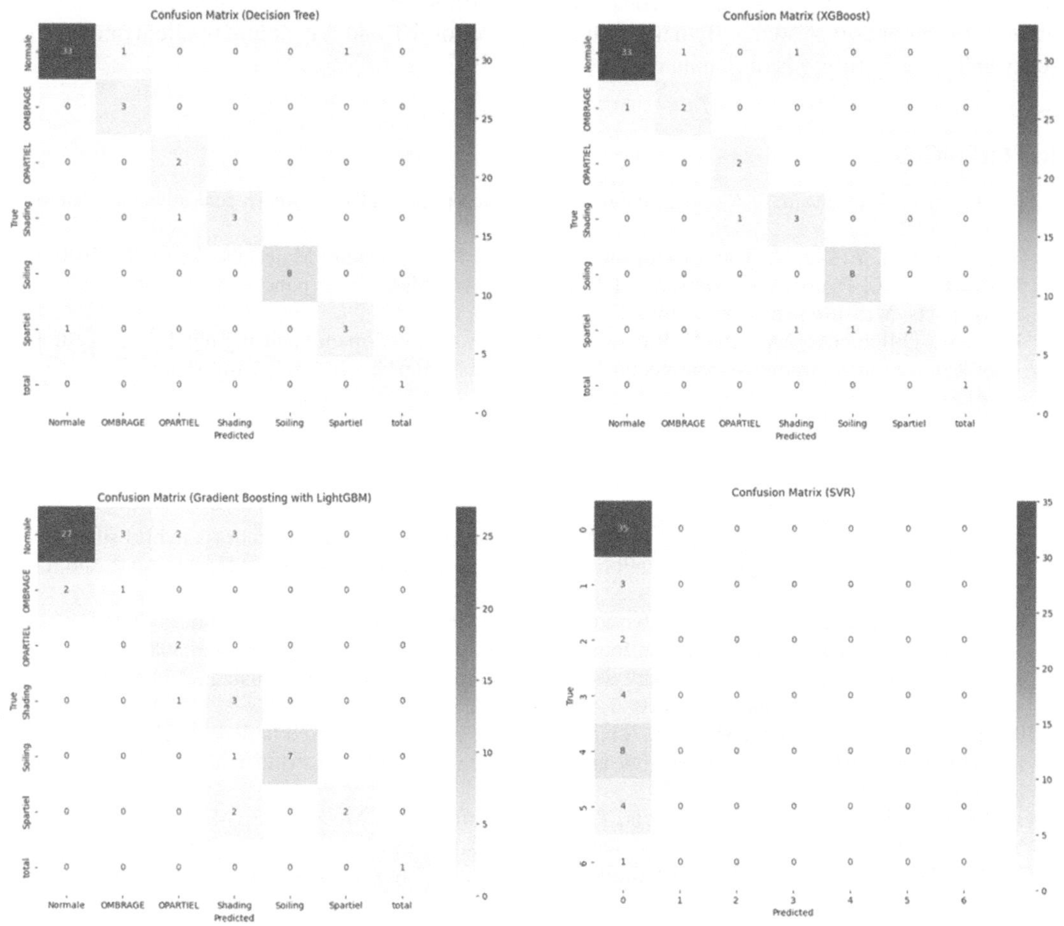

FIGURE 9.6 Assessing model performance: Confusion matrix analysis.

false positive (FP), and false negative (FN) forecasts produced by each model [19]. This matrix serves as a valuable tool for evaluating the accuracy and effectiveness of the models in distinguishing between faulty and non-faulty instances within the PV system. By examining the elements of the confusion matrix, such as precision, recall, and F1 score, insights can be gained into the strengths and limitations of each model in accurately identifying faults.

9.4 CONCLUSION

The integration of IoT technology and advanced ML approaches presents a promising avenue for enhancing fault detection and maintenance in solar panel systems by leveraging real-time data collected from IoT sensors, such as sunlight intensity, temperature, and humidity, alongside parameters specific to solar panel performance. Through the application of diverse ML techniques, effective fault detection within photovoltaic systems is achievable. Notably, DTR exhibited the highest precision of 0.94, recall of 0.92, F1 score of 0.93, and accuracy of 0.92 among the models evaluated. These results underscore the exceptional performance in fault detection. Additionally, XGBoost demonstrated considerable precision, recall, F1 score, and accuracy, with values of 0.89, 0.90, 0.89, and 0.89 respectively, showcasing its effectiveness as well. LightGBM and SVR, while showing lower performance compared to the top-performing models, still contribute valuable insights into fault detection within PV systems. Moving forward, continued research and development in this field will further refine these approaches, leading to improved reliability, efficiency, and sustainability of solar energy systems. By harnessing the power of IoT and AI, a more resilient and efficient renewable energy future is anticipated.

REFERENCES

1. Akpan, J., Olanrewaju, O.A.: Sustainable energy development: History and recent advances. Energies. 16, 7049 (2023).
2. Chu, S., Cui, Y., Liu, N.: The path towards sustainable energy. Nature Materials. 16, 16–22 (2016).
3. Meribout, M., Tiwari, V.K., Herrera, J.P.P., Baobaid, A.N.M.A.: Solar panel inspection techniques and prospects. Measurement. 209, 112466 (2023).
4. Yu, W., Dillon, T.S., Mostafa, F., Rahayu, W., Liu, Y.: A global manufacturing big data ecosystem for fault detection in predictive maintenance. IEEE Transactions on Industrial Informatics. 16, 183–192 (2020).
5. Abid, A., Khan, M.T., Iqbal, J.: A review on fault detection and diagnosis techniques: Basics and beyond. Artificial Intelligence Review. 54, 3639–3664 (2020).
6. Ahmad, T., Zhang, D.: Using the internet of things in smart energy systems and networks. Sustainable Cities and Society. 68, 102783 (2021).
7. Ahmed, Q.Z., Attar, H., Amer, A., Deif, M.A., Solyman, A.A.: Development of a hybrid support vector machine with grey wolf optimization algorithm for detection of the solar power plants anomalies. Systems. 11, 237 (2023).
8. Mohan, V., Senthilkumar, S.: IoT based fault identification in solar photovoltaic systems using an extreme learning machine technique. Journal of Intelligent and Fuzzy Systems. 43, 3087–3100 (2022).
9. Abbas, M.Z., Zhang, D.: A smart fault detection approach for PV modules using adaptive neuro-fuzzy inference framework. Energy Reports. 7, 2962–2975 (2021).
10. Mehmood, A., Sher, H.A., Murtaza, A.F., Al-Haddad, K.: A diode-based fault detection, classification, and localization method for photovoltaic array. IEEE Transactions on Instrumentation and Measurement. 70, 1–12 (2021).
11. Berghout, T., Benbouzid, M., Bentrcia, T., Ma, X., Djurović, S., Mouss, L.H.: Machine learning-based condition monitoring for PV systems: State of the art and future prospects. Energies. 14, 6316 (2021).
12. Mustafa, R.J., Gomaa, M.R., Al-Dhaifallah, M., Rezk, H.: Environmental impacts on the performance of solar photovoltaic systems. Sustainability. 12, 608 (2020).
13. Alhelou, H.H., Hamedani-Golshan, M.E., Njenda, T.C., Siano, P.: A survey on power system blackout and cascading events: Research motivations and challenges. Energies. 12, 682 (2019).

14. Khala, M., Abouzid, H., Teidj, S., Eloutassi, O., Messaoudi, C.: LSTM Deep Learning Method for Radiation Short and Long-Term Prediction. In: Ben Ahmed, M., Boudhir, A.A., Santos, D., Dionisio, R., Benaya, N. (eds) Innovations in Smart Cities Applications Volume 6. SCA 2022. Lecture Notes in Networks and Systems, vol 629. Springer, Cham. https://doi.org/10.1007/978-3-031-26852-6_63
15. Ghizlane, F., Mabrouki, J., Ghrissi, F., Azrour, M.: Proposal for a high-resolution particulate matter (PM10 and PM2. 5) capture system, comparable with hybrid system-based internet of things: Case of quarries in the Western Rif, Morocco. Pollution. 8(1), 169–180 (2022).
16. Contributors, phpMyAdmin: PhpMyAdmin, https://www.phpmyadmin.net/, last accessed February 13, 2024.
17. Sun, X., Liu, M., Sima, Z.: A novel cryptocurrency price trend forecasting model based on LightGBM. Finance Research Letters. 32, 101084 (2020).
18. Chicco, D., Jurman, G.: The advantages of the Matthews correlation coefficient (MCC) over F1 score and accuracy in binary classification evaluation. BMC Genomics. 21, 6 (2020). https://doi.org/10.1186/s12864-019-6413-7
19. Caelen, O.: A Bayesian interpretation of the confusion matrix. Annals of Mathematics and Artificial Intelligence. 81, 429–450 (2017).

10 Hybrid MHD-EHD Desalination

A Sustainable Approach Integrating Centrifugation and Crystallization for Zero Liquid Discharge

A. Karbal, D. Azdem, J. Mabrouki, M. Faraji, D. Bougmoum, M. Y. Elhafidi, A. El bouari, and R. Mghaiouini

10.1 INTRODUCTION

Water scarcity is becoming a more pressing worldwide issue, triggered by the rapid impact of increasing population, expansion of industries and climate changes (Abanyie, Amuah, and Nang 2025; Fahdi, Mabrouki, et al. 2024). Based on the United Nations, nearly 2.4 billion people reside in countries with water shortages and the number is predicted to increase dramatically (Fahdi, Azdem, et al. 2024; Panhwar et al. 2022). In this regard, seawater desalination is a crucial technology solution to the lack of freshwater supply particularly in arid or semi-arid regions (Azdem et al. 2024b, 2024c; Fatni et al. 2025). Desalination capacity around the world has been growing rapidly over the last two decades (Azdem, Mabrouki, and El Hajjaji 2025b, 2025c; Perumal, Sekar, and Carvalho 2024). By the year 2020, 15,906 desalination facilities were operating across the globe in 177 countries and had the total daily production capacity of 95.37 million cubic metres of freshwater (Azdem, Mabrouki, and El Hajjaji 2024a; Muteti 2024). These facilities are used by over 300 million individuals across the globe, particularly within the Middle East, North Africa and in some parts of Asia (Anastasi 2024; Azdem et al. 2024b, 2024c). Despite its growing importance within water systems however, traditional desalination techniques have a number of crucial challenges. Traditional processes are generally are divided into membrane-based techniques (reverse osmosis) and thermal-based processes (multi-stage flash distillation as well as multi-effect distillation) require a significant energy investment and result in higher maintenance costs and, most importantly, they produce huge quantities of brine that is highly concentrated (Azdem et al. 2025a; Tareemi and Sharshir 2023). Global production of brine is estimated at around 141.5 million cubic metres daily, which is more than 55% of it discharged into marine ecosystems without proper treatment (Azdem, Mabrouki, and El Hajjaji 2025b, 2025c; Pramanik, Shu, and Jegatheesan 2017). The uncontrolled release can have negative environmental impacts, such as small-scale increases in salinity, thermal pollutants and heavy metal accumulation that could seriously harm marine ecosystems on both the macro- and micro-scales.

To solve these operational and environmental challenges, zero liquid discharge (ZLD) systems have been recognized as an innovative alternative that is sustainable and long-lasting. ZLD technologies are designed to completely eliminate liquid waste completely by reclaiming both water and important minerals or salts in the stream of brine (Liang et al. 2024). Systems that integrate multi-effect distillation (MED) with the evaporative crystallization process have proven the capability to recover almost all the input water as well as the energy required for thermal processes ranging

 DOI: 10.1201/9781003647263-10

from 600 to 1100 kJ/kg. Another new ZLD technique, like the hybrid system which combines high-pressure reverse-osmosis (HPRO) brine concentrators and membrane-promoted crystallization (MPC), have achieved the rate of water recycling to more than 97%; however, they often come at the expense of complex operation and increased energy requirements (Panagopoulos et Giannika 2024). In this research, we have created a novel hybrid physical process that combines magnetico-hydrodynamic (MHD) as well as electrohydrodynamic (EHD) principles, as well as the process of crystallization and centrifugation in order to reach brine production as well as valuation. This method was developed to override the limitations of existing methods that rely only on physical processes, eliminating the requirement to use chemical additives, thereby decreasing their environmental footprint. Through computational fluid dynamics (CFD) simulation and modeling using COMSOL Multiphysics saltwater's behaviour in the presence of electrical and magnetic fields is thoroughly studied that allows for an immediate optimization of the parameters in the test cell for treatment (Chalá et al. 2024).

The mechanism that is at the heart of it involves the use of two important physical forces:

- EHD is a field in which Coulomb force is applied to dielectric fluids in order to cause an ion flow through the action of an electric field (Iranshahi et al. 2024).
- MHD, which utilizes Lorentz force to control conductors of electrical charge in the absence of an electromagnetic field encouraging ion movement and increasing the efficiency of desalination (Lin et al. 2020).

The combination with EHD as well as MHD fields allows for control of the charge-producing species in the saline matrix. The electromagnetic treatment will be followed up with centrifugation that separates solids on the basis of densities as well as crystallization that triggers solid-phase recovery of salts dissolved the end, the system permits the extraction of important and useful metals like bromine, lithium, arsenic and phosphorus copper, lead, silver, nickel and manganese by precipitation in the solid state. The concept for electromagnetic separation has been employed in microbiology, biomedicine agriculture, biotechnology, as well as in mining, wastewater treatment as well as the construction industry areas which benefit from precise control of the movement of particles and the ionic material. Through leveraging this established concept and extending the concept to desalination this system provides high selectivity, a low environmental footprint, and is compatible to circular economic models. This research contributes to the creation of an eco-friendly, energy efficient and scalable desalination process which not only generates potable water, but additionally transforms brine that was a waste product to an energy stream. In the next sections, we will discuss the physical model and simulation results, the experiments, and evaluation of performance. The goal of this research is to offer a feasible technological way to achieve zero discharge of liquids, resource recovery as well as ecologically efficient water management.

10.2 PHYSICAL MODEL

10.2.1 Magnetohydrodynamic (MHD) Processes

The magnetohydrodynamic (MHD) process is a key element in the physical simulation that is proposed for desalination process (Said 2024). It is based on the principles of plasma physics as well as fluid dynamics, particularly in the Dynamo theory that describes the method by the magnetic field is amplified and generated by the motion of an electrically conducting fluid. In the MHD scenario, the fluid that is conductive could include saline solutions, or plasma, in which free ions react to electric and magnetic fields. The Dynamo effect is a self-exciting phenomenon in which an initial magnetic field B0, when it interacts with a fluid conductive moving (characterized by the velocity field v) and generates the creation of an electromotive force (EMF). The EMF produces electric currents (i), which will in turn increase the magnetic field by using the law of Faraday's electromagnetic induction. The feedback cycle can be visualized schematically in the form of:

$$\mathbf{B0} \rightarrow \mathbf{e0} \rightarrow \mathbf{i0} \rightarrow \mathbf{B1} \rightarrow \mathbf{e1} \rightarrow \mathbf{i1} \ldots \quad (10.1)$$

$\mathbf{B_0}$: Initial magnetic field (seed magnetic field)
$\mathbf{e_0}$: EMF induced by the motion of the conductive fluid in the presence of B0
$\mathbf{i_0}$: The electric current generated due to e0
$\mathbf{B_1}$: A new (amplified) magnetic field generated by the current i0
$\mathbf{e_1}$: A new EMF induced by B1
$\mathbf{i_1}$: A new current generated by e1

As the induced magnetic field intensifies, it begins to influence the motion of the fluid itself through the **Lorentz force**, defined as:

$$F_{Lorentz} = \vec{J} \times \vec{\mathbf{B}} \quad (10.2)$$

J: Density of current
B: Field of magnetic energy

The feedback loop will continue to grow through an exponential increase in the energy of the magnetic field until it reaches saturation. Magnetic fields are powerful sufficient in order to create the force that's resistant to the motion of the fluid which results in the equilibrium dynamics. This result in the development of a constant-state model for MHD in which the magnetic field and the velocity of the fluid can be observed in the same space and create a balance of magnetic fields that are driven by the flow. In practical terms, MHD is a process that enhances the movement of ions by altering paths of charged particles in the liquid when it is used in the desalination process of seawater. The interaction between the magnetic field that is applied externally B_{ext} and Ionic stream in the saline waters creates Lorentz forces that create secondary flows that enhance the separation of ions in space. This process increases the overall effectiveness of the removal of ions without the requirement to use membranes, or any chemicals or reagents. The MHD subsystem of this model for desalination has two functions: to enhance an ion's separation by physical forces and is a part of the overall stableness of the process. Incorporating this principle into the model, the system becomes eco-friendly and energy efficient using natural physical laws for a more advanced fluid control.

10.2.2 Electromagnetic EMF and Electrohydrodynamic (EHD) Processes

This EHD method that is part of this desalination method relies on the use of an externally controlled rotating electromagnetic field which interacts directly with the ionic components of the solution of saline (Bai et al. 2025). In this case, it is the direction of the electromagnetic field that is aligned opposite to the direction of the primary flow of fluid which results in a deflection of charged particles away from their original streamlines. This deflection plays a significant function in dissolving the ion clustering and improving the separation of ions inside the fluid matrix. In parallel simultaneously, a parallel electric field is applied to the flow channels. The field serves in the role of the acceleration for ionic species and causes directional movement dependent on the charge polarity. The effect of the two electromagnetic interactions alters the electrokinetic properties of ions, improving their mobility and facilitating precise control of space of ion transport.

The dimension and geometrical configuration that is used to construct the electromagnetic separation, cells were specifically designed to take into account the movement time of ions and were dependent on the speed of fluid flow, ion charge and field intensity. The structure of the system makes sure that the time taken to move every ion that passes through the active zone remains long enough to ensure effective deflection and accelerated in the presence of applied fields. From a physicochemical point of view, the process could be compared to the electrodialysis (ED) system in which dissolved salts comprising positively charged cations as well as negatively charged anions moved towards electrodes with opposite charges by the force of an electrical field. The electrodes exert an electrostatic force on particles, which direct them along distinct pathways depending on their nature of polarity. This principle is widely used in traditional desalination processes, the processing of wastewater from industrial processes extracting valuable minerals from waste streams, as well as when it comes to making the processing and accumulation of salts. The inclusion to EHD strategies in this procedure is an important improvement over traditional ED methods, as it makes use of the potential of oscillating electrical fields as well as static electrical fields to provide more flexibility and control over the behaviour of ions. By increasing the field's parameters and integrating them with the flow's hydrodynamic pattern, it can maximize the efficiency of separation ions without the need for chemicals or component. It is also a durable flexible, adaptable, and efficient alternative to advanced desalination methods.

10.2.3 Design of an ED–CFG Crystallization and Centrifugal Extraction System

This article describes the concept and operation of a unique hybrid extraction system that combines electrodeposition (ED) along with force-driven crystallization (CFG) (Orellana-Palma, Guerra-Valle, and Zúñiga 2021). The combination of these two physical and chemical procedures provides a very efficient alternative to traditional desalination as well as material recovery processes. The most noteworthy innovation is the integration of electrochemical precipitation and the centrifugal separation method that enhances the effectiveness of separation via the use of a phase transition induced by the field that is later followed by mechanically controlled phase segregation. The ED mechanism promotes the nucleation of dissolved ions by applying a controlled electric field, leading to localized ion saturation and electro-crystallization. This is especially efficient to remove precious heavy metal ions in brines due to the usage of redox potentials which differ as well as electrochemical affinities for the various ionic species.

Then comes the centrifugation phase where the solution is pushed through an intense rotational force which generates the centrifugal force (Fc) that is much greater than the gravitational acceleration. With this force, particles with a greater density move outwards radially and are able to sediment quickly. The rate of sedimentation is controlled by the sedimentation coefficient that is determined by the size, shape, density as well as the viscosity of the medium. This separation by mechanical means allows for an extremely high degree of selectivity and throughput, specifically for crystallized salts as well as metal precipitates.

The ED-CFG module was developed with a single intake that is able to receive the saline water that has been partially treated through a downstream EMF (electromagnetic field) process, as well as three outlet ports that are devoted to:

- The primary outlet for the first outlet was designed in order to get rid of constantly crystallized Salts which are separated through concentration using centrifugal sedimentation.
- The second outlet can be used to draw the concentrated brine that is safe to be released or returned to the system to be processed further to increase the efficiency of the process. This allows for the ZLD operation.
- The third outlet is drinking water which has been filtered without the dissolving or suspended salts.

The modular structure does not just facilitate efficient desalination, it also increases brine's potential through the reclaiming of strategic minerals and cutting down on amounts of garbage by combining the effects of deposition by using electric fields as well as centrifugal separation, The ED-CFG system provides an extremely scalable and chemical-free and green alternatives to traditional brine management methods.

10.3 MATHEMATICAL MODEL

The mathematical framework used for this research is rooted in an analysis of the motion of the electrically conducting liquid, caused by the transportation charge particles (ions) through an electromagnetic field, which is linked (E B). The system is described by electromagnetic hydrodynamic (MHD) behaviour, in which the electromagnetic and fluid dynamics fields interact in large ways. The behaviour of the fluid flows is controlled through the incompressible Navier-Stokes equations. They are extended by electromagnetic force which is vital for understanding the effect on the Lorentz force. The general formulation of the equation of momentum conservation is defined as (10.3):

$$\rho\left(\frac{\partial \vec{v}}{\partial t}+(\vec{v}.\nabla)v\right)=-\nabla p+\mu\nabla^2\vec{v}+\vec{J}\times\vec{B} \quad (10.3)$$

ρ : fluid density
$\vec{\upsilon}$: velocity field
p : pressure
μ : dynamic viscosity
$\vec{J}\times\vec{B}$: Lorentz force, with:
$\vec{J}$: current density
$\vec{B}$: magnetic field vector

This formulation is a combination of the fundamentals of electromagnetism and fluid mechanics and is crucial to describe the dynamic of electroconductive fluids within environments where magnetic and electric forces are evident, for instance in desalination processes that employ electrokinetic processes, or in marine environments that are affected by electromagnetic sources anthropogenic to humans.

10.3.1 Electromagnetic Contribution and Fluid Coupling in MHD Systems

The behaviour of electromagnetic fields in conductive fluid exposed to external magnetic fields is governed by Maxwell's equations, which constitute the fundamental framework for understanding the behaviour of magnetic and electric fields. The equations, in differential form, are as follows:

- **Magnetic flux conservation**

$$\nabla\cdot\vec{B}=0 \quad (10.4)$$

- **Gauss's law (electrostatics)**

$$\nabla\cdot\vec{E}=\frac{\rho_0}{\varepsilon_0} \quad (10.5)$$

- **Faraday's law of induction**

$$\nabla\times\vec{E}=-\frac{\partial\vec{B}}{\partial t} \quad (10.6)$$

- **Ampère-Maxwell law**

$$\nabla\times\vec{B}=\mu_0\vec{j}+\mu_0\varepsilon_0\frac{\partial\vec{E}}{\partial t} \quad (10.7)$$

- **Charge conservation (continuity equation)**

$$\nabla \cdot \vec{j} = -\frac{\partial \rho}{\partial t} \tag{10.8}$$

Additionally, the transport of electric charge in the fluid medium is described by:

$$\frac{\partial \rho_0}{\partial t} = -\sigma \varepsilon_0 \rho_e \text{ with } \frac{d}{dt} = \frac{\partial}{\partial t} + \vec{\upsilon} \cdot \nabla \tag{10.9}$$

The electric field ($\vec{E}$) can also be expressed in terms of the scalar potential (ϕ) and the vector potential $\left(\overrightarrow{A}\right)$ as:

$$\vec{E} = -\nabla \phi - \frac{\partial \vec{A}}{\partial t} \tag{10.10}$$

10.3.2 Electromagnetic-Fluid Coupling

In magnetohydrodynamic systems it is the interplay between the fluid's velocity field along with its magnetic field that results in the creation of an electromagnetic force. These are described using an amalgamation of magnetic and electrostatic forces:

- **Electrostatic force:**

$$\vec{F}_E = q\vec{E}$$

- **Magnetic (Lorentz) force:**

$$\vec{F}_M = q\left(\vec{\upsilon} \times \vec{B}\right)$$

- **Total electromagnetic force:**

$$\vec{F}_{EM} = q\left[\vec{E} + \left(\vec{\upsilon} \times \vec{B}\right)\right] \tag{10.11}$$

$$\vec{j} = \sigma \vec{E}^* = \sigma\left(\vec{E} + \vec{\upsilon} \times \vec{B}\right) = \sigma\left(-\nabla V + \vec{\upsilon} \times \vec{B}\right) \tag{10.12}$$

The induced current density, $\vec{j}$, is governed by the generalized Ohm's law for a moving conductive fluid:

The corresponding Lorentz force acting on the fluid becomes:

$$\vec{F}_L = \vec{j} \times \vec{B} \tag{10.13}$$

Neglecting displacement currents and applying the divergence operator to Ohm's law yields the following Poisson-type equation for the electric potential, ϕ:

$$\Delta \phi = -\nabla \cdot \left(\frac{\downarrow \vec{A}}{\partial t} + \vec{\upsilon} \times \vec{B}\right) \tag{10.14}$$

Symbol Definitions:

- ρ_0 : volume charge density
- $\vec{j}$: electric current density
- ε_0 : vacuum permittivity
- μ_0 : magnetic permeability of free space
- σ : electrical conductivity
- q : electric charge
- $\vec{E}, \vec{B}$: electric and magnetic field vectors
- $\vec{\upsilon}$: fluid velocity
- $\phi, \vec{A}$: electric and magnetic potentials

10.3.3 Magnetohydrodynamic Mechanism

- **Magnetic induction:**

$$B_I \propto B_{0EX}$$

- **Voltage amplification:**

$$U_{0EX} = G_A \cdot B_I$$

- **Magnetic field control via coil voltage:**

$$B_{0EX} = G_K \cdot U_{0EX} \tag{10.15}$$

In MHD systems, interactions between a magnetic field as well as inducing the currents in the fluid produces secondary magnetic fields. These interactions are described by a variety of transfer equations:

10.3.4 Balance of Forces—Centrifugal Versus Gravitational

In rotating MHD systems, centrifugal and gravitational forces acting on particles can be compared as follows:

- **Centrifugal force:**

$$F_C = m\gamma = m \cdot r\omega^2 \text{ with } \omega = \frac{2\pi \mathrm{N}}{60}, \ \upsilon = r\omega \tag{10.16}$$

- **Gravitational force:**

$$F_g = mg \tag{10.17}$$

- **Ratio of forces:**

$$\frac{F_C}{F_g} = \frac{mr\omega^2}{mg} = \frac{r}{g}\left(\frac{2\pi \mathrm{N}}{60}\right)^2 = 0.001118 \cdot r \cdot N^2 \tag{10.18}$$

Where:

m : particle mass (kg)
F_C : centrifugal force
F_g : gravitational force

γ : centripetal acceleration (m/s^2)
r : radius (m)
ω : angular velocity (rad/s)
N : rotational speed (RPM)

This relationship shows that the centrifugal force increases quadratically with rotation speed, and can exceed gravitational forces depending on system parameters.

10.4 MODELING AND SIMULATION WITH COMSOL METAPHYSICS

This MHD model that is used in this research is built on the combined resolution of electromagnetic field equations. The term "governing force" in the Navier Stokes equation, which typically depicts convection, is changed to incorporate the Lorentz force, also called the Laplace force, which is the result of interactions between electric current ($\vec{J}$) that is injected into the electrodes and magnetic flux density ($\vec{B}$). This force is the one responsible for the induction of fluid motion, regardless of an absence in underlying differential pressure between the outlet and inlet of the channel.

10.4.1 Electromagnetic–Hydrodynamic Coupling

The hydrodynamic and electromagnetic effects are interdependent and have to be addressed simultaneously. The motion of the fluid is influenced by electromagnetic body forces while the induced density of current is controlled through the field of velocity local to it in accordance the Ohm's law. The coupling system comprises:

- **Electromagnetic equations**: Used to compute the induced current density ($\vec{J}$) considering both the electric field and motion-induced contributions (Xiao et al. 2025).
- **Hydrodynamic equations**: Govern the velocity field ($\vec{v}$) of the conductive fluid, accounting for the Lorentz force as a body force in the Navier–Stokes formulation (Murugaiyan 2024).

10.4.2 Electromagnetic Coupling and Ion Transport

Ion transport is modeled by incorporating various driving mechanisms acting simultaneously:

- **Diffusion**: Due to gradients in ionic concentration.
- **Migration**: Driven by the electric potential gradient.
- **Convection**: Both forced and natural convection contribute to the bulk movement of ions through the fluid medium.

The total molar flux of an ionic species (i), denoted ($\vec{N}_i$), is thus expressed as the sum of the diffusive, migrative, and convective fluxes.

10.4.3 Hydrodynamic–Ion Transport Coupling

The spatio-temporal evolution of species concentrations in the fluid domain is governed by a convection diffusion equation. The local ionic concentration (ci) of species (i) is coupled with the flow field ($\vec{u}_i$) via the convective flux term, $\vec{N}_i^{conv}$. The volumetric mass density (ρ\rhoρ) of the fluid is calculated from the sum of the contributions of all species:

$$\rho = \sum\nolimits_i \mathbf{M}_i \times \mathbf{c}_i. \tag{10.19}$$

Where M_i is the molar mass of species i.

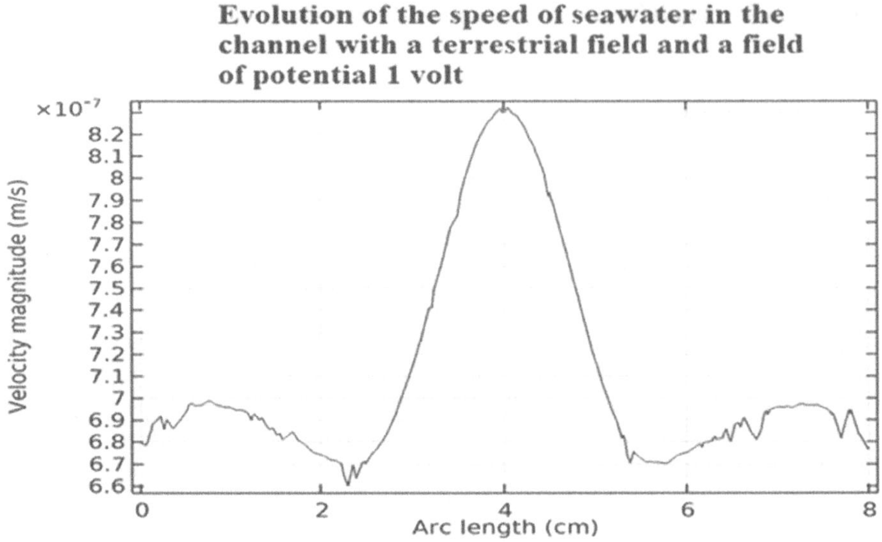

FIGURE 10.1 Effect of the earth's magnetic field on flow velocity using COMSOL Multiphysics software.

10.4.4 Velocity Field Distribution under Electromagnetic Actuation

Simulation results generated using COMSOL Multiphysics show the evolution of the velocity of flow of seawater inside a microchannel that is confined and subjected to a magnetic field of the terrestrial as well as an electrical potential that is 1 volt (Zhou et al. 2022). As seen in Figure 10.1, the velocity is at its highest near the middle of the channel as a result of the highest point in Lorentz strength, demonstrating the efficacy of electromagnetic action for purposes of pumping in without pressure gradients.

Figure 10.2 illustrates the variation in the velocity of seawater along the length of an arch channel, under different electromagnetic actuation scenarios, and illustrates the effects upon the intensity of the electric field, magnetic and electric fields. In the case of a magnetic field of 0.5 T and the electrical voltage 1V (Figure 10.2a), the velocity increases slowly and reaches a peak of around 0.0055 M/s. Then it drops a little. If it is enlarged to 1 T using similar electric voltage (Figure 10.2c), the peak velocity rises to 0.009 milliseconds. This is a clear increase in velocity of flow as a result of the higher Lorentz force. Furthermore, an increase in the magnetic field to 2 T, with no voltage (Figure 10.2b), can result in significant speed increases of up to 0.0155 milliseconds, which is three times quicker in contrast with 0.5 T in the case. It is important to know that one electric voltage, which is 1 V, is the same frequency as the magnetic one (Figure 10.2d) does not result in any significant increase in the speed of the fluid. It is 0.0155 milliseconds. This suggests a saturation effect that magnet field functions as the main driving force behind the acceleration of fluids and the contribution in the role of electric potentials is insignificant. The results demonstrate that the magnetic field acts as the primary driver of fluid motion in this scenario, while the electric potential serves as a secondary enhancer that is particularly efficient at lower magnetic intensities. The general pattern of behavior follows the basic principles that are underlying MHD, which is the area where the Lorentz force generated by the interaction between magnetic and electric forces in the seawater which conducts is accountable for the flow characteristics. These results are important for improving fluid control in electromagnetic systems, such as those employed in desalination channels as well as engineering of the marine environment.

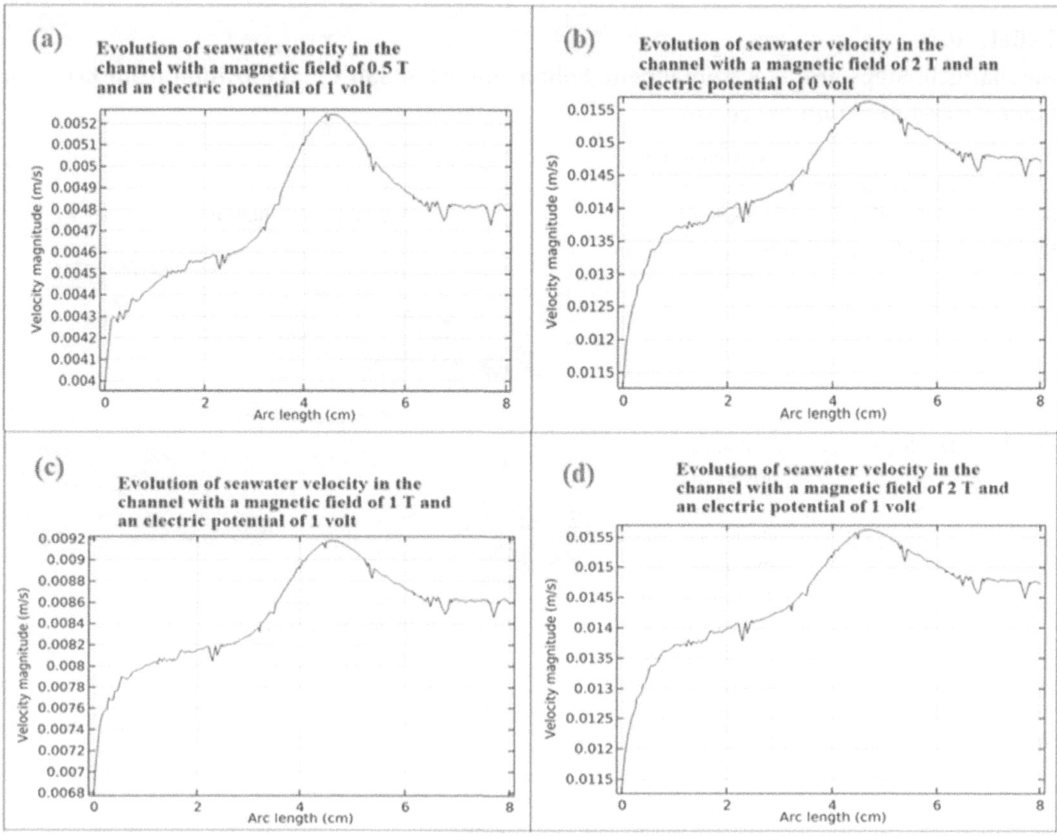

FIGURE 10.2 a–d MHD simulation of seawater velocity distribution under electromagnetic actuation using COMSOL Multiphysics.

10.5 CRYSTALLIZATION

To crystallize salt from a brine solution, two major energy barriers must be overcome. First is the enthalpy of hydration, the energy advantage ions gain when dissolved in water (Nikkhah et al. 2024; Zhao et al. 2024). Second, the system must overcome lattice enthalpy, the energy needed for those ions to come together and form a solid crystal structure. Modern technologies provide the energy required to break through these boundaries is provided by electromagnetic fields that are generated by MHD techniques. This technique is based on the principles of electro-crystallization, which is the process of removing salt crystals out of water-based substances. The application of the electric field is demonstrated to stimulate crystal nucleation by aligning molecules and increasing the density fluctuations. In addition, because crystallites are polar, applying an electric field decreases the energy they have to work with, which makes it more attainable for crystals to create. Once nucleation occurs, the crystallization process continues in the solid phase through this mechanism (Table 10.1).

Step 1: Stabilizing the aqueous solution.

At this stage, the aqueous solution is pre-balanced to ensure a stable chemical composition and uniformity of parameters such as ionic concentration, pH, and temperature. This creates the initial conditions required for the rest of the process.

$$F(n) = -n\Delta\,\mu + \gamma a n^{2/3} \tag{10.20}$$

TABLE 10.1
Mechanistic Steps and Electromagnetic Enhancement of Salt Crystallization from Brine via Electro-crystallization Processes

Step	Thermodynamics/ Kinetics	Mechanism or Principle of Operation	Visual Représentation
Step 1	**Equilibrium**	Chemical Potential in Equilibrium:	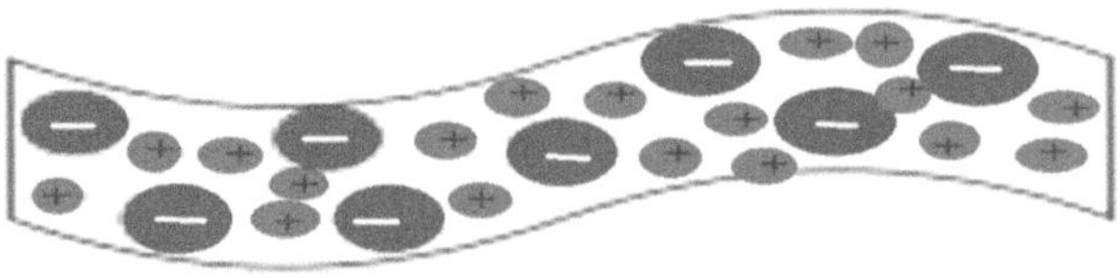 Ions in equilibrium
Step 2	**Driving Forces**	MHD affects the driving force by increasing the chemical potential and adding momentum to the particles in order to overcome the energy barriers to produce nuclei.	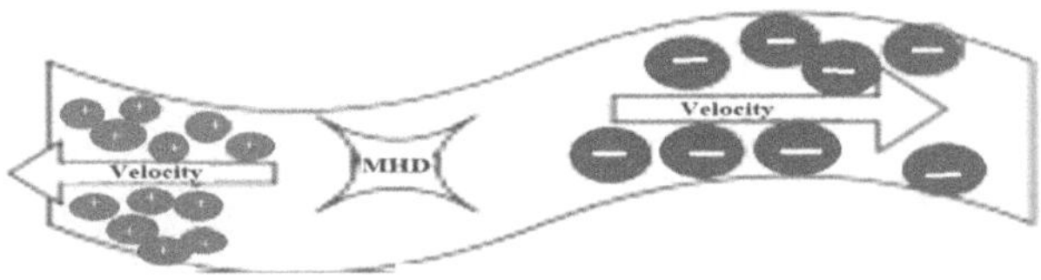 Magnetic field (MHD), ion motion and orientation
Step 3	**Seeding**	EMF use to promote the formation of the first seeds and to increase the nucleus growth velocity.	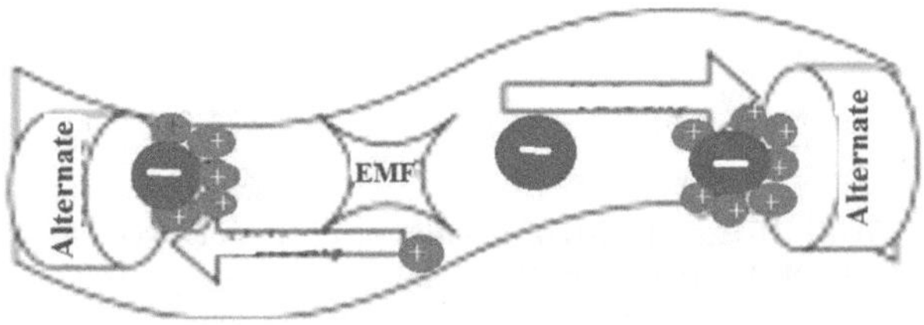 Electric field (EMF), ion flow, and seed formation
Step 4	**Nucleation**	Promoted by increasing the attachment frequency for nucleation by overcoming the two resistances: 1. Diffusion résistance. 2. Ion desolvation.	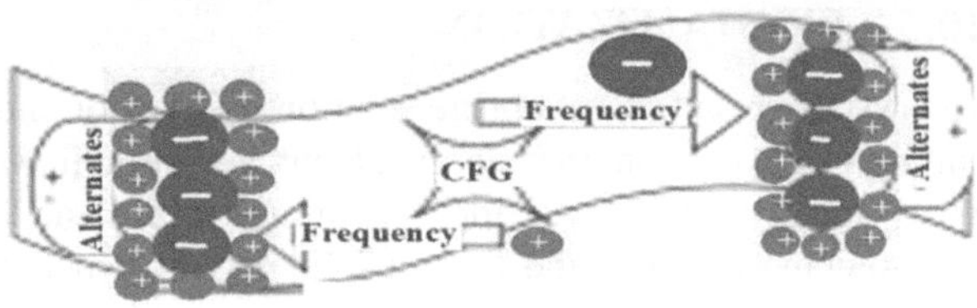 Attachment of ions and nucleation step
Step 5	**Nucleation Rate**	Assess the nucleation rate to measure how quickly new crystals form, a key factor in understanding and optimizing crystallization kinetics	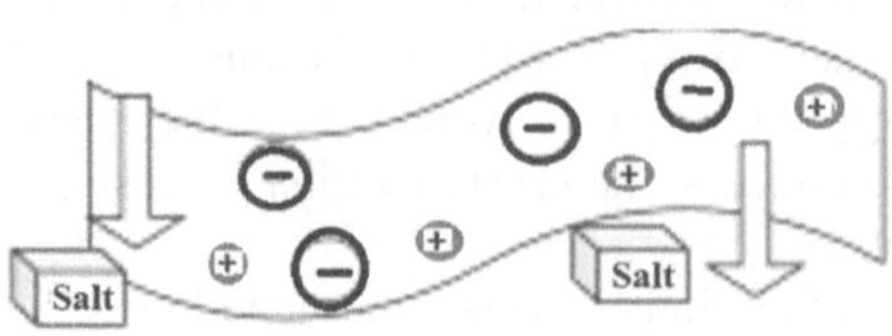

Step 2: Driving forces: Disturbance of the solution's equilibrium.

Once MHD is induced, significant changes in the driving forces of the solution occur. MHD, generated by the application of magnetic and electric fields, breaks the equilibrium of the solution, initiated by external forces.

$$\Delta\mu_{\varepsilon} = \Delta\mu + c_{\varepsilon}E^2 \tag{10.21}$$

$$C_{\epsilon} = \frac{3\epsilon_0\epsilon_m(\epsilon_c - \epsilon_m)V_0}{2(\epsilon_c - \epsilon_m)} \tag{10.22}$$

Step 3: Core growth with EMF for priming. EMF is used to stimulate the growth of crystalline nuclei, thereby initiating the crystallization process. This step is crucial to guarantee homogeneous and controlled crystallization.

$$\Delta\, t\; d(n)/dt \ll csatV(S-1) \tag{10.23}$$

Step 4: Crystallization: Fixation frequency for nucleation, followed by centrifugation.

Crystallization occurs at a frequency that is suitable for the nucleation of crystals. This process is followed by centrifugation, which is a mechanical process that breaks down the components based on their density, which allows crystals to form.

Step 5: Assessment of nucleation rate.

The final stage involves an exact assessment of the nucleation rate, which is measured by the rate at which crystals begin to form. This measurement is vital for understanding the mechanism of crystallization as well as optimizing the whole process.

10.6 MECHANICAL CENTRIFUGAL EXTRACTION

Centrifugation makes use of centrifugal acceleration to divide phases according to their density variations (Farshidfar et al. 2025). This technique is especially effective as the system reaches saturation. In the course of centrifugation, components that are denser move outward towards the outer edges of the bowl rotating. A fast settling speed is vital for getting a quick and effective separation. The effectiveness of separation depends on a variety of factors, including the density differences between particles, the size of the fluid, as well as viscosity and temperature. Most dissolved salts consist of positively or negatively charged ions, which tend to migrate toward electrodes of opposite charge (Figure 10.3). The electrodes apply an electrostatic force on these ions, drawing them toward their respective poles.

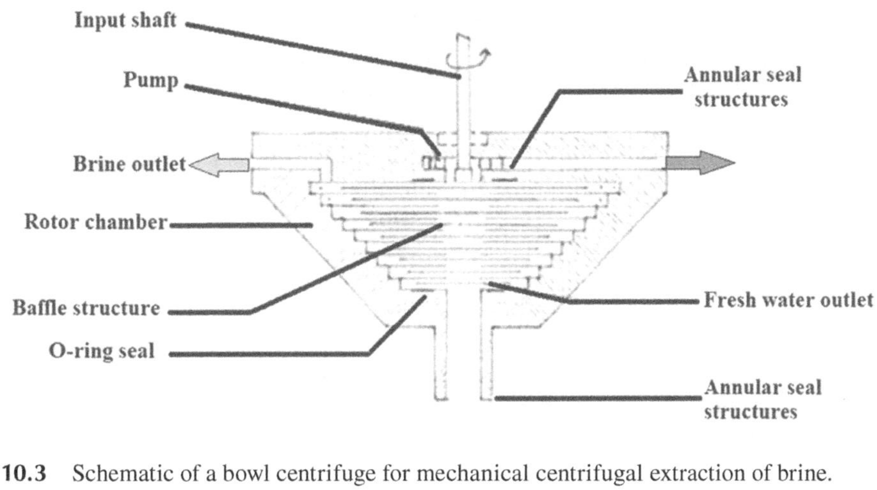

FIGURE 10.3 Schematic of a bowl centrifuge for mechanical centrifugal extraction of brine.

10.7 MATERIALS AND METHODS

The experimental system developed for this study integrates a hybrid magnetohydrodynamic–electrohydrodynamic (MHD–EHD) approach with mechanical centrifugation and crystallization to achieve ZLD desalination.

10.7.1 MHD Subsystem Configuration

The MHD subsystem is based on the dynamo effects, using an elongated channel (20 mm width 200 mm length) in which the brine flows in a continuous manner. A permanent parallel-pipe–shaped magnet (200 mm × 50 mm × 50 mm) is positioned to generate a stable magnetic field across the flow path, while a direct current (1 VDC) is applied perpendicular to the direction of flow to create an electric field that accelerates charged ions (Na^+ and Cl^-). The interaction of these fields increases ionic mobility and facilitates charge separation.

10.7.2 EHD Electrochemical Cell

The EHD subsystem consists of a cylindrical bifurcation cell made of a non-conductive polymer, tailored for continuous treatment of seawater and brackish water. The cell features a central axial inlet for saltwater and two concentric outlets: the inner axis discharges fresh water while the outer cylindrical wall collects concentrated brine. Two half-plate electrodes that are curved (anode as well as cathode) are placed in a symmetrical manner inside the cell, and are powered by an alternating current (AC) generator which enables ion adsorption as well as dynamic separation of charge through capacitive effects.

10.7.3 Centrifugation and Crystallization Unit

After electrochemical treatment, the solution is then transferred to a bowl centrifuge that is designed for mechanical separation at high speeds. The centrifugal force assists in the collection and precipitation of salt crystals on the edge of the rotating chamber and purified water is retrieved in the canter. The nucleation process is accelerated due to prior exposure to electromagnetic fields which lower the threshold of free energy for the formation of crystallites.

10.7.4 Process Monitoring and Data Analysis

Throughout the experiment, important variables like conductivity, temperature, voltage and flow rate, as well as rotational speed and ion concentration were observed with calibrated sensors. Every experiment was run three times in the same conditions to ensure that it was reproducible. Data were analyzed by using OriginPro 2023 software. the statistical significance was evaluated with the aid of ANOVA to determine the effect the operational variables on efficiency of desalination.

10.8 RESULTS AND DISCUSSION

A series of experiments were conducted to evaluate the performance of the hybrid MHD-EHD desalination system under identical operational conditions. A 5-litre volume of seawater from the natural source (EM_0) was tested in conjunction with three solutions of NaCl synthesized (P_1,

TABLE 10.2
Desalination Performance of Natural Seawater and Synthetic NaCl Solutions Under MHD–EHD Treatment Conditions

Sample	Initial Salinity (Si) g/L	Final Salinity (Sf) g/L	ΔS (g/L)	Reduction Rate (%)
EM_0	35.5	32.2	3.3	9.29
P_1	17.8	16.1	1.7	9.55
P_2	9.9	7.9	2.0	20.20
P_3	6.4	3.6	2.8	43.75

P_2 and P_3) that were each made with different salt concentrations (17.8 g/L, 9.9 g/L and 6.4 g/L, respectively), all of which were maintained at 20°C and with the same stream rate of 24 litres per minute.

10.8.1 Desalination Efficiency

The results (Table 10.2) revealed a clear trend: as initial salinity decreases, the system's desalination efficiency improved significantly. EM_0, with an initial salinity of 35.5 g/L, showed a modest reduction rate of 9.29%. Similarly, P_1 (17.8 g/L) achieved 9.55%, indicating limited performance at medium salinity. In contrast, P_2 (9.9 g/L) and P_3 (6.4 g/L) displayed much higher reduction rates 20.2% and 43.75%, respectively.

These findings confirm that the MHD-EHD treatment system performs best in low to moderately saline environments, as seen in Figure 10.4. In high-salinity waters, such as seawater, its efficiency is limited, and supplementary treatment steps or enhanced field intensities may be required. Conversely, the system shows strong potential for pre-treated brackish water or for polishing steps in multi-stage desalination systems.

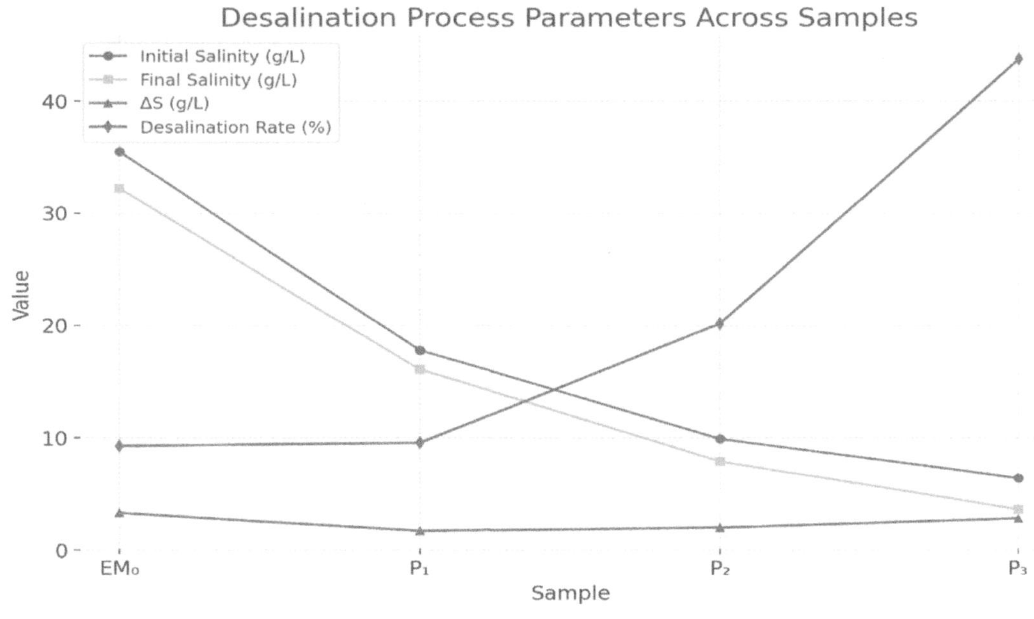

FIGURE 10.4 Evolution of desalination parameters across samples treated by MHD–EHD hybrid system.

TABLE 10.3
Salt Recovery Efficiency and Brine Concentration as a Function of Initial Salinity under MHD-EHD Treatment

Sample	Initial Salinity (Si, g/L)	Brine Salt Concentration (g/L)	Salt Recovery Rate (%)
EM_0	35.5	68	52.20
P_1	17.8	45	39.55
P_2	9.9	27	36.66
P_3	6.4	19	33.68

10.8.1.1 Salt Recovery from Brine

To complement desalination performance, salt recovery from residual brine was assessed. Table 10.3 shows that samples with higher initial salinity yielded greater salt concentrations in the brine. EM_0 produced 68 g/L of recoverable salt with a 52.20% recovery rate, while P_3, with the lowest initial salinity, yielded only 19 g/L and a 33.68% recovery. This trend reflects the greater availability of ions for recovery in highly saline samples.

The results suggest that feedwater with higher salinity contains a larger ionic load, providing an increased amount of material for extraction in the separation and concentration stages (Figure 10.5). The MHD-EHD technology is believed to aid in the process of concentrating salt better in areas that have high ions, perhaps due to the stronger interactions between charged particle magnetic fields. Furthermore, the current trends suggest that the chance to recover salt can be heavily dependent on the amount of free ions more frequent in salt-rich matrixes. The continuous decrease in salt concentration as well as speed of recovery ranging from EM_0 to P_3 demonstrates the limits in the extraction of salt in low-saline solutions.

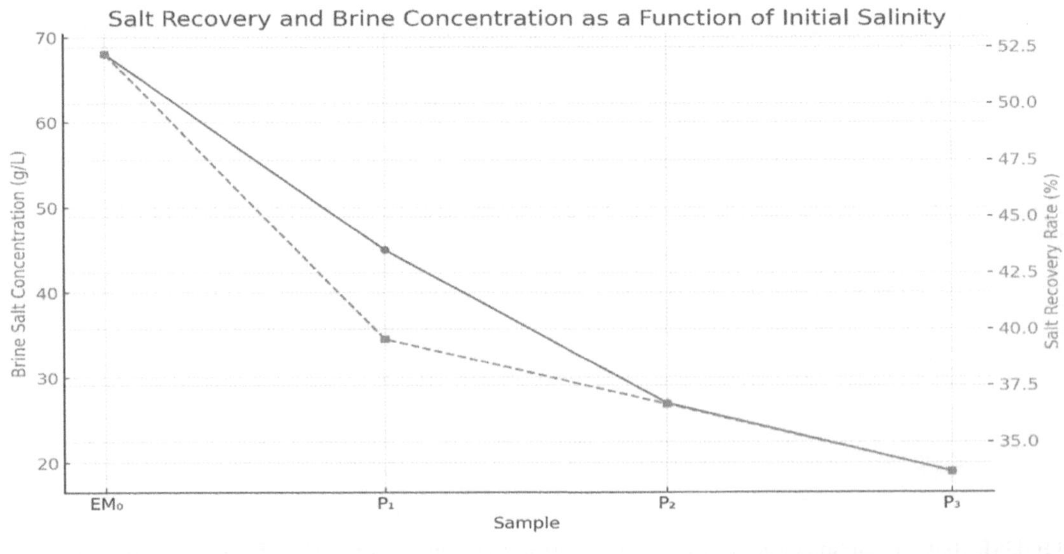

FIGURE 10.5 Evolution of brine salt concentration and salt recovery rate as a function of initial salinity.

10.9 CONCLUSION

This study presented a comprehensive approach to desalination through the use of hybrid MHD-EHD processes, pushing the boundaries of sustainable water treatment. Through the use of electromagnetic forces, specifically Lorentz as well as Laplace forces, this technique enables an effective separation of charged particles (Na^+ and Cl^-) from brackish and seawater water, without the need for the use of membranes or chemical compounds. The primary goal was to generate water that is fresh and also to extract the concentrated brine with a high concentration in heavy metals and ions which could be repurposed. Our research has confirmed that the effectiveness of this system was dependent on the strength that the magnetic field exerted as well as bias voltage that increased ions' mobility by increasing the velocity of flow and reducing the length of residence. Additionally, the flow rate was highest in the central part of the treatment channel. This was the area where ion separation was most effective. The joining of multiple magnets in series improved the flow rate and helped to stabilize it, which permitted for constant desalination across every part of the process.

What makes this method unique was the combination of crystallization and centrifugation that not only enhanced salt extraction, but tackled the urgent environmental issue of discharges from brine, which is a recognized environmental threat of marine ecosystems. Crystallized salts have liquid ions that can be recovered, such as nickel, lithium and a variety of other heavy metals. This adds an important dimension of circularity as well as resource recovery to this treatment process.

In practical terms, the system has a wide range of applications:

- As a pre-treatment unit for reverse osmosis, it can eliminate heavy metals, reduce scaling and fouling and prolong membrane life span.
- It can be deployed as a stand-alone, autonomous unit for potable water generation in remote or off-grid areas.
- It supports ion recovery processes, particularly for strategic elements like lithium, with ongoing development toward integrating solid-state centrifugation, electrodeposition and crystallization downstream.

REFERENCES

Abanyie, Samuel Kojo, Ebenezer Ebo Yahans Amuah, and Douti Biyogue Nang. 2025. "Water scarcity and its implications on sanitation: A perspective study in an emerging city in Northern Ghana". *Green Technologies and Sustainability* 3 (2): 100138.

Anastasi, Chris. 2024. *Desalination and Water Security*. Taylor & Francis.

Azdem, Driss, Abdelilah Fatni, Jamal Mabrouki, and Souad El Hajjaji. 2025a. "The desalination strategy and economic integration of Morocco: Water scarcity, prospects, and challenges". *Circular Economy Applications in Energy Policy*, 257–68.

Azdem, Driss, Jamal Mabrouki, and Souad El Hajjaji. 2025b. "The use of artificial intelligence in a desalination plant powered by renewable energy". https://Services.Igi-Global.Com/Resolvedoi/Resolve.Aspx?Doi=10.4018/979-8-3373-1220-0.Ch004. IGI Global. 1 janvier 1apr. J.-C. https://www.igi-global.com/chapter/the-use-of-artificial-intelligence-in-a-desalination-plant-powered-by-renewable-energy/373103.

Azdem, Driss, Jamal Mabrouki, and Souad El Hajjaji 2024a. "Methods and Technologies for the Treatment and Disposal of Brine in Seawater Desalination Plants". In *Advances in Environmental Engineering and Green Technologies*, édité par Jamal Mabrouki et Mourade Azrour, 79–87. IGI Global. https://doi.org/10.4018/979-8-3693-3807-0.ch007.

Azdem, Driss, Jamal Mabrouki, and Souad El Hajjaji. 2025c. "MarocFarm: A Digital Revolution for Agriculture in Morocco". In *Circular Economy Applications in Energy Policy*, 137–52. IGI Global Scientific Publishing.

Azdem, Driss, Jamal Mabrouki, Ahmed Moufti, Souad El Hajjaji, and Abdelilah Fatni. 2024b. "Assessment of heavy metal contamination in seawater in Agadir coastline, Morocco". *Desalination and Water Treatment* 317:100129. https://doi.org/10.1016/j.dwt.2024.100129.

Azdem, Driss, Jamal Mabrouki, Ahmed Moufti, Souad El Hajjaji, and Abdelilah Fatni. 2024c. "Present Situation and Environmental Impact of Seawater Desalination Plants in Morocco". In *Technical and Technological Solutions Towards a Sustainable Society and Circular Economy*, édité par Jamal Mabrouki et Azrour Mourade, 581–92. Cham: Springer Nature Switzerland. https://doi.org/10.1007/978-3-031-56292-1_46.

Bai, Long, Ziru Wang, Dong Ye, Hanghang Wei, Zihan Peng, Jiaying Ge, Siwei Tan, Tianxiang Li, and Hao Wu. 2025. "High efficient near-infrared sintering for electrohydrodynamic printed frequency selective surface". *Materials & Design* 252: 113774.

Chalá, Dayana Carolina, Claudia Castro-Faccetti, Edgar Quiñones-Bolaños, and Mehrab Mehrvar. 2024. "Salinity intrusion modeling using boundary conditions on a laboratory setup: Experimental analysis and CFD simulations". *Water* 16 (14): 1970.

Fahdi, G., D. Azdem, A. Lamchaimech, M. Benrhanem, J. Mabrouki, and S. El Hajjaji. 2024. "Analysis of the impact of climate change on water resources: Case of the Tensift basin (Morocco)". *Water Conservation and Management* 8 (4): 461–65. https://doi.org/10.26480/wcm.04.2024.461.465.

Fahdi, G., J. Mabrouki, A. Lamchaimech, D. Azdem, M. Benrhanem, and S.E. Hajjaji. 2024. "Impacts of climate change on groundwater quality and recharge in the Tensift basin, Morocco". *Water Conservation and Management* 8 (4): 466–74. https://doi.org/10.26480/wcm.04.2024.466.474.

Farshidfar, Nima, Karol Alí Apaza Alccayhuaman, Nathan E. Estrin, Paras Ahmad, Anton Sculean, Yufeng Zhang, and Richard J. Miron. 2025. "Advantages of horizontal centrifugation of platelet-rich fibrin in regenerative medicine and dentistry". *Periodontology 2000*.

Fatni, Abdelilah, Mourad Addich, Abdelilah Essekri, Noureddine EL Baraka, Driss Azdem, Abdelaziz Ait Taleb, and Abdellatif EL Hammadi. 2025. "Characterization and mitigation strategies for inorganic scaling in reverse osmosis system treating brackish groundwater". *Sustainable Chemistry One World* 5: 100035. https://doi.org/10.1016/j.scowo.2024.100035.

Iranshahi, Kamran, Thijs Defraeye, Rene M. Rossi, and Ulf Christian Müller. 2024. "Electrohydrodynamics and its applications: Recent advances and future perspectives". *International Journal of Heat and Mass Transfer* 232: 125895.

Liang, Ziyu, Wenjian Yang, Zheng Yin, Xin Wang, Yansheng He, Kui Yang, and Jinxing Ma. 2024. "Chlor-alkali membrane cell process for industrial waste salt utilization: fundamentals and challenges". *Desalination* 587: 117921.

Lin, Lu, Wenbin Jiang, Xuesong Xu, and Pei Xu. 2020. "A critical review of the application of electromagnetic fields for scaling control in water systems: mechanisms, characterization, and operation". *NPJ Clean Water* 3 (1): 25.

Murugaiyan, Suresh. 2024. "Modeling the dynamics of liquid metal in fusion liquid walls using Maxwell-Navier-Stokes equations".

Muteti, Christopher K. 2024. *Enhancing Regulatory Framework for Sustainable Seawater De-Salination: Lessons and Options for Kenya*. University of Nairobi.

Nikkhah, Hasan, Deniz Ipekçi, Wenjun Xiang, Zachary Stoll, Pei Xu, Baikun Li, Jeffrey R. McCutcheon, and Burcu Beykal. 2024. "Challenges and opportunities of recovering lithium from seawater, produced water, geothermal brines, and salt lakes using conventional and emerging technologies". *Chemical Engineering Journal* 498: 155349.

Orellana-Palma, Patricio, María Guerra-Valle, and Rommy N. Zúñiga. 2021. "Centrifugal filter-assisted block freeze crystallization applied to blueberry juice". *Processes* 9 (3): 421.

Panagopoulos, Argyris, and Vasiliki Giannika. 2024. "A comprehensive assessment of the economic and technical viability of a zero liquid discharge (ZLD) hybrid desalination system for water and salt recovery". *Journal of Environmental Management* 359: 121057.

Panhwar, Aijaz, Rashid Abro, Aftab Kandhro, Abdul Rauf Khaskheli, Nusrat Jalbani, Khadim Ali Gishkori, Atta Muhammad Mahar, and Sofia Qaisar. 2022. "Global Water Mapping, Requirements, and Concerns Over Water Quality Shortages". In *Water Quality-New Perspectives*. IntechOpen.

Perumal, Muthukumar, Selvam Sekar, and Paula CS Carvalho. 2024. "Global investigations of seawater intrusion (SWI) in coastal groundwaters in the last two decades (2000–2020): A bibliometric analysis". *Sustainability* 16 (3): 1266.

Pramanik, Biplob Kumar, Li Shu, and Veeriah Jegatheesan. 2017. "A review of the management and treatment of brine solutions". *Environmental Science: Water Research & Technology* 3 (4): 625–58.

Said, Drid. 2024. "Magnetohydrodynamic (MHD) energy harvesting in reverse osmosis desalination plants". *Journal of Electrical Systems* 20 (3): 5675–84.

Tareemi, Ahmad A., and Swellam W. Sharshir. 2023. "A state-of-art overview of multi-stage flash desalination and water treatment: Principles, challenges, and heat recovery in hybrid systems". *Solar Energy* 266:112157. https://doi.org/10.1016/j.solener.2023.112157.

Xiao, Yiru, Bo Feng, Jikai Zhang, Kun Mao, Kai Wang, Yini Song, and Yihua Kang. 2025. "Effect of excitation frequency on motion-induced eddy current and speed measurement". *IEEE Sensors Journal* 25 (4): 6151–59.

Zhao, Senyao, Rongjie Zhu, Jiatong Song, and Han Yuan. 2024. "An overview of sustainable desalination with freezing crystallization: Current development, future challenges, and prospects". *Sustainability* 16 (22): 10138.

Zhou, Peng, Chunyue Zhang, Jingkun Ai, Yongqiang Ge, Xiaoqing Peng, Qiaoling Gao, Wei Wang, Zhonghui Zhou, and Jiawang Chen. 2022. "The numerical simulation and experimental study of heat flow in seabed sediments based on COMSOL". *Journal of Marine Science and Engineering* 10 (10): 1356.

11 IoT for Green Hydrogen

A Smarter and More Sustainable Energy Transition

A. Ghailane, J. Mabrouki, K. El-Moustaqim, D. Azdem, C. Lamiri, and Y. Abrouki

11.1 INTRODUCTION

In a world faced with a climate emergency and the need to reduce greenhouse gas emissions, the energy transition has become a major challenge for the XXI century (Holdren 2006; Kabeyi and Olanrewaju 2022; Wiseman 2018). In this context, green hydrogen stands out as a clean, versatile energy carrier, produced from renewable energies (wind, solar, hydraulic) via processes such as water electrolysis. It represents a sustainable option to fossil fuels, particularly for sectors that are complex to decarbonize, such as heavy industry, long-distance transport, and large-scale energy storage (El-Moustaqim et al. 2024d, 2024e, 2024h). However, the large-scale development of green hydrogen faces a number of major challenges (El-Moustaqim et al. 2024f, 2024g), including high production costs, complex transport and storage logistics, and the need for optimized infrastructure management (Jayachandran et al. 2024; Taibi et al. 2020). In this context, Internet of Things (IoT) technologies appear to be an essential lever for overcoming these obstacles and improving the sector's efficiency (EL-Moustaqim et al. 2024d, 2024e, 2024h; Raj and Raman 2017; Vermesan and Friess 2014). IoT integration optimizes energy production, storage, and consumption through real-time monitoring and predictive analytics (Hajri et al. 2024). This intelligent connectivity facilitates more efficient resource management, preventive infrastructure maintenance, and optimal integration with renewable energy networks (El-Moustaqim, Mabrouki, and Hmouni 2024c; El-Moustaqim et al. 2024a, 2024d, 2024e, 2024h). IoT are in sectors such as smart homes, smart cities, and industrial systems (Chataut, Phoummalayvane, and Akl 2023; Tripathi et al. 2023; Umair et al. 2021). Today, IoT is at the heart of the fourth industrial revolution, revolutionizing various sectors thanks to a vast network of interconnected devices. This technology is transforming our daily lives by optimizing processes, improving energy efficiency (Munirathinam 2020). By combining automation, artificial intelligence and real-time data analysis, the IoT is transforming the green hydrogen ecosystem, making this technology safer, more competitive, and fully adapted to the needs of a carbon-free energy future (El-Moustaqim, Mabrouki, and Azrour 2023; El-Moustaqim et al. 2023).

This chapter explores how the synergy between IoT and green hydrogen paves the way for a decarbonized economy. In particular, it examines the impact of smart technologies on optimizing green hydrogen production, storage, and distribution processes, while highlighting the opportunities and challenges to be overcome to maximize their potential.

11.2 THE EMERGING IOT REVOLUTION

Over the past decade, the concept of the IoT has attracted attention by projecting the vision of a global infrastructure of networked physical objects, offering constant and universal connectivity for anything and anyone (Kosmatos, Tselikas, and Boucouvalas 2011). IoT refers to the encoding and networking of everyday objects so that they can be individually read and tracked by machines on

DOI: 10.1201/9781003647263-11

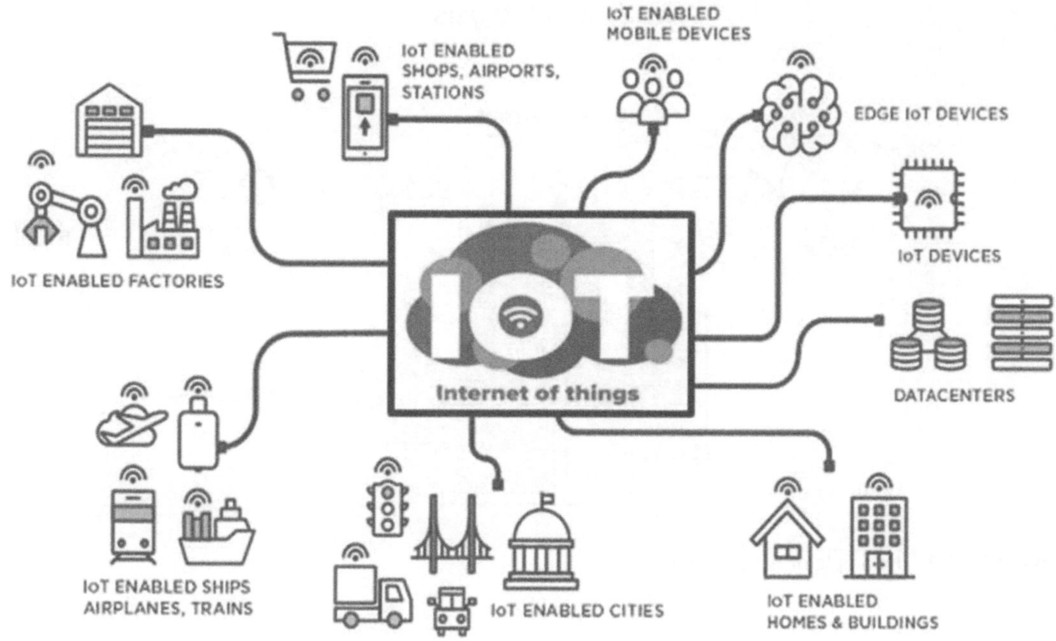

FIGURE 11.1 The main IoT applications.

the Internet (Borgia 2014; Madakam, Ramaswamy, and Tripathi 2015). Today, thanks to significant technological advances, this vision is becoming a concrete reality. The IoT is undergoing a rapid transformation, marking the start of a veritable revolution in a variety of fields (Figure 11.1). Thanks to advanced connectivity, intelligent sensors and real-time data analysis (Verma et al. 2017). The IoT can also be seen as a global network enabling communication between individuals, between individuals and objects, and between objects themselves (Madakam, Ramaswamy, and Tripathi 2015). Devices and objects equipped with embedded sensors are connected to the IoT network, which aggregates data from a variety of devices. This requires analysis to ensure that the most valuable information is shared for use in meeting specific requirements, enabling intelligent identification and management (El-Moustaqim et al. 2024b). The IoT is redefining the way we produce, manage, and consume resources (Zikria et al. 2021). In the industrial sector, it refines production processes and facilitates predictive machine maintenance, resulting in lower costs and fewer downtimes (Liu et al. 2022). In the energy sector, it enables better management of power grids and encourages the integration of renewable energy sources. Similarly, in smart cities, the IoT optimizes urban mobility, waste management, and infrastructure efficiency (Anagnostopoulos et al. 2017; Belli et al. 2020).

11.3 GREEN HYDROGEN AS AN ENERGY SOURCE

Green hydrogen is a promising and sustainable energy source, produced by water electrolysis as seen in Figure 11.2, which separates water (H_2O) into hydrogen (H_2) and oxygen (O_2) using electricity from renewable energies such as wind or solar power (Hassan et al. 2024a, 2024b). This process creates completely carbon-free hydrogen, helping to reduce greenhouse gas emissions (Du et al. 2024; Maka and Mehmood 2024; Nnabuife et al. 2025; Pradhan et al. 2024). As an energy carrier, green hydrogen offers many possible uses. It can be used as a clean fuel for the transport sector or industry, replacing the process with a high CO_2 footprint (Asiegbu, Kahn, and Almaktoof 2024; Ma et al. 2024; Muradov and Veziroğlu 2008; Reda et al. 2024). Green hydrogen is seen as an essential pillar in the global energy transition to a cleaner, more environmentally friendly future

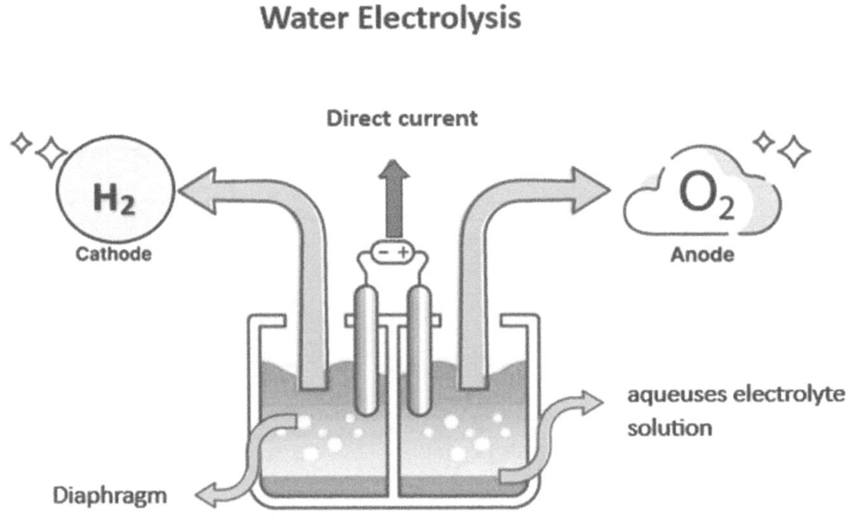

FIGURE 11.2 Principle of water electrolysis.

(Cheekatamarla 2024). What is more, global demand for green hydrogen and its applications are set to grow exponentially over the next decade.

11.4 IoT AND GREEN HYDROGEN TECHNOLOGIES

IoT technology enables physical objects to connect to the Internet to gather, share, and examine information in real time (Li, Xu, and Zhao 2015). This ability to manage information on a continuous basis creates new opportunities in a variety of fields, including energy (Hossein Motlagh et al. 2020; Siddique et al. 2021). In this context, IoT plays a key role in the development and optimization of green hydrogen production, storage, and distribution systems. Through the integration of smart sensors, advanced communication networks and artificial intelligence–based solutions, IoT enables instant monitoring of infrastructures, thus improving operational performance and consolidating process safety (Nwankwo et al. 2024; Tiwari 2022). In production facilities, these technologies enable detailed monitoring of electrolyzer performance and optimize the use of renewable energies. IoT platforms powered by artificial intelligence simplify the management of renewable resources by dynamically adapting hydrogen production according to the availability of solar and wind energy.

11.5 IOT APPLICATIONS FOR GREEN HYDROGEN

11.5.1 Optimizing Green Hydrogen Production

Optimizing green hydrogen production depends on efficient management of renewable energy resources and electrolysis systems (Alhussan et al. 2023; Park et al. 2024). The IoT offers the possibility of real-time monitoring of key parameters in the electrolysis process, such as temperature, pressure, and current intensity. Thanks to intelligent sensors and artificial intelligence algorithms, these parameters can be dynamically adjusted to maximize energy efficiency and reduce electricity consumption during the process. In addition, the IoT makes it possible to incorporate meteorological and renewable energy production data (solar, wind, hydro) to optimize the operation of electrolyzers according to the availability of resources (Ali et al. 2023). This approach not only minimizes operating costs, but also ensures continuous, stable production of green hydrogen in line with demand.

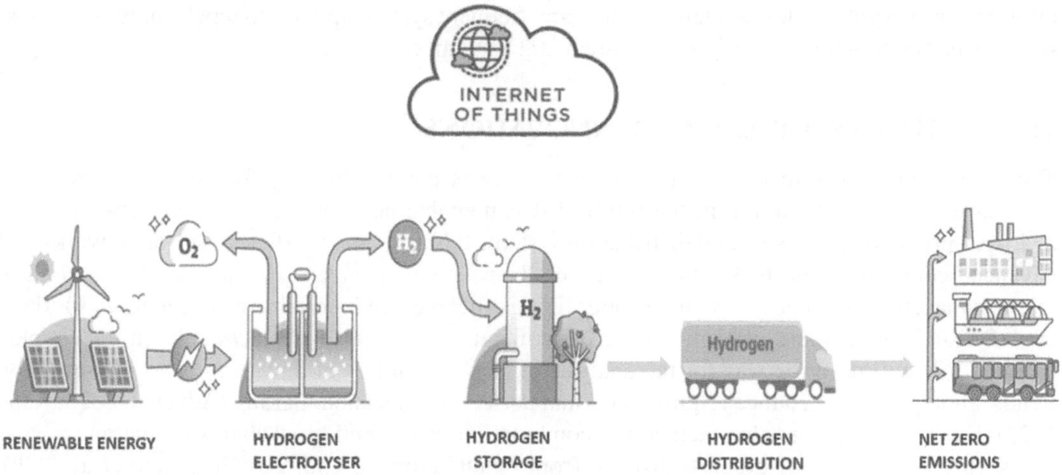

FIGURE 11.3 Integrating IoT into green hydrogen production and storage systems.

11.5.2 Intelligent Storage Management

The main challenge in storing green hydrogen is the extremely volatile and flammable nature of this gas (Ahad et al. 2023; Jansons et al. 2022). To overcome these challenges, the IoT offers live monitoring of storage tanks via intelligent sensors that evaluate key indicators (Figure 11.3). These devices immediately detect any anomalies, such as leaks or overpressure, and trigger notifications to anticipate incident-related hazards. The integration of these technologies with artificial intelligence systems also makes it possible to anticipate possible failures by analyzing trends in the data collected, thus facilitating proactive maintenance and reducing the costs associated with corrective interventions (Ben-Lhachemi et al. 2024; Gbadamosi et al. 2021). In addition, the IoT is helping to improve various hydrogen storage processes, whether in the gas phase at high pressure, in the liquid phase at a very low temperature or consigned in solid materials such as metal hydrides. Thanks to connected platforms, infrastructure managers can adjust storage conditions according to variations in hydrogen demand and supply, improving the flexibility and efficiency of the energy system.

11.6 CHALLENGES AND LIMITS OF IOT INTEGRATION FOR GREEN HYDROGEN

Integrating the IoT into the production and management of green hydrogen poses a number of technical, economic, and security challenges. System complexity remains a major challenge, and IoT requires a robust and reliable communications infrastructure, which can be difficult to set up in remote areas where renewable energies are often used (Frustaci et al. 2018; Hameed, Khan, and Hameed 2019; Ma et al. 2019). The reliability and robustness of sensors and IoT devices is also an issue (Moore et al. 2020; Sarkar 2016). The industrial environment in which hydrogen is produced and stored can be exposed to extreme conditions (high temperatures, air humidity, variable pressures) that can affect sensor durability and accuracy. It is therefore essential to ensure periodic maintenance and systematic redundancy to guarantee optimal operation when implementing IoT technologies. In addition, sensors, connected devices, and data management platforms have high initial costs, limiting the deployment of major standards (Sheng et al. 2015). In addition, data security is a critical issue as IoT systems are susceptible to cyber-attacks and can affect the stability of hydrogen production networks (Anawar et al. 2019; Bout, Loscri, and Gallais 2022). A malicious intrusion could lead to changes in production parameters, service interruptions, or even safety hazards. To deal with these dangers, it is essential to establish solid cybersecurity protocols, including information encryption, device identification, and frequent system updates to repel emerging aggressions (Altulaihan, Almaiah, and Aljughaiman 2022; Aslan et al. 2023).

11.7 FUTURE PROSPECTS AND INNOVATIONS

The integration of IoT in the green hydrogen sector is paving the way for promising prospects and transformative innovation. In the future, IoT can enable new interconnectivity between green hydrogen production, storage, and distribution systems thanks to more efficient sensor networks and advanced communication technologies. These advances will enable even more precise and automatic management of energy resources, optimizing efficiency and reducing costs (Hossein Motlagh et al. 2020). What is more, the emergence of artificial intelligence and machine learning, coupled with the IoT, will pave the way for the analysis of large quantities of data to forecast consumption trends, anticipate maintenance requirements and perfect live decision making (Adi et al. 2020; Rane 2023). Future innovations also include the combination of IoT and blockchains to guarantee clear and secure traceability of green hydrogen, from manufacture to end use (Grecuccio et al. 2020; Tsang et al. 2019).

11.8 CONCLUSION

Integrating the IoT into the production and management of green hydrogen represents a major step towards a more efficient and sustainable energy transition. With process optimization, real-time monitoring, and improved system security, the IoT is capable of making full use of this potential as a clean energy source. However, challenges remain in terms of cybersecurity costs, implementation standardization, and technology. Future innovations, such as artificial intelligence, blockchain, and smart sensors, provide the right solutions to overcome these obstacles and promote widespread acceptance. Green hydrogen, combined with IoT, thus embodies the hope of an energy revolution that brings progress and environmental responsibility.

REFERENCES

Adi, Erwin, Adnan Anwar, Zubair Baig, and Sherali Zeadally. 2020. "Machine learning and data analytics for the IoT". *Neural Computing and Applications* 32 (20): 16205–33. https://doi.org/10.1007/s00521-020-04874-y.

Ahad, Md Tanvir, Md Monjur Hossain Bhuiyan, Ahmed Nazmus Sakib, Alfredo Becerril Corral, and Zahed Siddique. 2023. "An overview of challenges for the future of hydrogen". *Materials* 16 (20): 6680. https://doi.org/10.3390/ma16206680.

Alhussan, Amel Ali, El-Sayed M. El-Kenawy, Mohammed A. Saeed, Abdelhameed Ibrahim, Abdelaziz A. Abdelhamid, Marwa M. Eid, M. El-Said, Doaa Sami Khafaga, Laith Abualigah, and Osama Elbaksawi. 2023. "Green hydrogen production ensemble forecasting based on hybrid dynamic optimization algorithm". *Frontiers in Energy Research* 11. https://doi.org/10.3389/fenrg.2023.1221006.

Ali, Awais, Tajamul Hussain, Noramon Tantashutikun, Nurda Hussain, and Giacomo Cocetta. 2023. "Application of smart techniques, internet of things and data mining for resource use efficient and sustainable crop production". *Agriculture* 13 (2): 397. https://doi.org/10.3390/agriculture13020397.

Altulaihan, Esra, Mohammed Amin Almaiah, and Ahmed Aljughaiman. 2022. "Cybersecurity threats, countermeasures and mitigation techniques on the IoT: Future research directions". *Electronics* 11 (20): 3330. https://doi.org/10.3390/electronics11203330.

Anagnostopoulos, Theodoros, Arkady Zaslavsky, Kostas Kolomvatsos, Alexey Medvedev, Pouria Amirian, Jeremy Morley, and Stathes Hadjieftymiades. 2017. "Challenges and opportunities of waste management in IoT-enabled smart cities: A survey". *IEEE Transactions on Sustainable Computing* 2 (3): 275–89. https://doi.org/10.1109/TSUSC.2017.2691049.

Anawar, Syarulnaziah, Nurul Zakaria, Zaki Masud, M. Zulkiflee, Norharyati Harum, and Rabiah Ahmad. 2019. "IoT technological development: Prospect And implication for cyberstability". *International Journal of Advanced Computer Science and Applications* 10. https://doi.org/10.14569/IJACSA.2019.0100256.

Asiegbu, A. D., M. T. E. Kahn, and A. M. Almaktoof. 2024. "Green hydrogen: A clean energy solution for Germany's transportation sector". *SAIEE Africa Research Journal* 115 (1): 16–23. https://doi.org/10.23919/SAIEE.2024.10463751.

Aslan, Ömer, Semih Serkant Aktuğ, Merve Ozkan-Okay, Abdullah Asim Yilmaz, and Erdal Akin. 2023. "A comprehensive review of cyber security vulnerabilities, threats, attacks, and solutions". *Electronics* 12 (6): 1333. https://doi.org/10.3390/electronics12061333.

Belli, Laura, Antonio Cilfone, Luca Davoli, Gianluigi Ferrari, Paolo Adorni, Francesco Di Nocera, Alessandro Dall'Olio, Cristina Pellegrini, Marco Mordacci, and Enzo Bertolotti. 2020. "IoT-enabled smart sustainable cities: Challenges and approaches". *Smart Cities* 3 (3): 1039–71. https://doi.org/10.3390/smartcities3030052.

Ben-Lhachemi, Najlaa, Souad Nasrdine, Jamal Mabrouki, and Miloudia Slaoui. 2024. "Integration of Bio-Sensors-IoT System for the Evaluation and Control of Soil Quality". In *Advanced Technology for Smart Environment and Energy*, édité par Jamal Mabrouki et Azrour Mourade, 221–29. Cham: Springer Nature Switzerland. https://doi.org/10.1007/978-3-031-50871-4_14.

Borgia, Eleonora. 2014. "The internet of things vision: Key features, applications and open issues". *Computer Communications* 54: 1–31. https://doi.org/10.1016/j.comcom.2014.09.008.

Bout, Emilie, Valeria Loscri, and Antoine Gallais. 2022. "How machine learning changes the nature of cyber-attacks on IoT networks: A survey". *IEEE Communications Surveys & Tutorials* 24 (1): 248–79. https://doi.org/10.1109/COMST.2021.3127267.

Chataut, Robin, Alex Phoummalayvane, and Robert Akl. 2023. "Unleashing the power of IoT: A comprehensive review of IoT applications and future prospects in healthcare, agriculture, smart homes, smart cities, and industry 4.0". *Sensors* 23 (16): 7194. https://doi.org/10.3390/s23167194.

Cheekatamarla, Praveen. 2024. "Hydrogen and the global energy transition—Path to sustainability and adoption across all economic sectors". *Energies* 17 (4): 807. https://doi.org/10.3390/en17040807.

Du, Lifeng, Yanmei Yang, Luli Zhou, and Min Liu. 2024. "Greenhouse gas reduction potential and economics of green hydrogen via water electrolysis: A systematic review of value-chain-wide decarbonization". *Sustainability* 16 (11): 4602. https://doi.org/10.3390/su16114602.

El-Moustaqim, Khadija, Benchrifa Mohammed, Jamal Mabrouki, and Driss Hmouni. 2024a. "Application of IoT-Based Intelligent Systems in Microalgae Biofuel Production". In *Advanced Technology for Smart Environment and Energy*, édité par Jamal Mabrouki et Azrour Mourade, 293–305. Environmental Science and Engineering. Cham: Springer Nature Switzerland. https://doi.org/10.1007/978-3-031-50871-4_20.

El-Moustaqim, Khadija, Benchrifa Mohammed, Jamal Mabrouki, and Driss Hmouni 2024b. "Application of IoT-Based Intelligent Systems in Microalgae Biofuel Production". In *Advanced Technology for Smart Environment and Energy*, édité par Jamal Mabrouki et Azrour Mourade, 293–305. Cham: Springer Nature Switzerland. https://doi.org/10.1007/978-3-031-50871-4_20.

El-Moustaqim, Khadija, Houria El Bakraoui, Jamal Mabrouki, Yasser Fouad, Miloudia Slaoui, Driss Hmouni, Mbadiwe S. Benyeogor, and Tobore Louis Igbigbi. 2023. "Combination of microalgae method, decantation, and filtration for domestic wastewater treatment". *Sustainability* 15 (22): 16110. https://doi.org/10.3390/su152216110.

El-Moustaqim, Khadija, Jamal Mabrouki, and Driss Hmouni. 2024c. "Applied Sciences Green Microalgae for Future Biomass Development". In *Advanced Systems for Environmental Monitoring, IoT and the application of Artificial Intelligence*, 203–11. Springer. https://doi.org/10.1007/978-3-031-50860-8_12.

El-Moustaqim, Khadija, Jamal Mabrouki, and Mourade Azrour. 2023. "Monitoring of Water Toxicity Through the Internet of Things to Protect the Health of the Population". In *IoT, Machine Learning and Data Analytics for Smart Healthcare*, 85–92. Taylor & Francis. https://doi.org/10.1201/9781003430735-7.

El-Moustaqim, Khadija, Jamal Mabrouki, Mourade Azrour, Mouhsine Hadine, and Driss Hmouni. 2024d. "Enabling Smart Agriculture Through Integrating the Internet of Things in Microalgae Farming for Sustainability". In *Smart Internet of Things for Environment and Healthcare*, édité par Mourade Azrour, Jamal Mabrouki, Abdulatif Alabdulatif, Azidine Guezzaz, et Fatima Amounas, 209–22. Cham: Springer Nature Switzerland. https://doi.org/10.1007/978-3-031-70102-3_15.

El-Moustaqim, Khadija, Toufik Rachiq, Jamal Mabrouki, Miloudia Slaoui, and Driss Hmouni. 2024e. "Application of Microalgae in Green Hydrogen Production: A Diverse Biomass for Multiple Applications". In *Advanced Technology for Smart Environment and Energy*, 231–40. Springer. https://doi.org/10.1007/978-3-031-50871-4_15.

El-Moustaqim, Khadija, Salah Eddine Sbai, Yassine El Yousfi, Jamal Mabrouki, and Driss Hmouni. 2024f. "New strategy for the advancement of modern agriculture through the use of microalgae as biofertilizers". *Euro-Mediterranean Journal for Environmental Integration*. https://doi.org/10.1007/s41207-024-00678-x.

El-Moustaqim, Khadija, Salah Eddine Sbai, Jamal Mabrouki, and Driss Hmouni. 2024g. "Critical review of green microalgal biomass development through CO2 sequestration". *Euro-Mediterranean Journal for Environmental Integration*. https://doi.org/10.1007/s41207-024-00548-6.

EL-Moustaqim, Khadija, Jamal Mabrouki, Mourade Azrour, and Driss Hmouni. 2024h. "How Can Green Energy Be Improved by Integrating IoT into the Microalgae Process?" In *Sustainable and Green Technologies for Water and Environmental Management*, édité par Mourade Azrour, Jamal Mabrouki, et Azidine Guezzaz, 239–52. Cham: Springer Nature Switzerland. https://doi.org/10.1007/978-3-031-52419-6_17.

Frustaci, Mario, Pasquale Pace, Gianluca Aloi, and Giancarlo Fortino. 2018. "Evaluating critical security issues of the IoT world: Present and future challenges". *IEEE Internet of Things Journal* 5 (4): 2483–95. https://doi.org/10.1109/JIOT.2017.2767291.

Gbadamosi, Abdul-Quayyum, Lukumon O. Oyedele, Juan Manuel Davila Delgado, Habeeb Kusimo, Lukman Akanbi, Oladimeji Olawale, and Naimah Muhammed-Yakubu. 2021. "IoT for predictive assets monitoring and maintenance: An implementation strategy for the UK rail industry". *Automation in Construction* 122: 103486. https://doi.org/10.1016/j.autcon.2020.103486.

Grecuccio, Jacopo, Edoardo Giusto, Fabio Fiori, and Maurizio Rebaudengo. 2020. "Combining blockchain and IoT: Food-chain traceability and beyond". *Energies* 13 (15): 3820. https://doi.org/10.3390/en13153820.

Hajri, Nawaf Hamed Al, Rahaf Naji Al Harthi, Gopi Krishna Pasam, and Rajababu Natarajan. 2024. "IoT and machine learning based green energy generation using hybrid renewable energy sources of solar, wind and hydrogen fuel cells". *E3S Web of Conferences* 472: 01008. https://doi.org/10.1051/e3sconf/202447201008.

Hameed, Sufian, Faraz Idris Khan, and Bilal Hameed. 2019. "Understanding security requirements and challenges in internet of things (IoT): A review". *Journal of Computer Networks and Communications* 2019 (1): 9629381. https://doi.org/10.1155/2019/9629381.

Hassan, N. S., A. A. Jalil, S. Rajendran, N. F. Khusnun, M. B. Bahari, A. Johari, M. J. Kamaruddin, and M. Ismail. 2024a. "Recent review and evaluation of green hydrogen production via water electrolysis for a sustainable and clean energy society". *International Journal of Hydrogen Energy* 52: 420–41. https://doi.org/10.1016/j.ijhydene.2023.09.068.

Hassan, Qusay, Vahid Sohrabi Tabar, Aws Zuhair Sameen, Hayder M. Salman, and Marek Jaszczur. 2024b. "A review of green hydrogen production based on solar energy; Techniques and methods". *Energy Harvesting and Systems* 11 (1). https://doi.org/10.1515/ehs-2022-0134.

Holdren, John P. 2006. "The energy innovation imperative: Addressing oil dependence, climate change, and other 21st century energy challenges". *Innovations: Technology, Governance, Globalization* 1 (2): 3–23. https://doi.org/10.1162/itgg.2006.1.2.3.

Hossein Motlagh, Naser, Mahsa Mohammadrezaei, Julian Hunt, and Behnam Zakeri. 2020. "Internet of things (IoT) and the energy sector". *Energies* 13 (2): 494. https://doi.org/10.3390/en13020494.

Jansons, L., L. Zemite, N. Zeltins, I. Bode, I. Geipele, and K. Kiesners. 2022. "The green hydrogen and the EU gaseous fuel diversification risks". *Latvian Journal of Physics and Technical Sciences* 59 (4): 53–70. https://doi.org/10.2478/lpts-2022-0033.

Jayachandran, M., Ranjith Kumar Gatla, Aymen Flah, Ahmad H. Milyani, Hisham M. Milyani, Vojtech Blazek, Lukas Prokop, and Habib Kraiem. 2024. "Challenges and opportunities in green hydrogen adoption for decarbonizing Hard-to-abate industries: A comprehensive review". *IEEE Access* 12: 23363–88. https://doi.org/10.1109/ACCESS.2024.3363869.

Kabeyi, Moses Jeremiah Barasa, and Oludolapo Akanni Olanrewaju. 2022. "Sustainable energy transition for renewable and low carbon grid electricity generation and supply". *Frontiers in Energy Research* 9. https://doi.org/10.3389/fenrg.2021.743114.

Kosmatos, Evangelos A., Nikolaos D. Tselikas, and Anthony C. Boucouvalas. 2011. "Integrating RFIDs and smart objects into a Unified internet of things architecture". *Advances in Internet of Things* 1 (1): 5–12. https://doi.org/10.4236/ait.2011.11002.

Li, Shancang, Li Da Xu, and Shanshan Zhao. 2015. "The internet of things: A survey". *Information Systems Frontiers* 17 (2): 243–59. https://doi.org/10.1007/s10796-014-9492-7.

Liu, Changchun, Haihua Zhu, Dunbing Tang, Qingwei Nie, Tong Zhou, Liping Wang, and Yejia Song. 2022. "Probing an intelligent predictive maintenance approach with deep learning and augmented reality for machine tools in IoT-enabled manufacturing". *Robotics and Computer-Integrated Manufacturing* 77:102357. https://doi.org/10.1016/j.rcim.2022.102357.

Ma, Nan, Weihua Zhao, Wenzhong Wang, Xiangrong Li, and Haiqin Zhou. 2024. "Large scale of green hydrogen storage: Opportunities and challenges". *International Journal of Hydrogen Energy* 50:379–96. https://doi.org/10.1016/j.ijhydene.2023.09.021.

Ma, Zheng, Ming Xiao, Yue Xiao, Zhibo Pang, H. Vincent Poor, and Branka Vucetic. 2019. "High-reliability and low-latency wireless communication for internet of things: Challenges, fundamentals, and enabling technologies". *IEEE Internet of Things Journal* 6 (5): 7946–70. https://doi.org/10.1109/JIOT.2019.2907245.

Madakam, Somayya, R. Ramaswamy, and Siddharth Tripathi. 2015. "Internet of things (IoT): A literature review". *Journal of Computer and Communications* 3 (5): 164–73. https://doi.org/10.4236/jcc.2015.35021.

Maka, Ali O. M., and Mubbashar Mehmood. 2024. "Green hydrogen energy production: Current status and potential". *Clean Energy* 8 (2): 1–7. https://dx.doi.org/10.1093/ce/zkae012.

Moore, Samuel J., Chris D. Nugent, Shuai Zhang, and Ian Cleland. 2020. "IoT reliability: A review leading to 5 key research directions". *CCF Transactions on Pervasive Computing and Interaction* 2 (3): 147–63. https://doi.org/10.1007/s42486-020-00037-z.

Munirathinam, Sathyan. 2020. "Chapter Six - Industry 4.0: Industrial Internet of Things (IIOT)". In *Advances in Computers*, édité par Pethuru Raj et Preetha Evangeline, 117:129–64. The Digital Twin Paradigm for Smarter Systems and Environments: The Industry Use Cases. Elsevier. https://doi.org/10.1016/bs.adcom.2019.10.010.

Muradov, Nazim Z., and T. Nejat Veziroğlu. 2008. "Green" path from fossil-based to hydrogen economy: An overview of carbon-neutral technologies". *International Journal of Hydrogen Energy* 33 (23): 6804–39. https://doi.org/10.1016/j.ijhydene.2008.08.054.

Nnabuife, Somtochukwu Godfrey, Abdulhammed K. Hamzat, James Whidborne, Boyu Kuang, and Karl W. Jenkins. 2025. "Integration of renewable energy sources in tandem with electrolysis: A technology review for green hydrogen production". *International Journal of Hydrogen Energy* 107: 218–40. https://doi.org/10.1016/j.ijhydene.2024.06.342.

Nwankwo, Samuel, Omolayo M. Ikumapayi, Imhade P. Okokpujie, Stella I. Monye, Ayodele Ajayi, Yinka Ajiboye, Boluwaji Jayesimi, and Sunday A. Afolalu. 2024. "Application and Optimization of IoT in Storage Devices for Hydrogen Energy". In *2024 International Conference on Science, Engineering and Business for Driving Sustainable Development Goals (SEB4SDG)*, 1–7. https://doi.org/10.1109/SEB4SDG60871.2024.10629891.

Park, Joungho, Sungho Kang, Sunwoo Kim, Hyun-Seok Cho, Seongmin Heo, and Jay H. Lee. 2024. "Techno-economic analysis of solar powered green hydrogen system based on multi-objective optimization of economics and productivity". *Energy Conversion and Management* 299: 117823. https://doi.org/10.1016/j.enconman.2023.117823.

Pradhan, Bijay B., Bundit Limmeechokchai, Achiraya Chaichaloempreecha, and Salony Rajbhandari. 2024. "Role of green hydrogen in the decarbonization of the energy system in Thailand". *Energy Strategy Reviews* 51:101311. https://doi.org/10.1016/j.esr.2024.101311.

Raj, Pethuru, and Anupama C. Raman. 2017. *The Internet of Things: Enabling Technologies, Platforms, and Use Cases*. New York: Auerbach Publications. https://doi.org/10.1201/9781315273095.

Rane, Nitin. 2023. "Integrating leading-edge artificial intelligence (AI), internet of things (IoT), and big data technologies for smart and sustainable architecture, engineering and construction (AEC) industry: Challenges and future directions". SSRN Scholarly Paper. Rochester, NY: Social Science Research Network. https://doi.org/10.2139/ssrn.4616049.

Reda, Bassma, Amr A. Elzamar, Shehab AlFazzani, and Shahira M. Ezzat. 2024. "Green hydrogen as a source of renewable energy: A step towards sustainability, an overview". *Environment, Development and Sustainability*. https://doi.org/10.1007/s10668-024-04892-z.

Sarkar, S. 2016. "Chapter 11 - Internet of Things—Robustness and Reliability". In *Internet of Things*, édité par Rajkumar Buyya et Amir Vahid Dastjerdi, 201–18. Morgan Kaufmann. https://doi.org/10.1016/B978-0-12-805395-9.00011-3.

Sheng, Zhengguo, Chinmaya Mahapatra, Chunsheng Zhu, and Victor C. M. Leung. 2015. "Recent advances in industrial wireless sensor networks toward efficient management in IoT". *IEEE Access* 3: 622–37. https://doi.org/10.1109/ACCESS.2015.2435000.

Siddique, Abdul Hasib, Sumaiya Tasnim, Fahim Shahriyar, Mehedi Hasan, and Khalid Rashid. 2021. "Renewable energy sector in Bangladesh: The current scenario, challenges and the role of IoT in building a smart distribution grid". *Energies* 14 (16): 5083. https://doi.org/10.3390/en14165083.

Taibi, Emanuele, Raul Miranda, Marcelo Carmo, and Herib Blanco. 2020. "Green hydrogen cost reduction". https://www.h2knowledgecentre.com/content/researchpaper1613.

Tiwari, Ashutosh. 2022. "Hydrogen leading the green energy future". *Advanced Materials Letters* 13 (2): 2202–1690. https://doi.org/10.5185/amlett.2022.021690.

Tripathi, Aditya, Nidhi Sindhwani, Rohit Anand, and Aman Dahiya. 2023. "Role of IoT in Smart Homes and Smart Cities: Challenges, Benefits, and Applications". In *IoT Based Smart Applications*, édité par Nidhi Sindhwani, Rohit Anand, M. Niranjanamurthy, Dinesh Chander Verma, et Emilia Balas Valentina, 199–217. Cham: Springer International Publishing. https://doi.org/10.1007/978-3-031-04524-0_12.

Tsang, Yung Po, King Lun Choy, Chun Ho Wu, George To Sum Ho, and Hoi Yan Lam. 2019. "Blockchain-driven IoT for food traceability with an integrated consensus mechanism". *IEEE Access* 7: 129000–17. https://doi.org/10.1109/ACCESS.2019.2940227.

Umair, Muhammad, Muhammad Aamir Cheema, Omer Cheema, Huan Li, and Hua Lu. 2021. "Impact of COVID-19 on IoT adoption in healthcare, smart homes, smart buildings, smart cities, transportation and industrial IoT". *Sensors* 21 (11): 3838. https://doi.org/10.3390/s21113838.

Verma, Shikhar, Yuichi Kawamoto, Zubair Md. Fadlullah, Hiroki Nishiyama, and Nei Kato. 2017. "A survey on network methodologies for real-time analytics of massive IoT data and open research issues". *IEEE Communications Surveys & Tutorials* 19 (3): 1457–77. https://doi.org/10.1109/COMST.2017.2694469.

Vermesan, Ovidiu, and Peter Friess, éd. 2014. *Internet of Things Applications - From Research and Innovation to Market Deployment*. Taylor & Francis. https://doi.org/10.1201/9781003338628.

Wiseman, John. 2018. "The great energy transition of the 21st century: The 2050 zero-carbon world oration". *Energy Research & Social Science*, Energy and the Future, 35:227–32. https://doi.org/10.1016/j.erss.2017.10.011.

Zikria, Yousaf Bin, Rashid Ali, Muhammad Khalil Afzal, and Sung Won Kim. 2021. "Next-generation internet of things (IoT): Opportunities, challenges, and solutions". *Sensors* 21 (4): 1174. https://doi.org/10.3390/s21041174.

12 Innovative Approaches in Water Treatment and Monitoring

The Role of AI, Machine Learning, and Smart Technologies

B. R. Vinod, M. Niranjanamurthy, and M. S. Sumathi

12.1 INTRODUCTION

Water security is one of the most urgent issues of the 21st century, with the United Nations estimating that by 2025 two-thirds of the world's population could be exposed to water stress conditions [1]. While traditional approaches to water treatment and monitoring approaches are fundamental, they are increasingly insufficient to address the complex and evolving challenges of water management, such as new impurities, the impact of climate change, and aging infrastructure [2]. The water sector is an important time when traditional technology alone cannot ensure sustainable water safety for future generations [3].

The convergence of artificial intelligence (AI), machine learning (ML), and smart technologies offers an unprecedented opportunity to change water management practices (Figure 12.1). These technologies allow for more accurate control of the treatment process, predictive capabilities for system optimization, and a wide range of surveillance networks that can recognize subtle changes in water quality parameters [4].

Integrating AI into water treatment represents a paradigm shift from reactive to aggressive management of water resources [5]. However, with the latest advances in computing power, sensory diseases, and algorithm development, AI and ML applications are more accessible to water suppliers and environmental authorities [6, 7]. Lower costs and increased reliability of intelligent technologies gradually overcome the historical unwillingness of innovation in the water sector [8].

The purpose of this study is to systematically assess the current state of AI, ML, and smart technology applications in water treatment and monitoring, identify key success factors and obstacles for implementation, and develop a strategic framework for the implementation of technology. By bridging the gap between theoretical possibilities and practical implementation, this study contributes to the increased knowledge of the interface between data science and water technology, thereby meeting the urgent need for innovative approaches for the management of water resources.

12.2 LITERATURE REVIEW

12.2.1 Evolution of Water Treatment Technologies

Water treatment has evolved significantly from basic filtration and disinfection to complex multi-barrier approaches incorporating advanced oxidation, membrane filtration, and biological processes [9]. Traced this evolution, noting that each technological advancement has been driven by increasing water quality standards and emerging contaminant concerns [10]. The most recent phase in

DOI: 10.1201/9781003647263-12

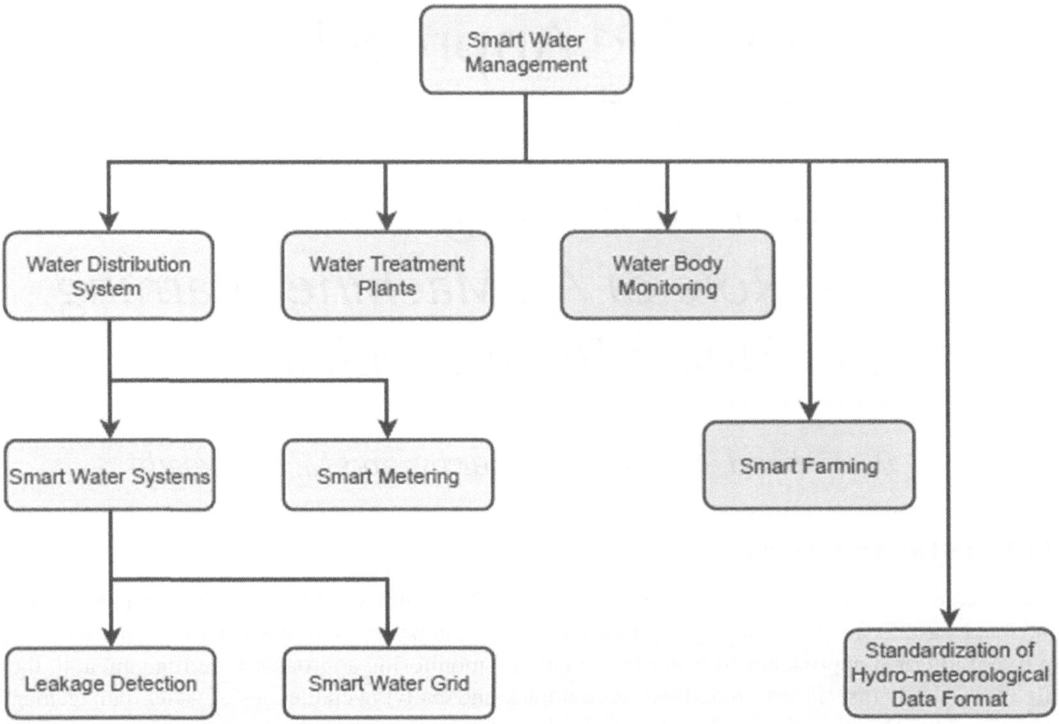

FIGURE 12.1 A flow chart outlining several organized aspects and developments in an intelligent resource management system [4].

this evolution involves the integration of digital technologies with physical and chemical treatment processes, created using cyber-physical water systems [11].

12.2.2 AI and ML Applications in Water Treatment

AI applications in water treatment have expanded rapidly over the past decade. The applications fall into three primary domains: process control optimization, predictive maintenance, and treatment performance forecasting [12]. Their comprehensive review identified neural networks and random forests as the most commonly applied ML algorithms in water treatment optimization, with reported energy savings between 10–25% across various case studies.

Deep learning approaches have shown particular promise in optimizing coagulant dosing in conventional treatment plants, demonstrating that a deep neural network model could predict optimal coagulant doses with 92% accuracy based on influent turbidity, pH, temperature, and alkalinity, reducing chemical usage by approximately 17% compared to conventional control approaches [13]. Similarly, reinforcement learning algorithms have been successfully applied to optimize membrane filtration processes; they reported a 15% reduction in membrane fouling and a 20% decrease in backwash frequency through adaptive control strategies [14].

12.2.3 Smart Monitoring Systems for Water Quality

Smart monitoring systems represent another significant area of innovation. Traditional water quality monitoring relies heavily on manual sampling and laboratory analysis, creating inevitable delays between contamination events and their detection [15]. Real-time monitoring networks employing in-situ sensors and wireless communication technologies have emerged as an alternative approach. After reviewing 45 smart monitoring implementations globally it was found that early detection of contamination events improved by an average of 8.4 hours compared to conventional monitoring programs [16].

The integration of IoT sensors with ML algorithms for anomaly detection has proven particularly effective. The system combined multiparameter sensor arrays with random forest classifiers to identify water quality anomalies with 86% sensitivity and 92% specificity [17]. Their approach enabled the detection of subtle changes in water quality that might indicate contamination events before they reached levels detectable by traditional monitoring methods.

The effectiveness of smart monitoring systems is fundamentally limited by sensor reliability, data quality, and appropriate interpretation algorithms [18]. Their critical analysis highlighted persistent challenges in sensor fouling, calibration drift, and false positives that continue to impede the widespread adoption of these technologies.

12.2.4 Predictive Analytics for Water Distribution Systems

Predictive analytics has emerged as a powerful approach for managing water distribution networks. Leaks, breaks, and water quality degradation in distribution systems represent significant challenges for water utilities, with infrastructure failures in the United States alone resulting in the loss of approximately 6 billion gallons of treated water daily [19]. ML-based predictive models have demonstrated the capacity to forecast pipe failures based on historical data, environmental factors, and infrastructure characteristics.

The gradient boosting model predicted pipe failures with 78% accuracy up to 14 days in advance, allowing for preventive maintenance and significantly reducing emergency repair costs [20]. Their model incorporated infrastructure age, material, pressure patterns, soil conditions, and weather variables to generate risk profiles for different segments of the distribution network.

12.2.5 Research Gaps and Opportunities

Despite these advances, significant research gaps remain. First, most AI applications in water treatment have focused on individual unit processes rather than holistic treatment trains, limiting their overall impact on system performance [12]. Second, the transferability of ML models between different treatment plants and water sources remains problematic, often requiring extensive retraining and calibration [13]. Third, the integration of multiple data streams from diverse sensors and monitoring systems presents substantial challenges in data standardization, quality control, and interpretation [17].

These gaps highlight the need for more comprehensive approaches that combine domain expertise with data science, address the unique characteristics of water systems, and develop robust frameworks for technology implementation that can be adapted to diverse contexts and requirements.

12.3 METHODOLOGY

12.3.1 Research Design

This study employed a mixed-methods approach combining quantitative analysis of performance data from AI and ML implementations in water treatment and monitoring with qualitative assessments of implementation challenges and success factors. The research was conducted in three phases:

1. Systematic review and meta-analysis of published case studies
2. Comparative analysis of technology implementations across different scales and contexts
3. Development and validation of an implementation framework through expert consultations

12.3.2 Data Collection

Data was collected from multiple sources to ensure comprehensive coverage of the field:

1. Peer-reviewed literature: 142 articles published between 2010–2024 were identified through structured database searches in Web of Science, Scopus, and Engineering Village
2. Industry reports: 37 technical reports from water utilities, technology providers, and research institutions
3. Case study interviews: 28 semi-structured interviews with water utility managers, technology developers, and regulatory officials
4. Performance data: Operational data from 15 water treatment facilities and 9 distribution networks that had implemented AI or smart monitoring systems

The inclusion criteria for case studies required documented implementations with at least one year of operational data, clear performance metrics, and detailed information on technology specifications and implementation processes.

12.3.3 Data Analysis

The quantitative analysis focused on key performance indicators including:

- Treatment efficiency (% removal of target contaminants)
- Energy consumption (kWh per million gallons treated)
- Chemical usage (kg per million gallons treated)
- Detection accuracy (sensitivity and specificity for monitoring systems)
- Cost savings (operational and capital expenditures)
- System reliability (downtime and maintenance requirements)

Statistical analysis was conducted using R (version 4.2.1), with meta-analysis performed using the "metafor" package following the methodology outlined [21]. Effect sizes were calculated as standardized mean differences between conventional and AI-enhanced systems, with random-effects models employed to account for heterogeneity across studies.

Qualitative data from interviews and reports were analyzed using thematic analysis, following the six-step approach [22]. Coding was performed independently by three researchers, with discrepancies resolved through consensus discussions. The resulting themes were organized into a conceptual framework addressing technological, organizational, and environmental factors influencing technology adoption.

12.3.4 Framework Development

The implementation framework was developed through an iterative process involving:

1. Initial framework development based on literature review and case study analysis
2. Refinement through expert workshops (n = 3) involving 42 participants representing water utilities, technology providers, research institutions, and regulatory agencies
3. Validation through application to three water utilities of varying sizes and technological sophistication

The framework evaluation employed a multi-criteria assessment approach incorporating technical feasibility, economic viability, organizational readiness, and regulatory compliance dimensions, with each dimension rated on a 1–5 scale based on predefined criteria.

TABLE 12.1
Distribution of AI and ML Applications in Water Treatment Processes

Treatment Process	Neural Networks	Random Forests	Support Vector Machines	Reinforcement Learning	Other ML Algorithms	Total Applications
Coagulation/Flocculation	28 (42.4%)	15 (22.7%)	11 (16.7%)	5 (7.6%)	7 (10.6%)	66 (24.5%)
Disinfection	19 (38.8%)	12 (24.5%)	8 (16.3%)	3 (6.1%)	7 (14.3%)	49 (18.2%)
Membrane Filtration	24 (53.3%)	8 (17.8%)	6 (13.3%)	5 (11.1%)	2 (4.4%)	45 (16.7%)
Biological Treatment	16 (37.2%)	13 (30.2%)	6 (14.0%)	2 (4.7%)	6 (14.0%)	43 (16.0%)
Sedimentation	12 (38.7%)	9 (29.0%)	5 (16.1%)	0 (0.0%)	5 (16.1%)	31 (11.5%)
Advanced Oxidation	8 (30.8%)	6 (23.1%)	5 (19.2%)	4 (15.4%)	3 (11.5%)	26 (9.7%)
Other Processes	5 (45.5%)	2 (18.2%)	1 (9.1%)	0 (0.0%)	3 (27.3%)	11 (4.1%)
Total	**112 (41.6%)**	**65 (24.2%)**	**42 (15.6%)**	**19 (7.1%)**	**33 (12.3%)**	**269 (100%)**

Note: Percentages in parentheses represent the proportion within each treatment process (for algorithm columns) or the overall distribution (for the Total row). Some case studies implemented multiple algorithms, resulting in totals exceeding the number of reviewed studies. Data synthesized from literature review of 142 articles published between 2010–2024 [23].

12.4 RESULTS AND DISCUSSION

12.4.1 Current State of AI and ML Applications in Water Treatment

The meta-analysis of case studies revealed significant variability in the types of AI and ML applications across different water treatment contexts. Table 12.1 summarizes the distribution of applications by treatment process and algorithm type based on the reviewed literature.

As shown in Table 12.1, neural networks were the most widely applied algorithm across all treatment processes, accounting for 41.6% of all applications. Coagulation/flocculation processes received the most attention from AI researchers (24.5% of applications), likely due to the significant chemical cost savings potential and the relatively straightforward implementation of control algorithms in this process. Advanced oxidation processes, despite their complexity and sensitivity to operating conditions, had the fewest AI applications, suggesting a potential area for future development.

12.4.2 Smart Monitoring Systems: Performance and Limitations

The analysis of smart monitoring implementations revealed significant advances in early detection capabilities, data resolution, and coverage compared to conventional monitoring approaches. Table 12.2 summarizes the key performance metrics for different types of smart monitoring systems based on the reviewed case studies.

The most effective monitoring approaches were AI-enhanced systems that combined multiple sensor types with advanced analytics, achieving detection time reductions of 80–95% compared to conventional laboratory testing while maintaining low false positive (2–6%) and false negative (1–4%) rates. These systems typically employed anomaly detection algorithms that could identify subtle deviations in water quality parameters before they reached threshold levels of concern (Figure 12.2).

However, the analysis also revealed significant challenges in implementing and maintaining smart monitoring networks. Sensor fouling and calibration drift were reported as persistent issues in 78% of the reviewed case studies, requiring frequent maintenance and periodic replacement. Data transmission reliability, particularly in remote or underground locations, also presented challenges, with reliability rates ranging from 60–95% depending on the communication technology employed.

TABLE 12.2
Performance Comparison of Smart Monitoring Systems for Water Quality

Monitoring Approach	Detection Time Reduction	False Positive Rate	False Negative Rate	Parameter Coverage	Operational Lifetime	Data Transmission Reliability
Conventional Lab Testing	Baseline	< 1%	5–15%	High	N/A	N/A
Fixed-Position Smart Sensors	65–85%	5–12%	2–8%	Medium	2–5 years	85–95%
Mobile Sensor Networks	75–90%	8–15%	3–10%	Medium-High	1–3 years	75–90%
Distributed Consumer-Grade Sensors	50–70%	15–25%	8–15%	Low	0.5–2 years	60–80%
Hybrid Systems	70–92%	3–8%	1–5%	High	2–4 years	80–95%
AI-Enhanced Monitoring	80–95%	2–6%	1–4%	Medium-High	2–5 years	85–95%

Note: Detection time reduction compared to conventional laboratory methods. Parameter coverage rated as Low (1–3 parameters), Medium (4–7 parameters), or High (8+ parameters). Data compiled from 37 case studies and technical reports [4].

12.4.3 Implementation Challenges and Success Factors

The thematic analysis of interview data and case reports identified several key challenges and success factors for AI and smart technology implementation in water utilities. The most frequently cited challenges included:

1. Data quality and availability (mentioned in 82% of interviews)
2. Integration with legacy systems (76%)

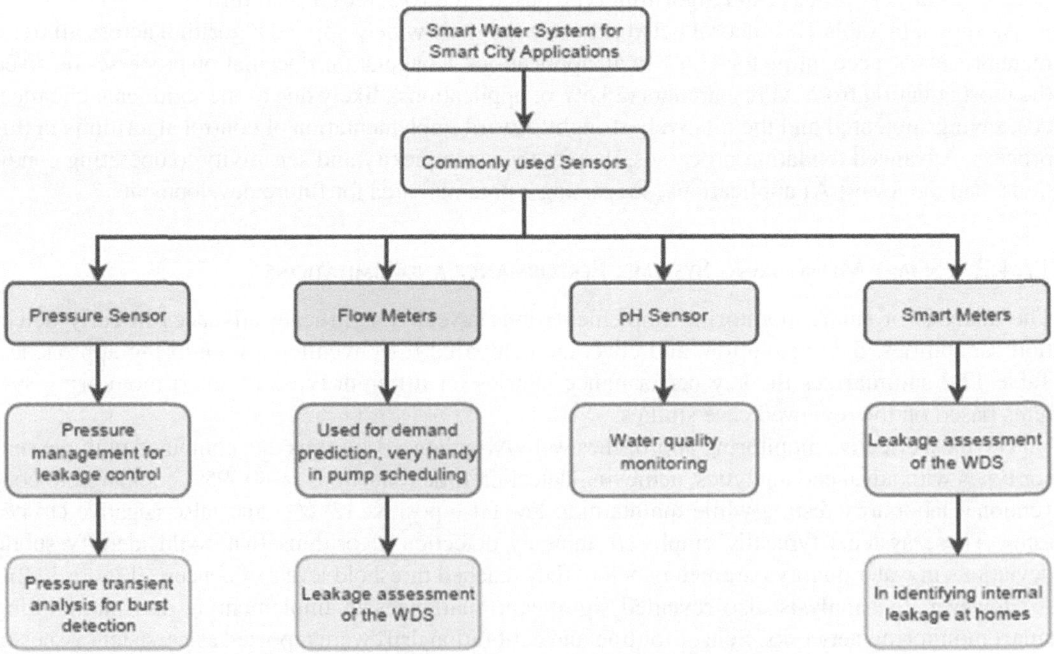

FIGURE 12.2 A flow chart presents an overview of several monitoring apparatuses and their definite roles in an innovative technological structure [4, 28].

3. Staff expertise and training requirements (71%)
4. Cost justification and return on investment uncertainty (68%)
5. Regulatory barriers and compliance concerns (59%)

Success factors that facilitated effective implementation included:

1. Strong leadership support and clear vision (mentioned in 88% of interviews)
2. Phased implementation approach (76%)
3. Collaborative partnerships with technology providers (72%)
4. Comprehensive staff training programs (69%)
5. Robust data management practices (65%)

The organizational context emerged as a critical factor influencing technology adoption success. Utilities with established innovation cultures, dedicated resources for technology evaluation, and collaborative relationships with research institutions demonstrated significantly higher success rates in AI implementation. The technological readiness of water utilities often matters less than their organizational readiness for innovation [22].

12.4.4 Implementation Framework

Based on the research findings, a comprehensive framework for AI and smart technology implementation in water utilities was developed. The framework consists of five key components:

1. **Technology Assessment**: Systematic evaluation of technology readiness, alignment with utility needs, and integration requirements
2. **Organizational Readiness**: Assessment of leadership support, staff capabilities, and innovation culture
3. **Implementation Planning**: Phased approach with clear milestones, resource requirements, and risk management strategies
4. **Performance Monitoring**: Development of key performance indicators, evaluation methodologies, and feedback mechanisms
5. **Scaling and Sustainability**: Long-term planning for technology maintenance, upgrades, and expansion

The framework was validated through application in three water utilities of varying sizes, with all three reporting improved clarity in decision making and more structured approaches to technology evaluation. The framework's flexibility allowed for customization based on utility-specific constraints and priorities, while maintaining a consistent methodology for technology assessment and implementation planning.

12.4.5 Future Directions and Emerging Technologies

Several emerging technologies show particular promise for future water treatment and monitoring applications:

1. **Digital Twins**: Virtual representations of physical water systems that can simulate different operational scenarios and optimize treatment strategies in real time. Early implementations have demonstrated potential for 15–25% operational cost reductions through optimized control strategies [24].
2. **Edge Computing**: Distributed computing architectures that process data near the point of collection, reducing latency and bandwidth requirements for remote monitoring systems. This approach has shown particular value in time-sensitive applications such as contamination event detection [25].

3. **Federated Learning**: Collaborative ML approaches that allow models to be trained across multiple utilities without sharing sensitive operational data, addressing privacy concerns while improving model robustness through diverse training data [26].
4. **Explainable AI**: Algorithms designed to provide transparent reasoning for their recommendations, addressing a key concern among water professionals regarding the "black box" nature of many AI systems. This approach has shown promise in increasing operator trust and adoption rates [27].

The convergence of these technologies with advances in sensor miniaturization and energy harvesting suggests a future where water systems are continuously monitored and autonomously optimized, with human operators focusing on strategic decision making rather than routine process control.

12.5 CONCLUSION

This research demonstrates that AI, ML, and smart technologies offer significant potential to transform water treatment and monitoring practices, with documented improvements in efficiency, accuracy, and sustainability. The meta-analysis of case studies revealed consistent performance enhancements across multiple applications, with the most substantial benefits observed in complex treatment processes and distributed monitoring systems.

The key conclusions from this study include:

1. AI-enhanced control systems can reduce operational costs by 15–30% and increase treatment efficiency by up to 25% compared to traditional approaches, as the greatest improvement in coagulation/aggregation and membrane filtration processes are observed.
2. Smart surveillance systems with ML-based anomaly recognition can be identified 4–8 hours earlier than traditional surveillance approaches, improving identification of 18–42% of subtle contamination events.
3. The effectiveness of these technologies is strongly influenced by organizational factors, such as management support, HR skills, and data management practices, which often outweigh technical considerations when determining successful implementation.
4. A structured implementation framework that takes into account both technical and organizational aspects can significantly improve recruitment outcomes, especially for suppliers with limited prior experience in implementing advanced technologies.
5. Estimation techniques such as digital twin, edge computing, and federated learning show special promises to treat current limitations and expand the application of AI in water systems.

Despite these promising results, scaling these technologies and ensuring accessibility in a variety of user contexts presents considerable challenges. Data quality issues, integration into legacy systems, and regulatory obstacles continue to hinder widespread acceptance, especially in small suppliers and developmental regions where potential benefits are maximized.

12.5.1 Future Research Directions

Several promising avenues for future research have been identified:

1. Development of a transmission learning approach to improve model tolerance between different water treatment systems and reduce data requirements and implementation costs for small supply companies.
2. Investigation of hybrid modeling approaches that combine physics-based process understanding with data-driven techniques to improve model robustness and generalizability.

3. Exploration of collaborative data sharing frameworks that enable smaller utilities to benefit from AI developments while addressing privacy and security concerns.
4. Assessment of the long-term reliability and maintenance requirements of smart monitoring networks under diverse environmental conditions and contamination scenarios.
5. Evaluation of the sustainability impacts of AI and smart technology implementation, including life-cycle energy consumption, resource usage, and environmental footprint.

As water utilities continue to face increasing challenges from climate change, emerging contaminants, and aging infrastructure, the integration of AI, ML, and smart technologies will likely play an increasingly important role in ensuring water security and sustainability. By addressing the implementation challenges identified in this research and building on the promising results demonstrated in early adopters, these technologies have the potential to fundamentally transform water management practices globally.

REFERENCES

1. Hellberg, S., & Söderbaum, F. Routledge Handbook of Water and Development. https://books.google.co.ma/books?hl=fr&lr=&id=RV3bEAAAQBAJ&oi=fnd&pg=PA1964&dq=Hellberg,+S.,+%26+S%C3%B6derbaum,+F.+Routledge+handbook+of+water+and+development&ots=ncoYodwRBw&sig=TQIMuOqjMRIiFiidKLj-2UqqGvI&redir_esc=y#v=onepage&q=Hellberg%2C%20S.%2C%20%26%20S%C3%B6derbaum%2C%20F.%20Routledge%20handbook%20of%20water%20and%20development&f=false
2. Smith, P., Adams, J., Beerling, D. J., Beringer, T., Calvin, K. V., Fuss, S., ... & Keesstra, S. (2019). Land-management options for greenhouse gas removal and their impacts on ecosystem services and the sustainable development goals. Annual Review of Environment and Resources, 44(1), 255–286.
3. Boyle, C., Ryan, G., Bhandari, P., Law, K. M., Gong, J., & Creighton, D. (2022). Digital transformation in water organizations. Journal of Water Resources Planning and Management, 148(7), 03122001.
4. Gupta, A. D., Pandey, P., Feijóo, A., Yaseen, Z. M., & Bokde, N. D. (2020). Smart water technology for efficient water resource management: A review. Energies, 13(23), 6268.
5. Li, X., Su, J., Wang, H., Boczkaj, G., Mahlknecht, J., Singh, S. V., & Wang, C. (2024). Bibliometric analysis of artificial intelligence in wastewater treatment: Current status, research progress, and future prospects. Journal of Environmental Chemical Engineering, 12, 113152.
6. Ahmed, F., Johnson, D., Hashaikeh, R., & Hilal, N. (2023). Barriers to innovation in water treatment. Water, 15(4), 773.
7. Srivastava, N., Sen, A., Srivastava, A., Ali, S. E., & Khare, S. (2025). Cost-Benefit Analysis of Automated Water Quality Management: Investment, Savings, and Strategic Advantages. In Computational Automation for Water Security (pp. 157–174). Elsevier.
8. Ghizlane, F., Mabrouki, J., Ghrissi, F., & Azrour, M. (2022). Proposal for a high-resolution particulate matter (PM10 and PM2. 5) capture system, comparable with hybrid system-based internet of things: Case of quarries in the Western Rif. Morocco Pollution, 8(1), 169–180.
9. Adesina, O. B., William, C., & Oke, E. I. (2024). Evolution in water treatment: Exploring traditional self-purification methods and emerging technologies for drinking water and wastewater treatment: A review. World News of Natural Sciences, 53, 169–185.
10. Ma, X., Xue, X., González-Mejía, A., Garland, J., & Cashdollar, J. (2015). Sustainable water systems for the city of tomorrow—A conceptual framework. Sustainability, 7(9), 12071–12105.
11. Hinojosa-Palafox, E. A., Rodríguez-Elías, O. M., Hoyo-Montaño, J. A., Pacheco-Ramírez, J. H., & Nieto-Jalil, J. M. (2021). An analytics environment architecture for industrial cyber-physical systems big data solutions. Sensors, 21(13), 4282.
12. Alam, G., Ihsanullah, I., Naushad, M., & Sillanpää, M. (2022). Applications of artificial intelligence in water treatment for optimization and automation of adsorption processes: Recent advances and prospects. Chemical Engineering Journal, 427, 130011.
13. Heddam, S. (2021). Extremely Randomized Tree: a New Machines Learning Method for Predicting Coagulant Dosage in Drinking Water Treatment Plant. In Water Engineering Modeling and Mathematic Tools (pp. 475–489). Elsevier.
14. Bagheri, M., Akbari, A., & Mirbagheri, S. A. (2019). Advanced control of membrane fouling in filtration systems using artificial intelligence and machine learning techniques: A critical review. Process Safety and Environmental Protection, 123, 229–252.

15. Wilson, J. L. (2023). Impacts of legacy mining on stream structural and functional processes in an urban stream (Doctoral dissertation, University of Missouri-Columbia).
16. Akhter, F., Siddiquei, H. R., Alahi, M. E. E., & Mukhopadhyay, S. C. (2021). Recent advancement of the sensors for monitoring the water quality parameters in smart fisheries farming. Computers, 10(3), 26.
17. Zheng, Y., Jin, M., Liu, Y., Chi, L., Phan, K. T., & Chen, Y. P. P. (2022). From unsupervised to few-shot graph anomaly detection: A multi-scale contrastive learning approach. arXiv preprint arXiv:2202.05525.
18. Torres-Martínez, J. A., Mahlknecht, J., Kumar, M., Loge, F. J., & Kaown, D. (2024). Advancing groundwater quality predictions: Machine learning challenges and solutions. Science of The Total Environment, 949, 174973.
19. Guo, C. (2024). Regulator Preferences and Underinvestment in Drinking Water Infrastructure. Electronic Theses & Dissertations (2024 - present). 19. https://scholarsarchive.library.albany.edu/etd/19
20. Chen, R., Wang, Q., & Javanmardi, A. (2025). A review of the application of machine learning for pipeline integrity predictive analysis in water distribution networks. Archives of Computational Methods in Engineering, 1–29.
21. Nickow, A., Oreopoulos, P., & Quan, V. (2024). The promise of tutoring for PreK–12 learning: A systematic review and meta-analysis of the experimental evidence. American Educational Research Journal, 61(1), 74–107.
22. Solís, P. (2022). The Routledge Handbook of Development and Environment. B. McCusker, W. Ahmed, M. Ramutsindela (Eds.). Routledge.
23. Tasnim, Z. (2024). Development and Application of Discrete Element Models for Granular Materials-From Rigid Spheres to Complex-Shaped Flexible Particles With Breakage (Doctoral dissertation, Clemson University).
24. Corrado, C. R., DeLong, S. M., Holt, E. G., Hua, E. Y., & Tolk, A. (2022). Combining green metrics and digital twins for sustainability planning and governance of smart buildings and cities. Sustainability, 14(20), 12988.
25. Dou, F., Ye, J., Yuan, G., Lu, Q., Niu, W., Sun, H., … Song, W. (2023). Towards artificial general intelligence (AGI) in the internet of things (IoT): Opportunities and challenges. arXiv preprint arXiv:2309.07438.
26. Rejula, M. A., Minija, S. J., Sophia, S. G., & Barakka, J. A. (2025). Decentralized water quality classification using federated learning with recurrent neural networks. Water Quality Research Journal, 60(1), 135–150.
27. Walker, C. M., Agarwal, V., Lin, L., Hall, A. C., Hill, R. A., Mortenson, T. J., & Lybeck, N. J. (2023). Explainable artificial intelligence technology for predictive maintenance (No. INL/RPT-23-74159-Rev000). Idaho National Laboratory (INL), Idaho Falls, ID (United States).
28. Manjunatha, M. B. (2017). A hybrid gesture recognition method for American sign language. Indian Journal of Science and Technology, 10(1), 1–12.

13 Hydrogen Energy Storage
Technologies, Challenges, and Opportunities

S. Boudraham, J. Mabrouki, I. Ait Oujallal, and Y. Abrouki

13.1 INTRODUCTION

Hydrogen is increasingly recognized as a key player in the transition to a sustainable energy future. As a clean and efficient energy carrier, hydrogen has the potential to address the growing demand for energy while reducing the environmental impact of traditional energy systems (Hossain Bhuiyan and Siddique 2025a). Unlike fossil fuels, hydrogen produces only water as a by-product when used in fuel cells, making it an ideal solution for decarbonizing sectors like transportation, industry, and power generation (El-Moustaqim, Rachiq, et al. 2024; El-Moustaqim, Sbai, El Yousfi, et al. 2024; El-Moustaqim, Sbai, Mabrouki, et al. 2024). As global efforts to combat climate change intensify, hydrogen demand is increasing across various sectors. In the transportation sector, hydrogen fuel cells are emerging as a clean alternative to fossil fuel–powered vehicles, especially for heavy-duty transportation such as trucks, buses, and trains. In the industrial sector, hydrogen is vital for processes such as refining, ammonia production, and steel manufacturing, where it can replace conventional carbon-intensive fuels (Gomonov, Permana and Handoko 2025). Additionally, hydrogen is gaining traction as an energy storage solution for renewable energy, helping to balance supply and demand by storing excess energy from intermittent sources like wind and solar. As a result, hydrogen is poised to become a key element in decarbonizing multiple industries, driving the demand for efficient and scalable storage technologies (Pathak, Yadav and Padmanaban 2023). The mounting interest in hydrogen as a sustainable energy vector has precipitated a substantial surge in research endeavors concerning hydrogen storage technologies. As demonstrated in Figure 13.1, the number of scientific publications in this field has increased exponentially over the past two decades, reflecting the global efforts to develop efficient, safe, and cost-effective storage solutions (Muhammed et al. 2023). This surge in research is driven by the pressing need to overcome the challenges associated with hydrogen storage, including its low energy density, the requirement of high pressure, and material limitations. The broad range of applications, from transportation to industrial processes and large-scale energy storage, underscores the critical importance of developing advanced hydrogen storage technologies for the successful establishment of a sustainable hydrogen economy.

One of the key challenges to hydrogen's widespread adoption, however, lies in the effective storage of this energy carrier. Hydrogen has a very low energy density in its natural form, necessitating advanced storage methods to enable its efficient use in various applications (Osman et al. 2024a). These storage systems must meet specific criteria such as high energy density, safety, cost-effectiveness, and the ability to store hydrogen under a wide range of conditions (El-Moustaqim, Mabrouki and Hmouni 2024; El-Moustaqim, Mabrouki, et al. 2024). Current storage technologies include compressed hydrogen, liquid hydrogen, and solid-state hydrogen storage using materials like metal hydrides, each with its own set of advantages and drawbacks (AlZohbi, Almoaikel and AlShuhail 2023a). This chapter aims to explore the various hydrogen storage technologies, the challenges that come with them, and the opportunities for overcoming these obstacles. By examining the current

DOI: 10.1201/9781003647263-13

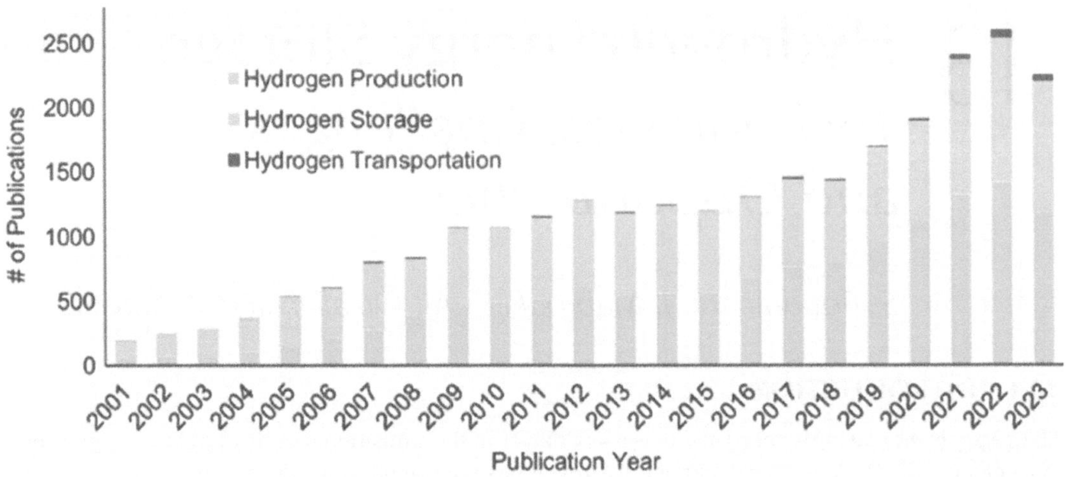

FIGURE 13.1 Growth of scientific publications on hydrogen production, storage, and transportation in recent years.

state of research and development in hydrogen storage, we aim to highlight the potential of hydrogen as a cornerstone of clean energy systems and provide insight into its future role in the global energy landscape.

13.2 PROPERTIES AND APPLICATIONS OF HYDROGEN

Hydrogen (H), the lightest and most abundant element in the universe, constitutes approximately 73% of the elemental mass of the cosmos ("Origin of the Elements" s. d.). In its fundamental state it is a flammable gas that is colorless, odorless, and tasteless. The simplest form of hydrogen, known as protium($^{1}_{1}H$), consists of a single proton and one electron, with no neutrons present in its nucleus. This makes it the smallest and lightweight element. Despite its simplicity, hydrogen exhibits remarkable versatility and plays a pivotal role in various scientific and industrial applications. Additionally, its naturally occurring isotopes, deuterium($^{2}_{1}H$) and tritium($^{3}_{1}H$), are integral to nuclear fusion research. Deuterium contains one proton and one neutron in its nucleus, while tritium possesses one proton and two neutrons, highlighting the multifaceted nature of hydrogen and its importance in advancing modern energy technologies (Abe et al. 2019). Hydrogen is among the most extensively utilized chemical elements globally, playing a critical role across a wide range of industries, including chemical manufacturing, oil refining, metallurgy, food production, glass-making, and electronics. Due to its low molecular weight and exceptional energy density, hydrogen is also a favored fuel for space exploration and serves as an environmentally friendly energy source in fuel cell technologies (Idriss, Scott and Subramani 2015). Hydrogen exhibits distinctive physical and chemical characteristics that position it as a promising contender in the pursuit of sustainable energy solutions. Table 13.1 presents a summary of the general properties of this element ("Characteristics and Safety of Hydrogen" s. d.).

As the lightest element, hydrogen offers an exceptional energy content of 120 MJ/kg, which is nearly three times that of gasoline on a mass basis. However, its lower volumetric energy density (0.0899 kg/m^3) presents challenges for storage and transport (Osman et al. 2022). To address these limitations, innovative solutions are being developed, including compressed gas storage, advanced pressure vessels, and materials-based technologies. Moreover, hydrogen stands out as a clean and environmentally friendly energy source. Its combustion results solely in the production of water vapor, making it highly appealing. However, the environmental impact of hydrogen depends significantly on the method used for its production. The most carbon-intensive is brown hydrogen,

TABLE 13.1
Properties of Hydrogen

Property	Hydrogen
Molecular Weight (g/mol)	2.016
Density (kg/m³ at 20°C and 1 atm)	0.08375
Normal Boiling point (°C)	–252.8
Flash Point (°C)	< –253
Flammability Limits in Air (Volume %)	4.0–75.0
CO_2 Production per Energy Unit	0
Auto-ignition Temperature in Air (°C)	585
Higher Heating Value (MJ/kg)	142.0
Lower Heating Value (MJ/kg)	120.0

which is derived from fossil fuels. In contrast, green hydrogen, which is produced from renewable energy sources, and blue hydrogen, which employs carbon capture technologies, is regarded as more environmentally friendly. Furthermore, hydrogen's versatility as an energy carrier enables its use in a multitude of applications across various industries (Figure 13.2) (Alptekin and Celiktas 2022). It serves as a power source for fuel cell vehicles, integrates with electricity grids for on-demand energy, supports industrial processes, and stores excess renewable energy, thereby ensuring a continuous supply. These characteristics illustrate the potential of hydrogen to play a pivotal role in achieving net-zero emissions and fostering a sustainable energy future.

13.3 HYDROGEN STORAGE TECHNOLOGIES

Prior to elucidating the subject of hydrogen storage technologies, it is first necessary to provide a concise overview of the principal methods of hydrogen production. As a vital energy vector for the transition to a low-carbon economy, hydrogen is produced through a range of methods, each with distinct environmental and economic implications. At present, the principal sources for hydrogen production are water, biomass, and fossil fuels (Taipabu et al. 2022). Fossil fuel–based methods include steam

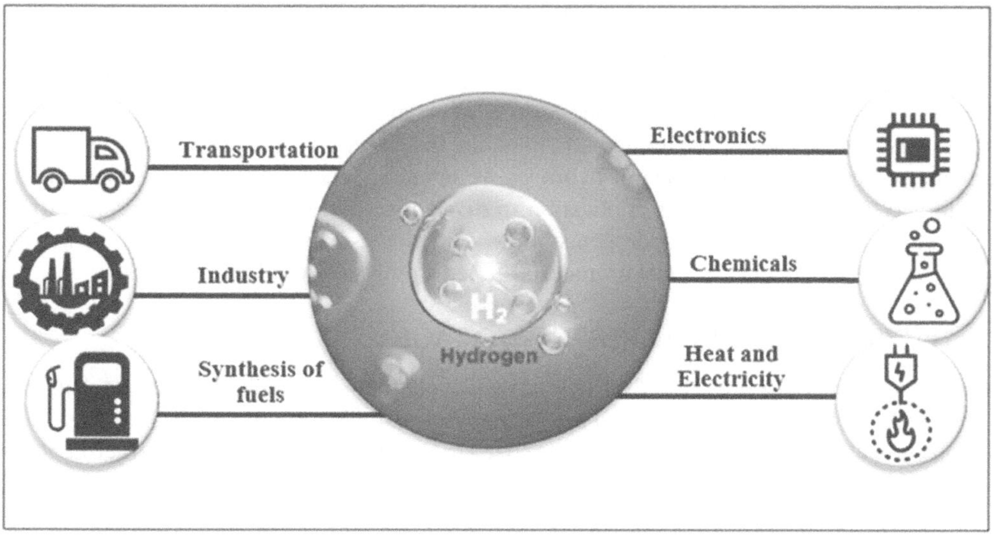

FIGURE 13.2 Applications of hydrogen.

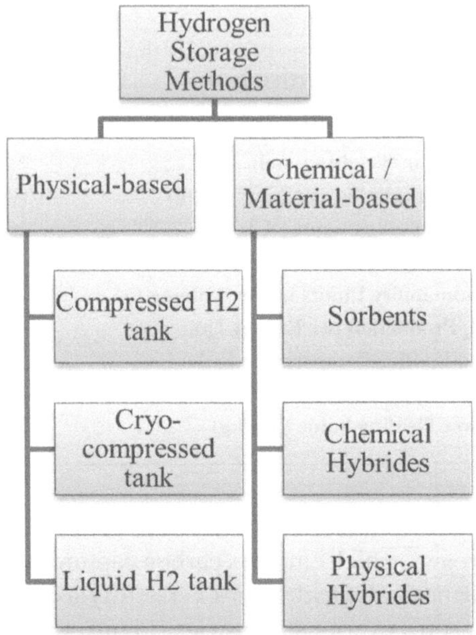

FIGURE 13.3 Basic hydrogen storage techniques.

reforming of natural gas, whereby methane reacts with steam at high temperatures to yield hydrogen and carbon monoxide; thermal cracking of natural gas, which decomposes methane into hydrogen and solid carbon; partial oxidation of heavier hydrocarbons, such as naphtha, producing hydrogen and carbon monoxide; and coal gasification, which converts coal into syngas, a mixture of hydrogen, carbon monoxide, and carbon dioxide (Afanasev et al. 2024). Another pathway for hydrogen production is through the use of biomass, a renewable resource. This can be achieved through pyrolysis or gasification, which are thermochemical processes that generate a gaseous mixture containing hydrogen. Furthermore, water can be utilized as a sustainable hydrogen source through various methods, including electrolysis, which employs electricity to split water into hydrogen and oxygen; photolysis, where solar energy drives water dissociation; thermochemical cycles involving high-temperature reactions with chemical intermediates; and thermolysis, which is the direct breakdown of water at extreme temperatures (Ishaq, Dincer and Crawford 2022). Finally, biological processes make use of microorganisms, such as algae or bacteria, to produce hydrogen via enzymatic or photosynthetic reactions, representing a promising avenue for renewable and low-impact hydrogen production (Xu, Zhou and Yu 2022). These diverse methods illustrate the versatility and potential of hydrogen as a clean energy carrier. The utilization of hydrogen as an energy necessitates the development of efficient storage methodologies for its application in diverse sectors, including industry, mobility, and energy production. The primary hydrogen storage technologies can be categorized into two distinct types: physical storage and chemical or material-based storage (Figure 13.3).

13.4 PHYSICAL-BASED HYDROGEN STORAGE

Compressed hydrogen: is stored at high pressures of 700 bars, which increases its density to 42 kg/m^3. Composite overwrapped pressure vessels (COPVs) are typically used for storage, offering a gravimetric density of up to 5.7%. This method allows for quick refueling and is widely used, especially in aviation, due to its commercial availability and high technology readiness. However, it has drawbacks, including a relatively low volumetric energy density and the risk of tank degradation due to the high pressures, which can lead to leaks or explosions (Massaro et al. 2023a).

Liquid hydrogen: is produced by cooling hydrogen to cryogenic temperatures (20 K°), which increases its density to 71 kg/m³. This method provides higher volumetric and gravimetric densities, as well as operates at much lower pressures, improving safety (AlZohbi, Almoaikel and AlShuhail 2023b). However, the main challenge is the boil-off phenomenon, where hydrogen evaporates if the tank temperature increases. To minimize this, tanks are insulated with multilayer systems. Continuous developments in insulation and materials like aerogels and carbon-fiber composites aim to further improve storage efficiency and reduce weight.

Cryogenic hydrogen: The cryogenic storage of hydrogen involves the cooling of the gas to a temperature below –252.87°C, which transforms it into a dense liquid for long-term storage (Afanasev et al. 2024). The process entails the purification of hydrogen, whereby impurities such as moisture and CO_2 are removed. This is followed by the cooling of the hydrogen using cryogenic systems, which employ liquid nitrogen or helium. The liquefied hydrogen is stored in insulated tanks constructed from materials such as stainless steel or aluminum, which are designed to maintain extremely low temperatures and minimize evaporation. This method ensures high-energy density and efficient storage (Nam et al. 2023).

Cryo-compressed gas: The process of cryo-compression of hydrogen involves the combination of high-pressure compression (50–700 bars) with cryogenic cooling (25–110 K) (Massaro et al. 2023b). This hybrid approach enhances both volumetric and gravimetric densities. A significant advantage is the reduction of boil-off losses in comparison to liquid hydrogen, as the tank is capable of withstanding minor pressure increases. This method supports rapid refueling and is based on existing technologies, although safety concerns persist due to the combination of high pressure and low temperatures. Ongoing research into lighter tank materials, such as advanced composites, may improve performance and reduce risks (Li et al. 2024).

13.5 CHEMICAL HYDROGEN STORAGE

Hydrogen sorption: Hydrogen sorption refers to the process by which hydrogen is stored in solid-state materials, classified into two mechanisms: physisorption and chemisorption (Figure 13.4) (Mulky et al. 2024). Both methods are central to addressing the challenges of efficient hydrogen storage, pivotal for the hydrogen economy.

- *Physisorption:* is a process that relies on weak van der Waals forces to bind hydrogen molecules to porous materials, including MOFs, COFs, and carbon compounds. These materials possess high surface areas and adjustable pore structures, which may facilitate the potential storage of hydrogen (Chen et al. 2022). Nevertheless, their performance at room temperature is constrained, with carbon-based materials demonstrating a storage capacity of less than 1% hydrogen (Xu et al. 2007). Significant challenges remain in improving the room-temperature capacity of this method.

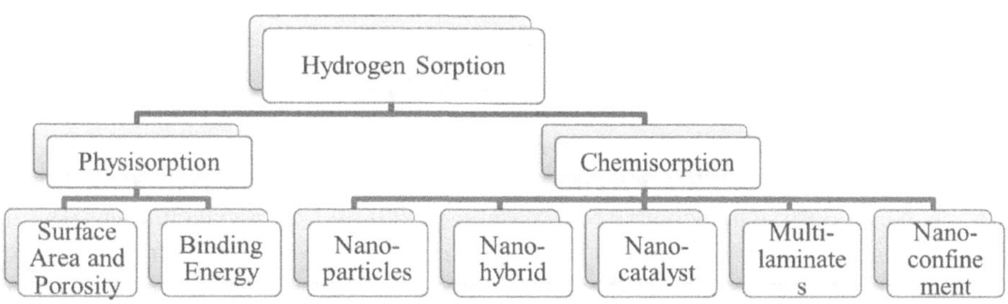

FIGURE 13.4 Hydrogen sorption methods.

- *Chemisorption:* In the process of chemisorption, hydrogen molecules are split into atoms that form chemical bonds with materials such as metals or complex hydrides. This process provides stable and efficient storage, as evidenced by the use of materials such as titanium or magnesium hydrides (Qureshi et al. 2024). Nevertheless, the necessity for elevated temperatures during desorption may prove to be a limiting factor in the practical applications of this process. Notwithstanding these challenges, chemisorption represents a promising avenue for reversible hydrogen storage and conversion in advanced systems.

Metal hybrids: Metal hydrides are compounds where hydrogen atoms are incorporated into the lattice structure of a metal, forming stable solid solutions. Hydrogenation can occur through two mechanisms: direct dissociative chemisorption or electrochemical water splitting. These materials provide superior volumetric storage due to lattice expansion during hydrogen uptake. Transition metal hydrides, such as vanadium hydrides, show promise for improved hydrogen capacity, particularly at low temperatures, although their reliance on heavy elements can reduce gravimetric efficiency (Scarpati et al. 2024).

Chemical Hydrides: Chemical hydrides are non-metallic substances capable of absorbing or releasing hydrogen through specific reactions or processes. They form ionic or covalent bonds with hydrogen, making them effective for storage. While they are typically more stable than metal hydrides, this stability often demands significant energy input to release the stored hydrogen. Although hydrogen storage in liquids such as ammonia and toluene is possible, ammonium borohydride (AB) is the primary non-metal solid used in slurry form to date (Osman et al. 2024b).

Liquid Organic Hydrogen Carriers (LOHCs): LOHCs offer an efficient and sustainable solution for large-scale hydrogen storage and transport. They utilize aromatic compounds, often derived from fossil fuels or sustainable lignocellulose biomass, particularly lignin. Through thermochemical processes, lignin can produce intermediates capable of reversible hydrogen storage. This approach is enhanced by optimizing catalysts and reaction conditions to improve efficiency and scalability (Sage et al. 2024).

Porous Materials: These materials, prized for their extensive surface area and adjustable pore sizes, are being investigated for efficient hydrogen storage. These materials can reversibly adsorb and desorb hydrogen under controlled conditions. At 77 K and pressures between 20–100 bar, their hydrogen storage capacity ranges from 1.49 to 7.0 t%, while at 298 K, it ranges from 0.15 to 0.7% (Firlej et al. 2021). Graphene oxide (GO) and activated carbon both perform excellently in hydrogen storage, demonstrating high adsorption capacities under various conditions. Moreover, certain metal-organic frameworks (MOFs) are gaining attention for their superior hydrogen storage capabilities (Chen et al. 2022).

Carbon Materials: Carbon-based materials such as graphene, carbon nanotubes (CNTs), and carbon nanofibers are widely studied for hydrogen storage due to their high porosity and lightweight nature. Graphene can store significant amounts of hydrogen but faces challenges in cycle stability, while CNTs offer strong structural support but require surface modifications to improve hydrogen uptake (Wen et al. 2025). Physisorption, the main hydrogen interaction mechanism, is essential for these materials storage capabilities. Despite their promising characteristics, carbon-based materials are less efficient than metals in hydrogen storage, though they offer advantages like lower cost and mass density. Overall, their potential in hydrogen storage requires further optimization.

Electrochemical Hydrogen Storage: this technology offers an efficient method for hydrogen storage, particularly when compared to gaseous and liquid storage techniques. In this process, aqueous solutions are used to break down and adsorb atomic hydrogen onto storage materials (Osman et al. 2024a). Various electrochemical techniques, such as voltammetry, amperometry, charge-discharge testing, and electrochemical impedance spectroscopy (EIS), help assess the hydrogen storage capacity of materials (Gholami and Pirsaheb 2021).

By analyzing oxidation peaks and discharge potentials, the quantity of stored hydrogen can be determined. The storage mechanism primarily involves two processes: hydrogen adsorption on the surface and its insertion into the material's bulk.

13.6 CHALLENGES AND OPPORTUNITIES

Hydrogen storage technologies face several critical challenges while also offering promising opportunities for advancement. One major challenge is ensuring safety, particularly for compressed hydrogen storage, where the risk of high-pressure tank explosions remains significant. Similarly, liquid hydrogen storage requires extremely low temperatures and advanced insulation, leading to high energy consumption and operational costs. Metal hydrides, although stable and capable of high storage capacity, face limitations due to their slow kinetics, high material costs, and the need for additional cooling systems (Hossain Bhuiyan and Siddique 2025b). Emerging technologies, such as carbon-based materials and hydrate-based hydrogen storage, are still in the early stages of development and suffer from technical uncertainties, such as low hydrogen density, complex manufacturing, and limited commercial scalability. High density hydrogen storage remains a significant challenge, particularly in the context of transportation. Current methodologies depend on large-volume compressed gas tanks, which are adequate for stationary applications but are impractical for fuel-cell vehicles requiring over 300 miles of range. While certain hydrogen-powered vehicles have been developed with high-pressure composite tanks that can withstand 700 bar, storage constraints persist, particularly for compact models. Hydrogen possesses a high gravimetric energy density (120 MJ/kg) but a low volumetric density (8 MJ/L for liquid hydrogen vs. 32 MJ/L for gasoline), necessitating 5–13 kg of onboard storage (Figure 13.5) ("Hydrogen Storage Materials Database – CKAN" 2025).

To address these challenges, research efforts are currently focused on two primary avenues: the near-term optimization of high-pressure composite tanks and the long-term exploration of advanced storage methods. These include cryo-compressed hydrogen and material-based solutions such as sorbents, chemical hydrogen carriers, and metal hydrides. These advancements are designed to

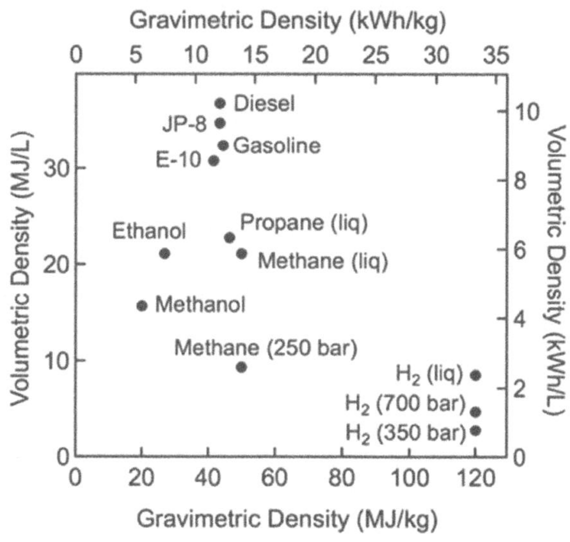

FIGURE 13.5 Comparison of gravimetric and volumetric energy densities of hydrogen and other fuels. (From "Hydrogen Storage". s. d. Energy.Gov. Consulté le 25 mars 2025. https://www.energy.gov/eere/fuelcells/hydrogen-storage.)

TABLE 13.2
Advantages and Challenges of Main Types of Hydrogen Methods

Technology	Advantages	Disadvantages
Compressed hydrogen	Well-established technology that is efficient and user-friendly.	High costs for storage cylinders and frequent refueling due to limited capacity.
Liquid hydrogen	Efficient storage system offering high-energy density.	Requires extremely low temperatures, leading to high-energy demands and significant costs.
Porous materials	High surface area and porosity enable significant hydrogen adsorption.	Reduced hydrogen absorption rates at lower temperatures.
Metal hybrids	Provides stable and safe storage with high hydrogen density and reversible cycling performance.	Impurities absorbed over time can reduce storage capacity and tank longevity.

Source: Mulky et al. (2024).

enhance hydrogen storage efficiency, thereby facilitating its widespread adoption in the energy and transportation sectors (Feki et al. 2025). Table 13.2 presents the advantages and disadvantages of some hydrogen storage technology.

On the other hand, the field presents several opportunities for innovation and integration. Advancements in tank fabrication and cryogenic technologies can significantly improve safety and cost efficiency for compressed and liquid hydrogen storage, respectively. The development of efficient catalysts and reaction conditions for chemical hydrides opens avenues for long-term hydrogen storage solutions. Moreover, the integration of hydrogen storage systems with renewable energy sources aligns with global efforts toward clean energy transition. Technologies like carbon-based storage and hydrates, with further research and optimization, could unlock new possibilities for large-scale, energy-efficient hydrogen storage. Support from government policies and industry investments further accelerates the potential for widespread adoption of these technologies in future energy systems (Ma et al. 2024).

13.7 CONCLUSION AND FUTURES PERSPECTIVES

In conclusion, the field of hydrogen energy storage presents a duality of significant opportunities and formidable challenges. The technologies currently available, including compressed hydrogen, liquid hydrogen, and metal hydrides, each possess distinctive advantages for specific applications. However, they also present challenges related to cost, efficiency, safety, and scalability. As the demand for clean energy continues to increase, the capacity to efficiently store and transport hydrogen will be crucial for its integration into global energy systems. In the future, research and innovation in hydrogen storage technologies are likely to focus on enhancing the efficiency, safety, and economic viability of these systems. Advances in materials science, including the development of novel carbon-based materials, metal hydrides, and porous frameworks, present promising avenues for achieving higher storage densities and enhancing the life span and performance of hydrogen storage tanks. Furthermore, innovations in cryogenic storage, solid-state storage, and even novel approaches, such as hydrogen storage in chemical compounds, could further enhance the versatility and reliability of hydrogen as an energy carrier. The future of hydrogen storage will depend not only on technological advancements but also on supportive policies, investments in infrastructure, and collaboration across sectors. As governments and industries align their efforts to meet ambitious carbon reduction targets, hydrogen storage systems will assume an increasingly pivotal role in enabling a clean, resilient, and sustainable energy future. By addressing the current limitations and seizing the opportunities for improvement, hydrogen storage has the potential to become a cornerstone of a global low-carbon economy, contributing to energy security and climate mitigation efforts worldwide.

REFERENCES

Abe, J.O., A.P.I. Popoola, E. Ajenifuja, and O.M. Popoola. 2019. "Hydrogen energy, economy and storage: Review and recommendation". *International Journal of Hydrogen Energy* 44 (29):15072–86. https://doi.org/10.1016/j.ijhydene.2019.04.068.

Afanasev, Pavel, Aysylu Askarova, Tatiana Alekhina, Evgeny Popov, Strahinja Markovic, Aliya Mukhametdinova, Alexey Cheremisin, and Elena Mukhina. 2024. "An overview of hydrogen production methods: Focus on hydrocarbon feedstock". *International Journal of Hydrogen Energy* 78:805–28. https://doi.org/10.1016/j.ijhydene.2024.06.369.

Alptekin, Fikret Muge, and Melih Soner Celiktas. 2022. "Review on catalytic biomass gasification for hydrogen production as a sustainable energy form and social, technological, economic, environmental, and political analysis of catalysts". *ACS Omega* 7 (29): 24918–41.

AlZohbi, G., A. Almoaikel, and L. AlShuhail. 2023a. "An overview on the technologies used to store hydrogen". *Energy Reports*, Proceedings of 2022 7th International Conference on Renewable Energy and Conservation, 9:28–34. https://doi.org/10.1016/j.egyr.2023.08.072.

AlZohbi, G., A. Almoaikel, and L. AlShuhail. 2023b. "An overview on the technologies used to store hydrogen". *Energy Reports* 9:28–34. https://doi.org/10.1016/j.egyr.2023.08.072.

"Characteristics and Safety of Hydrogen". s. d. Consulté le 21 novembre 2024. https://www.fuelcellstore.com/blog-section/component-information/hydrogen-storage-information/hydrogen-characteristics-safety.

Chen, Zhijie, Kent O. Kirlikovali, Karam B. Idrees, Megan C. Wasson, and Omar K. Farha. 2022. "Porous materials for hydrogen storage". *Chem* 8 (3):693–716. https://doi.org/10.1016/j.chempr.2022.01.012.

El-Moustaqim, Khadija, Jamal Mabrouki, Mourade Azrour, and Driss Hmouni. 2024. "How Can Green Energy Be Improved by Integrating IoT into the Microalgae Process?" In *Sustainable and Green Technologies for Water and Environmental Management*, 239–52. https://doi.org/10.1007/978-3-031-52419-6_17.

El-Moustaqim, Khadija, Jamal Mabrouki, and Driss Hmouni. 2024. "From microalgal biomass to biohydrogen: A novel approach to environmentally sustainable technology". *Advancements in Climate and Smart Environment Technology*, 29–38. https://www.igi-global.com/chapter/from-microalgal-biomass-to-biohydrogen/346691.

El-Moustaqim, Khadija, Salah Eddine Sbai, Yassine El Yousfi, Jamal Mabrouki, and Driss Hmouni. 2024. "New strategy for the advancement of modern agriculture through the use of microalgae as biofertilizers". *Euro-Mediterranean Journal for Environmental Integration*. https://doi.org/10.1007/s41207-024-00678-x.

El-Moustaqim, Khadija, Salah Eddine Sbai, Jamal Mabrouki, and Driss Hmouni. 2024. "Critical review of green microalgal biomass development through CO2 sequestration". *Euro-Mediterranean Journal for Environmental Integration*. https://doi.org/10.1007/s41207-024-00548-6.

El-Moustaqim, Khadija, Toufik Rachiq, Jamal Mabrouki, Miloudia Slaoui, and Driss Hmouni. 2024. "Application of Microalgae in Green Hydrogen Production: A Diverse Biomass for Multiple Applications". In *Advanced Technology for Smart Environment and Energy*, édité par Jamal Mabrouki and Azrour Mourade, 231–40. Environmental Science and Engineering. Cham: Springer Nature Switzerland. https://doi.org/10.1007/978-3-031-50871-4_15.

Feki, Imen, Mohammadali Shirinbayan, Samia Nouira, Robert Tie Bi, Jean-Baptiste Maeso, Cedric Thomas, and Joseph Fitoussi. 2025. "Composites in high-pressure hydrogen storage: A review of multiscale characterization and mechanical behavior". *Composites Part C: Open Access* 16:100555. https://doi.org/10.1016/j.jcomc.2024.100555.

Firlej, Lucyna, Bogdan Kuchta, Katarzyna Walczak, and Catherine Journet. 2021. "Hydrogen storage in pure and boron-substituted nanoporous carbons—Numerical and experimental perspective". *Nanomaterials* 11 (9):2173. https://doi.org/10.3390/nano11092173.

Gholami, Tahereh, and Meghdad Pirsaheb. 2021. "Review on effective parameters in electrochemical hydrogen storage". *International Journal of Hydrogen Energy* 46 (1):783–95. https://doi.org/10.1016/j.ijhydene.2020.10.003.

Gomonov, Konstantin, Chrisna T. Permana, and Chanel Tri Handoko. 2025. "The growing demand for hydrogen: Current trends, sectoral analysis, and future projections". *Unconventional Resources* 6:100176. https://doi.org/10.1016/j.uncres.2025.100176.

Hossain Bhuiyan, Md Monjur, and Zahed Siddique. 2025a. "Hydrogen as an alternative fuel: A comprehensive review of challenges and opportunities in production, storage, and transportation". *International Journal of Hydrogen Energy* 102:1026–44. https://doi.org/10.1016/j.ijhydene.2025.01.033.

Hossain Bhuiyan, Md Monjur, and Zahed Siddique. 2025b. "Hydrogen as an alternative fuel: A comprehensive review of challenges and opportunities in production, storage, and transportation". *International Journal of Hydrogen Energy* 102: 1026–44. https://doi.org/10.1016/j.ijhydene.2025.01.033.

"Hydrogen Storage". s. d. Energy.Gov. Consulté le 25 mars 2025. https://www.energy.gov/eere/fuelcells/hydrogen-storage.

"Hydrogen Storage Materials Database – CKAN". 2025. Consulté le 25 mars 2025. https://datahub.hymarc.org/dataset/hydrogen-storage-materials-db.

Idriss, H., M. Scott, and V. Subramani. 2015. "1 - Introduction to Hydrogen and Its Properties". In *Compendium of Hydrogen Energy*, édité par Velu Subramani, Angelo Basile, et T. Nejat Veziroğlu, 3–19. Woodhead Publishing Series in Energy. Oxford: Woodhead Publishing. https://doi.org/10.1016/B978-1-78242-361-4.00001-7.

Ishaq, Haris, Ibrahim Dincer, and Curran Crawford. 2022. "A review on hydrogen production and utilization: Challenges and opportunities". *International Journal of Hydrogen Energy*, SI: Progress in Hydrogen Production, Storage and Distribution (Ahmadi), 47 (62):26238–64. https://doi.org/10.1016/j.ijhydene.2021.11.149.

Li, Kexin, Xueqiang Dong, Haocheng Wang, and Maoqiong Gong. 2024. "Study on the influence factors of gravimetric hydrogen storage density of type III cryo-compressed hydrogen storage vessel". *International Journal of Hydrogen Energy* 96:680–91. https://doi.org/10.1016/j.ijhydene.2024.11.375.

Ma, Nan, Weihua Zhao, Wenzhong Wang, Xiangrong Li, and Haiqin Zhou. 2024. "Large scale of green hydrogen storage: Opportunities and challenges". *International Journal of Hydrogen Energy* 50:379–96. https://doi.org/10.1016/j.ijhydene.2023.09.021.

Massaro, Maria Chiara, Roberta Biga, Artem Kolisnichenko, Paolo Marocco, Alessandro Hugo Antonio Monteverde, and Massimo Santarelli. 2023a. "Potential and technical challenges of on-board hydrogen storage technologies coupled with fuel cell systems for aircraft electrification". *Journal of Power Sources* 555:232397. https://doi.org/10.1016/j.jpowsour.2022.232397.

Massaro, Maria Chiara, Roberta Biga, Artem Kolisnichenko, Paolo Marocco, Alessandro Hugo Antonio Monteverde, and Massimo Santarelli. 2023b. "Potential and technical challenges of on-board hydrogen storage technologies coupled with fuel cell systems for aircraft electrification". *Journal of Power Sources* 555:232397. https://doi.org/10.1016/j.jpowsour.2022.232397.

Muhammed, Nasiru S., Afeez O. Gbadamosi, Emmanuel I. Epelle, Abdulrahman A. Abdulrasheed, Bashirul Haq, Shirish Patil, Dhafer Al-Shehri, and Muhammad Shahzad Kamal. 2023. "Hydrogen production, transportation, utilization, and storage: Recent advances towards sustainable energy". *Journal of Energy Storage* 73:109207. https://doi.org/10.1016/j.est.2023.109207.

Mulky, Lavanya, Shashwat Srivastava, Thillai Lakshmi, Eashan Reddy Sandadi, Santusti Gour, Noel Abraham Thomas, S. Shanmuga Priya, and K. Sudhakar. 2024. "An overview of hydrogen storage technologies – Key challenges and opportunities". *Materials Chemistry and Physics* 325:129710. https://doi.org/10.1016/j.matchemphys.2024.129710.

Nam, Gi-Dong, Hae-Jin Sung, Dong-Woo Ha, Hyun-Woo No, Tea-Hyung Koo, Rock-Kil Ko, and Minwon Park. 2023. "Design and analysis of cryogenic cooling system for electric propulsion system using liquid hydrogen". *Energies* 16 (1):527. https://doi.org/10.3390/en16010527.

"Origin of the Elements". s. d. Consulté le 21 novembre 2024. https://www2.lbl.gov/abc/wallchart/chapters/10/0.html#toc.

Osman, Ahmed I., Mahmoud Nasr, Abdelazeem S. Eltaweil, Mohamed Hosny, Mohamed Farghali, Ahmed S. Al-Fatesh, David W. Rooney, and Eman M. Abd El-Monaem. 2024a. "Advances in hydrogen storage materials: Harnessing innovative technology, from machine learning to computational chemistry, for energy storage solutions". *International Journal of Hydrogen Energy* 67:1270–94. https://doi.org/10.1016/j.ijhydene.2024.03.223.

Osman, Ahmed I., Mahmoud Nasr, Abdelazeem S. Eltaweil, Mohamed Hosny, Mohamed Farghali, Ahmed S. Al-Fatesh, David W. Rooney, and Eman M. Abd El-Monaem. 2024b. "Advances in hydrogen storage materials: Harnessing innovative technology, from machine learning to computational chemistry, for energy storage solutions". *International Journal of Hydrogen Energy* 67:1270–94. https://doi.org/10.1016/j.ijhydene.2024.03.223.

Osman, Ahmed I., Neha Mehta, Ahmed M. Elgarahy, Mahmoud Hefny, Amer Al-Hinai, Ala'a H. Al-Muhtaseb, and David W. Rooney. 2022. "Hydrogen production, storage, utilisation and environmental impacts: A review". *Environmental Chemistry Letters* 20 (1):153–88. https://doi.org/10.1007/s10311-021-01322-8.

Pathak, Pawan Kumar, Anil Kumar Yadav, and Sanjeevikumar Padmanaban. 2023. "Transition toward emission-free energy systems by 2050: Potential role of hydrogen". *International Journal of Hydrogen Energy* 48 (26):9921–27. https://doi.org/10.1016/j.ijhydene.2022.12.058.

Qureshi, Fazil, Mohammad Yusuf, Salman Ahmed, Moinul Haq, Alhafez M. Alraih, Tarek Hidouri, Hesam Kamyab, Dai-Viet N. Vo, and Hussameldin Ibrahim. 2024. "Advancements in sorption-based materials for hydrogen storage and utilization: A comprehensive review". *Energy* 309:132855. https://doi.org/10.1016/j.energy.2024.132855.

Sage, Valerie, Jim Patel, Philip Hazewinkel, Qurat Ul Ain Yasin, Feng Wang, Yunxia Yang, Karen Kozielski, and Chao'en Li. 2024. "Recent progress and techno-economic analysis of liquid organic hydrogen carriers for Australian renewable energy export – A critical review". *International Journal of Hydrogen Energy* 56:1419–34. https://doi.org/10.1016/j.ijhydene.2023.12.269.

Scarpati, Gabriele, Emmanuele Frasci, Giovanni Di Ilio, and Elio Jannelli. 2024. "A comprehensive review on metal hydrides-based hydrogen storage systems for mobile applications". *Journal of Energy Storage* 102:113934. https://doi.org/10.1016/j.est.2024.113934.

Taipabu, Muhammad Ikhsan, Karthickeyan Viswanathan, Wei Wu, Nikmans Hattu, and A. E. Atabani. 2022. "A critical review of the hydrogen production from biomass-based feedstocks: Challenge, solution, and future prospect". *Process Safety and Environmental Protection* 164:384–407. https://doi.org/10.1016/j.psep.2022.06.006.

Wen, Yuhui, Xingzai Chai, Yunpeng Gu, Wenqi Wu, Wenhao Ma, Jubing Zhang, and Ting Zhang. 2025. "Advances in hydrogen storage materials for physical H2 adsorption". *International Journal of Hydrogen Energy* 97:1261–74. https://doi.org/10.1016/j.ijhydene.2024.11.459.

Xu, W. -C., K. Takahashi, Y. Matsuo, Y. Hattori, M. Kumagai, S. Ishiyama, K. Kaneko, and S. Iijima. 2007. "Investigation of hydrogen storage capacity of various carbon materials". *International Journal of Hydrogen Energy*, ICHS-2005, 32 (13):2504–12. https://doi.org/10.1016/j.ijhydene.2006.11.012.

Xu, Xianxian, Quan Zhou, and Dehai Yu. 2022. "The future of hydrogen energy: Bio-hydrogen production technology". *International Journal of Hydrogen Energy* 47 (79):33677–98. https://doi.org/10.1016/j.ijhydene.2022.07.261.

14 Optimization of Reagent Consumption at the CHICHAOUA Drinking Water Treatment Plant

T. Rachiq and J. Mabrouki

14.1 INTRODUCTION

Water covers 70% of the earth's surface, but unfortunately 97% of this water is salty and undrinkable and is not suitable for irrigation or consumption. Fresh water represents 3% of the total water on our planet. In this small percentage, rivers and lakes represent 0.3%, while all the rest is stored in the polar ice caps. One of the major factors governing the development of human societies is the concern to obtain and maintain an adequate supply of water. Having a sufficient quantity of water, societies dominated the early phases of development. Water is a vital resource, essential to human beings for their survival, health and food; for this, it must be protected, defended and treated. Water is of capital biological and economic importance. It is at the same time a food, possibly a medicine, an industrial, energy and agricultural raw material. Its uses are therefore multiple. The collective or individual supply of drinking water that can be used for food, but also for domestic and hygiene purposes [1]. The required level of water quality obviously depends on these uses and particular attention is paid to the quality of water intended for human consumption [1]. Quality limits concern parameters whose presence in water is likely to generate immediate or longer-term effects on the health of the consumer. They concern both microbiological parameters and chemical substances, such as nitrates, pesticides, certain metals and chlorinated solvents, and by-products of water disinfection. Quality limits are generally based on the recommendations of the World Health Organization (WHO) [2]. Based on our determinations of the quality of raw water from Abou El Abbas Sebti and the possibilities of their treatment by conventional physicochemical processes, omnipresent in the treatment chain for the purification of surface water, the objective of this work is therefore to determine the optimal conditions for treating these waters during the prechlorination, coagulation/flocculation and disinfection phases.

14.2 MATERIALS AND METHODS

14.2.1 Abou EL Abbas Sebti Water Treatment Plant

"TASKOURT", it is located in the territorial commune of Assif El Mal in the province of Chichaoua in the region of Marrakech-Safi, as illustrated in Figures 14.1–14.3.

This station was commissioned on January 25, 2019, with a water treatment capacity of 300 L/s, it mainly supplies the centers of Chichaoua, Imintanoute, Mejjat, Idwirane, Guemassa, Sidi Ghanem, Mzoudia, Mzouda and Sid Mokhtar in the province of Chichaoua and the center of Amizmiz in the province of Tahnaout.

14.2.2 Physicochemical Analysis of Drinking Water

Pour analyser l'eau on effectue plusieurs méthodes d'analyses tel que la mesure de la conductivité, la température, le titre hydrotimétrique, l'oxydabilité.

DOI: 10.1201/9781003647263-14

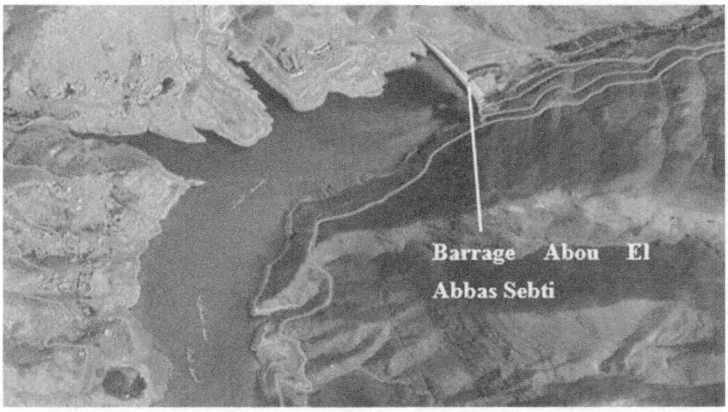

FIGURE 14.1 Satellite image of the location of the Abou El Abbas Sebti Dam.

FIGURE 14.2 Photo of Abou El Abbas Sebti Dam.

FIGURE 14.3 ABOU EL ABBAS SEBTI treatment plant.

14.2.2.1 Hydrogen Potential pH

The pH of water represents its acidity or alkalinity, that is to say the concentration of hydrogen ions (H^+). At pH 7, water is said to be neutral, at a pH lower than 7, water is said to be acidic, and at a pH higher than 7, it is said to be basic. The pH of natural water is linked to the nature of the terrain crossed; it usually varies between 7.2 and 7.6. Generally speaking, very calcareous water has a high pH, and water from terrain poor in limestone or siliceous water has a pH close to 7 and sometimes lower (around 6) [3].

14.2.2.2 Determination of pH

Determining the pH value is based on measuring the potential difference of an electrochemical cell using a suitable pH meter.

The pH of a sample also depends on temperature due to dissociation equilibrium. Therefore, the sample temperature is always indicated with the pH measurement.

14.2.2.3 Temperature

Temperature plays a key role in salt solubility; it influences the amount of dissolved oxygen, viscosity, density, etc. Temperature is measured using a pH meter [4].

The temperature of drinking water should be lower in summer and higher in winter than the air temperature. For drinking water to be thirst-quenching, its temperature must be between 8 and 15°C; between 20 and 25°C, it is not very thirst-quenching. In practice, water temperature has no direct impact on human health. However, a temperature above 15°C promotes the growth of microorganisms in pipes and can intensify odors and flavors. On the other hand, a temperature below 10°C slows down chemical reactions in various water treatments [2].

14.2.2.4 Turbidity

Turbidity is an indicator of the presence of suspended particles in water, either in coarse suspension or in colloidal suspension (colloidal turbidity). It is determined using a turbidimeter, the unit of measurement being the NTU. This device measures the light scattered by suspended particles at a 90° angle to the incident light beam [2].

Turbidity has the same consequences as suspended solids [5].

14.2.2.5 Determination of Turbidity

Turbidity is an indicator of the presence of suspended particles in water. It is determined using a nephelometer. This device measures the light scattered by suspended particles at a 90° angle to the incident light beam.

14.2.2.6 Determination of TA and TAC Alkalinity

Alkalinity is the measure of water's ability to neutralize acids. It corresponds to the presence of bicarbonates (HCO_3^-), carbonates (CO_3^{2-}) and hydroxides (HO^-). It is determined by calculating two titers:

14.2.2.7 Alkalimetric Title TA

The alkalimetric title TA makes it possible to know the content of carbonates and strong bases present in the water [6].

Total alkalimetric title TAC:

The total alkalimetric title TAC is the quantity used to measure the rate of hydroxide ions, carbonates and bicarbonates in water; its unit is the French degree (°f) [7].

14.2.2.8 Determination of Alkalinity (TA-TAC) by Titrimetry

Volumetric dosage by addition of dilute hydrochloric acid solution, in the presence of a colored indicator, necessary to neutralize the volume of water to be analyzed.

14.2.2.9 Determining the TA

- Take 100 mL of water to be analyzed into a flask;
- Add 1 to 2 drops of phenophthalein; a pink color should develop.

If no pink color develops, the pH is zero;

- Then gently pour the acid into the flask using a burette, stirring constantly, until the solution completely discolors.
- Let V' be the volume (in mL) of acid poured to obtain the color change.

If the phenolphthalein is added and no pink color is observed, then the pH is zero.

14.2.2.10 Determination of the TAC

- Use the previously processed sample or the original sample if no color was observed;
- Add 2 drops of the bromocresol green and methyl red solution;
- Titrate with 0.02 N HCl until the greenish-blue color disappears and the pink color appears.

Let V'' be the volume (in mL) of acid added since the beginning of the assay.

$$TA = (V' \times N \times 1000) / V = V' \times N \times 5(\text{in °F}) TAC = (V'' \times N \times 1000) / V = V'' \times N \times 5(\text{in °F}).$$

- V″: volume of HCl added.
- V′: volume of acid solution added.
- N: normality of HCl.
- V: volume of the sample

Hydrometric strength (hardness):

Water hardness is defined by the hydrotimetric strength, which corresponds to the sum of the concentrations of calcium and magnesium salts (Ca^{++} and Mg^{++}). The richer the water is in calcium and magnesium, the harder it is considered [8].

In most cases, hardness is mainly due to calcium and magnesium ions, sometimes with the addition of iron, aluminum, manganese and strontium ions.

$$TH = r\left[Ca^{2+}\right] + r\left[Mg^{2+}\right].$$

The importance of this determination lies in the fact that hard water is not directly usable; a high content is detrimental to cooking vegetables and the proper use of ordinary soaps. Low hardness prevents the formation of a carbonate layer, which often protects pipes against certain corrosion risks; however, high hardness poses a significant risk of scaling in pipes [9].

14.2.2.11 Determination of Hardness

The alkaline earth metals present in the water are caused to form a chelate-type complex by the disodium salt of ethylenediaminetetraacetic acid. The disappearance of the last traces of free elements to be measured is detected by the change in color of a specific indicator, eriochrome black. In a suitably buffered medium to prevent the precipitation of magnesium, the method allows the sum of calcium and magnesium ions to be measured.

Total calcium and magnesium concentration, expressed in milliequivalents per liter
Total hardness in millequivalents/L = 1000 × C V1/V2

C: concentration in millequivalents per liter of the EDTA solution.
V1: volume of EDTA added.
V2: sample volume.

14.2.2.12 Oxidability

Oxidability expresses the amount of oxygen required to oxidize most organic matter and oxidizable mineral salts [10].

Determination of oxidizability:

The permanganate index of water corresponds to the quantity of oxygen expressed in mg/l given up by the permanganate ions (MnO^{4-}) and consumed by the oxidizable materials contained in a liter of water under the conditions defined by this standard.

The permanganate index of the sample, expressed in mg/L of oxygen, is given by the following expression:

$$Ip = V \times 0.8 \left(mg / L\right).$$

Where V: The volume of $KMnO_4$ added.

14.2.2.13 Dissolved Oxygen

Dissolved oxygen concentration is an essential parameter for estimating the amount of organic matter present in water. Dissolved oxygen is measured using an oximeter [11].

The dissolved oxygen content in water rarely exceeds 10 mg/L. It depends on the water's origin: surface waters can contain relatively large quantities close to saturation, while deep waters most often contain only a few mg/L.

Low oxygen levels can lead to organoleptic alteration of the water.

In environments with a low renewal rate (lakes, dams, etc.), the dissolved oxygen content tends to decrease with depth, and anaerobic phenomena can develop at the bottom.

Natural self-purification phenomena in surface waters are a function of dissolved oxygen levels [12].

14.2.2.14 Determination of Dissolved Oxygen

Dissolved oxygen concentration is an essential parameter for estimating the amount of organic matter present in water.

Dissolved oxygen is measured using an oximeter.

14.2.2.15 Conductivity

Electrical conductivity (EC) is a numerical expression of a solution's ability to conduct electric current, which is measured using a conductivity meter. Conductivity measurement allows us to assess the quantity of salts dissolved in water. It is a function of the total ion concentration, their mobility, their valence, their relative concentration and the temperature. Its unit is µsiemens/cm [3]. The relationship between conductivity and mineralization [13]:

Conductivity < 100 µs/cm → Very low mineralization
100 µs/cm < Conductivity < 200 µs/cm → Low mineralization
200 µs/cm < Conductivity < 333 µs/cm → Medium mineralization
333 µs/cm < Conductivity < 666 µs/cm → High medium mineralization
666 µs/cm < Conductivity < 1000 µs/cm → Significant mineralization
Conductivity > 1000 µs/cm → High mineralization

14.2.2.16 Free Residual Chlorine

Free residual chlorine in treated water is determined by the DPD test, which is a test indicator using a visual comparator. This test consists of adding a reagent tablet to the sample, which gives a color that will then be compared to the colors of the comparator, and finally the quantity of residual chlorine present in the water in mg/L is determined [14].

14.2.3 Experimental Protocol for the Treatment of Drinking Water

14.2.3.1 Determining the Break Point

- Dilute the bleach solution with distilled water to obtain a 0.1 g/L solution;
- Prepare 10 brown glass bottles with a volume of 250 mL, which are numbered;
- Add 100 mL of water to be analyzed to each bottle, then add known quantities of chlorine solutions, increasing from bottle to bottle, to obtain precise concentrations of active chlorine;
- Leave the bottles in the dark for 30 minutes after being capped and shaken;
- After exactly 30 minutes (contact time), the residual chlorine is recorded using a comparator by introducing the usual colorimetric reagent DPD;
- Plot the curve of residual chlorine as a function of injected chlorine;
- We determine the break-point: In our test on a sample of raw water presented in Figure 14.4, we found that the dose of chlorine necessary to obtain the break-point is 3.00 mg Cl_2/L.

We can see that the curve is divided into four zones:

- A: Zone of chlorine destruction by mineral compounds (Fe, Mn);
- B: Zone of formation of organic chlorine compounds and chloramines, reduction of monochloramines and dichloramines, formation of trichloramines;
- C: Destruction of chloramines by the addition of additional chlorine, breakpoint (break point), mono-, di- and trichloramines have practically disappeared;
- D: Production of active chlorine, all the added chlorine will be in the form of hypochlorous acid (HClO), but some free chlorine remains.

14.2.3.2 Jar Test Operating Protocol

Coagulation is optimal for the introduction of positive electrical charges corresponding to the cancellation of the zeta potential of the colloids. The purpose of this test is to determine the nature and

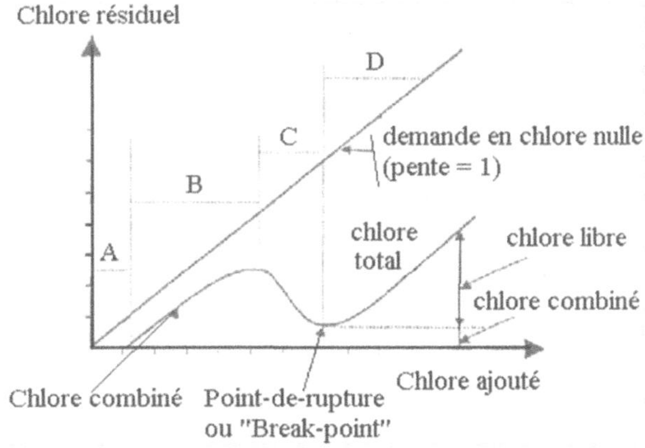

FIGURE 14.4 Break point determination curve.

doses of reagents that allow water to be treated in the best conditions: coagulants and flocculants, pH adjustment, etc. [8].

- Take six beakers containing 1 L of raw water with rapid stirring;
- In the six beakers, add the dose of bleach corresponding to the previously determined chlorine demand and start the stirrers at a speed of approximately 120 rpm;
- In our case, we will perform the Jar Test twice, first with the coagulant $Al_2(SO_4)_3$ and the second with $FeCl_3$; for each test, we will follow these steps:
- Add increasing doses of coagulant and maintain rapid stirring at 120 rpm for 2 minutes; □ For aluminum sulfate, the polyelectrolyte is then added and stirring is reduced to 40 rpm for 20 minutes;
- The time taken for floc to appear (in minutes and seconds) from the start of slow stirring is noted for both tests;
- After slow stirring, the appearance of the flocs is noted:
  0—no floc; 02—slight opalescence; 04—small dots; 06—medium-sized flocs; 08—good floc; 10—excellent;
- The propellers are raised and all portions are allowed to settle for 30 minutes in the beakers;
- During this settling time, the sedimentation rate of the flocs formed is estimated (L: slow; M: medium; R: fast);
- The pH, turbidity, oxidizability and residual chlorine of the supernatant are determined;
- The six beakers from each test are filtered into an Erlenmeyer flask using filter paper placed in the funnels to produce results that are as comparable as possible to those obtained by sand filtration;
- The turbidity and alkalinity are determined for each sample;
- The AL and Fe content are also determined using the kit:

14.2.3.3 Iron Determination by Kit

Method Reference:

This method is described in "Procedures for Spectrophotometric Analyses," January 2021 version.

Procedure:

- In two 20 mL tubes, A measuring tube and B blank;
- Place 6 mL of sample;
- Add 3 drops of Fe;
- Let stand for 3 minutes;
- Slide the colorimetric card to the left until the colors, viewed from the top through the two uncapped tubes, match as closely as possible.
- Read the result in mg/L of Al on the colorimetric card on the inside bottom of the box, directly to the right of the comparator block.

Aluminum determination by kit:

Method reference:

This method is described in "Procedures for Spectrophotometric Analyses", January 2021 version (Figure 14.5).

Procedure:

- Place 5 mL of sample in two tubes, A for measuring and B for blanks;
- Add 1 teaspoon (blue cap) of Al;
- 1 syringe (1.2 mL) of Al;
- 4 drops of Al, then let stand for 7 minutes;
- Slide the colorimetric card to the left until the colors, viewed from the top through the two uncapped tubes, match as closely as possible.

FIGURE 14.5 Jar Test.

- Read the result in mg/L of Al on the colorimetric card on the inside bottom of the box, directly to the right of the comparator block.

14.3 RESULTANTS AND DISCUSSIONS

14.3.1 Optimization of the Prechlorination Step

14.3.1.1 Experimental Concept

Injection Point:

Chlorine is induced in the form of chlorinated water from chlorine gas, according to the following reaction:

$$H2O + CL2 \rightarrow HCLO + HCL$$

The chlorinated water is injected using an injection pipe immersed in a raw water tank at a minimum depth of 3 m [15, 16].

Contact Time:

The desired contact time (raw water with the chlorine solution) should be around 30 minutes before the pre-chlorinated water reaches the coagulation system [17, 18].

Chlorine Demand:

This method determines the amount of chlorine required, also known as the breakpoint, to be injected into the raw water (Table 14.1). This is the breakpoint, or critical point, the point beyond which the NH_4^+ ion is completely oxidized by the chlorine. The introduced chlorine is then in the form of free chlorine [19].

Test Results:

According to the test results as seen in Figure 14.6, the break point is 3 mg/L.

TABLE 14.1
Chlorine Demand Test

N° Flacon	I	II	III	IV	V	VI	VII	VIII	IX	X
Chlore injected (mg/L)	0,5	1	1,5	2	2,5	3	3,5	4	4,5	5
Chlore residual (mg/L)	**0**	**0,2**	**0,7**	**1,2**	**1,5**	**1,5**	**2**	**2,5**	**3**	**3,5**

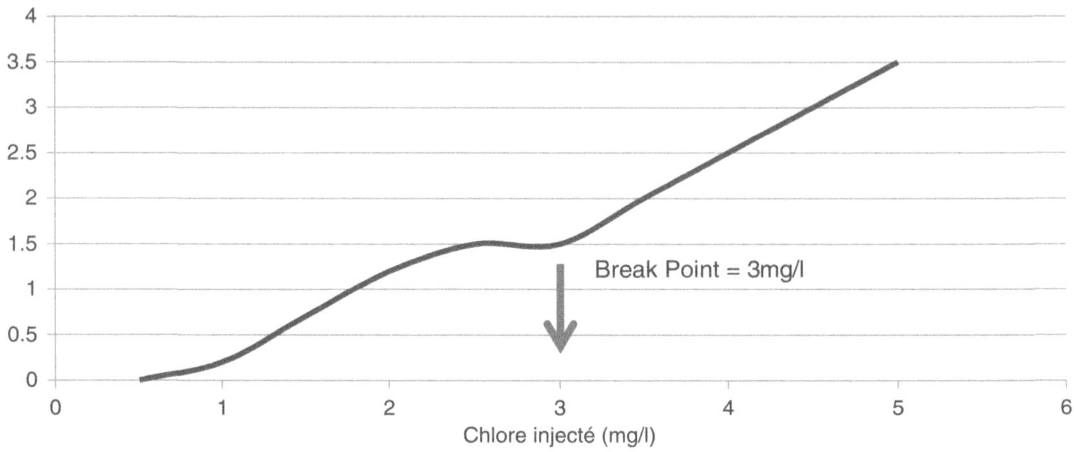

FIGURE 14.6 Curve of the results of the Break Point determination test.

14.3.2 Optimization of the Coagulation/Flocculation Stage

14.3.2.1 Experimental Design

To meet the objective of this study, the Jar Test was chosen as the experimental technique to determine the nature of the reagent and the optimal concentrations required for maximum reduction of water turbidity after simple decantation.

Preparation of Coagulant Solutions:

Preparation of Aluminum Sulfate Solution:

- Using laboratory beakers and a magnetic mixer, weigh 10 g of aluminum sulfate crystal, place it in a container and pour 1 L of water into it;
- Let the mixer run until saturation and the resulting solution is almost as clear as drinking water.

Preparation of Ferric Chloride Solution:

- Weigh 41 g of $FeCl_3$;
- Add to the volumetric flask and fill with distilled water to obtain 100 mL of solution. Characterization of raw water (Table 14.2):

Criteria for selecting the optimal reagent dose (best beaker):

TABLE 14.2
Characterization of Raw Water

Setting	Results
Room temperature (°C)	22
Water temperature (°C)	16
pH	8.03
Conductivity (μS/Cm)	460
Turbidity (NTU)	6.72
Colloidal turbidity (NTU)	2.04
Chlorine demand (mg/L)	**2.50**
TA (meq/L)	**0.00**
TAC (meq/L)	**14.50**

Note: The results of TAC (meq/L) are depicted in bold.

TABLE 14.3
Jar Test Result with $Al_2(SO4)_3$

Beaker No.	I	II	III	IV	V	VI
Aluminum sulfate (mg/L)	5	10	15	**20**	25	**30**
Polymer (mg/L)	0,10	0,10	0,10	**0,10**	0,10	**0,10**
Chlore (mg/L)	3,00	3,00	3,00	**3,00**	3,00	**3,00**
Turbidity of settled water (NTU)	1,62	1,23	0,70	**0,46**	0,48	**0,55**
Free residual chlorine (mg/L)	0,90	0,90	1,00	**1,00**	0,90	**0,90**
Turbidity of filtered water (NTU)	0,56	0,42	0,17	**0,10**	0,19	**0,22**
pH settled water	7,90	7,73	7,68	**7,60**	7,54	**7,45**
Conductivity (us/cm)	458	458	458	**458**	458	**458**
Residual aluminum (mg/L)	—	—	0,20	**0,00**	—	—
TAC (meq/L)	—	—	—	**14**	—	—
Appearance of the flakes	04	06	08	**10**	10	**08**
Sedimentation rate	L	M	M	R	R	M

The objective of the Jar Test is to determine the optimal doses for producing drinking water that exhibits the following characteristics [10]:

- Turbidity of the settled water < 5 NTU;
- Turbidity of the water filtered through the paper < 0.50 NTU;
- Settling pH between 6.50 and 8.50;
- Residual aluminum < 0.20 mg/L;
- Residual iron < 0.30;
- Conductivity < 2700 us/cm;
- Residual chlorine in the order of 1 mg/L;
- Minimum coagulant dose.

Jar Test Results with Aluminum Sulfate and Ferric Chloride (Table 14.3):

Results Obtained: Coagulant of Aluminum Sulfate with the Polymer.

Observation:

According to the criteria for choosing the best beaker, the one that perfectly meets these criteria is beaker No. IV.

Results obtained: Ferric chloride coagulant (Table 14.4):

TABLE 14.4
Jar Test Results with $FeCl_3$

Becher N°	I	II	III	IV	V	VI
Ferric chloride (mg/L)	15	17	**19**	21	23	**25**
Polymer (mg/L)	—	—	—	—	—	—
Chlore (mg/L)	3,00	3,00	**3,00**	3,00	3,00	**3,00**
Turbidity of filtered water (NTU)	0,96	0,72	**0,41**	0,48	0,56	**0,64**
Chlore résiduel libre (mg/L)	0,9	0,9	**0,9**	0,9	0,9	**0,9**
Turbidité des eaux filtrées (NTU)	0,54	0,32	**0,11**	0,15	0,17	**0,23**
pH settled water	7,69	7,66	**7,60**	7,56	7,58	**7,58**
Conductivity (us/cm)	458	458	**458**	458	458	**458**
Fer résiduel (mg/L)	—	—	**0,00**	0,10	—	—
TAC (méq/L)	—	—	**14**	—	—	—
Appearance of the flakes	06	08	**10**	10	08	**08**
Sedimentation rate	M	R	R	R	R	R

Observation:

According to the criteria for choosing the best beaker, the one that perfectly meets these criteria is beaker No. III.

14.3.2.2 Interpretation of the Results

Jar test with aluminum sulfate:

Turbidity is a parameter that influences the proper functioning of the coagulation process. Therefore, based on the results obtained previously, we note that the higher the aluminum sulfate dose, the greater the decrease in turbidity for both the settled water (from 1.62 to 0.46 NTU) and the filtered water (from 0.56 to 0.10).

However, the turbidity remains high for the first beakers due to the presence of suspended particles in the supernatant, and for the last beakers, the coagulant dose is excessive, resulting in destabilization of the colloids.

Therefore, based on the test, we observe that the best beaker is beaker No. IV:

- The sedimentation rate is rapid;
- The size of the flocs formed is larger when the water is less turbid;
- Conductivity < 2700 us/cm ;
- Decreased turbidity of filtered water <0.50 NTU;
- Lack of residual aluminum.

Jar test with ferric chloride:

According to the results of the test, we note that the higher the dose of ferric chloride, the greater the decrease in turbidity of both the settled and filtered water (from 0.96 to 0.41 NTU for the settled water and from 0.54 to 0.11 NTU for the filtered water).

The test therefore shows that the best beaker is beaker No. III:

- The sedimentation rate is fast;
- The size of the flocs formed is excellent;
- Decreased turbidity of filtered water < 0.50 NTU;
- Conductivity < 2700 us/cm ;
- Lack of residual iron.

Comparison between aluminum sulfate and ferric chloride:

From the tables of results obtained, we note:

- Turbidity decreases to a minimum, which is the optimal dose;
- The good condition of the flocs obtained during the tests on aluminum sulfate and ferric chloride, as they are almost the same size;
- The conductivity for both coagulants remains within the standards;
- The color of the water for the test with ferric chloride is more intense than that of the water with aluminum sulfate, due to iron residue, which gives a yellow or brown color that can pose a health risk;
- The values of residual aluminum and iron after filtration are acceptable and do not exceed the standard limits. (Aluminum < 0.2 mg/L and Iron < 0.3 mg/L);
- Ferric chloride is corrosive and can cause problems for plant equipment, wildlife and humans due to its toxicity.

Therefore, based on this study, we can conclude that, technically and qualitatively, aluminum sulfate is the most suitable coagulant for the coagulation/flocculation process.

14.4 CONCLUSION

This work, carried out at the water treatment plant of the "Abou El Abbas Sebti Dam", allowed me to both apply the scientific concepts we acquired during our university training and familiarize myself with the professional world. Indeed, the topic covered in this internship is a study on the optimization of the reagents used in the treatment of this water, specifically the prechlorination, coagulation/flocculation and disinfection processes, which were clearly explained given their importance in the latter. The choice of reagents is the first step in this study. For prechlorination and disinfection, chlorine gas was chosen, and for coagulation/flocculation, aluminum sulfate plus polymer was chosen. The next step was the quantification of the reagents, for chlorine it was based on the chlorine demand, which is determined by the break point, and for aluminum sulfate it was based on the Jar Test.

REFERENCES

1. Rachiq, T., Abrouki, Y., Mabrouki, J., Samghouli, N., Fersi, C., Rahal, S., & El Hajjaji, S. (2021). Evaluation of the efficiency of different materials to remove specific pollutants from landfill leachate. Desalination Water Treat, 238, 240–250.
2. Mabrouki, J., Azrour, M., Fattah, G., Dhiba, D., & El Hajjaji, S. (2021). Intelligent monitoring system for biogas detection based on the internet of things: Mohammedia, Morocco City Landfill case. Big Data Mining and Analytics, 4(1), 10–17.
3. Rachiq, T., Mabrouki, J., El Hajjaji, S., & Rahal, S. (2022). Simulation of the Treatment Performance of a Purification Plant for a Dairy Effluent. In IoT and Smart Devices for Sustainable Environment (pp. 19–27). Cham: Springer International Publishing.
4. Rachiq, T., Samghouli, N., Mabrouki, J., El-Moustaqim, K., Rahmani, M., Al-Jadabi, N., & Rahal, S. (2022). Study and modeling of dairy effluents treatment using design of experiments methodology. Desalination and Water Treatment, 261, 131–140.
5. Mabrouki, J., Azrour, M., & Hajjaji, S. E. (2021). Use of internet of things for monitoring and evaluating water's quality: A comparative study. International Journal of Cloud Computing, 10(5–6), 633–644.
6. El Alouani, M., Aouan, B., Rachdi, Y., Alehyen, S., El Herradi, E. H., Saufi, H., … Barka, N. (2024). Porous geopolymers as innovative adsorbents for the removal of organic and inorganic hazardous substances: A mini-review. International Journal of Environmental Analytical Chemistry, 104(16), 4784–4796.
7. Eloutassi, N., Louasté, B., Boudine, L., & Chaouch, M. (2013). Traitement du lactosérum pur issu de l'industrie laitière par la méthanisation. Rev ScienceLib Environnement, 5, 1–11.
8. Boguniewicz-Zablocka, J., Klosok-Bazan, I., & Naddeo, V. (2019). Water quality and resource management in the dairy industry. Environmental Science and Pollution Research, 26, 1208–1216.
9. Rachiq, T., Abrouki, Y., Mabrouki, J., Samghouli, N., Fersi, C., Rahal, S., & El Hajjaji, S. (2021). Evaluation of the efficiency of different materials to remove specific pollutants from landfill leachate. Desalination and Water Treatment, 238, 240–250.
10. Belghyti, D. et al. (2009). Caractérisation physico-chimique des eaux usées d'abattoir en vue de la mise en oeuvre d'un traitement adéquat: Cas de Kénitra au Maroc. Afrique Science: Revue Internationale des Sciences et Technologie, 5, 199–216.
11. Ghizlane, F., Mabrouki, J., Ghrissi, F., & Azrour, M. (2022). Proposal for a high-resolution particulate matter (PM10 and PM2. 5) capture system, comparable with hybrid system-based internet of things: Case of quarries in the Western Rif. Morocco. Pollution, 8(1), 169–180.
12. Mabrouki, J., Benbouzid, M., Dhiba, D., & El Hajjaji, S. (2022). Simulation of wastewater treatment processes with bioreactor membrane reactor (MBR) treatment versus conventional the adsorbent layer-based filtration system (LAFS). International Journal of Environmental Analytical Chemistry, 102(19), 7458–7468.
13. Benchrifa, M., Elouardi, M., Fattah, G., Mabrouki, J., & Tadili, R. (2023). Identification, Simulation and Modeling of the Main Power Losses of a Photovoltaic Installation and Use of the Internet of Things to Minimize System Losses. In Advanced Technology for Smart Environment and Energy (pp. 49–60). Cham: Springer International Publishing.
14. Le Moal M., et al. (2019). Eutrophication: A new wine in an old bottle? Science of the Total Environment, 651, 1–11.

15. Al-Jadabi, N., Laaouan, M., El Hajjaji, S., Mabrouki, J., Benbouzid, M., & Dhiba, D. (2023). The dual performance of Moringa oleifera seeds as eco-friendly natural coagulant and as an antimicrobial for wastewater treatment: A review. Sustainability, 15(5), 4280.
16. CCTG, version A. (Octobre 2012). Cahier des clauses techniques générales relatives au traitement eau potable.
17. Mgalaa, S., Mabrouki, J., Elouardi, M., El Azzouzi, L., Moufti, A., El Hajjaji, S., … El Belghiti, M. A. (2022). Study and evaluation of the degradation of procion blue dye by the ozonation method: Parametric and isothermal study. Nanotechnology for Environmental Engineering, 7(3), 691–697.
18. Qualité physico-chimique et chimique des eaux de surface. (Novembre 2005). Institut Bruxellois pour la Gestion de l'Environnement.
19. Méthodes d'analyse révision 5, 2016-02-03 centre expertise en analyse environnemental du Québec.

15 Seawater Desalination Driven by Renewable Energy

Toward a Zero-Waste Treatment Framework

D. Azdem, A. Fatni, J. Mabrouki, M. El Mouedden, A. Ghailane, and H. Taghlaoui

15.1 INTRODUCTION

Water has a crucial role to play in maintaining ecological well-being and balance of the modern world. Oceans account for 97% of the total amount of water available on Earth [1, 2]. Natural sources of water are generally classified into surface and groundwater. Surface water is the main source of fresh water; recent drought conditions have drastically decreased the availability for this resource [3, 4]. Furthermore, groundwater resources, generally considered more stable are also impacted because of the constant reduction in global rainfall [5, 6]. It is estimated that there are currently in excess of 1 billion people across the globe for whom the procurement of drinking water that is safe is not a viable option [7]. This leads to more than 5 million deaths per year caused by waterborne illness, 4 million of which are children [8]. The rapid increase in population and the rising demands for water has resulted in increasing concerns about water security [9, 10]. Although it is widely thought that fresh water is abundant, accessibility to freshwater is not easy, particularly in areas where there is a water shortage [11, 12]. Seawater, though abundant, requires extensive processing before it can be consumed. Desalination of seawater as well as brackish waters could provide potable drinking water to over 75 million people around the world [13, 14]. The World Health Organization (WHO) has found that 25% of the people live in areas without the infrastructure required for getting and removing waters from natural reservoirs and waterways [15]. This is especially the case in Africa and also the Arab world, where 7% of the population globally has access to only 1.4% of freshwater-based renewable resources. Desalination technology, especially ones that treat brackish and seawater, are increasingly essential to meet the ever-growing demand or water that is potable [16]. This study will establish the best solutions available to combat the worldwide water shortages, with a focus on improving seawater reverse osmosis (SWRO) systems that are powered by renewable energy sources that produce zero waste. Simulating the operational dynamics of SWRO plants is challenging due to the variability of conditions, including seasonal fluctuations and membrane degradation over time [17]. Engineers and investors need to be aware of the analysis of design parameters in addition to operating and capital costs because they are crucial to efficient execution and planning. One of the major challenges to SWRO desalination is the management of waste brine [11, 18], which has significant amounts of organic matter–based metallics and toxins [3]. A sustainable and environmentally friendly disposal of brine requires cost-effective, efficient technologies to limit ecological harm [19]. Additionally, the economic benefits can be derived from brine using advanced recovery methods that extract valuable minerals and salts, such as sodium chloride, bromine of industrial grade lithium, as well as potassium, boron and rubidium, in addition to magnesium sulfate and calcium [20]. The primary goal of this study is to examine the latest developments in desalination techniques based upon renewable energy sources, and the

DOI: 10.1201/9781003647263-15

probability that these advancements will result in zero waste. Particular attention will be paid to the efficiency of recycling and energy usage.

15.2 RESEARCH METHODS

The objective of this study was to establish the most significant issues, including worldwide water shortage, peak of seawater, the shortcomings of current desalination methods, and feasible solutions to these problems. A system was devised for the purpose of sourcing and evaluating academic literature, with the objective of ensuring the inclusion of quality studies. The evaluation process involves peer-reviewed, reputable databases, with particular emphasis on South African Department of Higher Education and Training (DHET) and Scopus indexes. A literature review was conducted within these databases, marking the initiation of the implementation process. The chosen data were cross-referenced to verify reliability and consistency. In this phase, established methodological frameworks were adhered to in order to ensure exact identification, classification and critical evaluation of the data. The information extracted was coded and compared to determine connections between the studies, which helped to provide a complete understanding of the issues and solutions. In the reporting phase, results were summarized and presented as well-structured qualitative summaries, accompanied with visual aids, such as tables and figures.

15.3 SCARCITY OF WATER RESOURCES

Water scarcity is caused by the complex interplay between factors that include increasing population and environmental pollution, climate change and inefficient resource management and the increasing rate of global usage and withdrawal [21, 22]. Additionally, the uneven geographic and social-economic distributive distribution of the fresh water resource causes this problem [23]. Rapid urbanization as well as industrialization and the increasing use of farming practices continue to put a significant strain on water supplies already strained. Without efficient strategies for water conservation and sustainable management of water resources, the issue of water scarcity will remain a major global issue. Around 80% of the world's population could be exposed to extreme water pressure, while approximately 25% of people already suffering from extreme water scarcity conditions [24]. In addition, climate-related events like drought and desertification, as well as global warming are likely to intensify the problem and could lead to water shortages in areas that were not previously confronted with similar problems. The increasing gap between the demand for water in the world and the supply available has reached the point that there are some regions where it poses a serious risk to the health of people and even survival. The depletion of freshwater resources is a growing concern due to the fact that only a small portion of Earth's fresh water can be used by humans. The remaining is mostly stored in groundwater reservoirs that are mostly brackish or saline. This is a sign of the dominance of marine water bodies within the earth's hydrological system, and just a tiny part of the water being available to human consumption. In response to the huge amount of saline water that accounts for nearly 90% of world's supply, desalination is emerging as an effective and widely used method, particularly in regions in which freshwater resources are being depleted or are being overused. Desalination has been recognized as a viable option for protecting water resources and has emerged as a top element in the global agenda for sustainable development [25].

15.4 SEAWATER AS A STRATEGIC WATER RESOURCE

Access to sufficient water resources is crucial to both economic growth and the health of communities [26]. While water covers a large part of the planet's surface, just 3% of the total is freshwater, and it can be used directly to meet the needs of humans and ecosystems [1]. This makes seawater inedible for consumption directly or the majority of practical applications, especially in areas that are already suffering from water stress. To overcome this issue, desalination technologies are vital, as they allow the conversion of saline waters into fresh water that can be used for drinking as well as other purposes [27].

15.5 DESALINATION TECHNOLOGIES

To increase the supply of potable water sources, many methods for desalinating seawater are in development and research [28, 29]. The methods are defined by the physical principles they use and include de-salination using thermal energy in addition to membrane-based desalination [30]. The thermal methods depend on the change in the phase of water, mostly by condensation and evaporation to remove salts that are dissolved [27], while membrane methods employ selective filtering to remove fresh water from sources of saline [31]. Thermal desalination includes a range of methods, including multiple-stage flash distillation (MSF) and the multiple-effect distillation (MED) as well as freeze separation and solar still distillation and systems for vapor compression [32]. However, membrane technologies typically comprise reverse osmosis (RO) and electrodialysis (ED), both made up of semi-permeable membranes for removing salts and impurities out of ocean water (Figure 15.1).

An exciting development in this area is forward the process of RO, offers potential gains in energy efficiency as well as operational costs [33]. The process involves the transfer of water over an impermeable membrane that is driven by an osmotic pressure gradient that allows extracting freshwater out of brackish or saline water. This method is attractive because of its low energy consumption compared to traditional methods; however, more research is required to create strong RO membranes as well as efficient methods for removing solutes before the

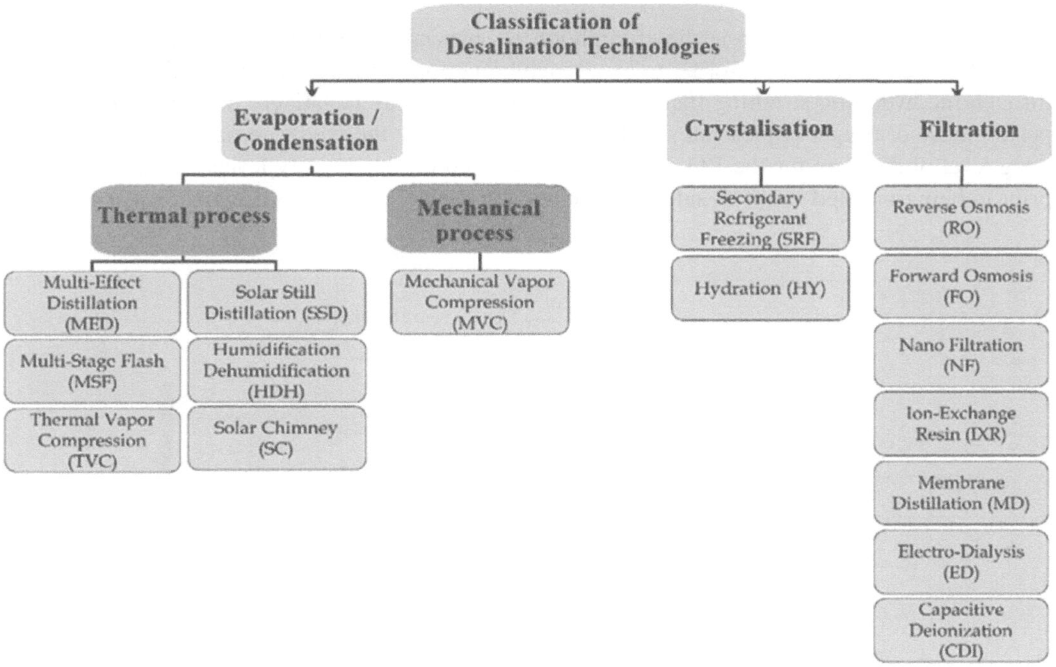

FIGURE 15.1 Classification of desalination processes according to functional principles.

technique can be adapted from research to practical applications. Research is currently focused on improving the efficiency of membrane flux and the development of innovative draw solutions that can be reused without requiring any energy input [34]. Parallel to this significant progress has been made in the integration of the solar power into desalination technology. Solar-powered systems, like solar stills, MSF, MED and humidification dehumidification (HDH) electrodialysis as well as adsorption-based systems are being studied to ensure sustainable production of fresh water [35]. In the list of the options are the solar photovoltaic (PV)-powered RO systems are gaining popularity because of the commercial viability and versatility of both RO and PV components [36].

15.5.1 Membrane-Based Desalination Methods

In the variety of methods employed for desalination of seawater, membrane-based processes are the most popular [37]. They are extremely adaptable, offering solutions to both brackish water treatment as well as seawater desalination in a variety of conditions. Desalination systems are classified by the type of components used and the type of energy required. In membrane desalination, the primary principal is particular permeability of a semi-permeable membrane that lets water molecules pass through, while also retaining some impurities, including salts. The two principal approaches within this class are forward osmosis (FO) and RO, both of which are the basis of classifying membrane desalination processes [38]. Osmosis is the term used to describe the natural flow of water from an area with a lower concentration of solute to one with greater concentration over the semi-permeable membrane (Figure 15.2). In the process of osmosis, the natural flow is reversed by applying pressure externally on the side with higher concentration of solutes and causing water molecules to travel to the other direction toward the side with lower concentration.

15.5.2 Reverse Osmosis Process

In RO systems, the first step is the elimination of suspended solids as well as particulates from brackish or seawater water in order to avoid membrane fouling [39]. This is usually accomplished through the hydraulic straining that filters out debris that may negatively impact the membrane's performance or trigger unpleasant smells. After this process, the feedwater is pumped to the operating level of pressure for the RO unit that varies based on the salinity levels of the source water prior to being injected in the desalination system. In the process of desalination, water molecules

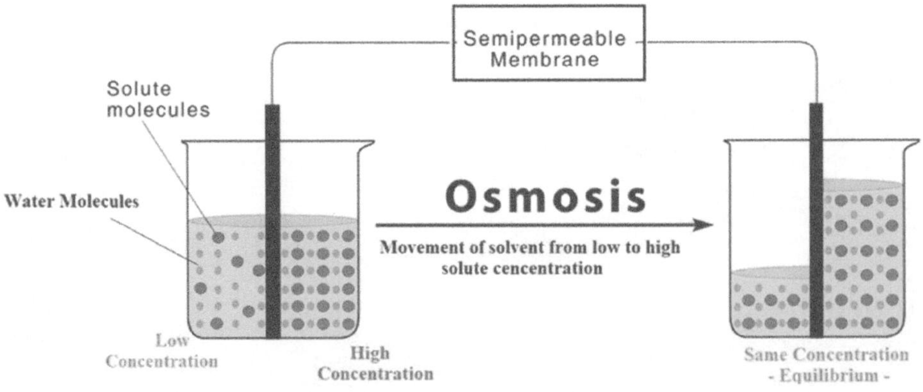

FIGURE 15.2 Schematic representation of osmosis across a semi-permeable membrane.

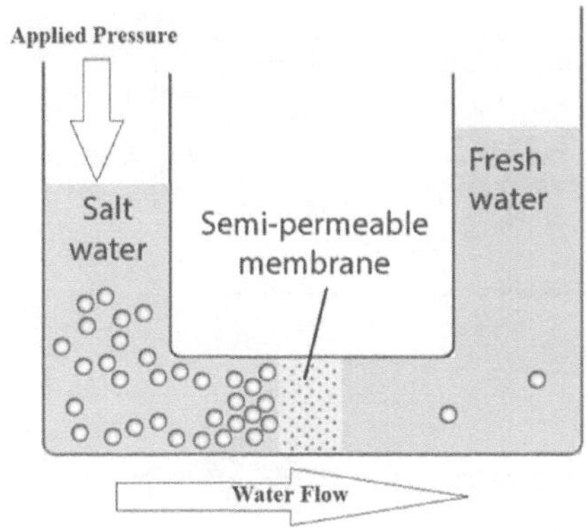

FIGURE 15.3 Representation of the reverse osmosis process.

pass through the semi-permeable membrane, creating a permeate flow down the other side whereas impurities and salts are kept (Figure 15.3).

15.5.3 Thermal Desalination Process

The process of thermal desalination involves applying the heat of saline water, usually seawater or other sources of brackish water to cause the evaporation which results in steam which then condenses into fresh water [40]. The process usually incorporates techniques to reduce pressure to reduce the temperature of boiling water and thus less energy is needed for evaporation. Through the manipulation of temperature and pressure conditions the thermal desalination system is capable of effectively getting rid of salts and other solids dissolved in water [41].

15.6 RENEWABLE ENERGY POWER SOLUTION

Desalination of seawater is extremely energy-intensive and expensive. This procedure poses challenges like high energy use as well as environmental pollution due to burning fossil fuels [42]. The use of fossil fuels for the supply of fresh water is ineffective and damaging to the climate due to the increasing concerns over fossil fuel–related environmental concerns in addition to carbon dioxide emissions. In addition, due to its energy-intensive nature desalination of water poses a serious risk to the environment of pollution. To reduce this risk and limit the environmental impact of desalination, plants require an energy source that produces low emissions and they must be economical [43]. This can be accomplished through the use of renewable sources of energy like solar photovoltaic, wind geothermal or thermal energy sources [44] (Figure 15.4). Just 3% of the desalination capacity in the world is currently accounted for through renewable-powered facilities and plants [45]. Photovoltaic (PV) systems transform sunlight in direct current (DC) electricity through solar cells made of semiconductors. They may either store the generated energy in batteries or transfer it to inverters to generate AC to be used in operation. RO units can be powered directly by PV systems that comprise components like battery storage devices as well as charge control devices to regulate the flow of energy.

One of the challenges for solar-powered systems is the variations in sunlight. This makes the creation of membranes that are battery-free is a crucial area for research [46]. Field tests have shown that solar-powered systems are able to provide superior quality desalinated water but the results can differ based upon the kind of membrane that is in use. The main benefit that comes with RO technology is the ability to alter the amount of water produced based on the energy available and allowing for the operation to be flexible in response to fluctuating energy supply.

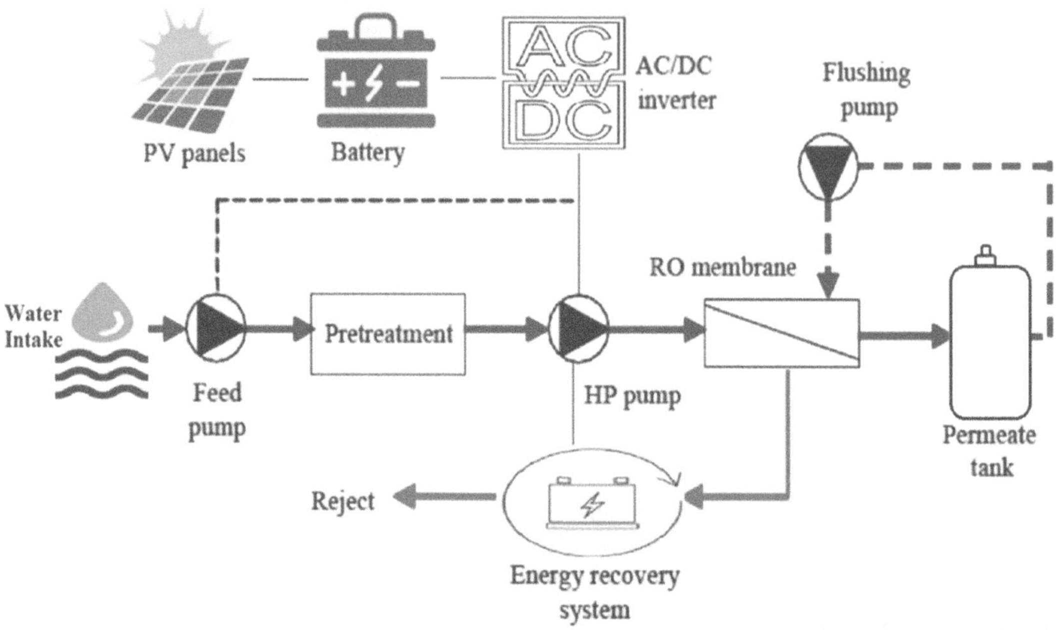

FIGURE 15.4 Schematic diagram of a solar-powered reverse osmosis desalination system with energy recovery.

15.7 SUSTAINABLE SEAWATER DESALINATION WITH ZERO LIQUID DISCHARGE

The technology of seawater desalination is classified as an investment with low returns because of the high operating expenses and low permeate recovery flux, and the high amount of system produced brine concentrate [47]. Separating salt concentrations and valuable metals from brine prior to disposal significantly reduces the cost of producing clean water through SWRO and also eliminates the negative environmental impacts of dumping brine concentrates with no salt and metals in the sea [27]. Brine has been managed with different strategies, such as evaporation ponds, deep-well injectors, discharge to the ocean/sea and the outflow of sewage. These methods are no longer practical because of rising costs for management as well as stricter discharge regulations and environmental hazards [11, 48]. Therefore, new solutions for managing the brine concentrates needed urgently to limit environmental impact while preserving economic viability.

To limit the harmful consequences of discharges from brine, managing brine concentrate is now an absolute necessity that requires creation of feasible and effective technologies. The search for environmentally technologically, economically and technically efficient methods to control brine concentration is becoming more complicated. The complexity and complexity in this problem increases, it is clear that traditional approaches to desalination and disposal of brine are not sustainable economically. Therefore, there is a demand for new techniques, like those that are based in minimal liquid discharge (MLD) and ZLD systems to tackle these issues with a method that reduces the environmental impact while increasing the efficiency of energy and resource recovery.

15.8 CONCLUSION

This study demonstrates the fact that to reverse the process of RO, when successfully integrated into a larger system, is an extremely efficient way to achieve ZLD in desalination processes. It is the most popular membrane-based technology to treat of both seawater and brackish water RO systems can effectively produce large amounts of potable water in the most cost-effective way. But, one of the major drawbacks is their large energy consumption. The study also examined the potential economic benefits of powering RO systems using sustainable energy sources in order to improve the sustainability of these systems and meet ZLD's goals. In the longer-term managing brine is a crucial aspect of sustainable managing of brine is crucial to safeguarding ecosystems and ensuring the integrity of the environment. Implementing technology that minimize the volume of brine and facilitate the recovery of resources is crucial for sustainable control of brine. While residual salts could need to be properly disposed of however, the ecological impact is much less if they are taken care of. The latest technologies show promising possibilities in reducing the volume of brine as well as enabling recovering of valuable metals and salts that can be re-sold or recycled and offset operational costs. Integrated method which combines water recovery and the extraction of chemicals is the key to turning brine, which is a waste product, into a beneficial resource. The end goal is to address the global water crisis requires an extensive strategy that includes conservation of water resources and sustainable management, as well as investment in infrastructure and strong policies. These measures can help address the root causes of problems with water and aid in the security of water for the future generation.

REFERENCES

1. G. Fahdi, D. Azdem, A. Lamchaimech, M. Benrhanem, J. Mabrouki, and S. El Hajjaji, "Analysis of the impact of climate change on water resources: Case of The Tensift Basin (Morocco)," *Water Conservation and Management*, vol. 8, no. 4, pp. 461–465, 2024. doi: 10.26480/wcm.04.2024.461.465.
2. M. M. Mekonnen, M. M. Kebede, B. W. Demeke, J. A. Carr, A. Chapagain, C. Dalin, P. Debaere, L. Marston, C. Ray, L. Rosa, and L. Zhuo, "Trends and environmental impacts of virtual water trade," *Nature Reviews Earth & Environment*, vol. 5, no. 12, pp. 890–905, 2024. doi: 10.1038/s43017-024-00605-2.
3. D. Azdem, J. Mabrouki, A. Moufti, S. El hajjaji, and A. Fatni, "Assessment of heavy metal contamination in seawater in Agadir coastline, Morocco," *Desalination and Water Treatment*, vol. 317, p. 100129, 2024. doi: 10.1016/j.dwt.2024.100129.
4. P. K. Singh *et al.*, "Critical review on toxic contaminants in surface water ecosystem: Sources, monitoring, and its impact on human health," *Environmental Science and Pollution Research*, vol. 31, no. 45, pp. 56428–56462, 2024.
5. A. Fatni *et al.*, "Characterization and mitigation strategies for inorganic scaling in reverse osmosis system treating brackish groundwater," *Sustainable Chemistry One World*, vol. 5, p. 100035, 2025. doi: 10.1016/j.scowo.2024.100035.
6. P. Huang, M. Hou, T. Sun, H. Xu, C. Ma, and A. Zhou, "Sustainable groundwater management in coastal cities: Insights from groundwater potential and vulnerability using ensemble learning and knowledge-driven models," *Journal of Cleaner Production*, vol. 442, p. 141152, 2024.
7. C. O. Okafor, U. I. Ude, F. N. Okoh, and B. O. Eromonsele, "Safe Drinking Water: The Need and Challenges in Developing Countries," in *Water Quality-New Perspectives*, IntechOpen, 2024.
8. H. Needs, "Interventions to mitigate burdens of waterborne and water-related disease," *Textbook of Children's Environmental Health*, vol. 378, 2024.
9. G. Fahdi, J. Mabrouki, A. Lamchaimech, D. Azdem, M. Benrhanem, and S. E. Hajjaji, "Impacts of climate change on groundwater quality and recharge in the Tensift Basin, Morocco," *Water Conservation and Management*, vol. 8, no. 4, pp. 466–474, 2024. doi: 10.26480/wcm.04.2024.466.474.
10. D. Azdem, J. Mabrouki, and S. El Hajjaji, "MarocFarm: A Digital Revolution for Agriculture in Morocco," in *Circular Economy Applications in Energy Policy*, IGI Global Scientific Publishing, 2025, pp. 137–152.
11. D. Azdem, J. Mabrouki, and S. El Hajjaji, "Methods and Technologies for the Treatment and Disposal of Brine in Seawater Desalination Plants," in *Advances in Environmental Engineering and Green Technologies*, J. Mabrouki, and M. Azrour, Eds., IGI Global, 2024, pp. 79–87. doi: 10.4018/979-8-3693-3807-0.ch007.
12. D. Azdem, A. Fatni, J. Mabrouki, and S. El Hajjaji, "The desalination strategy and economic integration of Morocco: Water scarcity, prospects, and challenges," *Circular Economy Applications in Energy Policy*, pp. 257–268, 2025.
13. K. El-Moustaqim, J. Mabrouki, M. Benchrifa, D. Azdem, and D. Hmouni, "Microalgae-Based Food Additives for Improved Shelf Life and Nutritional Value," in *Biology and Life Sciences Forum*, MDPI, 2025, p. 42. Consulté le: 23 mars 2025. [En ligne]. Disponible sur: https://www.mdpi.com/2673-9976/40/1/42.

14. D. Azdem, J. Mabrouki, A. Moufti, S. E. Hajjaji, and A. Fatni, "Present Situation and Environmental Impact of Seawater Desalination Plants in Morocco," in *Technical and Technological Solutions Towards a Sustainable Society and Circular Economy*, J. Mabrouki, and A. Mourade, Eds., Cham: Springer Nature Switzerland, 2024, pp. 581–592. doi: 10.1007/978-3-031-56292-1_46.
15. B. D. Richter *et al.*, "Lost in development's shadow: The downstream human consequences of dams," *Water Alternatives*, vol. 3, no. 2, p. 14, 2010.
16. Z. S. Ali, I. M. Abdul Hameed, and M. M. Al-Fiqi, "Environmental impacts of seawater desalination technologies in the gulf states - A review," *Zagazig Journal of Agricultural Research*, vol. 51, no. 5, pp. 1001–1019, 2024.
17. S. Helali, S. Albalawi, and N. Bel Hadj Ali, "Harnessing physics-informed neural networks for performance monitoring in SWRO desalination," *Water*, vol. 17, no. 3, p. 297, 2025.
18. J. Wu, and E. M. Hoek, "Current opportunities and challenges in membrane-based brine management," *Current Opinion in Chemical Engineering*, vol. 47, p. 101079, 2025.
19. L. Ebrahimi, and E. Ebrahimi Sarindizaj, and D. R. Arab, "Eco Environmentally Friendly Sustainable Desalinated Water Production: Turning a Challenge into an Opportunity," *Davood Reza, Eco Environmentally Friendly Sustainable Desalinated Water Production: Turning a Challenge into an Opportunity.*
20. P. Kumari, Y. S. Chang, G.-J. Witkamp, J. Vrouwenvelder, L. F. Vega, and L. F. Dumée, "Brine valorization through resource mining and CO2 utilization in the Middle East–A perspective," *Desalination*, p. 117598, 2024.
21. I. Tinazzi, "Water scarcity, migrations and climate change: An assessment of their Nexus," 2024. https://unitesi.unive.it/handle/20.500.14247/16549.
22. S. Pacheco-Treviño, and M. G. F. Manzano-Camarillo, "Review of water scarcity assessments: Highlights of Mexico's water situation," *Wiley Interdisciplinary Reviews: Water*, p. e1721, 2024.
23. M. Benchrifa, J. Mabrouki, M. Elouardi, M. Azrour, and R. Tadili, "Detailed study of dimensioning and simulating a grid-connected PV power station and analysis of its environmental and economic effect, case study," *Modeling Earth Systems and Environment*, vol. 9, no. 1, pp. 53–61, 2023.
24. M. M. Mekonnen, and A. Y. Hoekstra, "Four billion people facing severe water scarcity," *Science Advances*, vol. 2, no. 2, p. e1500323, 2016.
25. S. Jain, A. Srivastava, L. Khadke, U. Chatterjee, and A. Elbeltagi, "Global-scale water security and desertification management amidst climate change," *Environmental Science and Pollution Research*, vol. 31, no. 49, pp. 58720–58744, 2024.
26. P. H. Gleick, "Water in crisis: Paths to sustainable water use," *Ecological Applications*, vol. 8, no. 3, pp. 571–579, 1998.
27. S. M. J. S. Sabour, and B. Ghorashi, "A comprehensive review of major water desalination techniques and mineral extraction from saline water," *Separation and Purification Technology*, vol. 349, p. 127913, 2024.
28. R. K. Guduru, R. Singh, and R. K. Vij, "Desalination of Seawater and Related Technologies: a Step Towards Achieving Sustainable Development Goals," in *Water, The Environment, and the Sustainable Development Goals*, Elsevier, 2024, p. 305–325.
29. J. Mabrouki, M. Benbouzid, D. Dhiba, and S. El Hajjaji, "Simulation of wastewater treatment processes with bioreactor membrane reactor (MBR) treatment versus conventional the adsorbent layer-based filtration system (LAFS)," *International Journal of Environmental Analytical Chemistry*, vol. 102, no. 19, pp. 7458–7468, 2022.
30. S. Xu et J. F. Torres, "Thermodiffusive (de) salination in multichannel devices: A membrane less all-liquid thermal separation method confirmed with seawater and brine."
31. S. Aziz *et al.*, "A comprehensive review of membrane-based water filtration techniques," *Applied Water Science*, vol. 14, no. 8, p. 169, 2024.
32. I. Bencheikh, K. Azoulay, J. Mabrouki, S. El Hajjaji, A. Dahchour, A. Moufti, and D. Dhiba, "The adsorptive removal of MB using chemically treated artichoke leaves: Parametric, kinetic, isotherm and thermodynamic study," *Scientific African*, vol. 9, p. e00509, 2020.
33. I. Nurjanah, T.-T. Chang, S.-J. You, C.-Y. Huang, and W.-Y. Sean, "Reverse osmosis integrated with renewable energy as sustainable technology: A review," *Desalination*, vol. 581, p. 117590, 2024. doi: 10.1016/j.desal.2024.117590.
34. J. Mabrouki, M. Azrour, Y. Farhaoui, and S. El Hajjaji, "Intelligent System for Monitoring and Detecting Water Quality," in *Big Data and Networks Technologies*, Springer International Publishing, 202, vol. 3, pp. 172–182.

35. A. Unnarkat, A. Bhavsar, S. Ostwal, and P. Vashi, and S. Dharaskar, "Solar energy in water treatment TS." https://www.researchgate.net/profile/Ashish-Unnarkat-3/publication/360112282_Solar_Energy_in_Water_Treatment_Processes-An_Overview/links/637efdb52f4bca7fd0883dec/Solar-Energy-in-Water-Treatment-Processes-An-Overview.pdf.
36. Y. Aryanfar *et al.*, "Technologies Review for Solar Thermal Integrated Photovoltaic Desalination," in *Performance Enhancement and Control of Photovoltaic Systems*, Elsevier, 2024, p. 59–87.
37. J. Ramkumar, "Membrane Based Technology for Water Desalination," in *Advances in Separation Sciences*, Elsevier, 2025, p. 251–264.
38. H. Loukili, J. Mabrouki, A. Anouzla, Y. Kouzi, S. A. Younssi, K. Digua, and Y. Abrouki, "Pre-treated Moroccan natural clays: Application to the wastewater treatment of textile industry," *Desalination and Water Treatment*, vol. 240, pp. 124–136, 2021.
39. X. Chen, X. Meng, C. Si, and Y. Fu, "Recovery of wastewater from the pulp and paper industry by cellulose acetate reverse osmosis membrane," *International Journal of Biological Macromolecules*, vol. 297, p. 139862, 2025.
40. S. Rajesh, and C. Chiranjeevi, "Hybrid thermal desalination systems for sustainable development–A critical review," *Solar Energy*, vol. 270, p. 112364, 2024.
41. I. Nabil, A. M. Abdalla, T. M. Mansour, A. I. Shehata, and M. M. K. Dawood, "Salinity impacts on humidification dehumidification (HDH) desalination systems: Review," *Environmental Science and Pollution Research*, vol. 31, no. 2, pp. 1907–1925, 2023. doi: 10.1007/s11356-023-31327-5.
42. M. Khashehchi *et al.*, "Solar desalination techniques: Challenges and opportunities," *Highly Efficient Thermal Renewable Energy Systems*, pp. 305–329, 2024. https://www.taylorfrancis.com/chapters/edit/10.1201/9781003472629-19/solar-desalination-techniques-morteza-khashehchi-sivasakthivel-thangavel-pooyan-rahmanivahid-milad-heidari-taleb-moazzeni-vikas-verma-ashwani-kumar.
43. M. Benchrifa, and J. Mabrouki, "Simulation, sizing, economic evaluation and environmental impact assessment of a photovoltaic power plant for the electrification of an establishment," *Advances in Building Energy Research*, vol. 16, no. 6, pp. 736–753, 2022.
44. Z. M. Ghazi, S. W. F. Rizvi, W. M. Shahid, A. M. Abdulhameed, H. Saleem, and S. J. Zaidi, "An overview of water desalination systems integrated with renewable energy sources," *Desalination*, vol. 542, p. 116063, 2022.
45. S. Sarathe, P. V. Baredar, G. Dwivedi, S. Tapdiya, and A. Gaurav, "Review of various types of renewable-powered desalination technologies with economic analysis," *Materials Today: Proceedings*, vol. 56, pp. 326–335, 2022.
46. M. A. Dawoud, G. R. Sallam, M. A. Abdelrahman, and M. Emam, "The performance and feasibility of solar-powered desalination for brackish groundwater in Egypt," *Sustainability*, vol. 16, no. 4, p. 1630, 2024.
47. J. Mabrouki, M. Benchrifa, M. Ennouhi, K. Azoulay, I. Bencheikh, T. Rachiq, and S. El Hajjaji, "Geographic information system for the study of water resources in Chaâba El Hamra, Mohammedia (Morocco)," in The international conference on artificial intelligence and smart environment (pp. 469–474). Cham: Springer International Publishing, 2022, November.
48. J. Mabrouki, A. Moufti, I. Bencheikh, K. Azoulay, Y. El Hamdouni, and S. El Hajjaji "Optimization of the Coagulant Flocculation Process for Treatment of Leachate of the Controlled Discharge of the City Mohammedia (Morocco)," in *Advanced Intelligent Systems for Sustainable Development (AI2SD'2019) Volume 7-Advanced Intelligent Systems for Sustainable Development Applied in Energy and Electrical Engineering*, Springer International Publishing, 2020, pp. 200–212.

16 Advancing Environmental Sustainability

Intelligent Microalgae-Bacteria Wastewater Remediation

K. El-Moustaqim, J. Mabrouki, A. Ghailane, C. Lamiri, and D. Hmouni

16.1 INTRODUCTION

The dangerous substances that pollute the air, water, and land are now found everywhere; they have absorbed the entire planet, from the atmosphere to the ocean below (Mujtaba and Lee 2016; Nagabalaji et al. 2023). Along with organic trash (El-Moustaqim, Mabrouki, and Azrour 2023; El-Moustaqim et al. 2023), inorganic trace elements like phosphorus and heavy metals, nitrogen, and microplastics, the new pollutants, these chemicals can also harm the environment and the population of living things (El-Moustaqim, Benchrifa, et al. 2024; El-Moustaqim, Tallou, Mabrouki, Slaoui, et al. 2024). The necessity for freshwater and the issues with water body contamination have long been the world's greatest worries (Toth 2020; Virta and Hedberg 2024; Zhang et al. 2024). Wastewater from homes, businesses, and farms may be treated using a variety of methods (Azdem et al. 2024a, 2024b). Utilizing bioaugmentation and the symbiosis of bacteria, microalgae, fungus, and other microorganisms that reside in a certain biome, bioremediation is one of these techniques (El-Moustaqim, Mabrouki, and Hmouni 2024b, 2024a; El-Moustaqim, Rachiq, Mabrouki, Slaoui, et al. 2024). Previous studies have shown that using wastewater to cultivate microalgae while using the nutrients from wastewater can cut freshwater use by up to 90% (Sahu et al. 2023; Yu, Ong, and Zaman 2024). The most practical and realistic way to lower the cost of producing microalgal biofuel while reducing greenhouse gas emissions through microalgae growth is to reuse and recycle wastewater as the main source of water and nutrients (Deepika et al. 2025; El-Sheekh, El-Kassas, and Ali 2025). Due to its dual benefits of treating water and valuing biomass for biofuel or bioproducts, microalgae-based wastewater treatment technology is gaining increased research interest in light of the energy crisis (Abuhasheesh et al. 2025; Lu 2025; Younas et al. 2025). Furthermore, using microalgae for wastewater treatment can provide sustainable biomass that can be used to produce value-added compounds and biofuel. Through symbiotic relationships, microalgae usually coexist with a variety of heterotrophic bacteria (El-Moustaqim et al. 2025), particularly in wastewater (Perera et al. 2022). Technologies based on microalgae provide a viable path toward converting wastewater nutrients into biomass with additional value for further resource recovery (Hosny, Elshobary, and El-Sheekh 2025). The main goal of traditional algae growth techniques in photobioreactors (PBRs) and high-rate algal ponds (HRAPs) is to extract organic matter and nutrients from wastewater. However, wastewater treatment in conjunction with resource recovery has received a lot of attention from academics in recent years (Agoun and Avci 2025; Kumar et al. 2025). Microalgae are considered the third generation of feedstocks for biofuel production due to their carbon-neutral life cycle, lack of competition with food or crops, and potential for vertical and high-density cultivation systems (Sun

DOI: 10.1201/9781003647263-16

et al. 2011; Tam et al. 2010; Zhu, Yang, and Wei 2023). The symbiotic and mutualistic relationships made possible by the interdependence of bacteria and algae offer several advantages for the environment, energy, and other sectors (Sahu et al. 2023). The primary obstacles to the commercialization of microalgal technologies are the elevated harvesting costs and the requisite technical expertise for algae cultivation (Soranno et al. 2025). Algal systems exhibit significant sensitivity to environmental disturbances. This requires continuous oversight and administration of these systems to sustain ideal conditions. These persistent issues can be mitigated by adopting diverse, advanced, automated, and intelligent technologies grounded in real-time monitoring (Gong et al. 2025; X. Liu et al. 2025). AI-enabled intelligent systems and dynamic models can greatly enhance the efficiency of algae systems (El-Moustaqim, Mabrouki, Azrour, and Hmouni 2024; El-Moustaqim, Mohammed, Mabrouki, and Hmouni 2024; El-Moustaqim, Mabrouki, Azrour, Hadine, et al. 2024). AI can be effortlessly integrated to monitor, optimize, anticipate uncertainties, and identify real-time errors inside complex environmental systems (Mabrouki, Azoulay, et al. 2022; Mabrouki, Fattah, Al-Jadabi, et al. 2022; Mabrouki, Fattah, Kherraf, et al. 2022). Artificial intelligence can simulate algal wastewater treatment and enhance process parameters for resource recovery. Owing to its robustness and dependability, AI has been extensively utilized for automation, prediction, and decision making in complex systems (Arya 2025; Capodaglio and Callegari 2025). Furthermore, enhanced modeling and prediction in wastewater treatment are made possible by the application of AI approaches (Ramandani et al. 2025; Recio-Colmenares et al. 2025). Numerous research demonstrates AI's significant potential for improving wastewater treatment procedures. AI can forecast and lower treatment plant energy use through energy optimization (Xu et al. 2025). There are many obstacles to the use of AI in wastewater treatment procedures (Dalibard et al. 2025). There is little research on machine learning (ML) modeling in wastewater treatment facilities, and much of it concentrates on other facets of water quality (Padri et al. 2025). A lack of data hampers process optimization and precise prediction models (Pitawala, Scales, and Martin 2025). Hiring and keeping qualified employees is hampered by a talent shortage. The absence of a well-defined AI strategy hinders its execution. Adoption of real-time monitoring is very limited, despite its promise (Gotovtsev et al. 2025).

Based on the most recent research, this study offers a critical analysis of the use of AI in microalgal wastewater treatment. This review paper addresses the synergy between bacteria and microalgae in the biodegradation of pollutants. It looks at the importance of algal-bacterial biotechnology's environmental uses, such as bioremediation, and the production of biofuel from the biomass used in bioremediation. This research aims to provide a fresh viewpoint on upcoming developments by considering the technologies that might affect agriculture productivity and sustainability. One has to understand ecology to promote the usage of sustainable manufacturing processes in the green industry.

16.2 THE ADVANTAGES OF BIOLOGICAL CO-CULTURING BACTERIA AND MICROALGAE FOR IMPROVED WASTEWATER TREATMENT

The employment of photosynthetic bacteria (purple non-sulfur bacteria, or PSB) to absorb those soluble metabolites and transform the stored energy into a high energy density form, H_2, through further photo-fermentation, is a crucial method that can eliminate the high concentrations of organic chemicals (Han et al. 2016; Younas et al. 2025). Microalgae and bacteria (or cyanobacteria) co-cultivation holds great potential to replace traditional biological treatment systems that employ a single microorganism (microalgae/bacteria) (Soranno et al. 2025). This is because co-cultivation requires fewer chemicals as reducing agents or pH buffers and oxygen as an oxidant for the denitrification process (Qi et al. 2018; Xu et al. 2020). Reactive oxygen species (ROS) generation is increased and the high output of lipids in microalgae is facilitated by oxidative stress, which is brought on by nutrient (nitrogen and phosphate) shortage or other environmental stressors in wastewaters (Makut, Das, and Goswami 2019).

Microalgae and bacteria, which are vital to the flow of energy and the cycling of nutrients, are found in almost all aquatic environments (Joshi et al. 2025). By consuming the organic waste products that algae create and lowering the dissolved oxygen content, bacteria generally promote the development of microalgae (Li et al. 2021). As seen in Figure 16.1, microalgae then return the favor by giving bacteria oxygen and extracellular substances.

When compared to axenic culture, the primary advantage of co-culturing bacteria and microalgae is the increased biomass (Soranno et al. 2025). To address the problem that monocultures are usually preferred for a variety of reasons, which raise production costs, require more work, and increase the likelihood of contamination, co-culturing systems for microalgae have recently developed and drawn attention (D. Li et al. 2025; X. Liu et al. 2025). Co-culturing technology's main objectives are to expand the possible feedstock and boost the productivity of the target substrate or value-added bioproducts (P. Liu et al. 2025). To further enhance the efficiency and sustainability of wastewater treatment, a new wave of research has emerged, incorporating cutting-edge technologies such as ML and AI (Recio-Colmenares et al. 2025). These novel approaches leverage the power of data-driven decision making and advanced modeling to optimize various stages of the treatment process, paving the way for more effective and eco-friendly solutions (Han et al. 2016; Qi et al. 2018; Yu, Ong, and Zaman 2024). In fact, the inhibitory compounds generated by native microorganisms found in wastewater may have an impact on the benefits of microalgal–bacterial consortia, leading to intense competition (Joshi et al. 2025). However, only with a thorough grasp of their anticipated interactions with the native microorganisms could the potential of such strong consortia in the bioremediation of actual wastewaters be fully used (Dalibard et al. 2025; Gotovtsev et al. 2025; Kumar et al. 2025).

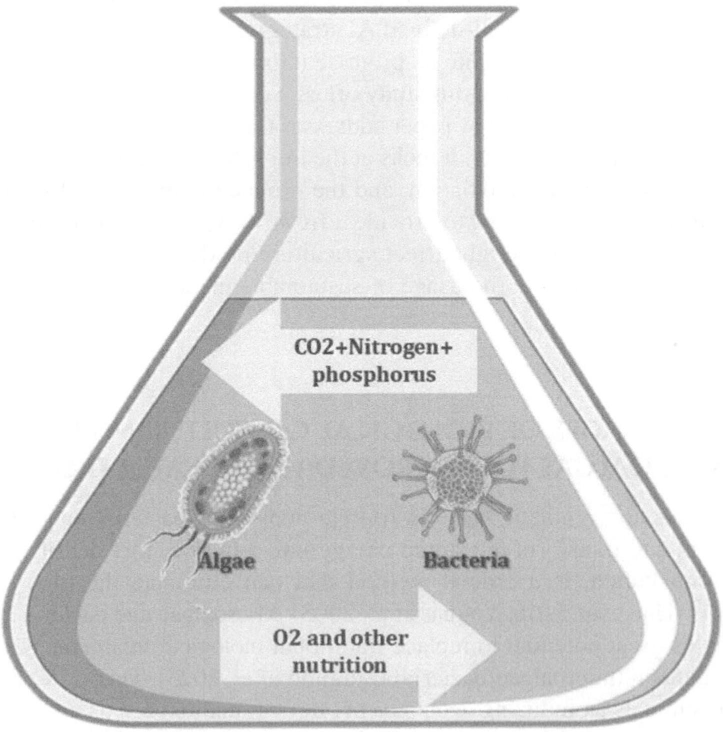

FIGURE 16.1 The relationship between algae and bacteria as co-culture.

16.3 THE EFFECTIVENESS OF MICROALGAE-BACTERIA IN WASTEWATER TREATMENT

In a reactor or pond, the bacteria and microalgae are grown together in a symbiotic interaction. An efficient method of treating wastewater is the coculture of microalgae with bacteria, which eliminates organic materials, nutrients, and pathogens (Xu et al. 2020; Xiu and Dong 2019; Yang et al. 2022). High organic matter elimination efficiency is one of the benefits of co-culturing microalgae and bacteria (Capodaglio and Callegari 2025). The data on the removal of inorganic nutrients from wastewater is inadequate, although the aerobic and anaerobic methods now in use are effective at removing organic carbon (Tian et al. 2024). Because of the high concentration of both organic and inorganic nutrients, wastewater can be a great substrate for microalgal development (Takahashi et al. 2024). Furthermore, using microalgae for wastewater treatment can provide sustainable biomass that can be used to produce value-added compounds and biofuel (Mujtaba et Lee 2016). The microalgae-bacteria wetland system's waste removal mechanism is a complicated series of interactions between algae and bacteria, and running the electrochemical reactions takes a lot of energy. This energy needs could not be financially viable for some applications, and it might raise the system's overall cost (Kumar et al. 2025; Li et al. 2021; P. Li et al. 2025).

Microalgae-bacteria co-culture has several benefits, such as high organic matter, nutrient, and pathogen removal efficiency; low energy consumption when compared to other wastewater treatment technologies; reduced sludge production when compared to traditional wastewater treatment technologies; and the potential to produce biofuel. The source of wastewater cleaned by microalgae-bacteria symbiosis and its primary components are displayed in Table 16.1.

Additionally, the system may be susceptible to variations in the wastewater and nutrient content, which may have an impact on the growth of microorganisms and the overall efficacy of the treatment process (Kumar et al. 2010; Pulz 2001; Wolf et al. 2016). In the bioflocs, the dual bioaugmentation approach was found to improve the establishment of an algal-bacterial symbiosis (Wang

TABLE 16.1
Types of Systems Based on Microalgae and Bacteria Co-Culture in Wastewater Treatment

Systems	Application	Performances	Refs
Algal-bacteria biofilm	Dairy manure wastewater (DMW) treatment.	• Efficacy in removing COD and ammonia nitrogen of 50% raw DMW wastewater.	Chu et al. (2025)
Co-cultures of *Desmodesmus* sp. and *Bacillus megaterium*	Nutrient removal, biomass production, and lipid accumulation.	• *Desmodesmus* sp. Inhibiting urea hydrolysis. • *B. megaterium* promoting urea utilization and NH_4^+-N assimilation.	D. Li et al. (2025)
Co-culture system of *C. sorokiniana* and *P. pantotrophus*	Lighting strategy and tetracycline hydrochloride (TCH) on ammonia nitrogen (NH_4^+-N) and phosphates removal.	• Removal of NH_4^+-N and phosphate, with the maximum removal efficiencies of 85.75% and 75.5%, respectively.	Zhang et al. (2025)
Aeration mode on the Microalgae Consortium	Two *Chlorella* strains and mixed cultures.	• The highest pigment accumulation in *Chlorella pyrenoidosa* (20.49 mg/L). • Nitrogen removal efficiency	P. Li et al. (2025)
Immobilizing microalgae onto coral velvet carriers, *Chlorella* sp., *Isochrysis galbana*, *Chaetoceros* sp., and *Nannochloropsis* sp.	Cultivating microalgae for the remediation of aquaculture wastewater.	• *Chlorella* sp., for a nitrate removal rate of 100% and an inorganic nitrogen removal rate of 79.13%. • Mono-cultures of *Chlorella* sp. and *Isochrysis galbana* showed removal rates of 77.76% and 45.72%, respectively.	Wang et al. (2025)

et al. 2025). The electroactive algae-bacteria wetland system combines electrochemical, microbial, and algae activities to improve the efficacy of cleaning micro-polluted wastewater. The system is divided into two sections: an electrochemical section and an algae-bacteria section (Carvalho, Meireles, and Malcata 2006). According to the microbial community study, heterotrophic bacteria, which were essential to the generation of bioflocs, predominated in the biofloc reactor whereas denitrifying bacteria dominated the denitrification reactor. Additionally, the microalgae supported the development of aerobic bacteria by assisting in the maintenance of dissolved oxygen levels in the biofloc reactor (P. Li et al. 2025; Makut, Das, and Goswami 2019). Furthermore, promising results for the elimination of pollutants from wastewater have been shown by the Chlorella-bacteria consortium and the nighttime aeration mode (El-Moustaqim et al. 2023). However, the effectiveness of these techniques may be affected by environmental conditions such as temperature, pH, and light availability, requiring the adoption of more advanced monitoring and control systems to optimize performance and reduce the risk of system failure (Zhang et al. 2025).

16.4 ARTIFICIAL INTELLIGENCE IN THE REMEDIATION OF WASTEWATER BY MICROALGAE

The use of AI techniques in the co-culturing of bacteria and microalgae has been proven in recent studies, providing innovative ways to improve the wastewater treatment process (Sahu et al. 2023). One research, for example, optimized the growth conditions for bacteria and microalgae in a co-culturing system using ML techniques (Sundaram et al. 2025). To optimize biomass production and pollutant elimination, the program dynamically modified the culture conditions by evaluating real-time data on ambient factors, nutrient levels, and microbial activity (Oruganti et al. 2023). These sensors delivered input into an AI model about dissolved oxygen, pH, and nutrition levels. Improved performance and energy efficiency resulted from the model's precise prediction of the ideal coculturing conditions utilizing deep learning techniques (Pitawala, Scales, and Martin 2025; Ramandani et al. 2025). To create a new phase with a focus on AI, automation, ML, big data, cloud technology, real-time data, and the Internet of Things (IoT), more informatics technology has begun to be incorporated into the current industrial applications in this era of the Fourth Industrial Revolution (IR 4.0) (Imamoglu 2024; Yu, Ong, and Zaman 2024).

Furthermore, using AI technology helps with complicated problem monitoring, prediction, and optimization across the microalgae culture and bioprocess system as seen in Table 16.2.

TABLE 16.2
Methods for Wastewater Treatment and Microalgal Production and Classification Using AI

AI Systems	Description of Application	Achievements	Refs
Neuronal network to classify different microalgae genera.	Allows the detection of contamination.	Obtaining a macro F1 score of 98.64%.	González-Hernández et al. (2025)
Reinforcement learning (RL)	Optimizing environmental conditions for microalgae growth for *Chlorella vulgaris* and *Scenedesmus obliquus*.	Absorption rates by 50% and 80%, respectively.	Al-Huqail et al. (2025)
Dry cell weight (DCW)	The influence of input features derived from food waste culture medium on the growth of *Chlorella vulgaris* FSP-E.	R^2 of 0.9897, MAE of 0.0632, MSE of 0.0098, and RMSE of 0.0992, respectively.	Ramandani et al. (2025)
Backpropagation Neural Network (BPNN)	Developing a ML approach, offering new insights for achieving efficient microalgal harvesting.	Higher testing R^2 value (0.923).	Xu et al. (2025)

Additionally, the co-culturing process may be monitored and controlled in real time with the help of AI and ML approaches. Researchers may collect ongoing data on important parameters and use AI algorithms to examine and improve the interactions between bacteria and microalgae by utilizing intelligent sensors and Internet of Things devices (Meenatchisundaram et al. 2024; Pavendan and Nagarajan 2022). By enabling prompt cultural condition modifications, this dynamic feedback loop may guarantee peak performance and efficient use of resources (El-Moustaqim, Mabrouki, and Azrour 2023). AI algorithms and modeling approaches, for instance, may improve some elements of microalgae culture, including temperature management, nutrient availability, contamination, and harvesting using sensors and real-time automated monitoring systems (Al-Huqail et al. 2025). The operator could readily make decisions based on the data gathered to optimize yields, productivity, and energy efficiency thanks to the real-time monitoring connected to the shifting environmental circumstances (Dalibard et al. 2025; Imamoglu 2024; Oruganti et al. 2023).

16.5 ENHANCING THE CO-CULTURE OF BACTERIA AND MICROALGAE IN WASTEWATER BIOREMEDIATION THROUGH THE INTEGRATION OF AI AND ML APPROACHES

As a form of aquatic photosynthetic microbial organism, microalgae produce organic compounds such as proteins, lipids, and carbohydrates in exchange for using carbon dioxide as a carbon source for photosynthesis (El-Moustaqim et al. 2023). AI and enormous volumes of data, especially image recognition and natural language processing, are processed when used to wastewater treatment and water recycling techniques. This method involves using massive data gathering, water quality, and water treatment plant operational factors to train AI models (Chang et al. 2024). The achievement of a high biomass yield is essential to setting up an economical microalgae production system. Microalgal production is influenced by several factors, including temperature, light levels, nutrition availability, pH, aeration, and fluid mechanics modulation (Tian et al. 2024; Xu et al. 2020). Microalgae have recently been used as a sustainable wastewater treatment option because of their great capacity to remove nutrients and their biomass, which can be used to produce value goods. ML algorithms have been used by many academics to forecast the optimal variable combinations for process optimization (Oruganti et al. 2023). Therefore, to obtain high biomass output and optimal treatment efficiency, it is crucial to supply the strain with ideal growing circumstances (Flores-Zambrano, Tapia, and Castillejo 2025). AI can be employed for the selection of appropriate strains for wastewater treatment and optimum parameters for cultivation. Because microalgae development is intermittent, optimizing its cultivation is a non-direct process that typically necessitates dynamic optimization strategies (Dalibard et al. 2025; Gotovtsev et al. 2025).

This technology allows for intelligent system control by analyzing the makeup of wastewater components and producing decision-making outcomes (Figure 16.2) (Roostaei et al. 2023; Yuliantini et al. 2019). Furthermore, AI technologies that are useful for water treatment and recovery techniques fall into three categories: natural language, picture recognition, and enormous data (Wang et al. 2025). AI also plays a key role in solving physical system problems, such as filter blockage prediction, membrane system blockage prevention, and overall system control optimization. These AI-powered tools improve responsiveness, accuracy, and efficiency in the administration of intricate wastewater treatment procedures (Semple and Pandhal 2025). The size, shape, cell density, community structure, color, and lipid content of microalgal growth systems are direct indicators that may be used to track the dynamics of the culture (Wang et al. 2025). Since the techniques used to measure these parameters are either too laborious, too time-consuming, or too harmful to microalgae, it is challenging to manage and monitor them continuously in real time (Amirian et al. 2025).

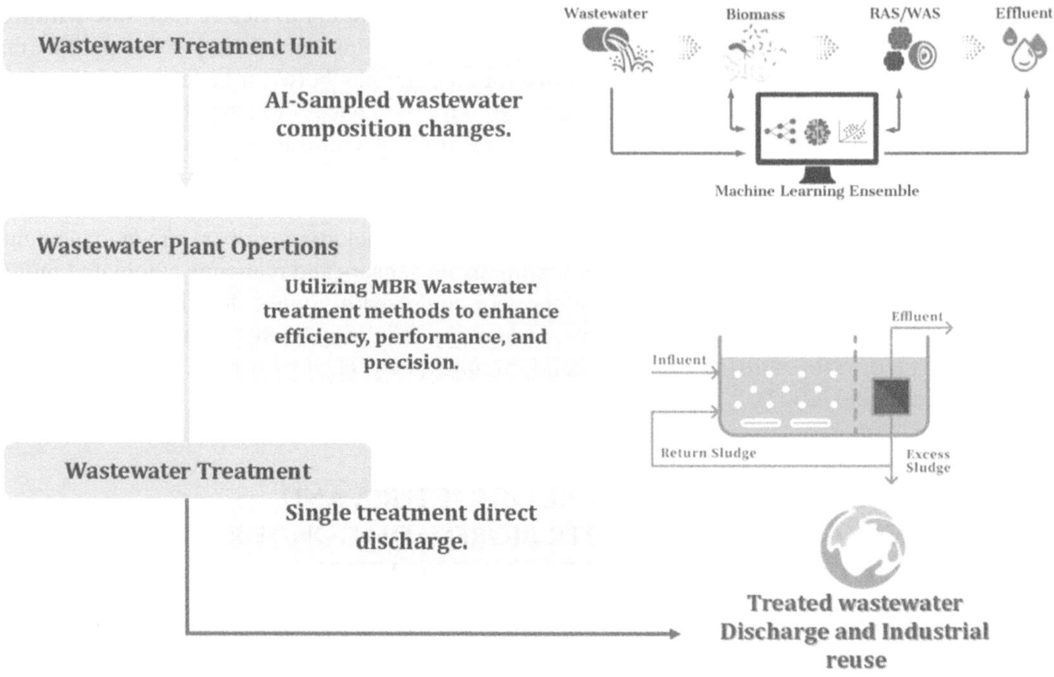

FIGURE 16.2 Combining artificial intelligence (AI) technology with membrane bioreactors (MBRs).

16.6 CHALLENGES AND FUTURE PERSPECTIVES

Incorporating AI into wastewater treatment using microalgae offers many opportunities, but also poses some challenges (Sahu et al. 2023). First of all, integrating artificial intelligence into wastewater treatment requires a detailed study of the costs and benefits (Baarimah et al. 2024). AI-based solution implementation frequently necessitates a large financial investment in operator training and technological infrastructure (Senthil Rathi et al. 2025). Furthermore, the quality and quantity of data accessible affect how accurate AI algorithms are, necessitating the usage of advanced sensors and well-structured databases. Accurate AI model training and refinement may be hampered by sensor data that contains mistakes, missing values, or uncertainty (Dalibard et al. 2025). Integrating different types of data from different sources raises interoperability issues (Khoiruddin et al. 2025). On the other hand, as wastewater treatment systems become increasingly digitalized and connected, the dangers of cyberattacks, data compromise, and system vulnerabilities multiply. It is therefore essential to adopt solid cybersecurity measures and robust data protection strategies (Agoun and Avci 2025; Capodaglio and Callegari 2025). But the future seems bright (Azdem, Mabrouki, and El Hajjaji 2025; Azdem and Mabrouki 2024; Azdem et al. 2025). Through the dynamic adaptation of culture conditions to changes in effluents, the combination of ML and neural networks will allow us to maximize the efficiency of microalgae (Alrashidi et al. 2025). Furthermore, developments in IoT systems and smart sensors will facilitate real-time data collection and analysis, increasing the precision of prediction models (Niju, Shruthi, et Priyadharshini 2025). Overcoming these obstacles and implementing intelligent, scalable solutions in microalgae wastewater treatment will need the creation of open-source solutions and strong cooperation between data scientists, technology suppliers, and wastewater treatment specialists (El-Sheekh, El-Kassas and Ali 2025).

16.7 CONCLUSION

Due to the complex and varied nature of microalgae, combining wastewater treatment with algal biorefineries requires a great deal of information and expertise from a variety of fields, including species selection, growth conditions, reactor design, and conversion technologies. A viable way to deal with the energy problems may be to use wastewater treatment technology based on microalgae to produce bioenergy. The creation of a sustainable environment is greatly aided by the numerous advantages of using microalgae-based systems for water treatment and biomass value conversion to biofuel or bioproducts. In conclusion, the symbiotic relationship between bacteria and microalgae offers a viable wastewater bioremediation approach that is more ecologically friendly than conventional techniques. With the help of open systems for cost-effectiveness and closed bioreactors for regulated operation, this co-culturing method speeds up biodegradation and eliminates hazardous chemicals from the environment, while contamination issues may reduce biomass output and efficiency. New developments in ML and AI provide approaches to lessen these difficulties. AI helps with biomass production, nutrition optimization, and strain selection. ML models improve nutrient removal and process efficiency, increasing the method's ecological viability and sustainability. A cleaner environment and a more environmentally friendly future are promised by the possible long-term wastewater treatment solution that microalgae-bacteria symbiosis offers. However, challenges like optimization and scalability still exist. Lastly, the combination between bacteria and microalgae has great potential for sustainable wastewater treatment. Its efficiency may be further increased by integrating AI and ML, which will support the objectives of ecological sustainability and a cleaner world.

REFERENCES

Abuhasheesh, Yazan, Aya Ghazal, Doris Ying Ying Tang, Fawzi Banat, Shadi W. Hasan, and Pau Loke Show. 2025. "Advances in chlorella microalgae for sustainable wastewater treatment and bioproduction". *Chemical Engineering Journal Advances*, 100715. https://www.sciencedirect.com/science/article/pii/S2666821125000122.

Agoun, Naila Amel, and Fatma Gizem Avci. 2025. "Phycoremediation: A path towards heavy metal bioremediation from wastewater". *Journal of Chemical Technology & Biotechnology* 100 (1):13–23. https://doi.org/10.1002/jctb.7745.

Al-Huqail, Arwa, Khidhair Jasim Mohammed, Meldi Suhatril, Hamad Almujibah, Sana Toghroli, Sultan Saleh Alnahdi, and Joffin Jose Ponnore. 2025. "Optimizing microalgal biomass conversion into carbon materials and their application in water treatment: A machine learning approach". *Carbon Letters*. https://doi.org/10.1007/s42823-024-00837-8.

Alrashidi, Wid, Safiah Alhazmi, Fotoon Sayegh, and Sherif Edris. 2025. "Microalga-based electricity production: A comprehensive review". *Energies* 18 (3):536. https://www.mdpi.com/1996-1073/18/3/536.

Amirian, Veghar, Mohammad Russel, Zetty Norhana Balia Yusof, Jit Ern Chen, Ali Movafeghi, Morteza Kosari-Nasab, Dayong Zhang, and Ewa Szpyrka. 2025. "Algae- and bacteria-based biodegradation of phthalic acid esters towards the sustainable green solution". *World Journal of Microbiology and Biotechnology* 41 (2):24. https://doi.org/10.1007/s11274-024-04243-0.

Arya, Adarsh Kumar. 2025. "Artificial Intelligence as a Sustainable Tool for Optimizing the Wastewater Treatment Plant". In *Practical Applications of Machine Learning and AI: Medicine, Environmental Science, Transportation, and Education*, 117–38. IGI Global Scientific Publishing. https://www.igi-global.com/chapter/artificial-intelligence-as-a-sustainable-tool-for-optimizing-the-wastewater-treatment-plant/370999.

Azdem, Driss, Abdelilah Fatni, Jamal Mabrouki, and Souad El Hajjaji. 2025. "The Desalination Strategy and Economic Integration of Morocco: Water Scarcity, Prospects, and Challenges". In *Circular Economy Applications in Energy Policy*, 265–76. IGI Global. https://doi.org/10.4018/979-8-3693-6680-6.ch018.

Azdem, Driss, and Jamal Mabrouki. 2024. "Methods and Technologies for the Treatment and Disposal of Brine in Seawater Desalination Plants". In *Advancements in Climate and Smart Environment Technology*. IGI Global. https://doi.org/10.4018/979-8-3693-3807-0.ch007.

Azdem, Driss, Jamal Mabrouki, and Souad El Hajjaji. 2025. "The Use of Artificial Intelligence in a Desalination Plant Powered by Renewable Energy". In *Advanced Computation Solutions for Energy Efficiency*, 81–94. IGI Global. https://doi.org/10.4018/979-8-3373-1220-0.ch004.

Azdem, Driss, Jamal Mabrouki, Ahmed Moufti, Souad El Hajjaji, and Abdelilah Fatni. 2024a. "Assessment of heavy metal contamination in seawater in Agadir coastline, Morocco". *Desalination and Water Treatment* 317. https://doi.org/10.1016/j.dwt.2024.100129.

Azdem, Driss, Jamal Mabrouki, Ahmed Moufti, Souad El Hajjaji, and Abdelilah Fatni. 2024b. "Present Situation and Environmental Impact of Seawater Desalination Plants in Morocco". In *Technical and Technological Solutions Towards a Sustainable Society and Circular Economy*, édité par Jamal Mabrouki and Azrour Mourade, 581–92. Cham: Springer Nature Switzerland. https://doi.org/10.1007/978-3-031-56292-1_46.

Baarimah, Abdullah O., Mahmood A. Bazel, Wesam Salah Alaloul, Motasem YD Alazaiza, Tharaa M. Al-Zghoul, Basheer Almuhaya, Arsalaan Khan, and Ahmed W. Mushtaha. 2024. "Artificial intelligence in wastewater treatment: Research trends and future perspectives through bibliometric analysis". *Case Studies in Chemical and Environmental Engineering*, 100926. https://www.sciencedirect.com/science/article/pii/S2666016424003207.

Capodaglio, Andrea G., and Arianna Callegari. 2025. "Use, potential, needs, and limits of AI in wastewater treatment applications". *Water* 17 (2):170. https://www.mdpi.com/2073-4441/17/2/170.

Carvalho, Ana P., Luís A. Meireles, and F. Xavier Malcata. 2006. "Microalgal reactors: A review of enclosed system designs and performances". *Biotechnology Progress* 22 (6):1490–1506.

Chang, Hung-Li, Yu-Lun Liu, Ching-Jui Keng, Han-Ling Jiang, and Jiayao Hu. 2024. "Challenges of industrial wastewater treatment: Utilizing membrane bioreactors (MBRs) in conjunction with artificial intelligence (AI) technology". *Journal of Industrial and Production Engineering* 41 (5):422–27. https://doi.org/10.1080/21681015.2024.2330401.

Chu, Ying, Xiaolin Jiao, Dan Li, Jiawen Yu, Changhai Wang, and Meilin He. 2025. "Integration of dairy manure wastewater treatment and biofuel production by microalgae-bacteria consortia in a biofilm reactor". *Algal Research* 86:103912. https://doi.org/10.1016/j.algal.2025.103912.

Dalibard, Antoine, Lukas Simon Kriem, Marc Beckett, Stephan Scherle, Yen-Cheng Yeh, and Ursula Schließmann. 2025. "Challenges and requirements of AI-based waste water treatment systems". *At - Automatisierungstechnik* 73 (1):40–49. https://doi.org/10.1515/auto-2024-0023.

Deepika, Charu, Gabriele Frascaroli, Anusha Gowri, and Alla Silkina. 2025. "Microalgae-Based Bioremediation: A Sustainable Approach for the Removal of Emergent Pollutants". In *Biotechnological Interventions in the Removal of Emerging Pollutants*, édité par Satarupa Dey and Sayan Bhattacharya, 101–39. Interdisciplinary Biotechnological Advances. Singapore: Springer Nature Singapore. https://doi.org/10.1007/978-981-97-9922-0_6.

El-Moustaqim, Khadija, Mohammed Benchrifa, Jamal Mabrouki, and Driss Hmouni. 2024. "Interaction between the Microalgae Cultivation System and Its Effects on Nutrient Removal from Wastewater". In *Artificial Intelligence Systems in Environmental Engineering*, 70–79. Taylor & Francis. https://doi.org/10.1201/9781003436218-8.

El-Moustaqim, Khadija, Houria El Bakraoui, Jamal Mabrouki, Yasser Fouad, Miloudia Slaoui, Driss Hmouni, Mbadiwe S. Benyeogor, and Tobore Louis Igbigbi. 2023. "Combination of microalgae method, decantation, and filtration for domestic wastewater treatment". *Sustainability* 15 (22):16110. https://doi.org/10.3390/su152216110.

El-Moustaqim, Khadija, Jamal Mabrouki, and Mourade Azrour. 2023. "Monitoring of Water Toxicity Through the Internet of Things to Protect the Health of the Population". In *IoT, Machine Learning and Data Analytics for Smart Healthcare*, 85–92. Taylor & Francis. https://doi.org/10.1201/9781003430735-7.

El-Moustaqim, Khadija, Jamal Mabrouki, Mourade Azrour, Mouhsine Hadine, and Driss Hmouni. 2024. "Enabling Smart Agriculture Through Integrating the Internet of Things in Microalgae Farming for Sustainability". In *Smart Internet of Things for Environment and Healthcare*, édité par Mourade Azrour, Jamal Mabrouki, Abdulatif Alabdulatif, Azidine Guezzaz, et Fatima Amounas, 209–22. Cham: Springer Nature Switzerland. https://doi.org/10.1007/978-3-031-70102-3_15.

El-Moustaqim, Khadija, Jamal Mabrouki, Mourade Azrour, and Driss Hmouni. 2024. "How Can Green Energy Be Improved by Integrating IoT into the Microalgae Process?" In *Sustainable and Green Technologies for Water and Environmental Management*, 239–52. Springer. https://doi.org/10.1007/978-3-031-52419-6_17.

El-Moustaqim, Khadija, Jamal Mabrouki, Mohammed Benchrifa, Driss Azdem, and Driss Hmouni. 2025. "Microalgae-based food additives for improved shelf life and nutritional value". *Biology and Life Sciences Forum* 40:42. https://www.mdpi.com/2673-9976/40/1/42.

El-Moustaqim, Khadija, Jamal Mabrouki, and Driss Hmouni. 2024a. "Applied Sciences Green Microalgae for Future Biomass Development". In *Advanced Systems for Environmental Monitoring, IoT and the application of Artificial Intelligence*, 203–11. Springer. https://doi.org/10.1007/978-3-031-50860-8_12.

El-Moustaqim, Khadija, Jamal Mabrouki, and Driss Hmouni 2024b. "From microalgal biomass to biohydrogen: A novel approach to environmentally sustainable technology". *Advancements in Climate and Smart Environment Technology*, 29–38. https://www.igi-global.com/chapter/from-microalgal-biomass-to-biohydrogen/346691.

El-Moustaqim, Khadija, Benchrifa Mohammed, Jamal Mabrouki, and Driss Hmouni. 2024. "Application of IoT-Based Intelligent Systems in Microalgae Biofuel Production". In *Advanced Technology for Smart Environment and Energy*, édité par Jamal Mabrouki et Azrour Mourade, 293–305. Environmental Science and Engineering. Cham: Springer Nature Switzerland. https://doi.org/10.1007/978-3-031-50871-4_20.

El-Moustaqim, Khadija, Toufik Rachiq, Jamal Mabrouki, Miloudia Slaoui, and Driss Hmouni. 2024. "Application of Microalgae in Green Hydrogen Production: A Diverse Biomass for Multiple Applications". In *Advanced Technology for Smart Environment and Energy*, édité par Jamal Mabrouki et Azrour Mourade, 231–40. Environmental Science and Engineering. Cham: Springer Nature Switzerland. https://doi.org/10.1007/978-3-031-50871-4_15.

El-Moustaqim, Khadija, Anas Tallou, Jamal Mabrouki, Miloudia Slaoui, and Driss Hmouni. 2024. "Phytoremediation Processes for the Removal of Heavy Metals". In *Integrated Solutions for Smart and Sustainable Environmental Conservation*, 33–46. Springer. https://doi.org/10.1007/978-3-031-55787-3_3.

El-Sheekh, Mostafa M., Hala Y. El-Kassas, and Sameh S. Ali. 2025. "Microalgae-based bioremediation of refractory pollutants: An approach towards environmental sustainability". *Microbial Cell Factories* 24 (1):19. https://doi.org/10.1186/s12934-024-02638-0.

Flores-Zambrano, Karla, Wilson Tapia, and Pablo Castillejo. 2025. "Microalgae strains isolated from piggery wastewater in Ecuador: Effective nitrogen compound removal and growth potential in extremophile conditions". *Biotechnology Reports*, e00883. https://www.sciencedirect.com/science/article/pii/S2215017X25000104.

Gong, Zichao, Jun Sha, Yunlu Jia, Jin Ma, Licheng Huang, Guofei Dai, and Lirong Song. 2025. "Differential photosynthetic and metabolic susceptibility of *Microcystis aeruginosa* and *Raphidocelis subcapitata* to environmentally relevant levels of diuron". *Journal of Environmental Sciences*. https://www.sciencedirect.com/science/article/pii/S1001074225001019.

González-Hernández, José, Martina Ciardi, José Luis Guzmán, José Carlos Moreno, and Francisco Gabriel Acién. 2025. "A low-cost methodology based on artificial intelligence for contamination detection in microalgae production systems". *Algal Research* 85:103849. https://doi.org/10.1016/j.algal.2024.103849.

Gotovtsev, Pavel, Yana Sergeeva, Daniil Sukhinov, Maria Petrova, Maria Vishnevskaya, Alexandra Shestakova, Anton Rogov, Alexander Romanov, Kirill Gorin, and Raif Vasilov. 2025. "Internet of things (IoT) use for remote monitoring of algal bioreactors". *Algal Bioreactors*, 205–14. Elsevier. https://www.sciencedirect.com/science/article/pii/B9780443140587000440.

Han, Jichang, Lin Zhang, Song Wang, Guanpin Yang, Lu Zhao, and Kehou Pan. 2016. "Co-culturing bacteria and microalgae in organic carbon containing medium". *Journal of Biological Research-Thessaloniki* 23 (1):8. https://doi.org/10.1186/s40709-016-0047-6.

Hosny, Shimaa, Mostafa E. Elshobary, and Mostafa M. El-Sheekh. 2025. "Unleashing the power of microalgae: A pioneering path to sustainability and achieving the sustainable development goals". *Environmental Science and Pollution Research*. https://doi.org/10.1007/s11356-025-35885-8.

Imamoglu, Esra. 2024. "Artificial intelligence and/or machine learning algorithms in microalgae bioprocesses". *Bioengineering* 11 (11):1143. https://www.mdpi.com/2306-5354/11/11/1143.

Joshi, Jnanada Shrikant, Laura Fladung, Olaf Kruse, and Anant Patel. 2025. "Novel co-cultivation bioprocess with immobilized paenibacillus polymyxa and scenedesmus obliquus for lipid and butanediol production". *Microorganisms* 13 (3):606. https://www.mdpi.com/2076-2607/13/3/606.

Khoiruddin, K., R. Boopathy, S. Kawi, and I. G. Wenten. 2025. "Towards next-generation membrane bioreactors: Innovations, challenges, and future directions". *Current Pollution Reports* 11 (1):15. https://doi.org/10.1007/s40726-025-00345-x.

Kumar, Amit, Sarina Ergas, Xin Yuan, Ashish Sahu, Qiong Zhang, Jo Dewulf, F. Xavier Malcata, and Herman Van Langenhove. 2010. "Enhanced CO2 fixation and biofuel production via microalgae: Recent developments and future directions". *Trends in Biotechnology* 28 (7):371–80.

Kumar, S. Sanjay, J. Arun, S. S. Dawn, and N. Nirmala. 2025. "Microalgae: A Novel Low-Cost Biotechnological Approach in Detoxification and Remediation of Radioactive Waste from Wastewater". In *Smart Waste and Wastewater Management by Biotechnological Approaches*, édité par Ali Mohd Yatoo et Mika Sillanpää, 223–39. Interdisciplinary Biotechnological Advances. Singapore: Springer Nature Singapore. https://doi.org/10.1007/978-981-97-8673-2_11.

Li, Dan, Ruiqing Liu, Ying Chu, Qiang Wang, Meilin He, and Changhai Wang. 2025. "Physiological and transcriptomic responses of microalgal-bacterial co-culture reveal nutrient removal and lipid production during biogas slurry treatment". *Bioresource Technology* 416:131810. https://www.sciencedirect.com/science/article/pii/S0960852424015141.

Li, Dan, Ruiqing Liu, Xiaoyu Cui, Meilin He, Shiyan Zheng, Weijie Du, Meng Gao, and Changhai Wang. 2021. "Co-culture of bacteria and microalgae for treatment of high concentration biogas slurry". *Journal of Water Process Engineering* 41:102014. https://www.sciencedirect.com/science/article/pii/S221471442100101X.

Li, Pan, Yajun Luo, Fei Ming, Jingxian Zheng, Zhicheng Pan, Rui Wang, and Yanqing He et al. 2025. "Aeration strategies for microalgae in wastewater treatment: Enhancing pollutant removal and community dynamics". *Journal of Environmental Management* 377:124629. https://doi.org/10.1016/j.jenvman.2025.124629.

Liu, Pengyang, Alitenai Tunuhe, Xiaoyu Zhang, Hongbo Yu, Fuying Ma, and Shangxian Xie. 2025. "The co-culturing of microalgae and fungi for synergistic valorization of alkali pretreatment liquor". *Journal of Water Process Engineering* 71:107380. https://www.sciencedirect.com/science/article/pii/S2214714425004520.

Liu, Xiao Xuan, Jun Mei Li, Ruo Bing Wang, Rongting Ni, Changqun Duan, Jiacheng Yu, Erik Jeppesen, and Ying Pan. 2025. "Antibiotics disrupt bacteria-phytoplankton symbioses: Unveiling ecological risks in aquatic ecosystems." *Oikos*, e11201. https://doi.org/10.1111/oik.11201.

Lu, Qian. 2025. "A state-of-the-art review of microalgae-based food processing wastewater treatment: Progress, problems, and prospects". *Water* 17 (4):536. https://www.mdpi.com/2073-4441/17/4/536.

Mabrouki, Jamal, Karima Azoulay, Saloua Elfanssi, Loubna Bouhachlaf, Fatimazahra Mousli, Mourade Azrour, and Souad El Hajjaji. 2022. "Smart System for Monitoring and Controlling of Agricultural Production by the IoT". In *IoT and Smart Devices for Sustainable Environment*, édité par Mourade Azrour, Azeem Irshad, and Rajasekhar Chaganti, 103–15. EAI/Springer Innovations in Communication and Computing. Cham: Springer International Publishing. https://doi.org/10.1007/978-3-030-90083-0_8.

Mabrouki, Jamal, Ghizlane Fattah, Naif Al-Jadabi, Younes Abrouki, Driss Dhiba, Mourade Azrour, and Souad El Hajjaji. 2022. "Study, simulation and modulation of solar thermal domestic hot water production systems". *Modeling Earth Systems and Environment* 8 (2):2853–62. https://doi.org/10.1007/s40808-021-01200-w.

Mabrouki, Jamal, Ghizlane Fattah, Siham Kherraf, Younes Abrouki, Mourade Azrour, and Souad El Hajjaji. 2022. "Artificial intelligence system for intelligent monitoring and management of water treatment plants". *Emerging Real-World Applications of Internet of Things*, 69–87.

Makut, Bidhu Bhusan, Debasish Das, and Gargi Goswami. 2019. "Production of microbial biomass feedstock via co-cultivation of microalgae-bacteria consortium coupled with effective wastewater treatment: A sustainable approach". *Algal Research* 37:228–39. https://www.sciencedirect.com/science/article/pii/S2211926418303217.

Meenatchisundaram, Karthikeyan, Sarath C. Gowd, Jintae Lee, Selvaraj Barathi, and Karthik Rajendran. 2024. "Data-driven model development for prediction and optimization of biomass yield of microalgae-based wastewater treatment". *Sustainable Energy Technologies and Assessments* 63:103670. https://www.sciencedirect.com/science/article/pii/S2213138824000663.

Mujtaba, Ghulam, and Kisay Lee. 2016. "Advanced treatment of wastewater using symbiotic co-culture of microalgae and bacteria". *Applied Chemistry for Engineering* 27 (1):1–9. https://doi.org/10.14478/ace.2016.1002.

Nagabalaji, Velmurugan, Pounsamy Maharaja, Rajendiran Nishanthi, Ganesan Sathish, Rangasamy Suthanthararajan, and Shanmugham Venkatachalam Srinivasan. 2023. "Effect of co-culturing bacteria and microalgae and influence of inoculum ratio during the biological treatment of tannery wastewater". *Journal of Environmental Management* 341:118008. https://doi.org/10.1016/j.jenvman.2023.118008.

Niju, Subramaniapillai, V. Shruthi, and K. Priyadharshini. 2025. "Comprehensive insights into biological and bio-electrochemical treatment of the sago industry wastewater: Challenges and future perspectives". *Sustainable Chemistry for the Environment*, 100242. https://www.sciencedirect.com/science/article/pii/S2949839225000379.

Oruganti, Raj Kumar, Alka Pulimoottil Biji, Tiamenla Lanuyanger, Pau Loke Show, Malinee Sriariyanun, Venkata KK Upadhyayula, Venkataramana Gadhamshetty, and Debraj Bhattacharyya. 2023. "Artificial intelligence and machine learning tools for high-performance microalgal wastewater treatment and algal biorefinery: A critical review". *Science of The Total Environment* 876:162797. https://www.sciencedirect.com/science/article/pii/S0048969723014134.

Padri, Mohamad, Nittaya Boontian, Maisarah Munirah Latief, Misdar Amdah, and Mohamed Sahrul Tamzil. 2025. "Application of syntrophic co-culture chlorella sorokiniana and streptomyces thermocarboxydus in semi-continuous system for cassava wastewater treatment and biomass production". *Green Technologies and Sustainability* 3 (3):100168. https://www.sciencedirect.com/science/article/pii/S2949736125000028.

Pavendan, K., and V. Nagarajan. 2022. "Modelling of wastewater treatment, microalgae growth and harvesting by flocculation inside photo bioreactor using machine learning technique". *Journal of Intelligent & Fuzzy Systems* 43 (5):5607–20. https://journals.sagepub.com/doi/abs/10.3233/JIFS-212676.

Perera, Isiri Adhiwarie, Sudharsanam Abinandan, Logeshwaran Panneerselvan, Suresh R. Subashchandrabose, Kadiyala Venkateswarlu, Ravi Naidu, and Mallavarapu Megharaj. 2022. "Co-culturing of microalgae and bacteria in real wastewaters alters indigenous bacterial communities enhancing effluent bioremediation". *Algal Research* 64:102705. https://doi.org/10.1016/j.algal.2022.102705.

Pitawala, Sulochana, Peter J. Scales, and Gregory J. O. Martin. 2025. "A new modeling approach for predicting the growth of filamentous algae in outdoor algae-Based wastewater treatment systems". *Biotechnology and Bioengineering*, bit.28941. https://doi.org/10.1002/bit.28941.

Pulz, Otto. 2001. "Photobioreactors: Production systems for phototrophic microorganisms". *Applied Microbiology and Biotechnology* 57:287–93.

Qi, Wenqiang, Shouhua Mei, Yizhong Yuan, Xuyang Li, Tao Tang, Quanyu Zhao, Minghong Wu, Wei Wei, and Yuhan Sun. 2018. "Enhancing fermentation wastewater treatment by co-culture of microalgae with volatile fatty acid-and alcohol-degrading bacteria". *Algal Research* 31:31–39. https://www.sciencedirect.com/science/article/pii/S2211926417307324.

Ramandani, Adityas Agung, Jun Wei Roy Chong, Sirasit Srinuanpan, Jun Wei Lim, Jheng-Jie Jiang, and Kuan Shiong Khoo. 2025. "Artificial intelligence-driven prediction models for the cultivation of Chlorella vulgaris FSP-E in food waste culture medium: A comparative analysis and validation of models". *Algal Research* 86:103935. https://www.sciencedirect.com/science/article/pii/S221192642500044X.

Recio-Colmenares, Carolina Livier, Jean Flores-Gómez, Juan Pablo Morales Rivera, Hasbleidy Palacios Hinestroza, and Belkis Sulbarán-Rangel. 2025. "Green materials for water and wastewater treatment: Mechanisms and artificial intelligence". *Processes* 13 (2):566. https://www.mdpi.com/2227-9717/13/2/566.

Roostaei, Javad, Yongli Z. Wager, Weisong Shi, Timothy Dittrich, Carol Miller, and Kishore Gopalakrishnan. 2023. "IoT-Based Edge Computing (IoTEC) for Improved Environmental Monitoring". *Sustainable Computing: Informatics and Systems* 38:100870. https://doi.org/10.1016/j.suscom.2023.100870.

Sahu, Sudarshan, Anupreet Kaur, Gursharan Singh, and Shailendra Kumar Arya. 2023. "Harnessing the potential of microalgae-bacteria interaction for eco-friendly wastewater treatment: A review on new strategies involving machine learning and artificial intelligence". *Journal of environmental management* 346:119004. https://www.sciencedirect.com/science/article/pii/S0301479723017929.

Semple, Austin, and Jagroop Pandhal. 2025. "Engineering the phycosphere: fundamental concepts and tools for the bottom-up design of microalgal-bacterial consortia". *Applied Phycology* 6 (1):21–51. https://doi.org/10.1080/26388081.2024.2435040.

Senthil Rathi, B., P. Senthil Kumar, S. Sanjay, M. Prem Kumar, and Gayathri Rangasamy. 2025. "Artificial intelligence integration in conventional wastewater treatment techniques: Techno-economic evaluation, recent progress and its future direction". *International Journal of Environmental Science and Technology* 22 (1):633–58. https://doi.org/10.1007/s13762-024-05725-2.

Soranno, Patricia A., Patrick J. Hanly, Katherine E. Webster, Tyler Wagner, Andrew McDonald, Arnab Shuvo, Erin M. Schliep, et al. 2025. "Abrupt changes in algal biomass of thousands of US lakes are related to climate and are more likely in low-disturbance watersheds". *Proceedings of the National Academy of Sciences* 122 (9):e2416172122. https://doi.org/10.1073/pnas.2416172122.

Sun, Jingxian, Xianzheng Yuan, Xiaoshuang Shi, Chunfeng Chu, Rongbo Guo, and Hainan Kong. 2011. "Fermentation of *Chlorella* sp. for anaerobic bio-hydrogen production: Influences of inoculum–substrate ratio, volatile fatty acids and NADH". *Bioresource technology* 102:10480–85. https://doi.org/10.1016/j.biortech.2011.09.016.

Sundaram, Karthikeyan Meenatchi, Deepak Kumar, Jintae Lee, Selvaraj Barathi, and Karthik Rajendran. 2025. "Time series forecasting of microalgae cultivation for a sustainable wastewater treatment". *Process Safety and Environmental Protection*, 106845. https://www.sciencedirect.com/science/article/pii/S0957582025001120.

Takahashi, Miiku, Ryosuke Yamada, Takuya Matsumoto, and Hiroyasu Ogino. 2024. "Co-culture systems of microalgae and heterotrophic microorganisms: Applications in bioproduction and wastewater treatment and elucidation of mutualistic interactions". *World Journal of Microbiology and Biotechnology* 40 (11):368. https://doi.org/10.1007/s11274-024-04173-x.

Tam, Do, Kyoung-Rok Kim, Minh-Thu Nguyen, Mi Kim, Donhue Kim, and Sang Sim. 2010. "Enhancement of fermentative hydrogen production from green algal biomass of Thermotoga neapolitana by various pretreatment methods". *International Journal of Hydrogen Energy* 35:13035–40. https://doi.org/10.1016/j.ijhydene.2010.04.062.

Tian, Lei, Yuhong Li, Jing Yang, Lin Qin, Hanglong Ma, Liandong Zhu, and Zhihong Yin. 2024. "Promoting the achievement of wastewater treatment by microalgae-based co-culture systems: From interactions mechanisms to pollution control performance". *Energy Conversion and Management* 320:118981. https://www.sciencedirect.com/science/article/pii/S0196890424009221.

Toth, Andras Jozsef. 2020. "Modelling and optimisation of multi-stage flash distillation and reverse osmosis for desalination of saline process wastewater sources". *Membranes* 10 (10):265. https://doi.org/10.3390/membranes10100265.

Virta, Leena, and Per Hedberg. 2024. "Declining salinity and increasing temperature reduce the diversity and resilience of benthic diatoms". *Environmental Microbiology* 26 (2):e16569. https://doi.org/10.1111/1462-2920.16569.

Wang, Baolong, Pengfei Ma, Mingyuan Liu, Ruiping Huang, Zhujun Qiu, Lanlan Pan, Jinghan Wang, Ying Liu, and Qian Zhang. 2025. "Enhancement of microalgae co-cultivation self-settling performance and water purification capacity of microalgae biofilm". *Environmental Research* 265:120342. https://doi.org/10.1016/j.envres.2024.120342.

Wolf, Juliane, Evan Stephens, Sven Steinbusch, Jennifer Yarnold, Ian Lawrence Ross, Christian Steinweg, Anja Doebbe, C. Krolovitsch, Steffen Müller, and Gisela Jakob. 2016. "Multifactorial comparison of photobioreactor geometries in parallel microalgae cultivations". *Algal research* 15:187–201.

Xiu, Chunbo, and Liying Dong. 2019. "Design of sewage treatment monitoring system based on internet of things". In *2019 Chinese Control and Decision Conference (CCDC)*, 960–64. IEEE.

Xu, Kaiwei, Zihan Zhu, Haining Yu, and Xiaotong Zou. 2025. "Prediction of microalgae harvesting efficiency and identification of important parameters for ballasted flotation using an optimized machine learning model". *Algal Research* 87:103985. https://www.sciencedirect.com/science/article/pii/S2211926425000943.

Xu, Ming, Zhaoxia Xue, Shiqing Sun, Chunzhi Zhao, Jinhua Liu, Juan Liu, and Yongjun Zhao. 2020. "Co-culturing microalgae with endophytic bacteria increases nutrient removal efficiency for biogas purification". *Bioresource Technology* 314:123766. https://www.sciencedirect.com/science/article/pii/S0960852420310385.

Yang, Jingya, Yingqi Huang, Gaojuan Zhao, Boqun Li, Xiangshi Qin, Jianchu Xu, and Xiong Li. 2022. "Phytoremediation potential evaluation of three rhubarb species and comparative analysis of their rhizosphere characteristics in a Cd- and Pb-contaminated soil". *Chemosphere* 296:134045. https://doi.org/10.1016/j.chemosphere.2022.134045.

Younas, Muhammad, Fahad Rehman, Sulaiman Al Zuhair, Faisal Ahmed, Muzamal Muzafar, Ali Awad, Maryam Asif, and Fahed Javed. 2025. "Synergistic approach to industrial wastewater treatment: Combining plasmolysis and microalgae cultivation". *Chemical Engineering and Processing-Process Intensification*, 110198. https://www.sciencedirect.com/science/article/pii/S0255270125000480.

Yu, Kai Ling, Hwai Chyuan Ong, and Halimah Badioze Zaman. 2024. "Integrated energy informatics technology on microalgae-based wastewater treatment to bioenergy production: A review". *Journal of Environmental Management* 368:122085. https://www.sciencedirect.com/science/article/pii/S0301479724020711.

Yuliantini, Emy, Jacky Chin, Tatyana Nikolaevna Tukhkanen, E. Laxmi Lydia, and K. Shankar. 2019. "The role of decision support system and risk management". *Journal of Critical Reviews* 6 (5):111–16.

Zhang, Jingshen, Shuangxi Li, Huiling Lu, Liandong Zhu, and Fengchang Wu. 2025. "Lighting strategy drives removal of ammonia nitrogen and phosphate in microalgae-bacteria consortia under tetracycline hydrochloride exposure". *Algal Research* 88:103989. https://doi.org/10.1016/j.algal.2025.103989.

Zhang, Yu, Huizeng Liu, Fei Wang, Ping Zhu, Zhengxin Zhang, Yanru Wang, Yongquan Wang, Guofeng Wu, and Qingquan Li. 2024. "Toward applicable retrieval models of oceanic particulate organic nitrogen concentrations for multiple ocean color satellite missions". *IEEE Transactions on Geoscience and Remote Sensing* 62:1–14. https://doi.org/10.1109/TGRS.2024.3447699.

Zhu, Defei, Runqing Yang, and Dong Wei. 2023. "Enhancing fucoxanthin production in *Phaeodactylum tricornutum* by photo-fermentation". *Sheng wu gong cheng xue bao = Chinese journal of biotechnology* 39:1070–82. https://doi.org/10.13345/j.cjb.220540.

17 Monthly Analysis of Oceanographic Conditions in the Western Mediterranean Sea

S. Souiri, M. Hakkou, H. Haddaoui, A. Aangri, S. Oubdil, and T. Chtioui

17.1 INTRODUCTION

The Western Mediterranean Sea, a semi-enclosed basin with complex hydrological dynamics, exhibits seasonal and interannual variations in physico-chemical parameters that profoundly influence its ecosystem. Understanding these variations is crucial for the sustainable management of marine resources and for assessing the impacts of climate change on this particularly sensitive system. Anthropogenic pressures, including coastal pollution, exacerbate this sensitivity (Salem et al., 2021).

Sea surface temperature (SST), salinity (SSS), and ocean currents are key indicators of the ocean's state (Johnson et al., 2012; Kug & An, 2010). SST influences water column stratification, the metabolism of marine organisms, and species distribution. SSS is affected by evaporation, precipitation, river inputs, and exchanges with the Atlantic Ocean. Ocean currents redistribute heat, salt, and nutrients, playing a role in the dispersion of pollutants and larvae. While SST variations are traditionally interpreted in terms of heat exchange with the atmosphere (Yu & Weller, 2007), recent studies highlight the role of dynamic ocean processes such as advection and upwelling (Johnson et al., 2012; Kug & An, 2010; Spiro Jaeger & Mahadevan, 2018).

In regions where stratification is dominated by salinity, surface cooling can induce enthalpy (T-S) compensation, where a decrease in temperature is offset by an increase in salinity, thus maintaining water column stability (Spiro Jaeger & Mahadevan, 2018). Nutrients, particularly nitrate, are essential for primary production by phytoplankton (Joo et al., 2018). Nitrate availability is influenced by water column stratification, vertical mixing, and freshwater inputs. Primary production, in turn, forms the base of the marine food web and influences biogeochemical cycles. Climate change, with its increased temperatures, altered precipitation patterns, and acidification, threatens to disrupt these processes and affect the oceanographic dynamics of the Mediterranean.

Within this context, this study aims to analyze the monthly variation of physico-chemical parameters (SST, SSS, ocean currents) and their impact on primary production in the Western Mediterranean Sea. Specifically, we seek to describe the seasonal cycles of SST, SSS, and ocean currents, quantify the relative contribution of dynamic and thermodynamic processes to SST variations, and evaluate the potential impact of climate change on the monthly and seasonal variations of these three parameters in the Western Mediterranean Sea. To address these questions, we used satellite data and numerical model outputs covering the period 1995–2019, and applied a methodology

DOI: 10.1201/9781003647263-17

combining GIS tools and statistical analyses. The study area, data, and methods are described in detail in the following sections.

17.2 MATERIALS AND METHODS

17.2.1 Study Area

This study focuses on the Western Mediterranean Sea, a region of significant economic and strategic importance characterized by its complex bathymetry and hydrographic features (Figure 17.1). The area is geographically defined by longitudes extending from 2°E to 34.5°E and latitudes from 36°N to 47°N. This encompasses key sub-basins and areas, including: the Strait of Gibraltar (a critical point of exchange with the Atlantic Ocean), the Algero-Provençal Basin (a region of intense deep-water formation), and the Tyrrhenian Sea. Due to its location at the crossroads of major shipping routes and its rich marine resources, this region is of immense international interest and significance.

The Western Mediterranean Sea experiences a microtidal regime, with tidal ranges generally less than 0.5 meters. These tidal ranges, computed using the FES2014 model data from AVISO,

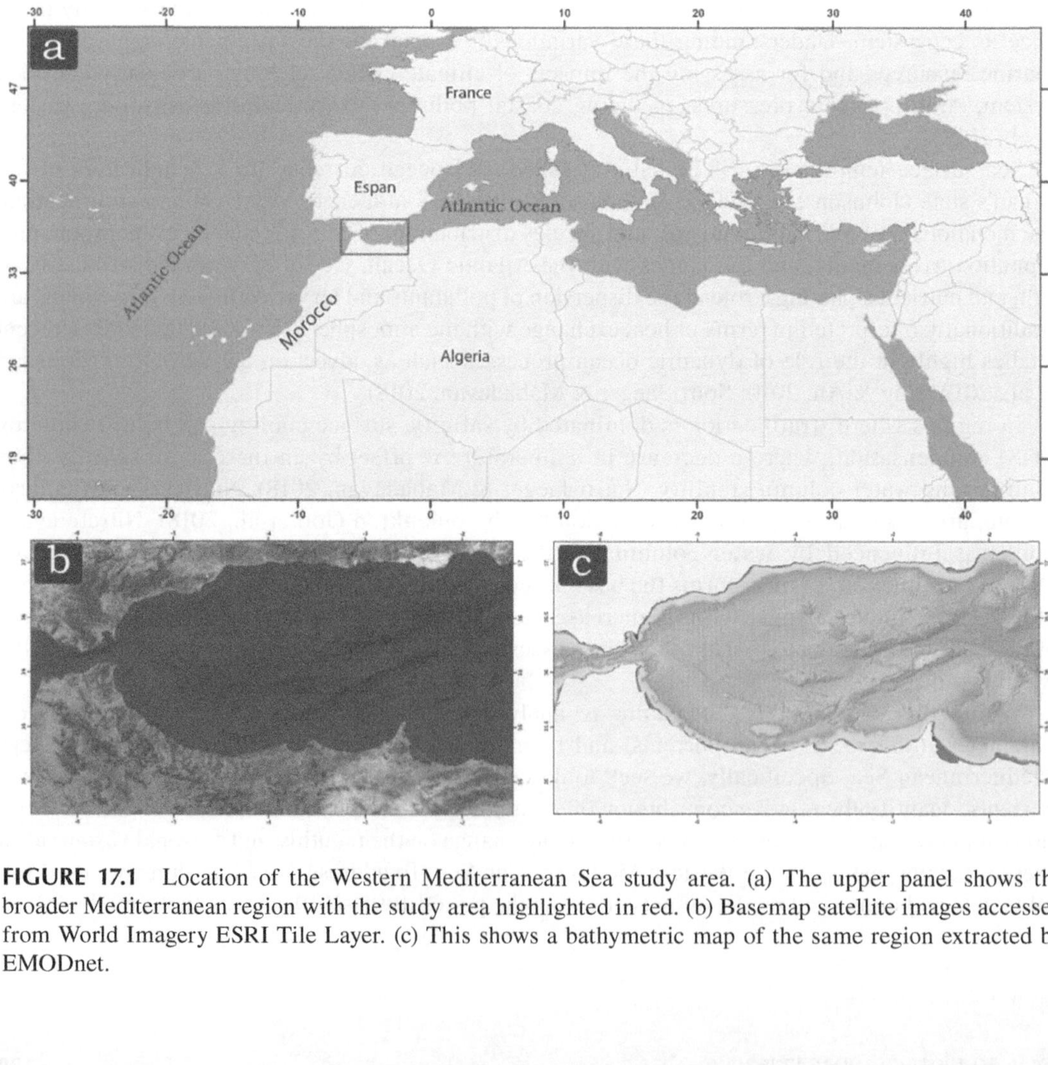

FIGURE 17.1 Location of the Western Mediterranean Sea study area. (a) The upper panel shows the broader Mediterranean region with the study area highlighted in red. (b) Basemap satellite images accessed from World Imagery ESRI Tile Layer. (c) This shows a bathymetric map of the same region extracted by EMODnet.

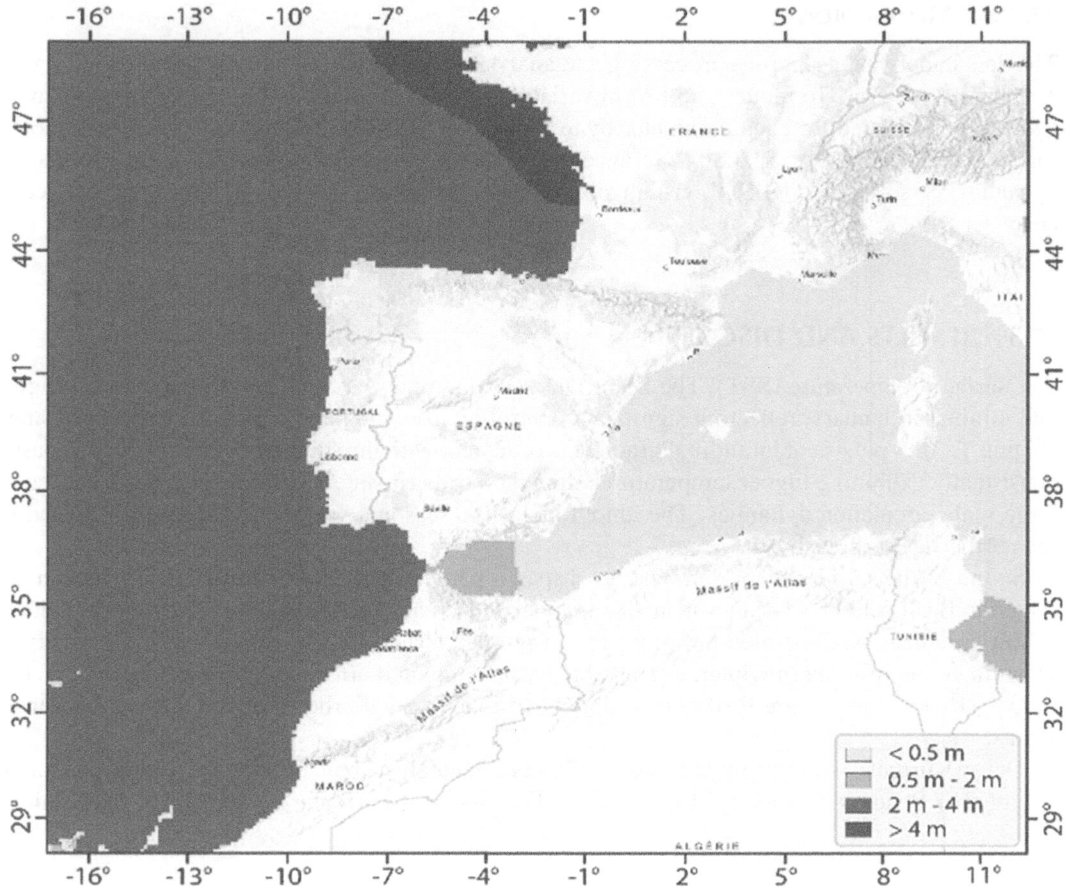

FIGURE 17.2 Tidal range along the coast of Europe. Data obtained from the Global Tidal Range data set computed by ESRI (https://www.arcgis.com/home/item.html?id=d5354dea41b14f0689860bf4b2cf5 e8a, accessed on 2 June 2023) using the FES2014 tide model produced by Noveltis, Legos and CLS and distributed by Aviso+, with support from Cnes (https://www.aviso.altimetry.fr/, accessed on 2 June 2023).

are based on lunar and solar semi-diurnal and diurnal harmonic amplitudes. Variations in the tidal range, as well as the complex bathymetry and coastline, influence local currents, stratification, and sediment transport patterns, as seen in Figure 17.2 (Souiri et al., 2025).

17.2.2 Data Used

Our present study utilizes satellite data and numerical model outputs spanning the period 1993–2019 to analyze the monthly variation of physico-chemical parameters in the Western Mediterranean Sea. The primary data source is the Copernicus Marine Service, providing access to high-resolution data sets for sea surface temperature (SST), sea surface salinity (SSS), and sea surface currents. The SST data, expressed in degrees Celsius (°C), provide critical information on surface water temperatures and their seasonal fluctuations. SSS data, measured in practical salinity units (psu), are essential for understanding freshwater inputs, evaporation patterns, and water mass mixing. Sea surface currents, quantified in meters per second (m/s), describe the horizontal movement of surface water, which influences heat and nutrient transport.

17.2.3 Methodology

The data underwent a rigorous processing and analysis procedure to ensure data quality and reveal meaningful patterns. To reduce short-term variability and highlight seasonal trends, monthly averages were calculated for each parameter by averaging all available data points within each month for the period 1993–2019. Spatially averaged data were then mapped using Geographic Information System (GIS) software (QGIS) to visualize spatial and temporal variations in SST, SSS, and ocean currents.

17.3 RESULTS AND DISCUSSION

Sea Surface Temperature (SST): The SST maps reveal strong seasonality, with maxima in August and minima in January, reflecting significant control by insolation and atmospheric heat exchange (Figure 17.3). A persistent latitudinal gradient is observed, with the southern waters (African coasts) consistently exhibiting higher temperatures than the northern waters (European coasts), reflecting large-scale circulation dynamics. The amplitude of the seasonal variation is more pronounced in the northern part of the basin.

Salinity: The salinity maps demonstrate a spatial pattern characterized by higher values in the Algerian Basin and lower values near the Spanish and French coasts, resulting in a south to north salinity gradient, as illustrated in Figure 17.4. Seasonal variability is also observed, with maxima in summer and minima in winter, attributable to the combined effects of evaporation and precipitation. These variations are further modulated by exchanges with the Atlantic Ocean and riverine inputs.

Ocean Currents: The current analysis highlights a general cyclonic circulation pattern, the intensity of which varies seasonally (Figure 17.5). The strongest currents are located in the Strait of

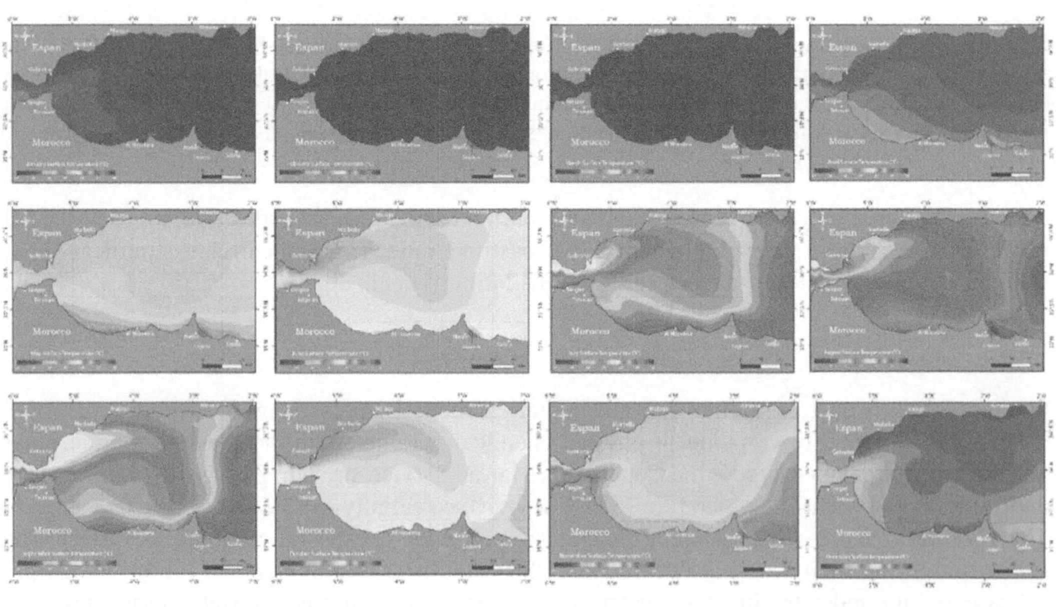

FIGURE 17.3 Climatological monthly sea surface temperature (SST) maps for the Western Mediterranean Sea between 1995 and 2019.

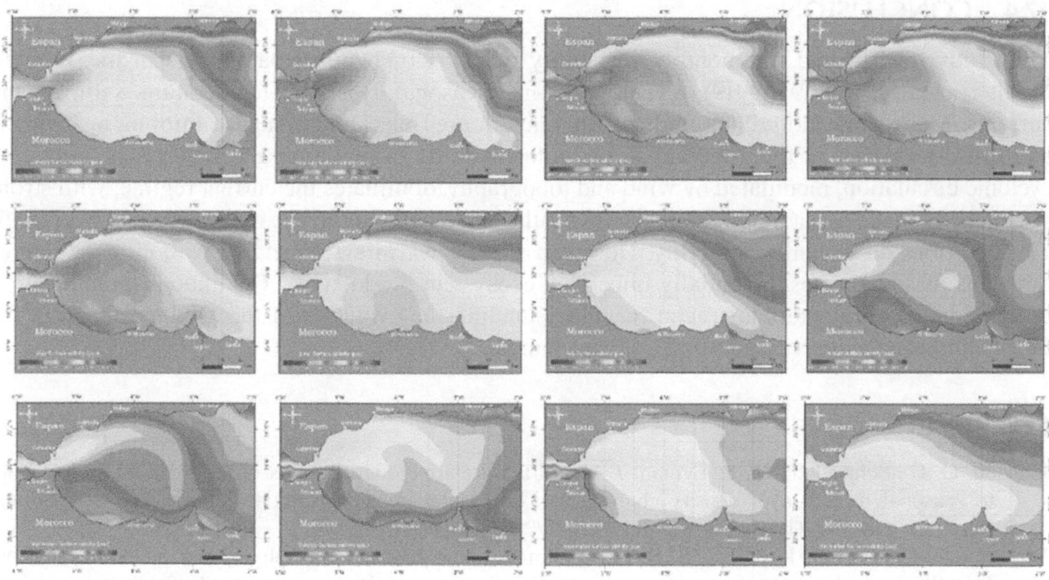

FIGURE 17.4 Climatological monthly sea surface salinity maps for the Western Mediterranean Sea between 1995 and 2019.

Gibraltar and along the North African coasts, while a weaker flow characterizes the center of the basin. This dynamic is modulated by wind regimes, with maximum intensities in winter and minimum intensities in summer. Bottom topography and the Coriolis force exert significant control over the configuration of this circulation, which plays a key role in the redistribution of hydrological properties.

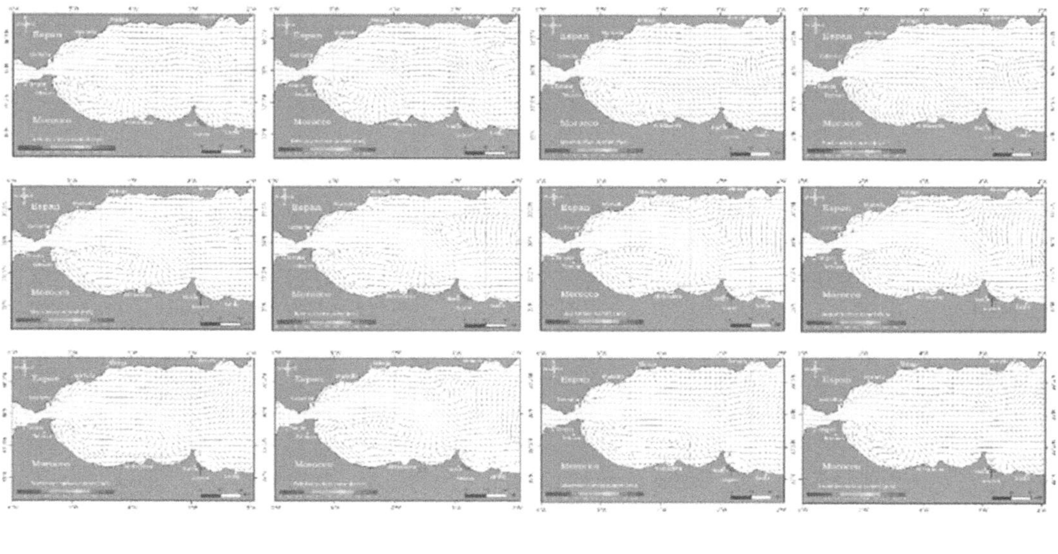

FIGURE 17.5 Climatological monthly surface current maps (vectors) for the Western Mediterranean Sea between 1995 and 2019.

17.4 CONCLUSION

This study investigated the monthly variability of physico-chemical parameters in the Western Mediterranean Sea (1995–2019), revealing distinct seasonal cycles for temperature, salinity, and currents. Maps show latitudinal gradients in temperature (maxima in August, minima in January) and salinity (higher in the Algerian Basin, influenced by evaporation and freshwater inputs). A cyclonic circulation, modulated by wind and topography, dominates the current regime, with stronger intensity in the Strait of Gibraltar. These results confirm the influence of climatic and hydrological conditions on oceanic variability. However, surface warming associated with climate change threatens these balances, potentially impacting ecosystem productivity. Continuous monitoring of oceanographic parameters is crucial for understanding the evolution of the Mediterranean in the face of anthropogenic pressures and climate change.

REFERENCES

Johnson, G. C., Schmidtko, S., & Lyman, J. M. (2012). Relative contributions of temperature and salinity to seasonal mixed layer density changes and horizontal density gradients. Journal of Geophysical Research: Oceans, 117(C4). https://doi:10.1029/2011JC007651.

Joo, H., Lee, D., Son, S. H., & Lee, S. H. (2018). Annual new production of phytoplankton estimated from MODIS-derived nitrate concentration in the East/Japan sea. Remote Sensing, 10(5), 806. https://doi.org/10.3390/rs10050806.

Kug, J.-S., & An, S.-I. (2010). Ocean dynamic processes responsible for the interannual variability of the Tropical Indian Ocean SST Associated with ENSO. Atmosphere, 20(2), 211–219.

Salem, D., Ibrahim, M., Mohamed, L., Hemaida, H., & El-shorbagi, E. (2021). Fluctuations of some physico-chemical parameters with depth in the Egyptian Mediterranean waters off Alexandria during winter 2016.

Souiri, S., Hakkou, M., Haddaoui, H., Oubdiln, S., & Lamrimed, N. (2025). Automatic coastline detection using CoastsatToolbox and GIS: A case study at Oued-LaouBeach, Morocco. E3S Web of Conferences, 607, 04024. https://doi.org/10.1051/e3sconf/202560704024.

Spiro Jaeger, G., & Mahadevan, A. (2018). Submesoscale-selective compensation of fronts in a salinity-stratified ocean. Science Advances, 4(2), e1701504.

Yu, L., & Weller, R. A. (2007). Objectively analyzed air–sea heat fluxes for the global ice-free oceans (1981–2005). Bulletin of the American Meteorological Society, 88(4), 527–539.

18 Self-Cleaning Solar Distillation System Using Piezoelectric Vibrations

S. Nasrdine, J. Mabrouki, and M. Slaoui

18.1 INTRODUCTION

Access to drinking water is a major problem in many parts of the world, particularly in arid zones where freshwater resources are limited (Ben-Lhachemi et al., 2024a; Nasrdine et al., 2024d, 2024g, 2024h). Solar distillation has emerged as an ecological solution to this problem, using the sun's energy to purify water (Ben-Lhachemi et al., 2024b, 2024c, 2024d; Nasrdine et al., 2024a). This process is based on the principle of evaporation and condensation using the heat emitted by the sun. However, in spite of its numerous advantages, this technology comes up against a major obstacle: the fouling of evaporation and condensation surfaces (Nasrdine et al., 2024c, 2024b, 2024e, 2024f). Over time, mineral deposits from salt water or suspended particles accumulate on these surfaces, reducing their efficiency and requiring frequent cleaning. This phenomenon prevents solar distillers from working independently and limits their use in many regions.

Integrating piezoelectric elements into solar distillers is an interesting new method for solving this problem (Zhang et al., 2023). These systems produce small vibrations under the effect of an electric current, which prevents impurities from adhering and facilitates their removal. This vibration technique has already been used in solar panel cleaning and membrane filtration, with good results in terms of efficiency in keeping surfaces clean and optimizing system performance (Manzo et al., 2025). Adding this technology to solar distillers will improve operation and reduce maintenance requirements. This chapter will present a prototype, focusing on individual components, analyze results and draw general conclusions.

18.2 LITERATURE REVIEW

The materials used, such as PZT (lead-zirconate-titanate), are highly efficient but not very environmentally friendly, which is why piezoelectric polymers are being explored as alternatives (Zhang et al., 2023). By adding improvements to piezoelectric composites, it is possible to optimize energy conversion, making these technologies more efficient and more suitable for industrial applications (Manzo et al., 2025). A new field of application for piezoelectric systems is their integration into self-cleaning solutions. By exploiting piezoelectric vibrations, unwanted deposits can be eliminated, limiting the accumulation of dust and contaminants on various surfaces (Mao et al., 2023). Recent studies have demonstrated that the combination of nanostructures and high-frequency vibrations significantly improves cleaning efficiency and reduces surface wear (Kar et al., 2019).

Many technologies have been employed in the field of self-cleaning systems to minimize particle adhesion and facilitate particle removal. Hydrophobic and superhydrophobic coatings based on nanometric structures prevent the attachment of dust and residues and improve the cleanliness of surfaces exposed to harsh environments (Austria et al., 2024). The same thing for photocatalytic coatings using titanium dioxide (TiO_2) exploit light to degrade impurities, these techniques are used in industry and the environment (Pu et al., 2021). Another concept is based on the use of ultrasonic waves generated by piezoelectric devices, which effectively loosen stuck particles and ensure

DOI: 10.1201/9781003647263-18

continuous cleaning for optical and solar equipment (Di et al., 2024). Recent studies have shown that integrating flexible microstructures with piezoelectric sensors can optimize these self-cleaning processes, improving both their efficiency and durability (Mao et al., 2019).

18.3 PROPOSED SYSTEM

18.3.1 System Principle

The proposed system is based on a conventional solar distiller, modified by the integration of piezoelectric modules at critical points:

On the evaporation surface: to prevent crystallization of salts by generating micro-vibrations.
On the condensation surface: to prevent the accumulation of droplets and dust, ensuring efficient flow of distilled water. These modules are connected to an excitation circuit, enabling precise control of the vibration frequency (100 Hz–10 kHz) to optimize their self-cleaning effect.

18.3.2 System Construction

The distiller consists of many important components, as illustrated in Figure 18.1:

A storage tank that is for saltwater exposed to the sun.
An absorber facilitates water evaporation.
A sloping glass cover allows steam to condense to form pure water.
A distilled water collector collects purified water in liquid form.

Piezoelectric modules are positioned on the evaporation and condensation surface, integrated into an electronic control circuit.

18.3.3 System Operation

Evaporation: salt water is heated by solar energy and then evaporates, leaving behind impurities and salts.
Activation of piezoelectric modules: Vibrations prevent salt crystallization on the evaporation surface and promote a smooth flow of steam.
Improved condensation: Water condenses on the glass surface, where vibrations prevent the accumulation of droplets and dust, improving system efficiency.
Distilled water collection: Purified water circulates freely in the recovery tank.

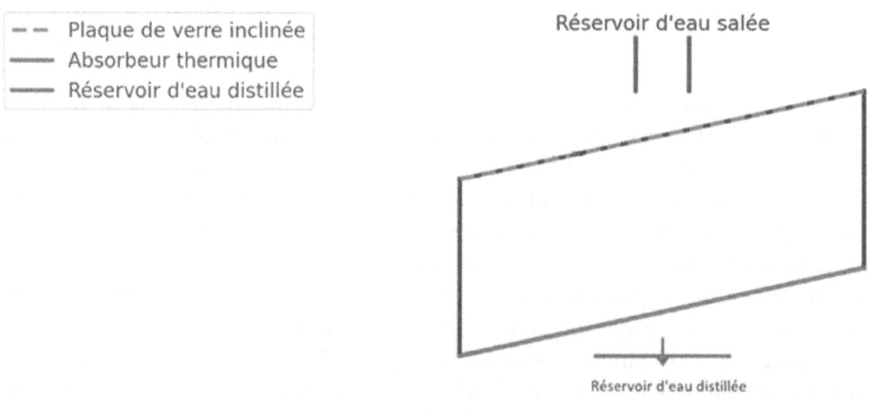

FIGURE 18.1 Simplified diagram of the conventional solar distiller.

18.3.4 System Advantages

Higher efficiency: thanks to clogging prevention, evaporation, and condensation are optimized.
Reduced maintenance requirements: self-cleaning limits the buildup of salt and dust deposits.
Energy autonomy: The piezoelectric modules do not consume much energy and can be powered by a small solar panel (Lan et al., 2022).

18.4 FEASIBILITY AND OPTIMIZATION PROSPECTS

18.4.1 Technical Feasibility

Mounting piezoelectric modules in a solar distiller is inspired by existing technologies, but their application to a self-cleaning system requires certain verifications:

Material compatibility: Piezoelectric modules must be resistant to wet and saline conditions to ensure their durability.
Optimization of vibration frequencies: Experimental studies are required to determine the optimum frequency to prevent salt crystallization and maximize droplet drainage.
Energy consumption: The addition of piezoelectric modules needs to be energy-efficient if it is to be powered autonomously via a solar panel (Li et al., 2023).

18.4.2 Optimization Potential

Many ways of improving the system's performance could be explored:

Combination with superhydrophobic coatings to reduce droplet adhesion and minimize vibration-related energy consumption.

Use of advanced piezoelectric materials, such as flexible polymers, for better adaptation to curved surfaces and optimized energy production.

Intelligent automation with salinity and fouling sensors to dynamically adjust vibration intensity as required (Ranasinghe et al., 2023).

18.5 CHALLENGES AND CONSTRAINTS

Despite its potential, the system presents some challenges:

Initial cost: The addition of piezoelectric modules and an excitation circuit may increase the price of the distiller.
Validation in real-life conditions: To assess the system's robustness, it is obligatory to test in different environments (arid, coastal, industrial).
Manufacturing complexity: Integration of the piezoelectric modules into the distiller structure must be straightforward to avoid compromising the commercial viability of the project (Huo et al., 2021).

18.6 ENVIRONMENTAL IMPACT AND DURABILITY

18.6.1 Reduction of the Ecological Footprint

The use of solar energy for water distillation represents an environmentally friendly alternative to traditional desalination methods that consume a lot of energy. The addition of piezoelectric modules further enhances this aspect, while reducing maintenance requirements and consuming a minimum of energy, contrary to conventional systems that require regular cleaning with chemicals or additional fresh water (Bagchi et al., 2020).

This technology enables:

- A reduction in water consumption for maintenance.
- A reduction in chemical waste associated with cleaning dirty surfaces.
- Greater energy efficiency, thanks to a renewable energy source that allows the system to operate autonomously.

18.6.2 The Durability of Materials

The challenge of piezoelectric technology is the durability of materials under extreme environmental conditions, notably prolonged exposure to humidity and temperature variations. To guarantee optimum system durability:

The use of corrosion-resistant materials, such as piezoelectric ceramics and advanced polymers.

The recycling of electronic components, in particular piezoelectric modules and excitation circuits, to limit the environmental impact of the system at the end of life (Pang et al., 2025).

18.7 CONCLUSION

The development of high-performance solar distillation systems is essential to meet the growing challenges of access to drinking water, particularly in arid regions. One of the main obstacles limiting the efficiency of conventional solar distillation systems is the fouling of evaporation and condensation surfaces. To resolve this problem, the piezoelectric vibrations are integrated to improve process efficiency and reduce maintenance requirements (Guptaa et al., 2019).

In this study, we proposed a system that combines energy efficiency, autonomy and sustainability, offering a viable alternative to conventional methods exploiting mechanical vibrations to prevent salt crystallization and promote efficient distilled water flow represents a significant advance. However, despite its many advantages, certain constraints remain, notably in terms of initial cost, optimization of vibration frequencies and material durability under real-life conditions. Prospects for improvement include integrating hydrophobic coatings, optimizing piezoelectric components and automating the system according to environmental conditions (Aydt & Nsofor, 2019).

REFERENCES

Austria, H. F. M., Sardome, R. P., Setiawan, O., Huang, T.-H., Lei, W.-C., Tian, X.-Y., Hu, C.-C., Lee, K.-R., Lai, J.-Y., Tayo, L. L., Caparanga, A. R., & Hung, W.-S. (2024). Efficient membrane fouling mitigation in self-cleaning piezoelectric PVDF-graphene loose nanofiltration membranes for sustainable textile wastewater treatment. *Separation and Purification Technology, 346*, 127317. https://doi.org/10.1016/j.seppur.2024.127317

Aydt, W. E., & Nsofor, E. C. (2019). Studies on solar-powered water purification with PCM energy storage. *6*(4). https://www.jmest.org/wp-content/uploads/JMESTN42352914.pdf

Bagchi, B., Hoque, N. A., Janowicz, N., Das, S., & Tiwari, M. K. (2020). Re-usable self-poled piezoelectric/piezocatalytic films with exceptional energy harvesting and water remediation capability. *Nano Energy, 78*, 105339. https://doi.org/10.1016/j.nanoen.2020.105339

Ben-Lhachemi, N., Benchrifa, M., Nasrdine, S., Mabrouki, J., & Slaoui, M. (2024a). Assessing the environmental effects associated with flood protection measures. In J. Mabrouki & M. Azrour (Éds.), *Integrated Solutions for Smart and Sustainable Environmental Conservation* (pp. 119–131). Springer Nature Switzerland. https://doi.org/10.1007/978-3-031-55787-3_9

Ben-Lhachemi, N., Benchrifa, M., Nasrdine, S., Mabrouki, J., Slaoui, M., & Azrour, M. ade. (2024b). Effect of IoT integration in agricultural greenhouses. In J. Mabrouki & A. Mourade (Éds.), *Technical and Technological Solutions Towards a Sustainable Society and Circular Economy* (pp. 435–445). Springer Nature Switzerland. https://doi.org/10.1007/978-3-031-56292-1_35

Ben-Lhachemi, N., Benchrifa, M., Nasrdine, S., Mabrouki, J., & Slaoui, M. (2024c). Intelligent temperature and humidity control system. In *Artificial Intelligence Systems in Environmental Engineering* (pp. 63–69). Springer. https://doi.org/10.1201/9781003436218-7

Ben-Lhachemi, N., Nasrdine, S., Mabrouki, J., & Slaoui, M. (2024d). Integration of bio-sensors-IoT system for the evaluation and control of soil quality. In J. Mabrouki & A. Mourade (Éds.), *Advanced Technology for Smart Environment and Energy* (pp. 221–229). Springer Nature Switzerland. https://doi.org/10.1007/978-3-031-50871-4_14

Di, B., Wang, Z., Wang, H., Gong, S., Zheng, L., Min, Y., & Li, H. (2024). Piezoelectric photocatalytic degradation of formaldehyde based on BFO@OCN heterojunctions. *Nano Energy*, *132*, 110384. https://doi.org/10.1016/j.nanoen.2024.110384

Guptaa, Y., Katakam, S. M., Aligireddy, S. R., Shameer, W., & Atheaya, D. (2019). Experimental study of self-sustainable hybrid solar photovoltaic cleaning mechanism coupled with water distillation unit. *Vibroengineering Procedia*, *29*, 213–218. https://doi.org/10.21595/vp.2019.21140

Huo, Z.-Y., Lee, D.-M., Kim, Y.-J., & Kim, S.-W. (2021). Solar-induced hybrid energy harvesters for advanced oxidation water treatment. *iScience*, *24*(7). https://doi.org/10.1016/j.isci.2021.102808

Kar, E., Bose, N., Dutta, B., Banerjee, S., Mukherjee, N., & Mukherjee, S. (2019). 2D SnO2 nanosheet/PVDF composite based flexible, self-cleaning piezoelectric energy harvester. *Energy Conversion and Management*, *184*, 600–608. https://doi.org/10.1016/j.enconman.2019.01.073

Lan, S., Jing, B., Yu, C., Yan, D., Li, Z., Ao, Z., & Zhu, M. (2022). Protrudent iron single-atom accelerated interfacial piezoelectric polarization for self-powered water motion triggered Fenton-like reaction. *Small*, *18*(2), 2105279. https://doi.org/10.1002/smll.202105279

Li, F., Miao, G., Hou, Z., Xu, L., Zhu, X., Miao, X., Song, Y., Ren, G., & Li, X. (2023). Tree-inspired aerogel with a radial and centrosymmetric structure for efficient solar-powered water purification. *ACS Applied Materials & Interfaces*, *15*(23), 27952–27962. https://doi.org/10.1021/acsami.3c02867

Manzo, M., Maymoun, M., Qamar, R., Rihani, Y., & Al-Jahran, A. (2025). A novel solar panel self-cleaning method based on piezoelectric films. *Journal of Renewable and Sustainable Energy*, *17*(1), 013501. https://doi.org/10.1063/5.0242347

Mao, H., Bu, J., Qiu, M., Ding, D., Chen, X., Verweij, H., & Fan, Y. (2019). PZT/Ti composite piezoceramic membranes for liquid filtration : Fabrication and self-cleaning properties. *Journal of Membrane Science*, *581*, 28–37. https://doi.org/10.1016/j.memsci.2019.03.022

Mao, H., Zhou, S., Qiu, M., Zhang, Z., Xue, A., Li, M., Zhao, Y., & Fan, Y. (2023). Piezoceramic membrane equipped with superwetting interface and in-situ ultrasound performance for efficient oil/water emulsion separation. *Desalination*, *555*, 116545. https://doi.org/10.1016/j.desal.2023.116545

Nasrdine, S., Benchrifa, M., Ben-Lhachemi, N., Mabrouki, J., & Slaoui, M. (2024a). Effect of various factors on the productivity of solar stills : Mini review. *Euro-Mediterranean Journal for Environmental Integration*, *9*(2), 977–987. https://doi.org/10.1007/s41207-024-00465-8

Nasrdine, S., Benchrifa, M., Ben-Lhachemi, N., Mabrouki, J., & Slaoui, M. (2024b). Remote monitoring of an inclined solar distiller. In *Artificial Intelligence Systems in Environmental Engineering* (pp. 56–62). Springer. https://doi.org/10.1201/9781003436218-6

Nasrdine, S., Benchrifa, M., Ben-Lhachemi, N., Mabrouki, J., Slaoui, M., & Azrour, M. (2024c). New design of an inclined solar distiller for freshwater production : Experimental study. In J. Mabrouki & A. Mourade (Éds.), *Technical and Technological Solutions Towards a Sustainable Society and Circular Economy* (pp. 447–453). Springer Nature Switzerland. https://doi.org/10.1007/978-3-031-56292-1_36

Nasrdine, S., Benchrifa, M., Mabrouki, J., Qisse, N., & Slaoui, M. (2024d). Comparative study of the quality of water produced by a solar distiller and by reverse osmosis. *Desalination and Water Treatment*, *317*, 100264. https://doi.org/10.1016/j.dwt.2024.100264

Nasrdine, S., Benchrifa, M., Mabrouki, J., & Slaoui, M. (2024e). The application of solar distillation technology as a sustainable solution for seawater desalination in the arid regions of the world. In J. Mabrouki & A. Mourade (Éds.), *Technical and Technological Solutions Towards a Sustainable Society and Circular Economy* (pp. 59–66). Springer Nature Switzerland. https://doi.org/10.1007/978-3-031-56292-1_4

Nasrdine, S., Benchrifa, M., Mabrouki, J., & Slaoui, M. (2024f). The use of solar stills in agriculture : Sustainable irrigation and food production. In M. Azrour, J. Mabrouki, A. Alabdulatif, A. Guezzaz, & F. Amounas (Éds.), *Smart Internet of Things for Environment and Healthcare* (pp. 241–249). Springer Nature Switzerland. https://doi.org/10.1007/978-3-031-70102-3_17

Nasrdine, S., Mabrouki, J., Ben-Lhachemi, N., & Slaoui, M. (2024g). Application of machine learning algorithms for air quality prediction. In J. Mabrouki & M. Azrour (Éds.), *Integrated Solutions for Smart and Sustainable Environmental Conservation* (pp. 95–104). Springer Nature Switzerland. https://doi.org/10.1007/978-3-031-55787-3_7

Nasrdine, S., Mabrouki, J., Ben-Lhachemi, N., & Slaoui, M. (2024h). Design and dimensioning of a green hydrogen production system. In J. Mabrouki & A. Mourade (Éds.), *Advanced Technology for Smart Environment and Energy* (pp. 43–51). Springer Nature Switzerland. https://doi.org/10.1007/978-3-031-50871-4_3

Pang, S. Y., Teh, W. J., Oh, T. H., & Chua, S. C. (2025). Design and development of a hemisphere flat base solar thermal water distillation system | *Journal of Engineering Technology and Applied Physics*. Consulté 22 mars 2025, à l'adresse https://journals.mmupress.com/index.php/jetap/article/view/767

Pu, L., Zhang, J., Wang, C., Pan, Y., Zhao, Y., Bu, Y., Zhang, Q., Pan, B., & Gao, G. (2021). Membrane cleaning strategy *via in situ* oscillation driven by piezoelectricity. *Journal of Membrane Science*, *638*, 119722. https://doi.org/10.1016/j.memsci.2021.119722

Ranasinghe, M. K. N. K., Kahandage, P. D., Kosgollegedara, E. J., & Weerasooriya, G. V. T. V. (2023). Design and implementation of an advanced solar water distillation system for the efficient purification of contaminated water into clean drinking water. *Journal of Dry Zone Agriculture*, *9*(2). https://doi.org/10.4038/jdza.v9i2.74

Zhang, L., Zheng, H., Zhou, X., Liu, B., Chen, S., & Li, K. (2023). A universal piezoelectric micropump with capabilities of self-cleaning, stable filtration and precise pumping. *IEEE Transactions on Industrial Electronics*, 1–9. https://doi.org/10.1109/TIE.2023.3241389

19 Environmental Implications of Anthropogenic Activities on Water Quality of Mohammedia Ramsar Site, Morocco

H. Jounaid, M. Taoufik, A. Moufti, E. El Hachimi, H. Noureddine, and M. Boutaib

19.1 INTRODUCTION

Wetlands offer many ecological services in both rural and urban contexts (Costanza et al., 1997). They improve water quality, provide flood storage, feed ground water, retain and remove nitrogen, host wildlife habitat and promote the general preservation of diminishing open space (Hammer, 1996; Mitsch and Gosselink, 1993; Richardson and Vepraskas, 2001). The numerous benefits of wetlands are now threatened by pollution and land use changes (Junhong et al., 2013). Because of this, many wetlands around the world are recognized by international treaties such as the Ramsar Convention on Wetlands (Toyra and Pietroniro, 2005) and their protection has been paid much more attention to by most governments such as Morocco, who ratified Ramsar Convention on Wetlands and set many plans and strategies to comply with the provisions of this convention. Despite this, natural wetlands in Morocco are in continuous loss and degradation. Losses on the only coastal wetlands are estimated, in Morocco, at about 34,000 hectares, which are covered for farming or urbanization (Alibou, 2002).

Mohammedia urban wetland is a Ramsar Site which corresponds to the estuary of El Maleh River and its floodplain (Dakki et al., 2011). According to Leopold (1968), in an urban environment every aspect of water is affected by the urban process. The latter impacts the ecosystems such as urban wetlands, leading to a potential breakdown of water services offered by such ecosystems. Within the current literature on Mohammedia wetlands, biotic and soil parameters are used dominantly in determining the ecosystem's health. Some studies are available on surface water characteristics but based on analysis carried out between 2000 and 2002 (Serghini, 2003; Serghini et al., 2010, 2013).

This article presents current water quality investigation with wider spatial coverage including Mohammedia wetland and a selected segment of El Maleh River which feeds the wetland. The study consists of determining the dominant contaminants and their spatial variation in the study area. The objective is to highlight the possible connections of El Maleh River water quality and ecological problems in Mohammedia wetland. This can help to better plan the protection and restoration of this important ecosystem. A field study was carried out to identify and map the pollution discharged into El Maleh River. The sampling points are selected to acquire representative data on the spatial variability of the impact of this pollution. To identify the dominant contaminants, physico-chemical and bacteriological analyses were carried out on 15 parameters. The results of the analyses will be compared to the threshold relating to fish life (Official Bulletin, 2004) to verify the ability of Mohammedia wetland to offer natural conditions for biodiversity.

DOI: 10.1201/9781003647263-19

19.2 MATERIALS AND METHODS

19.2.1 Description of the Study Area

Mohammedia Wetland City (MWC), located on the Atlantic coast in the west of Morocco. The geographical location of the MWC is between 33° 33'N and 07° 23' W. Average elevation of the MWC area is between –6 m and 6 m above sea level. The prevailing climate in the study area is semi-arid. The average annual rainfall is 430 mm. The maximum precipitation is recorded in December and the minimum in July. The mean temperature is 18°C. The average evaporation is 1400–1500 mm/year (Dakki et al., 2011). This wetland is extended over 3.4 km in a SW–NE direction; it is constituted mainly by some temporary marshes, fed by rainfall and El Maleh River floods, a permanent water body supplied by seawater during high tide and another permanent body of water fed by groundwater (Serghini et al., 2013). Mohammedia wetland is surrounded by residential areas, a golf covering 27 ha and several industries mainly an oil refinery and coal-fired power plant (Figure 19.1).

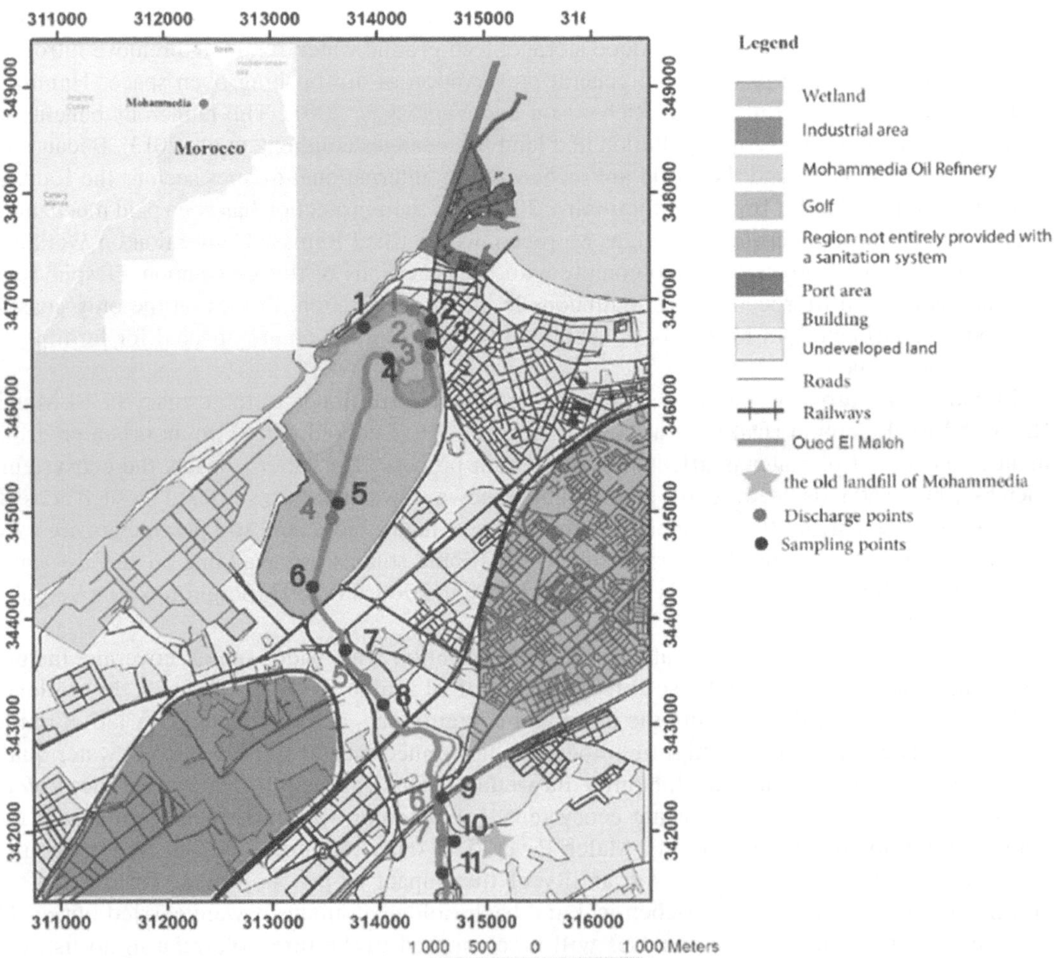

FIGURE 19.1 Study area and water quality sampling sites.

19.2.2 Sampling and Analysis

The location of the sampling sites was carried out with a Garmin 72 GPS. The surface water samples were in June 2019. Sampling sites included six sites in Mohammedia wetland (S1–S6) and five sites in selected segment of El Maleh River (S7–S11). The sampling network (Figure 19.1) was designed to cover sites, which reasonably represented the water quality of El Maleh River accounting discharges, identified during the field observation and which could impact Mohammedia wetland water quality. The sampling station 11 is chosen as the reference station, located upstream from urban activities observed along the El Maleh River. All samples were collected, in double, in new polypropylene bottles rinsed with deionized water and sterilized glass bottles (for bacteriology) [Rodier, 2009]. The parameters measured and recorded in situ were dissolved oxygen (DO), pH, temperature and electrical conductivity. All samples were stored at 4°C in the dark and transported immediately to the laboratory. BOD5 (20°) and Faecal Coliform were tested on the same day. The analyses were performed according Rodier, 2009) in GAIA laboratory in Hassan II University of Casablanca–Ain Chock Faculty of Sciences.

Five heavy metals (cadmium (Cd), manganese (Mn), lead (Pb), zinc (Zn) and chromium (Cr) were measured at seven selected points only (1–3, 5, 6, 9–11). The other parameters were measured in all the samples. The results of the analysis are presented in Table 19.1.

TABLE 19.1
Parameters Measured in the Study Area during June 2019

Station / Parameter	S1	S2	S3	S4	S5	S6	S7	S8	S9	S10	S11	Thresholds Tolerable for Fish life M.E.M.E. (2004)
Température	23.1	22.8	25.9	26.5	31.4	25	28	25.9	29.2	33.8	24.9	30
OD	8.25	7.46	6.2	3.41	3.16	5.94	5.23	6.67	7.14	5.04	7.7	3
pH	7.37	7.12	6.51	7.17	7.28	7.48	6.5	7.7	7.08	7.97	7.64	5–9
Condictivity (µS/cm)	7370	7120	6510	7170	7480	7280	6540	7700	7080	7970	7640	3000
DBO_5 (mg/L)	4	10	20	40	140	30	60	40	200	957	10	6
NO^{3-} (mg/L)	0.03	0.01	0.013	0.02	0.82	0.01	0.02	0.135	0.223	0.76	0.28	
NO^{2-} (mg/L)	0.03	0.002	0.035	0.12	0.03	0.001	0.02	1.495	1.47	1.37	0.47	0.5
$PO4^{3-}$(mg/L)	0.112	0.054	0.053	0.28	0.38	0.463	0.05	0.753	0.81	1.20	0.02	
FC (MPN / 100 mL)			43000	3000	51000	10000	45000	4000	25000	115000	15000	2000
SO_4^{2-} (mg/L)	1535.05	258.92	512.69	118.37	86.15	526.04	947.63	354.97	585.20	910.745	166	200
Chlorides (g/L)	10.36	10.26	8.46	7.9	3.46	0.3	0.3	0.15	0.4	9.2	0.28	
Cd (µg/L)	100.5		78		60	48.5			53	77.5	47.5	5
Mn (mg/L)	0.89		0.92		1.02	1.01			0.96	1.40	0.10	0.1
Pb (µg/L)	32.5		161.5		64.5	0.1			64.5	226	53.94	20
Zn (mg/L)	0.27		0.23		0.022	0.07			0.04	0.32	0.15	1.3

19.3 RESULTS AND DISCUSSION

19.3.1 Hydrogen Potential, Temperature, Conductivity and Dissolved Oxygen

The analysis of pH shows that surface water in the study area is slightly alkaline (Figure 19.2a). In the downstream part (S1–S3), the spatial distribution of pH ranging from 6.5 to 7.07 is characteristic of a tidal brackish water environment, as noted by Ajao and Fagade (2002). The alkalinity of water may also be due to the fact that this complex is located on a substrate containing carbonate and bicarbonates as reported by Serghini (2003). The lowest pH values recorded in station 7 may have resulted from industrial effluents discharged upstream from this sampling point. The highest value of pH (7.97) is recorded in sampling station 10, located close to a local landfill identified during the field study. This landfill was used earlier to dispose of urban solid waste high in organic content (Souabi et al., 2011) and its leachate is still flowing into El Maleh River. The mean pH value of the leached waste was 8.27 (Madinzi et al., 2011), which was compared to the value recorded in station 10. However, pH values recorded in all samples were well within the preferred pH of 5 to 9 recommended for fish life. Concerning the temperature, the highest value (33.8°C) was also recorded in station 10, attesting to a possible contamination by local landfill leachate. The lowest-value temperature was recorded in the downstream of the wetland (Sland S2) (Figure 19.2b). This may be due to the tidal action in this area, with cold incoming seawater.

Electrical conductivity was strongly related to Cl where both have the same distribution pattern along the study area (Figure 19.2c). It ranged from 51,970 mg/L to 4419. In all samples, EC exceeded the Moroccan standard for fish life (3000 us/c). These very high values were not only related to the high salinity of the predominantly maritime estuary in S1, S2 and S3, but could also be due to urban discharges, well expressed in station S10 located upstream.

The value of DO varied from 8.25 g/L (S2) to 3.16 g/L (S5) (Figure 19.2d). DO concentrations were within the permissible limits according to the Moroccan standards relating to fish life (DO should not be less than 3 mg/L). According to McNeely et al. (1979), the solubility of oxygen is dependent on temperature: at high temperature, the solubility of oxygen decreases and vice versa. Indeed, OD and temperature have an opposite distribution trend along the study area. The lowest levels of oxygen recorded near an offloading canal (S5) and at S4 indicated high organic matter and nutrients load (Davies et al., 2008; El-Gamel and Shafik, 1985). In the wetland (S3–S1), the higher concentrations of oxygen could be related to the brewing with seawater.

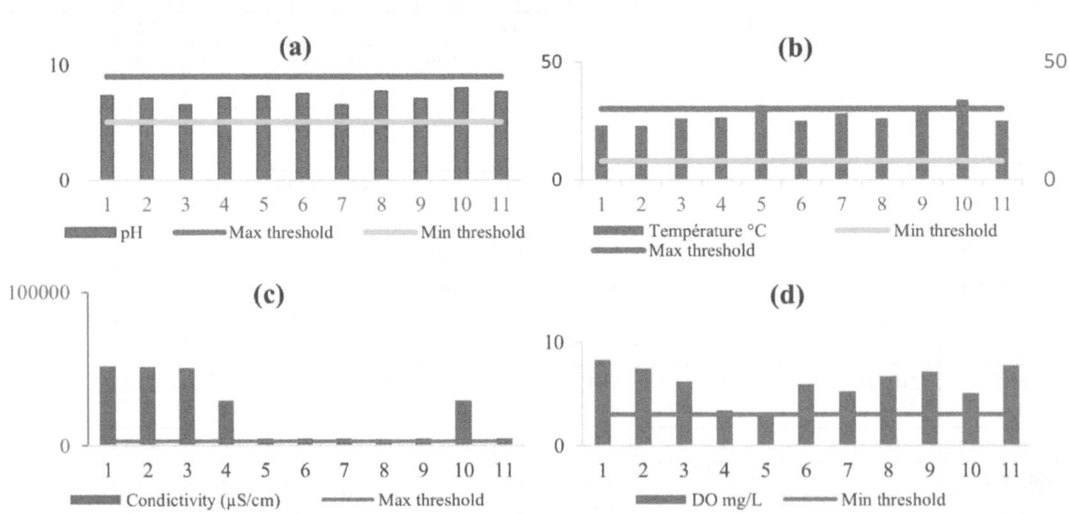

FIGURE 19.2 Spatial distribution of pH (a), temperature (b), electrical conductivity (c) and dissolved oxygen (d) in the study area during June 2019.

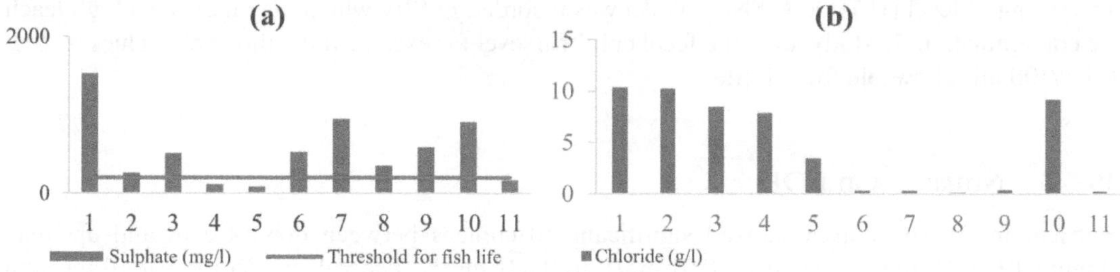

FIGURE 19.3 Spatial distribution of sulfate (a) and chloride (b) in the study area on June 2019.

19.3.2 Sulfate and Chloride

Sulfate varies between 86.15 g/L and 1535 g/L. The highest sulfate value was observed at sampling stations 1 (Figure 19.3a) located upstream the golf and near several industries, using fossil fuel, mainly coal-fired power plant. According to Bertolaccini and Gucci (1985), sulfate emission is due to the oxidation of sulphur dioxide. Delmas et al. (2010) noted that sulfate can precipitate from the atmosphere to the ground due to the rains. These very high values of sulfates are not only due to industrial activities, but could be related to urban discharges well expressed at stations S7 and S10 located upstream. The lower concentrations of sulfates recorded at sampling stations S5 and S4 could be related to a sulfate-reducing microorganisms' activity. These microorganisms might be highly active in wetlands and have a considerable stake in wetland sulfate reduction (Pester et al., 2012). At most sampling sites, sulfate concentrations were more than 200 mg/L, which is the safe limit for fish life.

Regarding chloride, except at sampling station 10, it increased from upstream to downstream along the segment studied of the watercourse (Figure 19.3b). The concentration of chloride ranged between 0.3 g (S6) and 0.15 g/L (S8) upstream and increased rapidly from station 3 to attain 10,361 g/L in station 1. This was related to the predominantly seawater in this area (Dakki et al., 2011). The highest chloride value in the upstream (9.2 g) recorded in station 10 could be mainly related to contamination by the old landfill. In fact, the waste contained significant levels of chlorides (Reitzel, 1990).

19.3.3 Fecal Coliforms

The presence of coliform bacteria in water is considered a principal indicator of water pollution (Geoffray and Vial, 1975). The bacteria and other infectious agents are often associated with sewage wastes (Kennish, 2002). Indeed, in the segment studied of El Maleh River, the higher concentrations of fecal coliforms were recorded in S5, S7 and S3 located near sewage effluent (Figure 19.4).

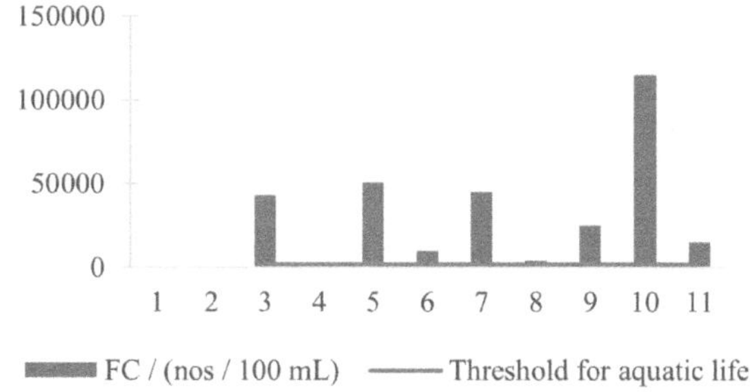

FIGURE 19.4 Spatial distribution of fecal coliform in the study area on June 2019.

The strongest level (115,000 MPN/100 mL) was recorded in S10, which may indicate a high leachate contamination. In study area, the fecal coliform level far exceeded the threshold values of 2000 MPN/100 mL allowable for fish life.

19.3.4 Nutrient and BOD

Nutrient in the study area showed significant differences between downstream and upstream (Figure 19.5). Nitrate ranges from 0.01 mg/L to 0.821 mg/L. The values recorded at all the sampling stations are lower than 25 mg/L, which is the maximum allowable for a good surface water according the Moroccan regulation (M.E.M.E., 2002). High nitrates levels were recorded upstream (Figure 19.5a). The nitrate peak recorded was at S5 (offloading canal) which drains runoff and domestic sewage discharged upstream. The lower concentrations recorded downstream (in the wetland) could be explained by the assimilation of nitrate by plants or using nitrates as a source of oxygen for degrading organic matter (denitrification). Haycock and Muscutt (1995) reported that the nitrogen cycle in wetlands was mainly characterized by two processes: uptake by plants and denitrification. Wetlands have been regarded as skins of nitrates and should be considered for this function (Harrison et al., 2011; Rutherford et al., 2009). A similar spatial variation pattern was recorded with regards to nitrite (Figure 19.5b) and phosphate (Figure 19.5c). According to Burt and Pinay (2005), wetlands bordering rivers can significantly reduce the flow of nitrogen and phosphorus through them. High nitrate and nitrite levels determined along the upstream, particularly, at S10, S9, S8 and S5 could be related to significant organic matter degradations by nitrification process through which ammonia is oxidized into nitrite and nitrate by aerobic bacteria (Abbott and Dawson, 1971). Nitrite and nitrate concentrations can explain the significant oxygen biologically demanded at S5, S9 and S10 (Figure 19.5d). BOD exceeds at all sampling sites, 6 mg/L, which is the tolerable threshold for fish life.

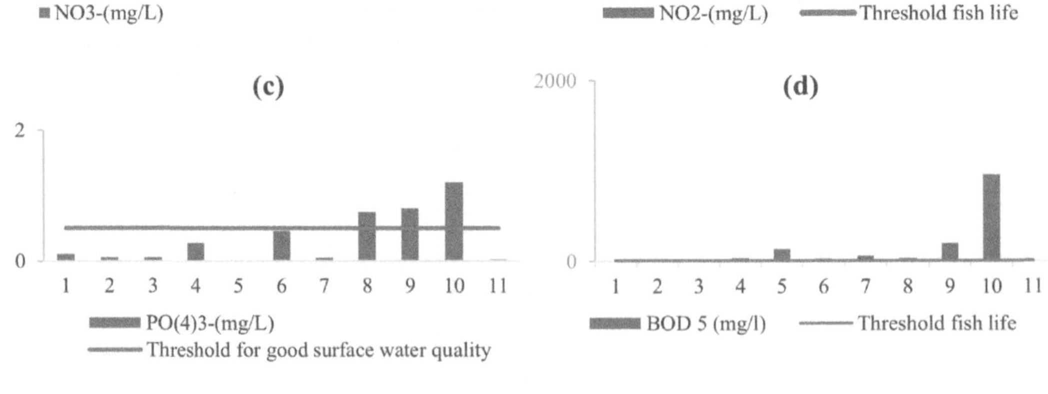

FIGURE 19.5 Spatial distribution of nitrate (a), nitrite (b), phosphate (c) and BOD_5 (d) in the study area during June 2019.

19.3.5 Heavy Metals

Chromium, copper, zinc (Zn), lead, manganese and cadmium (Cd) were measured in 7 selected sampling points: chromium was not detected in all samples and copper was detected in only three samples. The analyses show that apart from Zn (Figure 19.6a), all other heavy metals detected were found in levels exceeding the stipulated maximum allowable value for fish life (Official Bulletin, 2004) at all sampling stations. This indicates inorganic pollution, possibly by industrial sources (Hettiarachchi et al., 2011). The heavy metals may also originate from the domestic solid-waste dump. The typical urban domestic solid waste contains material such as dry batteries, paint containers, empty pesticide bottles and electronic waste which may leach heavy metals in the long run (Arceivala, 2004). Analyses carried out on the leachate of the Mohammedia old dump have shown lead concentrations of up to 690 µg/L (Souabi et al., 2010). The sampling station S10, located close to the landfill, recorded significant lead concentration of about 226 µg/L (Figure 19.6b), attesting to the contamination of surface water by the leachate. This station also shows the highest concentration of manganese (Figure 19.6c). Wastewater is also responsible for surface water contamination by manganese (WHO, 2004). Indeed, the sampling stations S9 and S5 located near wastewater discharges in El Maleh River show significant concentrations of manganese. Concerning cadmium, the values recorded downstream are more important than upstream (Figure 19.6d). Station S1, located in the estuary, showed the highest value (100.5 µg/L). This can be explained by the affinity of Cd with chlorine in a salty environment. This affinity resulted in a process of desorption of Cd from the particulates towards the dissolved phase (Bourg, 1987; Turner et al., 1993; Waeles et al., 2005). The significant contamination of sites S1 and S3 by Cd could also be related to phosphate fertilizer runoff (Brignon and Malherbe, 2005; N'Guessan, 2008) which could be used in Mohammedia golf covering 27 ha in this area. According to N'Gessman (2008), some phosphate fertilizers have Cd concentrations that can be about 300 times higher than the natural content of soils. We note that heavy metals are chemical contaminants which rank among the most dangerous in terms of potential impacts on organisms in estuaries such as the Mohammedia wetland (Kennish, 2002).

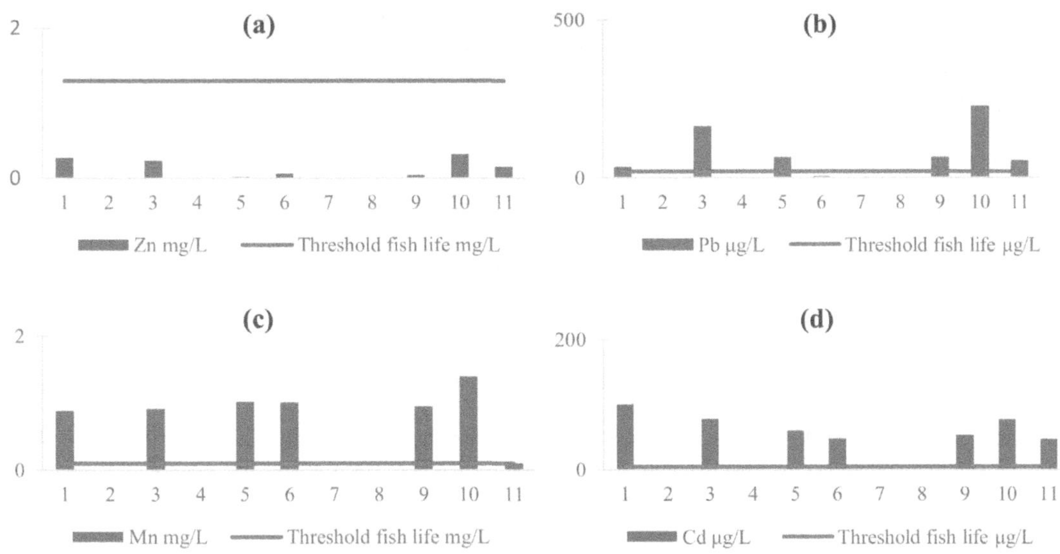

FIGURE 19.6 Spatial distribution of zinc (a), lead (b), manganese (c) and cadmium (d) in the study area during June 2019.

19.4 CONCLUSION

In this study, industrial activities, domestic sewage and the old landfill of Mohammedia city were identified as major pollution sources in the study area. Mn, Cd and Pb are the major heavy metal contaminants in surface water in the Mohammedia wetland. In the upstream, the concentrations of BOD, nitrite, phosphate and fecal coliform far exceeded the relevant threshold values. This is once more characteristic of any water source highly contaminated by organic pollutants (Metcalf and Eddy, 2001). The spatial variation of water quality parameters in the segment studied of El Maleh River showed that pollution generated upstream reaches the Mohammedia wetland and may lead to negative environmental impacts such as reduction of biodiversity and changes in composition of the wetland. The lower concentrations of nitrate in the downstream show that the Mohammedia wetland has the potential to reduce the level of nitrates and should be considered for this function. Wetlands often serve as nutrient sinks, filtering out bioavailable nutrients from receiving waters, storing them in sediments and converting them to organic forms, which may be stored or exported in less available forms (Brenner et al., 2001; Craft and Casey, 2000). It should be noted that, given the smaller volume and velocity of flow in El Maleh River during the sampling (hot dry season), the river system may have had a smaller capacity to mitigate pollution loads characteristics of the receiving water body (Chen et al., 2013). Serghini et al. (2010) found that some of the parameters studied had a lower concentration in the wetland during the high water period. Therefore, more frequent releases from the dam located upstream could help dilute the pollution reaching the wetland. These results provided useful help for decision makers to put in place effective measures to reduce pollution to acceptable levels in the Mohammedia wetland.

REFERENCES

Abbott W. and Dawson C.E. (1971) 'Physical, Chemical and Biological Characteristics of Estuaries', in *Water and Water Pollution*, 2nd edition. Ciaccio, New York, 2: 51–140.

Ajao E.A. and Fagade S.O. (2002) 'The ecology of Neritina glabrata in Lagos lagoon, Nigeria', *Archiv Hydrobiologie* **119** (3), pp. 339–350.

Alibou J. (2002) 'Impact of climate change on water resources and wetlands in Morocco', Available online: https://uicnmed.org/web2007/CDCambio_climatico/contenido/D/PDF/CC_d7a.pdf.

Arceivala S.J. (2004) '*Wastewater Treatment for Pollution Control*' 2 edition. Tata McGraw Hill.

Bertolaccini M.A. and Gucci P.M.B. (1985) 'Atmospheric pollution in the Venice area I. Sulphur dioxide, suspended particles and sulfates', *Science of The Total Environment* **43** (3), pp. 243–253.

Bourg A.C. (1987) 'Trace metal adsorption modelling and particle-water interactions in estuarine environments', *Continental Shelf Research* **7**, pp. 1319–1332.

Brenner M., Schelske C.L. and Keenan L.W. (2001) 'Historical rates of sediment and nutrient accumulation in marshes of the Upper St. Johns River Basin, Florida, USA', *Journal of Paleolimnology* **26**, pp. 241–257.

Brignon J.M. and Malherbe L. (2005) '*Cadmium Et Ses Dérivés. INERIS – Données Technicoéconomiques Sur Les Substances Chimiques En France*' Verneuil en Halatte.

Burt T.P. and Pinay G. (2005) 'Linking hydrology and biogeochemistry in complex landscapes', *Progress in Physical Geography: Earth and Environment* **29** (3), pp. 297–316.

Chen H., Teng Y. and Yue W. (2013) 'Characterization and source apportionment of water pollution in Jinjiang River, China', *Environ Monit Assess* **185**, pp. 9639–9650.

Costanza R., D'Arge R., De Groot R., Farber S., Grasso M., Hannon B., Limburg K., Naeem S., O'Neill R.V., Paruelo J., Raskin R.G., Sutton P. and van den Belt M. (1997) 'The value of the world's ecosystem services and natural capital', *Nature* **387**, pp. 253–260.

Craft C.B. and Casey W.P. (2000) 'Sediment and nutrient accumulation in floodplain and depressional freshwater wetlands of Georgia, USA', *Wetlands* **20** (2), pp. 323–332.

Dakki M., El Agbani M.A. and Qninba A. (2011) 'Zones humides du Maroc inscrites jusqu'en 2005 sur la liste de la convention Ramsar', *Bull. Ins. Scien. Rabat, Série Générale* 7, p. 226.

Davies O.A., Ugwumba A.A. and Abolude D.S. (2008) 'Physico-chemistry quality of trans-amadi (Woji) creek, Port Harcourt, Niger delta, Nigeria', *Journal of Fisheries International* **3** (3), pp. 91–97.

Delmas R., Baudet J. and Servant J. (2010) 'Mise en évidence des sources naturelles de sulfate en milieu tropical humide', *Inter Science* **30** (2), pp. 158–168.

El-Gamel A. and Shafik Y. (1985) 'A study on the monitoring of pollutants discharging to the River Nile and their effect on river water quality', *Water Quality Bull* **10**, pp. 111–115.

Geoffray C. and Vial J. (1975) *'Bacteriological analysis'* In Rodier, J (ed.). Analysis Fifth International Water Technology Conference IWTC 2000, Alexandria, Egypt.

Hammer D.A. (1996) *'Creating Freshwater Wetlands'*, 2nd edition. Boca Raton: CRC Press.

Harrison M.D., Groffman P.M., Mayer P.M., Kaushai S.S. and Newcomer T.A. (2011) 'Denitrification in alluvial wetlands in an urban landscape', *Journal of Environmental Quality* **40** (2), pp. 634–646.

Haycock N.E. and Muscutt A.D. (1995) 'Landscape management strategies for the control of diffuse pollution', *Landscape and Urban Planning* **31** (1–3), pp. 313–321.

Hettiarachchi M., Anurangi J. and De Alwis A. (2011) 'Characterisation and description of surface water quality in the threatened urban wetlands around the City of Colombo', *Journal of Wetlands Ecology* **5** (0), pp. 28–3.

Junhong B., Baoshan C., Huicong C., Ainong L. and Baiyu Z. (2013) 'Wetland degradation and ecological restoration', *The Scientific World Journal* **2013**, pp. 1–2.

Kennish M.J. (2002) 'Environmental threats and environmental future of estuaries', *Environmental Conservation* **29** (01), pp. 78–107.

Leopold L. (1968) *'Hydrology for Urban Land Planning – A Guidebook on the Hydrologic Effects of Urban Land Use'* US Geol Surv Circ.

M.E.M.E. (2002) http://www.environnement.gov.ma/fr/lois-et-reglementations/normes?id=1008 (accessed January 6th, 2020).

M.E.M.E. (2004). http://www.environnement.gov.ma/fr/lois-et-eglementations/normes?id=1010 (accessed January 6th, 2020).

Madinzi A., Aboulhassane M., Assou L., Fakid H., Bahlaoui M.A., Souabi S. and Hafidi M. (2011) 'Biological treatment of stabilized landfill leachate produced by the public discharge of the city Mohammedia (Morocco)', *Scientific Study & Research Chemistry & Chemical Engineering, Biotechnology, Food Industry* **15** (2), pp. 155–167.

McNeely R.N., Neimanis V.P. and Dwyer L. (1979) *'Water quality sources book – A guide to water quality parameters'*, pp. 112.

Metcalf and Eddy Inc. (2001) *'Wastewater Engineering-Treatment and Reuse'*, McGraw-Hill.

Mitsch W.J. and Gosselink J.G. (1993) *'Wetlands'* 2nd edition. Van Nostrand Reinhold.

N'Guessan Y.M. (2008) 'Dynamique des éléments traces dans les eaux de surface des bassins versants agricoles de Gascogne', National Polytechnique Institute of Toulouse, Toulouse, Ph. D. Thesis, 253 p.

Official Bulletin of the Kingdom of Morocco No. 5196. (2004) 'Joint order no. 2028-03 of 5th November 2003 defining the quality norms of fish water', *Moroccan Official Bulletin: Rabat, Morocco* **2004**, pp. 505–507.

Pester M., Knorr K.H., Friedrich M.W., Wagner M. and Loy A. (2012) 'Sulfate-reducing microorganisms in wetlands - fameless actors in carbon cycling and climate change', *Frontiers in Microbiology* **3**, p. 72.

Reitzel A. (1990) 'The temporal characterization of municipal solid waste leachate', Waterloo, Ontario, Univ. Waterloo, Masters of applied Science Thesis, 144 p.

Richardson J.L., and Vepraskas M.J. (2001) *'Wetland Soils: Genesis, Hydrology, Landscapes, and Classification'*, Boca Raton, CRC Press.

Rodier J. (2009) *'L'analyse De l'eau, Eaux Naturelles, Eaux Résiduelles, Eaux De Mer'* 9th edition. Dunod.

Rutherford J.C., Schrorer D. and Timpagny G. (2009) 'How muchrunoff do riparian wetlands affect?, *New Zealand Joutnal of Marine and Freshwater Research* **43** (5), pp. 1079–1094.

Serghini A. (2003). Diagnose du complexe zones humides de Mohammedia: étude qualitative et quantitative des composantes physique (eau – sédiment) et biologique de l'écosystème. Univ. Mohammed V, Rabat, PhD Thesis, 178 p.

Serghini A., El Abidi A., El Blidi S. and Fekhaoui M. (2013) 'Typologie spatio- temporelle et fonctionnement hydrologique du complexe de zones humides de Mohammedia (Maroc)', *Bulletin de l'Institut Scientifique, Rabat, Section Sciences de la Vie* **35**, pp. 61–65.

Serghini A., Fekhaoui M., El Abidi A., El Blidi S. and Ben Akkame R. (2010) 'Caractérisation hydrochimique d'un site Ramsar: le complexe de zones humides de Mohammedia (Maroc)', *Bull. Inst. Scien. Rabat, section Sciences de la Vie* **32**, pp. 133–145.

Souabi S., Touzar K., Chtioui H., Khalil F., Digua K. and Tahiri M. (2010) 'Problématiques du chrome et du plomb dans les lixiviats des décharges publiques des villes de Mohameddia et de Fès', *Déchets sciences et techniques* **58**, pp. 37–44.

Souabi S., Touzare K., Digua K., Chtioui H., Khalil F. and Tahiri M. (2011) 'Triage and valorization of solid waste in landfill city of Mohammedia', *les technologies de laboratoire* **6** (25), pp. 121–130.

Toyra J. and Pietroniro A. (2005) 'Towards operational monitoring of a northern wetland using geomatics-based techniques', *Remote Sensing of Environment* **97**, pp. 174–91.

Turner A., Millward G., Bale A. and Morris A. (1993) 'Application of the K D concept to the study of trace metal removal and desorption during estuarine mixing', *Estuar Coast Shelf Science* **36**, pp. 1–13.

Waeles M., Riso R.D. and Le Corre P. (2005) 'Seasonal variations of cadmium speciation in the Penzé estuary, NW France Estuar', *Coast Shelf Science* **65**, pp. 143–152.

World Health Organization. (2004) Manganese and its compounds: Environmental aspects. Tech. rep. World Health Organization.

20 Applications of Wireless Sensor Networks in Smart Agriculture

M. El Mouedden, J. Mabrouki, D. Azdem, and S. El Hajjaji

20.1 INTRODUCTION

Climate change, soil degradation, growing resource constraints, and a rapidly expanding population that is predicted to reach 10 billion by 2050 are all posing previously unheard-of challenges to the global agricultural landscape. A shift from conventional, labor-intensive farming to intelligent, data-driven agriculture is necessary to meet the growing demand for food, fiber, and fuel without further endangering the environment. In this regard, precision agriculture (PA) has become a game-changing method that uses digital technologies to track, evaluate, and improve farming operations on both a micro and macro level [1].

Wireless sensor networks (WSNs), a collection of widely dispersed sensor nodes equipped with wireless transceivers, microcontrollers, and environmental sensors, are at the core of precision agriculture. Real-time data on temperature, humidity, pH, light intensity, soil moisture, and plant health indicators are continuously gathered and transmitted by these nodes. To help farmers make timely and well-informed decisions, the collected data is either processed locally or sent to a central control system for advanced analytics. WSN enabled precision farming uses site-specific management, lowering input costs and increasing crop productivity, in contrast to traditional agricultural methods that apply uniform treatments across entire fields [2].

Usually incorporated into the Internet of Things (IoT) architecture, WSNs serve as the sensory layer, collecting fine-grained field-level data. WSNs offer smooth connectivity, remote monitoring capabilities, and data visualization tools when paired with IoT platforms and cloud services [3]. Additionally, communication protocols like Wi-Fi, SigFox, LoRaWAN, and ZigBee have been developed specifically for agricultural environments and provide low-power, long-range, and affordable solutions that work well in remote and infrastructure-constrained settings. These protocols are essential for guaranteeing dependable data transfer across a range of topographies and climates [4].

WSNs have a wide range of quickly growing applications in agriculture. These include early pest and disease detection systems based on microclimatic and biometric parameters, fertilizer optimization platforms that track soil nutrient content, smart irrigation systems that modify water delivery based on real-time soil moisture levels, and environmental monitoring tools that evaluate conditions in greenhouses or open fields. By facilitating accuracy in these areas, WSNs support sustainable practices that preserve soil health, water, and energy in addition to increasing crop yields and quality [5].

WSN deployment in agriculture is not without difficulties, despite their enormous potential. Careful system design and strong engineering solutions are required to address problems like short battery life, severe environmental conditions, sensor calibration drift, data loss, and network scalability. New technologies like edge computing, energy harvesting, AI-powered analytics, and blockchain-based data security are being investigated to overcome these constraints and improve WSN dependability in dynamic agricultural environments [6].

DOI: 10.1201/9781003647263-20

This chapter provides a comprehensive review of the current state and future potential of WSNs in precision agriculture. It begins with an overview of WSN architectures and their integration within IoT-based farming systems. It then examines key application areas such as irrigation, soil and crop monitoring, fertilization, and disease control highlighting both the technical frameworks and real-world implementations. The discussion also addresses critical limitations and emerging solutions, offering a road map for researchers, engineers, and policymakers seeking to advance the frontier of smart agriculture [7].

20.2 ARCHITECTURAL LAYERS AND WIRELESS SENSOR NETWORKS IN SMART AGRICULTURE

20.2.1 Structural Overview

WSNs are foundational components within the IoT ecosystem, particularly suited for gathering and transmitting real-time data from diverse physical and environmental conditions. In the domain of smart agriculture (SA), WSNs comprise an interconnected array of sensor nodes that communicate wirelessly and are capable of autonomous configuration, self-diagnosis, and dynamic adaptation to changing conditions. These networks are deployed in various agricultural settings, with terrestrial and subsurface (underground) WSNs being the most prevalent configurations [8]. A standard WSN architecture in agriculture typically consists of five functional layers (Figure 20.1):

- **Physical Layer**: This layer includes all hardware interfaces such as sensors, transceivers, and energy units. Sensors in this layer are engineered to detect critical agro environmental variables (e.g., soil moisture, air temperature, and radiation levels) [9].
- **Data Link Layer**: This layer governs node-to-node communication, ensuring data integrity and minimal transmission errors through mechanisms such as medium access control (MAC), framing, and error correction. Its reliability is vital for the accuracy of monitoring systems deployed across the farm landscape [10].
- **Network Layer**: Responsible for routing and data forwarding, this layer handles the transmission of packets across the sensor field to a central sink or server. Efficient routing algorithms are critical to ensure minimal energy consumption while maintaining robust data delivery, especially in geographically expansive or topographically complex farms.

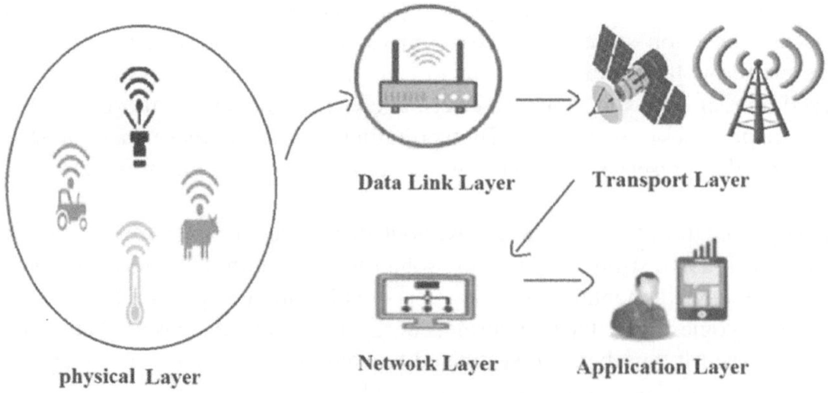

FIGURE 20.1 Architectural layers wireless sensor networks in smart agriculture.

- **Transport Layer**: This layer maintains end-to-end communication integrity and manages congestion control. It ensures that data transmitted from field sensors reaches the application server without loss, duplication, or delays, which is essential for time-sensitive agricultural operations like irrigation or pesticide application [11].
- **Application Layer**: The interface between the WSN and end users; this layer provides services such as data analytics, visualization dashboards, and decision-support tools. It enables farmers and agronomists to interpret sensor outputs and implement precision management practices based on actionable insights [12].

20.2.2 Wireless Communication Protocols in Smart Agriculture

Selecting the appropriate wireless communication protocol is a crucial design consideration in developing efficient WSNs for agricultural applications. Each protocol offers a unique combination of features that influence power consumption, data rate, range, and scalability. Understanding these differences allows system designers to tailor solutions that align with specific environmental and operational requirements [13].

In localized networks, like greenhouses or research plots, where mesh networking or high data throughput are desired, short-range protocols like ZigBee and Wi-Fi are frequently used. ZigBee is appropriate for low-power sensor clusters with moderate communication demands because it supports both star and mesh topologies and uses the IEEE 802.15.4 standard. Wi-Fi, which is based on the IEEE 802.11g standard, uses more energy but allows for high-speed connectivity over short distances [14].

On the other hand, long-range protocols such as LoRaWAN and SigFox are more appropriate for large-scale, open-field settings. SigFox employs an ultra-narrowband low-power wide-area network (LPWAN) technique that is highly energy efficient and optimized for small data transfers over long distances. With coverage spanning 5 to 45 kilometers and adjustable data rates, LoRaWAN provides a highly flexible deployment that can accommodate a range of use cases, from intermittent reporting to real-time monitoring. Adaptability based on energy and latency requirements is improved by its support for multiple device classes (A, B, and C) [15]. Table 20.1 summarizes key differences across these communication protocols:

TABLE 20.1
Role of Wireless Communication Technologies in Agricultural Sensor Networks

Attribute	ZigBee	Wi-Fi	SigFox	LoRaWAN	References
Topology	Star and mesh cluster	Star	Star	Star-of-stars	[16]
Data Rate	250 kbps	150 kbps	100–600 bps	0.3–27 kbps	[17]
Network Type	WPAN	WLAN	LPWAN	LPWAN	[17]
Frequency	2.4 GHz	2.4/60 GHz	868/902 MHz	Sub-GHz (varies by region)	[18]
Standard	IEEE 802.15.4	IEEE 802.11g	SigFox proprietary	LoRaWAN specification	[19]
Coverage Range	10–100 m	10–50 m	3–10 km (urban), up to 50 km (rural)	5–45 km	[19]
Key Features	Low-power mesh, IEEE MAC/PHY	High availability, web integration	Long-range, energy efficient	Low power, scalable and flexible	[20]

By leveraging the strengths of these technologies and matching them to the specific communication demands of different farming operations, WSNs can be made more robust, energy-aware, and scalable across a range of agricultural environments.

20.3 APPLICATIONS OF WSN IN AGRICULTURE

20.3.1 Smart Irrigation Management

Smart irrigation systems are among the most significant uses of WSNs in agriculture. Real-time soil moisture monitoring across various field zones is made possible by WSNs, which enables irrigation schedules to be dynamically adjusted in response to field conditions. Volumetric water content is measured by sensors and sent to a central processing unit, which uses actuators to start automated irrigation. This lowers water usage and avoids over-irrigation, which can cause nutrient leaching and root diseases [21].

The accuracy of irrigation decisions is further improved by integration with meteorological data (such as rainfall and evapotranspiration). Algorithms can predict the best times to water plants and estimate their water requirements, maximizing yield and minimizing water use (Figure 20.2) [21].

20.3.2 Soil Moisture and Nutrient Monitoring

Important agronomic factors like soil pH, salinity, nitrogen levels, and moisture content can be continuously monitored by soil sensors equipped with WSNs. By detecting nutrient toxicities or deficiencies, these measurements aid in the optimization of fertilization techniques. Root zone conditions can be accurately evaluated by placing multi-depth sensors, which offers information on water retention, infiltration rates, and the risks of nutrient leaching [22].

The creation of variable-rate application models and fertility maps is made possible by advanced data analytics. This results in higher crop quality, less impact on the environment, and significant fertilizer use savings [22].

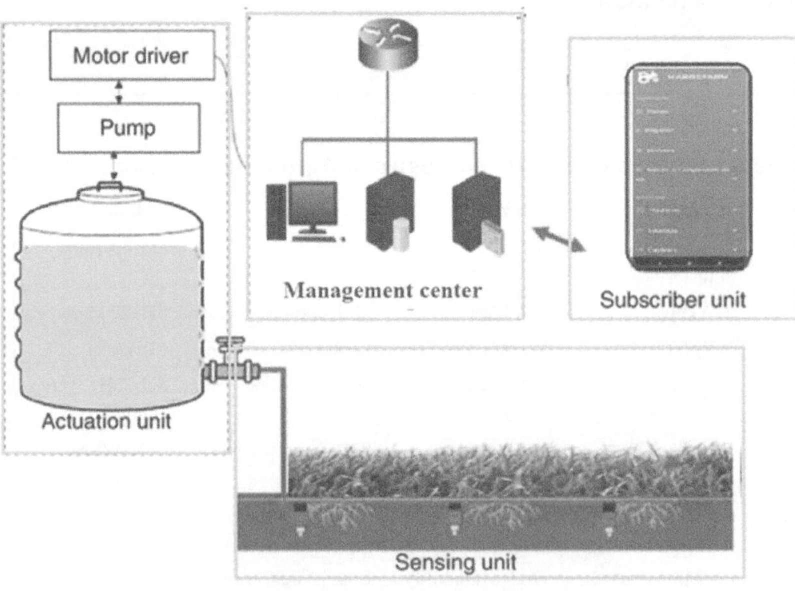

FIGURE 20.2 Architecture of a wireless sensor network-based smart irrigation system.

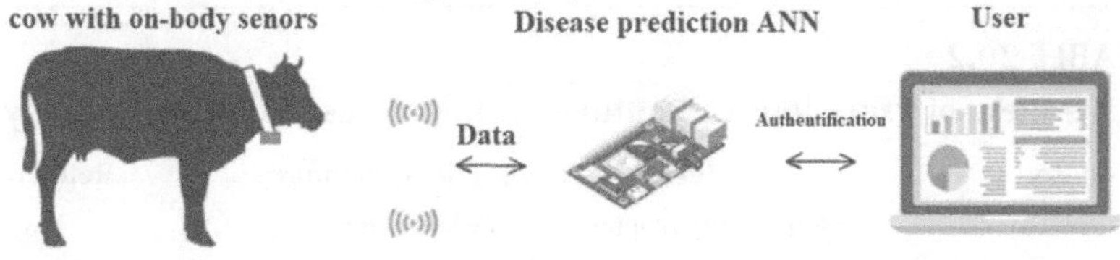

FIGURE 20.3 Livestock disease prediction system using on-body sensors.

20.3.3 Pest and Disease Detection

By keeping an eye on the microclimate that affects pathogen development, WSNs help detect crop diseases and pest infestations early. To measure disease pressure, variables like humidity, canopy temperature, and leaf wetness are gathered. To identify observable symptoms or variations in chlorophyll fluorescence, certain systems incorporate optical or spectral sensors [23].

Farmers can reduce pesticide use and increase crop protection effectiveness by using timely alerts to apply preventive treatments only when needed. High-value crops like fruits, vegetables, and vineyards benefit greatly from these systems [23].

20.3.4 Greenhouse Climate Control

WSNs are essential for preserving ideal growing conditions in controlled settings like greenhouses. In order to control HVAC systems, shading mechanisms, and misting units, sensors keep an eye on temperature, CO_2 concentration, humidity, and light intensity. They then communicate with actuators. Predictive model-based feedback loops minimize energy consumption and maximize productivity by maintaining stability within target thresholds [24].

20.3.5 Livestock and Equipment Monitoring

Although predominantly plant-focused, WSNs are increasingly applied in animal farming for health and movement tracking. Wearable sensors track heart rate, body temperature, and activity levels to detect illness or estrus cycles. In a similar vein, field machinery like tractors and irrigation pumps can be remotely monitored for predictive maintenance and fault identification (Figure 20.3) [25].

These apps are part of a larger ecosystem for digital agriculture, where linked systems enhance decision-making and resource use throughout the whole agricultural value chain [25].

20.4 PROBLEMS OR CHALLENGES IN WSNs

The integration of WSNs in PA has transformed traditional farming into an intelligent, automated, and data-driven process. WSNs are central to this evolution, enabling continuous monitoring and control of key agro-environmental variables such as temperature, soil pH, and moisture levels. These networks are particularly vital in greenhouses and large-scale open-field crops like sugarcane, where real-time alerts through Global System for Mobile (GSM) or IoT systems help optimize decisions. However, despite their numerous advantages, WSNs face a range of technical, economic, and operational challenges that hinder their widespread and efficient adoption [26].

20.4.1 Technological Adoption and Infrastructure Gaps

PA adoption is still restricted in many developing nations, especially in rural areas, because of high implementation costs, a lack of awareness, and restricted access to digital infrastructure. While nations like France, Morocco, Japan, and Turkey have embraced technologies like drones, yield monitors, neural networks, and soil mapping systems, areas with a large number of smallholder farmers frequently lack the resources or expertise necessary to take advantage of these tools (Table 20.2). In places where WSN-based solutions could be most useful, this digital divide slows their spread [27].

TABLE 20.2
Overview of Precision Agriculture Technologies in Use by Country

Country	Technologies Used in Agriculture	References
Morocco	Soil fertility mapping, GNSS/VRA equipment, smartphone-based DSS, remote sensing	[28]
Japan	Drones for crop spraying/monitoring (40% rice fields), AI-driven decision systems	[29]
Turkey	Yield monitors, GNSS-guided systems, automatic steering systems	[30]
France and Mexico	Artificial neural networks for fruit dehydration and quality estimation	[31]

20.4.2 Energy Efficiency and Power Limitations

One of the fundamental design constraints in WSNs is limited power availability. Most sensor nodes rely on low-capacity batteries (e.g., 0.5 Ah, 1.2 V), which are difficult to replace in remote or hazardous environments. Energy consumption in WSNs is driven primarily by three processes: sensing, wireless communication, and data processing. Additional factors such as communication range and the complexity of sensing events further impact power drain. The failure of even a few sensor nodes due to power loss can compromise the entire network topology, necessitating self-healing and re-routing protocols to restore connectivity [32].

20.4.3 Node Faults and System Reliability

Sensor nodes are susceptible to both hardware and software malfunctions, broadly classified into two categories [33]:

- **System Faults**: These include physical node failures, communication breakdowns, battery depletion, and packet loss.
- **Data Faults**: These occur when the sensor functions but delivers inaccurate or inconsistent readings, often due to environmental interference or calibration drift.

These faults can significantly degrade system performance, especially in mission-critical applications like disease detection or precision irrigation. Fault propagation across node tiers from the sensor node to the sink or gateway can escalate localized issues into network-wide failures.

20.4.4 Localization and Object Targeting

In agriculture, precise node localization is essential for targeted responses and spatial analysis. However, in large-scale deployments, it is prohibitively expensive to equip each sensor node with a GPS module. Positions are estimated using alternative methods like signal strength assessment, anchor-based triangulation, and proximity sensing. These techniques can be less accurate in areas with a lot of vegetation or hills, and they frequently call for extra infrastructure or calibration [34].

20.4.5 Communication Reliability and Topology Instability

WSNs function in outdoor settings that are vulnerable to atmospheric fluctuations, vegetation obstructions, and radio interference. These circumstances cause link instability, particularly in low-power protocols such as SigFox or ZigBee. Additionally, environmental events (such as storms or equipment collisions) and sensor node mobility (such as mounted on drones or livestock) can change network topology, requiring dynamic routing and recovery strategies [35].

20.4.6 Data Management and Security Concerns

As WSNs generate large volumes of real-time data, efficient data storage, filtering, and analysis become increasingly complex. Systems must handle noisy or incomplete data streams while providing actionable insights. In parallel, data transmitted across WSNs is vulnerable to interception, spoofing, or tampering. Implementing lightweight encryption, mutual authentication, and secure firmware updates are essential to ensure system integrity [36].

20.4.7 Market Integration and Value Chain Gaps

Even when technological systems like WSNs enhance crop productivity, farmers may not benefit economically unless their produce is effectively marketed. The agri-food value chain often involves intermediaries who capture a significant portion of profits. Hence, technologies must be accompanied by platforms that improve transparency, traceability, and market access—areas where blockchain and digital marketplaces offer promising solutions [37].

20.5 EMERGING TRENDS AND FUTURE PROSPECTS

The development of other cutting-edge technologies is intimately linked to the evolution of WSNs. The capabilities and reach of smart agriculture could be completely redefined by integrating WSNs with edge computing, blockchain, artificial intelligence (AI), and next-generation communication networks (5G/6G).

20.5.1 Integration with Artificial Intelligence

AI plays a transformative role in WSN-based precision agriculture by enabling intelligent data interpretation, anomaly detection, and predictive modeling. Machine learning (ML) algorithms process large volumes of sensor data to identify patterns in crop behavior, forecast weather-related risks, and optimize resource distribution. For example, AI models can analyze soil nutrient data to generate dynamic fertilization schedules or detect early signs of pest outbreaks using time-series analysis of microclimate data [38].

Deep learning architectures such as convolutional neural networks (CNNs) are also employed in image-based diagnostics, where camera-equipped sensor nodes monitor leaf color, plant morphology, or weed density in real time [38].

20.5.2 Edge and Fog Computing

Traditional cloud-based architectures can introduce latency and require stable internet connections, which may not be feasible in rural settings. Edge and fog computing mitigate this limitation by processing data near the source at the sensor node or local gateway. This decentralized model reduces bandwidth consumption and ensures faster response times for time-critical operations such as irrigation control or disease outbreak alerts [39].

20.5.3 Blockchain for Data Integrity

Data integrity and trust are critical in agricultural supply chains, particularly when WSNs are used for traceability and compliance monitoring. Blockchain technology offers a secure, decentralized ledger for recording sensor data, ensuring tamper-proof storage and transparent access for stakeholders. Smart contracts can automate quality checks, resource allocation, and payments based on sensor-verified conditions, fostering accountability and efficiency [40].

20.5.4 5G/6G Communication Networks

The rollout of 5G and anticipated emergence of 6G presents opportunities for ultra-low latency, high-bandwidth communication across agricultural landscapes. These networks enable real-time streaming of high-resolution video, rapid synchronization of distributed sensor nodes, and support for massive machine-type communication (MTC) [41].

20.6 CONCLUSION

One of the key technological enablers for the digital transformation of agriculture is a WSN. WSNs enable farmers to make data-driven decisions, maximize input use, and improve crop productivity by enabling continuous, real-time monitoring of vital agro-environmental parameters. These networks facilitate a wide range of applications, from pest detection and greenhouse automation to smart irrigation and nutrient monitoring, and are integrated with IoT platforms and supported by communication protocols like ZigBee, LoRaWAN, SigFox, and Wi-Fi. Notwithstanding their benefits, there are a number of technical and socioeconomic obstacles to WSN deployment in agriculture, such as resource-constrained areas' limited adoption, energy constraints, environmental robustness, node failures, and data security. To get around these restrictions and create intelligent, resilient, and self-adaptive agricultural systems, however, the integration of AI, edge computing, blockchain, and next-generation wireless technologies (5G/6G) presents encouraging avenues. The future of high-efficiency and sustainable farming will depend more and more on the effective integration of WSNs with other smart technologies as global agricultural demands continue to rise. To increase WSNs' affordability, scalability, and accessibility for farmers in a variety of contexts, interdisciplinary cooperation and investment are crucial.

REFERENCES

1. S. Getahun, H. Kefale, and Y. Gelaye, "Application of precision agriculture technologies for sustainable crop production and environmental sustainability: A systematic review," *Sci. World J.*, vol. 2024, no. 1, p. 2126734, 2024. doi: 10.1155/2024/2126734.
2. M. N. Mowla, N. Mowla, A. S. Shah, K. M. Rabie, and T. Shongwe, "Internet of things and wireless sensor networks for smart agriculture applications: A survey," *IEEE Access*, vol. 11, pp. 145813–145852, 2023.
3. M. Benchrifa, J. Mabrouki, M. Elouardi, M. Azrour, and R. Tadili, "Detailed study of dimensioning and simulating a grid-connected PV power station and analysis of its environmental and economic effect, case study," *Model. Earth Syst. Environ.*, vol. 9, no. 1, pp. 53–61, 2023.
4. L. Aldhaheri *et al.*, "LoRa communication for agriculture 4.0: Opportunities, challenges, and future directions," *IEEE Internet Things J.*, 2024, Consulté le: 25 mars 2025. [En ligne]. Disponible sur: https://ieeexplore.ieee.org/abstract/document/10735371/
5. I. Bencheikh, K. Azoulay, J. Mabrouki, S. El Hajjaji, A. Dahchour, A. Moufti, and D. Dhiba, "The adsorptive removal of MB using chemically treated artichoke leaves: Parametric, kinetic, isotherm and thermodynamic study," *Sci. Afr.*, vol. 9, p. e00509, 2020.
6. M. U. Mushtaq, H. Venter, A. Singh, and M. Owais, "Advances in energy harvesting for sustainable wireless sensor networks: Challenges and opportunities," *Hardware*, vol. 3, no. 1, p. 1, 2025.
7. J. Mabrouki, A. Moufti, I. Bencheikh, K. Azoulay, Y. El Hamdouni, and S. El Hajjaji, "Optimization of the Coagulant Flocculation Process for Treatment of Leachate of the Controlled Discharge of the City Mohammedia (Morocco)," in *Advanced Intelligent Systems for Sustainable Development (AI2SD'2019) Volume 7-Advanced Intelligent Systems for Sustainable Development Applied in Energy and Electrical Engineering*. Springer International Publishing, 2020, pp. 200–212.
8. M. A. R. Mushtaque, "Integration of wireless sensor networks, internet of things, artificial intelligence, and deep learning in smart agriculture: A comprehensive survey: Integration of wireless sensor networks, internet of things," *J. Innov. Intell. Comput. Emerg. Technol. JIICET*, vol. 1, no. 01, pp. 8–19, 2024.

9. M. Ren *et al.*, "Recent advances in self-powered sensors based on nanogenerators: From material and structural design to cutting-edge sensing applications," *ACS Appl. Electron. Mater.*, vol. 6, no. 6, pp. 3955–3997, 2024.
10. D. Jocil, and R. Vadivel, "Network Security Risks and Solutions Through Automated Toolkits in Underwater Sensor Network: A Survey," in *Leveraging Artificial Intelligence (AI) Competencies for Next-Generation Cybersecurity Solutions*, Apple Academic Press, 2024, pp. 1–37. Consulté le: 25 mars 2025. [En ligne]. Disponible sur: https://www.taylorfrancis.com/chapters/edit/10.1201/9781003478096-1/network-security-risks-solutions-automated-toolkits-underwater-sensor-network-survey-jocil-vadivel
11. Y. Wang, and Y. Yang, "A novel secure and energy-efficient routing method for the agricultural internet of things using whale optimization algorithm," *J. Cyber Secur. Mobil.*, vol. 13, no. 4, pp. 725–750, 2024.
12. K. Zangeneh, "Design of a graphic interface for precision agriculture," PhD Thesis, Politecnico di Torino, 2024.
13. E. Effah, O. Thiare, and A. M. Wyglinski, "Hardware evaluation of cluster-based agricultural IoT network," *IEEE Access*, vol. 12, pp. 33628–33651, 2024.
14. J. Mabrouki, M. Benchrifa, M. Ennouhi, K. Azoulay, I. Bencheikh, T. Rachiq, and S. El Hajjaji, "Geographic Information System for the Study of Water Resources in Chaâba El Hamra, Mohammedia (Morocco)," in *The International Conference on Artificial Intelligence and Smart Environment*, Cham, Springer International Publishing, 2022, November, pp. 469–474.
15. M. Benchrifa, and J. Mabrouki, "Simulation, sizing, economic evaluation and environmental impact assessment of a photovoltaic power plant for the electrification of an establishment," *Adv. Build. Energy Res.*, vol. 16, no. 6, pp. 736–753, 2022.
16. L. García, C. Cancimance, R. Asorey-Cacheda, C.-L. Zúñiga-Cañón, A.-J. Garcia-Sanchez, and J. Garcia-Haro, "Compliant and seamless hybrid (Star and mesh) network topology coexistence for LoRaWAN: A proof of concept," *Appl. Sci.*, vol. 15, no. 7, p. 3487, 2025.
17. R. Abdul Salam, N. Iqbal Ratyal, U. Ahmed, I. Aziz, M. Sajid, and A. Mahmood, "An overview of recent wireless technologies for IoT-enabled smart grids," *J. Electr. Comput. Eng.*, vol. 2024, no. 1, p. 2568751, 2024. doi: 10.1155/jece/2568751.
18. S. J. Al-Sofi, S. M. S. Atroshey, and I. A. Ali, "IEEE 802.15. 6 and LoRaWAN for WBAN in healthcare: A comparative study on communication efficiency and energy optimization," *Computers*, vol. 13, no. 12, p. 313, 2024.
19. H. Loukili, J. Mabrouki, A. Anouzla, Y. Kouzi, S. A. Younssi, K. Digua, and Y. Abrouki, "Pre-treated Moroccan natural clays: Application to the wastewater treatment of textile industry," *Desalin. Water Treat.*, vol. 240, pp. 124–136, 2021.
20. N. M. T. A. Ahmed, "Cross-layer design for IoT dedicated Healthcare," PhD Thesis, Université Polytechnique Hauts de France; University of Al Neelain Institute, 2024.
21. B. Nsoh *et al.*, "Internet of things-based automated solutions utilizing machine learning for smart and real-time irrigation management: A review," *Sensors*, vol. 24, no. 23, p. 7480, 2024.
22. J. Mabrouki, M. Azrour, Y. Farhaoui, and S. El Hajjaji, "Intelligent System for Monitoring and Detecting Water Quality," in *Big Data and Networks Technologies*, Springer International Publishing, 2020, vol. 3, pp. 172–182.
23. W. Q. Li, X. X. Han, Z. B. Lin, and A. Rahman, "Enhanced pest and disease detection in agriculture using deep learning-enabled drones," *Acadlore Trans. AI Mach. Learn*, vol. 3, no. 1, pp. 1–10, 2024.
24. S. Mallick, F. Airaldi, A. Dabiri, C. Sun, and B. De Schutter, "Reinforcement learning-based model predictive control for greenhouse climate control," *Smart Agric. Technol.*, vol. 10, p. 100751, 2025.
25. S. Nyamuryekung'e, "Transforming ranching: Precision livestock management in the internet of things era," *Rangelands*, vol. 46, no. 1, pp. 13–22, 2024.
26. M. Aarif K. O., A. Alam, and Y. Hotak, "Smart sensor technologies shaping the future of precision agriculture: Recent advances and future outlooks," *J. Sens.*, vol. 2025, no. 1, p. 2460098, 2025. doi: 10.1155/js/2460098.
27. S. Mgalaa, J. Mabrouki, M. Elouardi, L. El Azzouzi, A. Moufti, S. El Hajjaji, and M. A. El Belghiti, "Study and evaluation of the degradation of procion blue dye by the ozonation method: Parametric and isothermal study," *Nanotechnol. Environ. Eng.*, vol. 7, no. 3, pp. 691–697, 2022.
28. A. Hessane, A. Youssefi, Y. Farhaoui, and B. Aghoutane, "Precision agriculture in Morocco : Challenges and perspectives," 2021. doi: 10.13140/RG.2.2.15559.83361.
29. G. Phade, A. T. Kishore, S. Omkar, and M. Suresh Kumar, "IoT-Enabled Unmanned Aerial Vehicle: An Emerging Trend in Precision Farming," in *Drone Technology*, 1re éd., S. N. Mohanty, J. V. R. Ravindra, G. Surya Narayana, C. R. Pattnaik, and Y. Mohamed Sirajudeen, Éds., Wiley, 2023, pp. 301–324. doi: 10.1002/9781394168002.ch12.

30. F. Ghizlane, J. Mabrouki, F. Ghrissi, and M. Azrour, "Proposal for a high-resolution particulate matter (PM10 and PM2. 5) capture system, comparable with hybrid system-based internet of things: Case of quarries in the Western Rif, Morocco," *Pollution*, vol. 8, no. 1, pp. 169–180, 2022.
31. J. Mabrouki, M. Benbouzid, D. Dhiba, and S. El Hajjaji, "Simulation of wastewater treatment processes with bioreactor membrane reactor (MBR) treatment versus conventional the adsorbent layer-based filtration system (LAFS)," *Int. J. Environ. Anal. Chem.*, vol. 102, no. 19, pp. 7458–7468, 2022.
32. H. M. A. Fahmy, "WSNs Applications," in *Concepts, Applications, Experimentation and Analysis of Wireless Sensor Networks*, in Signals and Communication Technology., Cham: Springer Nature Switzerland, 2023, pp. 67–242. doi: 10.1007/978-3-031-20709-9_3.
33. M. El Alouani, B. Aouan, Y. Rachdi, S. Alehyen, E. H. El Herradi, H. Saufi, and N. Barka, "Porous geopolymers as innovative adsorbents for the removal of organic and inorganic hazardous substances: A mini-review," *Int. J. Environ. Anal. Chem.*, vol. 104, no. 16, pp. 4784–4796, 2024.
34. H. Ding, B. Zhang, J. Zhou, Y. Yan, G. Tian, and B. Gu, "Recent developments and applications of simultaneous localization and mapping in agriculture," *J. Field Robot.*, vol. 39, no. 6, pp. 956–983, 2022. doi: 10.1002/rob.22077.
35. M. A. Wase, and P. Siddaiah, "Enhanced sensor node lifetime based routing protocol in Mobile wireless sensor networks," *J. Electr. Syst.*, vol. 20, p. 302, 2024.
36. T. Rachiq, Y. Abrouki, J. Mabrouki, N. Samghouli, C. Fersi, S. Rahal, and S. El Hajjaji, "Evaluation of the efficiency of different materials to remove specific pollutants from landfill leachate," *Desalin. Water Treat.*, vol. 238, pp. 240–250, 2021.
37. A. Ahmed, I. Parveen, S. Abdullah, I. Ahmad, N. Alturki, and L. Jamel, "Optimized data fusion with scheduled rest periods for enhanced smart agriculture via blockchain integration," *IEEE Access*, vol. 12, pp. 15171–15193, 2024.
38. M. Benchrifa, M. Elouardi, G. Fattah, J. Mabrouki, and R. Tadili, "Identification, Simulation and Modeling of the Main Power Losses of a Photovoltaic Installation and Use of the Internet of Things to Minimize System Losses," in *Advanced Technology for Smart Environment and Energy*, Cham, Springer International Publishing, 2023, pp. 49–60.
39. N. Al-Jadabi, M. Laaouan, S. El Hajjaji, J. Mabrouki, M. Benbouzid, and D. Dhiba, "The dual performance of moringa oleifera seeds as eco-Friendly natural coagulant and as an antimicrobial for wastewater treatment: A review," *Sustainability*, vol. 15, no. 5, p. 4280, 2023.
40. M. I. Hossain, T. Steigner, M. I. Hussain, and A. Akther, "Enhancing data integrity and traceability in industry cyber physical systems (ICPS) through blockchain technology: A comprehensive approach," 2024, *arXiv*: arXiv:2405.04837.
41. M. N. A. Siddiky, M. E. Rahman, and M. S. Uzzal, "Beyond 5G: A comprehensive exploration of 6G wireless communication technologies," 2024.

21 AI for Smart Irrigation in Greenhouse Agriculture

Advancements, Challenges, and Future Prospects

F. Z. Khaled, J. Mabrouki, M. Slaoui, and Fatima Zahra Khaled

21.1 INTRODUCTION

The rising worldwide need for food, along with the difficulties brought about by climate change, has highlighted the necessity for more effective and sustainable farming methods. Greenhouse farming has become a feasible method to boost crop yields while reducing the negative impacts of environmental changes. Nonetheless, effective resource management, especially in water utilization, continues to be a vital issue in controlled environment agriculture (CEA) (Jones, 2004). Traditional irrigation methods, such as overhead sprinklers and flood irrigation, often lead to significant inefficiencies, including high water usage, nutrient depletion, and soil degradation. These conventional approaches lack the precision required to satisfy the evolving water needs of plants, resulting in either over-irrigation that wastes vital resources and increases production costs or under-irrigation that negatively impacts crop yields.

Artificial intelligence (AI) has become a revolutionary instrument for enhancing irrigation management through real-time decision making, predictive analytics, and automated control systems. Utilizing machine learning (ML) algorithms, sensor networks, and real-time data analysis, AI-driven irrigation systems can flexibly modify water delivery according to plant requirements, soil moisture content, and weather patterns (Shamshiri et al., 2018). These innovations facilitate accurate irrigation, decreasing water usage while preserving ideal plant health and enhancing productivity.

Recent research has examined different AI-based irrigation models, such as deep learning prediction systems, reinforcement learning for flexible irrigation timing, and smart irrigation networks integrated with IoT (Subeesh & Mehta, 2021). These technologies not only aid in conserving resources but also promote sustainable farming techniques by enhancing water use efficiency (WUE) and minimizing the ecological effects of over-irrigation.

Though it shows great promise, the adoption of AI-powered irrigation systems faces numerous challenges, such as elevated implementation expenses, the need for technical expertise, and worries about data privacy and energy use. Additionally, accessibility challenges, especially for small farmers in developing areas, could obstruct the broad adoption of these technologies. To overcome these obstacles, a comprehensive strategy is needed that includes technological innovations, policy backing, and capacity-building efforts to guarantee fair access to AI-driven irrigation solutions.

This chapter intends to offer a comprehensive examination of AI uses in enhancing irrigation in greenhouse settings. It will explore the main AI strategies used, the incorporation of sensors and IoT devices, as well as the advantages and difficulties related to AI-powered irrigation systems. By emphasizing contemporary research trends and practical applications, this discussion will aid

DOI: 10.1201/9781003647263-21

in understanding how AI can transform irrigation management and improve the sustainability of greenhouse agriculture.

21.2 CHALLENGES OF GREENHOUSE WATER MANAGEMENT

In greenhouse agriculture, conventional irrigation techniques (i.e., overhead and flood irrigation) can be very inefficient, therefore resulting in high water use, nutrient leaching, and environmental issues (García et al., 2020; Goldammer, 2024). These traditional systems are imprecise, which results in either over-irrigation that wastes water and encourages fertilizer runoff or under-irrigation, which leads to poor crop growth. As stated by Vallejo-Gómez et al. (2023), current irrigation systems have difficulty adapting to the time-variance of the soil moisture and the water requirements of the plant, leading to resource inefficiencies and increased operational costs. Moreover, smaller irrigating farms appear to use fewer modern irrigation technologies in comparison with larger farms, which are better positioned to invest heavily in capital (Schaible, 2017), and subsequently adopt new water-saving technologies more readily. In comparison, water consumption with sprinkler and drip irrigation—two significant efficient irrigation systems—can decrease down to 25% and 40% compared to traditional practices (Project Drawdown, 2024). Nevertheless, these improved methods still do not provide dynamic adaptability to climate variability and real-time changes in the environment, and thus are not enough to cope with the increasing water scarcity conditions that exists in productive systems of agriculture under controlled conditions. Climate change, as emphasized by García et al. (2020), has introduced elements, including extreme weather variability and rising evapotranspiration losses, that make accurate water use almost impossible without technology. Figure 21.1 illustrates the key problems associated with traditional irrigation techniques (overhead and flood irrigation) in greenhouse agriculture, including water waste, nutrient leaching, energy inefficiency, and poor adaptability to soil and plant needs.

Addressing these issues necessitates the incorporation of Internet of Things (IoT) sensors, AI, and ML algorithms for real-time monitoring and automated irrigation adjustments, which can significantly improve water efficiency and foster sustainable agricultural practices (Goldammer, 2024; Vallejo-Gómez et al., 2023).

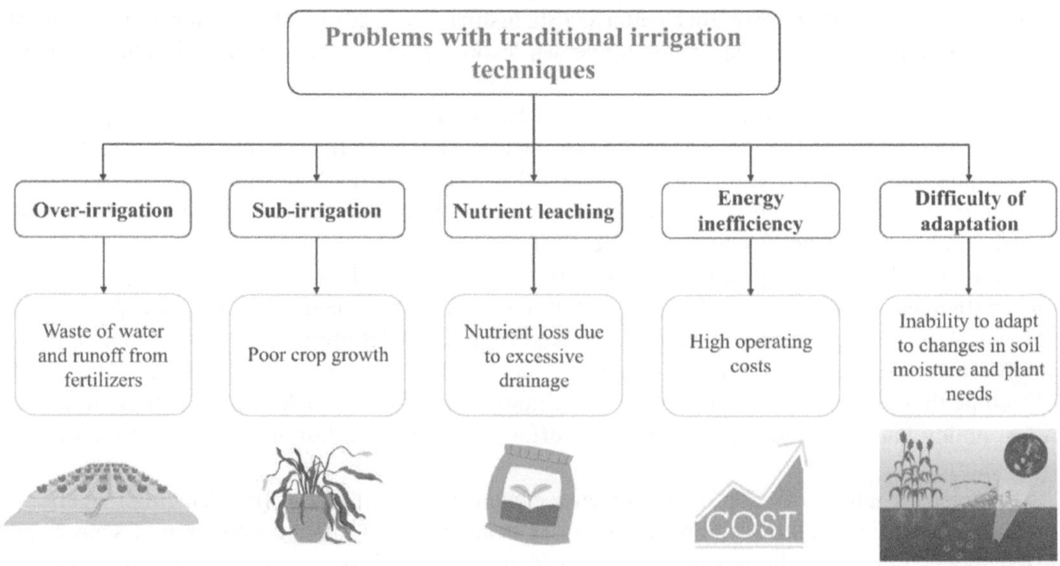

FIGURE 21.1 Problems of traditional irrigation techniques in greenhouse agriculture.

21.3 ROLE OF AI IN IRRIGATION OPTIMIZATION

AI has become a crucial element in changing conventional farming methods, especially in enhancing irrigation systems. By combining AI with IoT technologies, farmers can boost water efficiency, minimize waste, and enhance crop production. This examination explores the function of AI in enhancing irrigation efficiency, extracting knowledge from significant research and additional sources, while also offering a straightforward description of a smart irrigation system powered by AI.

AI algorithms are highly effective at handling large quantities of data instantly, allowing for adaptive modifications to irrigation timings. For example, Kumar et al. (2024) emphasize that AI is capable of examining data from different sources to enhance water efficiency in farming. Likewise, García et al. (2020) highlight that AI-powered systems can analyze sensor data to modify irrigation instantly, maintaining ideal soil moisture conditions. This feature enables accurate water distribution, minimizing waste and improving crop well-being.

The collaboration between AI and IoT is essential for contemporary irrigation systems. IoT sensors, including those that monitor soil moisture, temperature, and humidity, deliver ongoing data streams that AI models evaluate to make well-informed irrigation choices. Obaideen et al. (2022) explain that smart irrigation systems based on IoT employ sensors to gather real-time information, which is subsequently analyzed by AI algorithms to enhance water efficiency. This integration guarantees that irrigation meets the exact requirements of crops, enhancing sustainability and efficiency.

ML and deep learning (DL) algorithms play a crucial role in predicting the water needs of crops. By analyzing historical and real-time data, these models can forecast future irrigation requirements, enabling proactive water management. For instance, research featured in *Nature* by Ye et al. (2024) shows that AI systems can effectively forecast water needs, resulting in improved irrigation methods and lower water usage. Furthermore, Ding and Du (2023) created an irrigation system using deep reinforcement learning that demonstrated considerable water savings in field tests, highlighting AI's prospects in improving irrigation efficiency.

As depicted in Figure 21.2, an AI-based smart irrigation system works by combining various elements to ensure that water utilization occurs effectively. First, IoT sensors installed in

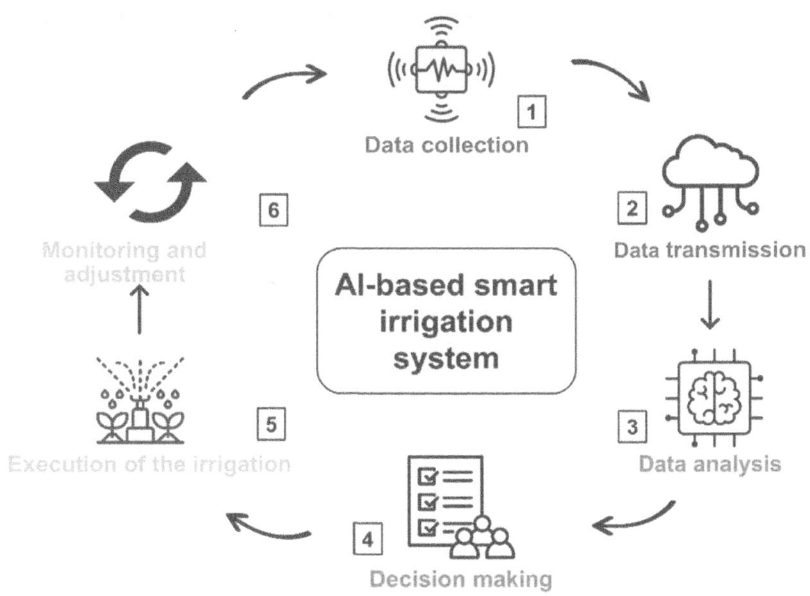

FIGURE 21.2 Operation of an AI-based smart irrigation system.

the agricultural field to monitor critical environmental parameters such as soil moisture, temperature, and humidity. This information is subsequently sent wirelessly to a central database for additional processing. Subsequently, AI algorithms examine the incoming data to assess existing soil and crop conditions, allowing precise forecasts of future water needs based on environmental elements and stages of crop development. According to this analysis, the system makes decisions by establishing the best irrigation timetable and the exact quantity of water required for optimal plant development. After a decision is reached, automated irrigation systems carry out the suggested measures, guaranteeing effective water allocation while reducing human involvement and the chances of over- or under-watering. Furthermore, the system utilizes ongoing monitoring and feedback processes, enabling it to adapt future irrigation plans dynamically based on shifting environmental factors. This closed-loop method guarantees sustainable water management along with enhanced agricultural productivity (García et al., 2020; Obaideen et al., 2022).

The incorporation of AI into irrigation systems signifies a major progress in farming methods. Through the utilization of real-time data analysis, integration of IoT sensors, and predictive modeling, AI facilitates accurate and effective water management. This not only preserves water resources but also boosts crop yield, aiding in sustainable farming. With the ongoing advancement of technology, the implementation of AI-powered irrigation systems is set to become a common practice, tackling worldwide issues related to water shortages and food safety.

21.4 EXAMPLES AND CASE STUDIES

AI is transforming agricultural irrigation, especially in greenhouse settings, by improving water efficiency and boosting crop production. Combining AI with IoT sensors allows for accurate monitoring and management, resulting in notable resource savings and enhanced productivity.

21.4.1 Case Study 1: Smart Irrigation Systems in Morocco

In the Souss-Massa region of Morocco, precision agriculture for development (PAD) has implemented smart irrigation technologies for date farms. Utilizing soil moisture sensors and automated irrigation resulted in a 30% reduction in water usage, while yields rose by 20%. This technology enhanced agricultural output and improved the economic stability of local farmers, aiding in poverty alleviation and food security (Bagalwa, 2024).

21.4.2 Case Study 2: Data Management Platforms in India

An agricultural cooperative in India implemented John Deere's Operations Center to oversee rice and wheat production. Consolidating data from IoT sensors and drones enabled the optimization of irrigation and fertilization plans, leading to a 25% boost in yield and a 15% decrease in production expenses (John Deere Operations Center, n.d.). This integration boosted profits and elevated living standards in rural areas (Bagalwa, 2024).

21.4.3 Case Study 3: Water Recycling Technologies in Spain

In Spain, a vineyard adopted Aquatech's water recycling systems, processing wastewater for reuse in irrigation (Aquatech-Home, n.d.). This method lowered fresh water requirements by 40%, safeguarded local water resources, and upheld high yields while decreasing operating expenses (Bagalwa, 2024).

TABLE 21.1
Comparative Table of the Case Studies

Case Study	Morocco	India	Spain	Canada	Tunisia
yield increase	20%	25%	not specified	not specified	15%
water consumption reduction	30%	not specified	40%	not specified	35%
production cost reduction	not specified	15%	not specified	not specified	20%
technology used	soil moisture sensors, automated irrigation systems	data management platform, IoT sensors, drones	water recycling systems	AI-driven greenhouse management (vent control, irrigation)	remote irrigation control system, mobile devices, AI
economic impact	boosted productivity, poverty reduction, food security	increased profitability, improved rural quality of life	reduced operational costs, improved water resource management	reduced water and energy costs, increased operational efficiency	increased production, reduced energy costs
environmental sustainability	conserved water resources, improved agricultural sustainability	optimized resource use, reduced environmental footprint	preserved fresh water, sustainable agriculture practices	reduced energy and water consumption, sustainable farming practices	reduced water and energy usage, improved resource sustainability

21.4.4 Case Study 4: AI-Driven Greenhouse Management in Canada

Nature Fresh Farms in Canada partnered with Intel to integrate AI into their greenhouse operations. The AI system automated operations like vent regulation and irrigation planning according to weather predictions, resulting in lower water and energy use. This proactive strategy enabled cultivators to concentrate more on monitoring crops instead of managing the system (Fresh Plaza, 2024).

21.4.5 Case Study 5: Remote Irrigation Control in Tunisia

In Tunisia, the start-up Irwise created a solution that allows farmers to oversee irrigation systems remotely using mobile devices. Through the examination of climate conditions, soil characteristics, and types of crops, the system improved water efficiency, lowering usage by 35%. Moreover, it lowered energy expenses by 20% and boosted agricultural output by 15%, showcasing AI's capacity to improve farming efficiency (L'Economiste Maghrébin, 2024).

In Table 21.1, we offer a comparative evaluation of the five case studies focusing on crucial indicators tied to agricultural efficiency, water management, and sustainability.

These cases illustrate the transformative impact of AI on agricultural irrigation, highlighting significant water conservation, cost savings, and yield enhancements in various geographical areas. Utilizing AI and IoT technologies, agriculture can enhance sustainability and productivity, tackling global food needs and resource issues.

21.5 BENEFITS AND LIMITATIONS OF AI IN IRRIGATION OPTIMIZATION

AI has greatly changed irrigation methods, providing remarkable benefits, encountering specific challenges, and showcasing encouraging future opportunities.

21.5.1 Benefits of AI in Enhancing Irrigation Efficiency

The incorporation of AI into irrigation systems has transformed water management in agriculture by allowing for accurate and data-informed decision making. These technologies assist farmers in maximizing water efficiency, enhancing crop production, and adjusting to shifting climate patterns. Here are several notable benefits of AI in irrigation:

- **Decreasing water usage:** Irrigation systems driven by AI enhance water efficiency by evaluating soil moisture and weather conditions in real time, minimizing excess water waste (Shahab et al., 2024).
- **Enhanced crop production:** Accurate irrigation guarantees that plants obtain ideal water quantities, boosting growth and efficiency (Vallejo-Gómez et al., 2023).
- **Climate adaptation:** AI facilitates real-time modifications in irrigation plans due to fluctuating climate conditions, enhancing resilience against droughts and severe weather phenomena (Obaideen et al., 2022).
- **Resource optimization:** AI combines data from IoT sensors to deliver suggestions on the application of fertilizers and pesticides, thereby lowering expenses and minimizing environmental effects (Wei et al., 2024).
- **Labor efficiency:** Automated irrigation systems minimize the necessity for manual oversight and intervention, enabling farmers to concentrate on other essential tasks.

21.5.2 Limitations of AI in Enhancing Irrigation Efficiency

Despite its many benefits, the use of AI in irrigation comes with its own set of challenges. Expensive prices, difficulties in access, and ethical dilemmas create major obstacles to broad implementation, especially for small farmers. Here are several of the key constraints:

- **Substantial initial expenses:** The deployment of AI-powered irrigation systems demands considerable financial investment in sensors, software, and infrastructure, which may be a barrier for small-scale farmers (Shahab et al., 2024; Wei et al., 2024).
- **Restricted access:** Small-scale farmers, especially in developing areas, might encounter obstacles in utilizing and embracing AI technologies because of financial and technical limitations (Vallejo-Gómez et al., 2023).
- **Requirement for technical knowledge:** Proper implementation of AI irrigation systems demands education and skills, which might not be easily accessible to every farmer (Obaideen et al., 2022).
- **Energy dependence:** AI models require substantial computational resources, potentially leading to increased energy use and carbon emissions if not paired with renewable energy sources (Okengwu et al., 2023).
- **Ethical and security issues:** Agriculture powered by AI depends on extensive data sets, provoking worries about data privacy, cybersecurity threats, and possible information abuse (Dara et al., 2022; Ryan, 2023).

Table 21.2 offers a comparative summary of the main advantages and disadvantages of AI-based irrigation, emphasizing both its possibilities and the challenges to its adoption.

In conclusion, although AI provides significant advantages for enhancing irrigation, it is crucial to address the related limitations and ethical issues to ensure sustainable and inclusive advancement in agriculture.

TABLE 21.2
Benefits and Limitations of AI in Irrigation Efficiency

Aspect	Description	References
Benefits		
reduced water usage	AI assesses soil moisture and weather patterns instantaneously, reducing unnecessary water waste.	Shahab et al. (2024)
improved crop production	accurate irrigation guarantees that plants obtain the ideal quantity of water, boosting growth and effectiveness.	Vallejo-Gómez et al. (2023)
climate adaptation	AI modifies irrigation plans based on evolving climate conditions, enhancing resilience to droughts and severe weather occurrences.	Obaideen et al. (2022)
resource optimization	AI combines IoT sensor information to offer advice on fertilizer and pesticide application, minimizing expenses and ecological effects.	Wei et al. (2024)
labor efficiency	automated irrigation systems minimize the requirement for manual involvement, enabling farmers to concentrate on other important activities.	Obaideen et al. (2022)
Limitations		
high initial costs	deploying AI-powered irrigation systems necessitates a substantial investment in sensors, software, and infrastructure.	Shahab et al., (2024) and Wei et al. (2024)
limited access	smallholder farmers, especially in emerging areas, encounter economic and technical obstacles to adopting AI.	Vallejo-Gómez et al. (2023)
need for technical expertise	effective use of AI irrigation systems necessitates knowledge and expertise, which might not be easily accessible to every farmer.	Obaideen et al. (2022)
energy dependence	AI models demand considerable computational power, potentially resulting in higher energy use and carbon emissions unless they are supported by renewable energy sources.	Okengwu et al. (2023)
ethical and security concerns	agriculture powered by AI depends on extensive datasets, raising issues regarding data privacy, cybersecurity risks, and possible information misuse	Dara et al. (2022) and Ryan (2023)

21.6 CONCLUSION

The incorporation of AI in greenhouse irrigation signifies a significant advancement in enhancing water efficiency, minimizing environmental effects, and boosting crop yields. Through the use of real-time data analysis, IoT sensor incorporation, and predictive modeling, AI-powered irrigation systems facilitate accurate water management, guaranteeing ideal growing conditions for plants while reducing waste. Case studies from different areas illustrate that AI can greatly decrease water usage, boost yields, and cut operational expenses, positioning it as a vital solution for tackling the issues of water shortage and sustainable farming.

Nevertheless, despite its many benefits, AI-driven irrigation systems encounter obstacles like significant upfront expenses, the need for specialized skills, and issues regarding data privacy and energy use. To break down these obstacles, it is essential for policymakers, researchers, and industry stakeholders to work together in creating AI solutions that are more affordable, accessible, and user-friendly for farmers globally.

Future advancements in AI for irrigation involve improving algorithm efficiency, incorporating renewable energy sources to sustainably power AI systems, and increasing the use of these technologies in various agricultural environments. Enhancing the cost-effectiveness and ease of use of AI tools can promote wider adoption among farmers. Furthermore, tackling ethical and economic issues, like guaranteeing data protection and offering fair access to technology, is essential for the responsible development of AI in agriculture.

As AI-powered irrigation advances, its significance in greenhouse farming is anticipated to grow, aiding in climate adaptability, resource preservation, and worldwide food safety. Through promoting interdisciplinary research and innovation, the agricultural industry can leverage AI's complete capabilities to develop more sustainable and efficient farming systems.

REFERENCES

Aquatech-Home. (n.d.). Aquatech. Retrieved March 13, 2025, from https://www.aquatech.com/

Bagalwa, D. P. (2024, September 27). Technologie de Gestion Efficace de l'Eau en Agriculture. *Kaziba Online*. https://kazibaonline.com/technologie-de-gestion-efficace-de-leau-en-agriculture/

Dara, R., Hazrati Fard, S. M., & Kaur, J. (2022). Recommendations for ethical and responsible use of artificial intelligence in digital agriculture. *Frontiers in Artificial Intelligence*, *5*, 884192. https://doi.org/10.3389/frai.2022.884192

Ding, X., & Du, W. (2023). *Optimizing irrigation efficiency using deep reinforcement learning in the field* (No. arXiv:2304.01435). arXiv. https://doi.org/10.48550/arXiv.2304.01435

Fresh Plaza. (2024, September 9). Comment l'IA rend l'agriculture sous serre plus efficace et plus rentable. https://www.freshplaza.fr/article/9656347/comment-l-ia-rend-l-agriculture-sous-serre-plus-efficace-et-plus-rentable/

García, L., Parra, L., Jimenez, J. M., Lloret, J., & Lorenz, P. (2020). IoT-based smart irrigation systems: An overview on the recent trends on sensors and IoT systems for irrigation in precision agriculture. *Sensors*, *20*(4), Article 4. https://doi.org/10.3390/s20041042

Goldammer, T. (2024). Chapitre 16 irrigating greenhouse crops-greenhouse irrigation systems-greenhouse management: A guide to operations and technology. APEX Publishers. https://www.greenhouse-management.com/greenhouse_management/irrigating_greenhouse_crops/greenhouse_irrigation_systems.htm?utm_source=chatgpt.com

John Deere Operations Center. (n.d.). John Deere. Retrieved March 13, 2025, from https://operationscenter.deere.com/

Jones, H. G. (2004). Irrigation scheduling: Advantages and pitfalls of plant-based methods. *Journal of Experimental Botany*, *55*(407), 2427–2436. https://doi.org/10.1093/jxb/erh213

Kumar, V., Sharma, K. V., Kedam, N., Patel, A., Kate, T. R., & Rathnayake, U. (2024). A comprehensive review on smart and sustainable agriculture using IoT technologies. *Smart Agricultural Technology*, *8*, 100487. https://doi.org/10.1016/j.atech.2024.100487

L'Economiste Maghrébin. (2024, May 9). Irrigation: L'IA Révolutionne l'Agriculture. *Leconomiste Maghrebin*. https://www.leconomistemaghrebin.com/2024/05/09/reprise-des-activites-de-ces-espaces-revoques-demeure-tributaire-de-regularisation-de-leur-statut-juridique-et-de-leur-respect-du-cahier-aux-dispositions-du-cahier-des-charges-en-vigueur/

Obaideen, K., Yousef, B. A. A., AlMallahi, M. N., Tan, Y. C., Mahmoud, M., Jaber, H., & Ramadan, M. (2022). An overview of smart irrigation systems using IoT. *Energy Nexus*, *7*, 100124. https://doi.org/10.1016/j.nexus.2022.100124

Okengwu, U. A., Onyejegbu, L. N., Oghenekaro, L. U., Musa, M. O., & Ugbari, A. O. (2023). Environmental and ethical negative implications of AI in agriculture and proposed mitigation measures. *Scientia Africana*, *22*(1), Article 1. https://doi.org/10.4314/sa.v22i1.13

Project Drawdown. (2024). Farm irrigation efficiency. https://drawdown.org/solutions/farm-irrigation-efficiency

Ryan, M. (2023). The social and ethical impacts of artificial intelligence in agriculture: Mapping the agricultural AI literature. *AI & Society*, *38*(6), 2473–2485. https://doi.org/10.1007/s00146-021-01377-9

Schaible, G. (2017). Understanding irrigated agriculture. USDA Economic Research Service. https://www.ers.usda.gov/amber-waves/2017/june/understanding-irrigated-agriculture?utm_source=chatgpt.com

Shahab, H., Iqbal, M., Sohaib, A., Ullah Khan, F., & Waqas, M. (2024). IoT-based agriculture management techniques for sustainable farming: A comprehensive review. *Computers and Electronics in Agriculture*, *220*, 108851. https://doi.org/10.1016/j.compag.2024.108851

Shamshiri, R., Kalantari, F., Ting, K. C., Thorp, K. R., Hameed, I. A., Weltzien, C., Ahmad, D., & Shad, Z. M. (2018). Advances in greenhouse automation and controlled environment agriculture: A transition to plant factories and urban agriculture. *International Journal of Agricultural and Biological Engineering*, *11*(1), 1–22. https://doi.org/10.25165/j.ijabe.20181101.3210

Subeesh, A., & Mehta, C. R. (2021). Automation and digitization of agriculture using artificial intelligence and internet of things. *Artificial Intelligence in Agriculture*, *5*, 278–291. https://doi.org/10.1016/j.aiia.2021.11.004

Vallejo-Gómez, D., Osorio, M., & Hincapié, C. A. (2023). Smart irrigation systems in agriculture: A systematic review. *Agronomy*, *13*(2), Article 2. https://doi.org/10.3390/agronomy13020342

Wei, H., Xu, W., Kang, B., Eisner, R., Muleke, A., Rodriguez, D., deVoil, P., Sadras, V., Monjardino, M., & Harrison, M. T. (2024). Irrigation with artificial intelligence: Problems, premises, promises. *Human-Centric Intelligent Systems*, *4*(2), 187–205. https://doi.org/10.1007/s44230-024-00072-4

Ye, Z., Yin, S., Cao, Y., & Wang, Y. (2024). AI-driven optimization of agricultural water management for enhanced sustainability. *Scientific Reports*, *14*(1), 25721. https://doi.org/10.1038/s41598-024-76915-8

22 Evolution of Digital Communication in Organizations

H. M. Driss, H. Kably, M. Elkouche, and A. O. Alaoui

22.1 INTRODUCTION TO DIGITAL COMMUNICATION IN ORGANIZATIONS

In recent years, digital communication has been a result of the adoption of digital tools, which has revolutionized organizational interaction, collaboration, and information sharing. Organizations can promptly adapt to market changes and stimulate innovation through real-time exchanges and enhanced decision-making tools (Kolasani, 2023). Digital communication now plays a central role in organizational life and has expanded throughout multiple industries, such as businesses, educational institutions, and non-profit organizations (Albanna et al., 2022). Digital communication encompasses information exchange via digital platforms, which includes both basic text messaging systems and sophisticated software solutions (Al-Jadabi et al., 2023). Management decisions directly shape organizational communication practices, which determine employee participation in information exchange. The emergence of digital communication brought about more effective processes that allow information to move more quickly and enable organizations to adapt faster despite existing challenges. Research remains dedicated to analyzing both the advantages and obstacles of digital communication while considering employee viewpoints and prospective developments (Grewal et al., 2022). This essay examines the evolution of digital communication its benefits and challenges for organizations and strategies for staying competitive in the digital age.

22.2 HISTORICAL CONTEXT OF DIGITAL COMMUNICATION

Insight into digital communication within organizations demands an investigation of its historical development. Organizational structures emerged from face-to-face discussions alongside printed materials and handwritten correspondence before the digital age took hold. The telegraph and typewriter emerged as essential inventions that established the groundwork for subsequent digital technologies (Seel, 2022). Digital age brought forth personal computers and email, which transformed communication methods, according to Nagesha (2024). Even as technology advances rapidly, organizations still frequently use non-digital communication methods. Organizational roles have been impacted by communication evolution from voice to text and further into digital formats (Johfre & Saperstein, 2023; Teytaut, 2023; Zhu, 2024).

22.2.1 Pre-digital Communication Methods

Organizations depended on face-to-face communication to build strong interpersonal relationships using verbal and non-verbal signals before digital communication tools arrived during the 1990s (Nguyen et al., 2022). Both personal and official paper-based communication served as the second primary method of interaction. Letters and notes served as personal communication methods, but official correspondence maintained organizational workflows despite being less accessible than digital solutions (Asadi, 2021).

DOI: 10.1201/9781003647263-22

22.2.2 Emergence of Digital Communication Technologies

Organizational efficiency improved during the late 20th century with the emergence of email and bulletin boards alongside web-based communication systems (Can, 2021). The transition from traditional paper memos to email and intranet systems in the late 1990s enabled organizations to share information more rapidly. Lotus Notes facilitated collaborative work, while social media became a dominant communication method during the 2000s (Ekeledirichukwu Onyejelem, 2018). New communication technologies have transformed organizational approaches worldwide, and digital tools keep advancing this evolution.

22.3 BENEFITS OF DIGITAL COMMUNICATION IN ORGANIZATIONS

Digital communication refers to the use of digital devices to communicate and share information. Designing, creating, sharing, or exchanging digital information through various devices like computers, smartphones, tablets, and the internet is digital communication. Over the past 20 years, digital communication has become essential to everyday life. Digital communication has increasingly become part of the organization's culture, policies, and practices. Using digital communication in organizations has changed the way people interact at work. Simply put, organizations put digital communication into practice in their culture, policies, and daily lives. Implementing digital communication brings many benefits to organizations. First and foremost, the efficiency and speed of information exchange are significantly enhanced. Teams can make quicker decisions and take quicker actions if they can share information anytime and anywhere. One of the great advantages of digital tools is that they can facilitate better collaboration by making it easier for employees to work together even if they are not physically in the same place. Teams can still generate ideas and work on common projects online, from brainstorming sessions to fully detailed reports or documents. Digital communication also helps organizations save costs because they can reduce physical resources, such as paper and postage for letters. In addition, the use of various digital communication tools improves record-keeping capability, and saving all emails, instant messages, or uploaded documents also helps organizations manage their knowledge better. The knowledge of previous business processes or deals is retained and can be easily accessed, even when an employee who previously worked on a specific task leaves the organization. Organizations can also enhance employee engagement through digital communication because they can create more opportunities for open dialogue, which can make employees feel more included in the decision-making processes of higher management. Thus, examining the benefits of digital communication in organizations is important to understand how they can use digital communication strategically.

22.4 CHALLENGES OF DIGITAL COMMUNICATION IN ORGANIZATIONS

Digital communication serves as a vital component in modern organizations by defining how team members, customers, and stakeholders connect with each other. A large portion of professional time goes into managing digital communication channels such as email and social media. Although these tools provide flexibility and advantages, they simultaneously create problems regarding staff well-being, operational efficiency, and communication. Research about digital communication has been extensive, yet organizations continue to face difficulties in comprehending its complete impact. A high volume of digital correspondence produces anxiety while breeding misunderstandings and reducing operational efficiency. Communication becomes harder when teams lack nonverbal cues, which poses a special challenge for culturally diverse teams. Data security breaches, along with privacy issues, continue to undermine employee trust. Communication faces obstacles due to unequal availability of digital devices alongside insufficient technological infrastructure and language differences. Organizations need to establish policies that focus on improving communication effectiveness while also ensuring security and inclusivity for all employees.

22.5 KEY TECHNOLOGIES AND TOOLS FOR DIGITAL COMMUNICATION

Digital communication technologies have become essential adoption elements for companies in every industry sector. Organizations started implementing these tools pre-pandemic. However, stay-at-home orders forced them to deploy these technologies quickly. The essential digital communication technologies encompass email, instant messaging platforms, videoconferencing solutions, webinars, and collaboration systems. Email serves as a fundamental tool for sending messages and files, whereas instant messaging provides users with real-time communication capabilities. The pandemic popularized videoconferencing for real-time visual communication, while webinars functioned as presentations with minimal interaction from participants. Through the integration of multiple managerial tools for project monitoring and secure file exchange, collaboration platforms boost team efficiency to enable successful collaborative efforts.

22.5.1 Email and Instant Messaging

Although organizations have access to numerous digital communication tools, email and instant messaging stand out as the primary methods. As Internet technology developed, both tools received new features that expanded their capabilities. The primary advantage of email lies in its capability to manage comprehensive messages and enable file sharing. The asynchronous communication feature of email makes it perfect for teams spread across different time zones. Professional communication relies on email as its standard platform, which includes productivity tools such as message organization, filtering options, and scheduling capabilities. Instant messaging works alongside email to facilitate real-time interactions while enabling faster response times. The platform, which started as a tool for casual conversations, has become a common fixture in professional environments. Instant messaging provides quicker response times than email while also enabling users to send files as email does. Users must exercise caution when using instant messaging to prevent disruption of work processes through off-hours communication and extended conversations. Email functions better than instant messaging when discussions require detailed information. After more than 20 years, both email and instant messaging have still played key roles in organizational communication. Organizational communication tools from the pre-digital era have evolved to integrate with modern platforms while maintaining their relevance. Organizations can maintain efficient communication by adhering to best practices even when facing email overload and miscommunication challenges. Email combined with instant messaging continues to play a fundamental role in organizational communication, according to Lee's 2022 research findings.

22.5.2 Videoconferencing and Webinars

Organizations now depend heavily on video conferencing and webinars to communicate with remote teams. Through videoconferencing, people can communicate face to face while breaking down geographical barriers and reducing travel time and expenses associated with physical meetings. Compared to audio-only meetings, videoconferencing creates more engaging interactions that help build stronger relationships between participants (Karl et al., 2022). Essential discussions become more difficult to maintain without videoconferencing because text-based communication does not adequately replace the depth of these exchanges. Webinar's function like videoconferencing platforms since they accommodate audiences larger than 20 people and work well for presentations where one-way communication is needed. Webinars enable participants to submit questions via chat while restricting alternative participation methods. The organization of webinars requires careful planning because teaching large audiences demands specific abilities. These events require a strong Internet connection and extensive preparation to handle possible technical problems. Remote webinars should operate without moderators because technical problems can be resolved better when participants are present face to face. The widespread phenomenon known as "Zoom fatigue" emerges as a drawback of video communications because people become exhausted from endless virtual meetings. Videoconferencing and webinars remain essential tools for team connectivity and the development of corporate culture despite facing multiple challenges.

22.5.3 Collaboration Platforms

Teams within organizations can improve their digital communication and teamwork capabilities through the various tools provided by collaboration platforms. These platforms, which originated from project management apps and document-sharing systems, fulfill different organizational functions through project management support and enable knowledge-sharing and social interaction. Teams can establish dedicated collaboration areas through web-based platforms where they can access shared document repositories along with real-time editing tools, calendar functions, and task management systems that include deadlines (Darics & Cristina Gatti, 2019). These platforms improve team collaboration and communication through the elimination of organizational barriers. These platforms provide basic wiki functions up to sophisticated systems that support instant messaging, videoconferencing capabilities, and document sharing. These platforms facilitate project workflows through dedicated spaces that enable team discussions and task clarification along with document updates. According to Razmerita and Tan (2017), to-do lists are a tool to monitor tasks by recording deadlines, assigned priorities, and designated discussion locations. The choice of platform presents significant challenges because its effectiveness relies on how well users adopt it and how it meets organizational requirements. Effective platform use requires proper training and support because different tools affect communication and knowledge sharing.

22.6 INTEGRATION OF DIGITAL COMMUNICATION WITH ORGANIZATIONAL PROCESSES

Organizational processes now utilize digital communication as an essential strategy that boosts efficiency and productivity beyond its original outreach purpose. By integrating digital communication into workflows, organizations achieve an uninterrupted flow of information between departments (Miller, 2011). Processes that incorporate ongoing feedback loops remain aligned with organizational objectives and remain responsive to technological advancements and client requirements. Digital communication has transformed into a strategic and relevant component of modern organizational communication. Integrating digital communication brings several benefits. Written communication formalization helps maintain consistency by establishing a clear understanding among all participants. The system increases transparency because information remains easily accessible while it reduces interpretation errors through automatic data transfer within integrated systems. This integration creates direct connections between communication and its supporting processes, which enables better tracking and adjustment of communication when processes undergo changes. Case studies show that integrating digital communication into organizational processes provides clear benefits despite challenges like resistance to change and required training (Vuchkovski et al., 2023).

22.7 IMPACT OF DIGITAL COMMUNICATION ON ORGANIZATIONAL CULTURE

The way organizations function culturally transforms through digital communication because it enhances reflective practices and promotes idea-sharing. The use of digital tools facilitates the development of an open, collaborative culture through simple feedback mechanisms and idea sharing, according to Bozkus (2023). Positive work environments depend on both formal and informal employee relationships and communication networks, while digital communication serves as a crucial element for trust-building and engagement. The blending of home and work responsibilities with digital communication creates worries about preserving organizational culture and work atmosphere (Tochukwu & Nonyelum, 2024). Leading a unified corporate culture throughout distant branches around the world presents significant difficulties. Studies into communication behavior demonstrate that organizational culture develops through the attitudes employees hold toward their work and their emotional bonds with their tasks (Bordi et al., 2018). Individual adaptation shapes organizational culture dynamically instead of being fully managed by leadership. Measuring digital communication's cultural effects presents difficulties, yet it remains essential for management to understand these effects in order to direct cultural progression.

22.8 DIGITAL COMMUNICATION AND EMPLOYEE ENGAGEMENT

Employee engagement represents the level of connection and satisfaction employees experience with their organization. Research indicates that multiple digital communication methods are used to connect with employees, including providing organizational updates to decrease rumors and gathering employee feedback through feedback systems (Estell & Davidson, 2019). Remote workers need transparent communication to build trust due to their potential sense of isolation. Organizations need to deliberately create clear and inclusive communication channels to effectively engage their hybrid workforce, according to Aaen Maagaard et al. (2019). Organizations can assess employee engagement by implementing digital platforms such as surveys and feedback sessions. The effectiveness of leadership communications plays a crucial role in building employee engagement because well-directed messages can inspire employees, but poorly constructed messages lead to confusion. Leadership communication enables organizational members to comprehend changes and implement them successfully across the entire organization.

22.9 DIGITAL COMMUNICATION AND KNOWLEDGE MANAGEMENT

Modern organizations rely strongly on digital communication and knowledge management, which became even more crucial after the pandemic. Organizations that utilize digital tools need to grasp how these tools enable the collection and distribution of knowledge along with its practical application (Lazarenko et al., 2018). Through knowledge management practices, organizations develop a sharing culture that ensures that information remains accessible and understandable for easy use. Organizations that depend on specialized knowledge must preserve competitive advantages through effective knowledge management regardless of employee turnover. Knowledge management benefits from digital communication tools, which transform knowledge into accessible digital formats such as written guides or video tutorials (Benchrifa et al., 2023). Digital platforms enable employees to exchange knowledge immediately via emails or chat and use specialized tools for joint problem-solving activities. Organizations need to plan carefully for knowledge management because digital tools primarily manage temporary information. Organizations need to create systems that make sure captured knowledge remains stored and accessible whenever required.

22.10 SECURITY AND PRIVACY CONSIDERATIONS IN DIGITAL COMMUNICATION

Digital communication has transformed the way organizations interact internally and externally. However, as communication increasingly moves online, security and privacy have become critical concerns (Khalil, 2023). Digital communication exposes organizations to various risks, including data breaches, cyber-attacks, and other malicious activities that compromise sensitive information. Protecting an organization's digital communication is essential for its credibility, reputation, and overall success (Hyland-Wood et al., 2021). Therefore, it is imperative to understand the security and privacy considerations that must be addressed in an organization's digital communication.

Email remains one of the most common forms of communication across organizations. However, it is also the most vulnerable to security breaches. These staggering figures underscore how vulnerable organizations are to security breaches in their digital communication (Rantao & Njenga, 2019). To combat this, organizations must establish robust security measures to protect their digital communication channels.

22.11 FUTURE TRENDS IN DIGITAL COMMUNICATION IN ORGANIZATIONS

The progression of digital communication began with basic written records before advancing through telephone technology to the Internet, which transformed interactions between individuals and organizations (Ghizlane et al., 2022). The widespread adoption of smartphones and computers has established digital communication as the standard while making video and text chats replace face-to-face interactions. Organizations now use digital tools beyond simple information sharing to develop connections and employee engagement. The rapid advancement of technology introduces new tools while upgrading existing ones, which requires organizations to modify their communication practices to stay effective. This text examines digital communication trends in organizations by looking at historical developments while addressing current trends and exploring future challenges.

REFERENCES

Aaen Maagaard, C., Jensen, A., & Wolff Lundholt, M. (2019). Making sense of the corporate philosophy: Dialogic employee engagement, and narrative positioning. [PDF]

Albanna, H., Alalwan, A. A., & Al-Emran, M. (2022). An integrated model for using social media applications in non-profit organizations. International Journal of Information Management, 63, 102452. [HTML]

Al-Jadabi, N., Laaouan, M., El Hajjaji, S., Mabrouki, J., Benbouzid, M., & Dhiba, D. (2023). The dual performance of *Moringa oleifera* seeds as eco-friendly natural coagulant and as an antimicrobial for wastewater treatment: A review. Sustainability, 15(5), 4280.

Asadi, A. R. (2021). Human-paper interaction in the digital era: Directions for human-information interaction design. EAI Endorsed Transactions on Creative Technologies, 8(29). semanticscholar.org

Benchrifa, M., Elouardi, M., Fattah, G., Mabrouki, J., & Tadili, R. (2023). Identification, Simulation and Modeling of the Main Power Losses of a Photovoltaic Installation and Use of the Internet of Things to Minimize System Losses. In Advanced Technology for Smart Environment and Energy (pp. 49–60). Cham: Springer International Publishing.

Bordi, L., Okkonen, J., Mäkiniemi, J. P., & Heikkilä-Tammi, K. (2018). Communication in the digital work environment: Implications for wellbeing at work. [PDF]

Bozkus, K. (2023). Organizational Culture Change and Technology: Navigating the Digital Transformation. In Organizational Culture-Cultural Change and Technology. IntechOpen. intechopen.com

Can, K. (2021). The evolution of communication technologies in Turkey's modern economy. Journal of Policy Options, 4(3). resdojournals.com

Darics, E., & Cristina Gatti, M. (2019). Talking a team into being in online workplace collaborations: The discourse of virtual work. [PDF]

Ekeledirichukwu Onyejelem, T. (2018). From evolution to revolution: Exploring the technological advances in mass communication. [PDF]

Estell, P., & Davidson, E. J. (2019). Employee engagement, voice mechanisms, and enterprise social network sites (ESNS). [PDF]

Ghizlane, F., Mabrouki, J., Ghrissi, F., & Azrour, M. (2022). Proposal for a high-resolution particulate matter (PM10 and PM2. 5) capture system, comparable with hybrid system-based internet of things: Case of quarries in the Western Rif, Morocco. Pollution, 8(1), 169–180.

Grewal, D., Herhausen, D., Ludwig, S., & Ordenes, F. V. (2022). The future of digital communication research: Considering dynamics and multimodality. Journal of Retailing. 98(2), 224–240. sciencedirect.com

Hyland-Wood, B., Gardner, J., Leask, J., & Ecker, U. K. (2021). Toward effective government communication strategies in the era of COVID-19. Humanities and Social Sciences Communications, 8(1), 1–11. nature.com

Johfre, S., & Saperstein, A. (2023). The social construction of age: Concepts and measurement. Annual Review of Sociology, 49, 339–358. annualreviews.org

Karl, K. A., Peluchette, J. V., & Aghakhani, N. (2022). Virtual work meetings during the COVID-19 pandemic: The good, bad, and ugly. ncbi.nlm.nih.gov

Khalil, J. F. (2023). Beyond revolutions, digital media economy in the Middle East: Continuing legacies and emerging disjunctures. The Sage Handbook of the digital media economy. [HTML]

Kolasani, S. (2023). Innovations in digital, enterprise, cloud, data transformation, and organizational change management using agile, lean, and data-driven methodologies. International Journal of Machine Learning and Artificial Intelligence, 4(4), 1–18. jmlai.in

Lazarenko, Y., Олександрівна Лазаренко, Юлія, & Александровна Лазаренко, Юлия (2018). Knowledge-sharing organizations in digital change. [PDF]

Miller, N. M. (2011). Communication in a digital age. [PDF]

Nagesha, H. (2024). Can non-digital marketing remain relevant in the digital era to achieve marketing goals? theseus.fi

Nguyen, M. H., Gruber, J., Marler, W., Hunsaker, A., Fuchs, J., & Hargittai, E. (2022). Staying connected while physically apart: Digital communication when face-to-face interactions are limited. New Media & Society, 24(9), 2046–2067. sagepub.com

Rantao, T., & Njenga, K. (2019). Communicating information systems security policies: Perception and determinism theories. [PDF]

Razmerita, L., & Tan, C. W. (2017). A mixed-methods study of a non-profit organization. [PDF]

Seel, P. B. (2022). Digital universe: The global telecommunication revolution. [HTML]

Teytaut, Y. (2023). On temporal constraints for deep neural voice alignment. hal.science

Tochukwu, I. C., & Nonyelum, O. F. (2024). Effectiveness of electronic meeting and video conferencing tools and techniques. IUP Journal of Computer Sciences, 18(4), 29–55. [HTML]

Vuchkovski, D., Zalaznik, M., Mitręga, M., & Pfajfar, G. (2023). A look at the future of work: The digital transformation of teams from conventional to virtual. Journal of Business Research, 163, 113912. sciencedirect.com

Zhu, X. (2024). Multiple phonological activation in writing: Evidence for cascadedness in Chinese written verb production. Frontiers in Psychology, 15, 1330522. https://doi.org/10.3389/fpsyg.2024.1330522. frontiersin.org

23 Improving English Communication Skills for Engineers in Morocco

A Comparative Analysis of the Private and Public Sectors in the Casablanca-Safi Region

M. D. Hanafi, H. Kably, M. Elkouche, and F. Z. A. Mahrez

23.1 INTRODUCTION

As globalization advances, professionals from all industries must develop English language skills due to their essential role in worldwide communication. Engineers in Morocco must navigate a complex linguistic environment that creates specific difficulties and possibilities, particularly in the Casablanca-Safi region. The Casablanca-Safi region stands as a dynamic center for engineering expertise and industries where English language knowledge boosts communication success rates and enables international business partnerships and progress. The educational system, where French and Arabic are dominant languages, creates significant language barriers for engineers. Engineers who learn English late develop difficulties with technical terminology and global team interactions and fail to comprehend specialized industry documentation. The industrially active region requires engineers to communicate with international clients while understanding detailed manuals and participating in global projects. Language barriers produce significant miscommunication and operational inefficiencies in Moroccan engineering workplaces. Engineers' struggles to convey ideas completely often restrict their potential for innovation and effective teamwork. Business leaders must identify these communication barriers and fund all-encompassing solutions to overcome language differences. Through the implementation of structured strategies and training programs that tackle current obstacles, Morocco stands to improve both the effectiveness and reach of its engineering industry in areas like Casablanca-Safi.

23.1.1 Background and Rationale

The present research proposal elaborates on the scientific study that is intended to be conducted by the author in the future on the theme of the improvement of the engineering profession by improving the English language communication skills using an ESP approach. The primary aim of the research is to develop an ESP course in English currently intended for engineering students in the public and private sectors in Morocco. The ambitious goals of the course are to lead to increased knowledge of engineering theory and practice; improved English language skills; improved employment opportunities and workplace integration for those who take part; and facilitate communication between engineers and foster international cooperation between engineering sectors in the Casablanca-Safi Region such as construction firms, energy companies, or aviation companies (Messaoudi and Hamzaoui-Elachachi, 2017). Concerning the matter of exportation and market competitiveness,

DOI: 10.1201/9781003647263-23

Morocco places 70th in the world with a GCI score of 4.14 out of 7. The country is particularly weak in innovation capability (108th), ICT Use (113th), and Company-Level Technology Development (106th). Thus, there is a need for the improvement of engineering professional practices to better prepare graduates for working in the field of complex system integration and high-tech industry.

23.1.2 Research Objectives

23.1.2.1 Research Problem

Moroccan employers indicate that workplace settings require more English language skills based on research findings. According to employers, recent graduates need strong English communication skills and technical expertise. Engineering schools and higher education institutions in Morocco face challenges in determining and examining English language requirements for their graduates. A needs analysis is essential to evaluate existing training provisions and determine curriculum development priorities.

Research exploring English language requirements exists for other countries in the Maghreb region, but the English needs of Moroccan private sector English have not been adequately studied. The study focused on the Casablanca-Safi region because of financial limitations despite its status as a hub for industrial enterprises. The study aims to analyze English language requirements for engineers from three different viewpoints—students, industrial professionals, and public service personnel—all of which impact their potential employment opportunities.

23.1.2.2 Research Rationale

The current market evolution demands skilled employees and requires English proficiency to reach international opportunities. Moroccan organizations need engineers who can handle English tasks due to the growing requirement for international market access. The engineering education system fails to teach essential soft skills needed for success in today's competitive labor market. The research intends to examine the essential English writing requirements for engineers by analyzing their language usage patterns and necessary communication abilities in the professional environment. The results will help employers and engineering educators design ESP courses that meet industry requirements for future training programs.

23.2 LITERATURE REVIEW

The subsection comprises evaluations of research papers and professional reports from academic dissertations to conference papers and analyses from Moroccan and international sources. The presentation order follows a chronological structure across four distinct parts. The initial section explores broad topics including the importance of English communication skills for engineering students, English's position in engineering education, and the difficulties these learners face. This section explores specific elements like teachers' assessments of student English skills while highlighting the significance of English discussions in academic curricula and the communication methods students employ in group work. Sections three and four widen their focus to investigate Morocco's broader context by examining how qualitative researchers and quantitative policymakers view English communication instruction while detailing how public and private sectors help engineering students develop these skills.

23.2.1 Importance of English Communication Skills in Engineering

The importance of English proficiency in engineering continues to grow because technical documentation exists mostly in English. Engineers must develop strong English skills across multiple areas to achieve success in the global engineering market (M. González Ardeo, 2010). Engineers should master English skills to analyze technical documents and write research articles while searching online for information and interacting with international colleagues. Engineers must maintain effective communication to appropriately address reviewer feedback and conduct successful presentations while actively engaging in workshops and team discussions. The increasing significance of international communication requires engineers to prioritize language education that combines communication abilities with professional mastery. Students in civil engineering at Universitas Muhammadiyah Malang (UMM) developed their communication capabilities by participating in English for specific purposes (ESP) courses according to Rif'ah (2018). The effectiveness of communication strategies relies heavily on social and sociolinguistic elements.

23.2.2 Challenges Faced by Engineers in Morocco

The analysis underlines fundamental distinctions between Morocco's private and public sectors while demonstrating that strong oral English abilities are essential for success in the private sector. Enhancing English proficiency through tertiary education leads to improved job opportunities for students. Morocco's present English education system fails to meet requirements, which creates obstacles for engineers who need English skills in private sector work. The job market faces increasing competition because more engineering graduates join the workforce every year. Stakeholders alongside educational institutions see engineering curricula as being excessively theoretical while lacking sufficient emphasis on practical capabilities such as English communication skills that are essential for professional growth. Civil engineering students believe professional English communication skills are essential for project proposals and discussions but find that their academic programs fail to provide adequate training, which forces new engineers to pursue extra English proficiency courses.

23.3 METHODOLOGY

The research applies a case study methodology to evaluate both English communication skills requirements and training delivery approaches in industrial sectors and higher education institutions across chosen regions. Researchers performed a needs analysis to determine representative input needs before comparing them to output needs through the application of weighted mean scores. Content and language integrated learning served as the basis for creating vocational/technical and crosscutting material while learners provided feedback through their perceptions and focus group discussions about content utility, relevance, and teaching methods. This research serves as a useful tool for developing educational programs and creating policy recommendations that benefit employers. The needs analysis creates an ESP course tailored to students' requirements. Essential goals involve selecting target industries and institutions before obtaining necessary permissions from authorities and gathering data through surveys and interviews as well as group discussions with key personnel in selected fields.

23.3.1 Research Design

This research investigates the English language demands of Moroccan engineering students who plan to work in Franco-Moroccan business studies. The primary research question investigates the language requirements students need to fulfill in their future professional roles. Our analysis of the English language requirements for Moroccan engineering students used triangulated research methods and instruments as documented in the Messaoudi and Hamzaoui-Elachachi (2017). The paper profiles the target situation by stressing English's role as the common language in engineering and its educational and research functions in Morocco. The theoretical framework adopts interpretivist constructivism to analyze students' difficulties in English-language computer-mediated communication (CMC) and the necessary skills for semi-formal educational discourse. The article examines how students' perceived needs match with professional specialist opinions and workplace documents utilized by graduate engineers.

23.3.2 Data Collection Methods

Researchers surveyed English learning obstacles for Casablanca-Safi English-medium instruction (EMI) engineering students and analyzed views from engineers working in both private and public sectors. Researchers gathered information from semi-structured student interviews as well as questionnaires from professionals and group discussions with both EMI teachers and English department professors. Morocco requires English in academic and professional fields, but French dominates engineering education. Universities can use the findings to understand student challenges and develop policies that serve both academic demands and job market requirements. Future professional students in Morocco will play a crucial role in developing the country's language use and educational approaches. Extensive research exists about EMI, but Moroccan-specific solutions in engineering for the Casablanca-Safi region have not been fully investigated.

23.4 THE PRIVATE SECTOR IN THE CASABLANCA-SAFI REGION

Engineering students from private Moroccan institutions show lower levels of English proficiency than their counterparts because their previous education was conducted in French and Arabic while they received minimal technical English training. Students who begin studying English during primary school and utilize digital platforms demonstrate notable progress. The universities these students attend use English as the primary language for teaching (M. González Ardeo, 2010). The swift expansion of companies in Morocco's private sector has generated a need for engineers with advanced technical skills. Engineering graduates from public schools often lack English fluency, which drives students to private institutions for superior instruction. The thesis investigates private sector approaches within the Casablanca-Safi region, which hosts most private sector companies and engineering schools throughout Morocco.

23.4.1 Overview of the Private Sector

Engineers working in private sector positions require both technical expertise and excellent communication abilities. Business engineers must articulate their ideas clearly while cooperating across multiple business sectors and directing project management. Business engineers play a vital role because they connect research with industry utilization and oversee both production processes and quality control as well as investment management (M. González Ardeo, 2010).

23.4.2 English Communication Requirements

Survey participants rated basic English communication skills as essential for engineering task completion. Writing reports and designing objects received high rankings, while collaboration with foreign companies emerged as another significant task. The results match studies that indicate English proficiency remains essential for engineers to stay competitive globally (Rif'ah, 2018).

23.5 THE PUBLIC SECTOR IN THE CASABLANCA-SAFI REGION

The Casablanca-Safi region serves as the basis for comparing private and public sectors while identifying the skills engineers should develop. Moroccan engineering students need to develop general abilities along with technical expertise. Technical skills are field-specific, while general abilities such as teamwork and English proficiency play vital roles in career progression. In international environments, English proficiency creates significant professional value for graduates, according to

Alaoui and Yassafi (2019). In higher education science and technology programs, French maintains its dominance, but English receives growing emphasis through programs such as European Region Action Scheme for the Mobility of University Students (ERASMUS) and Dubai's educational influence. Engineering students at their institutions must take ESP courses to improve their language abilities.

23.5.1 Overview of the Public Sector

Morocco's biggest city Casablanca plays a crucial economic role because engineers find work opportunities in both public and private domains. Engineers in the public sector generally work 35 to 40 hours weekly but must sometimes put in evening or weekend hours to meet project deadlines or meet government requirements. Service delivery companies require private sector engineers to manage more intense work schedules (M. González Ardeo, 2010).

23.5.2 English Communication Requirements

English communication skills are becoming essential in Morocco due to the country's growing openness to international investments. Engineers who wish to work with international companies need English proficiency as a critical asset. Businesses are now making English language capabilities a critical element of their hiring processes because they need graduates who can speak English. The private and public sectors show a clear trend in which English proficiency determines competitive success (Rif'ah, 2018).

23.6 COMPARATIVE ANALYSIS

This research compares its results with those of a prior investigation that examined master's students studying engineering in Algeria. Algerian students demonstrated high motivation for learning English because they understood its significance for their academic success and professional opportunities, similar to Moroccan students. The inside-out language learning method stood as the preferred choice for both groups, with a focus on grammar and vocabulary building alongside comprehension exercises. Algerians demonstrated no desire to write in English across various university departments, while Moroccan students showed this interest. Moroccan students expressed a desire to employ English in subject-related activities. The research examined English courses for students studying marketing, arts, and health in both Morocco and France.

23.6.1 Key Differences Between Private and Public Sectors

The engineering profession demands professionals to possess comprehensive knowledge and specialized technical skills along with general abilities. Procedural knowledge and knowledge organization serve as key skills while training outcomes require strong interpersonal communication and teamwork capabilities. Engineers maintain high self-evaluations regarding their expertise in their areas of training. This study used documentary analysis alongside employment inquiries and specialist interviews to collect data about social practices in engineering work. The study created a proposal to measure job readiness through fieldwork evaluations. The study investigated engineering practices in Spain because practical limitations dictated this choice. The theoretical framework discussed provided the basis for the survey design. The economy is heavily shaped by the construction sector which determines the necessary engineering qualifications. The research illustrates the facilities of site managers who manage both engineering and construction projects as prime examples. The study consists of situational needs assessments and theoretical components. The concluding remarks examine the methodological approach while identifying areas that could benefit from future enhancement.

23.7 FINDINGS AND DISCUSSION

In Africa, 65.5 million young people aged between 15 and 24 are out of school and do not participate in any form of training. A significant number of young people cannot go back to school because they finished only a few years of schooling. The need for skilled workers in a liberalized economy combined with technological changes has resulted in the expansion of non-traditional education and training programs that address diverse requirements. The market demands employees who possess modern skills, while people without work or those with insufficient work need training to enter self-employment or wage employment. The educational system must include non-traditional learning methods to address educational and socio-economic requirements through flexible literacy programs, basic education improvements, vocational training opportunities, and gender-sensitive practices.

23.7.1 English Proficiency Levels

This section analyzes the level of English communication abilities among engineers who work in the Casablanca-Safi region of Morocco. Among 335 engineers surveyed, it became evident that those in the private sector (210 respondents) and public sector (125 respondents) preferred receptive English skills such as reading and listening overwriting. Speaking and writing skills are the least practiced. Engineers regularly use English during professional meetings and training sessions, although confidentiality issues create limitations for some.

23.7.2 Training and Development Programs

This section explores training programs in Morocco. The public sector trains soft skills like language and computer proficiency on the job, but the private sector uses formal approaches to teach hard skills such as technical and science abilities. Technology and software dominate the field of hard skills, whereas soft skills encompass communication abilities, adaptability, negotiation competencies, and French language expertise (Hind & Soumia, 2017).

23.8 IMPLICATIONS FOR PRACTICE

The research examines the English communication requirements of engineering students and professionals in both private and public organizations located in Casablanca-Safi. Private sector organizations focus on technical English writing and speaking abilities, but engineering students receive insufficient training in technical English reading (Rif'ah, 2018). Morocco is becoming more aware of how essential English is for engineers who work in the worldwide market. An ESP course designed to meet these requirements is advised and can be adapted for use in multiple countries. Engineers' communication abilities could improve through partnerships among universities, government, and business organizations.

23.8.1 Recommendations for Engineers and Employers

Morocco shows clear differences between what universities teach and what the industry demands regarding English communication skills. The essential skills needed for professional success in international markets are often absent in engineering graduates, which restricts their career opportunities. Businesses require engineers with strong English communication skills, while higher education institutions fail to adequately prepare their students for these needs. The research recommends that enhanced emphasis on speaking and listening skills, together with practical experience, is essential to close this educational divide.

23.9 CONCLUSION

The research shows a pronounced difference in the level of English communication abilities between engineers working in Casablanca-Safi's public sector and those in the private sector. A joint initiative involving universities, industry stakeholders and government bodies presents the most effective solution. Future studies might investigate the effects of engineering skill gaps on career opportunities and ways to overcome these obstacles. The research emphasizes English proficiency as a key factor in international engineering while suggesting important consequences for professional growth and regulatory frameworks.

23.9.1 Summary of Key Findings

Engineering functions as a central element that enhances both living standards and industrial development in Morocco. National development relies heavily on engineering research and strategic planning. English proficiency stands out as a crucial factor in engineering according to this study, which considers globalization and technology developments. International conferences and strategic actions demonstrate this point by advancing the field.

23.9.2 Limitations and Future Research

Research shows that English proficiency plays a crucial role in career advancement in multinational corporations despite limited data on its impact. Engineering students understand English's significance yet encounter obstacles to obtaining proper training for speaking and comprehension skills. Upcoming studies should investigate how training deficiencies affect career achievements.

REFERENCES

Alaoui, M., & Yassafi, M. (2019). Communication marketing, culture et performances des entreprises marocaines dans les relations Sud-Sud: Etude exploratoire. [PDF]

Hind, C., & Soumia, B. (2017). Skills for employability: Identification of the soft skills required in engineering education. [PDF]

Messaoudi, N., & Hamzaoui- Elachachi, H. (2017). An analysis of Algerian manufacturing and engineering master's students' needs. Arab World English Journal.

M. González Ardeo, J. (2010). Communication skills and ESP courses: A Basque experience. [PDF]

Rif'ah, L. (2018). Study on English oral communication needs of ESP program for civil engineering students in university of Muhammadiyah Malang. [PDF]

24 The Impact of IoT Technologies on Enterprise Logistics Optimization

S. Elbouamri and L. Oubaouzine

24.1 INTRODUCTION

With the increasing complexity of global supply chains, companies face growing challenges in efficiency, cost management, and sustainability. To address these issues, businesses are turning to advanced technologies, with the Internet of Things (IoT) playing a crucial role in optimizing logistics operations. By integrating smart sensors, cloud computing, and artificial intelligence (AI)-driven analytics, IoT enhances real-time tracking, warehouse automation, and decision making in supply chain management [1]. Several industries, including retail, pharmaceuticals, and manufacturing like automotive and aeronautical, have already adopted IoT-based logistics solutions to improve inventory control, fleet management, and predictive maintenance [2,3]. The real-time capabilities of IoT-powered devices, such as radio frequency identification (RFID) tags and GPS sensors, provide greater visibility into supply chains, helping companies minimize losses and reduce inefficiencies [4,5]. Additionally, cloud-based logistics platforms enable businesses to process and analyze vast amounts of logistics data, leading to more informed decision making and streamlined operations [6].

However, despite these advantages, widespread IoT adoption in logistics is not without challenges. High infrastructure costs, cybersecurity vulnerabilities, and interoperability issues remain significant barriers [7,8]. Many companies still rely on legacy systems that are not fully compatible with IoT networks, further complicating the integration process [9]. To address these concerns, some researchers propose blockchain technology as a solution for securing IoT data and enhancing supply chain transparency, yet its implementation is still in early stages and requires further standardization [10,11].

Although IoT's role in logistics has been widely studied, most existing research focuses on specific applications, such as AI-based demand forecasting or warehouse robotics, rather than taking a holistic view of IoT's impact on logistics optimization and sustainability [12,13]. To address this research gap, this paper conducts a systematic literature review (SLR) to explore the following questions:

- How is IoT transforming logistics functions across different industries?
- What are the primary barriers to IoT adoption in logistics?
- Which emerging technologies (e.g., AI, blockchain, 5G) are expected to enhance IoT-driven logistics in the future?

Through an in-depth analysis of books, peer-reviewed journals, and industry reports, this study aims to provide valuable insights for researchers, logistics professionals, and policymakers, guiding the future adoption and development of IoT solutions in enterprise logistics.

The rest of the paper is structured as following. Section 24.2 explains the research method, following a systematic literature review (SLR) using PRISMA guidelines. Section 24.3 explores key IoT technologies such as real-time tracking, automation, and cloud-based analytics. Section 24.4 looks at the main challenges of IoT adoption, including cybersecurity risks, high costs, and integration issues. Section 24.5 discusses new trends shaping the future of IoT in logistics, such as AI-powered predictive analytics, 5G networks, and blockchain for supply chain security. Section 24.6 reviews

DOI: 10.1201/9781003647263-24

the key findings, highlighting research gaps and areas that need further study. Finally, Section 24.7 concludes with recommendations for industry professionals, policymakers, and researchers to support the growth of IoT-driven logistics.

24.2 METHODOLOGY

This study adopts a structured and transparent approach using the systematic literature review (SLR) method, guided by the PRISMA framework (Figure 24.1) [1]. The main objective is to explore how IoT technologies influence enterprise logistics in terms of efficiency, automation, and future trends.

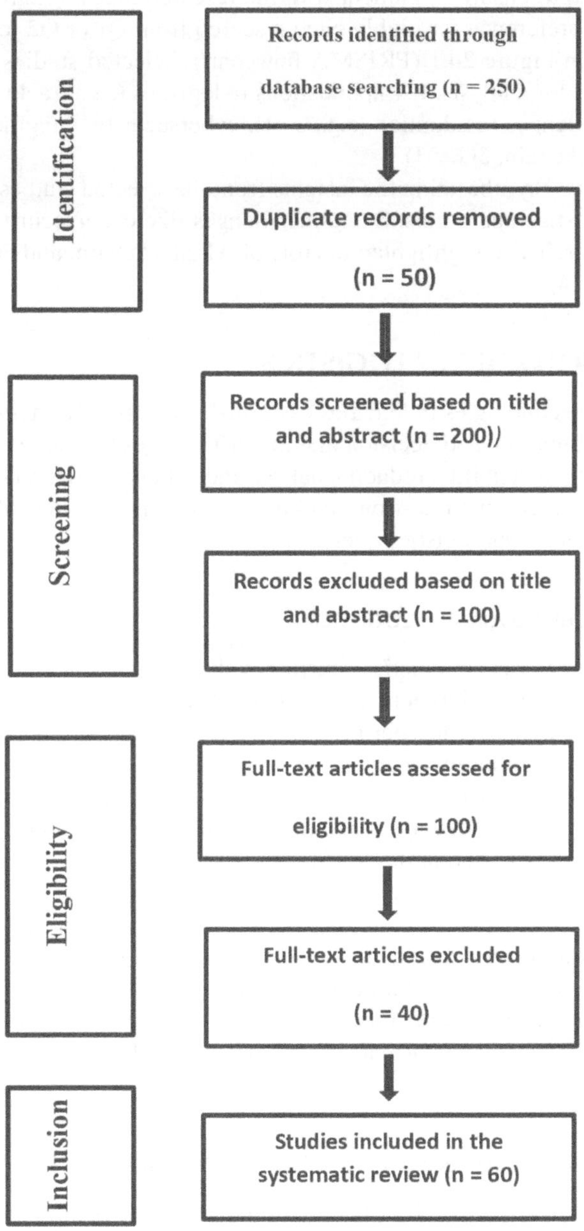

FIGURE 24.1 PRISMA-based study selection process.

To gather relevant and high-quality literature, I conducted a database search across IEEE Xplore, ScienceDirect, and Google Scholar. Boolean keywords such as "IoT in logistics", "supply chain optimization", and "real-time tracking in smart logistics" were used. The search focused on publications from 2018 to 2024 to ensure the inclusion of recent developments [2]. Selection criteria were applied to filter out irrelevant or low-quality sources. The inclusion criteria covered peer-reviewed journal articles and conference papers that focus on the use of IoT in logistics and supply chain operations. Studies addressing efficiency, automation, and implementation challenges were prioritized. Non-peer-reviewed papers, white papers, and duplicate studies were excluded.

The study selection process followed a three-step screening approach using the PRISMA methodology (Figure 24.1). First, titles and abstracts were reviewed to filter out unrelated papers. Next, full texts were assessed to ensure alignment with the research goals. Finally, a quality assessment was conducted with a preference for highly cited research from Q1 or Q2 journals. This process is visually summarized in Figure 24.1 (PRISMA flowchart). Selected studies were then categorized into key thematic areas for analysis: IoT applications in logistics (e.g., tracking, automation, predictive planning), [2] challenges to adoption (e.g., costs, cybersecurity, integration), and [3] emerging technologies (e.g., blockchain, 5G, AI).

The final step involved synthesizing the insights from the selected studies. IoT was shown to significantly improve logistics operations, though challenges like cybersecurity and integration costs persist [5, 9]. The research also highlighted the role of AI, blockchain, and 5G in shaping the future of smart logistics [10, 14]

24.3 IOT TECHNOLOGIES IN LOGISTICS

The integration of IoT technologies in logistics has transformed supply chain operations, enhancing real-time tracking, automation, and decision making. IoT enables businesses to optimize operations by improving supply chain visibility, reducing delays, and enhancing operational efficiency. Various IoT-based technologies, including real-time tracking, AI-driven analytics, blockchain security, and automation, are revolutionizing logistics operations.

24.3.1 Key IoT Technologies in Logistics

IoT is bringing major upgrades to logistics, starting with real-time tracking. Using GPS-connected devices, companies can follow shipments live, reduce delays, and respond quickly to disruptions. This not only improves transparency but also helps fleet managers make smarter route decisions and cut costs [8, 14]. In warehouses, automation powered by IoT sensors and robotics is changing the game (Table 24.1). From sorting to inventory control, connected robots are now handling tasks faster and with fewer errors. These systems boost efficiency while reducing labor costs, making supply chains more responsive and agile [5, 15].

AI takes it a step further by analyzing real-time data from IoT devices. This enables predictive logistics—helping businesses plan routes, forecast demand, and even predict equipment maintenance. As a result, downtime is reduced, and resources are used more effectively [12, 16,17].

Lastly, combining IoT with blockchain enhances supply chain security and transparency. These technologies work together to ensure that data is tamper-proof and products are traceable—an essential feature in industries like pharmaceuticals and food, where quality and compliance matter most [10, 18].

24.3.2 Industry-Specific Applications of IoT in Logistics

Although IoT is widely used across logistics, its impact often depends on the specific needs and challenges of each industry. The following examples highlight how different sectors are leveraging IoT to improve efficiency and reliability in their operations. In the pharmaceutical sector, cold-chain

TABLE 24.1
Overview of IoT Technologies in Logistics and Their Applications

IoT Technology	Application in Logistics	Real-World Example
RFID (Radio Frequency Identification)	Real-time inventory tracking and warehouse automation	Amazon uses RFID for automated warehouse tracking
GPS and Telematics	Fleet tracking and route optimization	DHL leverages GPS for real-time shipment tracking
Environmental Sensors	Cold-chain logistics for temperature-sensitive goods	Pfizer & Moderna use sensors to ensure vaccine integrity
Cloud Computing	Data storage, predictive analytics, and logistics automation	FedEx and UPS use cloud analytics for route optimization
Artificial Intelligence (AI)	Demand forecasting, route optimization, and real-time decision making	DHL employs AI-driven logistics planning to reduce delays
Blockchain	Supply chain security, fraud prevention, and traceability	Maersk's TradeLens blockchain enhances shipping security

logistics is a critical area where IoT makes a real difference. Companies like Pfizer and Moderna use temperature-sensitive IoT sensors and GPS tracking to monitor vaccine shipments in real time. These systems instantly alert managers to any temperature changes during transport, helping prevent spoilage and ensuring compliance with strict regulatory standards [12, 15, 19]. This level of control has become essential for safeguarding product quality and streamlining global distribution. In maritime logistics, smart ports are leading the transformation. The Port of Rotterdam, one of the world's busiest, has adopted IoT technologies combined with AI analytics to improve cargo tracking and ship docking schedules. This integration has helped reduce delays by 20% and optimize fuel usage, making operations not only faster but also more environmentally friendly [8, 14]. In the e-commerce sector, warehouse automation is where IoT truly shines. Ocado, a U.K.-based online grocery retailer, operates fully automated warehouses run by thousands of robots. These robots, guided by IoT sensors and a centralized cloud system, handle sorting and packing at incredible speeds. Predictive maintenance tools also keep the robots running smoothly, minimizing downtime and boosting overall efficiency [5, 18, 20,21].

24.4 CHALLENGES AND LIMITATIONS OF IOT IN LOGISTICS

This section identifies the major challenges of enterprises implementing IoT application in logistics and presents the possible solutions based on the latest research findings and practical applications (Table 24.2).

Despite the growing benefits of IoT in logistics, several key challenges continue to slow down its widespread adoption particularly for small and mid-sized companies. One of the most significant barriers is the high cost of implementation and maintenance. Rolling out IoT systems requires a large upfront investment in smart sensors, network infrastructure, and cloud storage platforms. On top of that, companies must manage ongoing maintenance, including regular software updates, cybersecurity protections, and real-time data processing [2]. These expenses are often out of reach for smaller logistics firms, which is why larger players like Amazon and DHL dominate early adoption [3]. To address this, companies are beginning to explore more affordable solutions such as AI-powered predictive maintenance and cloud-based IoT platforms, which help reduce operational costs and extend the life span of IoT systems [4].

Cybersecurity and data privacy are also major concerns. As IoT networks collect and transmit vast amounts of sensitive logistics data, they become prime targets for cyberattacks. With multiple

TABLE 24.2
Summary of IoT Challenges and Potential Solutions in Logistics

IoT Challenge	Description	Potential Solution
High Implementation and Maintenance Costs	IoT requires significant investment in infrastructure, hardware, and maintenance.	AI-driven predictive maintenance and cloud-based IoT platforms to reduce costs.
Cybersecurity and Data Privacy Risks	IoT networks are vulnerable to cyberattacks, data breaches, and hacking risks.	Blockchain security and AI-driven anomaly detection to enhance data protection.
Interoperability and Legacy System Integration	Many logistics firms struggle to integrate IoT with older legacy systems, causing inefficiencies.	Global IoT standards (e.g., ISO/IEC 30141) to improve system compatibility.
Regulatory and Compliance Issues	Different countries have varying IoT regulations, making global compliance difficult.	Harmonized international IoT regulations to simplify compliance.

entry points—from sensors and tracking devices to cloud-based control platforms—any vulnerability can expose the entire network. A real-world example came in 2021, when a cyberattack on a major global shipping company caused an estimated $300 million in damages, revealing serious weaknesses in IoT-enabled tracking systems [5]. In response, some companies are turning to blockchain-based security models and AI-driven anomaly detection tools to enhance protection and minimize risks [6]. Another challenge is the lack of interoperability and the difficulty of integrating IoT with legacy systems. Many logistics firms still rely on older technologies that don't easily communicate with new IoT platforms. The absence of global standards only complicates this further, making it hard to unify different systems within the same supply chain. For instance, Maersk experienced integration issues when attempting to align IoT tracking sensors with their older fleet management tools, leading to inconsistencies in data reporting [9]. To overcome this, logistics providers are beginning to adopt open standards like ISO/IEC 30141:2018, along with API-based system integration, to improve compatibility and create smoother adoption pathways [10].

Finally, regulatory and compliance issues present a growing concern, especially for companies operating across borders. Different countries have different rules, and that inconsistency makes global IoT deployment complex. The EU's GDPR, for example, has strict requirements for data handling, affecting how IoT devices track and store shipment data [11]. Without international agreement on data governance, many firms are left navigating a confusing legal landscape. As such, there is a strong call for harmonized international regulations and standardized compliance frameworks to support safe, efficient, and scalable IoT adoption worldwide [13, 22].

24.5 FUTURE RESEARCH AND EMERGING TRENDS IN IOT FOR LOGISTICS

As logistics become more digital and automated, IoT is taking on a bigger role in building smarter, faster, and more responsive supply chains. Combined with technologies like AI, 5G, blockchain, and sustainability-focused solutions, IoT is creating exciting new opportunities but also raising important questions that still need deeper research [2].

AI-Powered Predictive Logistics: AI is already making logistics more intelligent by helping companies anticipate disruptions, forecast demand, and optimize delivery routes. For example, DHL uses AI to cut fuel consumption and reduce delivery delays [5]. Still, real-time AI models that can adapt to fast-changing logistics scenarios are not fully developed yet and need more work [7]. 5G and Edge Computing: 5G is speeding up communication between IoT devices, while edge computing lets them process data locally instead of waiting for cloud servers—reducing delays and improving reliability. Amazon already applies these in its smart warehouses [6]. However, there is

TABLE 24.3
Summary of Emerging IoT Trends in Logistics and Future Research Areas

Emerging Trend	Description	Future Research Gap
AI-Powered Predictive Logistics	AI improves forecasting, route optimization, and predictive maintenance.	Developing real-time, adaptive AI models for dynamic logistics decision making.
5G and Edge Computing	5G enables fast communication, while edge computing processes data locally.	Addressing cybersecurity vulnerabilities in 5G-enabled IoT environments.
Blockchain for Secure Logistics	Blockchain secures and tracks logistics data transparently and tamper-proof.	Designing low-energy blockchain systems for sustainable, scalable logistics use.
IoT-Enabled Green Logistics	IoT sensors help reduce emissions and energy usage in vehicles and warehouses.	Measuring actual CO_2 reduction and long-term environmental impact across industries.
Hybrid AI-IoT Models	AI and IoT combined for enhanced real-time insight and decision making.	Exploring scalable implementations of hybrid systems in live supply chains.
Digital Twins in Logistics	Virtual models simulate logistics networks for planning and disruption testing.	Scaling digital twin systems for real-time, multi-site logistics operations.
Cyber-Physical Logistics Systems	IoT integrated with robotics enables autonomous logistics operations.	Evaluating technical, economic, and regulatory feasibility of full automation in logistics.

a growing need to address cybersecurity risks tied to these technologies, especially in large-scale IoT environments [10].

Blockchain for Supply Chain Security: Blockchain technology is gaining ground for securing logistics data and making supply chains more transparent. Maersk's TradeLens platform is a real-world case that proves how valuable blockchain can be [9]. But energy consumption remains a serious concern—future research should explore greener blockchain solutions [10]. Sustainable Logistics with IoT: As climate concerns grow, many logistics companies are turning to IoT to improve energy efficiency and reduce emissions. Smart sensors in electric vehicles and automated warehouses are already helping [12]. Still, we need better data on exactly how much CO_2 reduction IoT can bring across different industries [13]. Hybrid AI-IoT Models: Combining AI and IoT creates powerful tools for real-time logistics optimization. These hybrid models can enhance decision making by analyzing live data directly from the supply chain. More research is needed to understand how to implement these systems at scale across industries [4]. Digital Twins for Logistics Simulation: Digital twins—virtual copies of supply chain systems—are being used to simulate real operations, detect bottlenecks, and test improvements before making changes.

Future work should focus on how to scale digital twin technology for complex, multinational logistics networks [8]. Cyber-Physical Logistics Systems: As Table 24.3 illustrates, the integration of IoT with robotics and automation is paving the way for fully autonomous supply chains—think self-driving trucks, drone deliveries, and robot-run warehouses. To move forward, researchers need to explore not just technical feasibility, but also cost, legal, and regulatory aspects [14, 23,24].

24.6 RESULTS AND DISCUSSIONS

This study was guided by three main research questions, all focused on understanding the role of IoT in logistics. The first question explored how IoT is transforming logistics operations across different industries. Based on the literature, it is clear that IoT contributes significantly to improving efficiency through automation, real-time tracking, and better process optimization [25–27]. Companies like Amazon in retail, Maersk in maritime transport, and Pfizer in healthcare logistics are good examples of how IoT can help reduce costs, improve quality, and enhance visibility across the supply chain [28,29]. However, the scale of these benefits tends to vary depending on each company's level of digital maturity and the complexity of its operations. The second research question focused on the main barriers to IoT adoption. The findings show that high implementation costs are one of the most critical challenges, particularly for small and medium-sized enterprises [30]. Another major concern is cybersecurity, especially with the growing number of connected devices that can become access points for potential threats [4]. Additionally, the difficulty of integrating new IoT systems with existing legacy infrastructure slows down adoption and adds complexity [5]. The third research question looked at the emerging technologies that are expected to support and expand the impact of IoT in logistics. The review highlights how AI is being used to optimize logistics flows, while blockchain brings enhanced traceability and data security although it remains relatively expensive [6,7]. At the same time, technologies like 5G and edge computing are beginning to accelerate real-time data processing, making logistics systems more responsive and intelligent [8].

To better understand the combined impact of these technologies, a conceptual framework was developed and is presented in Figure 24.2. This framework shows how IoT functions as the backbone of modern logistics by integrating a range of tools that work together to create a smarter and more agile system.

The findings of this study are in line with those of other researchers. For example, Bidgoli et al. (2022) also noted the role of IoT in automating logistics and enhancing warehouse connectivity [9]. However, this study also goes beyond the benefits, by including the challenges and limitations of IoT implementation, which were not deeply explored in the work of Saxena (2021) [10]. Additionally, while Bowersox (2023) emphasized the impact of 5G, the current study broadens the scope by adding insights about regulatory and environmental issues [11]. These findings carry several practical implications for industry and public policy. Logistics companies will need to focus on improving data security, investing in more scalable cloud-based solutions, and providing better training for their teams. For governments and regulators, the key will be to push for IoT standardization, support smaller businesses with adoption incentives, and ensure that privacy regulations keep up with technological changes. From a research and innovation perspective, there is a clear need for

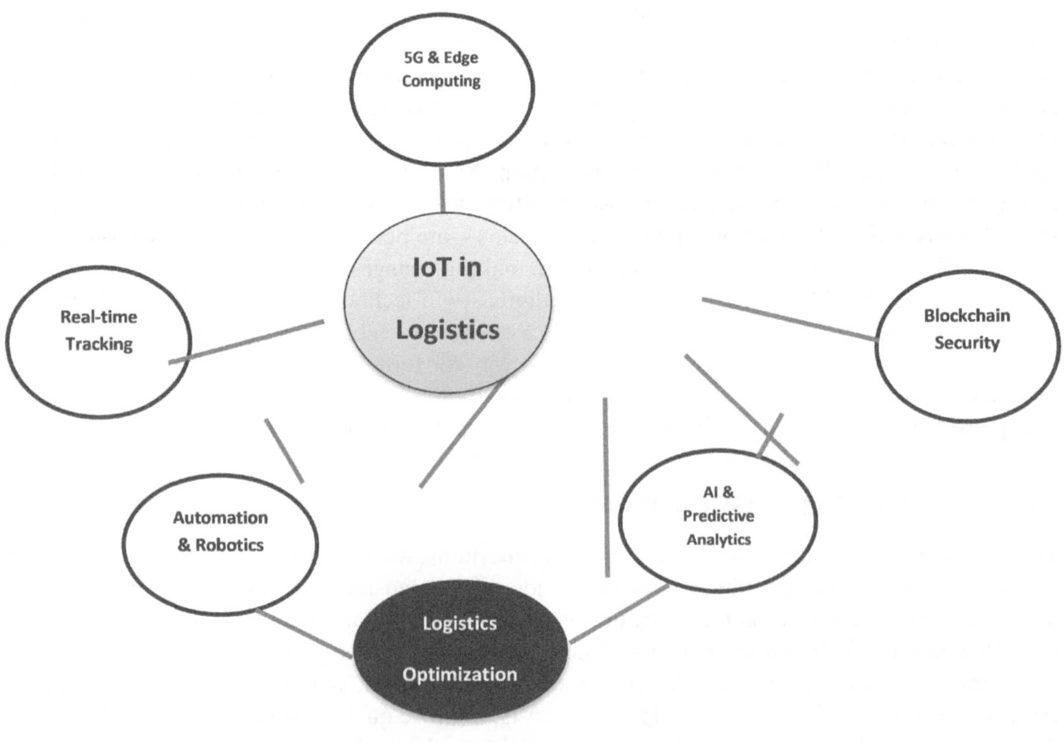

FIGURE 24.2 Conceptual framework for IoT-driven logistics optimization.

more advanced AI models, more sustainable blockchain platforms, and stronger evaluation of how IoT contributes to green and resilient supply chains.

In conclusion, this study confirms that IoT is already having a major impact on logistics but also highlights that several gaps still need to be addressed. As discussed earlier in Section 24.5, solutions such as hybrid AI-IoT systems, digital twins, and cyber-physical logistics networks are promising next steps. These technologies are worth further exploration as the logistics sector moves toward smarter, safer, and more sustainable operations.

24.7 CONCLUSION

This article presents a systematic literature review on the adoption of IoT technology in logistics, synthesizing major opportunities, key challenges, and emerging trends. The findings contribute to existing research by emphasizing technological advancements, security concerns, and future research directions. This study is particularly valuable for policymakers, industry leaders, and researchers aiming to enhance logistics efficiency and sustainability through IoT-driven innovations. IoT has revolutionized logistics by enabling real-time tracking, automation, and predictive analytics, significantly improving supply chain visibility, decision-making processes, and operational efficiency. However, several barriers still hinder widespread adoption, including high implementation costs, cybersecurity vulnerabilities, interoperability issues, and regulatory constraints. Overcoming these challenges is essential to fully unlock the potential of IoT in logistics operations.

To further facilitate the integration of IoT into logistics, future research should focus on:

- Advancing AI-powered IoT analytics to enhance real-time decision making and improve logistics efficiency.
- Strengthening security mechanisms for 5G and edge computing to ensure robust, low-latency IoT networks.
- Investigating sustainable IoT solutions that reduce carbon footprint, improve energy efficiency, and support eco-friendly logistics practices.
- Promoting the standardization of global IoT regulations to eliminate regulatory barriers and ensure seamless cross-border logistics.

As AI, blockchain, and IoT connectivity continue to evolve, logistics companies will have the opportunity to enhance efficiency, security, and sustainability. Future research must focus on eliminating technological barriers, reinforcing cybersecurity, and advancing environmental sustainability to ensure that IoT remains a transformative force in global supply chain management. By addressing these key issues, the logistics industry can fully harness the power of IoT to drive innovation, resilience, and long-term success.

REFERENCES

1. S. Ben-Daya, E. Hassini, and Z. Bahroun, "Internet of things and supply chain management: A literature review," *International Journal of Production Research*, vol. 57, no. 15–16, pp. 4719–4742, 2019.
2. S. Kamble, A. Gunasekaran, and H. Arha, "Understanding blockchain technology adoption in supply chains: Indian context," *International Journal of Production Research*, vol. 57, no. 7, pp. 2009–2033, 2020.
3. G. Büyüközkan, and F. Göçer, "Digital supply chain: Literature review and a proposed framework for future research," *Computers in Industry*, vol. 97, pp. 157–177, 2018.
4. M. Jabbar, J. Akhtar, and M. Dani, "Blockchain technology for enhancing supply chain resilience," *IEEE Access*, vol. 8, pp. 23632–23642, 2020.
5. S. Wamba, J. Queiroz, and R. Trinchera, "The role of IoT in sustainable supply chains: A systematic review and research agenda," *Journal of Cleaner Production*, vol. 310, p. 127591, 2022.
6. D. Ivanov, A. Dolgui, and B. Sokolov, "Digital supply chain twin: A simulation-based approach to supply chain resilience," *Omega*, vol. 102, p. 102414, 2021.

7. R. Ghadge, S. Dani, and H. Kalawsky, "Supply chain risk management: Present and future scope," *The International Journal of Logistics Management*, vol. 31, no. 3, pp. 813–839, 2020.
8. A. Dolgui, D. Ivanov, and B. Sokolov, "Reconfigurable supply chain: The X-network," *International Journal of Production Research*, vol. 58, no. 9, pp. 2823–2852, 2020.
9. H. Bidgoli, *Artificial Intelligence for Smart Logistics*, Springer, 2021.
10. S. Saxena, and M. Kumar, *Blockchain and IoT Integration: Supply Chain and Logistics*, Wiley, 2022.
11. D. Bowersox, and D. Closs, *Supply Chain Logistics Management*, McGraw-Hill, 2020.
12. S. Jeschke, and C. Brecher, *The Internet of Things in the Industrial Sector*, Elsevier, 2018.
13. M. Christopher, *Logistics & Supply Chain Management*, 5th ed., Pearson, 2021.
14. K. Lee, M. Patel, and P. Mehta, "AI-driven predictive analytics for IoT logistics optimization," *IEEE Transactions on Industrial Informatics*, vol. 18, no. 6, pp. 5124–5135, 2022.
15. P. Kumar, R. Shukla, and S. Tiwari, "Green IoT for sustainable supply chain management," *Sustainability*, vol. 14, no. 12, p. 7435, 2022.
16. H. Chan, C. Wong, and L. Zhang, "5G-enabled IoT applications in logistics," *IEEE Network*, vol. 34, no. 5, pp. 80–86, 2020.
17. J. Martin, and T. Nelson, "Edge computing in IoT logistics: Enhancing real-time processing," *Future Generation Computer Systems*, vol. 112, pp. 14–25, 2021.
18. S. Peters, "The role of AI in smart logistics: A systematic review," *Expert Systems With Applications*, vol. 202, p. 117435, 2022.
19. R. Maersk, and J. Thompson, "Blockchain-enhanced logistics transparency: The case of TradeLens," *Journal of Logistics Research*, vol. 9, no. 3, pp. 98–115, 2021.
20. L. Roberts, and J. Fernandez, "Big data analytics for logistics optimization," *Computers & Industrial Engineering*, vol. 153, p. 107084, 2021.
21. B. Edwards, and G. Peterson, *Digitalization in Logistics: The Role of AI and IoT*, Cambridge University Press, 2023.
22. A. Tiwari, J. R. Gupta, and S. Yadav, "Advances in IoT-based logistics and supply chain management," in *Proc. IEEE Int. Conf. on Industrial Informatics (INDIN)*, 2021, pp. 187–193.
23. X. Zhang, Y. Liu, and T. Wang, "Smart logistics enabled by 5G and AI: Challenges and opportunities," *IEEE Internet of Things Journal*, vol. 10, no. 4, pp. 2558–2572, 2023.
24. T. Wright, and H. Davis, "Cybersecurity risks in IoT logistics: A review," *Computers & Security*, vol. 108, p. 102329, 2021.
25. S. Mgalaa, J. Mabrouki, M. Elouardi, L. El Azzouzi, A. Moufti, S. El Hajjaji, and M. A. El Belghiti, "Study and evaluation of the degradation of procion blue dye by the ozonation method: Parametric and isothermal study," *Nanotechnology for Environmental Engineering*, vol. 7, no. 3, pp. 691–697, 2022.
26. C. Gonzalez, and F. Ramirez, "IoT security challenges in global logistics networks," *IEEE Security & Privacy*, vol. 18, no. 3, pp. 48–55, 2021.
27. J. Singh, and P. Verma, "Adoption of IoT for supply chain resilience: Challenges and strategies," *International Journal of Logistics Management*, vol. 34, no. 2, pp. 350–368, 2022.
28. X. Zhou, and H. Li, "Autonomous logistics with IoT and AI: A future roadmap," *IEEE Robotics and Automation Magazine*, vol. 29, no. 3, pp. 76–88, 2022.
29. A. Brown, and C. Scott, "AI-powered IoT networks for intelligent supply chain management," *Journal of Intelligent Manufacturing*, vol. 33, no. 7, pp. 1673–1690, 2023.
30. D. Richards, and M. Green, *Next-Generation Supply Chains: AI, IoT, and Blockchain Integration*, Routledge, 2023.

25 Use of Supercapacitors and Batteries for Efficient Energy Management in IoT

O. Enassiri, ELM. Benmalik, and O. Zytoune

25.1 INTRODUCTION

Today, the rapid development of the Internet of Things (IoT) opens up new possibilities in various fields such as health, home automation, or transportation through the use of connected objects. These objects, including sensors, smart home devices, and components of smart cities, have revolutionized the way we collect, share, and process data. However, this technological advancement also presents significant challenges, with energy autonomy being a cornerstone among them [1]. The continuous and reliable power supply for the multitude of connected objects has become a major concern, leading to the emergence of energy storage systems specifically tailored to meet the unique requirements of this era of connectivity.

Connected objects, known for compactness, low energy usage, and autonomous function, drive innovation in daily life. To enhance their longevity, we aim to explore advanced energy storage methods, evaluating their benefits, limitations, and proposing tailored solutions for sustained, reliable power [1, 2]. And to conduct a comprehensive literature review to gather recent advancements in energy storage for connected objects, supplemented by case studies to showcase varied applications. Addressing the perpetual energy demand from connected objects. Our objective is to ensure continuous and reliable power despite limitations in built-in battery capacity [3].

25.2 ENERGY STORAGE SYSTEM FOR CONNECTED OBJECTS

Developing energy storage systems tailored to the specific requirements of connected objects has emerged as a significant research and innovation challenge. Manufacturers and researchers are dedicated to devising solutions that address the demands for compactness, extended battery life, efficient power management, and convenient charging, all while ensuring durability and resilience to changing environmental conditions [4]. This chapter delves into the current state of energy storage systems for connected objects. We will discuss the technologies and approaches used for effective energy management and ensuring quality of service for these objects.

25.2.1 Energy Requirements for Connected Objects

The power needs of connected objects vary based on factors like size, function, and frequency of use, requiring tailored energy storage solutions. Key considerations include autonomy for independent operation, bridging the energy demand-production gap for seamless function, and addressing intermittency and fluctuation issues for consistent performance [3, 5, 6].

25.2.2 Energy Storage

Energy storage involves converting and storing energy for later use. This process includes three stages: charging, storage, and discharge [7]. To understand energy storage principles, we will explore commonly used components [8].

DOI: 10.1201/9781003647263-25

25.2.3 Batteries

Batteries store energy chemically and convert it to electrical energy when needed. They consist of cells containing electrodes, electrolytes, and separators. Chemical reactions inside the cells enable electron movement, generating a current to power devices or systems [9].

25.2.4 Supercapacitors

Supercapacitors are high-energy capacitors with significantly greater energy storage capacity compared to traditional ones. They consist of two porous electrodes in an electrolyte, storing energy electrostatically. Factors like electrode area, distance, and electrolyte's dielectric constant determine capacitance, similar to conventional capacitors, but supercapacitors utilize porous carbon-based electrodes for increased surface area [10].

25.2.5 Comparison between Energy Storage Technologies for Connected Objects

Batteries typically offer higher energy storage per unit mass/volume, making them well suited for long-term needs such as those of mobile devices. Supercapacitors, on the other hand, have lower storage capacity but can be charged and discharged rapidly, making them ideal for managing energy peaks and harvesting kinetic energy in transportation applications [11]. The Ragone diagram compares batteries, capacitors, and supercapacitors in terms of energy density and power density. Batteries excel in energy density for long-term storage, while capacitors have high power density for instant needs. Supercapacitors offer a balance between the two, making them suitable for applications needing rapid energy release over time [12, 13].

The Ragone diagram compares batteries, capacitors, and supercapacitors in terms of energy density and power density (Figure 25.1). Batteries excel in energy density for long-term storage, while capacitors have high power density for instant needs. Supercapacitors offer a balance between the two, making them suitable for applications needing rapid energy release over time [12, 13].

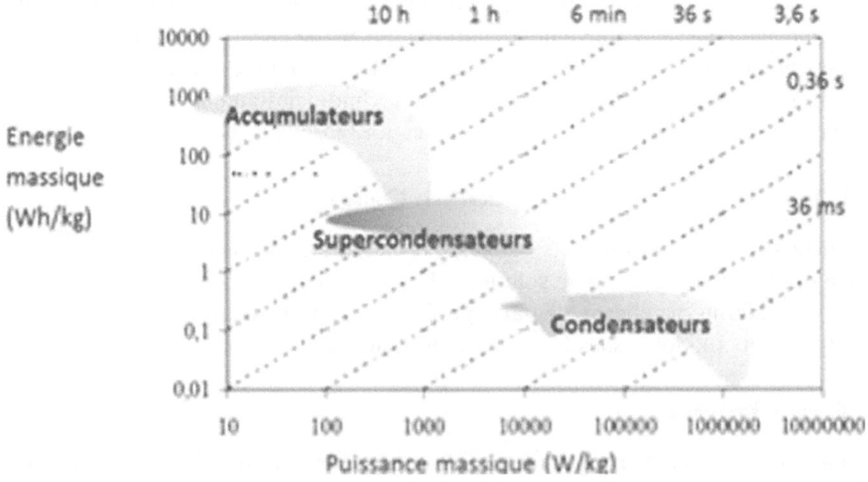

FIGURE 25.1 Ragone diagram for storage components.

25.3 ANALYSIS OF THE ADVANTAGES AND LIMITATIONS OF THE ENERGY STORAGE SYSTEM FOR CONNECTED OBJECTS

25.3.1 System Advantages

The advantages of lithium-ion batteries include high energy density, the durability where lithium-ion batteries are designed to be durable and can withstand multiple charge and discharge cycles without significant capacity loss and low self-discharge rate: Lithium-ion batteries can hold a charge for extended periods without frequent recharging [14]. The advantages of supercapacitors include their high capacity, rapid power delivery due to low internal resistance, making them ideal for surge applications. They also boast a long-life span, as their cycle count is infinite, and they offer fast charging times. Additionally, supercapacitors support a wide temperature range, further enhancing their versatility and suitability for various environments and applications [15, 16].

25.3.2 System Limitations

Lithium-ion batteries face limitations such as restricted range due to size and weight, along with a finite life cycle from repetitive charging. Temperature sensitivity poses risks of overheating, exacerbated by limited recharge rates [17, 18]. In contrast, supercapacitors have lower energy density and a narrow voltage range, impacting their long-term effectiveness. Rapid discharge and complex energy management challenge their utility, leading to performance degradation over time. Nonetheless, integrating supercapacitors with batteries presents a promising solution to address these limitations [19, 20].

25.4 SIMULATION OF ENERGY STORAGE TECHNOLOGIES

Simulation is pivotal for understanding system operations and functionalities. This chapter explores simulation outcomes using software tools to analyze switching mechanisms between batteries, supercapacitors, and capacitors for energy storage and power supply. The interaction between batteries and supercapacitors shows promise for efficient energy storage. Through simulation, we model these components and explore energy transition scenarios. This chapter aims to elucidate the significance of batteries and supercapacitors, introducing in-depth models, simulation conditions, and result analysis. The energy storage components chosen for powering connected objects or other systems are batteries and supercapacitors. As discussed in the previous chapter, batteries offer higher energy density, while supercapacitors excel in high power density. Therefore, for switching between the two, we will utilize batteries as backup power supplies, with power supplied by supercapacitors. Simulation using various software highlights the disparities between the two stacking units.

25.4.1 Simulation on MATLAB

In this simulation, we will explore the utilization of both batteries and supercapacitors for energy storage, as well as the switching between the two, considering scenarios involving low and high-power requirements. Before commencing the simulation, we will first select the components of the model and define their working characteristics.

The developed model comprises a battery, a supercapacitor, converters, and a constant load. The load's value was adjusted to investigate both low-power and high-power scenarios.

25.4.1.1 For Small Powers

In this case, we have adjusted the parameters of the battery and supercapacitor. The battery is set to be 80% charged with the following parameters: nominal voltage of 2.99 V, rated capacity of 0.66 Ah, initial state of charge (SOC) of 80%, maximum capacity of 0.66 Ah, and fully charged voltage of 3.07293 V. The supercapacitor has a capacity of 20 F, rate capacitance of 20 Farad, equivalent DC series resistance of $2.1*10^3\ \Omega$, rate voltage of 2.6 V, and a surge voltage of 2.7 V.

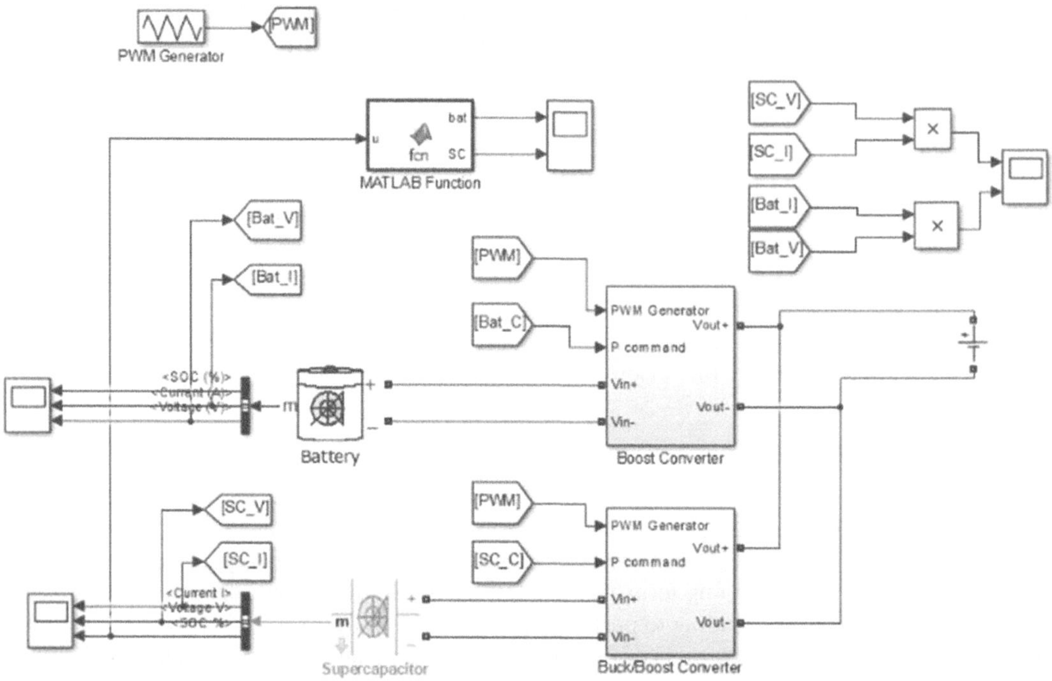

FIGURE 25.2 The general model of the battery and supercapacitor for low powers.

In Figure 25.2, we used the MATLAB Function block to create the function responsible for switching between the battery and the supercapacitor. This switching is determined by the state of charge (SOC) of the supercapacitor. If the SOC drops below the discharge threshold, the battery will be engaged.

25.4.1.2 For the Great Powers

For scenarios involving high power requirements, the characteristics of both the battery and supercapacitor will be adjusted accordingly in the model. This adjustment is necessary as we will be dealing with a higher constant charge value compared to the previous scenario.

The characteristics of the battery are summarized as follows: the nominal voltage is 26.4 V, the rated capacity is 6.6 Ah, and the initial SOC is 100%.

The supercapacitor characteristics for high power applications are as follows: rate capacitance of 500 Farad, equivalent DC series resistance of $2.1*10^3$ Ω, rate voltage of 16 V, 4 series capacitors, initial voltage of 24 V, and operating temperature of 25°C.

In this model, as illustrated in Figure 25.3, we will explore the difference between batteries and supercapacitors in terms of their capacity to power high-power systems.

25.4.2 Proteus 8 Professional

For this simulation (Figure 25.4), we will use Proteus 8 Professional to examine another type of energy storage capacitors in conjunction with batteries to power an LED continuously by switching between the two. The switching will be facilitated by relays:

Relay 1: closes to allow the capacitor to charge from source V1 and opens once the capacitor is fully charged.

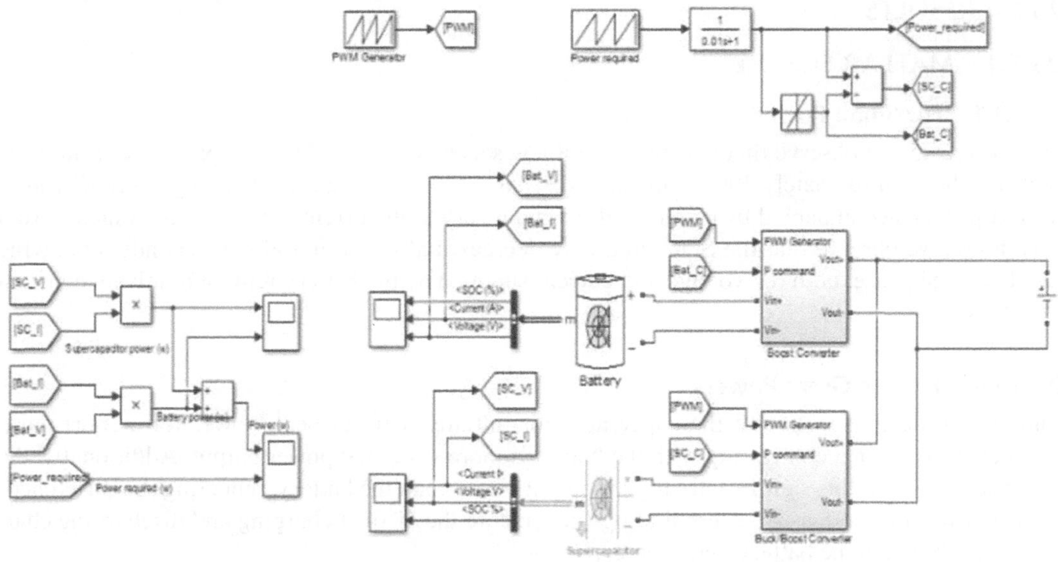

FIGURE 25.3 The general model of the battery and supercapacitor for high powers.

RL1 closes to power LED→ using the capacitor, RL2 closes to power LED→ using the battery and the capacitor is charging.

We will utilize an Arduino board to program these relays, measure the voltage of both the battery and capacitor, calculate the energy of each, and make a comparison between the two.

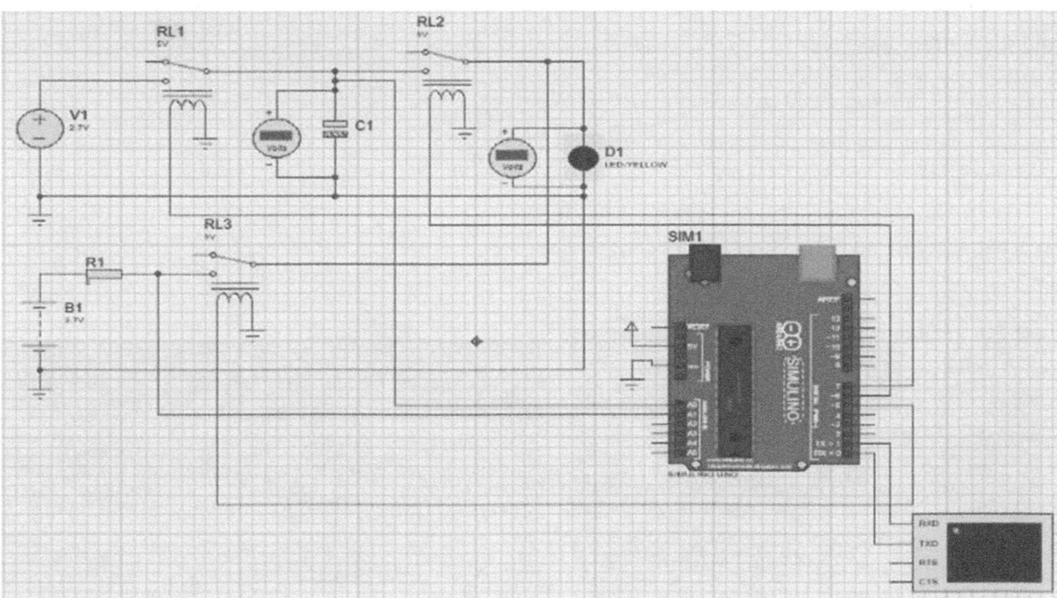

FIGURE 25.4 Battery and capacitor diagram in ISIS.

25.5 RESULTS

25.5.1 MATLAB Simulink

25.5.1.1 For Small Powers

In Figure 25.5, we observe that within the first few seconds, the capacitor supplies power until the SOC reaches approximately 40%. This rapid discharge leads to a significant change in SOC within a short period, accompanied by a voltage drop and a very high current at the initial moment. From this figure, we observe that the SOC gradually decreases after the initial few seconds. Following the discharge phase, both the voltage and current supplied by the battery remain relatively constant over time.

25.5.1.2 For the Great Powers

Figure 25.6 demonstrates that the supercapacitor initially delivers peak power for a short duration before declining, contrasting with the battery's more sustained power output. Additionally, the supercapacitor exhibits significantly higher output power than the battery, reflecting the differences in energy and power density. These findings underscore the distinct charging and discharging characteristics between the battery and supercapacitor.

25.5.2 Proteus 8 Professional

From Figures 25.7 and 25.8, we observe that the capacitor charges and discharges rapidly, enabling it to provide emergency power to keep the LED illuminated. The energy stored in the battery exceeds that of the capacitor, reflecting the battery's higher energy density.

By using these relationships, the energy stored in both the battery and the capacitor can be calculated: the capacitor energy:

$$E_C = \frac{1}{2} * C * U^2 \tag{25.1}$$

and the battery energy:

$$E_B = P * t = \frac{U^2}{R} * t \tag{25.2}$$

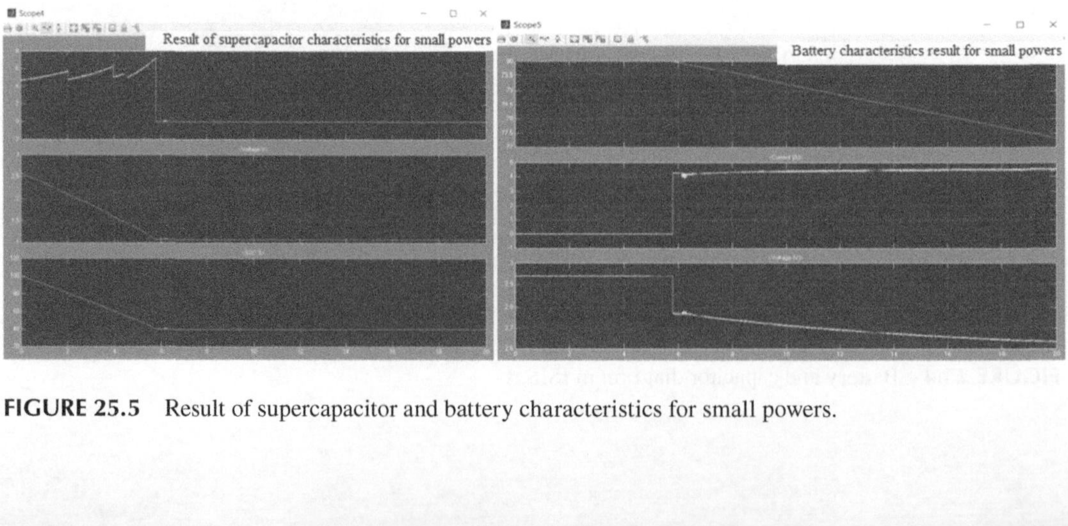

FIGURE 25.5 Result of supercapacitor and battery characteristics for small powers.

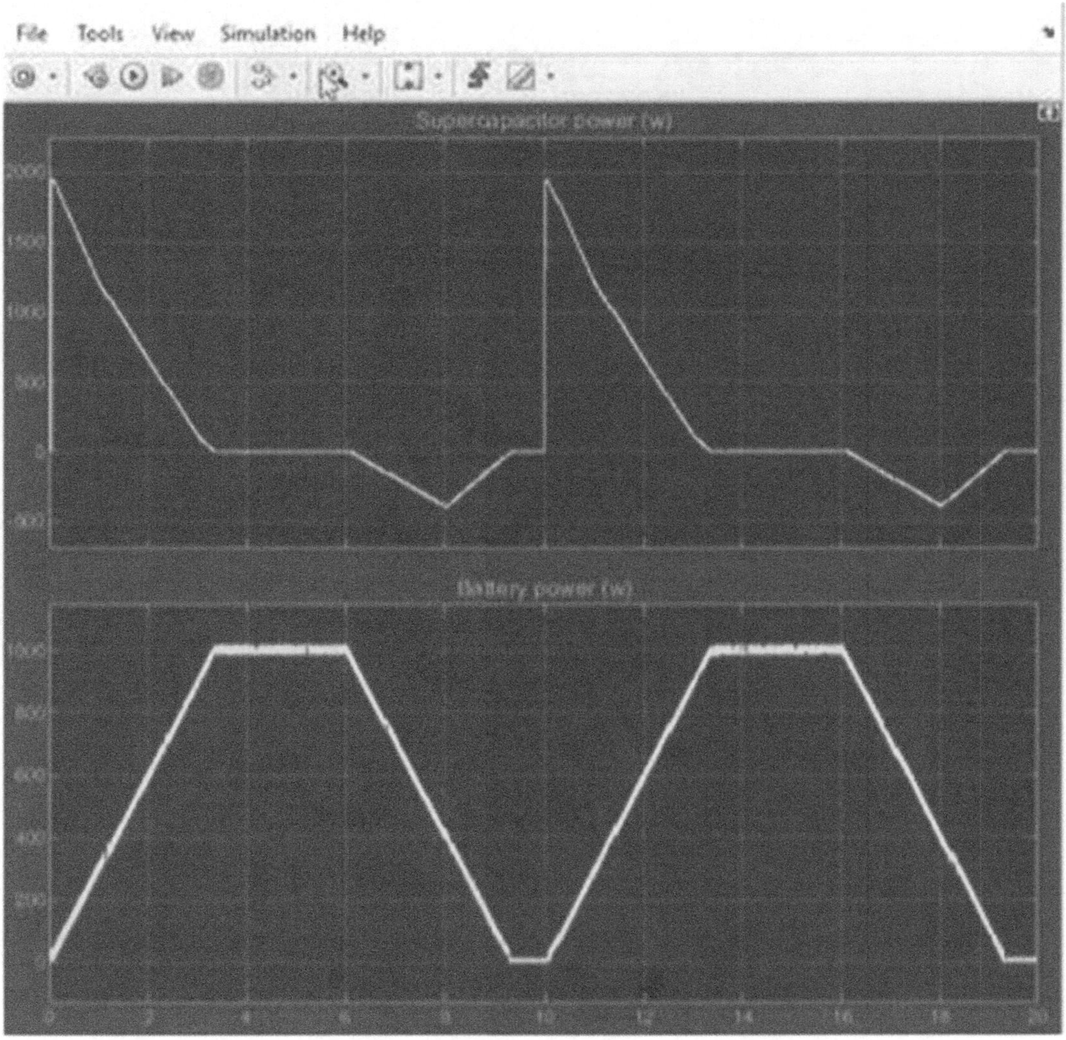

FIGURE 25.6 Battery power and supercapacitor for high powers.

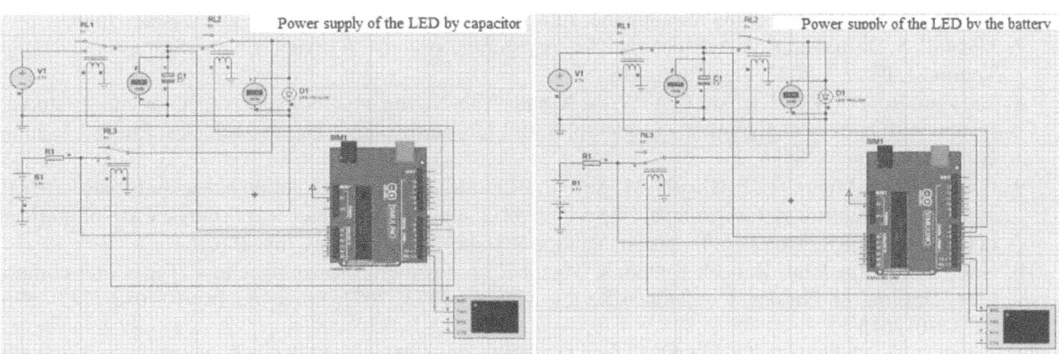

FIGURE 25.7 Power supply of the LED by capacitor or battery.

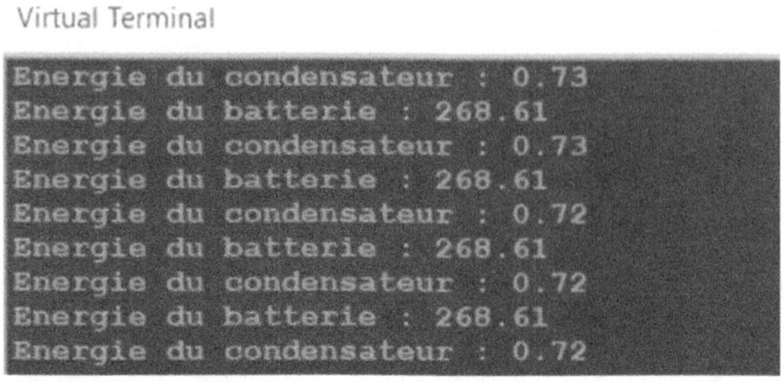

FIGURE 25.8 Capacitor and battery energies.

25.6 CONCLUSION

This work presents the outcomes of comprehensive research into the development of energy storage components, accentuating their variances. The emphasis on selecting suitable energy storage solutions for connected objects has underscored critical considerations within this domain. A comparative analysis between the two primary contenders, batteries and supercapacitors, has been conducted, elucidating the advantages and disadvantages of each through simulation. The batteries offer longer-term storage than supercapacitors, which discharge more slowly. Both degrade over time. Supercapacitors balance higher energy density than conventional capacitors but lower than batteries, with fast charging, suitable for connected objects.

REFERENCES

1. Seifeddine Messaoud, Abbas Bradai, Syed Hashim Raza Bukhari, Pham Tran Anh Quang, Olfa Ben Ahmed, Mohamed Atri, A survey on machine learning in Internet of Things: Algorithms, strategies, and applications, Internet of Things, Volume 12, 2020, Pages 100314.
2. A Michalski, Z. Watral, Problems of powering end devices in wireless networks of the internet of things, Energies, Volume 14, Issue 9, 2021, Pages 2417.
3. Sherali Zeadally, Faisal Karim Shaikh, Anum Talpur, Quan Z. Sheng, Design architectures for energy harvesting in the internet of things, Renewable and Sustainable Energy Reviews, Volume 128, 2020, Pages 109901.
4. M.R. Chakraborty, S. Dawn, P.K. Saha, J.B. Basu, T.S. Ustun, A comparative review on energy storage systems and their application in deregulated systems, Batteries, Volume 8, 2022, Pages 124.
5. D Zyukin, I. Khasambiev, IoT as a high degree of autonomy system, Journal of Physics: Conference Series, Volume 2176, 2022, Pages 012020. https://doi.org/10.1088/1742-6596/2176/1/012020
6. R. Kumar, S.K. Gupta, H.-C. Wang, C.S. Kumari, S.S.V.P. Korlam, From efficiency to sustainability: Exploring the potential of 6G for a greener future, Sustainability, Volume 15, 2023, Pages 16387.
7. J. Mitali, S. Dhinakaran, A.A. Mohamad, Energy storage systems: A review, Energy Storage and Saving, Volume 1, Issue 3, 2022, Pages 166–216.
8. Tao Zhu, Roberto Lot, Richard G.A. Wills, Xingda Yan, Sizing a battery-supercapacitor energy storage system with battery degradation consideration for high-performance electric vehicles, Energy, Volume 208, 2020, Pages 118336.
9. F. Ghizlane, J. Mabrouki, F. Ghrissi, M. Azrour, Proposal for a high-resolution particulate matter (PM10 and PM2. 5) capture system, comparable with hybrid system-based internet of things: Case of quarries in the Western Rif, Morocco, Pollution, Volume 8, Issue 1, 2022, Pages 169–180.
10. Noor I. Ibrahim Jalal, Raheek I. Oudah, Manal K, A review on supercapacitors: Types and components, Journal of Physics: Conference Series, Volume 1973, 2021. https://doi.org/10.1088/1742-6596/1973/1/012015
11. Amna Riaz, Mahidur R. Sarker, Mohamad Hanif Md Saad, Ramizi Mohamed, Review on comparison of different energy storage technologies used in micro-energy harvesting, WSNs, low-cost microelectronic devices: Challenges and recommendations, Sensors, Volume 21, Issue 15, 2021, Pages 5041.

12. Sehwan Kim, Pai Chou, Energy harvesting: Energy harvesting with supercapacitor-based energy storage. In Smart sensors and systems, 2015. Springer. https://doi.org/10.1007/978-3-319-14711-6_10.
13. K.C.S. Lakshmi, B. Vedhanarayanan, High-performance supercapacitors: A comprehensive review on paradigm shift of conventional energy storage devices, Batteries, Volume 9, 2023, Pages 202.
14. Chao Zhang, Yi-Li Wei, Peng-Fei Cao, Meng-Chang Lin, Energy storage system: Current studies on batteries and power condition system, Renewable and Sustainable Energy Reviews, Volume 82, Issue Part3, 2018, Pages 3091–3106.
15. Jie Zhang, Min Gu, Xi Chen, Supercapacitors for renewable energy applications: A review, Micro and-Nano-Engineering, Volume 21, 2023, Pages 100229.
16. M. Benchrifa, M. Elouardi, G. Fattah, J. Mabrouki, R. Tadili. Identification, simulation and modeling of the main power losses of a photovoltaic installation and use of the internet of things to minimize system losses. In Advanced technology for smart environment and energy (pp. 49–60), 2023. Cham: Springer International Publishing.
17. Michael.J. Lain, Emma Kendrick, Understanding the limitations of lithium ion batteries at high rates, Journal of Power Sources, Volume 493, 2021, Pages 229690.
18. R. de Fazio, D Cafagna, G Marcuccio, P. Visconti, Limitations and characterization of energy storage devices for harvesting applications, Energies, Volume 13, Issue 4, 2020, Pages 783.
19. Sambit Satpathy, Sanchali Das, Bidyut Kumar Bhattacharyya, How and where to use super-capacitors effectively, an integration of review of past and new characterization works on super-capacitors, Journal of Energy Storage, Volume 27, 2020, Pages 101044.
20. M.A. Hannan, S.B. Wali, P.J. Ker, M.S. Abd Rahman, M. Mansor, V.K. Ramachandaramurthy, K.M. Muttaqi, T.M.I. Mahlia, Z.Y. Dong, Battery energy-storage system: A review of technologies, optimization objectives, constraints, approaches, and outstanding issues, Journal of Energy Storage, Volume 42, 2021, Pages 103023.

26 Errachidia's Arid Climate Under AI Lens
14-Year Trends and Insights

A. Kabbori, A. Guezzaz, S. Benkirane, and M. Azrour

26.1 INTRODUCTION

Climate change poses an escalating threat to global ecosystems, with arid and semi-arid regions emerging as critical sentinels due to their heightened sensitivity to temperature increases, water scarcity, and extreme weather events. Projections indicate that North Africa may experience temperature rises of 1.5–2°C by 2050 under moderate emission scenarios, with amplified risks of desertification, heat stress, and disrupted agricultural systems. These trends highlight the urgent need for localized, high-resolution climate data to inform adaptation strategies in regions like the Maghreb, where semi-arid zones support millions through rain-fed farming and pastoral livelihoods. However, the scarcity of fine-scale meteorological records in such areas limits the precision of climate models and policy interventions, particularly in data-sparse environments like southeastern Morocco. Errachidia, a semi-arid city in Morocco's Draa-Tafilalet region, exemplifies this vulnerability. With an economy heavily reliant on date palm cultivation and a population exceeding 100,000, the region faces mounting pressures from climate variability, including prolonged droughts and episodic dust storms. Despite its ecological and socioeconomic significance, long-term, high-resolution climate analyses for Errachidia remain underexplored, with existing studies often relying on coarse satellite data or short-term observations. This gap is particularly critical as arid regions are expected to experience warming rates exceeding global averages, necessitating detailed insights into temperature trends, wind dynamics, and air quality to guide sustainable development. The absence of a comprehensive, multi-variable climate record spanning over a decade, combined with advanced analytical approaches, represents a significant barrier to understanding and mitigating climate impacts in this region. Global climate research has established that temperatures have risen by approximately 1.1°C since pre-industrial levels, with arid zones showing accelerated warming due to feedback mechanisms such as reduced surface reflectivity. In North Africa, this warming is accompanied by increased heatwave frequency and reduced precipitation, trends that exacerbate water stress and air quality challenges. The integration of artificial intelligence, particularly through techniques like deep learning, offers a promising avenue to enhance climate trend detection and forecasting, yet its application in arid climates—where high variability and episodic events like dust storms dominate—remains underexplored. This study addresses these challenges by analyzing a unique 14-year hourly weather dataset from Errachidia, encompassing temperature, wind, and employing advanced machine learning methods to uncover climate change signals. By providing a granular, multi-variable record, this research aims to enhance understanding of arid region climate dynamics and inform adaptation strategies for agriculture, urban planning, and public health. The remainder of this paper is organized as follows. "Literature Review" provides an overview of related works in the field of climate change analysis and forecasting using LSTM. In "Materials and Methods", the data set, data preprocessing process, and the model employed in this study are detailed. The "Results" section presents the observed climate trends and the performance of the models in predicting temperature and wind speed while "Conclusion and Future Work", is where the main contributions of the study and potential future research directions are summarized.

DOI: 10.1201/9781003647263-26

26.2 LITERATURE REVIEW

Climate change has precipitated a global temperature increase of approximately 1.1°C since pre-industrial levels, with arid and semi-arid regions experiencing amplified warming due to feedback mechanisms such as reduced albedo [1]. In North Africa, projections suggest temperature rises of 2–3°C by mid-century, accompanied by heightened heatwave frequency and diminished precipitation, intensifying desertification and water scarcity [2]. These trends underscore the vulnerability of arid ecosystems, where human-induced changes have notably reduced dry-season water availability [3]. Accelerated sea-level rise, driven by warming, further impacts coastal arid zones through salinity intrusion, amplifying regional climate challenges [4]. Regional climate dynamics in North Africa exhibit complex variability. Coupled climate models struggle to accurately depict the strengthening North Atlantic jet, a key driver of weather patterns [5], though improvements are noted in modeling Northern Hemisphere wintertime zonal winds [6]. Precipitation trends remain inconsistent, with recent wetting in East Africa from 1983–2021 contrasting with earlier drying linked to Indo-Pacific sea surface temperature variability [7, 8]. Greenhouse gas–driven warming in the Indo-West Pacific exacerbates these shifts, highlighting the need for high-resolution data to resolve discrepancies [9]. Such granularity has proven transformative, with hourly streamflow observations revealing trends obscured by coarse models [10]. In arid climates, high-resolution data sets have illuminated subtle dynamics. Middle Eastern studies using fine-scale data have identified seasonal anomalies tied to circulation changes [11], while Sahel research links dust storm intensification to desertification and temperature anomalies [12]. These findings underscore the value of localized, granular records for understanding arid climate evolution. The integration of artificial intelligence (AI) into climate science has further advanced trend detection and forecasting. Deep learning models, including long short-term memory (LSTM) networks, have enhanced temperature anomaly predictions in temperate regions [13, 14], yet their application in arid environments remains limited due to high variability and episodic events like dust storms [15]. AI's versatility extends beyond climate forecasting. It has reduced energy consumption by 30–50% in smart manufacturing [16], supported sustainable forest management by curbing deforestation [17], and informed resilient urban planning in climate-stressed cities [18]. In arid contexts, AI has improved El Niño predictions [19]. Globally, over 102,160 climate impact studies exist, yet an attribution gap persists, with evidence concentrated in high-income regions, leaving North Africa underrepresented [20]. In the Maghreb, research has focused on broad precipitation trends and climate variability [21], with agricultural vulnerability and thermal comfort in buildings receiving attention [22]. Weakening Northern Hemisphere summertime circulation, possibly driven by aerosol emissions [23], and challenges in modeling North Atlantic variability [24] highlight the need for localized, high-resolution analyses.

26.3 LITERATURE REVIEW

26.3.1 Data Collection

This study utilized a 14-year hourly weather dataset spanning 1 January 2010 to 31 December 2024, collected from a meteorological station in Errachidia, Morocco, operated by the National Moroccan Meteorological Service, known as the Direction Générale de la Météorologie (DMN). The station, located in a semi-arid region of the Draa-Tafilalet area (coordinates: Latitude 31-55-59N. Longitude 004-24-00W; elevation: 1,037 m), provided continuous measurements of key meteorological variables: temperature (°C), dew point (°C), wind speed (m/s), and wind direction (degrees). These variables were selected to capture a comprehensive profile of Errachidia's climate dynamics, with hourly resolution ensuring the detection of diurnal and seasonal variability. The data set comprised approximately 122,640 data points per variable over the study period.

26.3.2 Data Preprocessing

Raw data underwent systematic preprocessing to ensure quality and consistency for analysis. Initial cleaning involved identifying and addressing missing values and anomalies. Missing data points for all variables were interpolated using linear interpolation for gaps of less than 6 hours, preserving temporal continuity. and for longer gaps, monthly mean values were used to account for its cyclic nature. Outliers were identified as values exceeding three standard deviations from the monthly mean for each variable and were replaced with the median of the preceding and following 24-hour periods, affecting less than 0.5% of the data. Seasonal aggregation was performed by grouping hourly data into four seasons—Spring (March–May), Summer (June–August), Fall (September–November), and Winter (December–February)—to facilitate seasonal trend analysis. All preprocessing steps were conducted using Python 3.11 with the pandas and NumPy.

26.3.3 Statistical Analysis

Temporal trends in temperature, dew point, wind speed, and wind direction were assessed using linear regression on annual and seasonal means. Statistical significance was determined through two-tailed t-tests with a threshold of $\alpha = 0.05$, and 95% confidence intervals were calculated via bootstrap resampling with 1,000 iterations to ensure robustness. Time series decomposition was applied to temperature using the seasonal_decompose function in the statsmodels library, separating trends, seasonal components, and residuals to isolate long-term patterns. Extreme weather events were defined for temperature, with high events exceeding the 95th percentile and low events falling below the 5th percentile, computed annually to evaluate changes in frequency. Correlation between wind direction shifts and temperature anomalies was assessed using Pearson's correlation coefficient, with significance tested at $p < 0.05$.

26.3.4 Machine Learning Modeling

LSTM networks were employed to predict temperature and wind speed, leveraging their ability to model sequential data with temporal dependencies. Models were developed using pytorch on a CUDA-enabled GPU, featuring a single-layer architecture with 50 hidden units, a 24-hour input sequence, and 10 epochs of training (batch size: 32, Adam optimizer, learning rate: 0.001). Input features included temperature, dew point, wind speed, and wind direction, with missing values handled as described. The data set was split into 80% training and 20% testing sets, and performance was evaluated using root mean square error (RMSE) and mean absolute error (MAE) on normalized data (min-max scaling, 0–1 range). Predictions were validated over 100 time steps to assess short-term forecasting accuracy. Model robustness was ensured through k-fold cross-validation ($k = 5$).

26.4 RESULTS

26.4.1 Temperature Trends Prediction

The mean annual temperature exhibited a statistically significant increase of 0.1142°C per year ($p = 0.0003$, 95% confidence interval [CI]: 0.098–0.130°C), accumulating to approximately 1.6°C over the study period (Figure 26.1). The shaded standard deviation in this figure indicates rising variability, particularly after 2018, suggesting increased climatic instability. Seasonal decomposition further refines this trend, with fall temperatures rising at 0.1666°C per year ($p = 0.0002$, CI: 0.142–0.191°C), summer at 0.1155°C per year ($p = 0.0526$, CI: 0.102–0.129°C), winter at 0.0836°C per year ($p = 0.0391$, CI: 0.071–0.096°C), and spring at 0.0916°C per year ($p = 0.0998$,

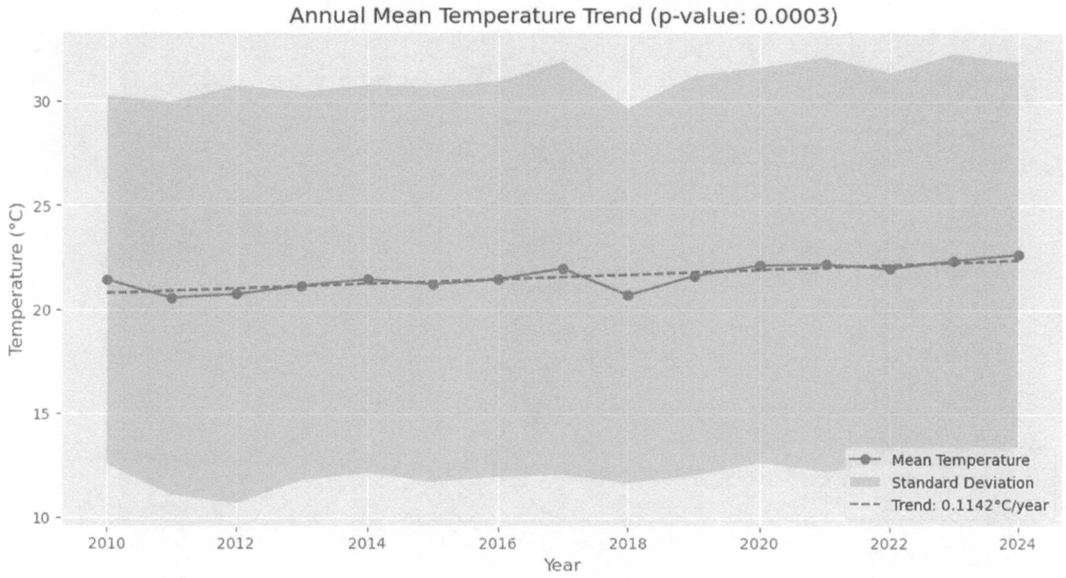

FIGURE 26.1 Annual mean temperature trend.

CI: 0.078–0.105°C) (Figure 26.2). These seasonal trends were derived by aggregating hourly data into seasonal means (Spring: March– May, Summer: June–August, Fall: September–November, Winter: December– February). A heatmap of monthly temperature patterns shows intensified summer heat (June–August) reaching ~35°C in 2022–2024, with winters warming from a baseline of ~10°C to ~15°C, reflecting a shift toward a longer hot season (Figure 26.3). Time series decomposition confirmed a robust trend component of 0.1290°C per year ($p = 0.0000$, CI: 0.115–0.143°C), reinforcing the annual increase across methodologies.

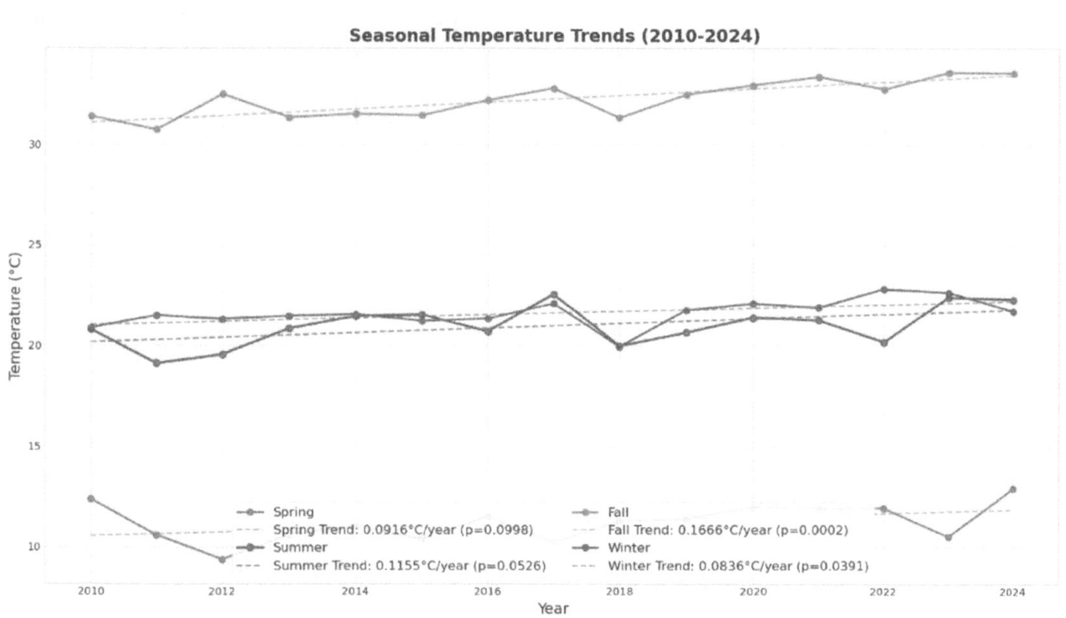

FIGURE 26.2 Seasonal temperature trends (2010–2024).

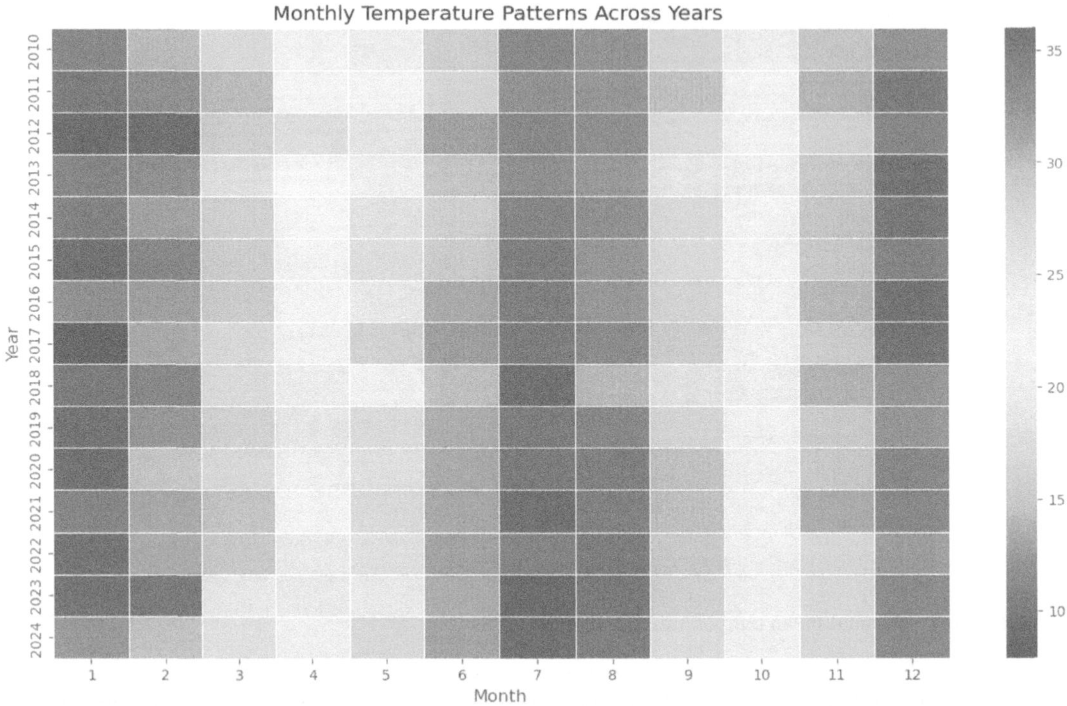

FIGURE 26.3 Monthly temperature patterns across the years.

26.4.2 Extreme Weather Events

Extreme weather events displayed divergent trajectories, underscoring a transition to a warmer climate. The frequency of extreme high-temperature events, defined as temperatures exceeding the 95th percentile, fluctuated with notable peaks at ~500 events in 2012 and 2024, reflecting persistent heat stress (Figure 26.4). Conversely, extreme low-temperature events, below the 5th percentile,

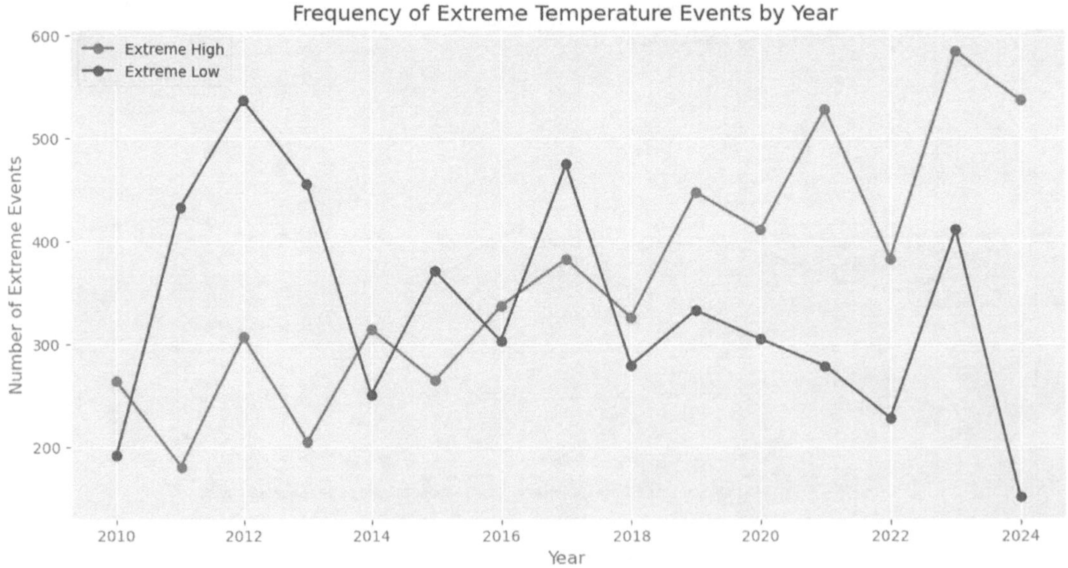

FIGURE 26.4 Frequency of extreme temperature events by year.

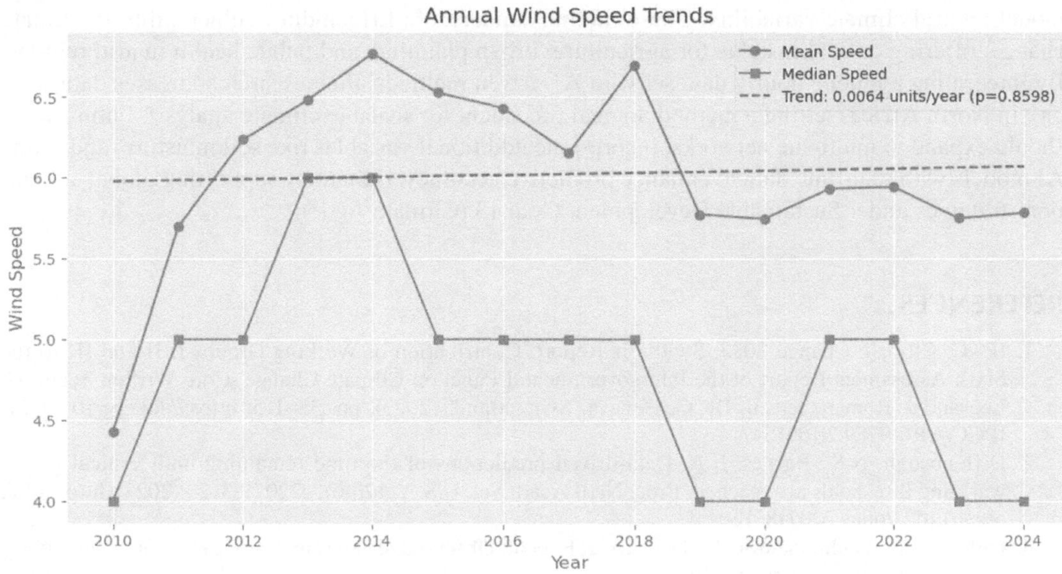

FIGURE 26.5 Annual wind speed trends.

peaked at ~500 in 2012 and 2020 but declined sharply to below 200 by 2024. This decline aligns with a non-significant reduction in cold spells (–9.29 events per year, p = 0.1688, CI: –19.5 to 0.9), while the earlier reported significant increase in heat events (23.75 per year, p = 0.0000, CI: 20.1–27.4) underscores the growing dominance of high-temperature extremes, a key indicator of climate change impact.

26.4.3 Wind Patterns

Wind pattern evolution revealed a directional shift without a corresponding change in intensity. Northern wind prevalence increased from 25.7% in 2011 to 30.4% in 2023, correlating with temperature anomalies (Pearson r = 0.4 in peak years), suggesting a potential link to altered atmospheric circulation (Figure 26.5). However, mean annual wind speed remained stable, with a negligible trend of 0.0064 m/s per year (p = 0.8589, CI: –0.015 to 0.028), and median wind speed showed similar stability, indicating that wind-driven phenomena like dust transport may be more influenced by direction than intensity.

26.4.4 LSTM Modeling

LSTM modeling evaluated the predictive capacity across variables. Temperature predictions failed to capture seasonal and diurnal variability, yielding a flat ~20°C forecast despite actual ranges of 5–35°C, with high errors (RMSE: 9.35°C, MAE: 7.89°C) (Figure 26.5). These results suggest that while LSTM models can detect long-term trends, their accuracy for extreme events is limited.

26.5 CONCLUSION

This study provides a high-resolution analysis of Errachidia's climate dynamics over 14 years (2010–2024), revealing a significant 1.6°C temperature increase, an extended hot season with fall warming at 0.1666°C per year, and intensified summer heat reaching ~35°C. Extreme high-temperature events persisted, while cold spells diminished. LSTM models, despite detecting trends, struggled with predictive accuracy (e.g., temperature RMSE: 9.35°C), highlighting challenges in modeling arid climate variability. These findings underscore Errachidia's vulnerability to climate change, offering critical insights for agriculture, urban planning, and public health in arid regions. By integrating a unique hourly data set with AI-driven methods, this research addresses data scarcity in North Africa, setting a methodological precedent for scalable climate analysis. Future work should expand to multi-site networks, incorporate additional variables like soil moisture, and refine AI models with real-time data to enhance predictive accuracy, ultimately supporting global adaptation strategies under Sustainable Development Goal 13 (Climate Action).

REFERENCES

1. IPCC. Climate Change 2023: Synthesis Report. Contribution of Working Groups I, II and III to the Sixth Assessment Report of the Intergovernmental Panel on Climate Change [Core Writing Team, H. Lee and J. Romero (eds.)]. IPCC, Geneva, Switzerland, (2023), pp. 35–115, https://doi.org/10.59327/IPCC/AR6-9789291691647
2. Diffenbaugh, N.S., Barnes, E.A. Data-driven predictions of the time remaining until critical global warming thresholds are reached. Proc. Natl. Acad. Sci. U.S.A. 120(6), e2207183120 (2023). https://doi.org/10.1073/pnas.2207183120
3. Padrón, R.S., Gudmundsson, L., Decharme, B. et al. Observed changes in dry- season water availability attributed to human-induced climate change. Nat. Geosci. 13, 477–481 (2020). https://doi.org/10.1038/s41561-020-0594-1
4. Nerem, R.S., Beckley, B.D., Fasullo, J.T., Hamlington, B.D., Masters, D., Mitchum, G.T. Climate-change–driven accelerated sea-level rise detected in the altimeter era. Proc. Natl. Acad. Sci. U.S.A. 115(9), 2022–2025 (2018). https://doi.org/10.1073/pnas.1717312115
5. Blackport, R., Fyfe, J.C., Climate models fail to capture strengthening wintertime North Atlantic jet and impacts on Europe. Sci. Adv. 8, eabn3112 (2022). https://doi.org/10.1126/sciadv.abn3112
6. Woollings, T., Drouard, M., O'Reilly, C.H. et al. Trends in the atmospheric jet streams are emerging in observations and could be linked to tropical warming. Commun Earth Environ. 4, 125 (2023). https://doi.org/10.1038/s43247-023-00792-8
7. Palmer, P.I., Wainwright, C.M., Dong, B. et al. Drivers and impacts of Eastern African rainfall variability. Nat Rev Earth Environ. 4, 254–270 (2023). https://doi.org/10.1038/s43017-023-00397-x
8. Yang, W., Seager, R., Cane, M.A., Lyon, B. The East African long rains in observations and models. J. Climate. 27, 7185–7202 (2014). https://doi.org/10.1175/JCLI-D-13-00447.1
9. Hoell, A., Funk, C., Zinke, J. et al. Modulation of the Southern Africa precipitation response to the El Niño Southern Oscillation by the subtropical Indian Ocean Dipole. Clim. Dyn. 48, 2529–2540 (2017). https://doi.org/10.1007/s00382-016-3220-6
10. Gudmundsson, L., Leonard, M., Do, H.X., Westra, S., Seneviratne, S.I. Observed trends in global indicators of mean and extreme streamflow. Geophys. Res. Lett. 46, 756–766 (2019). https://doi.org/10.1029/2018GL079725
11. Evans, J.P. Global warming impact on the dominant precipitation processes in the Middle East. Theor Appl. Climatol. 99, 389–402 (2010). https://doi.org/10.1007/s00704-009-0151-8
12. González Flórez, C., Klose, M., Alastuey, A., Dupont, S., Escribano, J., Etyemezian, V., González-Romero, A., Huang, Y., Kandler, K., Nikolich, G., Panta, A., Querol, X., Reche, C., Pérez García-Pando, C. Insights into the size-resolved dust emission from field measurements in the Moroccan Sahara. Atmos. Chem. Phys. 23, 7177–7212 (2023). https://doi.org/10.5194/acp-23-7177-2023
13. Reichstein, M., Camps-Valls, G., Stevens, B. et al. Deep learning and process understanding for data-driven earth system science. Nature. 566, 195–204 (2019). https://doi.org/10.1038/s41586-019-0912-1
14. Steininger, M., Abel, D., Ziegler, K., Krause, A., Paeth, H., Hotho, A. Deep learning for climate model output statistics. (2020). https://doi.org/10.48550/arXiv.2012.10394
15. Bochenek, B., Ustrnul, Z. Machine learning in weather prediction and climate analyses—Applications and perspectives. Atmosphere. 13, 180 (2022).
16. Saheb, T., Dehghani, M., Sahen, T. Artificial intelligence for sustainable energy: A contextual topic modeling and content analysis. Sustain. Comput.: Inform. Syst. 35, 100699 (2022). https://doi.org/10.1016/j.suscom.2022.100699
17. Liang, J., Gadow, K. Applications of artificial intelligence in forest research and management. (2023). https://www.researchgate.net/publication/376682757_Applications_of_Artificial_Intelligence_in_Forest_Research_and_Management

18. Chen, L., Chen, Z., Zhang, Y., Liu, Y., Osman, A., Farghali, M., Hua, J., Alfatesh, A., Ihara, I., Rooney, D., Yap, P.S. Artificial intelligence-based solutions for climate change: A review. Environ. Chem. Lett. 21 (2023). https://doi.org/10.1007/s10311-023-01617-y
19. Dijkstra, H., Petersik, P., Hernández-García, E., López, C. The application of machine learning techniques to improve El Niño prediction skill. Front. Phys. 7 (2019). https://doi.org/10.3389/fphy.2019.00153
20. Callaghan, M., Schleussner, C., Nath, S., Lejeune, Q., Knutson, T.R., Reichstein, M., Hansen, G., Theokritoff, E., Andrijevic, M., Brecha, R.J., Hegarty, M., Jones, C., Lee, K., Lucas, A., Maanen, N., Menke, I., Pfleiderer, P., Yesil, B., Minx, J. Machine-learning-based evidence and attribution mapping of 100,000 climate impact studies. Nat. Clim. Change. 11, 1–7 (2021). https://doi.org/10.1038/s41558-021-01168-6
21. Harraki, W., Ouazar, D., Bouziane, A., Hasnaoui, M. Climate change observations and trends overview: Focus on Morocco with a case-study of a future Reservoir's response to climate change. E3S Web Conf. 150, 01010 (2020). https://doi.org/10.1051/e3sconf/202015001010.
22. Schilling, J., Freier, K., Hertig, E., Scheffran, J. Climate change, vulnerability and adaptation in North Africa with focus on Morocco. Agric. Ecosyst. Environ. 156, 12–26 (2012). https://doi.org/10.1016/j.agee.2012.04.021
23. Coumou, D., Lehmann, J., Beckmann, J. Climate change. The weakening summer circulation in the Northern Hemisphere mid- latitudes. Science (New York, N.Y.). 348 (2015). https://doi.org/10.1126/science.1261768
24. Bonnet, R., McKenna, C., Maycock, A. Model spread in multidecadal North Atlantic Oscillation variability connected to stratosphere–troposphere coupling. Weather Clim. Dyn. 5, 913–926 (2024). https://doi.org/10.5194/wcd-5-913-2024

27 Renewable Energy
A Systematic Review and Sustainability Issues

I. I. Jouicha, A. Semmar, O. Jbaihi, H. Gziri, Wafaa Dachry, and H. Medromi

27.1 INTRODUCTION

The impact of energy sources on a country's strategic policies is essential, as energy plays a vital role in daily needs such as heating, cooking, manufacturing, and transport [1]. Nowadays, communities worldwide are increasingly concerned about relying on conventional resources to produce energy. This concern stems from the fact that the latter ones are finite, and their use causes significant greenhouse gas emissions. Communities have set themselves the goal of turning this concern into action towards a sustainable future. This target aligns with sustainable development goal (SDG) 13, which states that to ensure a livable climate, greenhouse gas emissions need to be reduced by 43% by 2030, a potential objective that can be achieved by increasing investments in renewable energies [2].

The latter was defined by the International Renewable Energy Agency (**IRENA**) as an energy that does not become depleted, meaning that it can be utilized without reducing its availability in the future [3]. There are many sources of renewable energy, such as wind, solar, geothermal, and biomass [4].

In this context, this paper examines the various renewable energy sources, their primary applications, limitations, and potential role in addressing environmental challenges. To reach our objectives, a systematic review was conducted combined with a content analysis, as these are currently the mainstream of research in the energy sector [5].

Otherwise, this document is structured as follows: The second section provides the background for this research, presenting concepts related to renewable energy sources and sustainable development. The third section describes the selection approach of the papers included in this review. The fourth section consists of a descriptive analysis of the collected papers, along with a presentation of various renewable energy applications. The fifth section discusses the challenges facing the field and its perspectives. The last section, the paper's conclusion, will highlight the key findings.

27.2 BACKGROUND

27.2.1 Renewable Energy Sources

The world's expanding energy needs are now being met by alternative energy sources, replacing conventional ones [6]. The most common renewable energy sources are: biomass, hydropower, geothermal, wind, and solar [7].

Solar energy: By harnessing the sun's rays, solar energy technologies generate electricity [8]. Solar energy is one of the best options, as it offers numerous advantages, such as the lowest

DOI: 10.1201/9781003647263-27

environmental impact, cost-effectiveness, reliability, and durability, as well as, adaptability to both small and large-scale energy needs [9]. The two leading solar energy technologies are photovoltaic "PV" and concentrating solar power "CSP" [10].

- **PV technology:** Thanks to its capacity to produce electricity directly from sunlight through semiconductor devices in the solid state, photovoltaic technology has become the driving force behind the solar revolution [11]. Photovoltaic cells have been developed over four generations. The fourth one is called hybrid inorganic cells. This generation is expected to be more flexible and cost-effective, despite the issues that remain [11].
- **CSP technology:** Using mirrors or lenses, CSP technology concentrates solar energy over a small area, thereby generating heat that can be used to generate electricity through conventional turbines. The most mature of the different forms of CSP technology are as follows: parabolic trough systems, solar power towers, linear Fresnel reflectors, and dish stirling systems [11].

A comparison between the main solar energy technologies in terms of capital costs, maintenance, maturity, and energy storage over the life cycle of the installation is shown in Table 27.1.

Wind energy: Using the kinetic energy of flowing air, wind energy is generated by wind turbines installed either on land (**onshore**) or in seawater or freshwater (**offshore**), land-based solutions have been successfully tested and are in widespread use [8]. Note that a typical 6 kW wind power system is expected to save 2.5 to 5 tons of CO_2 per year [12]. IRENA reports that in 2023, installed wind power capacity worldwide stood at 1017.39 GW, including 944.205 GW onshore and 73.185 GW offshore [13].

Hydropower: Hydropower exploits the potential energy of water as it flows from one high altitude to another [8]. Hydropower represents the world's largest renewable energy sources [12]. IRENA reports that installed hydropower capacity worldwide in 2023 was 1406.863 GW [13].

Geothermal: Geothermal energy is a renewable, carbon-free resource that harnesses the flow of heat from the earth's core. Currently, for electricity generating, deep geothermal resources based on hot water and steam are exploited, in contrast to shallow geothermal heat pump systems that provide cooling and heating without the use of heat stored in the ground [9]. In 2023, the installed capacity of geothermal energy worldwide stood at 15,026 GW [13].

Biomass: All organic substance, living or recently dead, originating from plants or animals, is considered as biomass. In order to improve the properties of biomass to match those of coal, torrefaction was born. It is a thermal pre-treatment process carried out at temperatures ranging from 200 to 300°C within an inert atmosphere to modify the chemical properties of the biomass [14].

TABLE 27.1
A Comparative Table of the Leading Solar Energy Technologies [11]

Parameter	PV Technology	CSP Technology
Low CAPEX	✓	×
Low maintenance	✓	×
High maturity	✓	×
Cost-effective long-term energy storage	×	✓

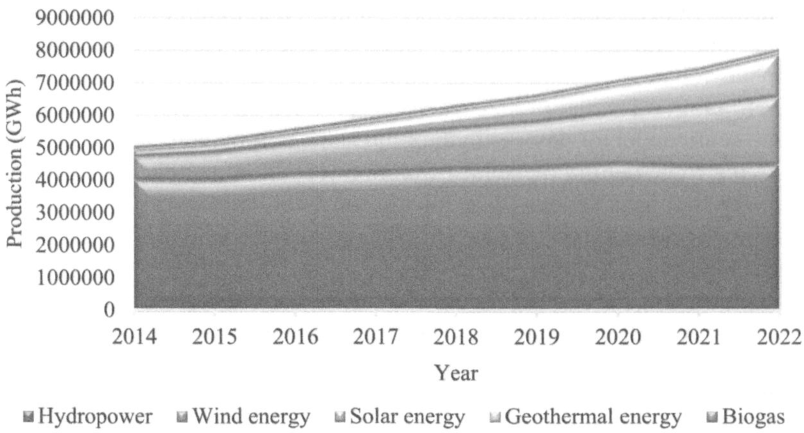

FIGURE 27.1 Energy generated by renewable energy sources seen in this section worldwide between 2014 and 2022.

Renewable energy statistics published in July 2024 by IRENA [13], were used to extract data relating to the contribution of the sources presented in this section to the global renewable energy production between 2014 and 2022, and Excel to graph them (Figures 27.1 and 27.2).

27.2.2 Sustainable Development

Sustainable development refers to development that meets current needs without compromising the ability to meet the needs of future generations. Sustainable development will not be possible without climate action, in large part because public health, food and water security, migration, peace, and security are all already impacted by climate change. In 2015, the world's leaders assembled and made a landmark commitment to ensure the rights and well-being of everyone on a healthy and prosperous planet as they adopted the 2030 Agenda for Sustainable Development [15, 16].

The sustainable development goals: The SDGs are a roadmap towards a more sustainable and better future for all. There are 17 interconnected goals [17].

Climate action represents the thirteenth goal [18].

- **Fact:** By 2030, emissions must be cut by almost half to limit global warming to 1.5 degrees Celsius above pre-industrial levels. However, we are still a long way from this target [18].

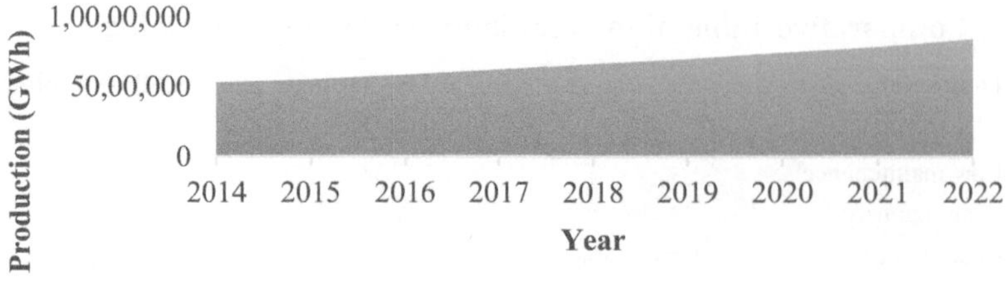

FIGURE 27.2 Total renewable energy produced worldwide between 2014 and 2022.

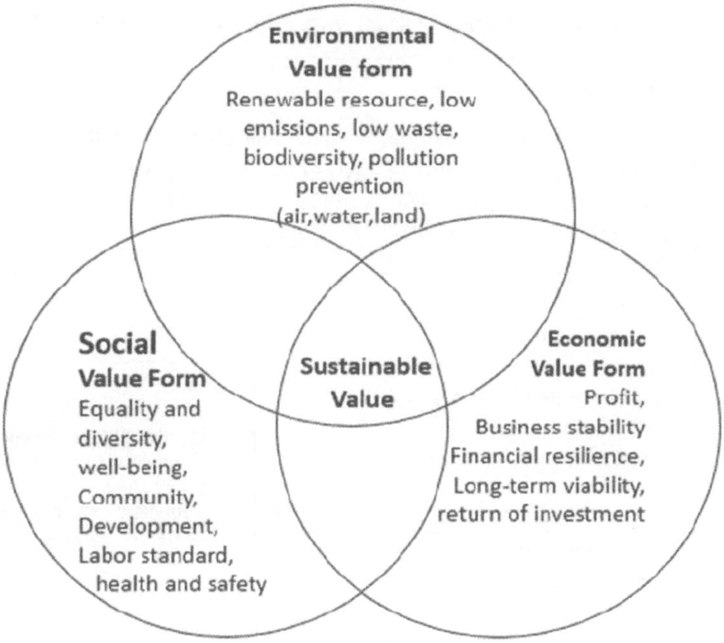

FIGURE 27.3 Pillars of sustainability [19].

- **Reaching the goal:** To limit the global temperature to under 2°C, or even 1.5°C, it is essential that the world transform its energy, industrial, transport, food, agricultural and forestry systems [18].

Sustainability pillars [19]: Sustainable development involves a mixture of environmental, social and economic considerations, which are often referred to as the pillars of sustainability, which are represented in Figure 27.3.

It is clear now that integrating renewable energies is a key factor in promoting sustainable development.

27.3 METHODOLOGY

A systematic review was conducted to investigate the role of renewable energies in addressing environmental issues and promoting sustainable development. The data used in this study were extracted from Scopus database. This database was selected because their publications in peer-reviewed journals are frequently regarded as exceptional in quality and credibility [20]. To collect data, the following query was used: ((("renewable energies") AND ("sustainability" OR ("sustainable development")) AND ((" comprehensive review ") OR " drivers " OR " enablers " OR " challenges " OR (" future directions") OR " applications"))), which was based on keywords only. In addition, to keep abreast of the latest developments in the field, the search was limited to documents published between 2015 and 2025. Next, the filters "Document type, open access and publication stage" were applied to refine the search results. After selecting reports based on their title and abstract, some of them were discarded. This method is called preferred reporting items for systematic reviews and meta-analyses (PRISMA) and it is illustrated in Figure 27.4.

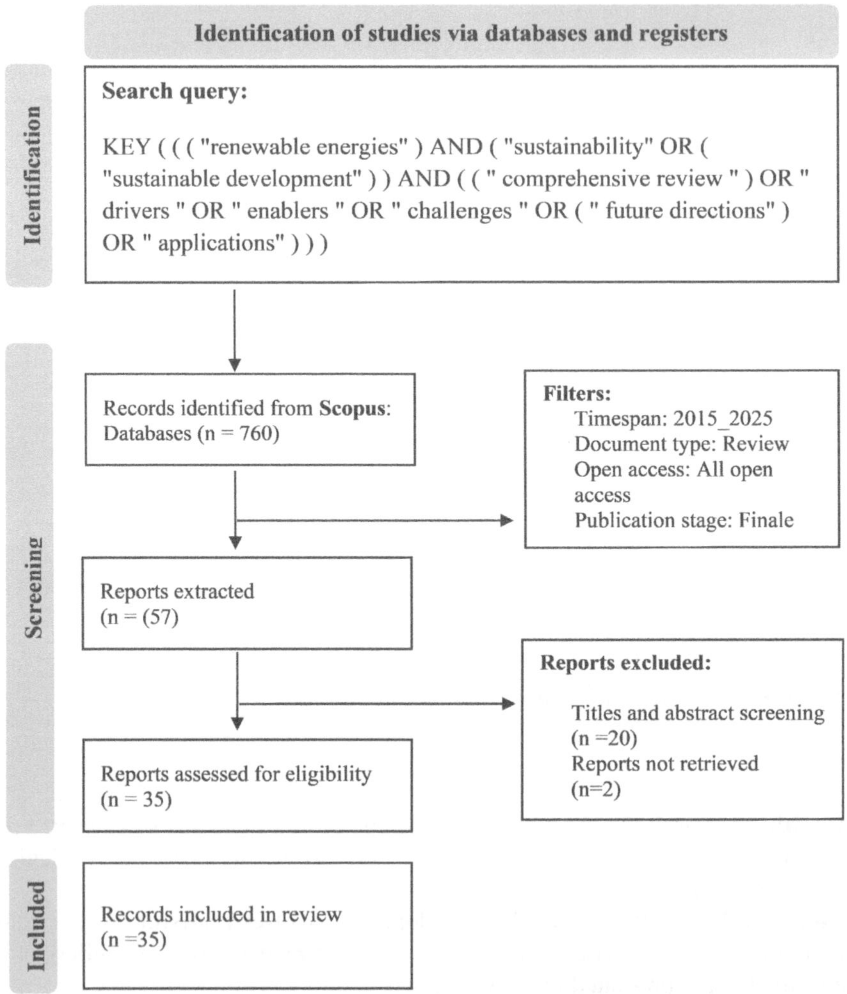

FIGURE 27.4 PRISMA diagram.

To evaluate the data set and present the results, we used the following tools: **Biblioshiny** for **Bibliometrix** [21] and **VOSviewer** [22].

27.4 RESULTS

27.4.1 Data Set Descriptive Analysis

Table 27.2 presents an overview of the primary information about our dataset, such as the 35 papers from 21 sources, the significant annual growth rate, and the number of authors. We reveal that 45.71% of the papers included in our data set have been internationally co-authored, highlighting the value of teamwork in scientific research.

Figure 27.5 generated by Biblioshiny and style-adjusted using MATLAB, illustrates country production over time. The figure indicates that 42.85% of the data collected was produced in India, which may indicate the country's energy supply security problems. It might also indicate that this country has obtained substantial subsidies to reduce its greenhouse gas emissions.

TABLE 27.2
Primary Information for the Data Set

Description	Results
MAIN INFORMATION ABOUT THE DATA	
Timespan	2015:2025
Sources (Journals, Books, etc.)	21
Documents	35
Annual Growth Rate %	20.85
DOCUMENT CONTENTS	
Author's Keywords (DE)	157
AUTHORS	
Authors	152
AUTHORS COLLABORATION	
Co-Authors per Doc	4.34
International Co-authorships %	45.71
DOCUMENT TYPES	
Review	35

Figure 27.6 presents the VOSviewer keywords co-occurrence network. To generate it, we uploaded our data set into VOSviewer in CSV format, selected "co-occurrence, all keywords," and three as the minimum number of occurrences of a keyword. To finalize and create the map, we relied on the following website [23]. So, we merged the synonyms into a single word and selected the most pertinent terms from the list offered by VOSviewer after choosing two as the minimum number of occurrences of a keyword this time. This gives us 11 terms, which are represented in Figure 27.5 by nodes of variable size indicating the keyword's frequency of appearance within our data set. The link between nodes reveals the number of times the keywords appear together.

Table 27.3 illustrates the most relevant sources. The majority of the papers included in our data set were published in journals with a significant H-index, which could indicate their meaningful contribution to the field.

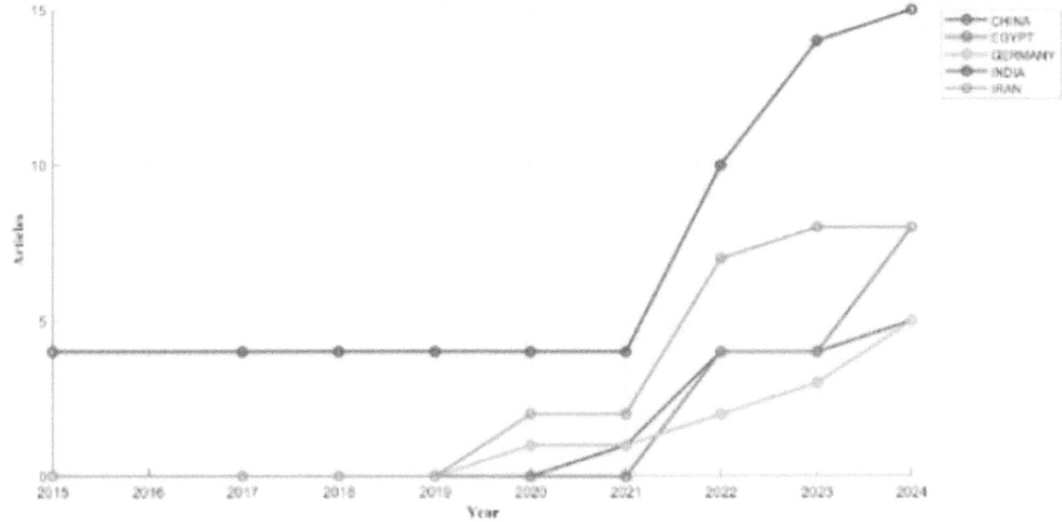

FIGURE 27.5 Countries' production over time.

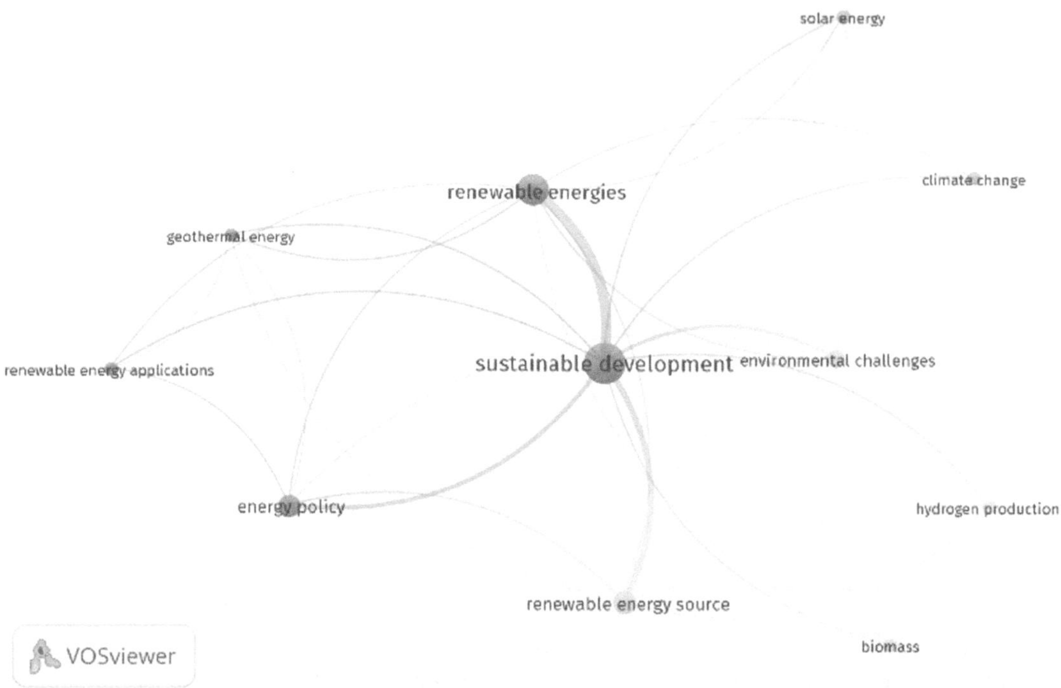

FIGURE 27.6 Co-occurrence network generated using VOSviewer.

27.4.2 Renewable Energy Applications

Table 27.4 shows the main applications of the renewable energy sources mentioned in Section 27.2.

27.5 DISCUSSION

Despite worldwide efforts to promote a sustainable future through various initiatives, partnerships, annual meetings, regular events, etc., we are still in the midst of a global crisis. It is easy to say that increasing the percentage of renewable energies in the global energy mix is a good solution as it creates jobs, reduces the world's greenhouse gas emissions, and so on. However, as with every solution, difficulties arise. One of these is the reality associated with the intermittence of renewable

TABLE 27.3
Most Relevant Sources

Sources	No. of Documents	H-Index
Energies	9	152
Sustainability (Switzerland)	3	169
Current sustainable/renewable energy reports	2	26
Energy strategy reviews	2	65
Environmental science and pollution research	2	179
Journal of environmental management	2	243
Deep underground science and engineering	1	5
Energy exploration and exploitation	1	43
Energy reports	1	88
Energy science and engineering	1	51

TABLE 27.4
Applications of Renewable Energy Sources: "Key Finding & Gaps"

Renewable Energy Source	Applications	Key Finding	Gaps
Solar energy [4, 24–30]	Solar electricity generation	Photovoltaic power plants generate lower carbon footprints than their conventional thermal counterparts; Solar collectors that can produce high-quality steam for electricity generation are complex and expensive. However, technical variants of linear Fresnel reflectors (LFR) show promise for meeting a variety of requirements.	Some environmental worries linked to the use of Pb, Sn, Cd, and Te may lead to adverse effects on human health
	Heating and cooling	Even if all solar cooling systems are potentially very attractive because of their environmental and energy benefits, the thermal choice was extensively studied, with a focus on adsorption, absorption, and desiccant cooling systems. Generally speaking, these systems perform well in meeting cooling demand. Solar thermal energy systems provide heat for water preheating or steam production, and may be incorporated easily into industrial systems already in place.	
	Desalination	At the heart of sustainable water management strategies, solar-powered reverse osmosis desalination will allow the MENA region to pave the way to a future where water resources are clean, sustainable, and accessible for the next generations. For higher efficiency solar desalination systems, the most attractive option seems to be direct solar desalination systems using solar thermal collectors. In fact, solar-powered desalination helps directly to achieve most of the SDGs: SDG 1, SDG 2, SDG 6, SDG 7, and SDG 13.	
Wind energy [31–33]	Transportation Grinding grain Pumping water Power generation Hydrogen production	When it comes to generating electricity from different renewable resources, wind power proved to have the lowest investment costs, followed by CSP and then PV. Offshore wind energy hydrogen production efficiently tackles the space and environmental constraints of onshore wind energy, and can be integrated into existing offshore wind installations.	High costs are involved in construction and maintenance, complexity of the technology, marine environment impact, and equipment must meet high standards of corrosion resistance
Hydropower [34, 35]	Electricity production	The use of micro-hydrokinetic river turbine technology, a low-tech solution for regions with difficult conditions, is viewed as a better option in South Africa.	Full development of this technology is still some way off
Geothermal [4, 36]	Geothermal electricity, heating and cooling	Geothermal energy plays an important role in the worldwide transition to sustainable, low-carbon energy systems, thanks to its capacity to provide regular, flexible electricity, especially for basic demand, and its competitive cost profile against fossil fuel–based energy options.	Higher dangers associated with drilling and unproductive wells
Biomass (biogas) [37]	Transportation Electricity generation Heat production Combined heat and power (CHP) systems Fuel cells	Biogas is currently used to produce electricity, heat, or a combination of both. Its energy quality, low production and operating costs, availability, ease of storage, and low greenhouse gas emissions (60–80% lower than gasoline) make biogas an attractive alternative, particularly for cogeneration. To improve its efficiency, it is important to focus on anaerobic digestion.	Management of biogas

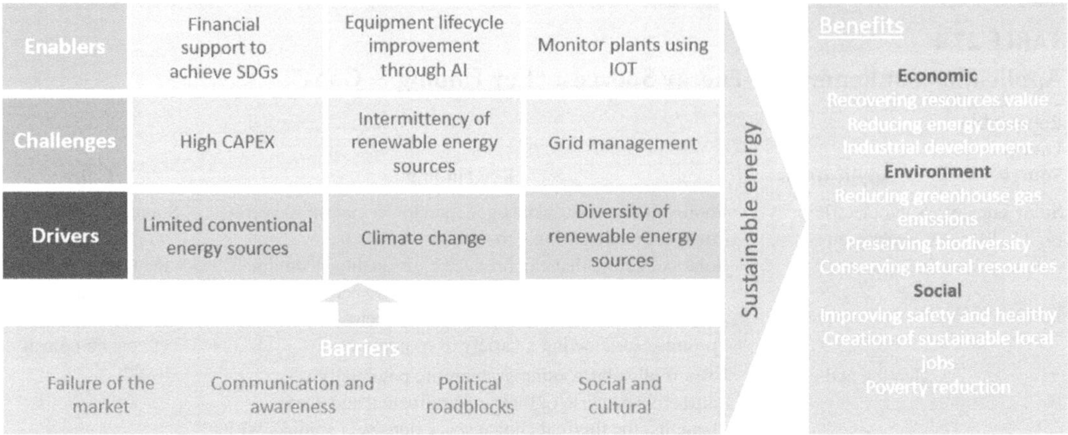

FIGURE 27.7 Conceptual framework of sustainable energy (Adapted to [19]).

energy sources. However, this problem could be solved thanks to the energy storage options. In this context, green hydrogen is becoming a trend, even if it must be produced using renewable energies that are being expanded. So, is this a profitable solution?

Grid-connected renewable energy facilities do not require storage; however, when these solutions are integrated into the electrical grid, they can introduce challenges in terms of management, adaptability, and grid stability.

IoT and AI have emerged as solutions to enhance the integration of renewable energy technologies, thereby promoting sustainability. For example, these technologies can be very valuable in estimating energy consumption, scheduling energy production, and improving equipment life cycles through predictive maintenance, among other applications. This sounds like a useful way to achieve our goals, but we need to ask ourselves the following questions: Do we have the resources to encourage the use of these solutions worldwide? Will the use of these solutions have a harmful effect on the public in terms of cybersecurity? Will these solutions be economically viable or not?

In this context, and based on our research and analysis, we propose a conceptual framework designed to enhance sustainability in the context of renewable energy. Thus, Figure 27.7 illustrates the "Sustainable Energy Conceptual Framework," highlighting key environmental, economic, and social benefits. This framework, adapted from the study of [19], provides insight into the drivers, enablers, and barriers. This conceptual framework attempts to propose a perspective for future studies.

27.6 CONCLUSION

In conclusion, this document highlights the urgent need for a livable climate, which can be achieved by promoting sustainable development by improving the integration of renewable energy technologies. However, although we found many studies on renewable energy and sustainability in the majority of databases, we need more, especially financial feasibility studies, suggestions for improving the international grid infrastructure, and the integration of renewable energy, IoT, and AI in buildings, industry, agriculture, and desalination; as well as studies related to the security of people's personal information.

REFERENCES

1. E. Ilbahar, S. Cebi, and C. Kahraman, "A state-of-the-art review on multi-attribute renewable energy decision making," *Energy Strategy Reviews*, vol. 25, pp. 18–33, 2019, Elsevier Ltd., doi: 10.1016/j.esr.2019.04.014.

2. "ACT NOW." [Online]. Available: www.un.org/en/actnow
3. "2. Overview of Renewable Energy IRENA Renewable Energy Statistics Training."
4. M. El Haj Assad, M. Alhuyi Nazari, and M. A. Rosen, "Applications of Renewable Energy Sources," in *Design and Performance Optimization of Renewable Energy Systems*, pp. 1–15, Jan. 2021, doi: 10.1016/B978-0-12-821602-6.00001-8.
5. M. Szarucki, R. Rybkowski, J. Bugaj, and K. Bracio, "A comprehensive review of research approaches in the energy sector: A management sciences perspective," *Energies*, vol. 15, no. 22, p. 8495, 2022, MDPI, doi: 10.3390/en15228495.
6. Z. Gareiou, E. Drimili, and E. Zervas, "Public Acceptance of Renewable Energy Sources," in *Low Carbon Energy Technologies in Sustainable Energy Systems*, pp. 309–327, Jan. 2021, doi: 10.1016/B978-0-12-822897-5.00012-2.
7. B. Zohuri, and P. McDaniel, "Other Electrical Power Generation Energy Sources," in *Introduction to Energy Essentials*, pp. 223–254, Jan. 2021, doi: 10.1016/B978-0-323-90152-9.00006-2.
8. M. Benchrifa, and J. Mabrouki, "Simulation, sizing, economic evaluation and environmental impact assessment of a photovoltaic power plant for the electrification of an establishment," *Advances in Building Energy Research*, vol. 16, no. 6, pp. 736–753, 2022.
9. F. Sher, N. Smječanin, H. Hrnjić, E. Bakunić, and J. Sulejmanović, "Prospects of renewable energy potentials and development in Bosnia and Herzegovina – A review," *Renewable and Sustainable Energy Reviews*, vol. 189, p. 113929, 2024, doi: 10.1016/J.RSER.2023.113929.
10. A. O. M. Maka, T. Ghalut, and E. Elsaye, "The pathway towards decarbonisation and net-zero emissions by 2050: The role of solar energy technology," *Green Technologies and Sustainability*, vol. 2, no. 3, p. 100107, 2024, doi: 10.1016/J.GRETS.2024.100107.
11. M. Osman, and I. Qureshi, "Review of photovoltaic and concentrated solar technologies including their performance, reliability, efficiency and storage," *Results in Engineering*, vol. 25, p. 104424, 2025, doi: 10.1016/J.RINENG.2025.104424.
12. M. Aien, and O. Mahdavi, "On the way of policy making to reduce the reliance of fossil fuels: Case study of Iran," *Sustainability*, vol. 12, no. 24, p. 10606, 2020, *MDPI*, doi: 10.3390/su122410606.
13. I. Renewable Energy Agency, Renewable Energy Statistics 2024 Statistiques D'énergie Renouvelable 2024 Estadísticas De Energía Renovable *2024*. 2024. [Online]. Available: www.irena.org
14. M. Benchrifa, M. Elouardi, G. Fattah, J. Mabrouki, and R. Tadili, "Identification, Simulation and Modeling of the Main Power Losses of a Photovoltaic Installation and Use of the Internet of Things to Minimize System Losses," in *Advanced Technology for Smart Environment and Energy*, Cham: Springer International Publishing, pp. 49–60, 2023.
15. "The Sustainable Development Agenda - United Nations Sustainable Development." Accessed: Mar. 15, 2025. [Online]. Available: https://www.un.org/sustainabledevelopment/development-agenda/
16. 2024 4th International Conference on Innovative Research in Applied Science, Engineering and Technology (IRASET). IEEE, 2024. https://www.iraset.org/2024/
17. "Take Action for the Sustainable Development Goals - United Nations Sustainable Development." Accessed: Mar. 15, 2025. [Online]. Available: https://www.un.org/sustainabledevelopment/sustainable-development-goals/
18. "Climate Change - United Nations Sustainable Development." Accessed: Mar. 15, 2025. [Online]. Available: https://www.un.org/sustainabledevelopment/climate-change/
19. A. Dachry, N. El Assad, H. Fouraiji, H. Gziri, W. Dachry, and N. Messaoudi, "Sustainability in the Era of Industry 4.0 and Circular Economy: Theoretical Background and Bibliometric Analysis," in *2024 4th International Conference on Innovative Research in Applied Science, Engineering and Technology, IRASET 2024*, 2024, doi: 10.1109/IRASET60544.2024.10549392.
20. "Maximising the Benefits of Publishing in Scopus - Blog spubl.kg." Accessed: Feb. 22, 2025. [Online]. Available: https://spubl.kg/en/blog/maximising-the-benefits-of-publishing-in-scopus
21. M. Aria, and C. Cuccurullo, "Bibliometrix: An R-tool for comprehensive science mapping analysis," *Journal of Informetrics*, vol. 11, no. 4, pp. 959–975, 2017, doi: 10.1016/J.JOI.2017.08.007.
22. N. J. van Eck, and L. Waltman, "Software survey: VOSviewer, a computer program for bibliometric mapping," *Scientometrics*, vol. 84, no. 2, pp. 523–538, 2010, doi: 10.1007/s11192-009-0146-3.
23. "Analyser des données bibliographiques avec VOSViewer | Recherche numérique." Accessed: Mar. 07, 2025. [Online]. Available: https://recherchenumerique.uqam.ca/scenarios/analyser-des-donnees-bibliographiques-avec-vosviewer/
24. M. Al-Addous *et al.*, "Innovations in solar-powered desalination: A comprehensive review of sustainable solutions for water scarcity in the Middle East and North Africa (MENA) Region," *Water*, vol. 16, no. 13, p. 1877, 2024, Multidisciplinary Digital Publishing Institute (MDPI), doi: 10.3390/w16131877.

25. J. H. Reif, and W. Alhalabi, "Solar-thermal powered desalination: Its significant challenges and potential," *Renewable and Sustainable Energy Reviews*, vol. 48, pp. 152–165, 2015, doi: 10.1016/J.RSER.2015.03.065.
26. R. V. S. Madhuri, Z. Said, I. Ihsanullah, and R. Sathyamurthy, "Solar energy-driven desalination: A renewable solution for climate change mitigation and advancing sustainable development goals," *Desalination*, vol. 602, p. 118575, 2025, doi: 10.1016/J.DESAL.2025.118575.
27. A. Mdallal, A. Yasin, M. Mahmoud, M. A. Abdelkareem, A. H. Alami, and A. G. Olabi, "A comprehensive review on solar photovoltaics: Navigating generational shifts, innovations, and sustainability," *Sustainable Horizons*, vol. 13, p. 100137, 2025, doi: 10.1016/J.HORIZ.2025.100137.
28. S. J. Thomas, S. M. Varghese, M. M. Awad, and S. S. Sahoo, "Solar thermal energy-based electricity and desalination in India: The impact of viability gap funding (VGF) to normalize levelized cost or production," *Journal of Thermal Analysis and Calorimetry*, vol. 149, no. 16, pp. 8857–8864, 2024, doi: 10.1007/s10973-024-13264-8.
29. A. Allouhi, T. Kousksou, A. Jamil, P. Bruel, Y. Mourad, and Y. Zeraouli, "Solar driven cooling systems: An updated review," *Renewable and Sustainable Energy Reviews*, vol. 44, pp. 159–181, 2015, doi: 10.1016/J.RSER.2014.12.014.
30. K. Ravi Kumar, N. V. V. Krishna Chaitanya, and N. Sendhil Kumar, "Solar thermal energy technologies and its applications for process heating and power generation – A review," *Journal of Cleaner Production*, vol. 282, p. 125296, 2021, doi: 10.1016/J.JCLEPRO.2020.125296.
31. M. Mahmoud, M. Ramadan, M. A. Abdelkareem, and A. G. Olabi, "Applications of Wind Energy," in *Renewable Energy - Volume 1: Solar, Wind, and Hydropower Definitions, Developments, Applications, Case Studies, and Modelling and Simulation*, vol. 1, pp. 331–345, 2023, doi: 10.1016/B978-0-323-99568-9.00010-8.
32. M. Dale, "A comparative analysis of energy costs of photovoltaic, solar thermal, and wind electricity generation technologies," *Applied Sciences (Switzerland)*, vol. 3, no. 2, pp. 325–337, 2013, doi: 10.3390/app3020325.
33. A. Tuluhong, Q. Chang, L. Xie, Z. Xu, and T. Song, "Current status of green hydrogen production technology: A review," *Sustainability*, vol. 16, no. 20, p. 9070, 2024, Multidisciplinary Digital Publishing Institute (MDPI), doi: 10.3390/su16209070.
34. E. Vassoney, A. Mammoliti Mochet, and C. Comoglio, "Use of multicriteria analysis (MCA) for sustainable hydropower planning and management," *Journal of Environmental Management*, vol. 196, pp. 48–55, 2017, Academic Press, doi: 10.1016/j.jenvman.2017.02.067.
35. W. Awandu, R. Ruff, J. U. Wiesemann, and B. Lehmann, "Status of micro-hydrokinetic river technology turbines application for rural electrification in Africa," *Energies*, vol. 15, no. 23, p. 9004, 2022, *MDPI*, doi: 10.3390/en15239004.
36. M. J. Barasa Kabeyi, and O. A. Olanrewaju, "Geothermal wellhead technology power plants in grid electricity generation: A review," *Energy Strategy Reviews*, vol. 39, p. 100735, 2022, doi: 10.1016/J.ESR.2021.100735.
37. S. Abanades *et al.*, "A conceptual review of sustainable electrical power generation from biogas," *Energy Science & Engineering*, vol. 10, no. 2, pp. 630–655, 2022, John Wiley and Sons Ltd, doi: 10.1002/ese3.1030.

28 Blockchain Architecture for Healthcare IoT Systems

An Architectural Comparison

Z. Elhadari, H. Zougagh, N. Idboufker, and M. Ech-Chebaby

28.1 INTRODUCTION

The Internet of Things (IoT) in health care represents a network of interconnected devices responsible for collecting data on the healthcare environment and storing it in a centralized cloud for later analysis. The security of this data is vulnerable to threats. Blockchain technology, operating on a decentralized system, addresses these vulnerabilities by eliminating the need for third-party intermediaries and ensuring real-time data protection (Islam et al. 2023). This decentralization offers security, privacy, integrity, and interoperability advantages, while also eliminating dependencies on third parties (J et al. 2023). Blockchain uses cryptography to establish trust without depending on a central authority, offering benefits like mobility and transparency. However, challenges arise due to resource constraints and the increasing number of devices in use. Trust mechanisms play a crucial role in ensuring data integrity for healthcare IoT applications (Thantharate and Thantharate 2023). Integrating blockchain technology into healthcare IoT ensures transparency, auditability, and energy efficiency, ensuring secure and efficient operation of healthcare IoT systems. The contributions of this paper offer an overview of key smart healthcare architectures from 2021 to 2023, specifically concentrating on the integration of blockchain technology within healthcare IoT settings. It conducts a comparative analysis of relevant studies concerning the integration of blockchain technology within healthcare IoT systems. Furthermore, the paper proposes a comprehensive multi-layered architecture based on blockchain technology and IoT, designed to effectively gather and safeguard patient data for accessibility by healthcare service providers and stakeholders. This paper is organized into five sections. Section 28.2 presents an architectural comparison of existing studies related to the integration of blockchain technology into Internet of Things systems in health care. Section 28.3 presents an exploration of the key findings and discussions from the research studies reviewed. Section 28.4 presents a new proposed blockchain architecture for the healthcare IoT and concludes the work with the main conclusions and some final remarks in Section 28.5.

28.2 OVERVIEW OF BLOCKCHAIN TECHNOLOGY IN HEALTHCARE IoT SYSTEMS

28.2.1 Blockchain Technology

Blockchain technology is a decentralized digital ledger system that securely records transactions across multiple computers using cryptographic techniques. It consists of blocks of data linked together, creating a chronological and immutable record of transactions (Alam et al. 2023; J et al. 2023). This technology reduces fraud risk, ensures transparency, and is used in various fields like supply chain management, health care, voting systems, etc.

DOI: 10.1201/9781003647263-28

Blockchain technology is characterized by (Abdelmaboud et al. 2022; Azbeg et al. 2022):

Distributed Ledger: Provides secure and efficient transaction recording, eliminating single points of failure and ensuring data integrity, opening new avenues for innovative solutions in security and trust across various domains.

Authentication: Relies on cryptographic keys to verify user identity and secure transactions, enhancing security and trust in blockchain-based systems.

Smart Contracts: Autonomous programs on the blockchain that execute transactions automatically upon meeting predefined conditions, ensuring high security, transparency, and efficiency.

Consensus Mechanism: Governs how new transactions are validated and added to the blockchain, ensuring data consistency and integrity by achieving agreement among network participants.

Hash Cryptography: Utilizes hashing algorithms like SHA256 to secure and validate transactions, providing one-way cryptography, deterministic computation, and collision resistance, thus safeguarding data integrity and security on the blockchain.

These unique properties make blockchain technology appealing to researchers, offering opportunities to enhance information exchange, data security, and privacy protection across IoT domains.

28.2.2 Why Integrate Blockchain in Healthcare IoT Systems

Healthcare institutions face challenges in managing medical data and information, often incompatible formats. The integration of IoT in healthcare systems offers promising solutions, such as real-time patient monitoring, remote medical data collection, and automated patient surveillance and follow-up processes. However, this requires robust infrastructure for stocking, managing, and securing sensitive medical information, this is where blockchain comes in. Blockchain technology allows healthcare organizations to manage and store patient medical information securely and unifiedly, providing a distributed and immutable ledger, ensuring data integrity and medical security, while maintaining patient privacy (Bataineh et al. 2022).

Blockchain enhances healthcare IoT systems in several key ways:

- **Data Security**: It ensures the utmost security and integrity of medical data, safeguarding patient records and monitoring data against unauthorized access and tampering.
- **Efficient Transactions**: Blockchain automates administrative processes like insurance claims and billing through smart contracts, reducing errors and fraud, and optimizing resource allocation.
- **Patient Empowerment**: Patients gain greater control over their health data, granting permission for data access, fostering patient engagement in healthcare decisions.

28.3 STUDY OF EXISTING ARCHITECTURES

28.3.1 Generic Blockchain Architecture for Healthcare Internet of Things

A generic blockchain architecture is being developed for healthcare IoT systems to create a robust distributed ledger system (Alam et al. 2022; Alkatheiri and Alghamdi 2023). This model aims to guarantee secure and transparent communication between the various IoT devices in health care. The medical data generated by each IoT device is securely stored and shared by blockchain's robust mechanisms. This improves security and ensures efficient information exchange within the medical IoT system. A generic blockchain architecture for healthcare IoT systems in five layers, shown in Figure 28.1, can be described as follows:

Healthcare Infrastructure Layer: Forms the backbone of the healthcare ecosystem, facilitating connectivity, data collection, and efficiency in healthcare operations.

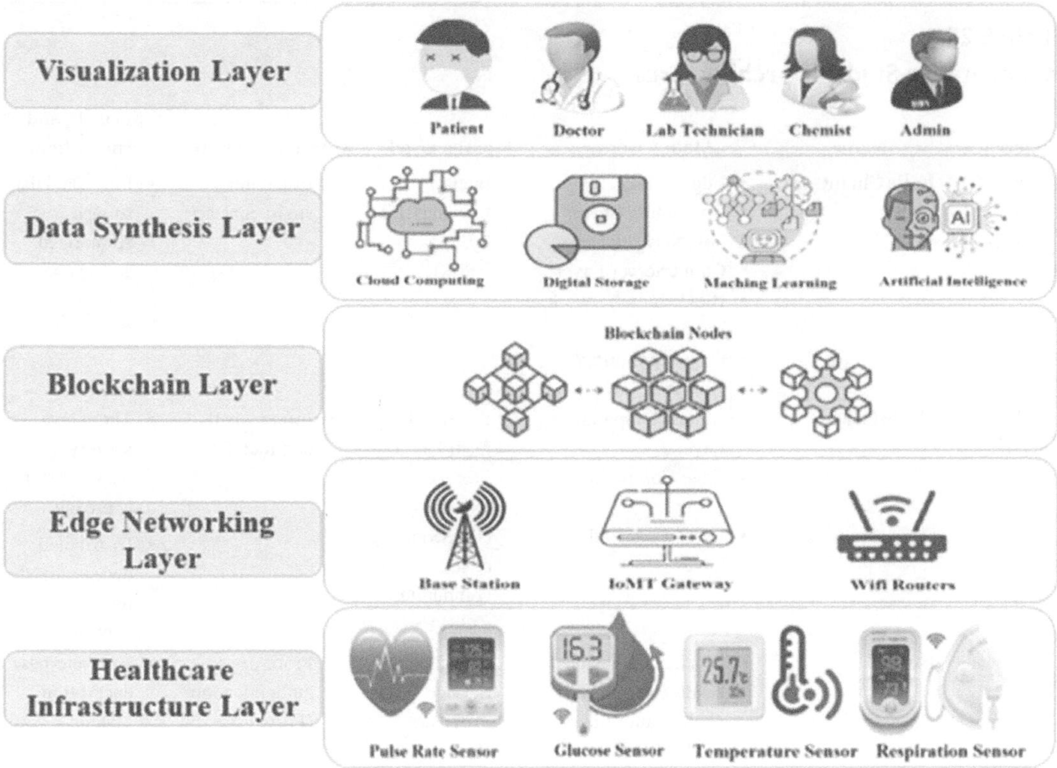

FIGURE 28.1 Generic blockchain architecture for healthcare IoT.

Edge Networking Layer: Focuses on localized data processing and communication to enhance speed and reduce latency.

Blockchain Layer: Utilizes distributed ledger technology to ensure transparency, immutability, and security in healthcare data management.

Data Synthesis Layer: Aggregates healthcare data from diverse sources to provide a comprehensive view for analysis and decision making.

Visualization Layer: Uses include user interfaces, dashboards, and visualization tools to present healthcare data in a meaningful, accessible manner, enhancing interpretability and decision making.

This model serves as a foundation for various other blockchain architectures in healthcare IoT systems, offering a structured and secure framework for decentralized registry systems.

28.3.2 Comparison of Architectures

The evaluation of blockchain architectures for healthcare IoT systems involves a thorough comparison of various criteria, including data storage mechanisms (Shaikh et al. 2021), consensus mechanisms (Auhl et al. 2022), and security level (Hireche et al. 2022). The storage mechanism, which can range from conventional databases to distributed storage solutions or hybrid combinations, is crucial. Consensus mechanisms, such as proof of work or proof of stake, ensure agreement on the distributed ledger's state. Security level integrate encryption methodologies, cryptographic algorithms, and safeguard data integrity, confidentiality, and access control. Table 28.1 presents a comparative analysis of the different architectures presented in the literature, taking into consideration several of the criteria described above.

TABLE 28.1
Comparative Study of Architectures

Reference	Name	Main Layers	Storage Used	Consensus Used	Security and Encryption
Dammak et al. (2022)	**LoRaChainCare**	• Edge/Sensor Computing Layer • Fog/Network Layer • Cloud/Server Layer • Blockchain Network Layer • Data Monitoring/ Analytics Layer	• Inter-Planetary File System (IPFS)	• Ethereum smart contract	• Encrypted data storage • Authorized access control
Egala et al. (2022)	**CoviBlock**	• Cloud Computing Layer • Edge Computing Layer • IoMT User End Layer	• Decentralised Distributed Storage System (DDSS) using edge computing and IPFS	• Infura smart contract	• Three-layer identity management approach • Distributed decentralized resource access control
Badri et al. (2023)	**BIoMT**	• Blockchain-Enabled Healthcare Architecture Layer • Medical Services Layer	• Cloud Service Providers (CSP)	• Proof of authentication (PoAh)	• Attribute-based encryption system
Karmakar et al. (2023)	**ChainSure**	• Application Layer • Blockchain Layer • Data Layer • Communication Layer	• Blockchain-Based Storage for electronic health records (EHRs)	• Ethereum smart contract	• Encryption layer based on tamper-proof encryption
Uppal et al. (2023)	**HealthDote**	• Data Layer • Network Layer • Consensus Layer • Incentive Layer • Application Layer	• Interplanetary File System (IPFS)	• Proof of stake (PoS)	• Encrypted storage before being hashed and stored in the ledger • Blocking unauthorized access
Ashraf et al. (2022)	**FIDChain**	• Healthcare Data Perception Layer • Edge-Based • Blockchain Layer • Network Layer • Cloud-Based Blockchain Layer • Application Layer • Business Layer	• Blockchain distributed ledger technology for storing	• Round robin	• The edge server encrypts collected data and generates corresponding signatures • Smart contracts control data privacy and integrity

(Continued)

TABLE 28.1 ***(Continued)***
Comparative Study of Architectures

Reference	Name	Main Layers	Storage Used	Consensus Used	Security and Encryption
Bathalapalli et al. (2022)	**PUFchain 2.0**	• Physical Layer • Communication Layer • Blockchain Layer • Puf-Edge Cloud Layer • Application Layer	• Blockchain distributed ledger technology	• Proof of PUF-enabled authentication	• Two-level authentication mechanism, combining MAC address and PUF key verification
Barka et al. (2021)	**STHM**	• User Layer • SDN Layer • Security Layer • Blockchain Layer	• Local databases connected to the software defined networking (SDN) and blockchain	• Proof of work (PoW)	• Authentication and Authorization modules • Trust module • Trust database • Access control decision module
Mayer et al. (2021)	**FogChain**	• IoHT Layer • Fog Layer • Blockchain Peers Layer • Smart Contracts Layer	• Decentralized blockchain storage with the option of off-chain information storage using IPFS	• Hyperledger fabric smart contract	• Use of cryptography within blockchain to ensure the security and privacy

28.4 DISCUSSION

The study assesses blockchain-integrated architectures for healthcare IoT systems, focusing on data storage, confidentiality, and blockchain goals. It explores diverse approaches, each tackling distinct challenges and proposing innovative solutions. Regarding data storage and confidentiality design, most healthcare IoT architectures favor centralized cloud storage, potentially compromising blockchain's integrity and security. However, a minority opt for distributed storage, which lacks robust confidentiality safeguards. The evaluation of current structures reveals a lack of universally viable solutions due to trade-offs between centralized and distributed storage. Centralized cloud storage compromises decentralization principles, while distributed storage lacks confidentiality safeguards. The analysis suggests blockchain's potential to revolutionize cloud computing in healthcare IoT by addressing issues like data security and interoperability. However, realizing this potential depends on addressing architecture shortcomings and balancing decentralized principles with robust data protection measures and system efficiency. The study emphasizes tailoring blockchain architecture to individual healthcare IoT applications, advocating for a flexible and adaptive approach. Balancing factors like data integrity, transparency, and security is crucial, considering economic and energy aspects. In conclusion, the comprehensive analysis highlights challenges, trade-offs, and the need for a nuanced approach. Achieving a universally viable solution necessitates reconciling conflicting principles, tailoring architectures, and harnessing blockchain's transformative potential.

28.5 PROPOSED ARCHITECTURE

Figure 28.2 illustrates a proposed architecture for integrating blockchain technology into healthcare IoT. This architecture is composed of eight layers, namely:

Healthcare perception layer: This layer collects data from healthcare IoT devices, such as medical sensors and wearable devices.

Edge-based blockchain layer: This includes the IoT gateways responsible for detecting multiple attacks. In this layer, blockchain operations are performed locally, close to the data source, to reduce latency and improve efficiency.

Fog/Network layer: Plays a crucial role in managing connectivity between healthcare IoT devices and the various layers of the blockchain. It is considered the connectivity layer that manages routing.

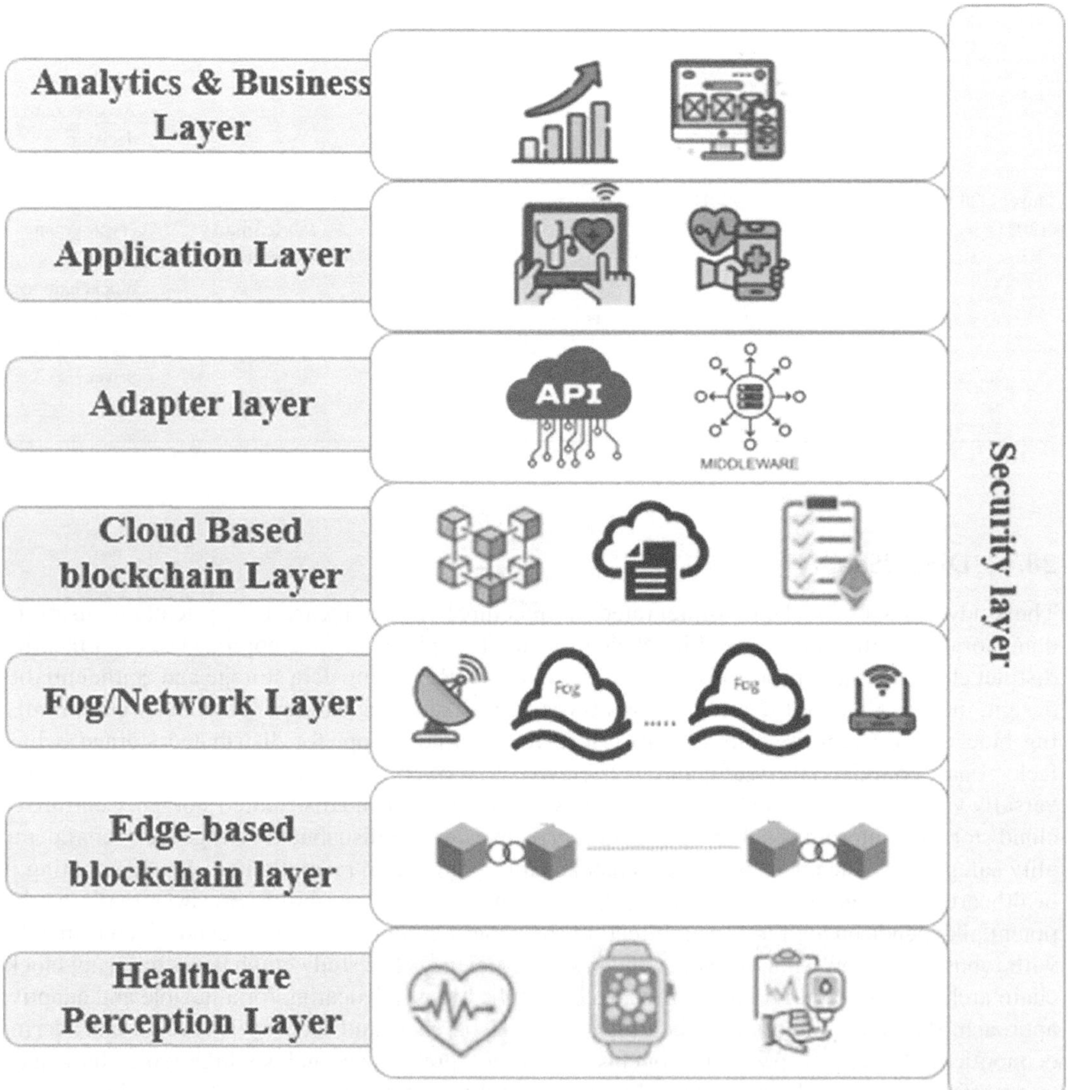

FIGURE 28.2 Proposed architecture for integrating blockchain technology into healthcare IoT systems.

Cloud-based blockchain layer: Offers decentralized data storage and processing on a cloud-based blockchain. This potentially enables greater scalability and accessibility from different locations.

Adapter layer: This layer is crucial for interoperability between different layers. Middleware and micro-services facilitate communication between heterogeneous systems, ensuring smooth integration of data across layers.

Application layer: Presents the user-oriented component that transforms the potential of healthcare IoT data stored on the blockchain into actionable information, providing a robust, user-friendly interface for efficient data management and visualization.

Analytics and business layer: Responsible for analyzing healthcare data, generating actionable insights, and creating business models based on these analyses.

Security layer: This is the layer transverse to all other previous layers.

28.6 CONCLUSION

The study of blockchain-based healthcare IoT architectures, including **LoRaChainCare, CoviBlock, BIoMT, ChainSure, HealthDote, FIDChain, PUFchain 2.0, STHM,** and **FogChain,** highlights the importance of data storage, security, and confidentiality. These architectures have both similarities and differences in addressing challenges posed by blockchain in healthcare IoT systems. The paper concludes that designing secure, and scalable solutions is crucial for implementing blockchain in healthcare IoT systems. The need for storage capabilities and fast computing power is also essential for digital signature verification efficient transaction management. The paper also highlights the complexities of integrating blockchain into healthcare IoT and the need for context-specific design choices. The survey calls for a comprehensive approach considering data storage, security, confidentiality, and scalability when implementing blockchain in healthcare IoT.

REFERENCES

Abdelmaboud A, Ahmed AIA, Abaker M et al (2022) Blockchain for IoT applications: Taxonomy, platforms, recent advances, challenges and future research directions. Electronics 11:630. https://doi.org/10.3390/electronics11040630

Alam S, Bhatia S, Shuaib M et al (2023) An overview of blockchain and IoT integration for secure and reliable health records monitoring. Sustainability 15:5660. https://doi.org/10.3390/su15075660

Alam S, Shuaib M, Ahmad S et al (2022) Blockchain-based solutions supporting reliable healthcare for fog computing and internet of medical things (IoMT) integration. Sustainability 14:15312. https://doi.org/10.3390/su142215312

Alkatheiri MS, Alghamdi AS (2023) Blockchain-assisted cybersecurity for the internet of medical things in the healthcare industry. Electronics 12:1801. https://doi.org/10.3390/electronics12081801

Ashraf E, Areed NFF, Salem H et al (2022) FIDChain: Federated intrusion detection system for blockchain-enabled IoT healthcare applications. Healthcare 10:1110. https://doi.org/10.3390/healthcare10061110

Auhl Z, Chilamkurti N, Alhadad R, Heyne W (2022) A comparative study of consensus mechanisms in blockchain for IoT networks. Electronics 11:2694. https://doi.org/10.3390/electronics11172694

Azbeg K, Ouchetto O, Jai Andaloussi S (2022) BlockMedCare: A healthcare system based on IoT, blockchain and IPFS for data management security. Egyptian Informatics Journal 23:329–343. https://doi.org/10.1016/j.eij.2022.02.004

Badri S, Ullah Jan S, Alghazzawi D et al (2023) BIoMT: A blockchain-enabled healthcare architecture for information security in the internet of medical things. Computer Systems Science and Engineering 46:3667–3684. https://doi.org/10.32604/csse.2023.037531

Barka E, Dahmane S, Kerrache CA et al (2021) STHM: A secured and trusted healthcare monitoring architecture using SDN and blockchain. Electronics 10:1787. https://doi.org/10.3390/electronics10151787

Bataineh MR, Mardini W, Khamayseh YM, Yassein MMB (2022) Novel and secure blockchain framework for health applications in IoT. IEEE Access 10:14914–14926. https://doi.org/10.1109/ACCESS.2022.3147795

Bathalapalli VKVV, Mohanty SP, Kougianos E et al (2022) PUFchain 2.0: Hardware-assisted robust blockchain for sustainable simultaneous device and data security in smart healthcare. SN Computer Science 3:344. https://doi.org/10.1007/s42979-022-01238-2

Dammak B, Turki M, Cheikhrouhou S et al (2022) LoRaChainCare: An IoT architecture integrating blockchain and LoRa network for personal health care data monitoring. Sensors 22:1497. https://doi.org/10.3390/s22041497

Egala BS, Pradhan AK, Gupta S, Sahoo KS, Bilal M, Kwak K-S (2022) CoviBlock: A secure blockchain-based smart healthcare assisting system. Sustainability 14:16844. https://doi.org/10.3390/su142416844

Hireche R, Mansouri H, Pathan A-SK (2022) Security and privacy management in internet of medical things (IoMT): A synthesis. JCP 2:640–661. https://doi.org/10.3390/jcp2030033

Islam MS, Ameedeen MAB, Rahman MA et al (2023) Healthcare-chain: Blockchain-enabled decentralized trustworthy system in healthcare management industry 4.0 with cyber safeguard. Computers 12:46. https://doi.org/10.3390/computers12020046

J A, Isravel DP, Sagayam KM et al (2023) Blockchain for healthcare systems: Architecture, security challenges, trends and future directions. Journal of Network and Computer Applications 215:103633. https://doi.org/10.1016/j.jnca.2023.103633

Karmakar A, Ghosh P, Banerjee PS, De D (2023) ChainSure: Agent free insurance system using blockchain for healthcare 4.0. Intelligent Systems with Applications 17:200177. https://doi.org/10.1016/j.iswa.2023.200177

Mayer AH, Rodrigues VF, Costa CAD et al (2021) FogChain: A fog computing architecture integrating blockchain and internet of things for personal health records. IEEE Access 9:122723–122737. https://doi.org/10.1109/ACCESS.2021.3109822

Shaikh M, Shibu C, Angeles E, Pavithran D (2021) Data Storage in Blockchain Based Architectures for Internet of Things (IoT). In: 2021 IEEE International IOT, Electronics and Mechatronics Conference (IEMTRONICS). IEEE, Toronto, ON, Canada, pp 1–5

Thantharate P, Thantharate A (2023) ZeroTrustBlock: Enhancing security, privacy, and interoperability of sensitive data through zero-trust permissioned blockchain. BDCC 7:165. https://doi.org/10.3390/bdcc7040165

Uppal S, Kansekar B, Mini S, Tosh D (2023) HealthDote: A blockchain-based model for continuous health monitoring using interplanetary file system. Healthcare Analytics 3:100175. https://doi.org/10.1016/j.health.2023.100175

29 Mobile Assistive Technology (MAT) Accessible for Persons with Visual Impairments

M. Aharchi and M. Ait Kbir

29.1 INTRODUCTION

Visual impairment (VI) refers to a significant decline in vision that cannot be fully corrected with glasses or medical treatments, ranging from partial sight to complete blindness, caused by various factors like eye diseases or injuries. It impacts daily tasks like navigation and object recognition, affecting approximately 2.2 billion people globally with expectations of further increase [1, 2].

Individuals with visual impairments (PVI) rely heavily on memory to navigate daily life, facing challenges in unfamiliar environments and object identification [3]. Emerging technologies like smartphones and smart glasses offer promising solutions through advanced software applications leveraging features like cameras and augmented reality (AR) [4]. These portable devices, connected to cloud services, facilitate access to information anytime, anywhere, enhancing daily engagements efficiently [5].

The present research focuses on exploring the existing options aimed at assisting PVIs in recognizing objects and their surroundings.

The primary objective of this study is to address the following research inquiries:

a. What are the various types of mobile assistive technology (MAT) accessible for PVIs?
b. What functionalities and capabilities are provided by the most advanced and effective solutions tailored for PVIs?
c. What are the advantages and disadvantages associated with the latest MAT solutions specifically designed for PVIs?

The second objective of this research is to explore and examine current solutions available in the field of MAT that specifically address the needs of individuals with visual impairments in terms of object detection and navigation in unfamiliar environments. Additionally, it explores the possible improvements these solutions can bring to PVIs and provides an overview of the difficulties and limitations associated with the proposed approaches.

The paper is organized into the following sections:

- Section 29.2: Research methodology outlines the study's approach, data collection methods, and procedures for investigating solutions for assisting PVIs.
- Section 29.3: Evaluation of solutions for assisting PVIs assesses various solutions, their features, and effectiveness in addressing the specific needs and challenges faced by PVIs.
- Section 29.4: Solutions advantages and research challenges discusses the benefits and limitations of identified solutions for PVIs.
- Section 29.5: Conclusion summarizes the study's findings, highlights key insights, implications, and recommendations, wrapping up the main arguments and contributions of the study.

DOI: 10.1201/9781003647263-29

29.2 RESEARCH METHODOLOGY

The paper follows the Kitchenham methodology, a systematic approach for conducting literature reviews in software engineering research. This method ensures reproducibility by employing consistent steps and keywords.

Using databases like IEEE Xplore and Science Direct, the study searched for MAT solutions for PVI, focusing on object recognition and navigation. The search spanned from 2015 to 2023 to capture recent findings.

From a pool of 244,307 papers on visual impairment, 231 were extracted based on specific keywords. These articles were further examined by reading their abstracts and applying the exclusion criteria specified below:

- Non-English language articles.
- Articles that do not directly talk about methods and techniques for object recognition.
- Articles that were only concerned on restricted object recognition.
- Duplicate articles.
- Articles lacking verifiable evidence.
- The full article text was not available
- Reviews and surveys.

After applying these exclusion criteria, a total of 191 articles remained. From this set, 21 articles were chosen for further analysis based on a detailed examination of their abstracts (Table 29.1).

29.3 MAT OBJECT DETECTION SOLUTIONS ACCESSIBLE FOR PVI

Visually impaired individuals encounter numerous challenges in object detection, stemming from their limited or complete absence of vision.

To address the challenge of object detection, we conducted a thorough search for existing solutions. Our aim was to identify technologies that assist in the easy detection of objects in the surrounding environment. As shown in Figure 29.1, we classified MATs for object detection into three categories: tag-based systems, hybrid systems, and systems based on computer vision (CV). By this classification, researchers and practitioners can gain a comprehensive understanding of the various technologies and methodologies utilized in MATs.

The next section gives an overview of recent technological advancements. We delve into the details of each technology, examining their features, functionalities and practical applications. Furthermore, we include research examples that showcase the effectiveness and potential of these solutions. By presenting this comprehensive examination, we aim to provide valuable insights into

TABLE 29.1
Keyword Database Searching Summary

SearchCriteria			Keywords			
Database	Field	Type	Object detection	Visual impairment	Navigation solution	OD+ VI+NS
IEE Xplore	All	All	35,629	1105	6559	10
Springer	All	All	263,864	91,977	80,703	204
Web of Science	All	All	19	68	23	13
Science Direct	All	All	120,645	151,160	48,208	4
Total			420,157	244,310	135,493	231

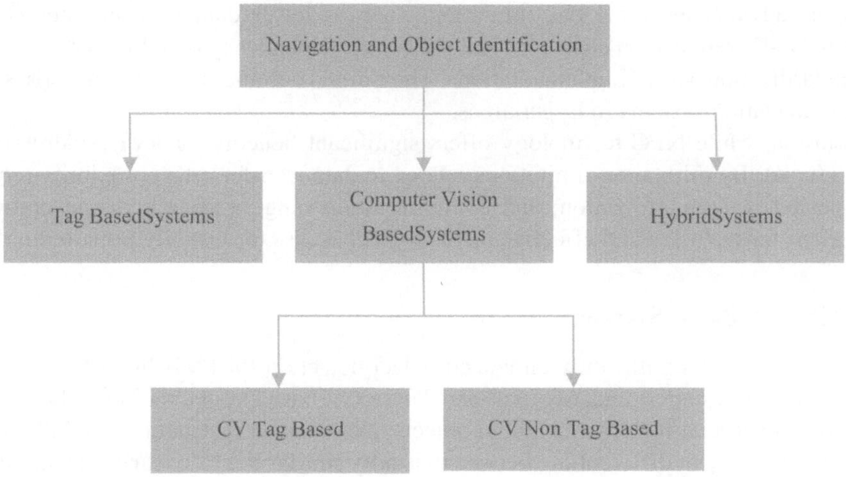

FIGURE 29.1 MAT solutions for the navigation and objects identification for PVIs.

the advancements in MATs and their role in facilitating object detection for visually impaired individuals.

29.3.1 Navigation and Object Identification

PVIs face not only the challenge of reaching unfamiliar places but also the complexity of navigating within these locations [6]. The task of navigating in new environments proves difficult for PVIs, and even more so is the identification of various objects present in these unfamiliar surroundings. As a result, PVIs often rely on the assistance of others, although this solution is not always readily available or easily accessible. Public institutions, for instance, are often ill-equipped to provide the necessary support [7]. They frequently lack appropriate signage and accessible information, which poses obstacles in terms of locating entrances, exits, specific areas, or facilities within the premises.

To address these challenges, researchers have developed numerous solutions aimed at aiding PVIs in navigation and object identification. These solutions are divided into three different categories: Tag-based systems, computer vision-based systems, Hybrid systems.

29.4 DISCUSSION

29.4.1 Tag-Based Systems

In this study, various strategies, methods, and tools have been discussed to aid a PVI in object recognition and navigation. Researchers have developed a wide range of solutions using tagging methods such as NFC and RFID.

RFID tags offer advantages such as vast signal coverage, ability to penetrate obstacles, and reuse of existing infrastructure, minimizing costs. They eliminate the need for precise pointing towards objects and enable embedding within objects for protection. However, drawbacks include the expense of installing numerous RFID tags, managing abundant information, limitations to labeled environments, and interference with metallic objects [8].

NFC technology has emerged as a valuable tool for enhancing PVI accessibility and independence. Through touch-based interaction with NFC-enabled devices, such as smartphones, PVIs can access object details, instructions, or navigation assistance effortlessly. NFC tags placed strategically indoors provide location-specific information and directions. Integration of NFC into assistive applications expands benefits, enabling object identification and audio descriptions.

Despite its advantages, NFC has limitations such as lower data transfer rates compared to Bluetooth or Wi-Fi, reliance on close physical proximity, dependency on NFC-enabled devices, and lack of standardization across implementations. The limited communication range poses challenges for outdoor navigation systems in larger areas.

In conclusion, while NFC technology offers significant benefits for PVIs, addressing its constraints is crucial. By mitigating limitations related to data transfer rates, proximity requirements, device dependency, standardization, and communication range, researchers can optimize NFC-based solutions for inclusive and effective technological access for visually impaired individuals.

29.4.2 CV Tag-Based Systems

CV tag technology has significantly enhanced object detection for PVIs by utilizing visual markers like barcodes, QR codes, and AR markers. These CV-based systems efficiently retrieve object details, providing crucial information about objects, locations, or instructions to PVIs. Particularly valuable is their ability to differentiate between visually similar items, ensuring accurate identification, which is essential in scenarios where item content may vary.

One key advantage of CV tag technology is its ability to provide object information with minimal computational power and storage requirements, leveraging existing visual tags like barcodes and QR codes. This cost-effective approach facilitates widespread adoption without significant financial barriers. However, challenges such as the need for precise tag placement and the potential for information overload when multiple tags are nearby must be considered. Additionally, the reliance on an unobstructed line of sight for tag detection and susceptibility to damage during transportation or adverse weather conditions are important factors to address in implementing CV tag technology effectively.

29.4.3 CV Nontag-Based Systems

Nontag-based computer vision (CV) systems significantly improve object detection for individuals with visual impairments, offering cost-effectiveness and easy integration with smartphones. These systems use advanced image processing techniques to help PVIs identify objects in their surroundings.

A major advantage of nontag-based CV systems is their ability to operate without specialized infrastructure or physical tags, making them adaptable in various environments. PVIs can simply use a smartphone to capture images and utilize the CV system to recognize objects.

However, nontag-based CV systems face challenges in real-world environments, including motion blur, varying lighting conditions, and changes in object orientation and scale, which affect detection accuracy and reliability. Overcoming these challenges requires substantial computational power for image processing and the capture of multiple high-quality photos.

Nevertheless, nontag-based CV systems represent a significant advancement in enhancing object detection for individuals with visual impairments. By leveraging computer vision, these systems empower PVIs to navigate independently and interact with objects confidently. Further advancements in algorithms and device capabilities will continue to improve their effectiveness.

29.4.4 Hybrid Systems

In recent years, researchers have developed hybrid systems to improve object detection by combining multiple technologies. These systems aim to enhance accuracy and usability by leveraging the strengths of different technologies while addressing their limitations. However, integrating these technologies requires significant infrastructure and expertise, posing challenges in coordination and maintenance. Despite these hurdles, hybrid systems offer promising advancements in accuracy and usability. Tables 29.2 and 29.3 provide an overview of the strengths, weaknesses, and evaluation criteria for these systems.

TABLE 29.2
Strengths, Challenges, and Drawbacks of Each Category

Type	Technology	Paper	Strengths	Challenges and Drawbacks
Tag-based systems	NFC	[9–12]	• **Simple Touch Activation**: PVIs can easily initiate services by touching NFC tags. • **Low-Cost Indoor Navigation**: NFC enables affordable indoor navigation systems for PVIs. • **Quick and Accurate**: NFC offers rapid response times and precise positioning. • **Robust in Unclean Environments**: NFC functions reliably even in dirty conditions. • **No Line of Sight Required**: NFC operates without a direct line of sight, enhancing accessibility for PVIs.	• **Proximity and Simplicity:** NFC is efficient for close-range interactions, making it simple to use. However, PVIs must be within the reading area and have an NFC-equipped mobile device. • **Specific Use Cases:** While not as versatile as technologies like Bluetooth or Wi-Fi Direct, NFC excels in specific applications where close proximity and ease of use are critical.
	RFID	[13, 14]	• **Non-Line-of-Sight Communication**: RFID can read tags even when they are not directly visible to the reader. • **High-Speed Data Capture**: RFID systems quickly capture data from tags. • **Multiple Tag Reading**: RFID readers can simultaneously process multiple tags within their range.	• **Cost**: RFID systems can be expensive to implement, particularly when considering the cost of RFID tags and readers. • **Interference**: RFID signals can be affected by interference from other electronic devices, metal, or liquids. • **Data Collision**: In scenarios with multiple RFID tags in close proximity, data collision can occur, leading to missed or erroneous readings. • **Environmental Factors**: RFID tags may not perform well in extreme environmental conditions, such as extreme temperatures, humidity, or exposure to chemicals. • **Tag Readability**: Some materials and items can interfere with RFID tag readability, such as liquids, metals, and densely packed objects. • **Maintenance**: RFID tags may require periodic maintenance, battery replacement (in active tags), or updates. • **Complex Integration**: Integrating RFID systems with existing infrastructure and software can be complex and may require custom solutions. • **Limited Data Capacity**: RFID tags have limited data storage capacity. • **Environmental Impact**: The disposal of RFID tags, especially if they contain batteries or hazardous materials, can pose environmental concerns.

(Continued)

TABLE 29.2 *(Continued)*
Strengths, Challenges, and Drawbacks of Each Category

Type	Technology	Paper	Strengths	Challenges and Drawbacks
Computer vision–based systems	CV	[15–21]	• **Cost-Effective**: CV methods are cost-effective as they often require little or no additional infrastructure to operate • **Ease of Implementation**: Many CV methods can be easily installed and utilized on smartphones and different platforms.	• **Inconsistent Real-World Performance**: CV methods can exhibit inconsistent performance in real-world environments due to factors like motion blur. • **High Computational Demands**: Many CV methods require extensive computational power, which can be a limitation for resource-constrained devices or applications. • **Photography Requirements**: PVIs may face challenges in capturing many good quality photos, which are often necessary for effective CV systems, potentially limiting their usability. • **Feedback Latency**: Reducing feedback latency is crucial to make CV systems more effective, as delays can hinder real-time applications and user interactions.
	QRcode, Barcode, Markers	[22, 23]	• **Efficiency and Resource-Friendly**: They are is resource-efficient, requiring low computational power and storage space, making it accessible for various devices. • **Utilizes Existing Visual Tags**: They complement existing visual tags like barcodes and QR codes, allowing PVIs to access information seamlessly without additional physical tags. • **Cost-Effective**: These tags can be generated and printed at a very low cost, making it an economical choice for augmenting accessibility and information retrieval.	• **Information Overload**: In areas with many tagged items, they can overwhelm PVI by providing information about all items at once. • **Line-of-Sight Requirement**: They must be in the direct line of sight of the camera, which can be challenging in some situations. • **Inaccessible Placement**: They cannot be placed inside items, limiting their use for internal item identification. • **Vulnerability to Damage**: They are susceptible to damage during movement across the supply chain or exposure to adverse weather conditions. • **Decreased Recognition with Distance**: There recognition may become challenging if PVI is moving rapidly, and the recognition rate decreases as the distance between the camera and the tag increases.

(Continued)

TABLE 29.2 *(Continued)*
Strengths, Challenges, and Drawbacks of Each Category

Type	Technology	Paper	Strengths	Challenges and Drawbacks
Hybrid systems		[8, 24, 25]	• **Balanced Trade-off:** Integrating multiple technologies allows for a balanced trade-off, leveraging the strengths of each to mitigate their individual limitations. • **Improved Accuracy**: Combining technologies can enhance accuracy by cross-verifying data or compensating for weaknesses in individual systems. • **Enhanced Robustness**: A combination of technologies can improve system robustness, making it more reliable in various conditions and scenarios. • **Usability**: By synergizing different technologies, usability can be improved, making systems more accessible and user-friendly. • **Enhanced Performance**: Combined technologies often lead to better overall system performance, addressing specific challenges and delivering more effective results.	• **Increased Infrastructure Usage**: Integrating multiple technologies can lead to higher infrastructure requirements, potentially causing congestion or resource allocation challenges. • **Increased Time Usage**: Managing multiple technologies may require additional time for setup, maintenance, and coordination, potentially impacting efficiency. • **Increased Complexity**: Combining technologies often results in increased system complexity, which can make troubleshooting and maintenance more challenging. • **Increased Cost**: Integrating multiple technologies may come with additional expenses, including equipment, software, and maintenance costs.

In Table 29.3, we have analyzed various criteria to compare the solutions discussed for assisting visually impaired individuals in object detection and identification. These criteria offer insights into the strengths and limitations of each solution.

a. **Cost**: Among the solutions, CV tag techniques stand out as the most cost-effective since they involve minimal expenses, mainly related to printing QR codes or AR markers. Barcodes are also budget-friendly as they are already on products. Tag-based techniques (RFID and NFC) are moderately priced due to the installation of RFID or NFC tags. In contrast, CV techniques can be costly, requiring high-quality equipment like cameras and powerful processors for optimal performance.
b. **Equipment Required**: In CV tag-based solutions, PVIs only need smartphones equipped with cameras for object detection. In CV techniques, some solutions are compatible with smartphones, while others demand high-resolution cameras and robust processors. For tag-based solutions, PVIs require RFID readers or smartphones supporting NFC technology.
c. **Scanning Multiple Items**: RFID readers, AR markers, and CV techniques have the capability to scan multiple items simultaneously, offering practicality, especially in scenarios such as shopping, where PVIs may want to identify and count items in their cart.

TABLE 29.3
Comparison of Objects Identification Technologies for PVIs

Technology	Number of Scanned Items	Range	RequiresLine of Sight	Equipment	Capacity	Cost
RFID	Multiple	3 cm	No	RFID Reader	8000 bytes	Low
NFC	1	10 cm	No	NFC Reader	1.6 MB	Low
QR code	1	Depends on the marker size	Yes	Camera	2953 bytes	Free
Markers	Multiple	Depends on the marker size	Yes	Camera	NA	Free
Barcode	1	Depends on the marker size	Yes	Camera	NA	Free
CV techniques	Multiple	Depends on the camera	Yes	Camera	–	High

d. **Line of Sight**: Tag-based solutions (RFID and NFC) offer flexibility as they do not necessitate PVIs to have a direct line of sight with objects, enabling object identification from various angles within their specified range (3 meters for RFID and 10 centimeters for NFC). Conversely, CV solutions rely on factors like tag size (for QR codes and barcodes) and camera parameters (for CV techniques).
e. **Storage Capacity**: Certain tags, such as RFID, NFC, and QR codes, possess storage capacity, allowing them to hold additional information. In contrast, AR markers and barcodes lack this capability.

These criteria provide valuable insights into the trade-offs and considerations when choosing a technology solution for assisting PVIs. Researchers and engineers should carefully assess these factors and the specific requirements of each application to design effective and accessible systems for PVIs. It is essential to consider accessibility guidelines and standards during the development process. Further research is warranted to continue enhancing technology solutions for the visually impaired.

29.5 CONCLUSION

PVIs face many problems during their navigation, needing efficient solutions for increased autonomy. The current approaches of efficiently assisting PVIs have been thoroughly examined in this study. It is obvious that each of these options has a certain set of benefits and cons. In order to find more complete and effective solutions, researchers have started designing and analyzing hybrid systems that cleverly combine the benefits of many technologies while minimizing the drawbacks of each. But it is important to note that these hybrid approaches could also need more effort, infrastructure, and complexity. The introduction provided by this study is crucial, providing guidance and motivation for future research projects. It emphasizes the significance of ongoing studies that could result in advanced and effective solutions, ultimately improving the navigating experience for people who are visually impaired.

REFERENCES

1. What is Visual Impairment? Available online: https://www.news-medical.net/health/What-is-visualimpairment.aspx (accessed on 23 December 2018).
2. Blindness and Vision Impairment. Available online: https://www.who.int/news-room/fact-sheets/detail/blindness-and-visual-impairment (accessed on 18 May 2023).
3. Capella-Mcdonnall, M. The need for health promotion for adults who are visually impaired. J. Vis. Impair. Blind. 2007, 2002, 133–146.
4. Kollmuss, A.; Agyeman, J. Mind the gap: Why do people act environmentally and what are the barriers to pro-environmental behavior. Environ. Educ. Res. 2002, 8, 239–260.

5. Giudice, N.A. Navigating without vision: principles of blind spatial cognition. In Handbook of Behavioral and Cognitive Geography; Montello, D.R., Ed.; Edward Elgar, UK, 2018; pp. 1–32.
6. Domingo, M.C. An overview of the internet of things for people with disabilities. J. Netw. Comput. Appl. 2012, 35, 584–596.
7. Kostyra, E.; Zakowska-Biemans, S.; Sniegocka, K.; Piotrowska, A. Food shopping, sensory determinants of ′ food choice and meal preparation by visually impaired people. Obstacles and expectations in daily food experiences. Appetite. 2017, 113, 14–22.
8. McDaniel, T.L.; Kahol, K.; Villanueva, D.; Panchanathan, S. Integration of RFID and computer vision for remote object perception for individuals who are blind. In Proceedings of the Ambi-Sys Workshop on Haptic User Interfaces in Ambient Media Systems, Quebec City, QC, Canada, 1–14 February 2008; p. 7.
9. Sakpere, W.E.; Mlitwa, N.B.W.; Oshin, M.A. Towards an efficient indoor navigation system: A near field communication approach. J. Eng. Des. Technol. 2017, 15, 505–527.
10. Virtualeyez, A.M. Developing NFC technology to enable the visually impaired to shop independently. Master Thesis, Dalhousie University, Halifax, Nova Scotia, July 2014.
11. Ozdenizci, B.; Coskun, V.; Ok, K. NFC internal: An indoor navigation system. Sensors 2015, 15, 7571–7595.
12. Ozdenizci, B.; Ok, K.; Coskun, V.; Aydin, M.N. Development of an indoor navigation system using NFC technology. In Proceedings of the Information and Computing (ICIC) International Conference, Phuket Island, Thailand, 25–27 April 2011; pp. 11–14.
13. Andò, B.; Baglio, S.; Marletta, V.; Crispino, R.; Pistorio, A. A measurement strategy to assess the optimal design of an RFID-based navigation aid. IEEE Trans. Instrum. Meas. 2018, 1–7. Available online: https://ieeexplore.ieee.org/abstract/document/8540063 (accessed on 21 November 2018).
14. Kornsingha, T.; Punyathep, P. A voice system, reading medicament label for visually impaired people. In Proceedings of the RFID SysTech 7th European Workshop, Smart Objects: Systems, Technologies and Applications, Dresden, Germany, 17–18 May 2011; pp. 1–6.
15. Zientara, P.A.; Lee, S.; Smith, G.H.; Brenner, R.; Itti, L.; Rosson, M.B.; Carroll, J.M.; Irick, K.M.; Narayanan, V. Third eye: A shopping assistant for the visually impaired. Computer (Long. Beach. Calif). 2017, 50, 16–24.
16. Jafri, R.; Ali, S.A. A multimodal tablet–based application for the visually impaired for detecting and recognizing objects in a home environment. In Proceedings of the International Conference on Computers for Handicapped Persons (ICCHP) Conference, Paris, France, 9–11 July 2014; pp. 356–359.
17. Tai, K.C.; Chen, C.H. Symbol detection in low-resolution images using a novel method. Int. J. Control Autom. 2014, 7, 143–154.
18. Li, Y. An object recognition method based on the improved convolutional neural network. J. Comput. Theor. Nanosci. 2016, 13, 870–877.
19. Kang, M.C.; Chae, S.H.; Sun, J.Y.; Lee, S.H.; Ko, S.J. An enhanced obstacle avoidance method for the visually impaired using deformable grid. IEEE Trans. Consum. Electron. 2017, 63, 169–177.
20. Kang, M.C.; Chae, S.H.; Sun, J.Y.; Yoo, J.W.; Ko, S.J. A novel obstacle detection method based on deformable grid for the visually impaired. IEEE Trans. Consum. Electron. 2015, 61, 376–383.
21. Yi, C.; Tian, Y.; Arditi, A. Portable camera-based assistive text and product label reading from hand-held objects for blind persons. IEEE/ASME Trans. Mechatron. 2014, 19, 808–817.
22. Azenkot, S.; Zhao, Y. Designing smartglasses applications for people with low vision. ACM SIGACCESS Access Comput. 2017, 119, 19–24.
23. Zhang, H.; Zhang, C.; Yang, W.; Chen, C.Y. Localization and navigation using QR code for mobile robot in indoor environment. In Proceedings of the IEEE International Conference on Robotics and Biomimetics (ROBIO), Zhuhai, China, 6–9 December 2015; pp. 2501–2506.
24. Fernandes, H.; Costa, P.; Paredes, H.; Filipe, V.; Barroso, J. Integrating computer vision object recognition with location based services for the blind. In Proceedings of the International Conference on Universal Access in Human-Computer Interaction, Crete, Greece, 22–27 June 2014; pp. 493–500.
25. Rashid, Z.; Melià-Seguí, J.; Pous, R.; Peig, E. Using augmented reality and internet of things to improve accessibility of people with motor disabilities in the context of smart cities. Future Gener. Comput. Syst. 2017, 76, 248–261.

30 Internal Audit Activities in the Health Sector in the Age of Artificial Intelligence

A. Hantem, A. Aarab, N. Youssoufi,
K. Outaaza, A. Mellouki, and A. Wahbi

30.1 INTRODUCTION

Improving performance remains a major challenge for healthcare establishments. To address this, public health establishments should adopt logic of innovation and modernization, regarding the conditions of success, effectiveness, and efficiency as paramount. These factors, viewed as two crucial vectors of hospital performance, are closely linked to the adoption of technological and managerial innovations (Hantem et al., 2024). These innovations facilitate the modernization of practices and the management of potential risks. Regulatory requirements, budget constraints, insufficient resources, and the growing demand for quality services pose significant challenges. In this context, internal control systems become strategic tools, providing systematic perspectives for assessing, adjusting, and perfecting hospital practices. They serve as an essential lever for optimizing overall performance. The healthcare sector is both sensitive and complex, where improving service quality and optimizing resource management represent significant challenges and priorities. Therefore, innovative internal audit and control systems play a vital role in enhancing the efficiency of hospital operations.

With the growth of activities and massive data, efficiently fulfilling the objectives of the audit and control mission and guaranteeing reasonable assurance as to the veracity of the financial statements is becoming increasingly challenging (Rodrigues et al., 2023). To cope with repetitive tasks and automate control processes, certain technologies have been developed to provide continuous assurance. The concept of continuous assurance in auditing is defined by as the application of modern information technologies to auditing activities and procedures. Artificial intelligence (AI) tools offer numerous advantages for the audit and control professions (Ait Lemqeddem and Chouay, 2020). Thanks to their algorithms and techniques, AI makes it possible to automate routine tasks and focus on high value-added activities. This automation provides advantages in terms of speed, massive data management, risk identification, fraud detection, and verification of compliance with rules, thereby improving the quality of audits in real time.

This study investigates the perspectives of auditors and internal controllers regarding the utilization of artificial intelligence in healthcare facility audits in Morocco. It examines the ways in which this technology enhances audit quality and increases the effectiveness of controls. Furthermore, our study assesses the perceptions of auditors and internal controllers concerning the integration of AI into their work. The methodological approach adopted is rigorous and objective to address these inquiries.

30.2 OVERVIEW

30.2.1 Artificial Intelligence in Auditing

AI comprises a set of methods and tools crafted to enable computing devices to operate. It incorporates advanced algorithms and techniques for processing data and information, aiming to simulate and enhance cognitive functions (Kok et al., 2009; Wahbi et al., 2023). As defined by McCarthy

DOI: 10.1201/9781003647263-30

(2007), AI is the science and engineering of creating machines capable of performing tasks typically carried out by humans, utilizing intelligence.

In the current era, marked by increasingly complex and repetitive tasks, optimizing the execution of audit and control activities demands the adoption of high-performance control devices and the strategic utilization of AI tools for analyzing vast and valuable data sets. This approach contributes to enhancing the quality of audits and controls. The application of AI tools in accounting and auditing exhibits varying degrees of adoption across sectors, offering a multitude of advantages. It provides a deeper understanding of data, enabling advanced statistical analysis and risk assessment, thereby prioritizing audit tasks. Moreover, AI aids in understanding and extracting textual data, including unstructured information, through natural language processing. This capability assists auditors in identifying potential inconsistencies and anomalies in documentation. AI's predictive analysis capabilities further empower the anticipation of future financial trends. By automating routine and repetitive tasks, AI liberates time, allowing professionals to focus on higher value-added activities (The Institute of Internal Auditors [IIA], 2017). Finally, it expedites the identification and assessment of risks, enhancing the overall efficiency of auditing processes.

The integration of AI into internal control and audit processes automates repetitive and time-consuming tasks (such as data entry and report generation), freeing auditors to focus on high-value activities. AI tools enhance risk identification and assessment, enable real-time data processing and analysis, and facilitate efficient resource allocation (INTOSAI, 2023; Mighiss, 2021). These capabilities collectively contribute to a significant improvement in the quality of the audit process.

While the utilization of AI tools presents numerous advantages and enhances the performance of auditors and controllers, the technology still depends on human judgment for its construction, requiring relevant, accurate, and unbiased data. Consequently, human auditors remain indispensable to the audit process. This underscores the importance of developing the skills of all individuals involved in control and audit processes, ensuring proficiency in technology use and analysis of the resulting data.

30.2.2 Certifying Accounts and Improving Performance

The primary objective of certifying the financial statements of public-sector healthcare establishments is to ensure the reliability and transparency of their accounts (Hantem, 2023). This process involves establishing robust control systems and, most importantly, certifying that the accounts are sincere, reliable, and provide a true and fair view of the financial situation of the establishment in question (Lizot and Laskawiec, 2014). The sincerity and reliability of the accounts, reflecting the patrimonial approach of an accounting system (Hantem, 2019), follow a change management approach. This approach entails analyzing the initial situation, identifying strengths and weaknesses, pinpointing risk areas, and implementing an action and correction plan (Lafon, 2016). Such a systematic approach not only provides assurance but also builds trust among various stakeholders. Another essential objective of certifying accounts is to enhance overall performance. The concept of performance, as defined by Pesqueux (2004), extends beyond the completion of a process or task; it encompasses the achieved results and the success that one can take pride in. According to Saulquin and Schier (2007), the interpretation of performance varies among individuals or groups. From a managerial perspective, it could denote the profitability or competitiveness of the organization. For an employee, it might involve the work climate, and for a customer, performance could be evaluated based on the quality of services rendered (Saulquin and Schier, 2007).

Hospital performance differs from other forms of organizational performance. It is characterized by its complex and specific nature (Etienne Minvielle et al., 2008). Hospital performance is often associated with elements such as financial stability, optimizing bed occupancy, meeting staffing standards, achieving a high percentage of patients satisfied with their stay, and quality assurance of services. According to the WHO (2007), "High hospital performance should be based on professional competence in the application of knowledge, technology, efficiency in the use of resources, patient satisfaction and health outcomes" (Markaoui and Hassine, 2018; WHO, 2007).

An effective internal audit system significantly contributes to enhancing hospital performance through its emphasis on internal control systems. However, the absence of controls can result in a substantial deterioration of a well-established management system, diminishing its effectiveness and consequently impacting the overall performance of the organization (Zarai and Bettebai, 2007; Zarhloul and Mrani Zentar, 2022).

30.3 MATERIAL AND METHOD

After collecting information, we proceeded to design our questionnaire, which we administered to internal auditors and controllers working in public health establishments in Morocco between December 15, 2023, and January 15, 2024. The questionnaire comprised firm, open-ended, and scaled questions. We processed and analyzed the 60 received responses using Excel and SPSS software. For the statistical processing of the data, we employed univariate, bivariate, and multivariate analyses. Additionally, to test our research hypotheses, we referred to the following tests. We would point out that the declare that their Institutional Ethics Committee confirmed that no ethical review was required for this study. Written informed consent for participation was not required because all participants' data was anonymized before the statistical analyses were done.

30.4 RESULTS

30.4.1 Descriptive Analysis

The table provides an overview of the distribution of participants by gender, occupation, and sector of activity. The demographic results reveal a predominance of women, representing 56.67% of participants, compared with men, who account for 43.33%. In addition, most respondents (56.7%) were auditors, showing a predominance of this profession over controllers. In addition, a higher proportion of respondents work in the public sector (56.7%) than in the private sector (43.3%).

30.4.2 Data Analysis

30.4.2.1 Reliability

We tested the reliability of our result by examining the four variables in our study. The table shows the results obtained. Cronbach's alpha is 0.867, while Cronbach's alpha based on standardized items is 0.865. These statistics indicate a good level of internal consistency, demonstrating a high degree of coherence in our measurements.

30.4.2.2 Correlation

Our results reveal significant correlations between the different variables studied. A moderate positive correlation is observed between the level of training on AI and preparation for its introduction ($r = 0.506$, $p < 0.05$). In addition, a strong positive correlation is found between preparation for the introduction of AI and the level of awareness of the implications of AI in the job ($r = 0.736$, $p < 0.01$), as well as with the level of awareness of the changes already induced by AI in the job ($r = 0.728$, $p < 0.01$). A significant but weaker correlation was also found between the level of AI training and the level of awareness of AI implications in the workplace ($r = 0.591$, $p < 0.05$), as well as with the level of awareness of changes already induced by AI in the workplace ($r = 0.418$, $p < 0.05$).

30.4.3 Analysis of Assumptions

30.4.3.1 Hypothesis 1

Our hypothesis, H1, that the perception of the use of AI in audit and control activities in the private sector is different from that in the public sector is confirmed. The results of students' t-tests,

TABLE 30.1
Results of Students' t-tests for Hypothesis 1

	T-Test		
Variable	**Private**	**Public**	**P Value**
Learn about AI	2,923	2,294	0.0234
Change my practices to better prepare for the introduction of AI.	2,846	2,412	0.0478
Be aware of what AI could change in my work.	3,154	2,647	0.0132
Be aware of what AI has already changed in my work.	2,923	2,471	0.0945

based on representative samples from the private and public sectors, show notable divergences in AI-related perceptions and actions.

In the private sector, a greater willingness to train in AI is observed compared to the public sector, with a significant difference ($p = 0.0234$). Similarly, a greater willingness to adapt their practices for the introduction of AI is found in the private sector compared to the public sector ($p = 0.0478$), as well as a greater awareness of the potential impacts of AI on their work ($p = 0.0132$). However, there was no significant difference between the two sectors in terms of recognition of the changes already brought about by AI in their work ($p = 0.0945$) (Table 30.1).

30.4.3.2 Hypothesis 2

The data collected significantly confirms the H2: hypothesis that experienced auditors and controllers, with more than 10 years' experience, tend to use AI tools in their activities. Out of the 60 auditors and controllers surveyed, a majority of individuals representing 60% of the sample, said they had used AI tools as part of their professional activities. Looking specifically at those with more than 10 years' experience, we observe 83% have used these tools. By contrast, among those with less than 10 years' experience, only 25% have used such tools.

30.4.3.3 Hypothesis 3

Hypothesis 3 suggests that awareness and prior understanding of AI lead to proactive adaptation by internal auditors and controllers is confirmed. Students' t-test results reveal significant differences between individuals who have already heard of AI and those who have never heard of AI. Participants who had already heard of AI showed a greater propensity to learn about AI compared to those who had never heard of AI ($p = 0.0004$). Similarly, they showed a greater willingness to change their practices to prepare for the introduction of AI compared to those who had never heard of AI, with a statistically significant difference ($p = 0.0012$) (Table 30.2).

TABLE 30.2
Results of Students' t-tests for Hypothesis 3

	T-Test		
Variable	**Heard about AI**	**Never Heard about AI**	**P Value**
Learn about AI	2,821	1,750	0.0004
Change my practices to prepare for the introduction of AI	2,771	1,500	0.0012

30.4.3.4 Hypothesis 4

Our hypothesis H4 that the barriers to the use of AI in audit and control work are similar in the private and public sectors is confirmed. The average scores for five specific obstacles were identical in both sectors. In the private sector, the lack of AI training and expertise was rated at 3.125, while in the public sector it was rated at 3.000. Similarly, the high cost of implementing AI was noted at 3,000 in the private sector and 3,125 in the public sector. Other barriers, such as concerns about AI reliability, lack of clarity on AI-related regulations and reluctance to change, showed similar scores in both sectors. Students' t-tests revealed no significant differences between barrier scores in the two sectors. These results suggest that the challenges faced in adopting AI are uniform, regardless of industry.

30.5 DISCUSSION

In our study, we aimed to explore various aspects of the use of AI in auditing and control activities. To achieve this, we employed a methodology based on data collected from 60 participants. Several key hypotheses were formulated to investigate aspects such as the difference in perception between the private and public sectors, the utilization of AI tools by experienced auditors and controllers, the impact of prior knowledge of AI on training and practice adaptation, and the similarity of obstacles encountered in both sectors. Our initial descriptive analysis unveiled significant demographic trends. Regarding the reliability of our measures, the results demonstrated a good level of internal consistency, with a Cronbach's alpha of 0.867. Correlation analyses further revealed significant links between different variables studied, including a positive correlation between the level of training on AI and readiness for its introduction, as well as between readiness for AI introduction and the level of awareness of AI's implications in the workplace.

The analysis of hypotheses in our study has provided valuable insights into various aspects of AI utilization in audit and control activities. The results confirm hypothesis H1, indicating a disparity in the perception of AI use between the private and public sectors. Professionals in the private sector exhibit greater openness and readiness to adopt AI in their audit and control practices compared to their public sector counterparts. This is evident in their increased willingness to learn about AI and their greater readiness to adapt their professional practices to incorporate this emerging technology. These differences in perception and readiness may be attributed to various factors, including organizational culture, available resources, and sector-specific regulatory requirements. Despite these distinctions, there is no significant divergence between the two sectors in recognizing the changes already brought about by AI in their work. This indicates that the challenges and opportunities associated with this technology are independent of the sector of activity.

Hypothesis 2 highlights a significant relationship between the professional experience of auditors and controllers and their propensity to use AI tools in their activities. The data show that professionals with more than 10 years' experience have a greater tendency to apply these tools than those with less than 10 years' experience. These results show that professional experience is a determining factor in the adoption of emerging technologies such as AI. Auditors and controllers who have accumulated significant experience are more likely to explore and integrate AI tools into their daily practices. On the contrary, less experienced professionals may be less inclined to adopt AI tools due to a lack of familiarity with the technology, concerns about their effective use, or a lesser perception of the benefits associated with these tools.

Regarding Hypothesis 3, our results highlight the significant impact of prior AI awareness on the proactive adaptation of internal auditors and controllers. This hypothesis postulated that individuals who had already heard of AI would be more inclined to train and change their professional practices to integrate this emerging technology. Our analyses, based on students' t-tests, revealed statistically significant differences between individuals informed about AI and those who were not. Specifically, participants who had prior knowledge of AI exhibited a stronger willingness to learn about this technology compared to those who had never been exposed to it, with a very high level of confidence. Furthermore, they demonstrated a significantly greater readiness to modify their professional practices to prepare for the introduction of AI into their work.

The results of our analysis confirm Hypothesis H4, which postulated that the barriers to using AI in audit and control work would be similar in the private and public sectors. These results indicate that audit and control professionals, whether working in the private or public sector, face similar challenges when it comes to integrating AI into their professional practice. We observed that the average scores attributed to the five specific obstacles were virtually identical in both sectors. Notably, the lack of AI training and expertise was rated at similar levels in both sectors. Similarly, concerns such as the high cost of AI implementation and worries about the reliability of this technology were present to similar degrees in both sectors. But despite the potential differences in contexts and practices between the private and public sectors, the challenges faced in adopting AI are uniform.

Our results provide valuable insights to guide efforts in promoting the use of AI in audit and control activities, highlighting the importance of awareness, professional experience, and understanding common barriers encountered in both the private and public sectors. Firstly, our results highlight a difference in perception between the private and public sectors regarding the use of AI. Professionals in the private sector appear to be more open to and better prepared for the adoption of AI in their audit and control practices than their public-sector counterparts. This finding underscores the need to tailor AI promotion and awareness strategies to the specificities of each sector. Secondly, our study reveals that professional experience plays a crucial role in the use of AI tools. Professionals with more than 10 years' experience tend to apply these tools more than those with less experience. Organizations should, therefore, consider their employees' level of experience when contemplating the integration of AI into their practices. Thirdly, our study highlights the importance of prior awareness of AI in promoting its adoption and use. Individuals informed about AI showed a stronger willingness to learn about this technology and a greater readiness to modify their professional practices to adapt to it. Therefore, efforts to promote the use of AI should include awareness-raising and training initiatives to ensure effective and widespread adoption of this technology. Finally, our results indicate that the barriers to the use of AI in audit and control work are similar in the private and public sectors. Although organizational contexts may vary, the challenges encountered in AI adoption remain uniform, such as the lack of AI training and expertise, the high cost of its implementation, and concerns about its reliability. To foster more widespread and effective use of AI in these fields, addressing these common challenges is essential.

30.6 CONCLUSION

Innovative control devices serve as security measures designed to manage risk, ensure compliance with laws and regulations, provide reliable information, and enhance performance. Our study reveals that the integration of AI into auditing and control significantly contributes to better risk management and improved performance, particularly in the healthcare sector. Examining various aspects of AI use in audit and control activities, our study identifies important trends. Firstly, professionals in the private sector appear to be more prepared to adopt AI compared to their public sector counterparts. This highlights the need for tailored AI promotion and awareness strategies for each sector. Secondly, professional experience emerges as a determining factor in the adoption of AI tools, with seasoned professionals exhibiting a stronger propensity to use them. Organizations should thus consider employees' experience levels when contemplating AI integration. Thirdly, prior awareness of AI proves crucial in fostering its adoption and use, as evidenced by the increased willingness of informed individuals to undergo training and modify their professional practices. Lastly, our findings indicate that barriers to AI use are similar in the private and public sectors, emphasizing the importance of overcoming common challenges such as the lack of AI training and expertise, as well as concerns about its reliability. To improve the use of AI in these areas, it is essential to invest in raising awareness, training and resolving the barriers identified. For future research, it would be interesting to investigate in more detail the specific awareness-raising and training strategies that are most effective, as well as to examine the long-term impacts of AI adoption on audit and control practices.

REFERENCES

Ait Lemqeddem, H. & Chouay, J. Le rôle de la digitalisation dans l'efficacité de l'audit interne. Revue Française d'Economie et de Gestion 1(5), 48–65 (2020).

Etienne Minvielle, C., Andre-Pierre, C., Marine, J., Nathalie, P., Annie, B., & Christian, R. Hospital performance: Competing or shared values? Health Policy 87(1), 8–19 (2008).

Hantem, A. Thèse de doctorat: introduction d'une comptabilité d'exercice au sein des CHU au Maroc: caractéristique et analyse des facteurs, sous la direction de Mohammed abou eljaouad, Université hassan II, Casablanca (2019).

Hantem, A. Le recours aux nouveaux outils du management public: vers un système comptable fiable et transparent. Revue D'Etudes en Management et Finance D'Organisation 8(1), 1–19 (2023).

Hantem, A., Emzi, KH., Youssoufi, N., Emzi, K., Mkik, M. Factors influencing patient satisfaction with the use of teleconsultation in Morocco. International Conference on Computing and Machine Learning. Book Series Lecture Notes in Networks and Systems. Springer (2024).

INTOSAI. The use of artificial intelligence (AI) in the execution of audits. Science and Technology in Audit 50(2) (2023). https://intosaijournal.org/journal-entry/the-use-of-artificial-intelligence-ai-in-the-execution-of-audits/, last accessed 2024/01/05

Kok, J. N., Boers, E. J., Kosters, W. A., Van der Putten, P., & Poel, M. Artificial intelligence: Definition, trends, techniques, and cases. Artificial intelligence 1, 270–299 (2009).

Lafon, L. Mazars. L'audit légal des établissements publics de santé. Gestion et management (2016).

Lizot, J., Laskawiec, A. Fiabilisation et certification des comptes des EPS. Gestion et Finances publiques, 3–4 (2014).

Markaoui, K., & Hassine, M. Définition de la performance des services hospitaliers: enquête au sein du chu mohammed vi d'oujda. Revue du Contrôle de la Comptabilité et de l'Audit 2(3) (2018).

McCarthy, J. From here to human-level AI. Artificial Intelligence 171(18), 1174–1182 (2007). http://wwwformal.stanford.edu/jmc/

Mighiss, S. L'audit à l'ère de la transformation digitale. Revue du contrôle, de la comptabilité et de l'audit 5(1), 193–209 (2021). https://doi.org/10.5281/zenodo.5595403

Pesqueux, Y. La notion de performance globale. 5ème forum international sur la performance Globale de l'entreprise. Université Tunis Carthage (2004).

Rodrigues, L., Pereira, J., da Silva, A. F., & Ribeiro, H. The impact of artificial intelligence on audit profession. Journal of Information Systems Engineering and Management 8(1), 19002 (2023).

Saulquin, J., & Schier, G. Responsabilité sociale des entreprises et performance: complémentarité ou substituabilité? La Revue des Sciences de Gestion 1(223), 57–65 (2007).

The Institute of Internal Auditors (IIA). Intelligence artificielle: le futur de l'audit interne. tone at the top, N°85 (2017). https://docs.ifaci.com/wp-content/uploads/2018/03/tone-at-the-top-85-ia.pdf

Wahbi, A., Khaddouj, K., & Lahlimi, N. Study of the relationship between chatbot technology and customer experience and satisfaction. International Journal of Accounting, Finance, Auditing, Management and Economics 4(6–1), 758–771 (2023). https://doi.org/10.5281/zenodo.10442472

World Health Organization Europe. Performance assessment tool for quality improvement in hospitals (2007).

Zarai, M. A., & Bettabai, W. Impact de l'efficacité des comités d'audit sur la qualité des bénéfices comptables divulgués. Revue Gouvernance 4(1), 1–23 (2007).

Zarhloul, Z., & Mrani Zentar, S. La contribution de l'audit interne dans la performance du système de contrôle interne: cas du secteur bancaire marocain. International Journal of Accounting, Finance, Auditing, Manag 4(1) (2022).

31 LSTM-Based Analysis of Emotions and Participation in Online Learning

J. E. Rafiq, Z. Abdelali, S. Nouh, and M. Amraouy

31.1 INTRODUCTION

In the current digital age, education has transcended traditional boundaries, expanding into virtual spaces through online learning platforms. Social interaction, a cornerstone of learning, as emphasized by psychologist Vygotsky, plays a pivotal role in cognitive development [1].

Online discussion forums have emerged as vital platforms for collaborative learning, fostering critical thinking and knowledge acquisition through diverse perspectives [2]. These forums also promote written expression and communication skills, essential in the digital age.

The level of student engagement in online forums significantly impacts academic outcomes [3]. This study explores the influence of engagement on academic performance, particularly in computer science education at RCETP.

E-learning platforms like Moodle have revolutionized education, providing dynamic environments for collaboration and knowledge access [4]. Moodle, the focus of this study, encourages learner engagement through its effective support of online interactions [4].

This research investigates the correlation between student emotional involvement and academic results, particularly in computer science teacher training. Analyzing involvement in virtual discussions has the potential to enhance digital learning environments, fostering profound interactions among learners [5]. Active participation in discussions positively impacts academic performance [6], enabling students to deepen their understanding and exchange ideas with peers. Factors such as motivation and clarity of instructions influence participation levels, highlighting the importance of consistent engagement for improved educational outcomes. Consequently, promoting active involvement in online discussions not only encourages meaningful interactions among learners but also leads to higher levels of academic achievement.

31.2 METHODOLOGY

In this section, we will outline our approach to investigating the relationship between student engagement on a Moodle discussion platform and their academic performance at RCETP.

31.2.1 Learning Environment

Using the Moodle learning environment, the course "Assessment of Learning in a Competency-Based Approach" was taught. In order to promote interactive learning, this open-source platform provides a wide range of tools and features, such as discussion boards, activity submissions, and grade tracking [7].

31.2.2 Selection Criteria for Participants

Future specialized computer science teachers for secondary education are the study's participants. Active enrollment was restricted to students who participated in course activities on a regular basis

DOI: 10.1201/9781003647263-31

for the duration of the 12-week study period. Data anonymization and confidentiality were guaranteed by securely storing and anonymizing all participant data, including grades and personal information, to preserve privacy.

31.2.3 Data Collection

Two primary sources of data were gathered for this study: first, learner performance was evaluated using anonymous final summative evaluation grades. Secondly, a collection of messages encompassing thread discussions, individual messages, and participant comments were gathered from the Moodle course discussion forums.

31.2.4 Data Set Description

An inactive learner (one who left no messages or comments) was removed from the data set, which had information on 30 active students who completed the course with success. There were 632 comments out of the total 735 forum entries in this corpus. For the purpose of this study, no distinction was made between postings and comments in order to fully evaluate the effect of active learner participation on academic success.

31.2.5 Analysis of Learner Sentiments

Various NLP techniques were employed to analyze learners' sentiments in their forum messages [8–11]. The study [12] focused on Arabic text analysis, [13] utilized Text Blob for socio-affective component detection, [14] used the BERT model to analyze emotional states, and [15] studied LSTM's role in sentiment analysis. We chose LSTM for its capability to capture long-term dependencies in textual data. Before LSTM application, messages underwent preprocessing, including punctuation removal, lowercase conversion, stop word removal, and normalization. LSTM predicted sentiment scores for each comment, which were then associated with learner messages. Results were presented in a DataFrame with sentiment scores categorized as positive, neutral, or negative based on a threshold of 0.5, facilitating interpretation.

31.2.6 Correlation Between Learner Performance and Participation Factors

We analyzed correlations between various learner participation metrics on the discussion forum and their academic performance. Measures included post frequency, response count, connection time, and sentiment analysis (detailed in Table 31.1). Using SPSS 24, we correlated these metrics with anonymized grades from learner assessments. Results revealed significant correlation between active forum engagement and academic performance in the "Assessment of Learning in a Competency-Based Approach" course for Moroccan secondary school computer science teacher trainees. These findings substantially addressed our investigation into how involvement affects academic results, particularly in this specialized program.

TABLE 31.1
Factors Affecting Participation and Their Descriptions

Factors of Participation	Description
NDP	Participation of the learners as measured by the quantity of messages submitted on the discussion board.
NRP	Participation of learners is measured by the quantity of replies to messages that are posted.
CT	The overall duration of each learner's activity on the platform.
ES	The emotional state detected from the discussions of each learner.

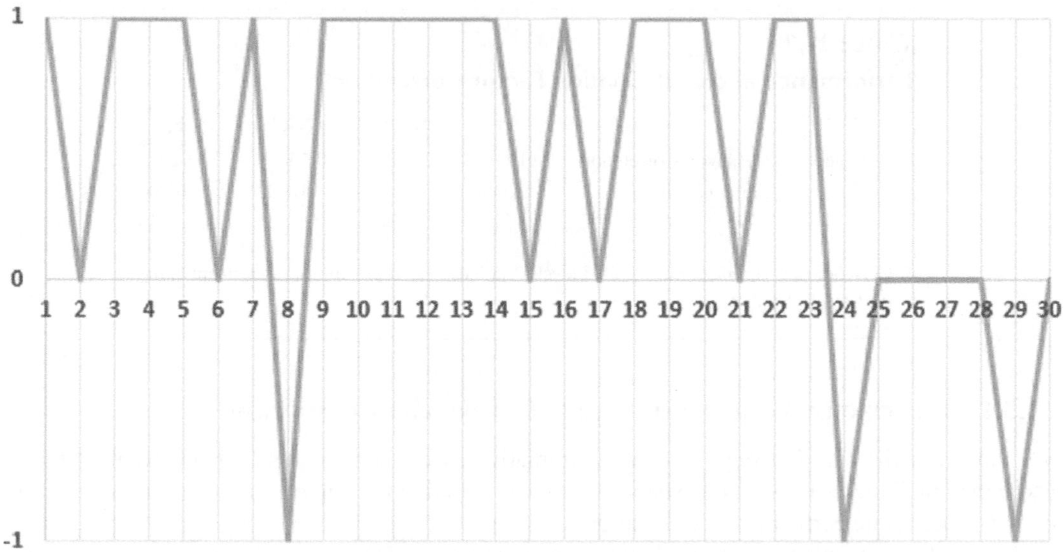

FIGURE 31.1 Sentimental state of students.

31.3 RESULTS

Here, we provide a detailed summary of the conclusions we reached after examining how student participation in an improved Moodle discussion forum affected their academic achievement as prospective computer science teachers at RCETP, Oujda, Morocco. Discussion participation, frequency of responses, length of platform usage, and emotional expression in student contributions were all taken into account in our analysis (Figure 31.1).

After thorough analysis, we categorized participants' emotional states: 17% negative (s = –1), 23% neutral (s = 0), and 60% positive (s = 1). This diversity reflects varied emotional reactions within the forum, providing insights into participant interactions and their potential impact on academic performance. Examples of these emotional states are provided in Table 31.2.

Our results highlight a significant correlation between student forum activity, emotional engagement, and academic success. Students who actively contributed to discussions with positive emotions achieved higher final assessment scores. This underscores the importance of fostering a vibrant and positive online discussion environment to enhance academic outcomes for future computer science teachers at RCETP, Oujda, Morocco.

TABLE 31.2
Examples of Learners' Sentimental State in Their Messages

Posts Examples	Sentiment Scores	Sentimental State
I really appreciate how competency-based learning emphasizes in-depth understanding of subjects. It encourages students to go beyond mere memorization and apply their knowledge meaningfully in real-life situations.	0.995	1
I find that competency-based learning can sometimes be too demanding for students. Skill-based assessments require a lot of work, and some students may feel overwhelmed.	0.109	–1
Competency-based assessment has sparked a lot of debates recently. There are clear advantages, but also challenges to overcome. I'm curious to see how this will evolve in the future.	0.545	0

TABLE 31.3
Performance and Participation Factor Correlation

		CT	NDP	NRP	ES
Performance	Pearson Correlation	0.316	0.398*	0.539**	0.610**
	Sig. (2-tailed)	0.089	0.029	0.002	0.000
	N	30	30	30	30

* Correlation is significant at the 0.05 level (2-tailed); ** correlation is significant at the 0.01 level (2-tailed).

31.3.1 Participation Factors and Learner Performance Correlation

We looked at the link between students' summative assessment scores and involvement factors (explained in Table 31.1) using the SPSS 24 software. Finding the correlation between these parameters and academic performance was our goal.

The analysis revealed intriguing insights, as depicted in Table 31.3. Time spent by connected learners on the platform showed a moderately positive correlation (0.316) with course success (Figure 31.2a). This suggests that increased interaction with online content and activities correlates with higher grades.

Moreover, active participation in discussions, as evidenced by the number of posts (Figure 31.2b) and responses (Figure 31.2c), displayed significant positive correlations with summative assessment scores. This emphasizes how crucial participation in online discussions is for achieving better academic results.

Notably, there was a strong positive association found between the emotional state of learners during online interactions and their academic achievement, suggesting that positive emotions may have an impact on learning.

In summary, a happy emotional state, active engagement in conversations, and connection time all have a big impact on academic performance. These results highlight the significance of emotionally healthy and active participation in virtual learning settings for improved academic outcomes.

31.4 DISCUSSION

Here, we will delve into the finding of our study and their implications in the context of training teachers at the RCETP in Oujda, Morocco. We will also address the limitations of our research and provide suggestions for future studies.

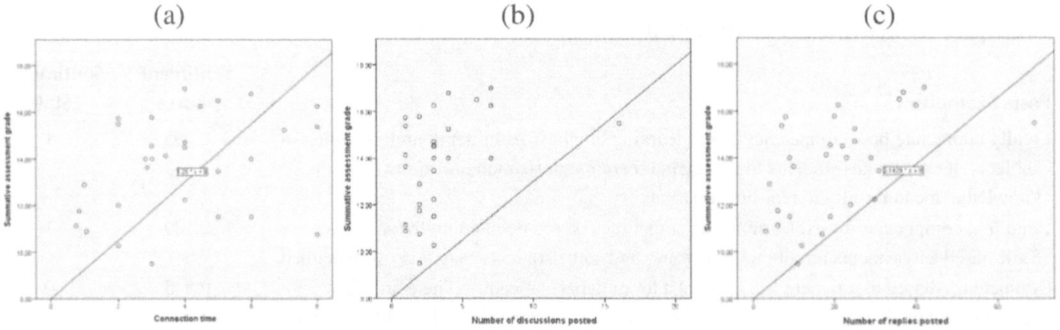

FIGURE 31.2 (a–c) Correlation between factors of participation and academic performance.

We found a significant positive correlation between summative assessment scores and participation levels (e.g., number of posted discussions and responses) that supports the critical relationship between active learner participation in online discussion forums and academic outcomes [6, 16–18]. This underscores the importance of fostering active engagement to create an optimal learning environment conducive to success. Moreover, our analysis of learners' emotional states revealed a positive correlation between positive emotions and increased participation [19, 20], emphasizing the need to consider emotional aspects in online learning activities to enhance engagement and potentially improve academic outcomes. These results imply useful ramifications for educators and course designers, highlighting the importance of creating stimulating online learning environments that promote interaction, participation, and positive emotional experiences [21, 22]. Utilizing sentiment analysis tools can aid educators in monitoring learners' emotions and providing timely support [23, 24]. To better understand participation dynamics and learning outcomes, larger samples and longitudinal studies are needed, as our study's shortcomings include a limited sample size and an emphasis on correlation rather than causation. Subsequent studies could investigate the relationship between online participation, emotions, and academic performance, as well as investigating specific pedagogical strategies aimed at promoting active participation and fostering positive emotional states in online learning contexts.

31.5 CONCLUSION

This study investigated the influence of student participation in an enhanced online discussion forum on academic performance within the context of training future computer science teachers at RCETP, Oujda, Morocco (Figure 31.1). Leveraging natural language processing (NLP) techniques, including LSTM, our analysis aimed to elucidate the impact of student engagement in discussion forums on academic outcomes. Our results underscore the significance of active learner participation, correlating with higher scores in final assessments, emphasizing the need for dynamic online discussion environments to enhance academic performance. Additionally, learners' positive emotional expressions correlated with increased participation and better academic results, highlighting the importance of considering emotional aspects in online learning activities. Despite limitations such as sample size constraints and the absence of causal relationships, this study offers valuable insights into optimizing online learning environments, fostering meaningful interactions, and promoting academic success for future computer science teachers.

REFERENCES

1. V. V. Rubtsov, "Two approaches to the problem of development in the context of social interactions: L.S. Vygotsky vs J. Piaget," *Cult.-Hist. Psychol.*, vol. 16, no. 3, pp. 5–14, 2020.
2. S. A. Aderibigbe, A. R. A. AbdelRahman, and H. Al Othman, "Using online discussion forums to enhance and document Students' workplace learning experiences: A semi-private Emirati University's context," *Educ. Sci.*, vol. 13, no. 5, 2023, doi: 10.3390/educsci13050458.
3. J.-S. Wong, B. Pursel, A. Divinsky, and B. J. Jansen, "An analysis of MOOC discussion forum interactions from the most active users," *Springer Int. Publ. Switz.*, vol. LNCS 9021, pp. 452–457, 2015, doi: 10.1007/978-3-319-16268-3_58.
4. R. A. Salas-Rueda, A. L. Eslava-Cervantes, and E. Prieto-Larios, "Teachers' perceptions about the impact of Moodle in the educational field considering data science," *Online J. Commun. Media Technol.*, vol. 10, no. 4, pp. 1–10, 2020, doi: 10.30935/ojcmt/8498.
5. E. C. Onyema, A. O. Deborah, Q. N. Alsayed, S. Naveed, and Sanober, "Online discussion forum as a tool for interactive learning and communication," *Int. J. Recent Technol. Eng. IJRTE*, vol. 8, no. 4, pp. 4852–4868, 2019, doi: 10.35940/ijrte.d8062.118419.
6. B. Yang, H. Tang, L. Hao, and J. R. Rose, "Untangling chaos in discussion forums: A temporal analysis of topic-relevant forum posts in MOOCs," *Comput. Educ.*, vol. 178, p. 104402, 2022, doi: 10.1016/j.compedu.2021.104402.

7. A. R. Vargas-Murillo, A. P. N. M. D. L. A. Pari-Bedoya, and A. P. L. Ypanaque-Pereira, "Analyzing the impact of Moodle as an interactive tool to improve students' engagement and motivation in the learning process: A systematic literature review," *ICSLT 23 Proc. 2023 9th Int. Conf. E-Soc. E-Learn. E-Technol.*, pp. 46–53, 2023, doi: https://doi.org/10.1145/3613944.3613952.
8. P. Wang, Z. Zhu, Q. Chen, and W. Dai, "Text reasoning chain extraction for multi-hop question answering," *Tsinghua Sci. Technol.*, vol. 29, no. 4, pp. 959–970, 2024, doi: 10.26599/TST.2023.9010060.
9. C. Hazman, A. Guezzaz, S. Benkirane, and M. Azrour, "Enhanced IDS with deep learning for IoT-based smart cities security," *Tsinghua Sci. Technol.*, vol. 29, no. 4, pp. 929–947, 2024, doi: 10.26599/TST.2023.9010033.
10. S. Li, T. Cui, and W. Viriyasitavat, "Edge device fault probability based intelligent calculations for fault probability of smart systems," *Tsinghua Sci. Technol.*, vol. 29, no. 4, pp. 1023–1036, 2024, doi: 10.26599/TST.2023.9010085.
11. N. Meenakshi *et al.*, "Efficient communication in wireless sensor networks using optimized energy efficient engroove leach clustering protocol," *Tsinghua Sci. Technol.*, vol. 29, no. 4, pp. 985–1001, 2024, doi: 10.26599/TST.2023.9010056.
12. T. Sabri, O. El Beggar, and M. Kissi, "Comparative study of Arabic text classification using feature vectorization methods," *Procedia Comput. Sci.*, vol. 198, no. 2021, pp. 269–275, 2021, doi: 10.1016/j.procs.2021.12.239.
13. M. Amraouy, M. Bellafkih, A. Bennane, and J. Talaghzi, "Sentiment Analysis for Competence-Based e-Assessment Using Machine Learning and Lexicon Approach," *3rd Int. Conf. Artif. Intell. Comput. Vis. AICV2023 March 5–7 2023 AICV 2023*, 2023, doi: 10.1007/978-3-031-27762-7_31.
14. J. E. Rafiq, A. Zakrani, M. Amraouy, A. Namir, and A. Bennane, "Optimizing Learning Performance through AI-Enhanced Discussion Forums," *2023 14th Int. Conf. Intell. Syst. Theor. Appl. SITA*, 2023, doi: 10.1109/SITA60746.2023.10373689.
15. Y. Zahidi, Y. Al-Amrani, and Y. El Younoussi, "Improving Arabic sentiment analysis using LSTM based on word embedding models," *Vietnam J. Comput. Sci.*, vol. 10, no. 3, pp. 391–407, 2023, doi: 10.1142/S2196888823500069.
16. T. Xu, Q. Wu, and Z. Xu, "The Impact of Online Learners' Social Interaction on Learning Achievement Based on Social Network Analysis," *2021 9th Int. Conf. Inf. Educ. Technol. ICIET 2021*, pp. 232–241, 2021, doi: 10.1109/ICIET51873.2021.9419643.
17. F. Ghizlane, J. Mabrouki, F. Ghrissi, and M. Azrour, "Proposal for a high-resolution particulate matter (PM10 and PM2. 5) capture system, comparable with hybrid system-based internet of things: Case of quarries in the Western Rif, Morocco," *Pollution*, vol. 8, no. 1, pp. 169–180, 2022.
18. A. Ntourmas, S. Daskalaki, Y. Dimitriadis, and N. Avouris, "Classifying MOOC forum posts using corpora semantic similarities: A study on transferability across different courses," *Neural Comput. Appl.*, vol. 35, no. 1, pp. 161–175, 2023, doi: 10.1007/s00521-021-05750-z.
19. J. E. Peng, and Y. Jiang, "Mining opinions on LMOOCs: Sentiment and content analyses of Chinese students' comments in discussion forums," *System*, vol. 109, p. 102879, 2022. doi: 10.1016/j.system.2022.102879.
20. F. Dalipi, K. Zdravkova, and F. Ahlgren, "Sentiment analysis of Students' feedback in MOOCs: A systematic literature review," *Front. Artif. Intell.*, vol. 4, pp. 1–13, 2021, doi: 10.3389/frai.2021.728708.
21. Z. Du, F. Wang, S. Wang, and X. Xiao, "Enhancing learner participation in online discussion forums in massive open online courses: The role of mandatory participation," *Front. Psychol.*, vol. 13, 2022, doi: 10.3389/fpsyg.2022.819640.
22. Ö Demir, M. Cinar, and S. Keskin, "Participation style and social anxiety as predictors of active participation in asynchronous discussion forums and academic achievement," *Educ. Inf. Technol. (Dordr)*, pp. 1–22, 2023, doi: 10.1007/s10639-022-11517-3.
23. J. Zhou, and J. Ye, "Sentiment analysis in education research: A review of journal publications," *Interact. Learn. Environ.*, vol. 31, no. 3, pp. 1252–1264, 2020, doi: 10.1080/10494820.2020.1826985.
24. Q. Tul *et al*, "Sentiment analysis using deep learning techniques: A review," *Int. J. Adv. Comput. Sci. Appl.*, vol. 8, no. 6, 2017, doi: 10.14569/ijacsa.2017.080657.

32 Urban Air Pollution Evaluations of Nitrogen Dioxide NO_2

The Case of the City of Kénitra–Morocco

M. Hadine, J. Mabrouki, and F. Z. Ennasri

32.1 INTRODUCTION

Anthropogenic activities are responsible for environmental degradation worldwide [1–3]. Indeed, the atmosphere, considered an essential component of our ecosystem, is constantly threatened and human beings are the main victims [1, 4–9]. Air pollution has become an increasingly important global problem. It affects both developed and developing countries. Although many measures to protect water and soil have been taken in major international cities, measures to protect the air are still very limited and require a great deal of effort. More and more potentially harmful gases and particles are entering the atmosphere, harming human health and the environment [10]. Due to the damage caused by various forms of air pollution, authorities in many countries have conducted surveys and research to better manage the problem and find appropriate solutions. In Morocco, several cities are increasingly exposed to this risk [11–13]. Due to its rapidly expanding industrial structure and its main cross-country road linking north and south, the city of Kénitra is frequently used by various modes of transport (on average 20,160 units/12 hours during the day and 7,100 units/12 hours at night), which can have an impact on air quality. Our aim is to set up a permanent monitoring system to control this scourge and to propose solutions and suggestions for potential problems [12, 14, 15]. Thus, a diagnosis of the city's air quality seems necessary to determine whether nitrogen dioxide could present a risk to residents.

32.2 STUDY AREA, DATA AND METHODS

32.2.1 Sampling Sites and Campaign

The study focuses on the city of Kénitra, which is located on the Atlantic coast in the north-west of the country. It is located about 40 kilometers north of Rabat, 131 kilometers from the economic capital Casablanca and 238 kilometers from Tangier, the gateway to Europe. It extends over 106 km^2. It is located in a slightly open depression to the northeast, with altitudes ranging from 2 meters near the river to 25 meters towards the Maâmora plateau. The city enjoys a semi-arid climate with temperate winters due to the presence of a dune cordon that separates it from the ocean, thus mitigating the oceanic influence on the region. Our study covered eight different sites (Figure 32.1) during the year 2018 with four campaigns, in order to identify spatiotemporal variations. The sampling points we selected were identified in a preliminary study and were distributed throughout the city, varying in pollution intensity and key meteorological parameters, in this case moisture. This spatial distribution was motivated by the desire to use it as a basis for considering the possibility of establishing a permanent network of measures in light of the results obtained.

DOI: 10.1201/9781003647263-32

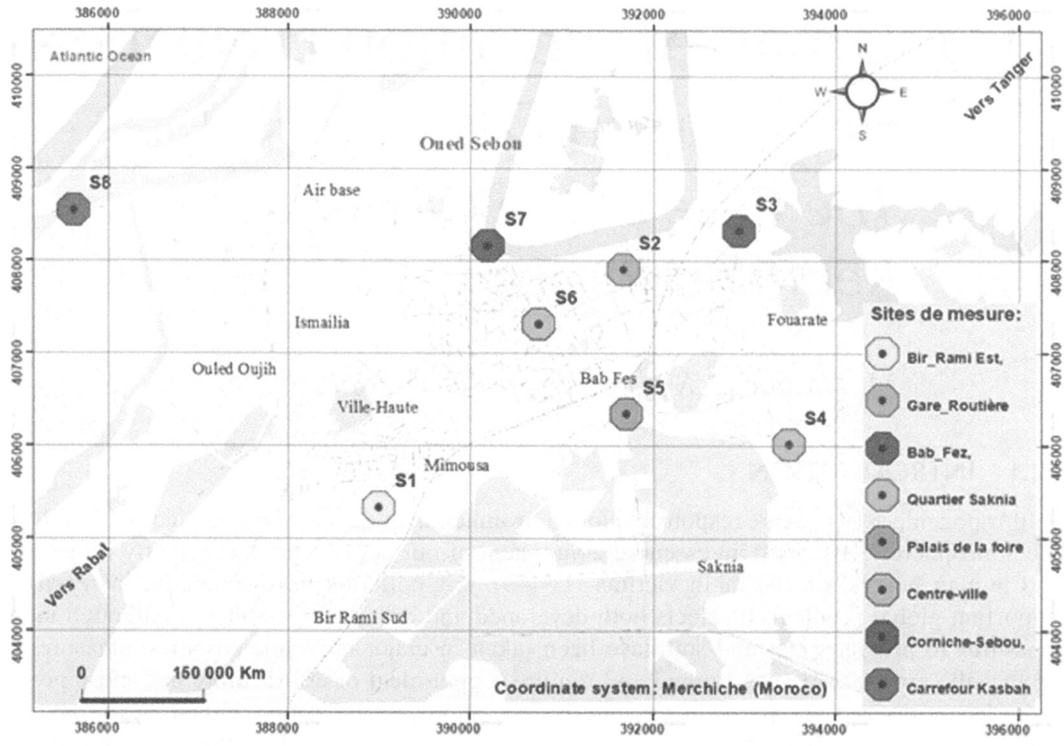

FIGURE 32.1 Location of sites in the study area.

32.2.2 Sampling Techniques and Protocols

Four sampling campaigns were conducted in 2018, four times a day. The parameter to be analyzed was chosen according to the types of discharges from the dominant industries and the emissions from the vehicle fleet. Thus, we opted for NO_2 (nitrogen dioxide). For most pollutants with cumulative and summative effects, continuous sampling is used, rather than instantaneous sampling for short periods of time. The NO_2 is sampled using a bubbling system [16] for 30 minutes; the sampling rate is maintained at 2 L/min. The sampling pump is a gas pump with a constant flow rate.

32.2.3 Analysis Methods

The analytical method used is Griess-Saltzman spectrophotometry [17]. It consists of drawing air into a solution of N-(1-Naphthyl)-ethylineglacial. The intensity of "pink" coloration is measured with a spectrophotometer. NO_2 concentration was determined from a calibration curve based on known sodium nitrite solutions. The nitrogen dioxide content is calculated taking into account the initial air volume sampled, the color developed and the calibration performed.

Calculation: The mass concentration of NO_2 is given by the following formula:

$$\mathbf{MNO_2 = \left(\left(C * Vs\right) / Va\right)\left(\mu g / m^3\right)}$$

where MNO_2: NO_2 mass concentration in µg/m^3; C: NO_2 concentration in µg/m^3; Vs: volume of sampling solution in m^3; Va: sampled air volume, in m^3.

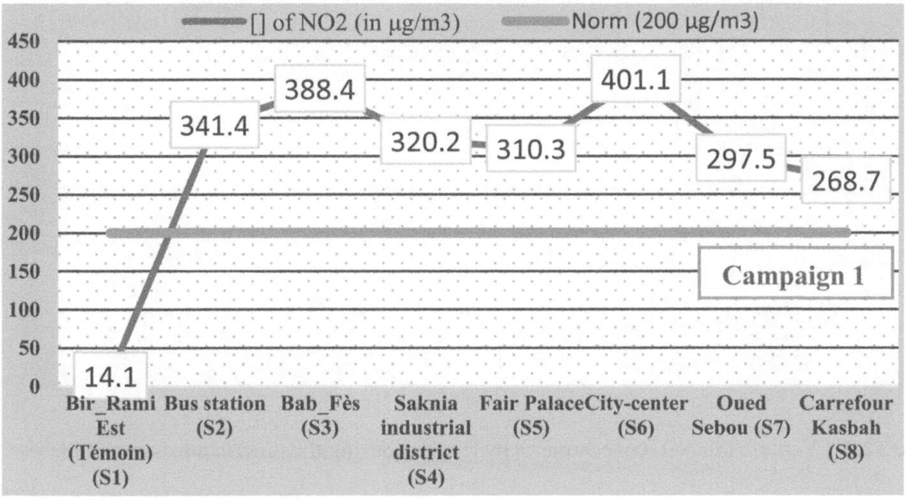

FIGURE 32.2 Variation in NO_2 concentration by location during the first campaign in 2018.

32.3 RESULTS AND DISCUSSION

During all the sampling campaigns, the control site of Bir_Rami_Est displayed nitrogen dioxide (NO_2) levels below the limit value of 100 µg/m^3 established by the Moroccan law on the fight against pollution [18], as demonstrated by the results presented in Figures 32.2–32.5. On the other hand, the data collected on the sites (S2), (S3), (S4), (S5), (S6), (S7) and (S8) exceeded all the standards established by Moroccan legislation of 200 µg/m^3 [18] as well as those imposed by the World Health Organization [19] and the European Union [20, 21], which were also 200 µg/m^3. The fluctuations in the NO_2 concentration per campaign and per site are illustrated in Table 32.1.

The high levels can also be attributed to phenomena such as the acceleration and deceleration phases, as well as the duration of the stops at the traffic lights (engine idling) on these sites of practically all these types of vehicles [22]. In addition, the proximity of these sites to junctions that are strongly impacted by traffic on the surrounding traffic lanes may also explain the high levels. Indeed, nitrogen dioxide is a very dispersive primary pollutant, which means that its concentrations

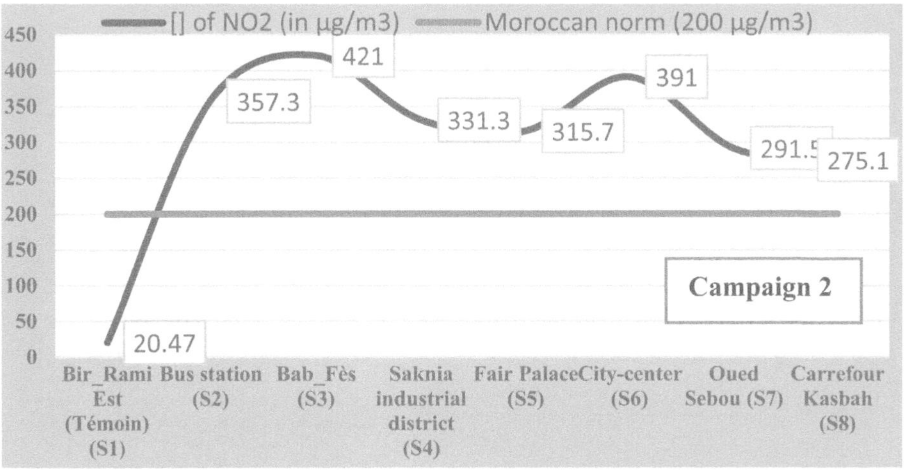

FIGURE 32.3 Variation in NO_2 concentration by location during the second campaign in 2018.

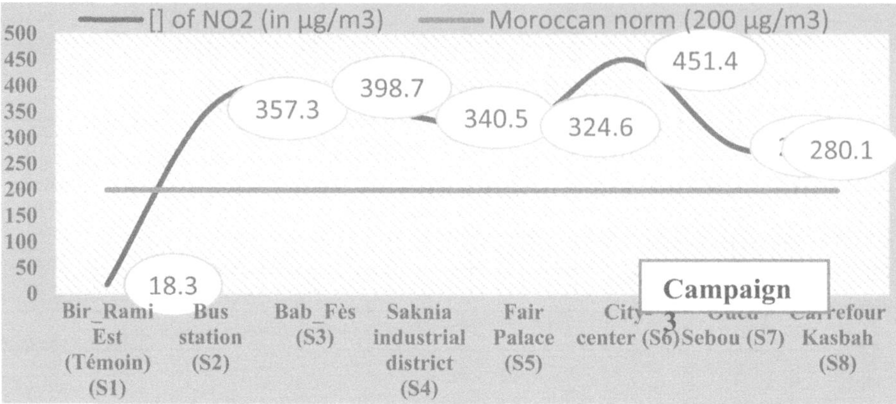

FIGURE 32.4 Variation in NO_2 concentration by location during the third campaign in 2018.

were higher on the traffic routes. Moving away from the axis, the levels decreased until they reached an ambient background level. The decrease was exponential [23].

In a similar way, research conducted by A. Venin and colleagues showed that the mode of driving could also lead to increased emissions of this pollutant [24]. Emissions increased significantly when the vehicle was in a cold state (when starting). This is a crucial parameter, especially since in France, for example, more than 25% of the total mileage was done with the cold engine [25, 26].

The results we have obtained are also in agreement with studies carried out in Switzerland, which were able to confirm that 74% of NO_2 emissions were attributable to transport and only 22% to industry [27]. They also agree with the studies carried out by Schutz in 1995, CITEPA in 1996, Karioun in 1997 and Venin and his collaborator Ecolivet in 2000 [13, 24, 28, 29] which demonstrated that the concentration of NO_2 increased depending on the number of vehicles [30–32].

In addition, the concentrations of NO_2 at sites (S2), (S3), (S4) and (S6) exceeded 300 µg/m^3, suggesting that air pollution may represent a serious danger to the health of the local population, particularly at the Bab-Fès and Centre-Ville sites, where it may cause symptoms such as irritation of the respiratory tract and the onset or worsening of asthma. ERPURS highlighted these signs and symptoms in a study conducted in Ile-de-France over five years, from 1991 to 1995 [13, 33–36].

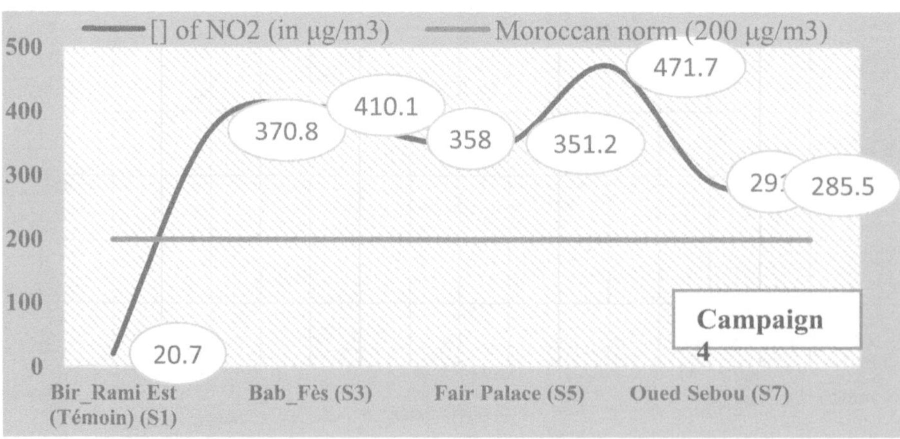

FIGURE 32.5 Variation in NO_2 concentration by location during the fourth campaign in 2018.

TABLE 32.1
Measured Concentrations of NO_2 per Site and per Campaign (in µg/m³)

Campaigns	Days	Sites	[] of NO_2 (in µg/m³)
1	02/01/2018	Bir_Rami Est (Témoin) (S1)	14.10
	03/01/2018	Bus station (S2)	341.40
	04/01/2018	Bab_Fès (S3)	388.40
	05/01/2018	Saknia industrial district (S4)	320.20
	06/01/2018	Fair Palace (S5)	310.30
	07/01/2018	City-center (S6)	401.10
	08/01/2018	Oued Sebou (S7)	297.50
	09/01/2018	Carrefour Kasbah (S8)	268.70
2	16/05/2018	Bir_Rami Est (Témoin) (S1)	20.47
	17/05/2018	Bus station (S2)	357.30
	18/05/2018	Bab_Fès (S3)	421.00
	19/05/2018	Saknia industrial district (S4)	331.30
	20/05/2018	Fair Palace (S5)	315.70
	21/05/2018	City-center (S6)	391.00
	22/05/2018	Oued Sebou (S7)	291.50
	23/05/2018	Carrefour Kasbah (S8)	275.10
	16/05/2018	Bir_Rami Est (Témoin) (S1)	20.47
3	02/08/2018	Bir_Rami Est (Témoin) (S1)	18.30
	03/08/2018	Bus station (S2)	357.30
	04/08/2018	Bab_Fès (S3)	398.70
	05/08/2018	Saknia industrial district (S4)	340.50
	06/08/2018	Fair Palace (S5)	324.60
	07/08/2018	City-center (S6)	451.40
	08/08/2018	Oued Sebou (S7)	285.70
	09/08/2018	Carrefour Kasbah (S8)	280.10
4	02/09/2018	Bir_Rami Est (Témoin) (S1)	20.70
	03/09/2018	Bus station (S2)	370.80
	04/09/2018	Bab_Fès (S3)	410.10
	05/09/2018	Saknia industrial district (S4)	358.00
	06/09/2018	Fair Palace (S5)	351.20
	07/09/2018	City-center (S6)	471.70
	08/09/2018	Oued Sebou (S7)	291.00
	09/09/2018	Carrefour Kasbah (S8)	285.50

32.4 CONCLUSION

Several research studies have been carried out in Africa and the Maghreb, in particular to monitor and evaluate the concentrations of atmospheric pollutants. Few of them, however, have been long enough to be interested in the characterization of air quality. In Morocco, even if some agglomerations have monitoring networks, there is very little data on air pollution, and even less is accessible to the public. The study in this city therefore seems essential, both on the levels of pollution and on their consequences on human health and its environment. Various campaigns have shown that road traffic can be perceived as a possible source of pollution. This could have an impact on the air quality of this city due to the high levels of NO_2. According to our results, the most affected neighborhoods are Bab-Fès (S3) and the City-Center (S6) which can be qualified as being black spots in the city of Kénitra. In addition, the industrial district of Saknia and the bus station are also affected, with fairly high pollution levels that exceed national standards, the World Health Organization and the European Union. In addition, the spatio-temporal variations in the levels of the pollutant mentioned could be linked to the influence of climatic conditions during the four campaigns considered. The study by J. Martin et al. showed that meteorological conditions have an impact on the distribution and dilution of pollutants [37–39]. In terms of health, high concentrations of pollutants can cause considerable damage in the areas surrounding their sources (asthma attacks, bronchial irritation, etc.). Residents of the region have confirmed this (personal survey, in progress).

REFERENCES

1. Ghizlane, F., Mabrouki, J., Ghrissi, F., & Azrour, M. (2022). Proposal for a high-resolution particulate matter (PM10 and PM2. 5) capture system, comparable with hybrid system-based internet of things: Case of quarries in the Western Rif, Morocco. Pollution, 8(1), 169–180.
2. Kampa, M., & Castanas, E. (2008). Human health effects of air pollution. Environmental Pollution, 151(2), 362–367. doi: 10.1016/j.envpol.2007.06.012.
3. Sleiman, H. (2016). Air pollution: Blood lead and road traffic in Beirut, 2268–3798.
4. Adams, M. D., & Kanaroglou, P. S. (2016). Mapping real-time air pollution health risk for environmental management: Combining mobile and stationary air pollution monitoring with neural network models. Journal of Environmental Management, 168, 133–141.
5. Brook, R. D., Franklin, B., Cascio, W., Hong, Y. L., Howard, G., Lipsett, M., Luepker, R., Mittleman, M., Samet, J., Smith, S. C., & Tager, I. (2004). Air pollution and cardiovascular disease - A statement for healthcare professionals from the expert panel on population and prevention science of the American Heart Association. Circulation, 109(21), 2655–2671. doi: 10.1161/01.CIR.0000128587.30041.C8.
6. Goldemberg, J., Martinez-Gomez, J., Sagar, A., & Smith, K. (2018). House-hold air pollution, health, and climate change–clearing the air. Environmental Research Letters, 13(3).
7. Khafaie, M. A., Yajnik, C. S., Salvi, S. S., & Ojha, A. (2016). Critical review of air pollution health effects with special concern on respiratory health. Journal of Air Pollution and Health, 1(2), 123–136.
8. Pope, C. A. III, & Dockery, D. W. (2006). Health effects of fine particulate air pollution: Lines that connect. Journal of the Air & Waste Management Association, 56(6), 709–742. doi:10.1080/10473289.2006.10464485.
9. WHO. (2006). Air Quality Guidelines: Global Update 2005: Particulate Mat-Ter, Ozone, Nitrogen Dioxide and sulfur Dioxide. World Health Organization, Regional Office for Europe, Copenhagen, Danmark.
10. Azrour, M., Mabrouki, J., Guezzaz, A., & Farhaoui, Y. (2021). Survey of high school students' usage of smartphone in Moroccan rural areas. International Journal of Cloud Computing, 10(3), 265–274.
11. Mabrouki, J., Azoulay, K., Elfanssi, S., Bouhachlaf, L., Mousli, F., Azrour, M., & El Hajjaji, S. (2022). Smart System for Monitoring and Controlling of Agricultural Production by the IoT. In IoT and Smart Devices for Sustainable Environment (pp. 103–115). Cham: Springer International Publishing.
12. Loukili, H., Anouzla, A., Jioui, I., Achiou, B., AlamiYounssi, S., Azoulay, K., & Riadi, Y. (2022). Combining multiple regression and principal component analysis to evaluate the effects of ambient air pollution on children's respiratory diseases. International Journal of Information Technology, 14(3), 1305–1310
13. Karioun, A. (1997). Moroccan review of civil engineering, special environment. 42–43, No. 68 – April/May.
14. Fattah, G., Mabrouki, J., Ghrissi, F., Azrour, M., & Elouardi, M. (2023). Application of Internet of Things Systems for Aerosol Monitoring of Quarries in Morocco. In Integrating Blockchain and Artificial Intelligence for Industry 4.0 Innovations (pp. 223–232). Cham: Springer International Publishing.
15. Mabrouki, J., Mourade, A., Irshad, A., & Chaudhry, S. A. (Eds.). (2023). Advanced Technology for Smart Environment and Energy. Springer Nature.
16. Mgalaa, S., Mabrouki, J., Elouardi, M., El Azzouzi, L., Moufti, A., El Hajjaji, S., & El Belghiti, M. A. (2022). Study and evaluation of the degradation of procion blue dye by the ozonation method: Parametric and isothermal study. Nanotechnology for Environmental Engineering, 7(3), 691–697.
17. Al-Jadabi, N., Laaouan, M., El Hajjaji, S., Mabrouki, J., Benbouzid, M., & Dhiba, D. (2023). The dual performance of moringa oleifera seeds as eco-friendly natural coagulant and as an antimicrobial for wastewater treatment: A review. Sustainability, 15(5), 4280.
18. Ministry of Energy, Mines, Water and Environment. (2010). Decree No. 2-09-286 of 20 Hija 1430 (8 December 2009) setting air quality standards and air monitoring procedures. Official bulletin no. 5806 of 21 January 2010.

19. Squilbin, S., & Davesne. (2015). Air quality guidelines of the World Health Organization. Brussels Environment-IBGE, collection of fact sheets, air theme, p 4, 8.
20. Vennin, A., & Ecolivet, F. (2000). Air quality: Standards and procedures. Pages 16–20.
21. CERTU, Ministry of Equipment, Transport, Spatial Planning, Tourism and the Sea. (2005). Methodological note on the evaluation of the health effects of air pollution in road impact studies. Technical guide, 52 pages
22. Benchrifa, M., Elouardi, M., Fattah, G., Mabrouki, J., & Tadili, R. (2023). Identification, Simulation and Modeling of the Main Power Losses of a Photovoltaic Installation and Use of the Internet of Things to Minimize System Losses. In Advanced Technology for Smart Environment and Energy (pp. 49–60). Cham: Springer International Publishing.
23. Air PACA. (2015). Air quality assessment of the municipality of Saint-Saturnin-Lès-Avignon – 25 March. page 9.
24. Rachiq, T., Abrouki, Y., Mabrouki, J., Samghouli, N., Fersi, C., Rahal, S., & El Hajjaji, S. (2021). Evaluation of the efficiency of different materials to remove specific pollutants from landfill leachate. Desalination and Water Treatment, 238, 240–250.
25. INRETS-LEN. (1990). Unit emissions of pollutants from light-duty vehicles. Report 116, Bron, 120 pages, 65 pages.
26. INRETS. (1995). Evolution of pollutant emissions from passenger cars during cold engine start. 1 sheet. Report resultsheet197.
27. OECD. (1995). Motor vehicle pollution - Reduction strategies beyond 2010, The Editions of OECD, Paris, 148 pages
28. Loukili, H., Mabrouki, J., Anouzla, A., Kouzi, Y., Younssi, S. A., Digua, K., & Abrouki, Y. (2021). Pre-treated Moroccan natural clays: Application to the wastewater treatment of textile industry. Desalination and Water Treatment, 240, 124–136.
29. Schutz, M. (1995). Environment and pollution how to measure? How to react? 13–86 pages.
30. CITEPA. (1996). Estimation of atmospheric pollutant emissions in France. 15–17–18–20–22 pages.
31. El Alouani, M., Aouan, B., Rachdi, Y., Alehyen, S., El Herradi, E. H., Saufi, H., & Barka, N. (2024). Porous geopolymers as innovative adsorbents for the removal of organic and inorganic hazardous substances: A mini-review. International Journal of Environmental Analytical Chemistry, 104(16), 4784–4796.
32. Khedidji, S., Ladji, R., & Yassaa, N. (2013). A wintertime study of polycyclic aromatic hydrocarbons (PAHs) in indoor and outdoor air in a big student residence in Algiers, Algeria. Environmental Science and Pollution Research, 20(7), 4906–4919.
33. Mabrouki, J., Azrour, M., Farhaoui, Y., & El Hajjaji, S. (2020). Intelligent System for Monitoring and Detecting Water Quality. In Big Data and Networks Technologies 3 (pp. 172–182). Springer International Publishing.
34. Mabrouki, J., Benchrifa, M., Ennouhi, M., Azoulay, K., Bencheikh, I., Rachiq, T., & El Hajjaji, S. (2022, November). Geographic Information System for the Study of Water Resources in Chaâba El Hamra, Mohammedia (Morocco). In The International Conference on Artificial Intelligence and Smart Environment (pp. 469–474). Cham: Springer International Publishing.
35. Padilla, C., Deguen, S., Lalloué, B., Blanchard, O., Beaugard, C., Troude, F., Zmirou-Navier, D., & Vieira, V. M. (2013). Cluster analysis of social and environment inequalities of infant mortality: A spatial study in small areas revealed by local disease mapping in France. Science of the Total Environment, 454–5, 433–441.
36. Mabrouki, J., Benbouzid, M., Dhiba, D., & El Hajjaji, S. (2022). Simulation of wastewater treatment processes with bioreactor membrane reactor (MBR) treatment versus conventional the adsorbent layer-based filtration system (LAFS). International Journal of Environmental Analytical Chemistry, 102(19), 7458–7468.
37. Benchrifa, M., & Mabrouki, J. (2022). Simulation, sizing, economic evaluation and environmental impact assessment of a photovoltaic power plant for the electrification of an establishment. Advances in Building Energy Research, 16(6), 736–753.
38. Pascal, L., Pascal, M., Stempfelet, M., Goria, S., & Declercq, C. (2013). Ecological study on hospitalizations for cancer, cardiovascular and respiratory diseases in the industrial area of Etang-de-Berre in the South of France. Journal of Environmental and Public Health, 2013, article IF 328737.
39. Wroblewski, A., Riffault, V., et al. (2007). Adaptation des plans d'échantillonnage aux objectifs des compagnes d'échantillonnage spatial. Guide de recommandation, Ecole d'Ingénieur Centre deRecherche, Mines de Douai.

33 Smart Farming
Revolutionizing Aquaponics and Ecosystems with Tiny ML Innovations

K. El Azzouzi, A. Hsain, K. Chawqi, and Y. Chtouki

33.1 INTRODUCTION

In the past few years, Morocco has been facing climatic challenges in different aspects: water scarcity, extreme temperature variations, and flow in its ecosystem became a threat to farmers. The filling rate in February 2022 stood at 33%, with water reserves equivalent to 5.3 billion cubic meters. In 2022, cereal production went down 60% from average, due largely to the drought [1]. As a "cherry on the cake", the outbreaks of COVID-19 had worsened the economic situation and directly affected farming production and agricultural yields. In this context, researchers suggested alternative solutions, among them aquaponics, recently embraced by some Moroccan farmers. Aquaponics is a sustainable farming system that combines aquaculture and hydroponics, also known as soilless culture, in a closed water recirculating system [1]. Fish effluents are converted by bacteria into fertilizers; water is then filtered and returned to the fish tank. This system was first developed by [2], from Auburn University, Alabama, in the 1970s. Aquaponics uses 90% less water than traditional agriculture, is highly productive, pesticide-free, and offers a promising solution for sustainable food production. However, this type of farming is known for its complexity; farmers struggle in monitoring water's chemical balances and so on. This project aims to develop smart aquaponics that are easy to use and can be adopted by Moroccan farmers by implementing censoring and Tiny ML devices, and creating easy tracking and monitoring solution.

33.2 THE ROLE OF AQUAPONICS IN SUSTAINABLE AGRICULTURE

33.2.1 Definition and Overview of Aquaponics

Aquaponics is a constructed ecosystem with the idea of growing crops with fish in a closed loop system [3] developed it based on combining RAS, short for recirculation aquaculture system, which is a solution for fish farming due to its low water consumption by aquaculture, with hydroponics. A hydroponic system is a biological plant-based system that purifies water before it is sent back to the fish tanks in a quasi-closed cycle, saving a considerable amount of water, 90%, as previously mentioned.

33.2.2 Sustainable Agricultural Solutions: Advantages of Aquaponics in Today's Farming

In the context of sustainable food production in a world where climate change and water scarcity are a menace, aquaponics holds a significant role for an improved food security. In an ecological level, aquaponics requires 10% of renewed water and less land area compared to traditional culture for a similar yield. According to a study by Rakocy (2006), the need for synthetic fertilizers is

DOI: 10.1201/9781003647263-33

minimized, which reduces the risk of nutrient runoff into surrounding ecosystems, mitigating surface and underground water pollution, by including rainwater harvesting and water recycling mechanisms in its functionality. To limit its impact on the local ecosystem, aquaponics prioritizes the use of native and non-invasive species. In the region of Ifrane in Morocco, Al Akhawayn University, where our study is conducted, we chose trout fish and lettuce for their efficient overlap and abundance in the region.

Accessibility to a healthier, more sustainable way of life is often expensive. A study conducted at FreshDirect revealed that organic food tends to cost approximately 22% more than non-organic food. To address this issue, aquaponics gives the opportunity to produce healthful, sustainable, diverse, and commercially viable seafood and terrestrial food that is both completely organic and affordable.

In a societal level, and in the region of Morocco in particular, the implementation of aquaponics encourages societal development by allowing farmers, local consumers, and other stakeholders to form strong ties and foster a sense of community around a common project.

33.2.3 Enhancing Aquaculture and Agriculture Through Technology

Enhancing aquaculture and agriculture mainly involves maintaining the plants, fish, and nutrients in an aquaponics system in a way that creates a harmonious and interdependent interaction is essential to its success. Integrated farming recycles the leftovers and by-products of one type of cultivation for another. Closing the loop between crops and animals is consequently shown to be one of the best strategies to increase fertilizer and water efficiency while minimizing waste [3, 4]. Therefore, to guarantee a consistent supply of food, it is necessary to investigate improved fresh food supply potential, modern technology, and integrated systems. Considering that it conserves water, solves soil issues, and addresses other issues with conventional farming, aquaponics presents itself as a potentially sustainable method in this context [2].

33.2.4 Enhancing Beneficial Water Quality in Aquaponics for Enhancing Plant and Fish Growth

Numerous factors contribute to pollution in the aquatic. Ecosystem and the eventual extinction of aquatic organisms. These include a fast-growing population, industrialization, urbanization, deforestation, environmental carelessness, lack of awareness, and inconsistent application of policies and regulations. These are all factors in contaminating water sources. Urbanization and population growth are linked to increased pollution loads for aquatic ecosystems [5, 6]. These contaminants change physicochemical, hydrological, and faunal features, creating serious risks to aquatic ecosystems [6]. The most dangerous issues facing emerging countries today are related to improper handling of the enormous amounts of waste material created by different human activities. The most difficult element is that these waste products are being wrongfully disposed of in the open air. Freshwater bodies, such as rivers and reservoirs, are most impacted by these activities and are rendered unsuitable for primary and secondary use as natural resources [7].

33.2.5 Tiny ML: Elevating Quality in Aquaponics and Natural Waters

Given the aforementioned, aquaponic systems and natural waters today have different fish and plant quality conditions. Fish and plant quality may be maximized in aquaponics where conditions are regulated by carefully controlling elements like nutrition, water quality, and environmental parameters. This frequently yields premium products with minimal impact on the environment. However, problems like nutrient imbalances and disease outbreaks might happen, requiring proactive control.

Numerous factors, such as habitat deterioration, pollution, overfishing, and climate change, have an impact on the quality of fish and plants in natural waters. Hence, the health and endurance of plant and fish populations in both aquatic ecosystems depend on efforts to protect and restore natural habitats. Thus, applying tiny machine learning (Tiny ML) to aquaponics and ecosystems presents a novel way to improve fish and plant quality, address environmental problems, and promote sustainable practices.

33.3 THE ADVENT OF SMART AQUAPONICS/THE ROLE OF TINY ML IN AQUAPONICS AND ECOSYSTEMS

33.3.1 Advantages of Smart Aquaponics

Smart aquaponics excels conventional hydroponics in water efficiency while lowering the need for water quality monitoring through the mutually beneficial convergence of hydroponics and aquaculture. The use of medium beds eliminates the requirement for soil in plant growth because they are also an oasis for bacteria that fix nitrogen. These bacteria convert fish waste into nutrients, which lowers the requirement for extra nutrients. The technology requires less chemicals, does not require cleaning up, and is less prone to disease than hydroponics. Its versatility allows for year-round regulated production and installation in a range of locations. This strategy provides an efficient and ecologically beneficial agricultural alternative for producers who care about the environment and wish to adopt sustainable farming methods. Consequently, the benefits of smart aquaponics include lower maintenance needs, preservation of resources, and a seamless shift from fish to plant culture [8].

33.3.1.1 Real-Time Monitoring and Management

In order to monitor vital parameters, AI algorithms may examine real-time data from IoT devices. characteristics examined include nutrient levels, plant growth indices, fish health, and water quality [9, 10, 11]. This makes it possible to identify any problems early on, such as imbalances, pollution, or disease outbreaks, and to take preventative action to avoid disruptions and preserve high system performance [12].

33.3.1.2 Predictive Analytics for Optimal Growth Conditions

Aquaponics is being revolutionized by predictive analytics, which uses real-time data to optimize plant and fish environments. Through the regulation of temperature, bacteria, nutrients, and water quality, this technology promotes healthier ecosystems and more effective growth [8].

- Water Quality Monitoring: The need of preserving water quality is underscored for the well-being of fish and vegetation equally. As the necessity of monitoring dissolved oxygen concentration, flushing chlorine out of water sources, and testing pH levels. With the use of real-time sensor data, predictive analytics can proactively modify environmental factors to maximize fish and plant development and predict changes in water quality.
- Temperature Control: In aquaponics systems, it is important to monitor the temperature of the water as well as the air. This emphasizes the significance of keeping fish and microorganisms at optimal temperatures. In order to anticipate how changes in environmental circumstances would affect plant and fish growth, predictive analytics can examine temperature trends over time. This enables proactive modifications to be performed.
- Bacteria Management: In aquaponics systems, bacteria are essential because they transform ammonia into nitrates, which are vital nutrients for plants. The conditions required for nitrifying bacteria to survive and function well are explained in the book. By tracking variables like temperature, pH, and oxygen saturation, predictive analytics can forecast bacterial health and activity, maximizing nutrient conversion for plant growth.

- Nutrient Management: The way that bacteria and fish waste work together to provide nutrients to plants. Usually, it is advised to start with plants that have lesser nutrient requirements because the system's nutrient levels may take some time to build up. Better planning for crop selection and rotation could be made possible by predictive analytics, which could measure nutrient levels over time and forecast when they would reach optimal levels for different plant species.

33.3.2 Added Value of Tiny ML in Aquaponics

33.3.2.1 Enhanced Efficiency and Productivity

Growing in popularity as a sustainable method of producing food, aquaponics uses a closed-loop system to grow both plants and fish. Soil or other environmental elements are not necessary for aquaponics to function. It can save up to 90–95% of water and fertilize plants with fish waste. Another innovative method for producing food is aquaponics, which has certain difficulties but is also very promising. It is a complicated ecosystem that calls for interdisciplinary expertise, careful monitoring of all important factors, and substantial initial and ongoing maintenance costs to establish the system. The Internet of Things (IoT) and artificial intelligence (AI) are important technologies that can help us overcome these obstacles. As a result, the application of AI and the IoT to automate processes, boost productivity and dependability, offer better management, and lower operational costs has been the subject of numerous recent research [13].

33.3.2.2 Potential Ways in Which the Tiny ML Results in an Improved Quality of Fish and Plants

Aquaponics is undergoing a revolution due to the IoT and AI integration, which is also improving productivity, sustainability, and system efficiency. When these technologies come together, operators may use real-time data to optimize performance, automate control, and make well-informed decisions. This collaboration makes it possible to have remote management capabilities, predictive analytics for early problem detection, and continuous monitoring, all of which help to guarantee optimal system performance [12].

Advanced aquaponic parameter monitoring and management, including nutrient levels, fish health, and water quality, is made possible by AI and IoT technologies. IoT sensor data in real-time enables the early detection of possible problems, leading to automated corrections to preserve ideal circumstances. AI-powered predictive maintenance avoids system outages by predicting component failures, and optimal nutrient management maintains balanced water chemistry for the health of fish and plants [14–16].

One more important benefit associated with combining AI and IoT is the optimization of fish health and feeding procedures. Precise feeding plans are informed by continuous monitoring of fish behavior and water quality using sensors and machine vision, which lowers waste and fosters a healthy aquatic ecosystem. AI-driven fish growth and feeding data analysis allow for a dynamic feed plan and formulation modification, assuring economical and effective use of resources [17, 18].

33.3.3 Overview of Existing Smart Aquaponic Systems

Smart aquaponics systems incorporate IoT technologies and data analytics to enhance hydroponics and aquaculture and facilitate efficient, sustainable food production. These systems provide climate control, automatic feeding, and real-time monitoring to optimum conditions for aquatic life and plants. Below is an overview of an existing smart aquaponic system that combines both the integration of IoT technologies and machine learning within an aquaponics system. The following technique uses an offline-first edge and fog computing approach to promote data privacy and reduce dependence on the cloud for an IoT aquaponics system. It features a ML model that monitors and sounds for anomalies in the aquaponics water pump's operation. Students at the St Peters Mbare IoT Makerspace developed this model, which uses the Edge Impulse platform and the Keras Neural Network library to compare the sounds of a pump with and without electricity. Hence, if an anomaly is discovered, the owner is informed. This approach minimizes bandwidth consumption, safeguards data privacy, and only moves data to the cloud, when necessary, by processing it locally. It is an illustration of a community-driven strategy for food security that makes use of safe technology to maintain the dependability and effectiveness of the system [19].

33.3.4 Limitations and Challenges

The technique mentioned above might have issues with accuracy due to background noise and acquiring real-time global data updates since it relies on local processing. There are technological and economical obstacles to establishing edge computing infrastructures, particularly in underserved areas. Furthermore, one of the important concerns is keeping the system reliable. Also, environmental problems arise from the energy requirements of edge and fog computing and the potential for greater electronic waste due to required upgrades or hardware failures. These characteristics demonstrate how important it is to consider ecologically friendly design and disposal methods in order to manage the environmental impact of such modern systems and ensure that technology and ecology both benefit from them.

33.4 CASE STUDY: ADDRESSING AQUATIC CHALLENGES WITH TINY ML

By utilizing the available data, Tiny ML enables intelligent decision making at the device level, enabling offline real-time responses to changes in the aquaponics environment. This reduces the amount of data that is processed and transmitted over the cloud, enhancing system dependability, and potentially reducing operational costs.

33.4.1 Overview on the Current Prototype We Have

The Ifrane Eco-village Aquaponics prototype is a soilless food production system suitable for cold climates that combines fish management and plant cultivation. It makes use of a 15×7 m^2 greenhouse with a $1.6 \times 1.5 \times 0.8$ m^3 fish tank and a $7.5 \times 1.2 \times 0.4$ m^3 plant unit with water recycling. Water level and temperature are monitored by sensors that are managed by an Arduino UNO for separate uses. Yet, there are certain limitations to the system, such as the potential for environmental factors to lead to inaccurate sensor measurements, the need for constant electricity for the water pump, and the necessity of manual data processing. Therefore, we are looking at Tiny ML enhancements to automate and enhance data processing to solve these problems and boost productivity. Finally, we aim to solve these issues with food security and the environment.

33.4.2 Data Collection

a. Usage of Tiny ML: Our enhanced aquaponics system employs two Tiny ML models, specifically optimized for microcontrollers, with a focus on the Arduino Nano 3 platform.
b. First Model: Anomaly Detection: The inaugural model functions as an anomaly detector, leveraging time series data for variables such as temperature, pH, turbidity, dissolved oxygen, ammonia, and nitrate. It predicts the forthcoming state of the aquaponics system after 10 minutes, utilizing data from the preceding 30 minutes. This predictive capability facilitates the identification of parameters deviating from normal ranges. In instances of anticipated anomalies or malfunctions, the system proactively notifies the user, thereby enhancing monitoring efficacy and response timeliness.

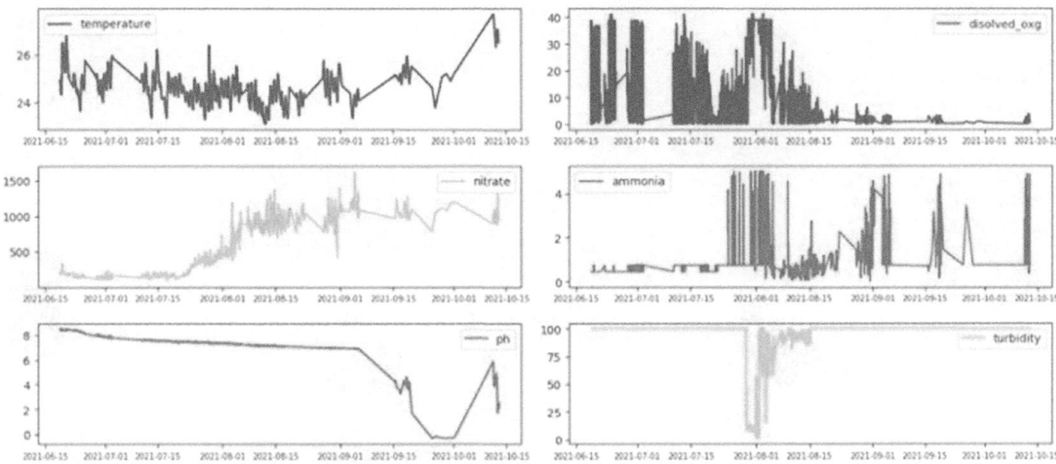

FIGURE 33.1 Visualizing the first pond from the data set.

c. Second Model: Growth Prediction: A subsequent model, utilizing the same environmental features, forecasts the expected average length and weight of fish within the system. This metric serves as an indicator of the system's health. Users receive weekly updates on fish growth projections, with alerts issued for any detected deterioration in health metrics.
d. Model Development and Deployment: Both models are developed using compact fully connected neural networks (Figure 33.1), employing Keras and Edge Impulse, and are deployed on an Arduino Nano. The models are trained on the data set "Sensor Based Aquaponics Fishpond Datasets" [20] comprising 11 CSV files from various aquaponic ponds, detailing relevant data and features. Given the data set's complexity, preliminary data cleaning and preprocessing are requisite steps prior to model training.
e. Hardware Specifications: The deployment hardware of choice is an Arduino Nano 33 BLE Sense, selected for its capability to run Tiny ML models locally [21, 22], thus obviating the necessity for continuous cloud connectivity or data transfer and storage costs. The Arduino Nano 33 BLE Sense's memory capacity is 256 KB SRAM, 1 MB flash, dictating that our neural networks should not surpass at max 100 K parameters, assuming the use of 16-bit floating-point numbers.
f. Operational Workflow: Predictive analyses occur every 10 minutes, based on data collected over the previous 30 minutes. Should predicted values fall outside predefined ranges for three consecutive predictions, an alert is dispatched to the user. This workflow ensures timely and effective anomaly detection and intervention without having many false alerts.
g. Data Processing and Retraining: Data collection occurs at designated intervals, feeding into the neural network for predictive analysis (Figure 33.2). Upon detection of a potential anomaly or a significant deviation in fish growth metrics, the system alerts the owner for manual verification. Subsequent to owner verification, actual data can be reintegrated into the system for model retraining or fine-tuning, enhancing overall system performance.
h. Evaluation Metrics: For the anomaly detection model, both precision and recall are deemed critical; hence, the F1 Score is utilized to comprehensively evaluate performance. The second model, focused on growth prediction, is assessed using mean square error, aligning with application requirements.

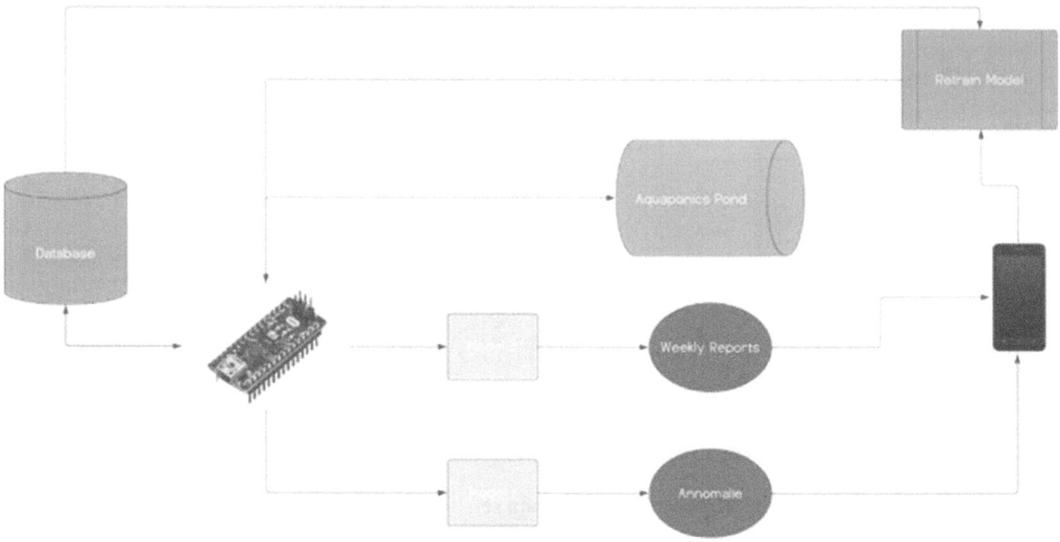

FIGURE 33.2 Data flow of the aquaponics system.

33.5 CONCLUSION

In the light of what has been said so far, the integration of Tiny ML into aquaponics is a step towards sustainable food production, as it addresses challenges in both hydroponics and aquaculture. The Ecovillage Aquaponics system in Ifrane, which integrates these two methods inside an IoT framework and leverages local edge, fog computing, and Tiny ML is a concrete example to the potential of community-driven, sustainable projects. Despite technological difficulties and environmental concerns, such as sensor mistakes and the energy requirements of computing infrastructure, the application of Tiny ML and smart technologies enables increased operational efficiency, water conservation, and a reduction in manual monitoring. With further advancements, this intelligent technology may become much more a part of bigger ecosystems, such as rivers and seas. We could look forward to a period when real-time aquatic environment monitoring and management leads to improved fish and plant quality, more efficient nutrient management, and healthier ecosystems through the use of Tiny ML. This proactive approach has the ability to preserve and restore our natural water bodies in addition to laying the groundwork for successful aquaponics systems. With additional technology advancements and careful policy implications, Tiny ML has the potential to transform sustainable aquaculture and ecosystem management, making it an essential part of the hunt for a sustainable and balanced relationship with the world's resources.

33.6 FUTURE WORK

In the present paper, we have introduced a conceptual framework for a Tiny ML-based system aimed at enhancing aquaponics monitoring. Future work will delve deeper into these aspects, focusing on achieving a comprehensive understanding of the system's practical implications. Future work will prioritize the realization of a physical implementation of the proposed Tiny ML-based system within an aquaponics environment. This entails the integration of Tiny ML-enabled hardware and sensors into existing aquaponics setups, ensuring seamless compatibility and functionality within the operational context. Furthermore, subsequent research will be dedicated to the rigorous evaluation of the performance of ML models integrated into the system. Through systematic experimentation and data analysis, the efficacy and reliability of the ML algorithms in accurately monitoring and optimizing aquaponic parameters will be thoroughly assessed. An essential aspect of future investigations will involve conducting a comparative analysis between the proposed Tiny ML-based system and conventional monitoring approaches prevalent in aquaponics systems. By juxtaposing the performance, efficiency, and practicality of the proposed system against traditional methodologies, valuable insights into the transformative potential of Tiny ML in aquaponics management will be gleaned.

REFERENCES

1. Aquaponics Italia, "Aquaponics: History and Benefits," Acquacoltura Italia, pp. 1–5. Available: www.acquacolturaitalia.com.
2. E. Joshi et al., "Aquaponics: An Innovative Sustainable Food Production Farming System," pp. 73–78, January 2022. Available: https://www.researchgate.net/publication/358021106
3. R. W. Marklin et al, "Aquaponics: A Sustainable Food Production System That Provides Research Projects for Undergraduate Engineering Students," Mechanical Engineering Faculty Research Publications, vol. 12, 2013.
4. R. P. Mishra, S. K. Das, N. Kumar, and J. P. Mishra, "Sustainable Aquaponics System and Its Challenges: A Review," International Journal of Current Microbiology and Applied Sciences, vol. 9, no. 5, pp. 577–588, 2020.
5. C. A. Edokpayi, and J. A. Nkwoji, "Annual Changes in the Physicochemical and Microbenthic Invertebrate Characteristics of the Lagos Lagoon Sewage Dump Site at Iddo, Southern Nigeria," Ecology, Environment and Conservation, vol. 13, pp. 13–18, 2000.
6. J. A. Nkwoji, A. Yakub, G. E. Ajani, K. J. Balogun, K. O. Renner, J. K. Igbo, A. A. Ariyo, and B. O. Bello, "Seasonal Variations in the Water Chemistry and Benthic Macroinvertebrates of a South Western Lagoon, Lagos, Nigeria," Journal of Animal Science, vol. 6, pp. 85–92, 2010.
7. D. S. Malik, A. K. Sharma, A. K. Sharma, R. Thakur, and M. Sharma, "A Review on Impact of Water Pollution on Freshwater Fish Species and Their Aquatic Environment," in Advances in Environmental Pollution Management: Wastewater Impacts and Treatment Technologies, vol. 1, V. Kumar, N. Kamboj, T. Payum, J. Singh, and P. Kumar, Eds. December 2020, pp. 10–28.
8. J. Underwood, and B. Dunn, "Aquaponics," Oklahoma Cooperative Extension Service, HLA-6721, Jan. 2016.
9. D. Karimanzira, and T. Rauschenbach, "An Intelligent Management System for Aquaponics," At-Automatisierungstechnik, vol. 69, pp. 345–350, 2021.
10. R. Murakami, and H. Yamamoto, "Growth Estimation Sensor Network System for Aquaponics using Multiple Types of Depth Cameras," in *Proc. of the 2022 International Conference on Artificial Intelligence in Information and Communication (ICAIIC)*, Jeju Island, Republic of Korea, Feb. 21–24, 2022, pp. 33–38.
11. Cameras. *Proceedings of the 2022 International Conference on Artificial Intelligence in Information and Communication (ICAIIC)*, Jeju Island, Republic of Korea, 21–24 February 2022, pp. 33–38.
12. A. A. Channa, K. Munir, M. Hansen, and M. F. Tariq, "Optimisation of Small-Scale Aquaponics Systems Using Artificial Intelligence and the IoT: Current Status, Challenges, and Opportunities," Encyclopedia, vol. 4, pp. 313–336, Feb. 2024. Available: https://doi.org/10.3390/encyclopedia4010023
13. A. Mandal, and A. R. Ghosh, "Role of Artificial Intelligence (AI) in Fish Growth and Health Status Monitoring: A Review on Sustainable Aquaculture," Aquaculture International, vol. 10, pp. 1–30, 2023.
14. J. John, and P. R. Mahalingam, "Automated Fish Feed Detection in IoT Based Aquaponics System," in *Proceedings of the 2021 8th International Conference on Smart Computing and Communications (ICSCC)*, Kochi, Kerala, India, 1–3 July 2021, pp. 286–290.
15. R. Murakami, and H. Yamamoto, "Growth Estimation Sensor Network System for Aquaponics using Multiple Types of Depth Cameras," in *Proceedings of the 2022 International Conference on Artificial Intelligence in Information and Communication (ICAIIC)*, Jeju Island, Republic of Korea, 21–24 February 2022; pp. 033–038.
16. M. F. Taha, et al, "Using Deep Convolutional Neural Network for Image-Based Diagnosis of Nutrient Deficiencies in Plants Grown in Aquaponics," Chemosensors, vol. 10, p. 45, 2022.
17. Z. Jie Ong, A. Keong Ng, and T. Ya Kyaw, "Intelligent Outdoor Aquaponics with Automated Grow Lights and Internet of Things," in *Proceedings of the 2019 IEEE International Conference on Mechatronics and Automation (ICMA)*, Tianjin, China, 4–7 August 2019; pp. 1778–1783.

18. "Water Quality in Aquaponics—Go Green Aquaponics," Available: https://www.gogreenaquaponics.com/blogs/news/water-quality-in-aquaponics (accessed on 5 February 2024).
19. F. Ghizlane, J. Mabrouki, F. Ghrissi, and M. Azrour, "Proposal for a High-Resolution Particulate Matter (PM10 and PM2. 5) Capture System, Comparable With Hybrid System-Based Internet of Things: Case of Quarries in the Western Rif, Morocco," Pollution, vol. 8, no. 1, pp. 169–180, 2022.
20. B. Ogbuokiri, "Sensor-Based Aquaponics Fish Pond Datasets," Kaggle, Available: https://www.kaggle.com/datasets/ogbuokiriblessing/sensor-based-aquaponics-fish-pond-datasets.
21. Arduino, "Arduino Nano 33 BLE Sense," Arduino Store, Available: https://store.arduino.cc/products/arduino-nano-33-ble-sense?selectedStore=eu
22. J. Rakocy, "The Godfather of Aquaponics: James Rakocy Creating Ecosystems," The Fish Site, [Online], Available: https://thefishsite.com/articles/the-godfather-of-aquaponics-james-rakocy-creating-ecosystems

34 Application of Deep Learning in Mammography Image Classification for Breast Cancer Prevention

A. Saydtahiri, H. Anas, A. Elmajidi, M. A. Bourkha, and J. Hicham

34.1 INTRODUCTION

Subsequent paragraphs, however, are indented. Worldwide, cancer has been one of the main causes of death for many years. Breast cancer is the most common form of cancer in women, with 10,136 new cases diagnosed in Morocco in 2018, and this figure is set to rise to 15,000 by 2040, according to forecasts by the Greater Casablanca Cancer Registry 2013–2017. In this context, early and accurate diagnosis plays a crucial role in improving prognosis. As early detection visibly reduces the fatal consequences, Morocco has undertaken to launch national cancer prevention and control plans 2020–2029, as well as implementing a number of programs in the package of health activities in care establishments. However, mammography has been proven to be one of the most effective and widely recognized screening tools, with observational studies showing a reduction in mortality of around 40% after mammography screening (1). According to the North American Society of Radiology (RSNA) (2), with the technological progression of image acquisition systems, the number of images produced has increased rapidly. Consequently, there is a growing need to standardize the interpretation of images by humans and explore new approaches to our work. The emergence of algorithms based on Deep Learning is notable in the field of health, and according to the RSNA, these systems could provide a partial response to this problem. And as Currie et al. (3), if deep learning is the engine that will drive forward the future of technology, data is the essence on which it relies. The algorithms used indeed demand a vast amount of data to function effectively. The RSNA claims that deep learning algorithms are able to assist the image interpretation process, speeding up the diagnostic process, reducing diagnostic errors and improving quantitative evaluation (4). Deep learning, as one of the most important and widely used branches of machine learning, is grounded in learning and experience acquisition. Recently, Trister et al. report that new algorithms based on convolutional neural networks (CNNs) improve mammography performance and increase the efficiency of breast imaging (5). Deep learning shows promise in a wide variety of applications, including tumor classification, segmentation and survival prediction. Automated analysis tools based on deep learning can significantly enhance diagnostic accuracy, especially in cases involving subtle pathological changes that may go undetected through visual assessment. Although deep learning has been the subject of extensive study for many years, significant breakthroughs in the field of medical image processing have emerged recently, making this discipline a fast-growing field of research. Substantial efforts have been made in research into computer aided design (CAD)-based systems for automated breast cancer detection (6). Kulkarni et al. have used a deep learning architecture based on U-Net, proposed for the detection and classification of benign and malignant breast masses, on two databases, INbreast and DDSM, achieving a sensitivity of 99.64% at 0.25 false positives per image for the INbreast data set, while the sensitivity of DDSM

DOI: 10.1201/9781003647263-34

is 97.36% and 0.38 FPs/I (7). Another study for Pesapane et al. suggested a data set included 1,000 patients aged 21–73 years, with 389 clusters of malignant and 611 benign microcalcifications, the selected patient data are heterogeneous, including women with different breast densities (8). Three expert radiologists annotated a mammography data set in order to locate microcalcifications in the image in a breast imaging department of a national reference center for breast cancer care. Three neural networks (Alex-Net, Res-Net18 and Res-Net34) were trained and evaluated using specific measures. Alex-Net performed best with a sensitivity of 0.85, specificity of 0.89 and AUC of 0.94 for microcalcification classification.

Meanwhile, researchers have used ensemble learning methods in breast cancer detection to improve classification performance. Adyasha et al. have worked on an set classifier based on deep learning is proposed for breast cancer detection (9), by integrating three transfer learning models (Alex-Net, Res-Net and Mobile-NetV2) using residual learning, depth-separable convolution and inverted residual bottleneck structure to make the system faster, as well as skipped connection to facilitate optimization and also, the use of Gaussian Laplacian (LoG) and modified high amplification to improve performance. The results achieve an accuracy of 99.17% for detecting abnormalities and 97.75% for detecting malignancies on the mini-DDSM data set. Keila et al. present an approach for improving the accuracy of cancer screening through the application of deep cross-training (10). The approach was evaluated by classifying mammography images from the INbreast and CBIS-DDSM databases, developing binary and multiclass models based on the Res-Net-18 architecture. The experiments showed that the DCTr approach increased classification accuracy and precision by 39% and 36%, respectively. In this work for Saber et al., tumor regions are automatically extracted using segmentation techniques prior to the learning process in order to reduce computation time (11). Image quality can be improved and segmentation results can become more accurate through noise removal, histogram equalization and morphological analysis before segmentation. For the proposed model, features are extracted from the Mammographic Image Analysis-Society (MIAS) data set using a pre-trained CNN architecture such as Inception V3, Res-net50, VGG-19, VGG-16 and Inception-V2 Res-Net. Six evaluation parameters were chosen to assess the performance of the proposed model in terms of accuracy, sensitivity, specificity, precision, F-score and area under the ROC curve (AUC). The experimental results showed that the TL of the VGG16 model is powerful for breast cancer diagnosis by classifying mammographic breast images with accuracy, of 98.96%.The study of Li Hua et al. proposes a two-view mammography classification model, composed of a CNN and a recurrent neural network (RNN), which is used to classify benign and malignant breast masses (12). The model is composed of two branch networks, and two modified ResNets are used to extract breast mass features from mammograms from the craniocaudal (CC) and mediolateral oblique (MLO) views, respectively. In order to make efficient use of the spatial relationship of the two-view mammograms, RNN's Grid Recurrent Unit (GRU) structures are used to merge the breast mass features from the two views. The Digital Screening Mammography Database (DDSM) will be used for training. Experimental results show that the classification accuracy, recall and AUC of our method reach 0.947, 0.941 and 0.968, respectively.

34.2 MATERIALS AND METHODS

34.2.1 Data

34.2.1.1 Data Description

The images used in this study were collected inside the Mohammed VI University Hospital in Marrakech. The data set is composed of two classes of images, with the objective of distinguishing between mammographic images indicative of a benign diagnosis and those confirming the presence of malignant tumors. This classification is based on the outcomes of mammography examinations and supplementary ultrasound assessments. The data set underwent a random partition into two segments: 80% allocated for training and the remaining 20% for testing. The data collection process involved the selection of mammography exams conducted between the inception of 2017 and the conclusion of 2019. The data contains 1,800 images, 900 of which belong to the malignant class and 900 to the benign class, taken in two standard views: bilateral craniocaudal (CC) views and mediolateral oblique (MLO) views. The average age of the subjects was 54 years, ranging from 32 to 83 years. Notably, the predominant age range, between 40 and 68, represents 86% of the entire database. Throughout the data collection process, radiological reports incorporated specific terminology concepts, utilizing the BI-RADS (Breast Imaging Reporting and Data System) (13, 14). This system aims to standardize descriptions by proposing an international classification that predicts the severity of images, grouped into seven categories from Bi-rads 0 to Bi-rads 6, with management coded according to BI-RADS group (13) where Bi-rads 3 images predominate.

34.2.1.2 Data Preparation

The image data undergoes several preprocessing steps to enhance classification efficiency and reduce training duration. Initially, the images are cropped to eliminate significant black spaces and remove patient markings and identification labels, potential interferences in the learning process (15).

Subsequently, using the Python Imaging Library (PILLOW), the cropped images are resized to a standardized 160 × 160 pixels. To simplify the database, data normalization is applied by scaling the original image values (ranging from 0 to 255) to a more manageable range between 0 and 1, achieved by dividing them by 255. A data generator is then created to automate the retrieval of data from folders and facilitate its integration into Keras. This generator efficiently loads the necessary data from the source folder, converting it into training data and targets seamlessly. For effective labeling, malignant mammographic images are assigned the label 1, while benign images receive the label 0. To diversify the data set and enhance the network's learning, data augmentation techniques are employed. This involves transformations such as rotations, translations and zooms, implemented through the keras preprocessing Image Data Generator class in Keras. These augmentations contribute to increased accuracy, prevent overfitting and promote better generalization of learning networks (16, 17).

34.2.2 Approach

34.2.2.1 General Overview

The aim of this study is to provide insight into the application of deep learning (DL) for binary classification of mammography images (benign and malignant) using pre-trained neural networks such as Res-net 50, Dense-Net 121 and Inception-Net V3. The data set comprises 1,800 images, with 80% allocated for training and 20% for testing. The initial phase involves collecting, sorting and classifying mammography images, followed by utilizing these data for automated diagnostic purposes. Finally, to assess the performance of the employed pre-trained models, a set of evaluation metrics is applied, facilitating a comparative analysis of the results. All these aspects are comprehensively illustrated in Figure 34.1.

34.2.2.2 Neural Network

Using a pre-trained model instead of using a model from scratch is called transfer learning. Training a neural network from scratch requires considerable data and computing power, whereas using this interface a set of models is created with their pre-trained weights provided. Transfer learning is implemented using network models loaded via the Keras DL library. The neural network

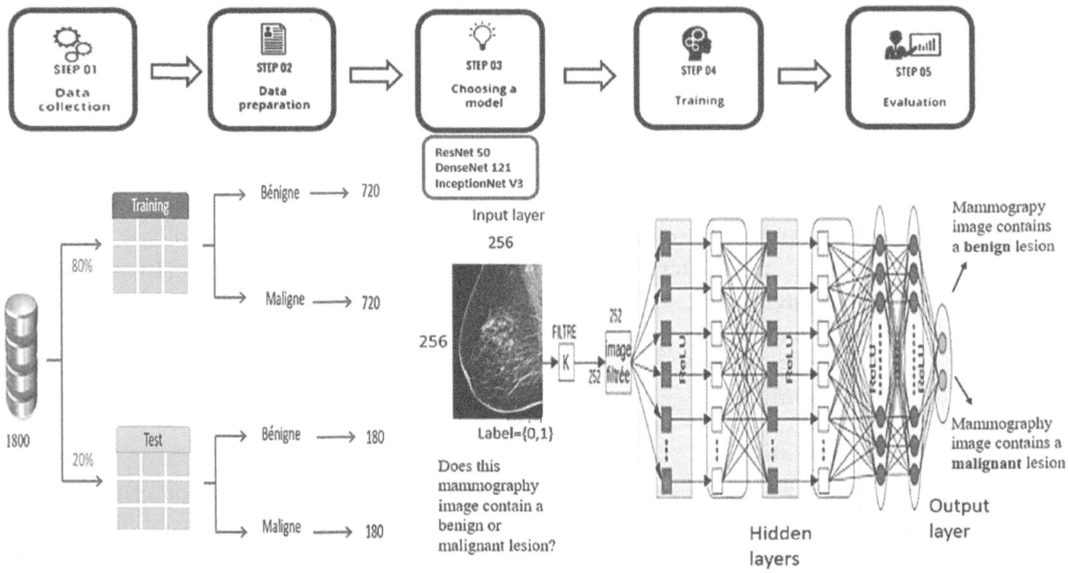

FIGURE 34.1 Flowchart of the proposed method.

architecture is based on the CNN, which is used in image processing and comprises two main layers: the convolutional layer and the pooling layer (18).

Res-Net50 (19): This model contains residual networks, which contain residual blocks linked by shortcut connections, these connections facilitate input to propagate information to future layers, by skipping weighted activations. The skip connection allows researchers to avoid costly computational gradients for these layers during backpropagation, which significantly improves performance. Res-Net50 is a variant of the 50-layer depth model and contains the following element: five convolution layers, a max pooling layer with a step size of 2 and a medium pooling with a fully connected layer containing 1,000 nodes.

Dense-Net-121 (20): The Dense-Net connects each convolutional layer to all other layers retroactively. Dense blocks differ from residual blocks in that they involve tensor concatenation rather than tensor addition. Not only are the feature maps of the previous layer used as input to the next layer, but the feature maps of all previous layers are also used as input. The densely connected convolutional network Dense-Net 121, with 120 convolution layers and four medium pooling layers. Dense-Nets are divided into dense blocks and layers between these blocks are called transition layers. Each dense block has a variable number of repeat layers with two convolutions, and each transition layer has a 1 × 1 convolution layer and a 2 × 2 pooling medium layer.

Inception-Net-V3: Inception V3 is a DL model based on CNNs, an advanced and optimized version of the basic Inception V1 model (21). Inception V3 is a model with a total of 42 layers with a stack of Inception, where the output size of each module is the input size of the following module.

34.3 RESULT AND DISCUSSION

These different evaluation metrics are used to compare the performance of each DL network model:

$$\text{Accuracy} = \frac{\text{TP} + \text{TN}}{TP + FP + FN + TN} \tag{34.1}$$

TABLE 34.1
Overall Performance Results of Proposed Models Applied

Model	Accuracy %	Precision %	Recall %	Score F1 %
Res-Net-50	84.44	84.06	85.00	84.54
Dense-Net-121	79.44	77.04	83.88	80.32
InceptionNet-V3	90.55	90.10	91.11	90.60

$$\text{Precision} = \frac{\text{TP}}{TP + FP} \tag{34.2}$$

$$\text{Recall} = \frac{\text{TP}}{TP + FN} \tag{34.3}$$

$$\text{F1} - \text{Score} = 2.\frac{\text{precision} \times \text{Recall}}{\text{precision} + \text{Recall}} \tag{34.4}$$

According to Table 34.1, the accuracy of the InceptionNet-V3 model is around 91%, which can be considered a good prediction result. It can also be seen that the prediction is more accurate for the InceptionNet-V3 model: 90.10%, compared to the Res-Net-50 and Dense-Net-121 networks. In addition, the Res-Net-50 network achieves better classification accuracy (84.06%) than the Dense-Net-121 classifier (77.04%). The other results obtained from the three approaches are compared, as shown in Table 34.1.

To further explore the results, we proceed with an examination of the confusion matrices derived from the pre-trained models. As illustrated in Figure 34.2a, the Res-Net50 model correctly identified 153 benign cases and 151 malignant images. Similarly, in Figure 34.2b, the Dense-Net 121 model achieved predictions of 151 benign cases and 135 malignant images. Furthermore, Figure 34.2c presents the superior results obtained by the InceptionNet-V3 neural network. The model predicted 164 benign cases, with 16 cases exhibiting poor predictions. For the malignant class, the model correctly predicted 162 cases, with 18 cases eluding accurate identification. During the training of each model, our emphasis is on managing the loss function to prevent both overfitting and underfitting also to optimizing the accuracy curve. For the DenseNet-121 network, the achieved accuracy after 16 epochs stands at 79.44%, as illustrated in Figure 34.3a. Similarly, the Res-Net50 model, when applied to the validation database, achieves an accuracy of 84.44%, as depicted in Figure 34.3b. The accuracy tends to increase with the number of epochs, and after six epochs,

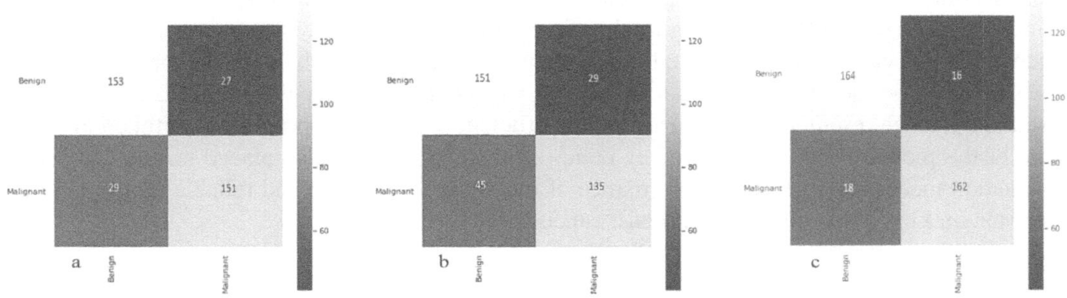

FIGURE 34.2 Confusion matrix of Res-Net50 (a), Dense-Net-121 (b) and InceptionNet-V3 (c).

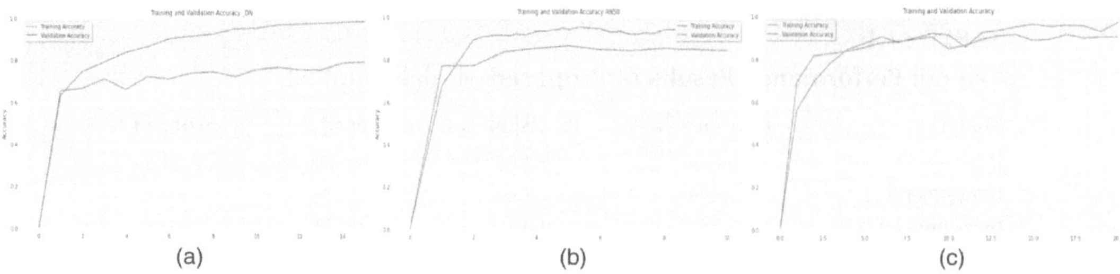

FIGURE 34.3 Accuracy curve (training and testing) of Dense-Net-121 (a), Res-Net-50 (b) and InceptionNet V3 (c).

the curve stabilizes, prompting us to halt training to prevent overfitting. The integration of the ReduceLrOnPlateau function has proven effective in enhancing classifier accuracy. In addition, the InceptionNet-V3 network demonstrates a commendable accuracy score of 90.55% after 20 epochs, as showcased in Figure 34.3c.

Focusing on the InceptionV3 Net model, we note substantial alignment in accuracy, sensitivity and precision metrics. Critical parameters such as learning rate, the number of epochs, batch size, model structure, activation function, optimization method and kernel size impact the model's ability to discern intricate variations within mammography images. Additionally, image quality, artifact absence, appropriate lighting and optimal image contrast emerge as influential factors in model performance. The results obtained from these architectures demonstrate a promise through the control of several parameters when using a Dense Block Architecture composed of several convolution, normalization and activation layers, which promotes a rich and continuous flow of information. It also includes transition layers used to reduce the spatial dimension of features and the global average pooling (GAP) technique at the output of the last dense block to produce a one-dimensional representation that can be used as an output vector. The Residual Block is widely used in various computer vision tasks that facilitate gradient flow during backpropagation and reduce the spatial resolution of features, enabling the strategy of subsampling convolution layers with strides (steps) of 2. The use of the Inception V3 model, known for its good performance, is distinguished by the convolution factorization technique, asymmetric convolutions and dimensional reduction.

34.4 CONCLUSION

This paper explores the growing influence of deep learning, made possible by the increasing power of computers and the abundance of data. The main objective of this study is to provide a practical introduction on the classification of mammography images. This classification is based on a database collected at the Mohammed VI University Hospital Center in Marrakech in Morocco, in a clinical context specific to this hospital establishment. The use of digital tools to automate detection and diagnosis is intended to support radiologists in their decision making. The choice of breast cancer is based on statistics showing its significant impact, particularly in Morocco, where it is the second-leading cause of cancer mortality among women. The approach adopted is based on CNNs, considered to be the most effective for detecting objects in computer vision. Transfer learning is used to optimize resources and increase model efficiency. This work proposes a detailed architecture of the models, highlighting the key components. Emphasis is also placed on the analysis of evaluation metrics to assess the performance of the models as a rapid and reliable tool to aid final decision making in the diagnosis of breast cancer.

REFERENCES

1. Lauby-Secretan B, Scoccianti C, Loomis D, Benbrahim-Tallaa L, Bouvard V, Bianchini F, et al. Breast-cancer screening—Viewpoint of the IARC working group. N Engl J Med. 2015;372(24):2353–8.

2. Van Ginneken B, Schaefer-Prokop CM, Prokop M. Computer-aided diagnosis: How to move from the laboratory to the clinic. Radiology. 2011;261(3):719–32.
3. Currie G, Hawk KE, Rohren E, Vial A, Klein R. Machine learning and deep learning in medical imaging: Intelligent imaging. J Med Imaging Radiat Sci. 2019;50(4):477–87.
4. Li H, Zhuang S, Li D, Zhao J, Ma Y. Benign and malignant classification of mammogram images based on deep learning. Biomed Signal Process Control. 2019;51:347–54.
5. Trister AD, Buist DS, Lee CI. Will machine learning tip the balance in breast cancer screening? JAMA Oncol. 2017;3(11):1463–4.
6. Anom BY. Ethics of big data and artificial intelligence in medicine. Ethics Med Public Health. 2020;15:100568.
7. Kulkarni S, Rabidas R. Fully convolutional network for automated detection and diagnosis of mammographic masses. Multimed Tools Appl. 2023;1–22.
8. Pesapane F, Trentin C, Ferrari F, Signorelli G, Tantrige P, Montesano M et al. Deep learning performance for detection and classification of microcalcifications on mammography. EurRadiol Exp. 2023;7(1):69.
9. Sahu A, Das PK, Meher S. An efficient deep learning scheme to detect breast cancer using mammogram and ultrasound breast images. Biomed Signal Process Control. 2024;87:105377.
10. dos Santos KL, dos Santos Silva MP. deep cross-training: An approach to improve deep neural network classification on mammographic images. Expert Syst Appl. 2024;238:122142.
11. Saber A, Sakr M, Abo-Seida OM, Keshk A, Chen H. A novel deep-learning model for automatic detection and classification of breast cancer using the transfer-learning technique. IEEE Access. 2021;9:71194–209.
12. Li H, Niu J, Li D, Zhang C. Classification of breast mass in two-view mammograms via deep learning. IET Image Process. 2021;15(2):454–67.
13. Jairam MP, Ha R. A review of artificial intelligence in mammography. Clin Imaging. 2022;88:36–44.
14. Shen L End-to-end training for whole image breast cancer diagnosis using an all convolutional design. ArXivPrepr ArXiv171105775. 2017;
15. Arevalo J, González FA, Ramos-Pollán R, Oliveira JL, Lopez MAG. Representation learning for mammography mass lesion classification with convolutional neural networks. Comput Methods Programs Biomed. 2016;127:248–57.
16. Kallenberg M, Petersen K, Nielsen M, Ng AY, Diao P, Igel C et al. Unsupervised deep learning applied to breast density segmentation and mammographic risk scoring. IEEE Trans Med Imaging. 2016;35(5):1322–31.
17. Kooi T, Litjens G, Van Ginneken B, Gubern-Mérida A, Sánchez CI, Mann R et al. Large scale deep learning for computer aided detection of mammographic lesions. Med Image Anal. 2017;35:303–12.
18. Huynh BQ, Li H, Giger ML. Digital mammographic tumor classification using transfer learning from deep convolutional neural networks. J Med Imaging. 2016;3(3):034501–034501.
19. Hasanah SA, Pravitasari AA, Abdullah AS, Yulita IN, Asnawi MH. A deep learning review of ResNet architecture for lung disease identification in CXR image. Appl Sci. 2023;13(24):13111.
20. Liao T, Li L, Ouyang R, Lin X, Lai X, Cheng G et al. Classification of asymmetry in mammography via the DenseNet convolutional neural network. Eur J Radiol Open. 2023;11:100502.
21. Jiang F, Liu H, Yu S, Xie Y. Breast mass lesion classification in mammograms by transfer learning. In: Proceedings of the 5th international conference on bioinformatics and computational biology. 2017. pp. 59–62.

35 Artificial Intelligence to Diagnosis and Measurement of Ocular Strabismus

F. Elkaoumi and M. Machkour

35.1 INTRODUCTION

Strabismus is an ophthalmologic disease in which eyes cannot be aligned at the same location [1–3]. Strabismus is caused by the problems occurring on optic nerve, brain or extraocular muscle [4], it appears in various forms: the affected eye can deviate horizontally, either inward or outward, and can be deviate vertically upward or downward, the horizontal and vertical deviations can appear to be combined in the same patient, and the misalignment may always affect the same eye (unilateral strabismus), or both eyes may be out of alignment in turn (alternating strabismus), it can be intermittent, appearing only when tired, even if good fusion is present at other times.

Strabismus has a serious impact on human life, as it can prevent the brain from fusing the two images received from the two eyes, leading to amblyopia [5], and again, strabismus has been shown to have serious psychosocial consequences for both children and adults [6], it affects personality traits and employment opportunities [7], these adverse consequences include education [8]. In general, the aesthetic aspect of strabismus can be treated at any age; however, the sensory disorders are more serious and can only be treated for a short period, which is why treatment of strabismus should begin as early as possible to ensure the best possible visual acuity [9]. Early treatment of strabismus in childhood can reduce the number of surgical interventions and the risk of developing amblyopia [10], strabismus can be treated in several ways, such as wearing glasses, vision therapy and surgery.

In this article, we present information on the related work on various methods of detection strabismus using artificial intelligence (AI).

35.2 METHOD

To diagnose strabismus, the following exams are performed: visual acuity, eye background, external examination of the eyes and eye movement exam, obtained by means of the Cover test and the Hirschberg test [1]. Most doctors use the Hirschberg test [11] to determine whether the patient has strabismus; all these methods are carried out manually by professional ophthalmologists, who make decisions based on their experience, and so the diagnostic results are subjective, and also despite the importance of strabismus screening, there are only a limited number of ophthalmologists capable of carrying out this task in many countries, such as Morocco [12].

Eye motion research laboratories use ocular trackers or magnetic devices to measure yaws and eye movements, but, despite the precision of these devices, these methods are costly and difficult to use in real-life situations [13].

These methods of detecting strabismus have the following disadvantages:

- Clinical assessment of strabismus, such as the Hirschberg and Krimsky tests, is labor intensive and should be performed by experts [14].
- These methods are often dependent on the skill of the examiner; a qualified observer is required to diagnose and measure strabismus.

DOI: 10.1201/9781003647263-35

- Strabismus measurements can vary within the same patient, even between experienced examiners.
- These methods are very approximate and do not detect microtropis.

These problems can be solved through automatic strabismus detection, and some AI experts have proposed automated solutions. AI has made great progress in various areas of ophthalmology, providing ophthalmologists with effective solutions to diagnose pathologies and obtain the desired results, and sometimes provide appropriate treatment for the diagnosed disease, allowing rapid eye care well developed away from conventional solutions, cheap and able to replace specialists. Compared with other ophthalmic disciplines, strabology has not until recently seen significant developments in the field of AI. In this research we summarize and discuss advances using AI for the detection of strabismus [15].

Before selecting a method to automate strabismus detection and measurement, we reviewed all published research articles on strabismus detection to analyze the results and identify the weaknesses and strengths of each method for strabismus detection, and can define the focus of our research (Table 35.1).

TABLE 35.1
Research into Strabismus Using AI

Author	Date	Method Used
Jolson and colleagues	1992	Jolson and colleagues [5] developed a device called the trophorometer, which is used to measure the position and movement of the eye using computerized image processing to aid the diagnosis of phoria and tropia, the moving window thinning technique was used to detect the edges of the pupil and limbus, and the Hough transformation was applied to locate the pupil.
Chandna and colleagues	2007	Using AI, Chandna and colleagues [13] measured vertical strabismus using measurements taken during the prism coverage test. Deviations are represented in the form of optical powers in prism dioptres, using the graphical representation of strabismus.
Chandna and colleagues	2009	Chandna and colleagues [16] have developed another accessible expert system (StrabNet) based on an artificial neural network architecture using deep learning, with a model based on 10 prism coverage test measurements. StrabNet is a web-based implementation, freely available on the Internet, for discovering and classifying vertical strabismus, and is clearly useful for diagnosing common vertical discrepancies in prism coverage test measurements.
Almeida and colleagues	2011	Dallyson Sousa de Almeida and colleagues [1] presented a methodology for the automatic detection of strabismus in digital images using the Hirschberg test. The work was organized into four stages: detection of the eye region, determination of the precise location of the eyes, localization of the limbus and luminosity and identification of the strabismus.
Khudmat	2013	Khumdat and Phukpattaranont [10] have created a method for automatic detection of strabismus processed by the Hirschberg test in digital images based on the central corneal light reflectance ratio (CCLRR).
Valente and colleagues	2017	Valente and colleagues [17] used large concept models (LCM) to detect the presence and classification of strabismus using digital video of eye movements and the coverage test, which analyzes the movement of the eye when strabismus is present and measures the extent of misalignment.
Lu and colleagues	2018	Lu and colleagues [18] used the DL to propose a method using RF-CNN framework to detect strabismus in a telemedicine setting using photographs of the eyes taken by the patients themselves.
Zheng and colleagues	2021	Zheng and colleagues [19] and colleagues have proposed a deep learning approach to screening for horizontal strabismus in children referable to primary gaze using face photographs from the analysis of the region of interest localized using Faster R-CNN, which is a DL model validated in two different external data sets. According to Zheng, this is the first DL model for screening strabismus in children.

35.3 RESULTS AND DISCUSSIONS

Several studies have used classical machine learning and deep learning on eye movement disorders of face photographs to detect strabismus with different techniques. The attempt to automate the detection of strabismus began in 1992 when Jolson and colleagues [5] developed a device called a trophorometer to measure the position and movement of the eyes. This method did not simplify the examination for strabismus because it required equipment and was more technical and costly given the equipment used at the time. This had no impact on hospital use, but in terms of research, it was an evolution in the 1990s. Since then many techniques have been used to diagnose strabismus, and some authors have suggested the use of photographic and videographic methods to automatically assess strabismus. Chandna and colleagues [13] have used the PCT to diagnose and classify vertical strabismus in order to develop an accessible expert system. The approach adopted is based on a parametric biomechanical model and is not based on real clinical data in the form of a study of strabismus. According to its authors, a diagnostic accuracy of 100% was obtained during the tests. The StrabNet expert system is simple and can be implemented using computer resources accessible to the user without requiring a local programming environment other than a browser. After two years, Chandna and colleagues [16] worked on another project to develop an expert system based on an accessible artificial neural network architecture (StrabNet), which will help in the clinical diagnosis of vertical strabismus. The expert system was found to be highly accurate at diagnosing vertical strabismus. Valente and colleagues [17] used LCM to detect and classify vertical strabismus using digital video of eye movements and the cover test. This methodology achieved 80% of the sensitivity specification, 100% of the specificity requirement and 93.33% of the accuracy requirement. Dallyson Sousa de Almeida and colleagues [1] recommended the use of image processing techniques, geostatistical functions and support vector machines for the automatic detection of strabismus in digital images using the Hirschberg method. A similar study presented by Khumdat and Phukpattaranont [10], which automated screening using the Hirschberg test on clinical photographs of 103 adult patients (age: 14 ± 3 years), the methodology produced results in the range of 94.17% accuracy, 97.23% sensitivity and 73.08% specificity. However, such an algorithm, if used in clinical practice, could trigger a high rate of false negatives leading to amblyopia. In an attempt to promote automated self-testing, Zheng and his colleagues [19] proposed a deep learning algorithm able to detect referable horizontal strabismus in children of images taken during clinical assessment. A total of 7,026 images were used to train the model and 277 images from an independent data set were tested. The algorithm achieved an accuracy of 95%, a better performance than the resident ophthalmologists (accuracy ranging from 81% to 85%). Lu and colleagues [18] proposed an automated strabismus detection system for telemedicine. The system achieved an accuracy of 93.9%, a sensitivity of 93.3% and a specificity of 96.2% for detecting strabismus. All these studies did not provide a solution for accurate detection and measurement, and the studies that obtained good results concentrated on detecting horizontal strabismus, even though vertical strabismus was also present.

35.4 CONCLUSION

Strabismus is an influential disease in human life, and its detection plays an important role in the prognosis and treatment of strabismus. Various traditional exams, such as the Hirschberg and Krimsky test, require a lot of work and must be carried out by experts, despite the relevance of this assessment. These measurements have a low degree of reproducibility, due to examiner-dependent factors. In the longer term, ophthalmologists could benefit from AI-assisted prognosis and disease monitoring that could enable personalized treatments. Non-ophthalmologists could also benefit from automated detection of disorders. AI offers a number of solutions for the detection and screening of strabismus, enabling rapid detection of strabismus, the application can potentially be used as an easy-to-apply tool to reduce time and increase diagnostic accuracy. In the future, we will try to publish established versions of the realization of an automatic diagnostic and classification system with AI for strabismus using the Hirshberg reflex, calculating the corneal light reflex ratio based on image processing.

REFERENCES

1. João Dallyson Sousa de Almeida, Aristófanes Corrêa Silva, Jorge Antonio Meireles Teixeira, Anselmo Cardoso de Paiva, Jorge Antonio Meireles Teixeira, "Computational methodology for automatic detection of strabismus in digital images through Hirschberg test," Elsevier.
2. M.W. Quick, R.G. Boothe, "A photographic technique for measuring horizontal and vertical eye alignment throughout the field of gaze," Investigative Ophthalmology & Visual Science 33 (1) (1992) 234.
3. Jiewei Lu, Zhun Fan, Ce Zheng, Jingan Feng, Longtao Huang, Wenji Li, Erik D. Goodman, "Automated strabismus detection for telemedicine applications," ArXiv, n° %1180902940 Cs (2018).
4. R. P. Rutstein, M. S. Cogen, S. A. Cotter, O. K. M. Daum, J. F. Amos, O. Barry Barresi, K. L. Beebe, O. J. Cavallerano, "Optometric clinical practice guideline care of the patient with strabismus: Esotropia and exotropia," Lindbergh Blvd. St. Louis: American Optometric Association.
5. Jolson et al., "Apparatus for evaluating eye," (1992).
6. Zenghai Chen, Hong Fu, Wai-Lun Lo, Zheru Chi, "Strabismus recognition using eye-tracking data and convolutional neural networks," Journal of Healthcare Engineering (2018) 9.
7. S. R. Hatt, D. A. Leske, P. A. Kirgis, E. A. Bradley, J. M. Holmes, "The effects of strabismus on quality of life in adults," American Journal of Ophthalmology 144 (5) (2007) 643–647.
8. O. Uretmen, S. Egrilmez, S. Kose, K. Pamukcu, C. Akkin, M. Palamar, "Negative social bias against children with strabismus," Acta Ophthalmologica Scandinavica 81 (2) (2003) 138–142.
9. E. T. Cunningham, P. Riordan-Eva, Vaughan & Asbury's general ophthalmology, 18th ed., (2011).
10. N. Khumdat, P. Phukpattaranont, "Development of a Computer System," Biomedical Engineering International Conference (2013)
11. J. B. Eskridge, B. Wick, D. Perrigin, "The Hirschberg test: A double-masked clinical evaluation," American Journal of Optometry and Physiological Optics 65 (9) (1988) 745–750.
12. Robert P. Rutstein, Martin S. Cogen, Susan A. Cotter, Kent M. Daum, Rochelle L. Mozlin, Julie M. Ryan, "Care of the patient with strabismus: Esotropia and exotropia," American Optometric Association.
13. Anthony C. Fisher, Arvind Chandna, Ian P. Cunningham, "The differential diagnosis of vertical strabismus from prism cover test data using an artificially intelligent expert system," Medical & Biological Engineering & Computing 45 (2007) 689–693. https://doi.org/10.1007/s11517-007-0212-z
14. B J Kushner, RY Choi, "The accuracy of experienced strabismologists using the Hirschberg and Krimsky test," Elsevier (1998).
15. Anthony C. Fisher, Arvind Chandna, Ian P. Cunningham, "The differential diagnosis of vertical strabismus from prism cover test data using an artificially intelligent expert system," (2007).
16. Arvind Chandna, Anthony C. Fisher, Ian P. Cunningham, Deborah Stone, Maureen Mitchell, "Pattern recognition of vertical strabismus using an artificial neural network (StrabNet©)," (2009) 131–138 | Received 6 Nov 2008, Accepted 19 Jul 2009, Published online: 10 Dec 2009.
17. Thales Levi Azevedo Valente, João Dallyson Sousa de Almeida, Aristófanes Corrêa Silva, Jorge Antonio Meireles Teixeira, Marcelo Gattass, "Automatic diagnosis of strabismus in digital videos through cover test," Computer Methods and Programs in Biomedicine (2017).
18. Jiewei Lu, Zhun Fan, Ce Zheng, Jingan Feng, Longtao Huang, Wenji Li, Erik D. Goodman, "Automated strabismus detection for telemedicine applications," arxiv, (2018).
19. Ce Zheng, Qian Yao, Jiewei Lu, Xiaolin Xie, Shibin Lin, Zilei Wang, Siyin Wang, Zhun Fan, Tong Qiao, "Detection of referable horizontal strabismus in children's," Translational Vision Science & Technology 10 (2021) 33.

36 An Analysis Evaluation for Spam Detection Using Optimized Machine Learning and Deep Learning Techniques

Applying a New Approach Based on NLP Features Extraction

I. Ennejjai, A. Ariss, J. Mabrouki, and S. Ziti

36.1 INTRODUCTION

Nowadays, the continuous development of the internet and mobile technology has made email and instant communication more practical media for human everyday conversation [1, 2]. Nevertheless, spam, commonly referred to as unwanted emails or messages, is increasingly seen as a serious challenge to the internet basis. Besides, it has badly affected individuals' normal communication, whether in the work or private environments [3–9]. Hence, the high volume of spam poses security risks for network-based computer systems whenever it is used to carry out virus and malware attacks. Accordingly, detecting and processing spam is extremely costly and time consuming, thus greatly diminishing the effectiveness of the system. In fact, numerous works that address the spam filtering problem using machine learning methods previously exist in the literature [10–14]. Text-based spam filtering can be viewed as an instance of text classification which basically entails the allocation of various textual materials to a predefined set of classes. The purpose of a classification scheme is to achieve the classification assignment with a reasonable level of efficiency.

In this paper, we report on the latest research in the area of machine learning for spam email classification. The goal of this work is, firstly, to consider and compare a set of machine learning methods that we have discovered during our study. Secondly, we intend to prepare the data set by performing new feature extraction to deal with this issue. Finally, we will conduct some experiments using different spam features for comparing machine learning methods, followed by a discussion of the outcomes obtained.

The remainder of this paper is organized as follows. The second section presents detailed information about published related works in the literature. Background materials are detailed in the third section. The fourth section illustrates our proposed model. The obtained results are discussed in the fifth section. Finally, the paper is concluded in the last section.

36.2 LITERATURE SURVEY

In this section, we will discuss some proposed works in the literature. As directed in Table 36.1, all proposed works are based on various machine learning algorithms and achieved different accuracies.

A spam detection method with usage forest algorithms and active learning to narrow down the clustering was further developed and reported on by DeBarr and Wechsler [15]. They utilized email

DOI: 10.1201/9781003647263-36

TABLE 36.1
The Accuracy Results of Some Published Works

Ref	Most Used Algorithms	Accuracy (%)
[15]	RF	95.20
[17]	SVM/NB	90.00
[17]	NB/RF/DT	99.30
[20]	hidden Markov	89.00
[18]	CART/REP and LAD tree/NBT	95.10
[21]	SVM/DL	93.00
[22]	ELM/SVM	94.06
[19]	SVM/KNN/DT/LR	92.10

data from RFC 822 [16] and segmented every email message into two groups. Furthermore, on behalf of a training data set, the authors chose a cluster-enabled email collection to tag the actual data and the inverse document frequency (TF/IDF) for all attributes of each email. Once the clustered email prototypes are considered for training, they perform experiments with the use of support vector machine (SVM), random forest (RF), Naive-Bayes (NB) and K-nearest neighbors (KNN) algorithms. According to the findings of experiments, the random forest algorithm is more efficient in classifying data with an overall accuracy of 95.

In 2017, Faris et al. [17] employed the NB network in spam detection. They achieved less than exceptional performance when using the assassin spam data set since the accuracy was only 89%. Yet, the goal of algorithms based on decision trees is to make predictions about a target variable's value over a specified set of inputs.

On the other hand, in 2018, an approach for spam filtering that is based on various decision tree algorithms was introduced by Subasi et al. [18] for comparing the accuracy and identifying the most suitable one for their data set. Hence, authors carry out clustering and regression tree (CART), REP and LAD tree, RF and rotation forest algorithms on the data set to classify the emails. The obtained outcomes show that the proposed modified RF model achieved the most significant accuracy compared to other decision tree methods for public data sets.

Jamil et al. [19] proposed a new framework based on blockchain technology that supports IoT and machine learning techniques. The model they suggest consists of at least two blocks. The first module is a blockchain-based network used for sensor device security and a relationship based on a smart contract and an inference engine that discovers hidden and actionable information from Internet of Things (IoT) data and user devices.

36.3 BACKGROUNDS (MACHINE LEARNING)

Machine learning is a branch of artificial intelligence (AI) and computer science that mainly uses data and algorithms to affect the way humans learn, gradually improving its accuracy. Within machine learning, we can identify three different types of machine learning: supervised learning, unsupervised learning and reinforcement learning.

Supervised learning focuses on previously categorized data. Hence, algorithms of this kind of learning are designed to accurately predict the correct class of a labeled data based on insights gathered from the initial observations and clusters. Besides, there are two categories of supervised learning algorithms: a classification and regression. In our case, we are just interested in classification.

The goal of support vector machines (SVMs) is to determine a decision-making threshold that partitions the area into two regions, in order to find the hyperplane that properly classifies the data

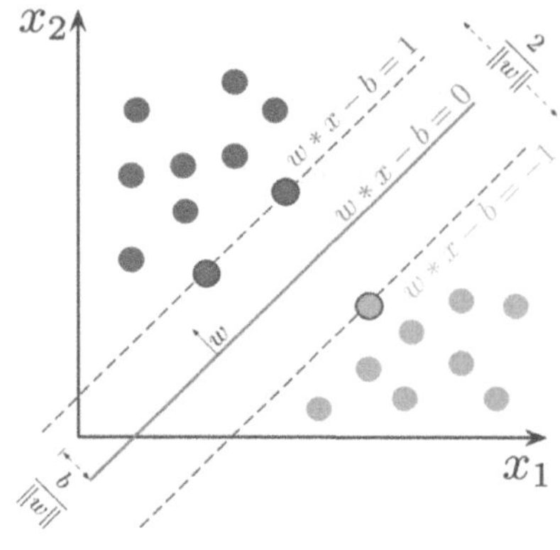

FIGURE 36.1 SVM algorithm [23].

and which is furthest away from all the data (Figure 36.1). One says that to maximize the margin, means the distance from the nearest point to the hyperplane. In this paper, we are going to consider that the entrances are textual documents and the outputs are categories.

In opposition to supervised learning, unsupervised learning addresses the issue where only inputs {X} are provided and no outputs are available. The aim of unsupervised learning or "clustering" is to constrict a heterogeneous set of objects into groups (clusters) of similar objects. There are several clustering algorithms. However in this paper, we are only interested in K-means algorithms.

According to Figure 36.2, K-means is an algorithm that allows the partitioning of data sets into K groups or clusters. Therefore, each item will only be assigned to one cluster. The amount of K is determined by the user. The purpose of this algorithm consists in grouping several sets of data according to their similarities.

Long-term memory (LSTM) is a kind of recurrent neural network (RNN). However, it performs even better than other conventional recurrent neural networks with regard to memory. Similar to other RNNs, the LSTM may have multiple hidden layers and since it runs through every layer, all pertinent information is retained and all non-pertinent information is eliminated in each cell (Figure 36.3).

36.4 SPAM

The question of what constitutes spam remains somewhat unclear, as everyone has a different opinion on the subject. Nowadays, electronic spam is a phenomenon that attracts everyone's attention. Email spam usually consists of special, spontaneous messages which are repeatedly being sent by

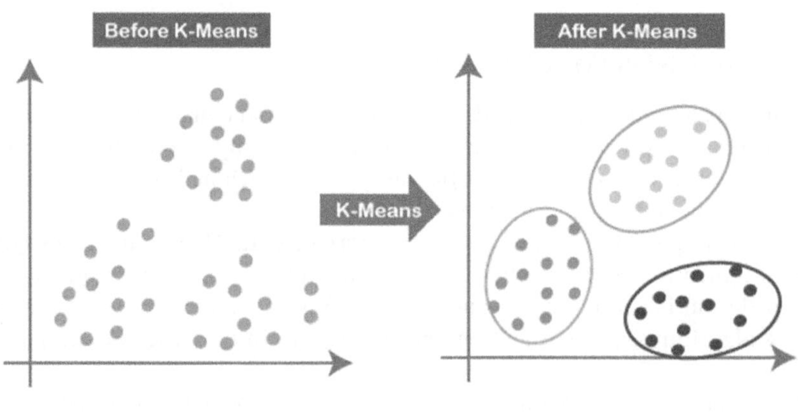

FIGURE 36.2 K-means algorithm [24].

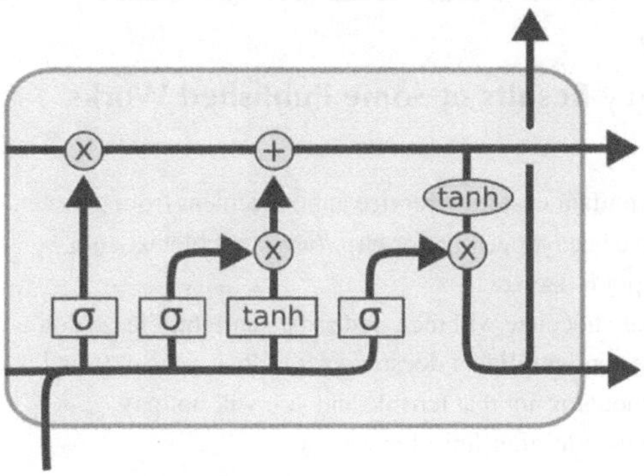

FIGURE 36.3 Long-term memory algorithm [25].

people you do not know. The notion of spam originated in a Monty Python sketch [26], where the Hormel boxed meat products have many annoying insertions. Even though the term "spam" was initially employed from 1978 to refer to unsolicited email, the term gained considerable momentum in the mid-1990s as it became increasingly common outside of the academic and research communities [27, 28].

36.5 PROPOSED MODEL

Our proposed framework for SPAM detect can be summarized through the pattern revealed in Figure 36.4. As exemplified in this figure, the suggested system comprises many phases containing data acquisition as the initial phase followed by the pre-processing of collected corpus. Afterwards, annotation was made through manual work. Next, tokenization, stop word elimination and steamining were performed. Eventually, our model was trained and tested with the chosen algorithms,

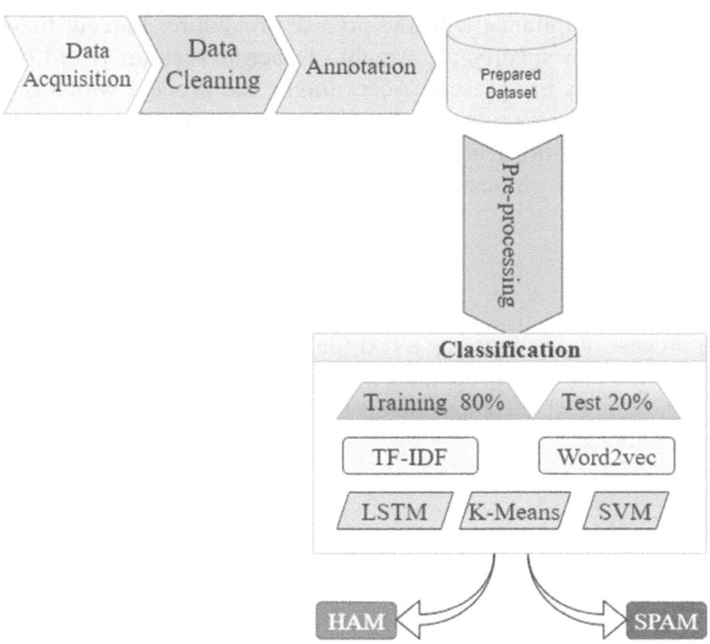

FIGURE 36.4 The proposed model architecture.

TABLE 36.2
The Accuracy Results of Some Published Works

Message	Category
You have an important customer service announcement from premier.	Spam
Your credits have been topped up for http://www.bubbletext.com. Your renewal pin is Tgxxrz	Spam
Science tells that chocolate will melt under the sunlight. Please don't walk under the sunlight. Baozi doesn't want to lose a sweet friend.	Ham
The monthly amount is not that terrible and you will not pay anything till 6 months after finishing school.	Ham

K-means, SVM and LSTM, by using the corpus with TF-IDF and word2vec techniques. Further details about each phase are provided in subsequent sub-sections.

36.6 DATA SET DESCRIPTION

The data set that we have used contains is downloaded from Kaggle website [29]. It contains about 57,208 rows. As shown in Table 36.2, the descriptive feature consists of email texts that are sent, but the targeted feature consists of two classes: "Ham" and "Spam". The classes are labeled for each message of the data set and represent the targeted functionality with a binary alphabet of type string {Ham; Spam}. The classes are then mapped to the integer 0 (Ham) and 1 (Spam).

36.7 PRE-PROCESSING AND ANNOTATION

In order to be able to create and train any classifier from any given text corpus, it is imperative that the text documents be converted into valid inputs which can be understood by any classification algorithm, in compliance with the precise procedure. Indeed, these valuable inputs are vectors or matrices that specify the weight of each descriptor (word or group of words) for each text file where they are present. Accordingly, the process which is performed on the corpus before returning it as a vector document is known as pre-processing, while the transformation from a text document (unstructured) to a vector document (structured data) is known as feature extraction. Hence, the phases of the corpus pre-processing that we follow are described in Figure 36.5.

36.7.1 Tokenization

Tokenization is a process of decomposing a text file into a collection of words and a collection of sentences. Its objective is to properly compute the data sets in these forms, in order to enhance the accuracy of the classifiers. Text decomposition is the process of rendering an input text into a set of words, including repetitions.

FIGURE 36.5 Email text pre-processing and transformation process.

36.7.2 Delete Stop Words

There are some words, called stop words, which occur very frequently regardless of the type of text. Due to this particularity, their appearance does not give any relevant indication for the classification of the text. On the other hand, the presence of this type of words can generate noise that impairs the system's performance. For this reason, it is advisable to eliminate these words in order to improve the classification capacity of the model that will later be exploited. Among this type of words, we find connectors, conjunctions and determiners, as well as verbs that frequently appear in all the headings of the classification (for instance "he, your, hat, have, be, so, how…").

36.7.3 Stemming

In the field of linguistic morphology and retrieval, extracting the root of a given term means the process of restoring deviated words to its original stem. The stem does not always have to be exactly the same thing as the morphological root of the word. It is generally enough if it enables words with a similar root and a similar sense to be clustered together, although the root might not actually be a legitimate root.

36.7.4 Vector Text Representation

We can consider the original text as a sequence of words. Currently, this kind of representation is impossible to interpret by machine learning algorithms for which it is necessary to provide numerical vector representations of the entities to be classified. The vector representation of the text consists of converting each part into a series of numbers, each of which represents a word of the dictionary of the series of documents or the corpus.

In order to be able to convert the texts into vectors, we first create a directory containing all the terms appearing in the texts of the learning set. Then, we obtain a numerical matrix where each row corresponds to one of the texts in the corpus and each column corresponds to a word in the corpus directory. If the word does not appear in the document, the number 0 is assigned to it; on the contrary, if it appears in the text, the number 1, or the number that represents the total number of occurrences of the word in the document, is assigned to it. This matrix is known as the frequency matrix.

36.8 DESCRIPTOR EXTRACTION

The term frequency-inverse document frequency (TF-IDF) is a restraint method that is commonly used for information research and in the field of text mining in particular. This statistical measure enables the evaluation of the significance of a term included in a given document, with respect to a whole collection or corpus. Its dimension is the rate of indexing terms in the document collection. The coordination of the vector of a document corresponds to the weights of the indexing words of this document. We give a greater weight to the terms characterizing a document presented as $d = (w_1, w_2, w_3, \ldots, w_n)$. Therefore, TF-IDF is determined by Equation 36.1.

$$TF_i - IDF_{i,j} = TF_{i,j} * IDF_i \tag{36.1}$$

With

$$TFi, j = \frac{n_{i,j}}{\sum_k n_{k,j}} \tag{36.2}$$

And

$$IDF_i = \log(\frac{D}{d_{j:t_i \in d_j}}) \tag{36.3}$$

On the other hands, the Word2vec method, which has been in use since 2013, is designed to efficiently create word embedding. However, beyond its benefits as a method for embedding words, many of its concepts have proven to be effective in creating recommendation tools and making sense of data sets, even for non-linguistic and professional tasks.

36.9 RESULTS AND DISCUSSIONS

Once we have prepared the pre-processed corpus to be used by the classifiers using the TF-IDF and word2vec methods, we begin the learning and evaluation phases. For this last phase, a series of computations on **precision**, **accuracy** and **F1-score** are computed by using Equations 36.4–36.6.

$$\text{Precision} = \frac{\text{True Positive}}{(\text{True Positive} + \text{False Positive})} \tag{36.4}$$

$$\text{F1-Score} = 2 \times \frac{(\text{Sensitivity} \times \text{Precision})}{(\text{Sensitivity} + \text{Precision})} \tag{36.5}$$

$$\text{Accuracy} = \frac{(\text{True Positive} + \text{True Negative})}{(\text{True Positive} + \text{False Positive} + \text{True Negative} + \text{False Negative})} \tag{36.6}$$

The results of the experimental classification evaluation of a conventional classifiers scheme together with a TF-IDF or Word2vec methods are shown in Table 36.3. From this table, one could easily observe that the LSTM classifier achieves those values 98.01%, 97.79% and 98.36% in precision, accuracy and F1-score, respectively, for our corpus with the usage of TF-IDF. Furthermore, when we replace TF-IDF with Word2vec, the obtained values are 94.84%, 95.02% and 96.24% in precision, accuracy and F1-score, respectively. On the other hand, the SVM classifier reaches 91.03% in precision, 93.26% in accuracy and 93.39% in F1-score when TF-IDF is used. However, the same algorithm attains 90.36% in precision, 92.65% in accuracy and 92.91% in F1-score through the usage of Word2vec. However, when TF-IDF is chosen, K-means results are 96.73%, 95.90% and 96.94% in precision, accuracy and F1-score, respectively. Besides, it has 94.79%, 92.54% and 94.26% in precision, accuracy and F1-score, respectively, with the usage of Word2vec.

Figure 36.6 depicts the performance of the three commonly used classifiers during the experiment where we have used TF-IDF as a characteristic extracting mechanism. Besides, Figure 36.7 illustrates the reached results after the usage of Word2vec. The details of the obtained outcomes are provided below. Hence, we can notice that LSTM achieves the high precision (98.01–94.84%), accuracy (97.79–95.02%) and F1-score (98.36–96.24%) with usage TF-IDF or Word2vec, respectively. Hence, they are high that the same values obtained when we use K-means or SVP wither the usage of

TABLE 36.3
Results Overview of the Classifiers, Including Pre-Processing

Algorithm		Precision (%)	Accuracy (%)	F1-Score (%)
K-MEANS	TF-IDF	96.73	95.90	96.94
	Word2vec	94.79	92.54	94.26
LSTM	**TF-IDF**	**98.01**	**97.79**	**98.36**
	Word2vec	94.84	95.02	96.24
SVM	TF-IDF	91.03	93.26	93.39
	Word2vec	90.36	92.65	92.91

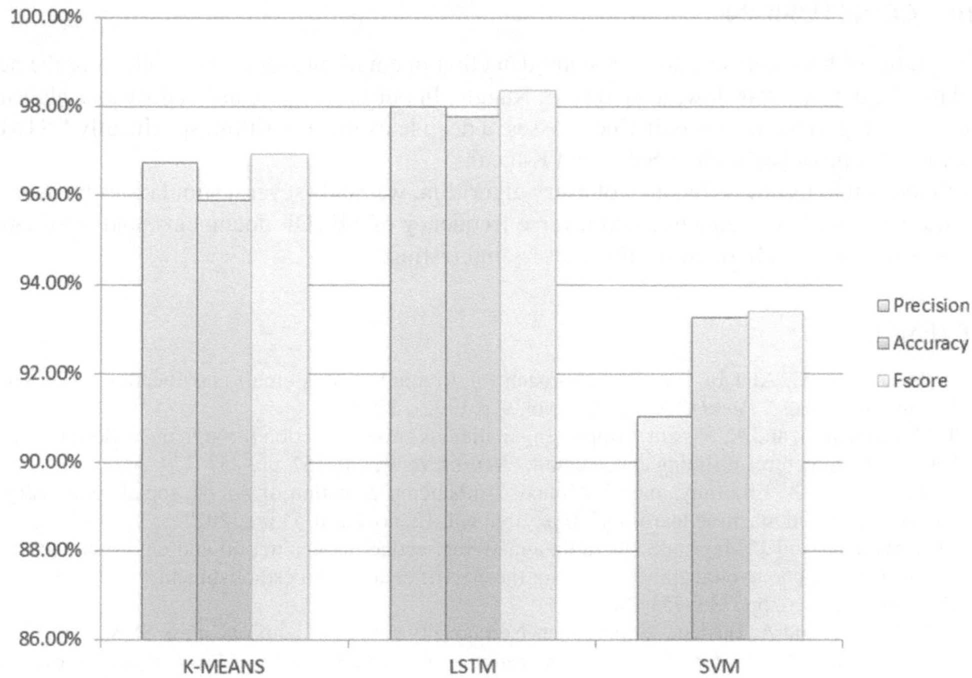

FIGURE 36.6 Classification results with TF-IDF.

TF-IDF or Word2vec. Moreover, we can remark that the performance of all algorithms is enhanced when TF-IDF is selected as a feature extraction technique instead of Word2vec.

These obtained outcomes demonstrate that the most suitable classifier to our proposal is the long and short-term memory recurrent network (LSTM), since it gives a high precision in comparison with the other methods, as well as a very high accuracy and a higher F1-score. We can also notice that the K-means classifier is equally useful for this situation, but this does not imply that the SVM is not useful.

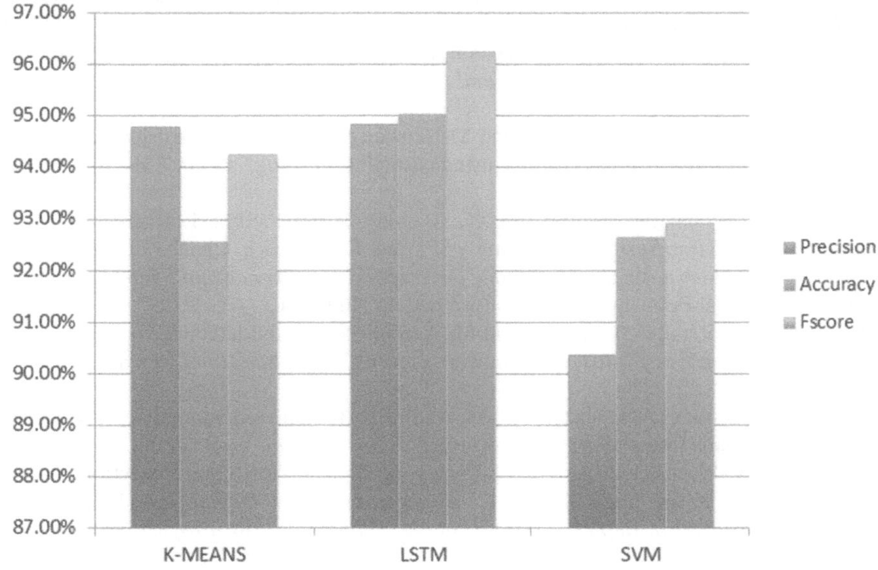

FIGURE 36.7 Classification results with TF-IDF.

36.10 CONCLUSION

In this study, we have concentrated on spam detection in email messages. For collecting the necessary data, the data set was downloaded from Kaggle. In our project, we utilized various algorithms that can be categorized into two distinct classes: a deep learning algorithm, specifically LSTM and conventional approaches such as SVM and K-means.

With the aim to test the efficiency of every algorithm, we made several simulations by using several parameters such as frequency and inverse frequency of TF-IDF documents, and word embedding (Word2vec). The obtained results are very interesting.

REFERENCES

1. A. Masri, and M. Al-Jabi, "A novel approach for Arabic business email classification based on deep learning machines," *PeerJ Comput. Sci.*, vol. 9, p. e1221, 2023.
2. T. Muralidharan, and N. Nissim, "Improving malicious email detection through novel designated deep-learning architectures utilizing entire email," *Neural Netw.*, vol. 157, pp. 257–279, 2023.
3. A. M. Alkadri, A. Elkorany, and C. Ahmed, "Enhancing detection of Arabic social spam using data augmentation and machine learning," *Appl. Sci.*, vol. 12, no. 22, p. 11388, 2022.
4. M. Benchrifa, and J. Mabrouki, "Simulation, sizing, economic evaluation and environmental impact assessment of a photovoltaic power plant for the electrification of an establishment," *Adv. Build. Energy Res.*, vol. 16, no. 6, pp. 736–753, 2022.
5. S. Y. Yerima, and A. Bashar, "Semi-supervised novelty detection with one class SVM for SMS spam detection," in *2022 29th International Conference on Systems, Signals and Image Processing (IWSSIP)*, 2022, pp. 1–4.
6. A. P. Rodrigues, R. Fernandes, A. Shetty, K. Lakshmanna, and R. M. Shafi, "Real-time twitter spam detection and sentiment analysis using machine learning and deep learning techniques," *Comput. Intell. Neurosci.*, vol. 2022, 2022.
7. N. Ahmed, R. Amin, H. Aldabbas, D. Koundal, B. Alouffi, and T. Shah, "Machine learning techniques for spam detection in email and IoT platforms: Analysis and research challenges," *Secur. Commun. Netw.*, vol. 2022, pp. 1–19, 2022.
8. M. Azrour, J. Mabrouki, Y. Farhaoui, and A. Guezzaz, "Experimental Evaluation of Proposed Algorithm for Identifying Abnormal Messages in SIP Network," in *Intelligent Systems in Big Data, Semantic Web and Machine Learning*, vol. 1344, N. Gherabi and J. Kacprzyk, Eds. Cham: Springer International Publishing, 2021, pp. 1–10. doi: 10.1007/978-3-030-72588-4_1.
9. M. Azrour, M. Ouanan, and Y. Farhaoui, "Survey of detection SIP malformed messages," *Indones. J. Electr. Eng. Comput. Sci.*, vol. 7, no. 2, pp. 457–465, 2017. doi: 10.11591/ijeecs.v7.i2.pp457-465.
10. A. Guezzaz, A. Asimi, A. Mourade, Z. Tbatou, and Y. Asimi, "A Multilayer Perceptron Classifier for Monitoring Network Traffic," in *Big Data and Networks Technologies*, Cham, 2020, pp. 262–270. doi: 10.1007/978-3-030-23672-4_19.
11. C. Hazman, A. Guezzaz, S. Benkirane, and M. Azrour, "lIDS-SIoEL: Intrusion detection framework for IoT-based smart environments security using ensemble learning," *Clust. Comput.*, vol. 26, pp. 1–15, 2022.
12. J. Mabrouki, M. Benchrifa, M. Ennouhi, K. Azoulay, I. Bencheikh, T. Rachiq, and S. El Hajjaji, "Geographic Information System for the Study of Water Resources in Chaâba El Hamra, Mohammedia (Morocco)," in *The International Conference on Artificial Intelligence and Smart Environment*. Cham: Springer International Publishing, 2022, November, pp. 469–474.
13. M. Douiba, S. Benkirane, A. Guezzaz, and M. Azrour, "Anomaly detection model based on gradient boosting and decision tree for IoT environments security," *J. Reliab. Intell. Environ.*, vol. 9, pp. 1–12, 2022.
14. M. Douiba, S. Benkirane, A. Guezzaz, and M. Azrour, "An improved anomaly detection model for IoT security using decision tree and gradient boosting," *J. Supercomput.*, vol. 79, pp. 1–20, 2022.
15. D. DeBarr, and H. Wechsler, "Using Social Network Analysis for Spam Detection," in *Advances in Social Computing: Third International Conference on Social Computing, Behavioral Modeling, and Prediction, SBP 2010, Bethesda, MD, USA, March 30-31, 2010. Proceedings*. Springer, *3*, 2010, pp. 62–69.
16. D. Crocker, "Standard for the format of ARPA Internet text messages," 1982.

17. H. Faris, J. Alqatawna, A.-Z. Ala'M, and I. Aljarah, "Improving email spam detection using content based feature engineering approach," in *2017 IEEE Jordan Conference on Applied Electrical Engineering and Computing Technologies (AEECT)*, 2017, pp. 1–6.
18. A. Subasi, S. Alzahrani, A. Aljuhani, and M. Aljedani, "Comparison of decision tree algorithms for spam e-mail filtering," in *2018 1st International Conference on Computer Applications & Information Security (ICCAIS)*, 2018, pp. 1–5.
19. F. Jamil, H. K. Kahng, S. Kim, and D.-H. Kim, "Towards secure fitness framework based on IoT-enabled blockchain network integrated with machine learning algorithms," *Sensors*, vol. 21, no. 5, p. 1640, 2021.
20. M. Verma, D. Divya, and S. Sofat, "Techniques to detect spammers in twitter- A survey," *Int. J. Comput. Appl.*, vol. 85, no. 10, pp. 27–32, 2014. doi: 10.5120/14877-3279.
21. M. Zavvar, M. Rezaei, and S. Garavand, "Email spam detection using combination of particle swarm optimization and artificial neural network and support vector machine," *Int. J. Mod. Educ. Comput. Sci.*, vol. 8, no. 7, p. 68, 2016.
22. S. O. Olatunji, "Extreme Learning machines and Support Vector Machines models for email spam detection," in *2017 IEEE 30th Canadian Conference on Electrical and Computer Engineering (CCECE)*, 2017, pp. 1–6.
23. M. A. Chandra, and S. S. Bedi, "Survey on SVM and their application in image classification," *Int. J. Inf. Technol.*, vol. 13, no. 5, pp. 1–11, 2021. doi: 10.1007/s41870-017-0080-1.
24. K. Krishna, and M. Narasimha Murty, "Genetic K-means algorithm," *IEEE Trans. Syst. Man Cybern. Part B Cybern.*, vol. 29, no. 3, pp. 433–439, 1999. doi: 10.1109/3477.764879.
25. K. Smagulova, and A. P. James, "A survey on LSTM memristive neural network architectures and applications," *Eur. Phys. J. Spec. Top.*, vol. 228, no. 10, pp. 2313–2324, 2019. doi: 10.1140/epjst/e2019-900046-x.
26. L. N. Petersen, "The ageing body in Monty python live (mostly)," *Eur. J. Cult. Stud.*, vol. 21, no. 3, pp. 382–394, 2018.
27. L. Zhuang, J. Dunagan, D. R. Simon, H. J. Wang, I. Osipkov, and J. D. Tygar, "Characterizing botnets from email spam records," *LEET*, vol. 8, no. 1, pp. 1–9, 2008.
28. B. Nelson *et al.*, "Exploiting machine learning to subvert your spam filter," *LEET*, vol. 8, no. 1–9, pp. 16–17, 2008.
29. "Find Open Datasets | Kaggle." https://www.kaggle.com/datasets (accessed Feb. 18, 2023).

37 Evaluation of Single Slope Solar Still Integrated with Flat Plate Collector

R. Aftiss, M. Najim, T. Tbatou, and M. Hissouf

37.1 INTRODUCTION

The evolution of energy and water demand is directly linked to population increases. The escalating demand has resulted in a crisis in many countries grappling with water scarcity, attributed to factors such as decreased rainfall and the impact of climate change. Consequently, there is a pressing need to explore eco-friendly solutions based on renewable energy sources to address the shortage of potable water. Thus, implementing the solar desalination system namely solar still could offer a practical solution for this issue. In this context, Figure 37.1 provides a comprehensive depiction of the conventional solar still. As illustrated in the figure, its operational principle closely resembles that of the water cycle. It primarily relies on the processes of evaporation and condensation of water to generate clean, potable water. The primary drawback of conventional solar stills (CSS) is their relatively low productivity in generating clean water. Thus, the coupling of conventional solar still with other external energy sources is recommended to enhance daily productivity. Badran et al. [1] conducted an experimental study on a solar still coupled with a flat plate collector. Their findings revealed a significant enhancement in productivity. Specifically, they observed a remarkable increase of approximately 231% when tap water was used as a feed. Additionally, an improvement of 52% in productivity was noted when saline water was used in the solar still. Dwivedi and Tiwari [2] numerically validated the experimental study of a double slope solar still coupled to an evacuated tube collector (ETC) under natural mode. The thermal modeling exhibited a strong agreement with the experimental study, showing an increase in daily productivity from 1.838 to 2.79 kg/m^2. Singh et al. [3] performed a numerical analysis of a single slope solar still integrated with an ETC operating under natural mode. Their study focused on investigating and optimizing the number of tubes and water depth. The findings indicated that the highest daily productivity was achieved with

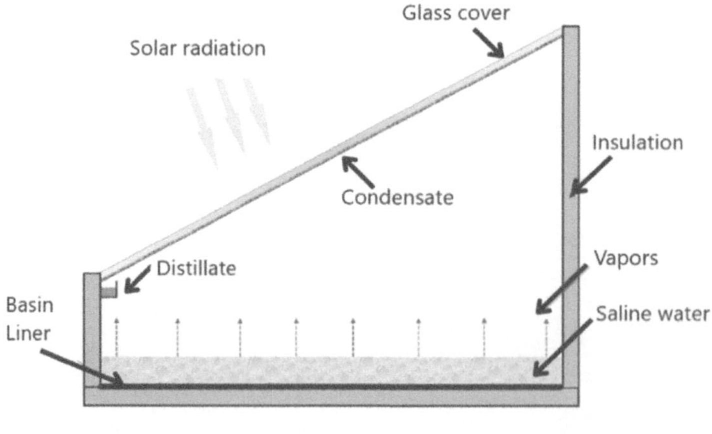

FIGURE 37.1 Single slope solar still.

DOI: 10.1201/9781003647263-37

a combination of 0.03 m water depth and 10 evacuated tubes. Tiwari et al. [4] conducted a comparative study of various active solar still configurations. They performed a numerical analysis comparing solar stills integrated with flat plate collector (FPC), concentrating tube collector, ETC, and ETC with heat pipes. The results revealed daily productivities of 3.19, 4.08, 4.00, and 4.24 kg/m^2/day, respectively, for the solar stills coupled to FPC, concentrating tube collector, ETC, and ETC with heat pipes. Moustafa et al. [5] enhanced the productivity of a tubular solar still by integrating it with a solar-powered electric heater. They developed a hybrid artificial intelligence (AI) model to predict the system's productivity. The results demonstrated an increase in accumulated productivity from 2.58 L/m^2 to 3.41 L/m^2/day. Furthermore, the AI model exhibited strong agreement when compared to the experimental findings. Kabeel et al. [6] experimentally improved the efficiency of a photovoltaic (PV) panel by integrating a solar reflector and employing air and water cooling techniques. They used the hot water output from the PV panel to enhance the evaporation rate in a solar still. This integration led to an observed improvement in productivity by approximately 40.08%. Alawee et al. [7] employed a novel technique based on incorporating a rotating cylinder on the absorber to decrease the depth of water as giving rise to maximizing the output productivity of pyramid solar still. Furthermore, three electric heaters of 30 W powered by the PV panel are installed to optimize the efficiency. The results revealed that the productivity of pyramid solar still, pyramid with cylinder, and pyramid with cylinder and heater are 2900, 7300, and 9100 mL/m^2.

The present paper presents a comprehensive study of active solar still. To analyze the effect of a FPC integrated on a single slope solar still, two configurations are studied and compared numerically: CSS, and solar still with flat plate collector (SSP).

37.2 SYSTEM DESCRIPTION

This paper presents a numerical study of a solar still coupled with a FPC. The thermal energy absorbed by the FPC is subsequently transferred to the solar still, thereby elevating the temperature of the saline water. This increase in temperature contributes to an enhancement in water productivity. The mass flow rate from the FPC to the solar still is regulated by a pump powered by a PV panel. The geometrical parameters of the studied solar still are considered consistent throughout the analysis. Figure 37.2 presents a schematic diagram of a single slope connected to a FPC (Table 37.1).

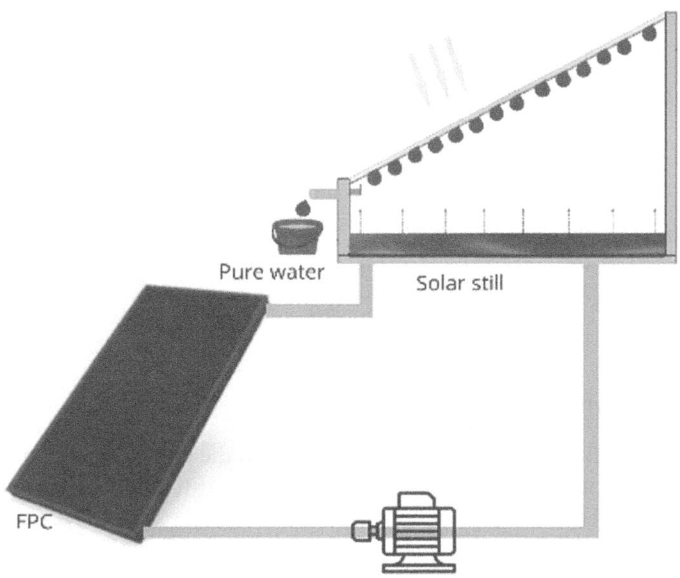

FIGURE 37.2 Single slope solar still with flat plate collector.

TABLE 37.1
Presents the Major Parameters Used in the Present Work for the Solar Still and Solar Collector

Parameter	Value	Parameter	Value
Ag, Aw, Ab	1 m²	Lback	0.04 m
Dc	0.008 m	Ledge	0.04 m
hwt	300 W/k.m²	Lins	0.05 m
F	0.91	Kins	0.035 W/m.k
mh	0.1 kg/s	Ksci	0.045 W/m.k
mw	40 kg	W	0.1 m

37.3 MATHEMATICAL MODEL

The thermal modeling of the solar still relies on the energy balance equation, which describes the mass and heat transfer occurring within the solar still. Consequently, the modeling of the energy balance equation can be viewed as a predictive model that plays a critical role in minimizing the time and materials required for experiments. The energy balance equations are resolved based on the following assumptions [8]:

- The sides and the bottom of the solar basin are well insulated.
- The water mass in the still basin is constant.
- The temperatures of the absorber, water, and glass cover are uniform.
- The solar still is impermeable to vapor leakage.

37.3.1 Conventional Solar Still

The governing equation of water, absorber, and glass cover can be written as follows:

- Energy balance for the basin plate [8]

$$\frac{m_b C_b}{A_b}\frac{dT_b}{dt} = q_{ib} - q_{cwb} - q_{pb} \tag{37.1}$$

- Energy balance for the saline water [8]

$$\frac{m_w C_w}{A_w}\frac{dT_w}{dt} = q_{iw} + q_{cwb} - q_{rwg} - q_{ewg} - q_{cwg} \tag{37.2}$$

- Energy balance for the glass cover [8]

$$\frac{m_g C_g}{A_g}\frac{dT_g}{dt} = q_{ig} + q_{cwg} + q_{rwg} + q_{ewg} - q_{cga} - q_{rga} \tag{37.3}$$

37.3.2 Solar Still with an FPC

The energy balance equation of an SSP system is similar to the conventional one, with the exception of the saline water equation. This equation undergoes a modification by incorporating additional terms representing the thermal energy supplied by the FPC [9].

$$\frac{m_w C_w}{A_w}\frac{dT_w}{dt} = q_{iw} + q_{cwb} + q_u - q_{rwg} - q_{ewg} - q_{cwg} \tag{37.4}$$

The heat flux delivered by the FPC can be written as follows [9]:

$$q_u = A_c \times Fr\left(\alpha_p \tau_G \times G - U_{collector}\left(T_w - T_a\right)\right) \quad (37.5)$$

Where:

$$Fr = \frac{\dot{m}_h \times C_p}{A_c \times U_{collector}}\left(1 - exp\left(\frac{-A_c \times U_{collector} \times F'}{\dot{m}_h \times C_p}\right)\right) \quad (37.6)$$

$$F' = \frac{\dfrac{1}{U_{collecteur}}}{w\left(\dfrac{1}{U_{collecteur}(D_c + (w - D_c) \times F)} + \dfrac{1}{\pi \times D_c \times h_{wt}}\right)} \quad (37.7)$$

37.3.3 Productivity and Energy Efficiency

The hourly productivity of a solar still can be calculated as follows:

$$\dot{m} = \frac{h_{e,w-g} \times (T_w - T_g) \times 3600}{L_{ev}} \quad (37.8)$$

Where $\dot{m}$ is the hourly productivity of freshwater and L_{ev} is the latent heat of vaporization of water.

The average energy efficiency for a CSS can be written as follows:

$$\eta = \frac{\sum \dot{m} \times L_{ev}}{3600 \times A_b \times \sum I(t)} \quad (37.9)$$

in the case of a solar still equipped with an FPC. Besides considering the solar intensity received by the absorber plate of the solar still, we also factor in the power delivered by the FPC and the power consumed by the pump.

$$\eta = \frac{\sum \dot{m} \times L_{ev}}{3600(A_b \times \sum I(t) + Q_u + P_{pum})} \quad (37.10)$$

37.4 NUMERICAL SIMULATION

37.4.1 Numerical Method

To solve the energy balance equations "(37.1)–(37.4)" the fourth-order Rung-Kutta method is adopted. This method is renowned for its accuracy and stability in numerical resolution. The implementation of this method is executed using a FORTRAN programming language program. The numerical calculations of the different parameters (temperature, productivity, efficiency...) were performed for each second as a time step. The flow chart (Figure 37.3) presents the different steps done during the resolution of the system.

37.4.2 Model Validation

To verify the accuracy and validity of the simulation, the present results are validated using both experimental [10] and theoretical [11] data reported in the literature. The findings demonstrate a good agreement with the literature results. Figure 37.4 illustrates a comparison of productivity and

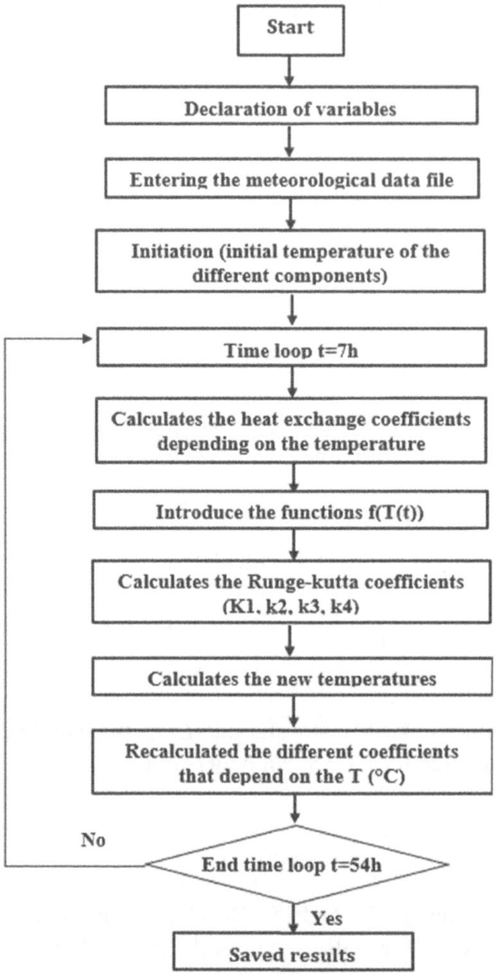

FIGURE 37.3 Flow chart diagram representing the numerical modeling steps.

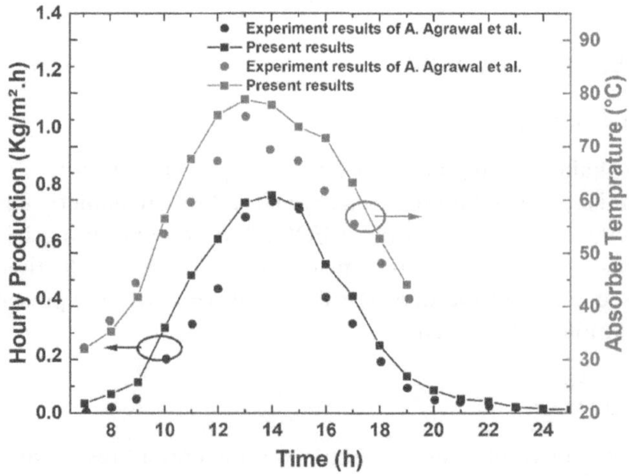

FIGURE 37.4 Validation of the numerical simulation with experimental data [10].

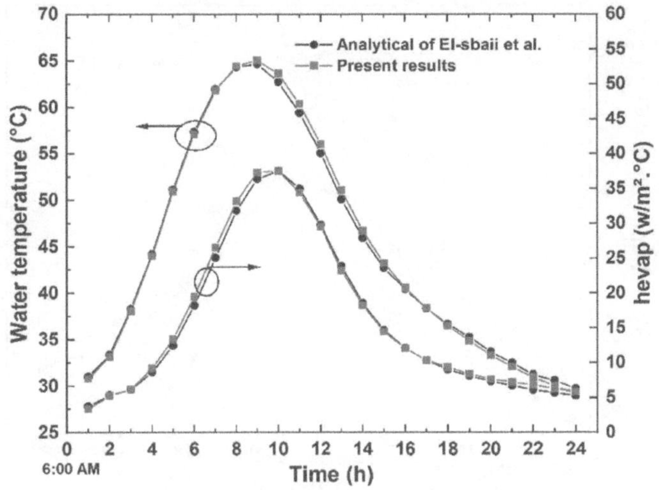

FIGURE 37.5 Validation of the numerical simulation with analytical data [11].

absorber temperature with the experimental results [10], while Figure 37.5 presents the validation results against the theoretical findings [11, 12]. It is evident that the present results behave similarly to those of the reference.

37.5 RESULTS AND DISCUSSION

The present study involves a numerical investigation of a CSS coupled with a FPC. The study consists of comparing two systems: CSS and a solar still with an FPC. To simplify the system's complexity, we focus on solving and analyzing the energy equation of the solar still. Meanwhile, the output energy of the FPC is estimated using a simplified Equation (37.5), derived from the energy balance equation of an FPC [8, 13]. This adopted approach proves to be valuable, demonstrating promising results under the climate conditions of Morocco. The same methodology is followed but in the steady state and under Jordan weather conditions [9]. Thus, the absence of an analysis of the FPC introduces certain limitations to the present work, particularly in evaluating the components of the FPC that could optimize its physical parameters (temperature distribution, thickness, length, etc…). This optimization could significantly enhance the overall productivity of the system. Hence, the mentioned limitation did not have an effect on the aim of the present work, which investigates the effect of integrating an FPC on the efficiency and productivity of a solar still system.

37.5.1 Temperature Profile

The study of the temperature variation of the solar still components, including the glass cover, saline water, and absorber, is essential to determine the energy efficiency and productivity of the system. Figure 37.6 presents the hourly variation of the glass cover temperature for the examined cases. The temperatures of solar still components increase until they reach the maximum value around noon. Subsequently, the curves begin to decline, following the solar intensity profile, until the end of the day. It is evident from the figure that the glass temperature in the SSP case is higher than in a CSS. This disparity is primarily attributed to the additional energy supplied by the FPC. The maximum temperature value for a CSS and SSP is about 50.91 and 68.17°C, respectively. Following the same trend as the glass cover, the temperature of water and absorber exhibit similar behavior. The maximum temperature is attained in the absorber plate of an SSP, distinguished by

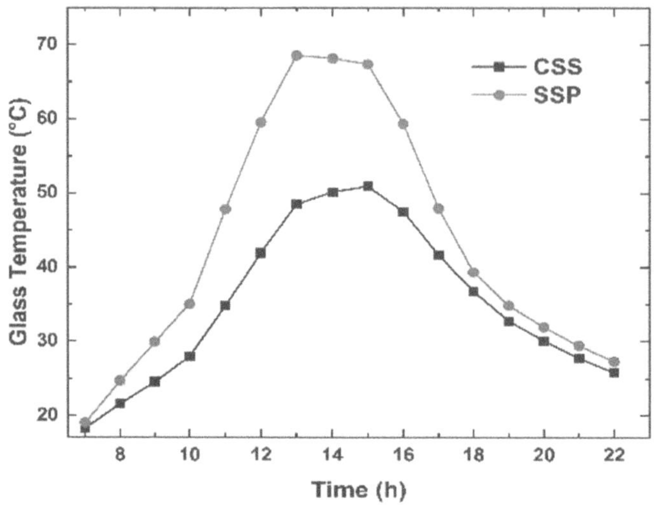

FIGURE 37.6 Hourly variation of glass cover temperature.

its high absorptivity and energy received by the FPC. Moreover, it is remarked that the water temperature closely approaches that of the absorber, which can be explained by the high convection heat transfer from the absorber to the saline water (Figures 37.7 and 37.8). In the evening, after 5:00 PM, we observe that the temperature of the solar still components in both cases are close to each other, indicating that the effect of the FPC diminishes due to the weakening solar radiation. Hourly and accumulated productivity

The productivity of solar still systems is considered an important indicator for evaluating their performance. Figure 37.9 presents the hourly productivity of a CSS and solar still equipped with an FPC. During the initial hour of computation, the productivity of the studied cases shows a slight similarity, suggesting that the water temperature is low and the heating process is in its nascent stage. Likewise, during the sunset hours when solar radiation is weak and the process of dissipating energy stored in water is commencing, we observe a comparable productivity between both systems. Additionally, the graph shows that the productivity of the solar still with the flat plate collector (SSP)

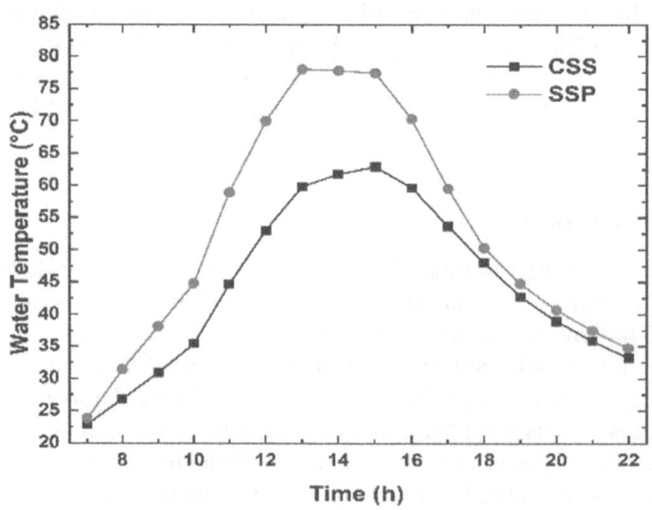

FIGURE 37.7 Hourly variation of water temperature.

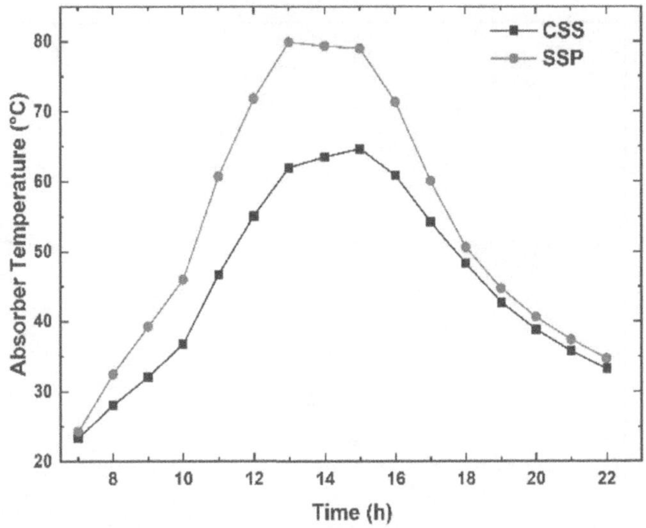

FIGURE 37.8 Hourly variation of absorber temperature.

is higher than that of the conventional system. This difference can be attributed to the additional energy transferred from the FPC, resulting in significant enhancement. The maximum production value for the studied systems is attained around noontime, approximately at 14:00, reaching about 0.61 and 1.03 kg/m^2 for the conventional and SSP, respectively.

Figure 37.10 displays the cumulative productivity of the CSS and SSP. The total freshwater productivity is 4.07 and 6.78 kg/m^2 for the CSS and SSP setups, respectively. From the current results presented in Figures 37.9 and 37.10, it is evident that the integration of FPC has a significant impact on enhancing system productivity. Thus, a remarkable improvement in the productivity of the CSS of about 66% is registered.

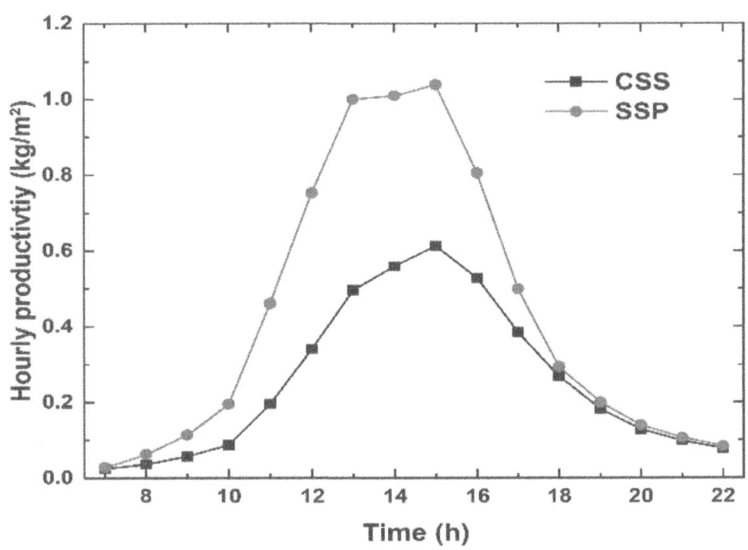

FIGURE 37.9 Hourly variation of freshwater productivity.

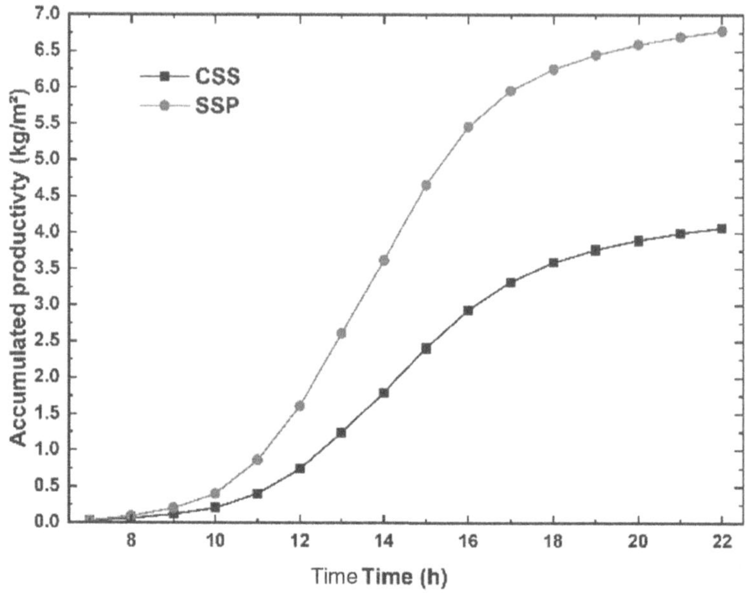

FIGURE 37.10 Accumulated productivity of the studied solar still.

37.5.2 Energy Efficiency

The energy efficiency of the studied solar stills is assessed based on the daily productivity as well as the energy being supplied to the system. Figure 37.11 compares the average energy efficiency of the studied solar stills. It is noted that the energy efficiency of the CSS and SSP is about 47.65% and 54.31%, respectively. An enhancement of 12.26% in the energy efficiency is obtained. By comparing the percentage enhancements for productivity and energy efficiency, we can observe a significant difference. This discrepancy primarily arises from the fact that the calculation of energy efficiency considers both internal and external FPC energy sources.

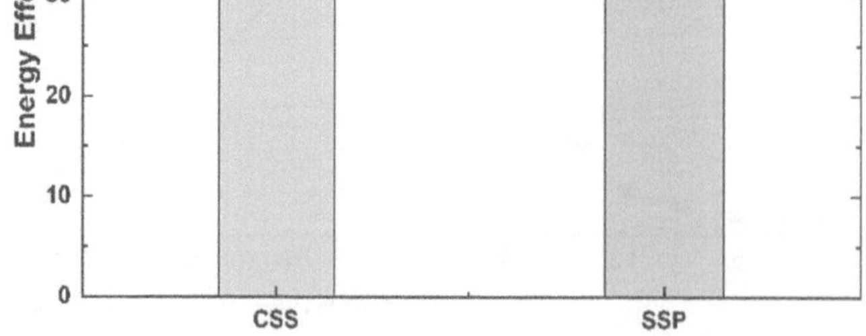

FIGURE 37.11 Energy efficiency of the studied solar still.

37.6 CONCLUSION

The present investigation involved a numerical study of the impact of coupling the FPC with a CSS. The results confirmed that incorporating an FPC as an external energy source enhances the productivity and energy efficiency of the conventional type. The primary conclusion drawn from this study is as follows:

- An improvement of 66% for the CSS productivity is attained.
- The total productivity of CSS and SSP is 4.07 and 6.78 kg/day, respectively.
- Energy efficiency is 47.65% for the CSS case and increased to 54.31% for the one with FPC.

NOMENCLATURE

$\dot{m}_h$ Mass flow rate of the hot fluid coming from the solar collector. (kg/s)
A Area. (m^2)
$\mathbf{D_c}$ Diameter of solar collector tubes. (m)
$\mathbf{F_r}$ Heat removal factor for the solar collector. (dimensionless)
$\mathbf{K_w}$ Thermal conductivity of brackish water. ($W.m^{-1}.°C^{-1}$)
$\mathbf{L_{ev}}$ Latent heat of vaporization of water. ($J.kg^{-1}$)
$\mathbf{P_g}$ Partial saturated vapour pressures at glass cover temperature. ($N.m^{-1}$)
$\mathbf{P_w}$ Partial saturated vapour pressure at a basin water temperature. ($N.m^{-1}$)
$\mathbf{m_w}$ Weight of water. (kg)
$\mathbf{U}_{collector}$ Overall heat loss coefficient of the solar collector. (W/m² K)
$\mathbf{h_{c,g-a}}$ Convective heat transfer coefficient from glass cover to ambient. ($W.m^{-2}.°C^{-1}$)
$\mathbf{h_{c,w-b}}$ Convective heat transfer coefficient from basin liner to water. ($W.m^{-2}.°C^{-1}$)
$\mathbf{h_{c,w-g}}$ Convective heat transfer coefficient from basin water to glass cover. ($W.m^{-2}.°C^{-1}$)
$\mathbf{h_{e,w-g}}$ Evaporative heat transfer coefficient from basin water to glass cover. ($W.m^{-2}.°C^{-1}$)
$\mathbf{h_{r,g-a}}$ Radiative heat transfer coefficient from glass cover to ambient. ($W.m^{-2}.°C^{-1}$)
$\mathbf{h_{r,w-g}}$ Radiative heat transfer coefficient from basin water to glass cover. ($W.m^{-2}.°C^{-1}$)
$\mathbf{h_{wt}}$ Heat transfer coefficient for water inside the collector tubes. (W/m² K)
F Fin efficiency. (dimensionless)
F' Fin efficiency factor for the solar collector. (dimensionless)
w Spacing between solar collector tubes. (m)

GREEK SYMBOLS

v_w Wind velocity. ($m.s^{-1}$)
μ Dynamic viscosity. ($kg.m^{-1}.s^{-1}$)
α Absorptivity
β Thermal expansion coefficient. (K^{-1})
ϵ Emissivity.
η Energy efficiency of solar still
ρ Density. ($kg.m^{-3}$)

SUBSCRIPTS

a Ambient
b Absorber
c Convection
e Evaporation

g Glass Insulation
r Radiation
w Water

REFERENCES

1. Badran, A. A., Al-Hallaq, I. A., Salman, I. E., & Odat, M. Z. A solar still augmented with a flat-plate collector. *Desalination*, *172*(3), 227–234 (2005).
2. Dwivedi, V. K., & Tiwari, G. Experimental validation of thermal model of a double slope active solar still under natural circulation mode. *Desalination*, *250*(1), 49–55 (2010).
3. Singh, R. V., Kumar, S., Hasan, M. M., Khan, M. T., & Tiwari, G. Performance of a solar still integrated with evacuated tube collector in natural mode. *Desalination*, *318*, 25–33 (2013).
4. Tiwari, G., Dimri, V., Singh, U., Chel, A., & Sarkar, B. Comparative thermal performance evaluation of an active solar distillation system. *International Journal of Energy Research*, *31*(15) (2007).
5. Moustafa, E. B., Hammad, A. H., & Elsheikh, A. H. A new optimized artificial neural network model to predict thermal efficiency and water yield of tubular solar still. *Case Studies in Thermal Engineering*, *30*, 101750 (2022).
6. Kabeel, A., & Abdelgaied, M. Performance enhancement of a photovoltaic panel with reflectors and cooling coupled to a solar still with air injection. *Journal of Cleaner Production*, *224*, 40–49 (2019).
7. Alawee, W. H., Mohammed, S. A., Dhahad, H. A., Abdullah, A., Omara, Z., & Essa, F. Improving the performance of pyramid solar still using rotating four cylinders and three electric heaters. *Process Safety and Environmental Protection*, *148*, 950–958 (2021).
8. Aftiss, R., Najim, M., & Hissouf, M. Numerical study of PCM-integrated solar still efficiency enhancement. *International Journal of Low-Carbon Technologies*, *19*, 443–454 (2024).
9. Abu-Arabi, M., Ahmad, M., & Mousa, H. Theoretical modeling of a glass-cooled solar still incorporating PCM and coupled to flat plate solar collector. *Journal of Energy Storage*, *29*, 101372 (2020).
10. Agrawal, A., Rana, R., & Srivastava, P. K. Heat transfer coefficients and productivity of a single slope single basin solar still in Indian climatic condition: Experimental and theoretical comparison. *Resource-Efficient Technologies*, *3*(4), 466–482 (2017).
11. Faidah, A. S., El-Sebaii, A. A., Al-Hazmi, F. S., & Al-Ghamdi, A. A. Thermal performance of a single basin solar still with PCM as a storage medium. *Applied Energy*, *86*(7–8), 1187–1195 (2009).
12. Duffie, J. A., & Beckman, W. A., Solar engineering of thermal processes, 4th ed., p. 936 (2013).
13. Benchrifa, M., Elouardi, M., Fattah, G., Mabrouki, J., & Tadili, R. (2023). Identification, simulation and modeling of the main power losses of a photovoltaic installation and use of the internet of things to minimize system losses. In Advanced technology for smart environment and energy (pp. 49–60). Cham: Springer International Publishing.

38 Sustainable Reuse of Harbor Dredged Sediment in Concrete Maritime Blocks

M. Bortali, M. Rabouli, M. Yessari, and A. Hajjaji

38.1 INTRODUCTION

Dredging of ports is necessary to ensure that they remain safe for navigation and access by ships on a permanent basis [1–4]. Since sediment continuously accumulates at the bottom of harbor channels, dredging operations are performed regularly [5, 6]. These dredging operations have a negative environmental impact. Dredged material is composed of organic and inorganic contaminants [7–9]. Once a tolerance threshold is exceeded, certain components of marine sediment are considered toxic [10]. Thus, the current management method of dumping dredged material at sea has harmful effects on the aquatic environment and human health [11, 12]. The use of dredged sediment in civil engineering is a potential alternative to natural resources [13]. This can address both the environmental impacts of dredging and the scarcity of non-renewable materials used in the construction industry [14]. Dredging operations generate significant quantities of sediment, exceeding 1 billion m^3 per year worldwide [15]. Much of this dredged sediment can be utilized in civil engineering, which is facing a growing scarcity of natural aggregates [16]. To address this issue, various research efforts have focused on sustainable solutions for waste valorization, including the beneficial reuse of dredged sediment [14, 15]. Studies have confirmed the technical feasibility of using dredged sediment as construction aggregate [4, 17–19]. Applications have been studied in concrete and mortar works [18, 20–23], cement production [24–27], bricks [1, 28], and road construction techniques [29–32]. Previous research efforts have rarely investigated case studies of reusing dredged sediment as concrete aggregate in a marine environment, which has specific challenges. To fill this gap and encourage the sustainable reuse of harbor dredged sediment in maritime construction, this case study was conducted. The objective of this paper is to produce concrete for manufacturing blocks used to protect maritime infrastructures, based on harbor dredged sediment. These blocks are exposed to an aggressive marine environment, which can accelerate concrete deterioration [33]. To assess the feasibility of valorizing dredged sediment in maritime works, four types of concrete were formulated with sand dredged from Safi port in Morocco, partially substituting conventional sand at rates of 25%, 50%, 75%, and 100%. The properties of fresh and hardened concretes were investigated and compared with those of reference concrete. Tests conducted on the manufactured concrete included slump test, air content, density, compressive strength, tensile splitting strength, flexural strength, and modulus of elasticity. Additionally, samples of grooved cubical blocks were manufactured and inspected. This study provides reliable technical data on the properties of concrete made with dredged sediment for potential use in maritime construction. The remainder of the paper is structured as follows: Section 38.2 describes the proposed methodology for the research. Sections 38.3 and 38.4 provide an analysis of the results obtained and discuss future research directions. The final section concludes the paper.

DOI: 10.1201/9781003647263-38

FIGURE 38.1 Location of Safi Port and its channel access.

38.2 MATERIALS AND METHODS

38.2.1 Study Area

For this case study, sediment was collected from the port of Safi, located in the west center of Morocco on the Atlantic coastline. The main traffics in transit through Safi harbor are phosphates, fertilizers, ores, cereals, phosphoric acid, sulfur, and fishing products. The port of Safi requires regular dredging of about 250,000 m^3 per year, due to its high level of silting [34, 35]. Representative samples were extracted, by a diver, from the access channel to the port. Figure 38.1 represents the location of Safi port and its channel access. Samples were placed in isothermal bags and kept in an opaque and cold room. Previous research work has focused on the physical, geotechnical, chemical, environmental, and mineralogical characterization of dredged sediment from the port of Safi, for sustainable reuse as construction material [36]. The results of the preliminary studies show that sediment dredged from the port access channel is identified as sand with medium to fine grains (about 97% between 0.63 mm and 1.25 mm), with a mean grain size of about 0.36 mm. Dredged sediment is sensitive to water and generally clean with a low percentage of clayey fines. While sulfate content is negligible, important water content (about 58%) and high concentration of chloride (between 0.15% and 0.18%) are explained by the salty marine origin. The main mineralogical phases of sediment dredged from Safi port are quartz and calcite.

For the remainder of this study, the feasibility of valorization was evaluated for a mixture of samples collected from the access channel to the port of Safi. A pretreatment step consisting on washing with fresh water then dehydrating the sediment mixture was performed, in order to reduce water and chloride contents.

38.2.2 Experimental Program

An experimental program was implemented using sediment dredged from Safi port for concrete production. The purpose was to investigate the change in concrete properties in relation to the optimal ratio of substitution of conventional sand by dredged sediment. In addition to reference concrete, four different mixtures incorporated dredged sediment as a partial substitute for sand. The percentage varied from 25% to 100% of the aggregate mass. Table 38.1 summarizes the nomenclature of the concrete formulas studied.

TABLE 38.1
Concrete Formulas Nomenclature

Concrete Formula	Nomenclature
Reference concrete	RC
Concrete incorporating 25% of dredged sediment	C25%
Concrete incorporating 50% of dredged sediment	C50%
Concrete incorporating 75% of dredged sediment	C75%
Concrete incorporating 100% of dredged sediment	C100%

38.2.2.1 Concrete Composition

The choice of concrete adapted to the type of work and environment was determined according to the recommendations of:

- The French Association of Civil Engineering (AFGC) on "Design of concrete for a given service life – Durability indicators" [37];
- The Central Laboratory of Bridges and Roads (LCPC) on "Control of the durability of concrete structures" [38];
- The European standard EN 206+A2 on "Concrete – Specification, performance, production and conformity" [39].

According to the performance approach of concrete durability (blocks designed for a 100-year life span), the exposure classes corresponding to the marine environment have been determined. We have opted for concrete B30 used in maritime works, with a minimum cement dosage of 300 kg/m^3 and a maximum water-cement ratio W/C of 0.45. The optimal quantities of materials necessary to the confection of this concrete were determined based on the method of formulation of Dreux-Gorisse [40].

The following components were used in concrete formulation:

- Dredged sediment (DS) excavated from the access channel of Safi port;
- Three aggregate fractions: A1 0/4 mm, A2 5/16 mm, and A3 16/40 mm, and dune sand A4 0/0.63 mm;
- Portland Cement CPJ-CEMII for exposure to marine environment according to the European standard EN 197-1 [41];
- Admixture of a new generation high water-reducing super plasticizer;
- Potable water for mixing.

The reference concrete (RC) was made using the three aggregate fractions: A1, A2, and A3 and the dune sand A4. Concrete C25%, C50%, C75%, and C100% were produced using, respectively, 25%, 50%, 75%, and 100% of dredged sediment in replacement of dune sand A4. Indeed, dredged sediment and dune sand A4 0/0.63 mm has similar particle size properties. Table 38.2 gives the nominal composition to produce 1 m^3 of concrete.

38.2.2.2 Concrete Properties

In order to assess the performance of produced concretes, laboratory tests were carried out to determine properties of concrete at both fresh and hardened states.

38.2.2.2.1 Properties of Fresh Concrete

- Workability, defined as the ease of fully compacting the concrete without segregation or bleeding, was measured by the slump test using an Abrams cone. For each concrete, the slump was measured at time T0 just after mixing the concrete, at T30 after 30 minutes, and at T60 after 60 minutes;

TABLE 38.2
Concrete Mixture Proportions

Component	RC	C25%	C50%	C75%	C100%
DS (kg)	–	70	150	220	290
A1 0/4 mm (kg)	600	600	600	600	600
A2 5/16 mm (kg)	400	400	400	400	400
A3 16/40 mm (kg)	690	690	690	690	690
A4 0/0.63 mm (kg)	290	220	140	70	–
Cement (kg)	300	300	300	300	300
Water (kg)	140	140	140	140	140
Admixture (kg)	3.6	3.6	3.6	3.6	3.6

- The air content was determined on the basis of the air-entrained test during mixing. It is interesting to measure the air content of the fresh concrete before placing it in order to give a first indication of the compactness of the mix;
- The physical properties of fresh concrete were determined by measuring the density.

38.2.2.2.2 Properties of Hardened Concrete

- The physical properties of hardened concrete were determined by measuring their density at the age of 28 days;
- The mechanical properties were assessed based on the measurement of compressive strength (at the ages of 7 days and 28 days), the tensile splitting strength (at the age of 28 days), the flexural strength (at the age of 28 days), as well as the modulus of elasticity (at the age of 28 days).

Table 38.3 summarizes the standards followed for each test.

All tests presented in this section were performed at least three times to confirm the results. The values considered in the following sections are the average of the three measurements obtained for each test.

38.2.3 In-situ Application

As an in-situ application of the studied concrete, we opted for artificial concrete blocks in the shape of a grooved cube. It is one of the types of blocks, of massive form, used on a large scale for important maritime projects [50].

TABLE 38.3
Summary of Tests Carried Out and Corresponding Standards

State	Property	Test	Standard
Fresh concrete	Air content	Air-entrained	EN 12350-7 [42]
	Physical properties	Density	EN 12350-6 [43]
	Workability	Slump test (T0)	EN 12350-2 [44]
		Slump test (T30)	
		Slump test (T60)	
Hardened concrete	Physical properties	Density	EN 12390-7 [45]
	Mechanical properties	Compressive strength (7 days)	EN 12390-3 [46]
		Compressive strength (28 days)	
		Tensile splitting strength	EN 12390-6 [47]
		Flexural strength	EN 12390-5 [48]
		Modulus of elasticity	EN 12390-13 [49]

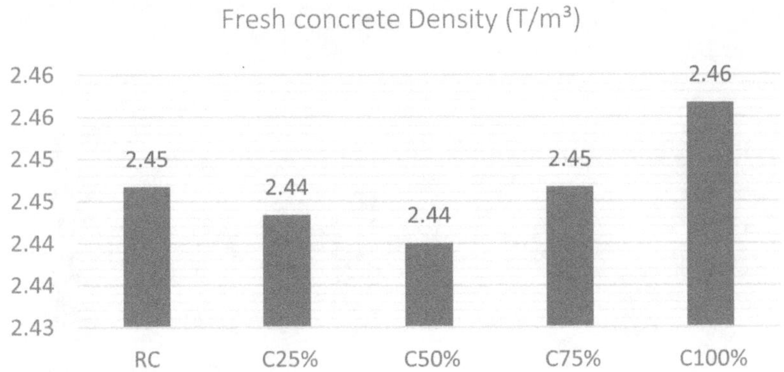

FIGURE 38.2 Density of fresh concrete.

A grooved cubic block, also called an Antifer block, is a simple cubic block with a semi-circular groove on all four non-parallel sides. The grooves allow a better circulation of water inside the block and a reduction of overpressure. In addition, they facilitate a stowage of the blocks between them, which improves their holding and gives them the behavior of a group [51, 52].

The following modalities have been respected for the prefabrication of the concrete blocks:

- Ensure perfect flatness of the block manufacturing area;
- The concrete casting for each block started is completed without interruption;
- The prefabricated blocks are unmolded 24 hours after casting;
- The prefabricated blocks are put in place 15 days after casting.

38.3 RESULTS

38.3.1 Fresh Concrete

38.3.1.1 Physical Properties (Density)

Figure 38.2 shows the density of fresh concretes containing dredged sediment as sand replacement. The findings highlight the similarity of density values between concretes formulated with different percentages of dredged sediment, and in comparison with reference concrete density. The density of fresh concrete is 2.45 T/m^3 on average.

The obtained results are explained by the fact that dredged sediment has the same specific gravity as dune sand; the use of dredged sediment as conventional sand replacement does not show any effect on the unit weight of formulated concrete [53].

38.3.1.2 Air Content

The air content in the five types of formulated concrete was determined based on the air-entrained test. Figure 38.3 shows the results of performed tests.

The obtained percentages of occluded air in the four concretes containing dredged sediment (C25%, C50%, C75%, and C100%) are almost the same in comparison with the RC. The average value of air entrained is 2.4%. These findings are explained by the fact that the content air is only linked to the finest particles of concrete aggregates [54]. Hence, the substitution of conventional sand by dredged sediment for concrete production does not have a significant impact on air content in fresh concrete.

38.3.1.3 Workability

Slump test provided the subsidence in millimeters observed when the frustum of Abrams cone filled with concrete is removed. The slump will be higher, the more fluid concrete is. Figure 38.4 shows the slump test results for the five manufactured concretes, at 0, 30, and 60 minutes from mixing the concrete.

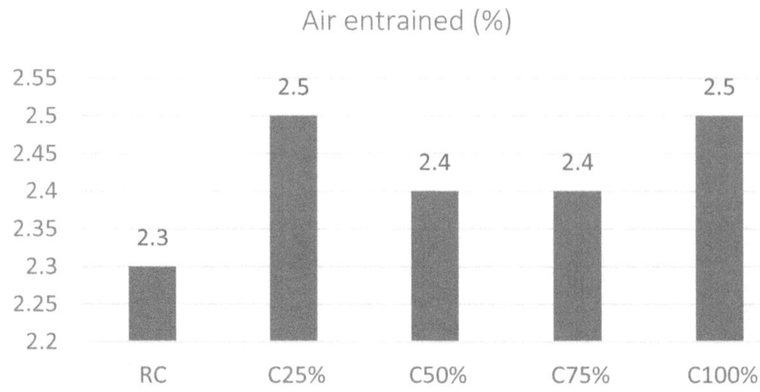

FIGURE 38.3 Air content in fresh concrete.

Results of the performed tests indicate that the slump values of the concrete mixtures containing 25%, 50%, and 75% of dredged sediment increase by 7%, 21%, and 29%, respectively, for T0, compared with RC. For T30, the same values increase by 9%, 18%, and 27%, respectively. The values relative to slump-test T60 increase by 17%, 50%, and 50%, respectively. A decrease in slump values was observed for concrete C100%, with results close to those of RC.

Slump-test results are mainly explained by the direct effect of particle size distribution on the workability of concrete [55]. As workability is linked to particle packing [56], the resulting slumps depend on the shape of particles [57], mainly the aspects related to roundness, surface roughness, and angularity of edges [58]. For the specific case of dredged sediment, concrete workability depends on the actual size fraction of the deposit [59]. In conclusion, concrete produced with dredged sediment, up to 75% incorporation rate, has the most workable mixes.

38.3.2 Hardened Concrete

38.3.2.1 Physical Properties (Density)

Figure 38.5 summarizes the density of all concretes fabricated, at their hardened state. According to the obtained results, the density of hardened concrete does not indicate a significant variation from one type of concrete to another. The average value of hardened concrete density is 2.44 T/m^3.

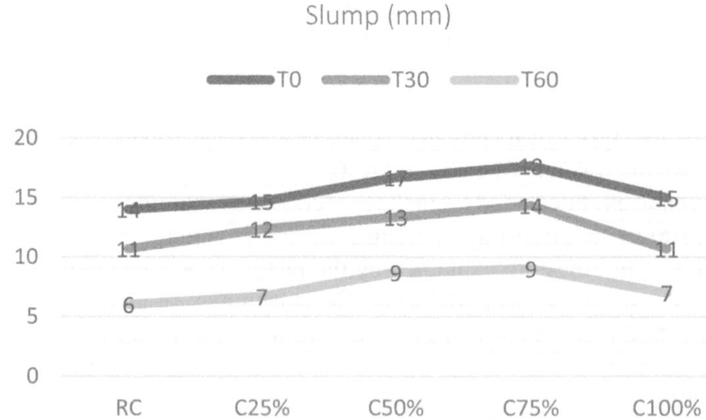

FIGURE 38.4 Slump-test results for fresh concrete.

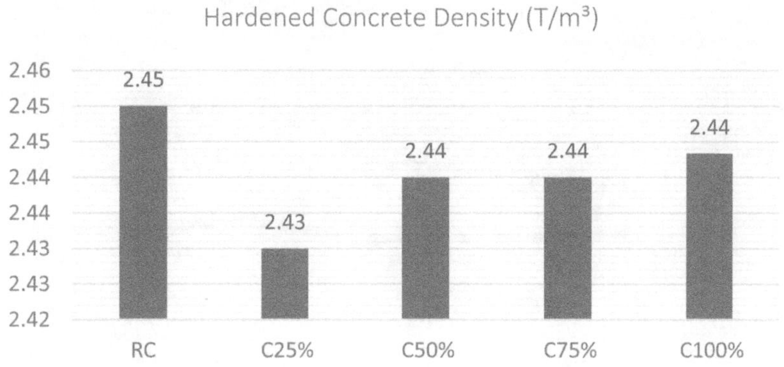

FIGURE 38.5 Density of hardened concrete.

As for fresh concrete, the results obtained for hardened concrete density confirm that the use of dredged sediment as conventional sand replacement does not have a significant impact on the density of concrete at its hardened state.

38.3.2.2 Mechanical Properties

Mechanical tests were performed for different formulas of concrete (RC, C25%, C50%, C75%, and C100%). Compressive strength was measured at 7 and 28 days, while tensile splitting strength and flexural strength were measured after 28 days. In addition, the modulus of elasticity was investigated to determine the mechanical performance of concrete mixtures.

38.3.2.2.1 Compressive Strength

Compressive strength is the most common measure of concrete strength, as it is the main criteria to decide whether the studied concrete will meet the needs of a specific project [60]; such as concrete maritime blocks in our case study. It is used to evaluate the performance of a given concrete mix, by measuring its ability to resist loads that will decrease its size [61]. Compressive strength test consisted of breaking cylindrical concrete specimens in a special machine (uniaxial compression testing setup) designed to measure this type of strength. The compressive strength values at 7, and 28 days of three batches were measured for each concrete mixture. Figure 38.6 shows the results of average compressive strength of the three obtained values.

According to the results obtained for the five types of concrete, an increase in the compressive strength over time was observed. A variation of 33%, 49%, 56%, 42%, and 33% was highlighted between 7 and 28 days for concretes RC, C25%, C50%, C75%, and C100%, respectively. In addition, the substitution of conventional sand by dredged sediment improved the compressive strength of the concretes mixture. The most significant improvement was observed for C75%. These findings can be explained by the fact that the fine particles of dredged sediment enhance the particle packing density of the cementitious system, which has a direct effect on the compressive strength of concrete made with it [62].

In order for concrete to withstand environmental aggression, the result fCE of the compression test at 28 days (average of measurements over the three batches) must satisfy the following two conditions [39]:

- Condition 1: $\text{fCE} > \text{fck} + (\text{CE} - \text{Cmin})$
- Condition 2: $\text{fCE} > \text{fck} + 3$

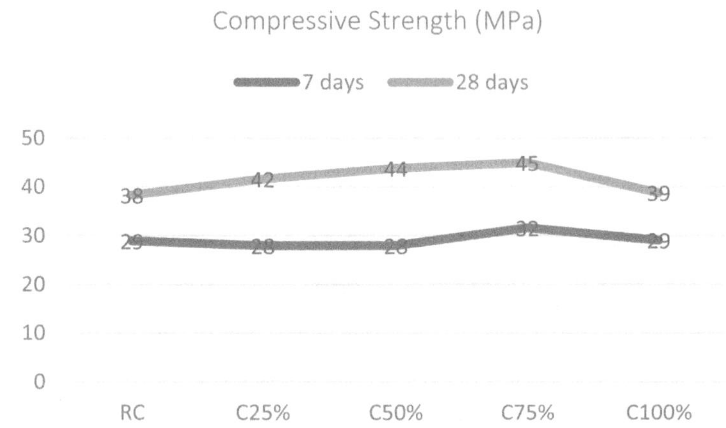

FIGURE 38.6 Compressive strength results for hardened concrete.

With:

- fck: the specified characteristic strength (30 MPa for B30);
- CE: the 28-day compressive strength of the cement used (48.3 MPa in our case study);
- Cmin: the minimum 28-day compressive strength value that can be met for the selected cement (42.5 MPa in our case study).

All the calculations done show that both conditions are satisfied for the five concrete mixtures. Therefore, the substitution of conventional sand by dredged sediment has no adverse impact on the compressive strength of concrete made with it.

38.3.2.2.2 Tensile Strength

Tensile strength reflects the ability of a given concrete to resist failure or cracking under tension. Direct testing of the tensile strength of concrete is not practical, so indirect methods are used. The most commonly used methods are splitting tensile strength and flexural strength [60].

- Splitting tensile strength: determined using a splitting tensile test on concrete cylinders.
- Flexural strength: determined by placing concrete samples between two points and then applying a load using a third point or two points.

Figure 38.7 shows the results of mechanical properties related to tensile strength.

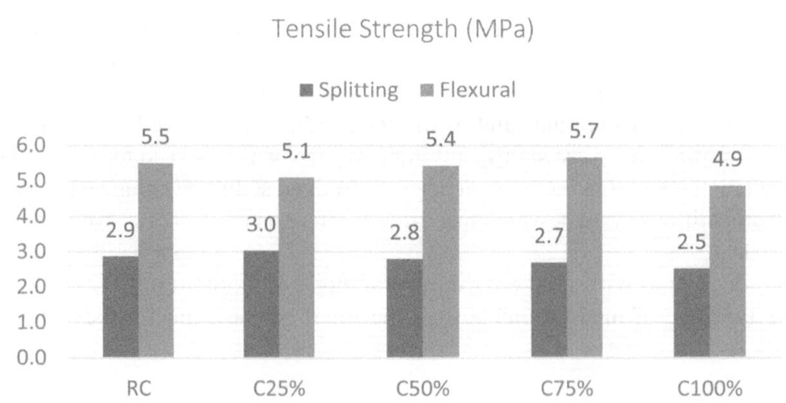

FIGURE 38.7 Tensile strength results for hardened concrete.

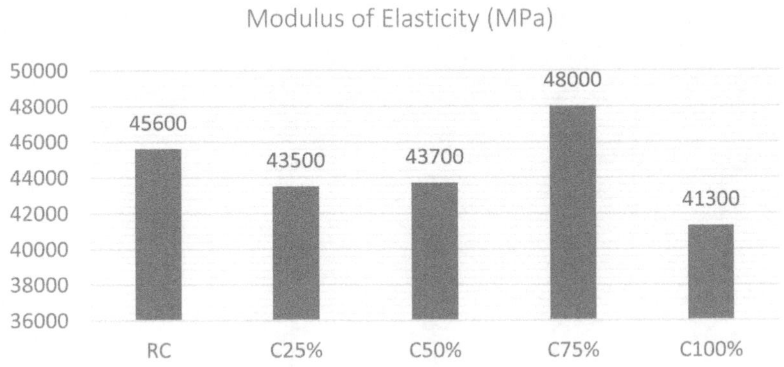

FIGURE 38.8 Modulus of elasticity for hardened concrete.

Results show that splitting tensile strength of concretes containing dredged sediment present a variation between 3% and 14% comparing to reference concrete, while the same variation for flexural strength is between 2% and 11%. Flexural strength of all types of concrete have values between 15% and 20% of their compressive strength at 28 days [63]. In a general way, concrete manufactured with dredged sediment show higher splitting tensile and flexural strengths at lower percentages of incorporated sediment (up to 25% for splitting tensile strength and up to 75% for flexural strength). These findings may be explained by the size fraction and the fineness of dredged sediment particles that improve the mortar phase [60, 62, 64].

38.3.2.2.3 Modulus of Elasticity

The modulus of elasticity of concrete is defined by the ratio of a strain to the elastic deformation caused by that strain. In order to calculate this mechanical property, cylindrical specimens of the five concrete mixtures were subjected to compression tests. The deformation of the specimens in relation to applied loads were recorded and analyzed during the tests. The modulus of elasticity of studied concrete was then determined from the load-deflection graph. Figure 38.8 shows the results of the modulus of elasticity for the five types of concrete.

The results of the modulus of elasticity of the concrete containing dredged material show a slight decrease compared to the result of 45,600 MPa for the reference concrete. A variation of 14%, 20%, 12%, and 4% was observed for concrete mixtures C25%, C50%, C75%, and C100%, respectively, in comparison with RC. It is almost the same variation observed for the compressive strength results. According to previous studies, the modulus of elasticity of concrete is influenced by three parameters; first, the curing conditions in the case of high strength concrete [60]; second, the fine aggregate properties at early age, such as the shape and the texture [65]; third, the properties of coarse aggregate, mainly regarding the stiffness and the relative volume [66].

38.3.3 In-situ Application

The in-situ application of the present case study was performed through the manufacturing of grooved cubical blocks employed to protect port infrastructures. We opted for Antifer blocks of unit volume 1 cubic meter. Figure 38.9 shows the shape and dimensions of the manufactured cubes.

Three samples of artificial concrete blocks were produced. Grooved cubical blocks were manufactured from concrete C50% and C75% (representing the best performances), as well as RC. A visual inspection of blocks obtained was performed after removal from the mold (24 hours after casting), and just before installation (15 days after casting). The verified aspects are color and

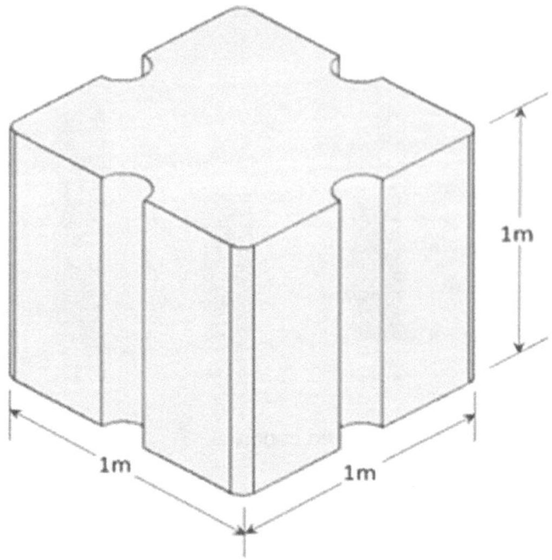

FIGURE 38.9 Manufactured Antifer blocks.

texture. Table 38.4 presents the visual aspects of inspected blocks. The results show that manufactured blocks do not present significant differences in terms of color. A slight decrease in porosity was observed for concrete blocks containing dredged sediment compared to concrete blocks made with conventional sand. This can be explained by the presence of fine particles in dredged sediment that fills the observed pores.

38.4 DISCUSSION

38.4.1 Feasibility of Dredged Sediment Reuse

This study explored the potential of using dredged sediment from the port of Safi as a substitute for conventional sand in the production of concrete B30, aimed at resisting marine environmental aggressions. The research involved three main phases: determining the optimal material mix based on the Dreux-Gorisse formulation method, assessing the physical and mechanical properties of both fresh and hardened concrete, and conducting in-situ application tests using Antifer concrete blocks.

Four concrete mixes were created with varying proportions of dredged sediment replacing conventional sand (25%, 50%, 75%, and 100%). The study found that concretes with dredged sediment exhibited physical and mechanical properties that met or exceeded the minimum standards required for maritime works. Among the tested mixes, concrete with 75% sediment substitution (C75%) showed the best performance, followed by the mix with 50% sediment (C50%). Concrete with 100% sediment (C100%) showed a significant drop in performance.

TABLE 38.4
Visual Inspection of Grooved Cubical Blocks

Concrete Type	24 Hours Age	15 Days of Age
RC	Dark gray, fairly porous	Light gray, fairly porous
C50%	Dark gray, fairly smooth surface	Light gray, fairly smooth surface
C75%	Dark gray, fairly smooth surface	Light gray, fairly smooth surface

The favorable results for concrete with up to 75% dredged sediment are attributed to the sediment's particle size, which is similar to that of conventional sand, allowing it to act effectively as a fine aggregate. Additionally, the fine particles in the dredged sediment helped reduce porosity by filling in pores. The visual inspection of the concrete blocks revealed no significant differences between those made with dredged sediment and those made with conventional sand. This suggests that dredged sediment can be effectively used without compromising the aesthetic quality of the concrete. Overall, the study confirms the feasibility of reusing dredged sediment as a partial substitute for conventional sand in the production of maritime concrete blocks. This approach not only provides a sustainable solution for managing dredged materials but also meets the necessary standards for use in marine environments.

38.4.2 Benefits of Dredged Sediment Reuse

From a general viewpoint, reuse of port dredged sediment as concrete aggregate offers a solution to many problems, mainly the adverse impacts of sea disposal of dredged material, and the scarcity of conventional sand linked to its over-exploitation. It presents a potential alternative to the increasing demand for construction materials, in the current context of rapid urbanization. From a perspective specific to the present case study, reuse of dredged material in maritime works presents supplementary advantages. First, the proximity of dredging zone to the construction sites enables to avoid the transportation of dredged material. Thus, benefits in terms of cost saving and CO_2 reduction will be achieved. Second, the reuse of marine sediment by port managers seems to be a valuable option to increase the acceptability of dredged waste by end users, and encourage the market demand.

In conclusion, the valorization of dredged sediment as aggregate of concrete subjected to marine environmental exposure is a technically feasible, economically viable and environmentally acceptable approach.

38.4.3 Limitations and Further Work

The properties of concrete produced using dredged sediment as a partial substitute for conventional sand were studied, focusing on its application in concrete maritime blocks used to protect harbor infrastructures. The results suggest that reusing dredged material in the manufacturing of these blocks does not significantly impact the concrete's properties. This practice can help mitigate environmental and ecological issues caused by sand extraction. However, the study has certain limitations. While the feasibility of using dredged sediment as concrete aggregate is assessed based on the properties of fresh and hardened concrete according to current standards, further research on the durability of the manufactured maritime blocks is necessary. Additionally, the economic feasibility and environmental impact of dredged sediment valorization should be evaluated before large-scale implementation.

Future research should also explore the use of dredged sediment in other types of maritime structures, both reinforced and unreinforced concrete. Among the study's limitations, it is essential to note that SEM (scanning electron microscopy) micrographs are highly requested to clarify changes in microstructures and investigate the effects of including dredged sand on concrete block characteristics. This detailed microstructural analysis will provide deeper insights into the performance and longevity of the concrete containing dredged sediment.

38.5 CONCLUSION

Waste valorization is a sustainable approach to address the scarcity of non-renewable resources in the construction sector. This study explores the potential of using dredged marine sediment from the Moroccan port of Safi as a substitute for construction aggregate in concrete, particularly for concrete blocks used in harbor infrastructure protection.

The research involved geotechnical, physical, chemical, environmental, and mineralogical characterization of the dredged sediment. The experimental program consisted of three main steps: determining the optimal material quantities for B30 concrete using the Dreux-Gorisse formulation method; testing the properties of fresh and hardened concrete with varying substitution rates of conventional sand with dredged sediment (25%, 50%, 75%, and 100%); and manufacturing and testing concrete maritime blocks using these formulations.

Results indicated that substituting conventional sand with dredged sediment did not negatively impact concrete properties. The best performance was seen in concrete with 75% and 50% sediment substitution. The properties of these concrete mixes were comparable to RC made from conventional materials, meeting standard requirements for workability, air content, and physical and mechanical properties. The presence of fine particles in the dredged sediment slightly decreased porosity by filling pores.

This study demonstrates that concrete made with dredged material has similar properties to concrete made with conventional aggregates, making it suitable for marine environmental exposure. The findings support the feasibility of valorizing dredged sediment as a construction material and provide a basis for further research on its use in other maritime structures.

REFERENCES

1. M. Hussain, D. Levacher, N. Leblanc, H. Zmamou, I. Djeran-Maigre, A. Razakamanantsoa, and L. Saouti, "Reuse of harbour and river dredged sediments in adobe bricks," *Clean. Mater.*, vol. 3, 2022, doi: 10.1016/j.clema.2022.100046.
2. H. Beddaa, I. Ouazi, A. Ben Fraj, F. Lavergne, and J. Torrenti, "Reuse potential of dredged river sediments in concrete : Effect of sediment variability," *J. Clean. Prod.*, vol. 265, p. 121665, 2020, doi: 10.1016/j.jclepro.2020.121665.
3. C. Sheehan, and J. Harrington, "Management of dredged material in the Republic of Ireland – A review," *Waste Manag.*, vol. 32, no. 5, pp. 1031–1044, 2012, doi: 10.1016/j.wasman.2011.11.014.
4. Y. C. Lim, Y. J. Shih, K. C. Tsai, W. D. Yang, C. W. Chen, and C. Di Dong, "Recycling dredged harbor sediment to construction materials by sintering with steel slag and waste glass: Characteristics, alkali-silica reactivity and metals stability," *J. Environ. Manage*, vol. 270, p. 110869, 2020, doi: 10.1016/j.jenvman.2020.110869.
5. G. H. P. Campmans, P. C. Roos, N. R. Van der Sleen, and S. J. M. H. Hulscher, "Modeling tidal sand wave recovery after dredging: Effect of different types of dredging strategies," *Coast. Eng.*, vol. 165, p. 103862, 2021, doi: 10.1016/j.coastaleng.2021.103862.
6. M. Zikra, S. Salsabila, and K. Sambodho, "Toward a better understanding of sediment dynamics as a basis for maintenance dredging in Nagan Raya port, Indonesia," *Fluids*, vol. 6, no. 11, 2021, doi: 10.3390/fluids6110397.
7. R. Bocchetti, D. Fattoreni, B. Pisanelli, S. Macchia, L. Oliviero, F. Pilato, D. Pelligrini, and F. Regoli, "Contaminant accumulation and biomarker responses in caged mussels, Mytilus galloprovincialis, to evaluate bioavailability and toxicological effects of remobilized chemicals during dredging and disposal operations in harbour areas," *Aquat Toxicol.*, vol. 89, pp. 257–266, 2008, doi: 10.1016/j.aquatox.2008.07.011.
8. J. Couvidat, V. Chatain, H. Bouzahzah, and M. Benzaazoua, "Characterization of how contaminants arise in a dredged marine sediment and analysis of the effect of natural weathering," *Sci. Total Environ.*, vol. 624, pp. 323–332, 2018, doi: 10.1016/j.scitotenv.2017.12.130.
9. N. Montero, M. J. Belzunce-segarra, J. Gonzalez, I. Menchaca, J. M. Garmendia, and N. Etxebarria, "Application of toxicity identification evaluation (TIE) procedures for the characterization and management of dredged harbor sediments," *Mar. Pollut. Bull.*, vol. 71, no. 1–2, pp. 259–268, 2013, doi: 10.1016/j.marpolbul.2013.01.038.
10. I. Bel Hadj Ali, Z. Lafhaj, M. Bouassida, and I. Said, "Characterization of Tunisian marine sediments in Rades and Gabes harbors," *Int. J. Sediment Res.*, vol. 29, no. 3, pp. 391–401, 2015, doi: 10.1016/S1001-6279(14)60053-6.
11. G. Barjoveanu, S. De Gisi, R. Casale, F. Todaro, M. Notarnicola, and C. Teodosiu, "A life cycle assessment study on the stabilization/solidification treatment processes for contaminated marine sediments," *J. Clean. Prod.*, vol. 201, pp. 391–402, 2018, doi: 10.1016/j.jclepro.2018.08.053.

12. M. Martins, P. M. Costa, J. Raimundo, C. Vale, A. M. Ferreira, and M. H. Costa, "Ecotoxicology and environmental safety impact of remobilized contaminants in Mytilus edulis during dredging operations in a harbour area : Bioaccumulation and biomarker responses," *Ecotoxicol. Environ. Saf.*, vol. 85, pp. 96–103, 2012, doi: 10.1016/j.ecoenv.2012.08.008.
13. M. Manni, F. F. de Albuquerque Landi, T. Giannoni, A. Petrozzi, A. Nicolini, and F. Cotana, "A comparative study on opto-thermal properties of natural clay bricks incorporating dredged sediments," *Energies*, vol. 14, no. 15, pp. 1–17, 2021, doi: 10.3390/en14154575.
14. S. Athira, and E. Subaida, "A Review on sustainable reuse of dredged sediments in earthwork infrastructures," *Proc. Int. Web Conf. Civ. Eng. a Sustain. Planet*, pp. 102–108, 2021, doi: 10.21467/proceedings.112.13.
15. B. P. Bose, and M. Dhar, "Dredged sediments are one of the valuable resources : A review," *Int. J. Earth Sci. Knowl. Appl.*, vol. 4, no. 2, pp. 324–331, 2022.
16. B. C. Guerra, and F. Leite, "Circular economy in the construction industry: An overview of United States stakeholders' awareness, major challenges, and enablers," *Resour. Conserv. Recycl.*, vol. 170, p. 105617, 2021, doi: 10.1016/j.resconrec.2021.105617.
17. M. Bagarani, A. De Vincenzo, C. Ievoli, and B. Molino, "The reuse of sediments dredged from artificial reservoirs for beach nourishment: Technical and economic feasibility," *Sustain*, vol. 12, no. 17, pp. 1–20, 2020, doi: 10.3390/SU12176820.
18. F. K. Aoual-Benslafa, D. Kerdal, M. Ameur, B. Mekerta, and A. Semcha, "Durability of mortars made with dredged sediments," *Procedia Eng.*, vol. 118, pp. 240–250, 2015, doi: 10.1016/j.proeng.2015.08.423.
19. H. Slimanou, D. Eliche-quesada, S. Kherbache, N. Bouzidi, and A. K. Tahakourt, "Harbor dredged sediment as raw material in fired clay brick production : Characterization and properties," *J. Build. Eng.*, vol. 28, p. 101085, 2020, doi: 10.1016/j.jobe.2019.101085.
20. R. Achour, R. Zentar, N. Abriak, P. Rivard, and P. Gregoire, "Case studies in construction materials," *Case Stud. Constr. Mater.*, vol. 11, p. e00244, 2019, doi: 10.1016/j.cscm.2019.e00244.
21. H. Beddaa, A. Ben Fraj, F. Lavergne, and J. M. Torrenti, "Reuse of untreated fine sediments as filler: Is it more beneficial than incorporating them as sand?" *Buildings*, vol. 12, no. 2, pp. 1–13, 2022, doi: 10.3390/buildings12020211.
22. S. H. Chu, and J. J. Yao, "A strength model for concrete made with marine dredged sediment," *J. Clean. Prod.*, vol. 274, p. 122673, 2020, doi: 10.1016/j.jclepro.2020.122673.
23. W. Maherzi, M. Benzerzour, Y. Mamindy-Pajany, E. van Veen, M. Boutouil, and N. E. Abriak, "Beneficial reuse of Brest-harbor (France)-dredged sediment as alternative material in road building: Laboratory investigations," *Environ. Technol.*, vol. 39, no. 5, pp. 566–580, 2018, doi: 10.1080/09593330.2017.1308440.
24. M. Amar, M. Benzerzour, J. Kleib, and N. E. Abriak, "From dredged sediment to supplementary cementitious material: Characterization, treatment, and reuse," *Int. J. Sediment Res.*, vol. 36, no. 1, pp. 92–109, 2021, doi: 10.1016/j.ijsrc.2020.06.002.
25. N. Junakova, and J. Junak, "Sustainable use of reservoir sediment through partial application in building material," *Sustain*, vol. 9, no. 5, pp. 1–13, 2017, doi: 10.3390/su9050852.
26. B. T. Smith, I. L. Howard, and F. Vahedifard, "Lightly cemented dredged sediments for sustainable reuse," *Environ. Geotech.*, vol. 5, no. 6, pp. 324–335, 2018, doi: 10.1680/jenge.16.00019.
27. Z. Zhao, M. Benzerzour, N. E. Abriak, D. Damidot, L. Courard, and D. Wang, "Use of uncontaminated marine sediments in mortar and concrete by partial substitution of cement," *Cem. Concr. Compos.*, vol. 93, no. June, pp. 155–162, 2018, doi: 10.1016/j.cemconcomp.2018.07.010.
28. H. Slimanou, D. Eliche-Quesada, S. Kherbache, N. Bouzidi, and A. K. Tahakourt, "Harbor dredged sediment as raw material in fired clay brick production: Characterization and properties," *J. Build. Eng.*, vol. 28, p. 101085, 2020, doi: 10.1016/j.jobe.2019.101085.
29. A. Loudini, M. Ibnoussina, O. Witam, A. Limam, and O. Turchanina, "Valorisation of dredged marine sediments for use as road material," *Case Stud. Constr. Mater.*, vol. 13, p. e00455, 2020, doi: 10.1016/j.cscm.2020.e00455.
30. M. Miraoui, R. Zentar, and N. Abriak, "Road material basis in dredged sediment and basic oxygen furnace steel slag," *Constr. Build. Mater.*, vol. 30, pp. 309–319, 2012, doi: 10.1016/j.conbuildmat.2011.11.032.
31. M. Balkaya, "Beneficial Use of Dredged Materials in Geotechnical Engineering BT - Recycling and Reuse Approaches for Better Sustainability," in *Recycling and Reuse Approaches for Better Sustainability*, N. Balkaya, and S. Guneysu, Eds., Cham: Springer International Publishing, 2019, pp. 21–38.
32. F. Ben Abdelghani, W. Maherezi, and M. Boutouil, "Geotechnical characterization of marine dredging sediments for use in road techniques," *Environ. Eng. Dev.*, vol. 66-mar, 2014, doi: 10.4267/dechets-sciences-techniques.269.

33. A. Kashi, A. A. Ramezanianpour, F. Moodi, and H. Malekitabar, "Effect of aggressive marine environment on strain efficiency factor of FRP-confined concrete," *Constr. Build. Mater.*, vol. 222, pp. 882–891, 2019, doi: 10.1016/j.conbuildmat.2019.07.055.
34. A. Minoubi, K. Elkhalidi, and M. Chaibi, "Impact of port works on the evolution of the coastline of the bay of Safi (Atlantic coast-Morocco)," *Moroccan Rev. Geomorphol.*, pp. 18–35, 2018, [Online]. Available: http://revues.imist.ma/?journal=remageom.
35. L. Charrouf, "Problems of silting of Moroccan ports on the Atlantic coast: Their sedimentological impact on the coast.," *La Houille Blanche.*, vol. 1, pp. 49–71, 1991, doi: 10.1051/lhb/1991004.
36. M. Bortali, M. Rabouli, M. Yessari, and A. Hajjaji, "Characterizing harbor dredged sediment for sustainable reuse as construction material," *Sustainability*, vol. 15, pp. 1–27, 2023.
37. AFGC, "Design of concrete for a given service life - Durability indicators," 2004. [Online]. Available: https://www.afgc.asso.fr/app/uploads/2007/10/Indicateurs-durabilite.pdf.
38. LCPC, "Control of the durability of concrete structures," 2010.
39. European Committee for Standardization, "EN 206+A2 - Concrete: Specification, performance, production and conformity," 2021.
40. C. Hamza, S. Bouchra, B. Mostapha, and B. Mohamed, "Formulation of ordinary concrete using the Dreux-Gorisse method," *Procedia Struct. Integr.*, vol. 28, pp. 430–439, 2020, doi: 10.1016/j.prostr.2020.10.050.
41. European Committee for Standardization, "EN 197-1 - Cement - Part 1: Composition, specifications and conformity criteria for common cements," 2012.
42. European Committee for Standardization, "EN 12350-7 - Testing fresh concrete - Part 7: Air content," 2019.
43. European Committee for Standardization, "EN 12350-6 - Testing fresh concrete - Part 6: Density," 1999.
44. European Committee for Standardization, "EN 12350-2 - Testing fresh concrete - Part 2: Slump test," 1999.
45. European Committee for Standardization, "EN 12390-7 - Testing hardened concrete - Part 7: Hardened concrete density," 2019.
46. European Committee for Standardization, "En 12390-3 - Testing hardened concrete - Part 3: Compressive strength," 2019.
47. European Committee for Standardization, "En 12390-6 - Testing hardened concrete - Part 6: Tensile splitting strength," 2009.
48. European Committee for Standardization, "EN 12390-5 - Testing hardened concrete - Part 5: Flexural stregth," 2019.
49. European Committee for Standardization, "En 12390-13 - Testing hardened concrete - Part 13: Determination of secant modulus of elasticity in compression," 2021.
50. I. Safari, F. Ropert, D. Mouaze, and A. Ezersky, "History of the artificial block used in embankment dikes," *Rev. Paralia.*, 2013, doi: 10.5150/revue-paralia.2013.004.
51. S. H. Mousavi, M. R. Kavianpour, and O. Aminoroayaie Yamini, "Experimental analysis of breakwater stability with antifer concrete block," *Mar. Georesources Geotechnol.*, vol. 35, no. 3, pp. 426–434, 2017, doi: 10.1080/1064119X.2016.1190432.
52. M. Castellino, M. Moroni, C. Cimorelli, M. Di Risio, and P. De Girolamo, "Riverbed protection downstream of an undersized stilling basin by means of antifer artificial blocks," *Water (Switzerland)*, vol. 13, no. 5, 2021, doi: 10.3390/w13050619.
53. K. Afshinnia, and P. R. Rangaraju, "Impact of combined use of ground glass powder and crushed glass aggregate on selected properties of Portland cement concrete," *Constr. Build. Mater.*, vol. 117, pp. 263–272, 2016, doi: 10.1016/j.conbuildmat.2016.04.072.
54. A. Richardson, K. Coventry, V. Edmondson, and E. Dias, "Crumb rubber used in concrete to provide freeze-thaw protection (optimal particle size)," *J. Clean. Prod.*, vol. 112, pp. 599–606, 2016, doi: 10.1016/j.jclepro.2015.08.028.
55. C. M. Cruz, E. M. Morais, R. H. F. Grillo, C. D. Roveri, and S. C. Maestrelli, "New mathematical modelling for grain size distribution adjustment in concrete," *Ceramica*, vol. 65, pp. 54–57, 2019, doi: 10.1590/0366-6913201965S12567.
56. W. Meng, M. Valipour, and K. H. Khayat, "Optimization and performance of cost-effective ultra-high performance concrete," *Mater. Struct.*, vol. 50, no. 1, p. 29, 2016, doi: 10.1617/s11527-016-0896-3.
57. B. M. Aïssoun, S.-D. Hwang, and K. H. Khayat, "Influence of aggregate characteristics on workability of super workable concrete," *Mater. Struct.*, vol. 49, no. 1, pp. 597–609, 2016, doi: 10.1617/s11527-015-0522-9.

58. M. PBenchrifa, M. Elouardi, G. Fattah, J. Mabrouki, and R. Tadili, "Identification, Simulation and Modeling of the Main Power Losses of a Photovoltaic Installation and Use of the Internet of Things to Minimize System Losses," in *Advanced Technology for Smart Environment and Energy*. Cham: Springer International Publishing, 2023, pp. 49–60.
59. J. Xiao, C. Qiang, A. Nanni, and K. Zhang, "Use of sea-sand and seawater in concrete construction: Current status and future opportunities," *Constr. Build. Mater.*, vol. 155, pp. 1101–1111, 2017, doi: 10.1016/j.conbuildmat.2017.08.130.
60. H. Constantinescu, O. Gherman, C. Negrutiu, and S. P. Ioan, "Mechanical properties of hardened high strength concrete," *Procedia Technol.*, vol. 22, pp. 219–226, 2016, doi: 10.1016/j.protcy.2016.01.047.
61. R. V. Silva, J. De Brito, and R. K. Dhir, "The influence of the use of recycled aggregates on the compressive strength of concrete: A review," *Eur. J. Environ. Civ. Eng.*, vol. 19, no. 7, pp. 825–849, 2015, doi: 10.1080/19648189.2014.974831.
62. J. Limeira, M. Etxeberria, L. Agulló, and D. Molina, "Mechanical and durability properties of concrete made with dredged marine sand," *Constr. Build. Mater.*, vol. 25, no. 11, pp. 4165–4174, 2011, doi: 10.1016/j.conbuildmat.2011.04.053.
63. F. Ghizlane, J. Mabrouki, F. Ghrissi, and M. Azrour, "Proposal for a high-resolution particulate matter (PM10 and PM2. 5) capture system, comparable with hybrid system-based internet of things: Case of quarries in the Western Rif, Morocco," *Pollution*, vol. 8, no. 1, pp. 169–180, 2022.
64. M. Ahmed, J. Mallick, and M. Abul Hasan, "A study of factors affecting the flexural tensile strength of concrete," *J. King Saud Univ. - Eng. Sci.*, vol. 28, no. 2, pp. 147–156, 2016, doi: 10.1016/j.jksues.2014.04.001.
65. H. Donza, O. Cabrera, and E. F. Irassar, "High-strength concrete with different fine aggregate," *Cem. Concr. Res.*, vol. 32, no. 11, pp. 1755–1761, 2002, doi: 10.1016/S0008-8846(02)00860-8.
66. F. P. Zhou, F. D. Lydon, and B. I. G. Barr, "Effect of coarse aggregate on elastic modulus and compressive strength of high performance concrete," *Cem. Concr. Res.*, vol. 25, no. 1, pp. 177–186, 1995, doi: 10.1016/0008-8846(94)00125-I.

39 Hydrogeochemical Characterization, Quality Assessment, and Vulnerability Mapping of the Tamouda Aquifer (Internal Rif, Morocco)

M. Chahid, J. Eddine Stitou El Messari, O. Kaibi, M. R. A. Mansour, A. Alitane, A. Oumou, A. Essahlaoui, A. V. Rompaey, I. Hilal, and A. El Ouali

39.1 INTRODUCTION

In Morocco, water resources face increasing pressures due to population growth, economic development, and reduced rainfall patterns caused by climate change [1]. This situation is exacerbated by the overexploitation of aquifers and their increased mineralization [2], which affects aquatic ecosystems [3]. Several factors lead to the deterioration of groundwater quality and an increase in its salinity, such as intensive agricultural practices [4], tourism development [5], high population density [6], and climate changes [7–9].

The excessive use of groundwater for agriculture, industry, and tourism has created a critical situation. Our research focuses on the Tamouda aquifer, located upstream of the Martil-Alila plain, which is important for supplying drinking water to Tetouan and nearby regions. This important resource is endangered by population growth, tourism, and other human activities. Understanding the dynamics within an aquifer is essential for maintaining groundwater quality. Our hydrogeochemical study of the Tamouda aquifer aims to achieve several goals, including the hydrogeochemical characterization of the water to prevent contamination that could impair the aquifer's capacity and ensure a reliable water supply. The study of hydrogeochemical speciation provides insights into the water's origin, its path, interactions between water and rock, and the impacts of human activities. We assess the groundwater quality in this aquifer using quality indices.

The vulnerability of aquifers to pollution is a complex issue. Vulnerability maps subjectively depict the subsurface's ability to protect groundwater resources. These maps aim to provide clear and accurate information about sensitivity to impacts, whether actual or potential, and facilitate comparisons between different aquifers or areas within the same aquifer. The creation of these maps typically involves combining or overlaying various thematic maps that detail selected physical factors to evaluate vulnerability. The study of groundwater vulnerability to pollution involves compiling data from multiple disciplines such as terrain topography, geology, hydrology, hydrogeology, and climatology [10].

DOI: 10.1201/9781003647263-39

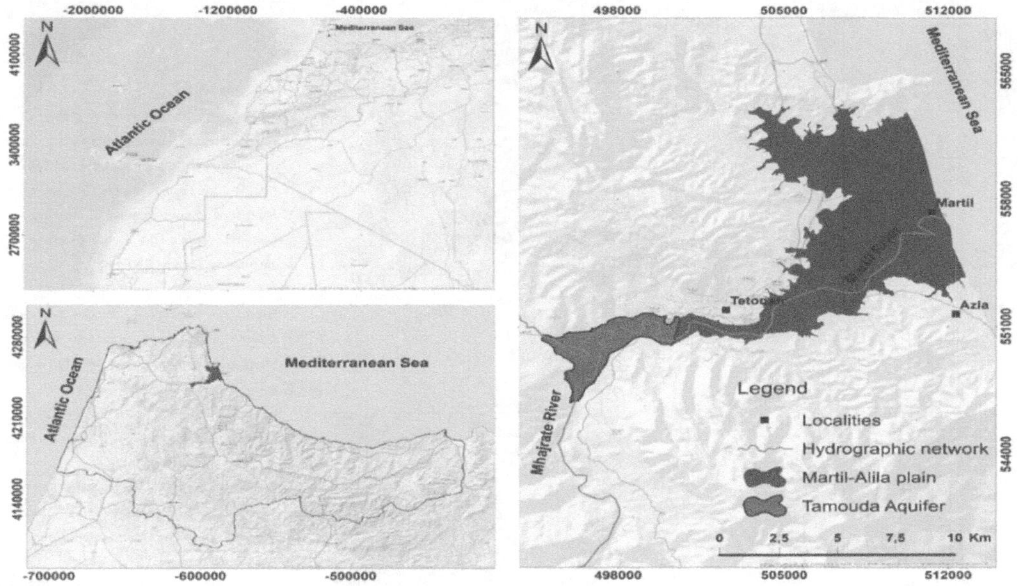

FIGURE 39.1 Geographic location of the Martil-Alila aquifer.

39.2 STUDY AREA

39.2.1 Geographical Situation

The Tamouda Plain, situated in the northwestern part of Morocco within the Tanger-Tetouan-Al Hoceima region, extends upstream from the Martil-Alila plain. It is intersected by the watercourses of the Martil and Mhajrate wadis, which are important for its water supply. This geographical region encompasses a total area of 16.40 km^2 (see Figure 39.1).

39.2.2 Geological Setting

The Tamouda region is characterized by its diverse geological formations, contributing to the complexity of its geological environment. To the north, carbonate layers from the Hettangian-Lias period are prevalent, accompanied by marls and limestone strata from the Turonian-Senonian period, as well as the blue marl of the lower Pliocene. Towards the south and west, similar geological formations are observed, with the boundaries of the Tamouda region delineated by late Quaternary river terraces (see Figure 39.2). Additionally, this area is characterized by the presence of sedimentary rocks, such as mudstones from the Lower Cretaceous. In the central part of the region, deposits of highly clayey silt predominate (see Figure 39.2). This specific sediment composition in central Tamouda plays an essential role in regulating the region's water resources.

39.2.3 Lithostratigraphy of Tamouda

The geological study of the Martil-Alila Plain, based on lithostratigraphic columns from mechanical boreholes conducted by the Loukkos Hydraulic Basin Agency, as well as vertical electrical soundings [11–14] and electrical tomography profiles, reveals that this plain comprises Plio-Quaternary alluvial deposits filling a geological depression. This depression resulted from the Mediterranean's collapse during a north-south extensional phase.

The primary stratigraphic and tectonic characteristics of these Plio-Quaternary alluvial formations are detailed in subsequent paragraphs. The Tamouda aquifer system comprises a sequence of detrital materials, including sediments from the Pliocene and Quaternary periods (see Figure 39.3).

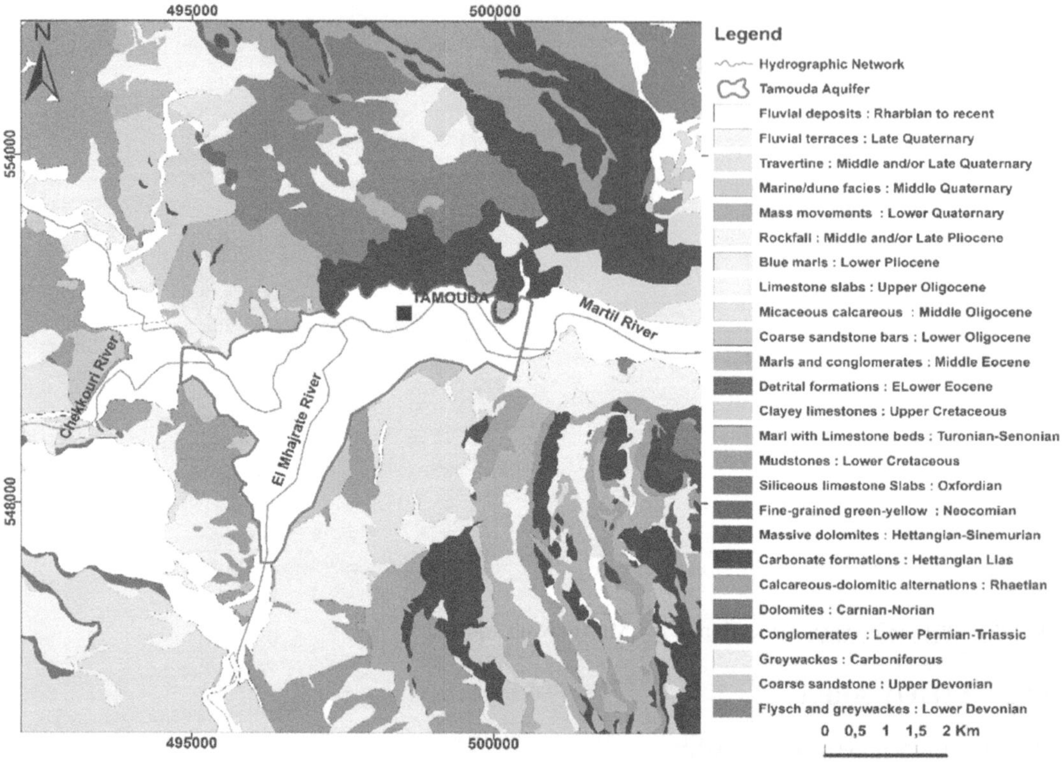

FIGURE 39.2 Geological map of the Tamouda aquifer (Modified and redesigned according to the 1/50000 geological map of Tetouan-Ras Mazari).

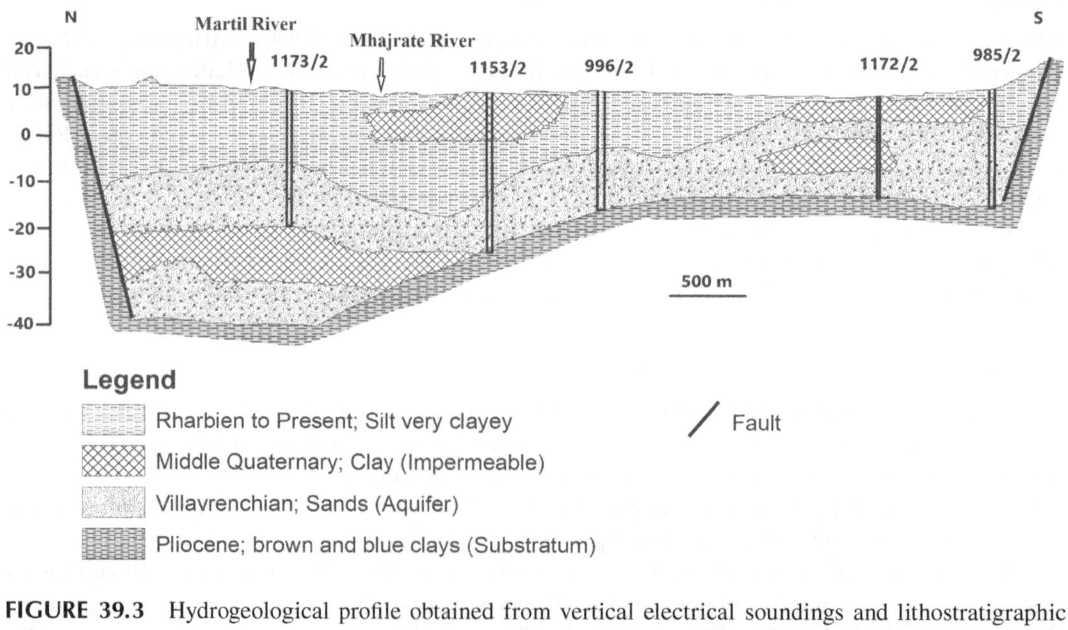

FIGURE 39.3 Hydrogeological profile obtained from vertical electrical soundings and lithostratigraphic columns.

The hydrogeological behavior of these differentiated materials in the study area is categorized as follows (see Figure 39.3):

Upper Quaternary. This layer contains detrital materials with sand at the base and highly clayey silt at the top. These formations, characterized by a distinctive yellow color and the presence of shell fragments, suggest a marine origin. The interspersed sand levels within this unit are highly permeable and constitute a significant focus of our study.

Middle Quaternary. This level is impermeable and divided into two parts: blue clays at the base and brown clays at the top.

Lower Quaternary. Composed of detrital materials from the Villafranchian age, this level includes coarse sands at the base and gravel at the top. The thickness of this layer varies between 26 and 35 m, functioning as a confined aquifer [15].

Pliocene. Characterized by a single lithological layer of brown and blue clays, the Pliocene acts as an aquiclude.

39.2.4 Hydrogeological Context

From a hydrogeological perspective, the flow moves eastward towards the Mediterranean Sea (see Figure 39.4). Various studies confirm depressions below sea level [5, 13]. In the Martil-Alila aquifer, depressions occur in Tamouda due to public water extraction (see Figure 39.4).

The piezometric levels fluctuate within the year, staying below 1.5 meters, correlating with precipitation patterns and runoff [11, 13]. Historical data shows a marked decline in these levels, referenced in [5, 13, 16, 17], attributed to prolonged drought and increased pumping withdrawals [15].

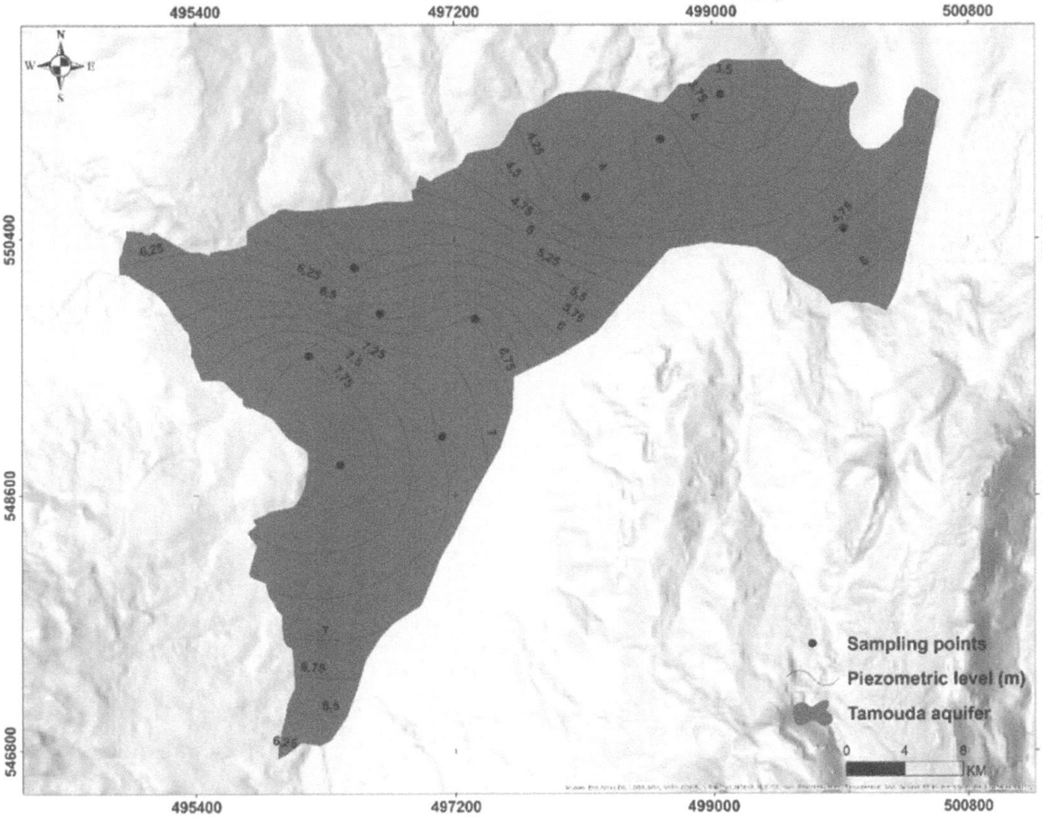

FIGURE 39.4 Piezometric map of the Tamouda aquifer.

39.3 MATERIAL AND METHODS

39.3.1 Selection of Sampling Points

The sampling point selection was meticulously planned after a reconnaissance mission on January 26, 2023. The aim was to achieve a strategically distributed coverage throughout the area, ensuring a thorough and balanced representation of the diverse hydrogeological conditions in the Tamouda aquifer.

Groundwater samples were collected from 10 wells over an area of 16 km^2, spanning various sectors of the study zone. Each location was selected from both banks of the Martil and Mhajrate rivers (see Figure 39.5). At every sampling site, data were gathered on water levels, well depths, and the types of fertilizers and pesticides used.

39.3.2 Field Sampling and Laboratory Analysis

On February 22 and 23, 2023, our team conducted a groundwater sampling mission in Tamouda. We recorded various in-situ parameters, including temperature, pH, electrical conductivity, and turbidity. For sample collection, we utilized a clean plastic bucket, ensuring its purity by rinsing it with the water under study beforehand. The collected samples were then transferred into 1-liter plastic bottles, sterilized with acetic acid and distilled water, tightly sealed, and labeled with collection details. Microbiological analyses were carried out in sterilized glass bottles, stored at 4°C in a cooler during transport to maintain integrity. Bacteriological tests were completed within 8 hours of collection to ensure accuracy.

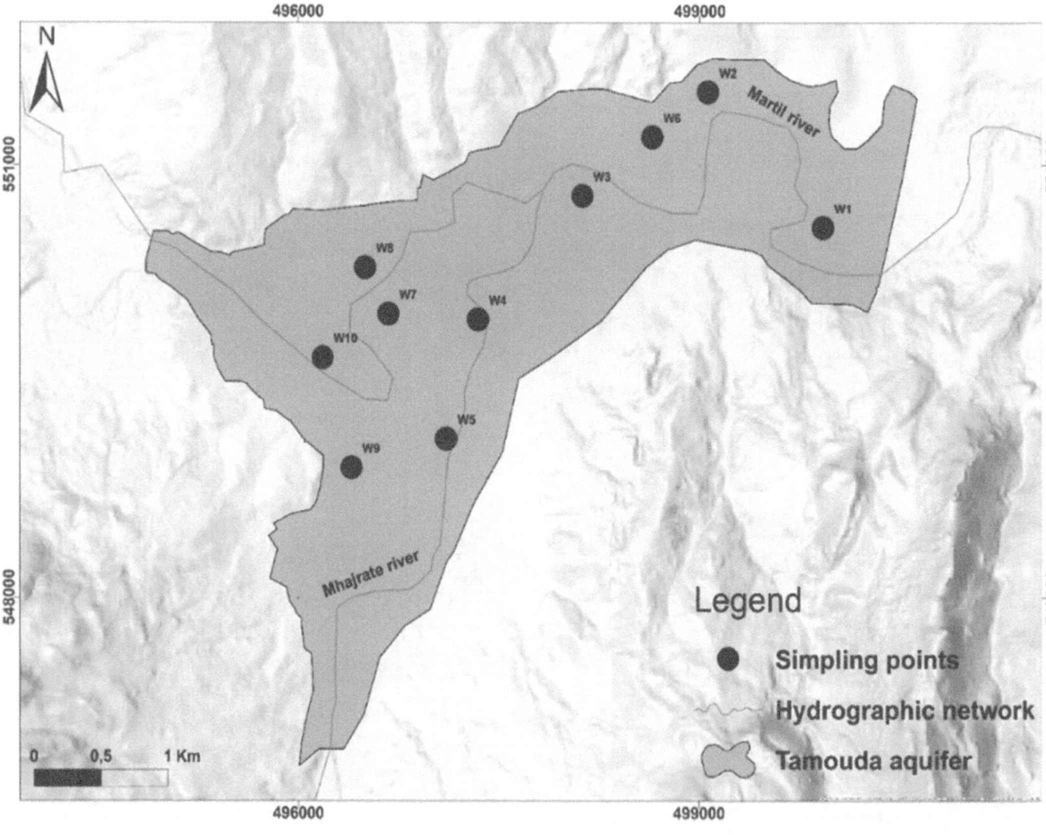

FIGURE 39.5 Localization of sampled wells in the study area.

Physicochemical analyses took place at the Loukkos Hydraulic Basin Agency laboratory in Tetouan, Morocco. These analyses encompassed various substances such as nitrites, nitrates, calcium, magnesium, total dissolved solids, organic matter, alkalinity, chlorides, sulfates, and bacterial content.

Potassium, sodium, and other trace element analyses were conducted at the laboratory of the National Center for Scientific and Technical Research in Rabat, Morocco.

39.3.3 Water Quality Index (WQI)

The water quality index (WQI) serves as a tool to categorize water quality by comparing various parameters against Moroccan or international standards.

Fundamentally, it streamlines the comprehension of voluminous water quality data by organizing it into distinct classes, including "Excellent," "Good," "Average," "Poor," and "Very Poor." Originally introduced by [18], this methodology utilizes the approach outlined by [19] for its computation (see Tables 39.1 and 39.2).

The procedure for calculating this index is outlined below:

39.3.3.1 Calculation of Relative Weights (Wi)

The relative weight (Wi) is calculated using the following equation:

$$\mathbf{Wi} = \sum_{i=1}^{n} \boldsymbol{qi} \times \boldsymbol{wi} \tag{39.1}$$

Wi: the relative weight
wi: the weight of each parameter
n: the number of parameters and qi is the quality rating scale (qi)

39.3.3.2 Calculation of the Quality Rating Scale (Qi)

The quality rating scale (qi) for each parameter is calculated by dividing the concentration (Ci) of each parameter by the Moroccan or WHO standard (Si) and then multiplying by 100.

$$qi = \left(\frac{Ci}{Si}\right) \times 100 \tag{39.2}$$

qi: Quality rating scale
Ci: Concentration of each parameter in mg/L
Si: Moroccan or WHO standard for each parameter in mg/L

39.3.3.3 Calculation of the Water Quality Index (WQI)

To calculate the WQI, the relative weight (Wi = 1/Si) is the first index to be determined by summing the weights of each parameter.

Then, the WQI for each sample is determined:

$$\mathrm{Wi} = \frac{\mathrm{K}}{\mathrm{Si}} \tag{39.3}$$

$$\mathrm{K}\left[\mathrm{constant}\right] = 1/\left[\left(1/\mathrm{S1}\right) + \left(1/\mathrm{S2}\right) + \left(1/\mathrm{S3}\right) + \ldots + \left(1/\mathrm{Sn}\right)\right.$$

The overall WQI is thus calculated by the following equation:

$$\mathrm{WQI} = \sum_{i=1}^{n} \mathrm{qi} \times \mathrm{Wi} \tag{39.4}$$

39.3.3.4 Water Classification According to the WQI

TABLE 39.1
Classification According to WQI [20]

Water Quality Index Classes	Water Quality
WQI < 50	Excellent quality
50 < WQI < 100	Good quality
100 < WQI < 200	Poor quality
200 < WQI < 300	Very poor quality
WQI > 300	Non-potable water

39.3.4 Pollution Index (PI)

The degree of water pollution was assessed and compared using the pollution index [21].

The calculation procedure of this index is as follows:

39.3.4.1 Calculation of Relative Weights (Rw)

A numerical value called the relative weight (Rw), ranging from 1 to 5, is assigned to each parameter, reflecting its degree of influence on water quality [22, 23].

A maximum weight of 5 has been assigned to NO_3^-, 4 to SO_4^{2-}, Cl^-, Na^+, and pH, 3 to HCO_3^-, σ, and a minimum weight of 1 to T °C and K^+.

39.3.4.2 Calculation of the Weights for Each Parameter (Wp)

The weight of each parameter (Wp) is the relative weight ratio (Rw) of each parameter divided by the total sum of these relative weights.

$$\text{Wp} = \frac{\text{Rw}}{\sum_{i=1}^{n}\text{Rw}} \tag{39.5}$$

Rw: Relative weight
Wp: Weight of each parameter

39.3.4.3 Calculation of the Quality Rating Scale (Qi)

The quality rating scale (qi) for each parameter is calculated by dividing the concentration of each parameter (Ci) by its corresponding standard for drinking water quality (Si):

$$\text{qi} = \left(\frac{\text{Ci}}{\text{Si}}\right) \tag{39.6}$$

The weight of each parameter (Wp) is multiplied by its corresponding qi to quantify the overall water quality (Qw). It reflects how many times Wp is increased based on the concentration of each parameter.

$$\text{Qw} = \text{qi} \times \text{Wp} \tag{39.7}$$

The water pollution index is calculated by adding up all the Qw values derived from the concentrations of each parameter.

$$\text{IP} = \sum_{i=1}^{n}\text{Qw} \tag{39.8}$$

39.3.4.4 Classification of PI

TABLE 39.2
Pollution Index Classification [24]

Pollution Index Classes	Water Quality
PI < 1	No pollution
1 < PI < 1.5	Low pollution
1.5 < PI < 2.0	Moderate pollution
2.0 < PI < 2.5	High pollution
PI > 2.5	Very high pollution

39.3.5 Mapping of Vulnerability to Pollution

The SINTACS methodology innovates by segmenting seven variables and assigning relative weights for groundwater quality assessment.

Each letter in the acronym corresponds to a specific factor: S for groundwater depth, I for effective infiltration, N for unsaturated zone attenuation capacity, T for soil type, A for saturated zone characteristics, C for hydraulic conductivity, and S for surface topographic slope.

This process leads to the derivation of a vulnerability index through Equation (39.9):

$$\text{The SINTACS index} = \sum_{i=1}^{7}(\text{Wf.Sf}) \tag{39.9}$$

With,

W: Weight of the factor (f)
S: Score of the factor (f)

The SINTACS method differentiates between normal and high-impact scenarios.

In the normal impact zone, with low topographic slope and sedimentary rock dominance, the focus lies on unsaturated zone parameters like water table depth and lithological composition.

Conversely, the high-impact zone, featuring intense human activities like agriculture and industry, emphasizes pollutant transfer to the aquifer. Key factors include water table depth, infiltration, and topography, weighted more heavily (Table 39.3).

TABLE 39.3
Weights of the SINTACS Factors [25]

Factors	Source	Normal Impact	High Impact
S: water table depth	Piezometric measurements	5	5
I: effective infiltration	Climatic data	4	5
N: unsaturated zone attenuation capacity	Drilling data	5	4
T: soil type	Drilling data	4	5
A: characteristics of the saturated zone	Drilling data	3	3
C: hydraulic conductivity	Drilling data	3	2
S: surface topographic slope	Digital elevation model	2	2

TABLE 39.4
SINTACS Vulnerability Classes

SINTACS Index	Vulnerability Class
[26 ; 80]	Very Low
[80 ; 105]	Low
[105 ; 140]	Moderate
[140 ; 186]	High
[186 ; 210]	Very High

39.3.5.1 SINTACS Index

Once the vulnerability map is generated, the SINTACS method offers a specific scale ranging from 26 to 260 to categorize pollution risk levels.

This scale is established based of 500 tests carried out across different regions of Italy (Table 39.4) [25].

39.4 RESULTS AND DISCUSSION

39.4.1 Spatial Distribution of Parameters

In the Tamouda region, analyses of water samples reveal total dissolved solids (TDS) concentrations ranging from 200 mg/L to 28,600 mg/L, with an average of 7,620 mg/L (see Figure 39.6). According

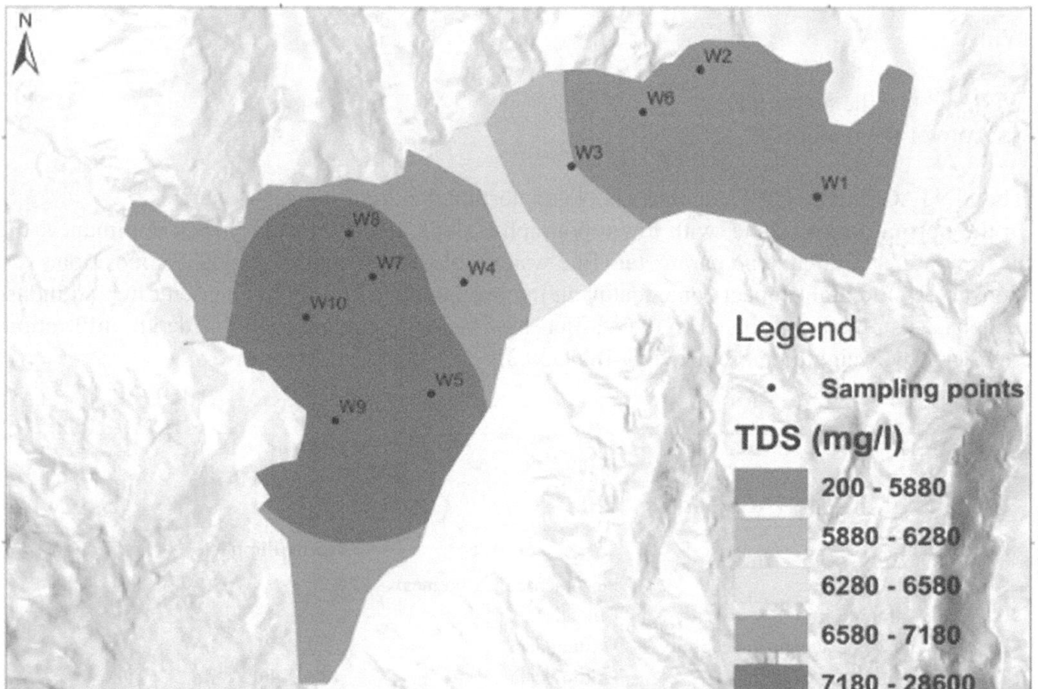

FIGURE 39.6 Maps of the spatial distribution of TDS.

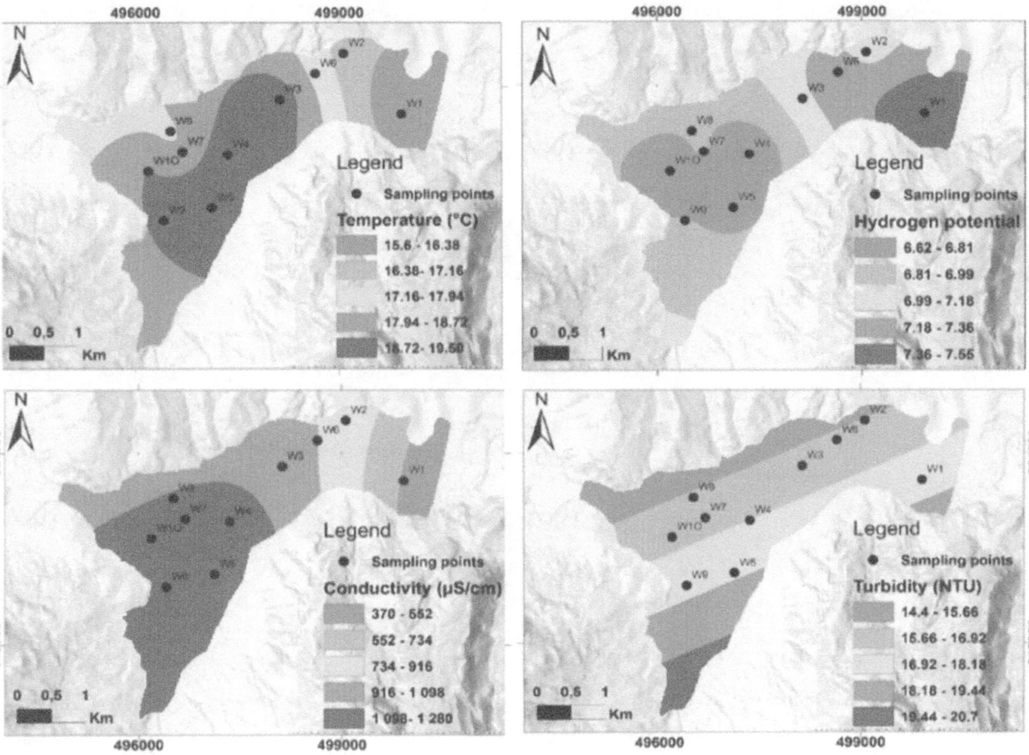

FIGURE 39.7 (a–d) Maps of the spatial distribution of in-situ parameters.

to the classification cited in reference [26], samples with TDS of 200 mg/L are considered freshwater, those between 1,000 mg/L and 3,000 mg/L as slightly salty, those between 3,000 mg/L and 10,000 mg/L (including the average of 7,620 mg/L) as moderately salty, and finally those between 10,000 mg/L and 28,600 mg/L as very salty.

Salinity increases as you move westwards. This trend is attributed to the geological characteristics of the region, which is mainly composed of silts and clays, and to human activities such as intensive agriculture. Irrigation in the west is more intensive than in the east.

Temperatures range between 15.6°C and 19.5°C, consistently below the maximum threshold for human consumption. A gradient from east to west correlates with increasing well depth (see Figure 39.7a). The pH levels fall within the range of 6.62 to 7.55, meeting WHO drinking water standards [27, 28]. This pH gradient decreases from east to west (see Figure 39.7b), indicating a geological interplay between acidic Tamouda sediments and the dissolution of alkaline limestone formations.

Electrical conductivity (σ) fluctuates between 370 µS/cm and 1,202 µS/cm, averaging 966 µS/cm ± 0.4 µS/cm (see Figure 39.7c), providing a precise measure of water mineralization well below the WHO (2011) standard of 2,700 µS/cm for groundwater.

Turbidity readings range from 14.4 to 20.7 NTU, below the Moroccan standard of 25 NTU. An increase in turbidity from north to south is attributed to silt and clay accumulation (see Figure 39.7d). The map displays elevated sodium levels at well 2 near the Martil River, likely due to contamination or interactions with sodium-rich rocks like clay or silt. Nearby points W5 and W4 along the Mhajrate River also exhibit high sodium content, probably influenced by surface water impacting groundwater (see Figure 39.8a).

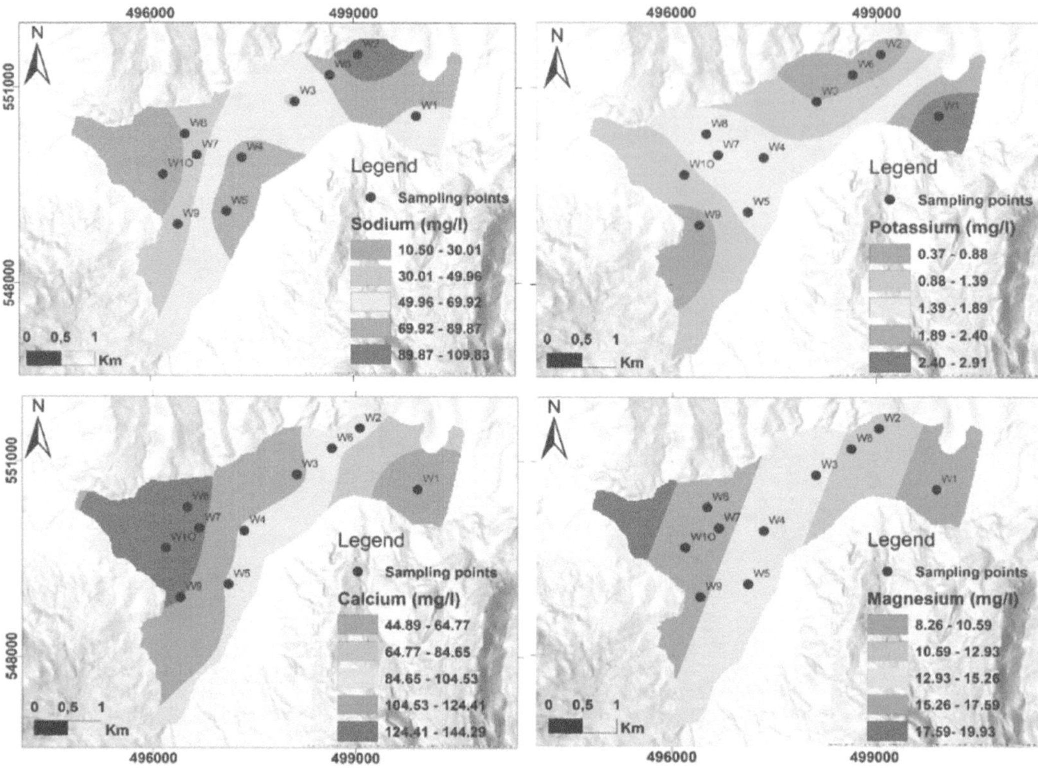

FIGURE 39.8 (a–d) Maps of the spatial distribution of major cations.

Potassium levels, ranging from 0.77 mg/L to 2.91 mg/L, remain well below the safe limit of 200 mg/L. Notably, higher concentrations are observed in the eastern regions, with peaks at well 1 and diminishing levels towards wells 2, 6, and 3, before increasing again at wells 4, 5, 7, and 8, and then declining towards the west (see Figure 39.8b), indicating a geological influence on potassium distribution.

Calcium concentrations span from 44.89 mg/L to 144.29 mg/L, peaking at well 10 but still below the 200 mg/L drinking water limit (see Figure 39.8c). This distribution shows an increase from east to west, with average levels in the central area, primarily sourced from limestone rocks influenced by rainwater and bicarbonate-rich limestone. Magnesium levels vary between 8.26 mg/L and 19.93 mg/L, peaking at well P9 yet remaining below Morocco's 50 mg/L drinking water limit (see Figure 39.8d). The lower levels near well P5 suggest that certain rocks retain magnesium north of the Mhajrate River, while the high concentrations at well 9 may result from the dissolution of magnesium-rich minerals.

The analysis of the Tamouda aquifer water unveils significant findings. Bicarbonate levels, exceeding the 180 mg/L drinking water standard, peak at points W5, W9, and W8 near the Martil River, decreasing towards the east and rising towards the central area (see Figure 39.9a).

Chloride concentrations, meeting drinking water standards, display a west-to-east gradient, with lower levels along the Mhajrate River and higher concentrations near wells 5 and 9, averaging 136 mg/L in the center (see Figure 39.9b), possibly influenced by agricultural practices.

Sulfate levels, remaining below the Moroccan standard limit of 400 mg/L, increase from east to west, reflecting human activities such as agriculture, industry, and urbanization, with average levels in the central area (see Figure 39.9c). Nitrate levels fluctuate, with most points adhering to the 50 mg/L limit, except for well 5, and elevated levels observed in the Mhajrate stream, south of the

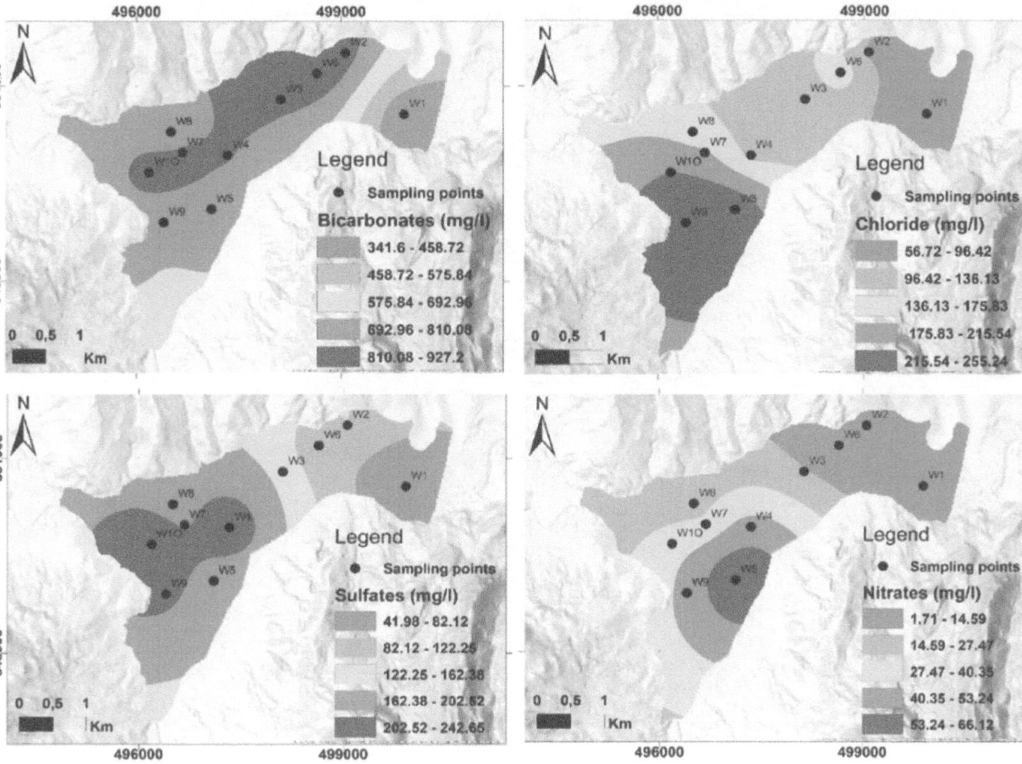

FIGURE 39.9 (a–d) Maps of the spatial distribution of major anions.

study area, due to fertilizer use and wastewater pollution, particularly near the Mhajrate River area (see Figure 39.9d).

39.4.2 Groundwater Hydrochemical Facies Evolution

Hydrochemical facies delineate the geochemical attributes of groundwater. Within the Tamouda aquifer, the Piper diagram illustrates a predominance of calcic and magnesian bicarbonate facies (see Figure 39.10).

This water, abundant in minerals such as calcite and dolomite, originates from the dissolution of carbonate rocks found in the limestone ridge.

39.4.3 Geochemical Speciation

The waters of Tamouda exhibit oversaturation with the following carbonate minerals: calcite, dolomite, and aragonite (see Figure 39.11).

39.4.4 Calculation of the Water Quality Index (WQI)

The WQI is a simple tool for assessing and determining water quality for drinking purposes [19, 20, 24]. We have observed that the majority of the obtained results were favorable. Indeed, 40% of the samples exhibited excellent quality according to the WQI, while 60% of them displayed good quality (see Figure 39.12a).

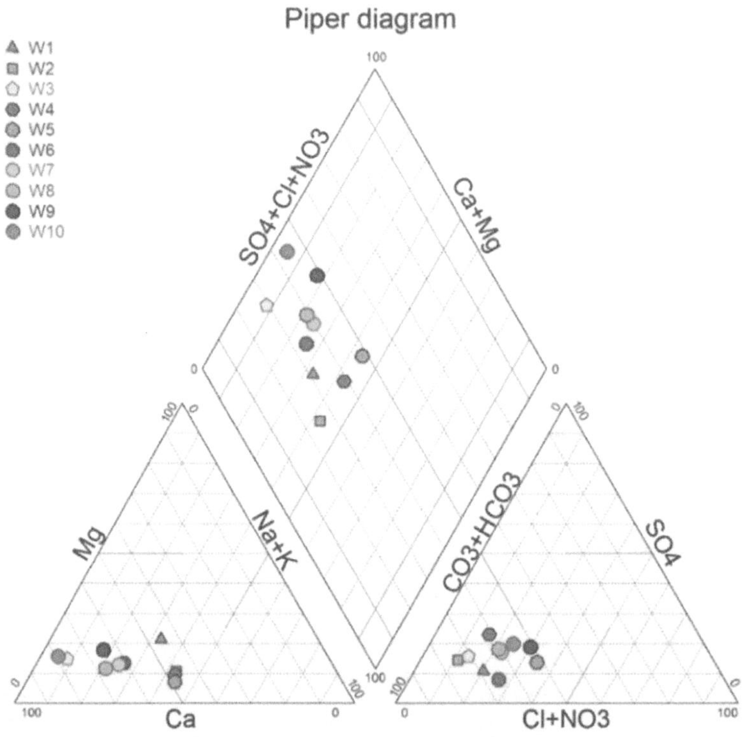

FIGURE 39.10 Hydrochemical facies of well water from the Tamouda Aquifer.

This indicates that the waters of Tamouda are generally of good quality, with a significant proportion meeting the criteria for excellence. It should be emphasized that, in our sampling, no sample was categorized as having average or poor quality according to the WQI. This suggests an absence of major issues or significant contamination for the evaluated parameters.

39.4.5 Calculation of the Pollution Index (PI)

All sampled wells had a pollution index below 1, indicating excellent water quality (see Figure 39.12b). This suggests effective protective measures in maintaining groundwater quality in the Tamouda

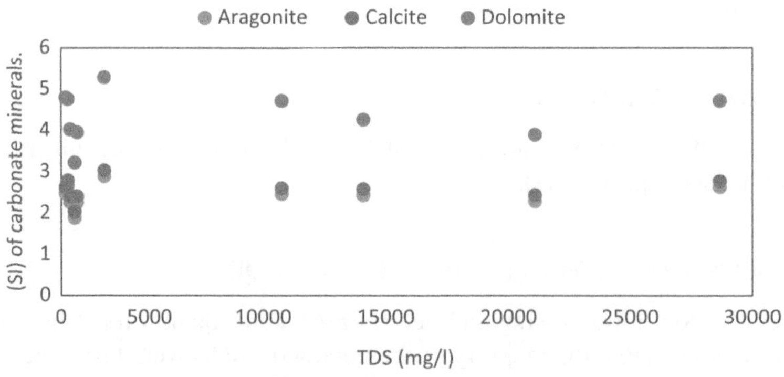

FIGURE 39.11 Saturation of the studied waters with respect to calcite, aragonite, and dolomite.

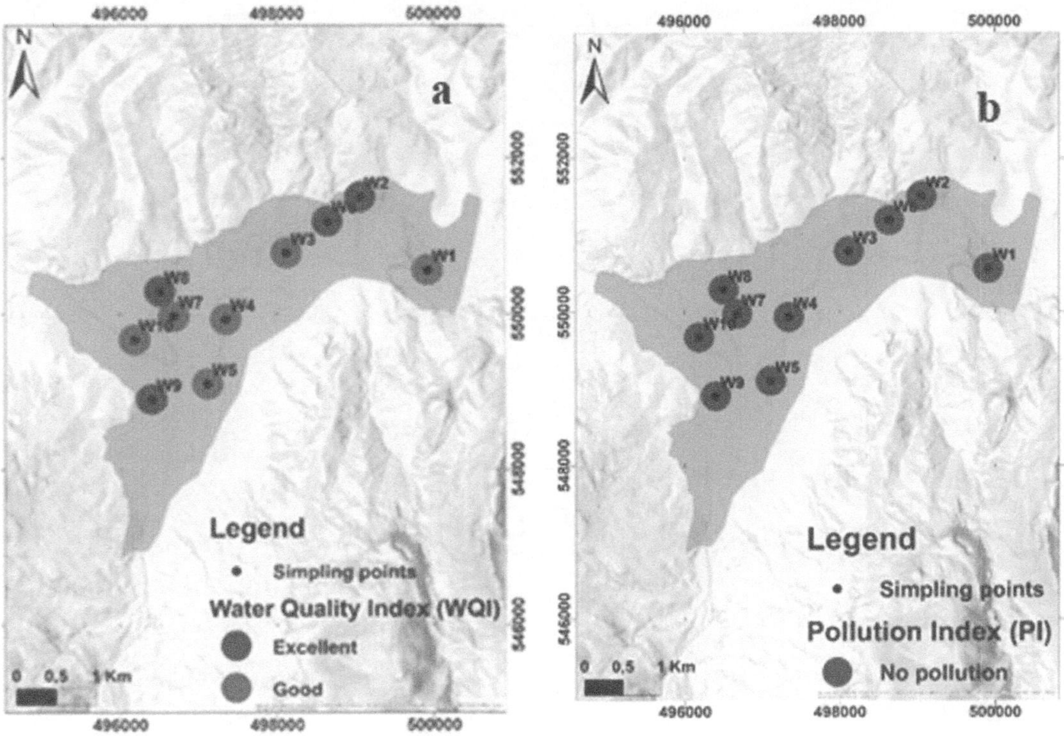

FIGURE 39.12 (a, b) Maps of the spatial distribution (PI) and (WQI).

aquifer. No exceptions were found among the sampled wells, enhancing result reliability. The findings are highly encouraging, suggesting minimal pollution in the Tamouda aquifer. Protective measures appear effective, ensuring groundwater quality. No exceptions among sampled wells bolster result reliability.

39.4.6 SINTACS Method

The SINTACS method customizes weights for hydrogeological specifics and human impacts. In the Tamouda Plain, two vulnerability maps were created due to increasing human activities.

The first map depicts the current regional conditions and typical human impacts. The second map anticipates a medium-term increase in human activities, highlighting areas with high anthropogenic impact.

39.4.6.1 SINTACS Normal Impact Zone

When combining the maps of the seven parameters with the weights of this zone, the aquifer demonstrates low to moderate vulnerability, being low in the east and moderate in the west (see Figure 39.13).

39.4.6.2 SINTACS High-Impact Zone

This scenario represents future conditions in our study area accurately. The vulnerability map indicates the disappearance of the low class, replaced by the moderate and high classes.

The region's socio-economic importance is increasing, leading to heightened resource pressure. In the "high-impact" scenario, the vulnerability map suggests greater vulnerability in the western part due to a shallower water table compared to the east (see Figure 39.14).

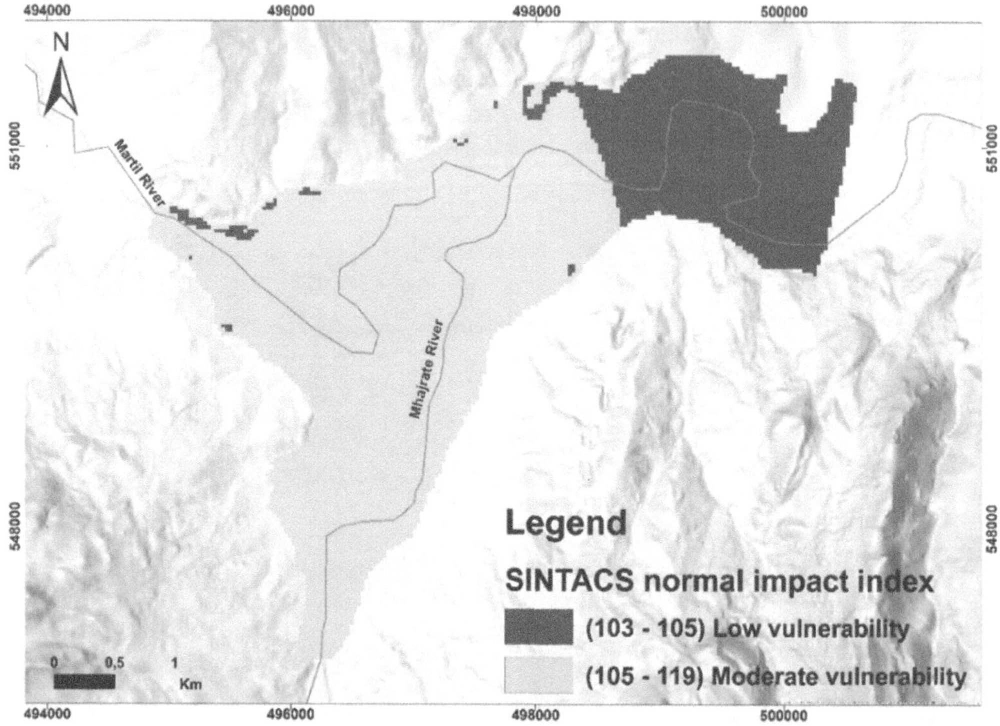

FIGURE 39.13 SINTACS index map with normal impact.

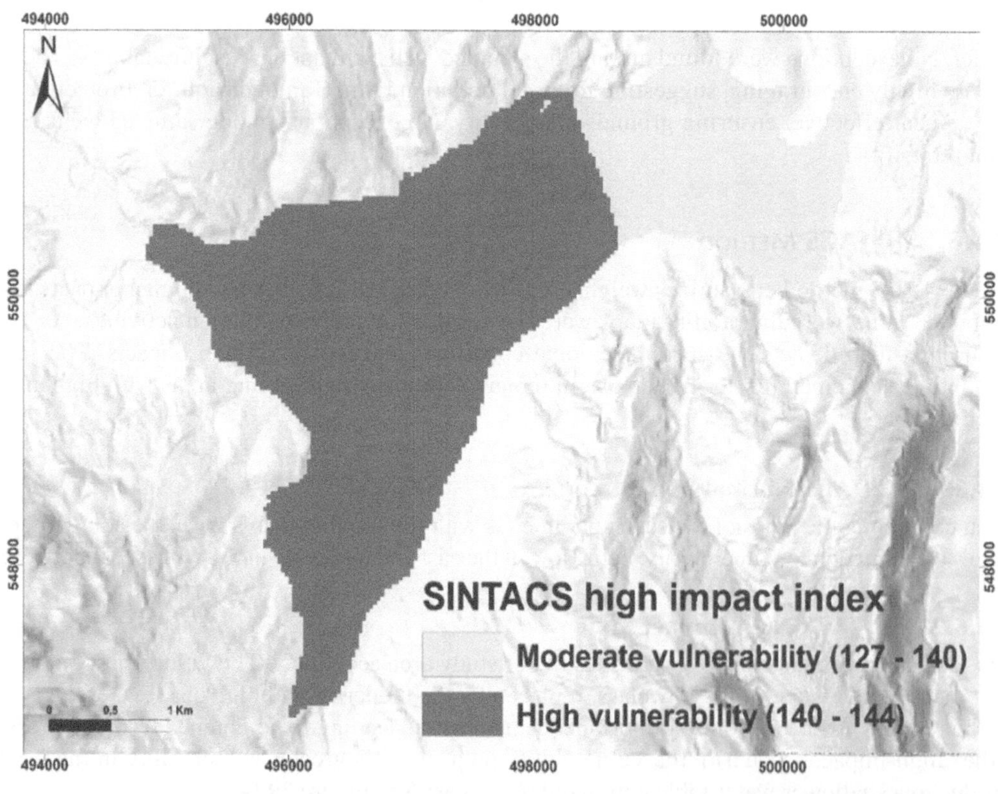

FIGURE 39.14 SINTACS index map with high impact.

39.5 CONCLUSION

The study analyzes the chemical composition of the groundwater in the Tamouda aquifer, focusing on its quality and vulnerability to contamination. Data collected from 10 wells in the region reveal that these waters are enriched with carbonate minerals, derived from the dissolution of limestones. The WQI, used to assess its suitability for human consumption, shows that 40% of the samples have excellent quality and 60% are of good quality. This indicates low pollution and confirms the high quality of this water. Furthermore, the SINTACS method was applied to assess the vulnerability of this aquifer to pollution. Two maps were produced, one reflecting the current situation and the other projecting medium-term forecasts with an expected increase in human activity. The first map shows a low to moderate vulnerability, while the second indicates a higher vulnerability, especially in the western areas where the aquifer is shallow. These results underline the importance of an integrated water resource management (IWRM) plan to protect this valuable aquifer from contamination. This need is particularly acute in its western part, where the aquifer is very close to the surface and thus requires enhanced protection.

REFERENCES

1. Ez-Zaouy Y, Bouchaou L, Saad A, Hssaisoune M, Brouziyne Y, Dhiba D, Chehbouni A. Morocco's coastal aquifers: Recent observations, evolution, and perspectives towards sustainability. Environ Pollut. 2022;293:118498. doi: 10.1016/j.envpol.2021.118498
2. Alabjah B, Amraoui F, Chibout M, Slimani M. Assessment of saltwater contamination extent in the coastal aquifers of Chaouia (Morocco) using the electric recognition. J Hydrol. 2018;566:363–376. doi: 10.1016/j.jhydrol.2018.09.003
3. Mirzavand M, Ghasemieh H, Sadatinejad SJ, Bagheri R. An overview on source, mechanism and investigation approaches in groundwater salinization studies. Int J Environ Sci Technol. 2020;17(4):2463–2476. doi:10.1007/s13762-020-02647-7
4. Abdelkarim B. Assessing and delineation of groundwater recharge areas in coastal arid area in Tunisia using GIS. Remote Sensing Techniques Self-Organizing Map. 2022:0–34. doi:10.21203/rs.3.rs-1160687/v1
5. Draoui J, Vias J, Andreo B, Targuisti K, El Messari JS. A comparative study of four vulnerability mapping methods in detritic aquifers under Mediterranean climatic conditions. Environ Geol. 2008;54.
6. Calabrese EJ, Tuthill RW. The influence of elevated levels of sodium in drinking water on elementary and high school students in Massachusetts. Sci Total Environ. 1981;18(C):117–133.
7. Bahir M, Ouhamdouch S, Ouazar D, El Moçayd N. Climate change effect on groundwater characteristics within semi-arid zones from western Morocco. Groundw Sustain Dev. 2020;11:100380. doi: 10.1016/j.gsd.2020.100380
8. Mongbo V, Ouendo EM, De Brouwere V, Alexander S, Dujardin B, Makoutodé M, Zhang WH. La césarienne de qualité: étude transversale dans 12 hôpitaux au Bénin. Rev Epidemiol Santé Publique. 2016;64(4):281–293.
9. Carol E, Galliari MJ, Santucci L, Nuñez F, Faleschini M. Assessment of groundwater-driven dissolved nutrient inputs to coastal wetlands associated with marsh-coastal lagoon systems of the littoral of the outer Río de la Plata estuary. Sci Total Environ. 2023;885.
10. Hmaidi A El, Ouali A El, Azzi Y, Essahlaoui A, Abdelhadi A, Ouali EL, Ayşe Muhammetoğlu, Habib Muhammetoğlu. Vulnerability assessment to pollution of groundwater of the Moroccan Atlas Causse (Region El Hajeb-Ifrane) by several methods using GIS. Am J Innov Res Appl Sci. 2016:2429–5396. www.american-jiras.com
11. El Morabiti K. Contribution à la connaissance hydrogéologique de l'aquifère de Martil-Alila (Tétouan, Maroc). Thesis, Université Abd Elmalek Essaâdi, Tetouan, Morocco. 1991.
12. Stitou El Messari J. Contribution à la connaissance hydrogéochimique des aquifères côtiers Martil-Alila, Oued Laou et Smir. Thèse de 3° cycle, Univ. Tétouan. 1995.
13. Stitou El Messari J. Etude de la salinité des eaux souterraines des aquifères côtiers Martil-Alila et Smir intégration des méthodes hydrogéochimiques, géophysiques et isotopiques. Thèse d'état, Univ. Tétouan. 2002.
14. Benabdelouahab S, Draoui M. Application de la prospection géophysique et des SIG pour étudier l'intrusion marine au sein de l'aquifère de Martil-Alila (Maroc septentrional). Publicaciones del Inst Geológico y Min España. 2012;29.

15. Jarar H. Etude hydrogéologique de la plaine de Tamouda (Aux environs de la ville de Tétouan). Proj fin d'étude, Maitrise Sci Tech géologie - hydrogéologie, Univ Tétouan, Maroc. 2001.
16. Manzali Y, El Far M, Chahhou M, Elmohajir M. Enhancing weak nodes in decision tree algorithm using data augmentation. Cybern Inf Technol. 2022;22(2):50–65.
17. Benabdelouahab S, Salhi A, Himi M, Eddine J, El S, Casas A, Mesmoudi H, Ben abdelfadel A. Using resistivity methods to characterize the geometry and assess groundwater vulnerability of a Moroccan coastal aquifer. Groundw Sustain Dev. 2018;7:293–304. doi: 10.1016/j.gsd.2018.07.004
18. Herschy RW. Water quality for drinking: WHO guidelines. Encycl Earth Sci Ser. 2012;876–883.
19. Alitane A, Essahlaoui A, Van Griensven A, Eisenreich S, Essahlaoui N, Elaaraj A, Kassou A, Essaied A. Assessment of drinking water quality based on trace elements concentrations in the semi-arid region, Morocco. E3S Web Conf. 2024;489.
20. El Yousfi Y, Himi M, El Ouarghi H, Aqnouy M, Benyoussef S, Gueddari H, Ait Hmeid H, Alitane A, Chaibi M, Zahid M, Essahlaoui N, Hitouri S, Essahlaoui A, Elaaraj A. Assessment and prediction of the water quality index for the groundwater of the Ghiss-Nekkor (Al Hoceima, Northeastern Morocco). Sustain. 2023;15(1).
21. De Jonge R, Van Esse HP, Maruthachalam K, Bolton MD, Santhanam P, Saber MK, Zhang Z, Usami T, Lievens B, Subbarao KV, Thomma BPHJ. Tomato immune receptor Ve1 recognizes effector of multiple fungal pathogens uncovered by genome and RNA sequencing. Proc Natl Acad Sci U S A. 2012;109(13):5110–5115.
22. Dhanasekarapandian M, Chandran S, Devi DS, Kumar V. Spatial and temporal variation of groundwater quality and its suitability for irrigation and drinking purpose using GIS and WQI in an urban fringe. J African Earth Sci. 2016;124:270–288. doi: 10.1016/j.jafrearsci.2016.08.015
23. Varol S, Davraz A. Evaluation of the groundwater quality with WQI (Water Quality Index) and multivariate analysis: a case study of the Tefenni plain (Burdur/Turkey). Environ Earth Sci. 2015;73(4):1725–1744.
24. Chahban M, Akodad M, Skalli A, Gueddari H, El Yousfi Y, Ait Hmeid H, Alitane A, Tabassum S, Saud Alam S, Husain Khan A, Sabi E, Zahmatkesh S. Hydrogeochemical assessment of aquifer salinization in north-eastern Morocco's Gueroaou coastal plain using statistical methods. Environ Res. 2024;244:117939. doi: 10.1016/j.envres.2023.117939
25. Civita MV. The combined approach when assessing and mapping groundwater vulnerability to contamination. J Water Resour Prot. 2010;2(1):14–28.
26. Robinove CJ, Langford RH, Brookhart JW. Saline-water resources of North Dakota. US Geol Surv Water Supply. 1958; 1428:1–77.doi: http://pubs.er.usgs.gov/publication/wsp1428
27. Dimane F, Haboubi K, Hanafi I, El Himri A, Andaloussi K. Impact Des Facteurs De Pollution Sur La Qualité Des Eaux De La Zone Aval De La Vallée De L'oued Nekor (Al-Hoceima, Maroc). Eur Sci J. 2017;13(3).
28. Barrocu G, Secci R, Uras G. The SINTACS method for evaluating aquifer vulnerability to pollution and saltwater intrusion. 2010;367–370.

40 Physical and Thermal Performance of Building Materials Based on *Washingtonia robusta* Fibers

H. Elmoudnia, P. Faria, R. Jalal, M. Waqif, and L. Saâdi

40.1 INTRODUCTION

The thermal properties of building materials play a pivotal role in determining their performance, especially in the context of sustainable and energy-efficient construction practices. Compressed earth bricks (CEBs) have emerged as a promising alternative in the construction industry due to their eco-friendly nature and abundant raw material source. Understanding the thermal behavior of CEBs is crucial for optimizing their application in various climates and building designs. This study delves into the thermal properties of compressed earth bricks, exploring factors such as conductivity, heat capacity, and insulation, to enhance our knowledge and inform the sustainable evolution of construction materials. The investigation into these thermal characteristics is instrumental in advancing the development of energy-efficient and environmentally conscious building solutions. The use of natural fibers in the construction sector has gained significant attention in recent years due to their sustainability, cost-effectiveness, flexibility, and mechanical properties [1]. Giroudon et al. [2] manufactured mud bricks using two different types of straw to gain a better understanding of the thermomechanical properties and durability of these materials. The results obtained showed that the highest thermal conductivity was obtained by barley straw blends, while lavender straw blends offered the best resistance to compression, erosion, dry abrasion, and resistance to mold growth. Caponetto et al. [3] evaluate the thermal and mechanical properties of cement–reed and lime–reed composites for energy-efficient buildings. It has been shown that thermal conductivities between 0.1709 and 0.2455 W/m.K are suitable for meeting non-bearing and thermal building block requirements. Hussien et al. [4] performed a study to evaluate the thermal and mechanical properties of wood pellets mixed with lime, sand, and clay for indoor insulation. It was observed that the mechanical properties of the samples reduced when the proportions of wood pellet shaving increased. Additionally, the results of thermal conductivity showed that the addition of fibers reduced the values from 0.37 W/m.K to 0.26 W/m.K. However, challenges such as moisture absorption and the need for chemical treatments must be addressed to fully harness their potential in various construction applications. This chapter presents an experimental investigation into the effect of *WRP* fiber content and compaction pressure on the bulk density of CEBs and subsequent impact on their thermal performance.

40.2 MATERIALS AND METHODS

40.2.1 Raw Materials

Washingtonia robusta fibers, as seen in Figure 40.1, derived from the petioles of the *W. robusta* palm, offer a unique and versatile material for various applications. These fibers exhibit notable strength and durability, making them suitable for use in diverse industries [5].

DOI: 10.1201/9781003647263-40

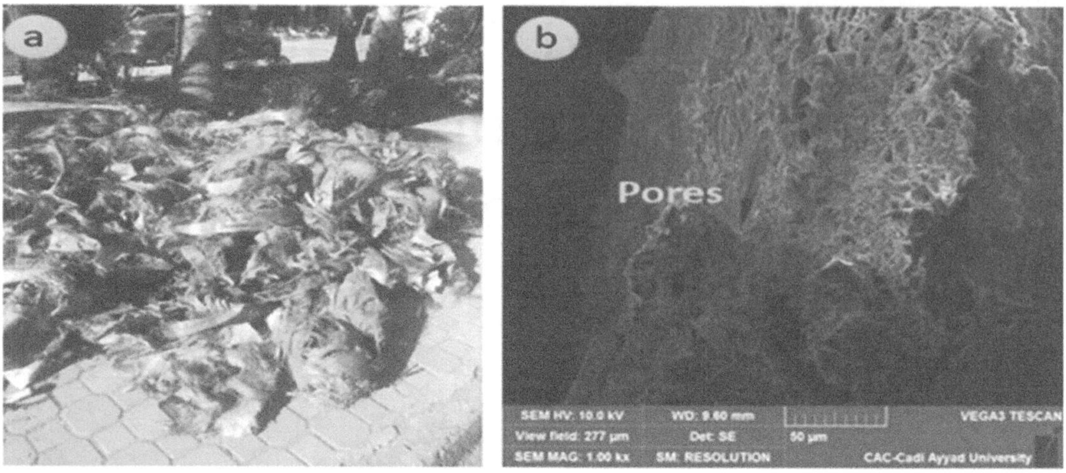

FIGURE 40.1 *Washingtonia robusta* fiber: (a) Waste fiber, (b) cross-section view.

The *W. robusta* fibers were collected from green spaces in the city of Marrakech, Morocco. The fibers are extracted from the petiole (renewable part of the palm tree). After drying under normal conditions, the petiole stems are cut into small pieces and then crushed. Granulometry and the separation into fibers of different diameters were achieved through a sieve belt (Figure 40.2). The selected fibers are in the range of 1–10 mm (Table 40.1).

In this study, a soil sample from the Tamaslouht area in the municipality of Marrakech (Figure 40.3) was used as the matrix due to its availability and abundance in the region. The soil was characterized according to the relevant Moroccan standards (Table 40.2).

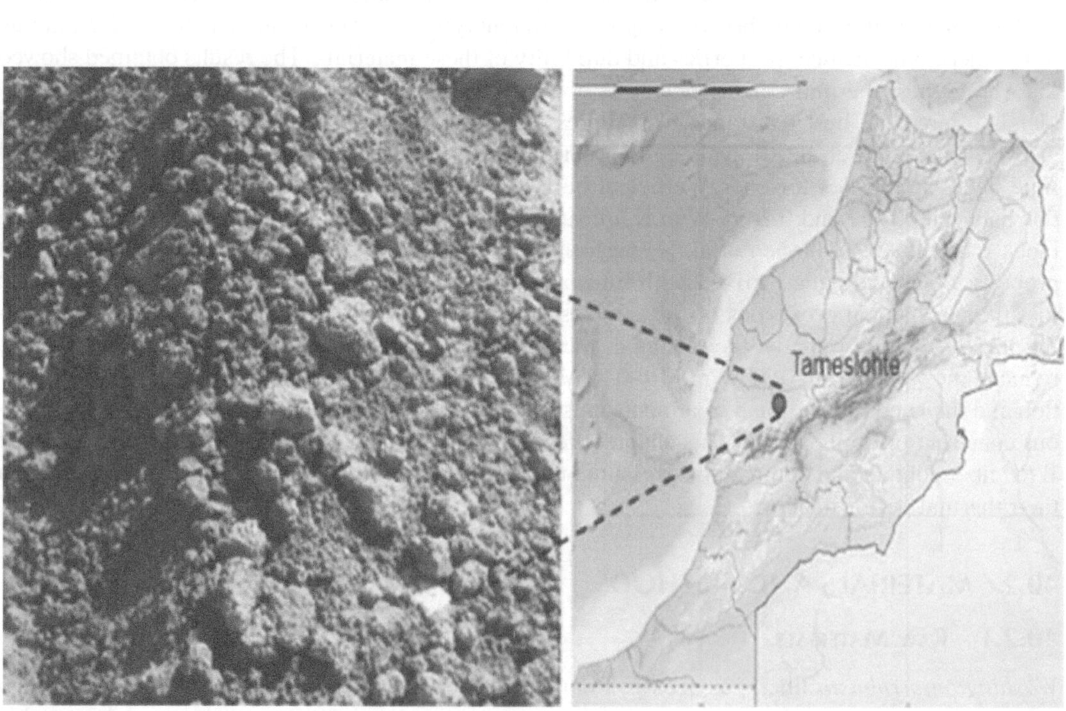

FIGURE 40.2 Schematic representation of chemical, physical, and geotechnical tests conducted on soil.

TABLE 40.1
Chemical Composition and Density of *Washingtonia robusta* with Other Natural Fibers

Fiber	Cellulose (%)	Hemicellulose (%)	Lignin (%)	Ash (%)	Density (g/cm³)	References
Washingtonia robusta	40	19.34	23.5	5	1.2	*Current study*
Pennisetum alopecuroides	40	29.5	19.66	6.76	0.87	[1]
Alfa	39.53	27.63	19.53	–	–	[6]
Phoenix sp.	61.13	12.56	19.91	7.79	1.25	[7]
Areca palm	57.49	18.34	7.26	1.43	1.090	[8]
Hibiscus vitifolius	75.09	13.34	10.42	0.94	1.530	[9]
Oil palm	49.6	18	21.2	–	1.55	[10]
Tunisian palm	43	8	35	–	1.33	[11]
Agave americana	68.54	18.41	6.08	–	1.035	[12]
Sansevieria	40.91	19.44	14.82	–	1.159	[13]

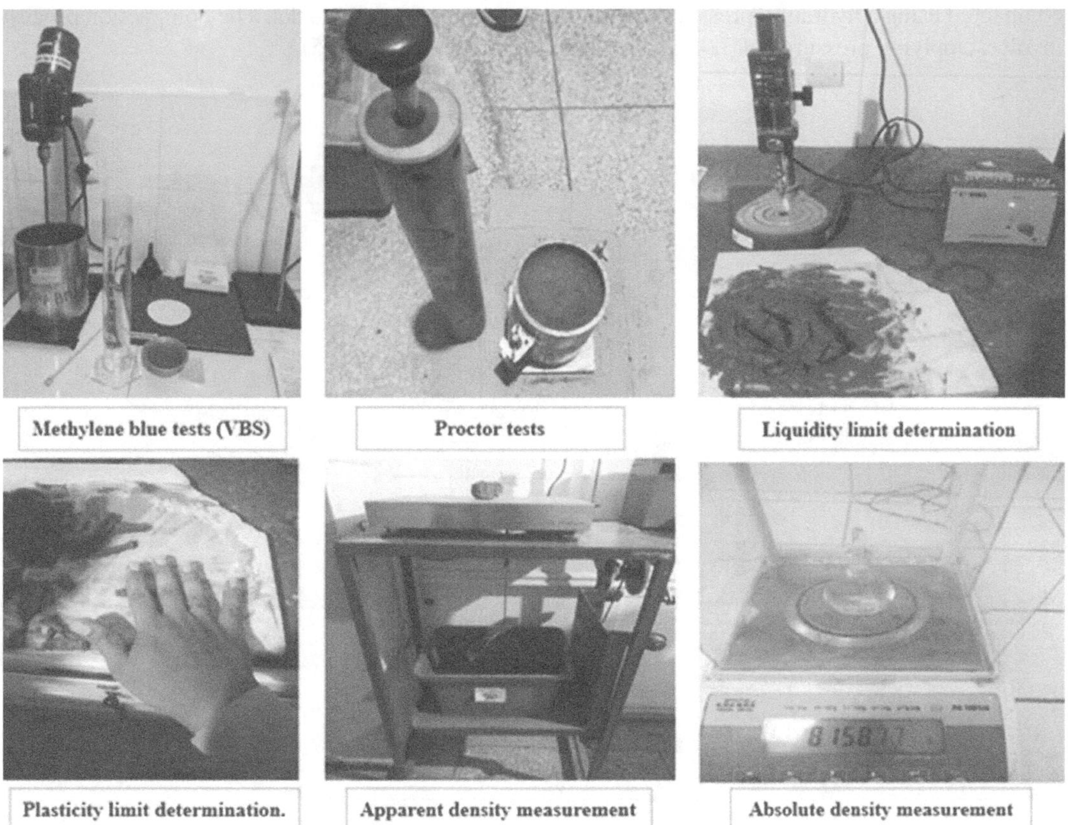

FIGURE 40.3 Presentation of the soil sampling site studied (ST).

TABLE 40.2
Chemical, Physical, and Geotechnical Characteristics of the Soil [14–21]

Property	Value	Standards
Liquid limit %	34	NM 13.1.012 2015
Plastic limit %	18	NM 13.1.012 2015
Plasticity index, PI	16	NM 13.1.012 2015
Absolute density (kg/m^3)	2580	NM 00.08.098 2015
Apparent density (kg/m^3)	1800	NM 13.1.119 2009
Maximum dry density (ρ_{dopt})	1690	NM 13.1.023 2019
The optimum moisture content (W_{opt})	14%	NM 13.1.023 2019
VBS	2.7	NM 13.1.178 2018
pH	8	NM ISO 10390 2013
Organic matter content	1.7	NFP 94-047 AFNOR 1998

40.2.2 Preparation of the Sample

The soil undergoes crushing and sieving through a 2 mm sieve, after which it is blended with *WRP* fiber ranging from 1 mm to 10 mm in length at varying percentages (0.5%, 1.5%, 2.5%, 3.5%, and 4.5%). Manual mixing is employed, incorporating the optimal water percentage (18%). The mixture was then introduced into a parallelepiped mold (80 × 60 × 40 mm³). Subsequently, they are compacted using a hydraulic press to manufacture compressed soil blocks. The compaction process involves applying pressures of 2, 4, and 6 MPa (Figures 40.4 and 40.5).

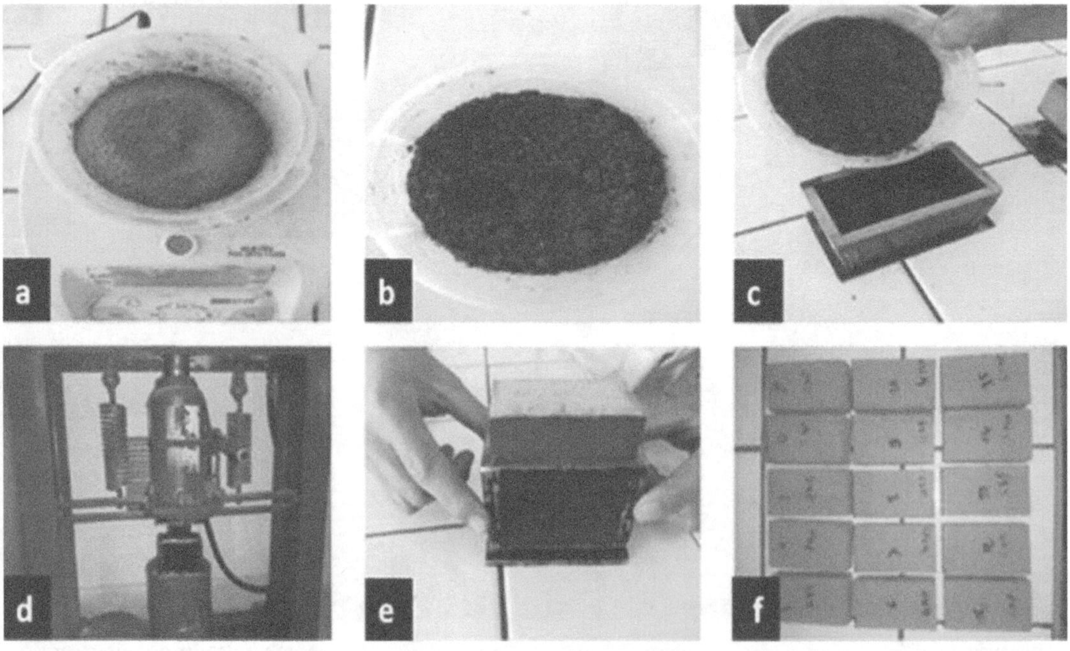

FIGURE 40.4 Preparation of test specimens: (a) weighed mixture; (b) the prepared dough; (c) put the mixture into the mold; (d) compression step; (e) demolding; (f) the samples prepared after drying.

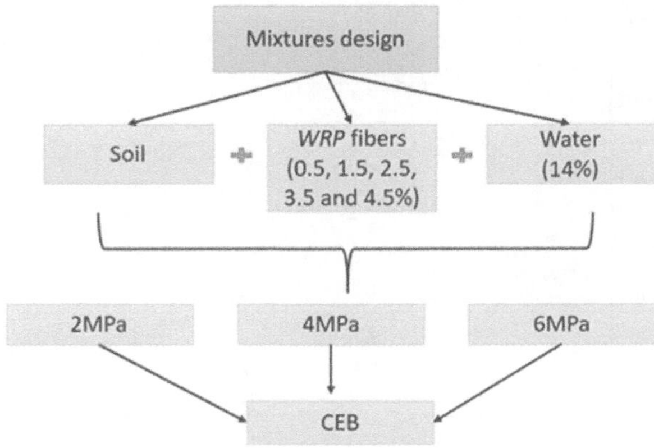

FIGURE 40.5 Mixture design of CEB.

40.2.3 Density Measurement

The test specimens were kept in the oven (60 ± 5°C) for 24 hours. After being removed from the oven, they were weighed and the following calculations were carried out:

$$\text{Density}\left(g/cm^3\right) = M/V \tag{40.1}$$

Where:

M = the mass of the test piece (g);
V = the volume of the test specimens (cm^3).

40.2.4 Thermal Conductivity Measurement

The hot disk method (Figures 40.6 and 40.7), derived from the transient plane source (TPS) technique, is a sophisticated approach to measuring thermal conductivity, thermal diffusivity and specific heat by volume. Following the standard (ISO 22007-2, 2015), this method can provide several thermal

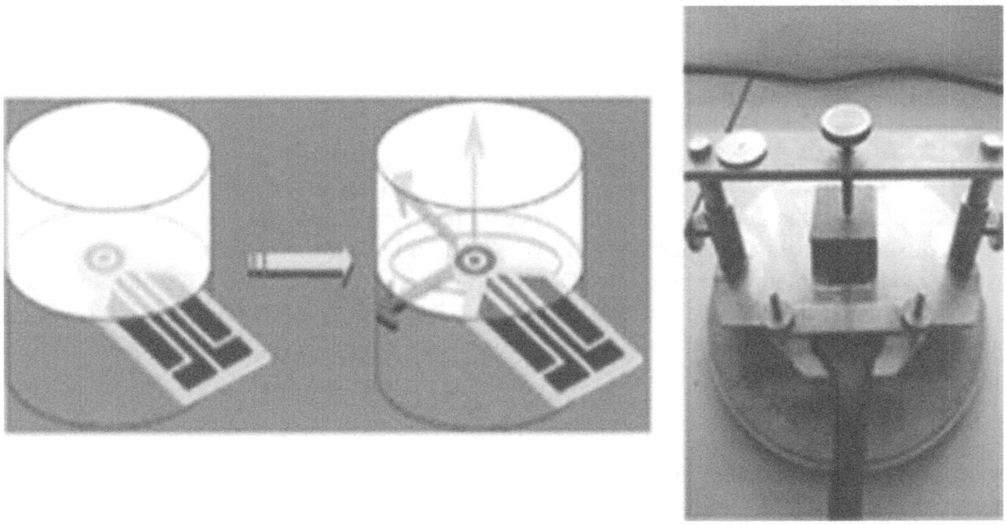

FIGURE 40.6 Diagram illustrating the principle of conductivity measurement.

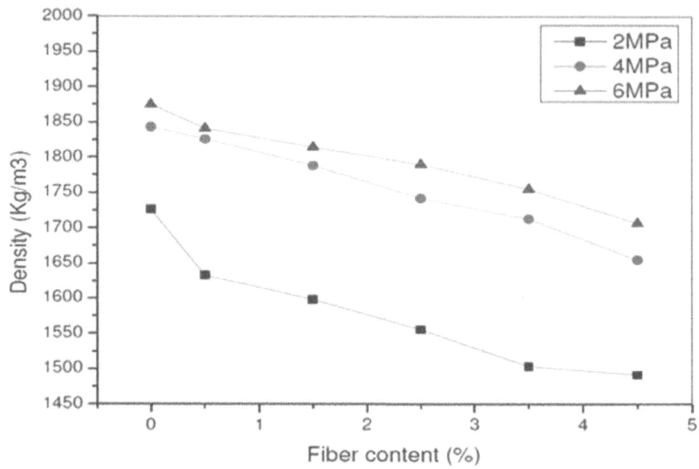

FIGURE 40.7 Density as a function of fiber content.

properties simultaneously and quickly by exploiting the transient phase of conductive heat transfer. The probe is positioned between two parallel-piped specimens slated for characterization. The probes come in various diameters, allowing for the analysis of samples of any size. Before the test, the specimens are dried at 60°C and placed in a desiccator to cool without reabsorbing moisture. Upon leaving the desiccator, they are carefully wrapped in plastic film to prevent any moisture absorption during the relatively lengthy measurement process. For a more in-depth understanding of the theoretical framework and the equations governing the hot disk method, we recommend consulting the work of [22].

40.3 RESULTS AND DISCUSSION

40.3.1 Physical and Chemical Composition of *WRP* Fibers

Several publications have discussed the chemical composition of materials derived from plants, but the processes used are not the same. Table 40.1 lists the composition of *W. robusta* fiber compared with other natural fibers. *WRP* fiber *contains 40% cellulose, 19.34% hemicellulose, and 23.5% lignin.* The physicochemical characteristics of natural fibers differ based on the geographical locations and climatic conditions in which the plants are cultivated [23]. The density of *WRP* fibers was 1.2 g/cm^3, thus indicating their suitability for lightweight composites.

40.3.2 Geotechnical Characterization of the Soil

Table 40.2 presents the geotechnical and physical characteristics of the soil analyzed. The composition of soil according to Atterberg limit is 34% liquidity limit, 18% plasticity limit, and 16% plasticity index. The methylene blue value (VBS) has been determined to be 2.7, which suggests that our clay falls under the classification of silty clay according to the GTR (Guide des Terrassements Routiers) classification system. The absolute density is determined to be around 2.58 g/cm^3.

40.3.3 Density

The density was identified as a significant indicator of block strength and durability. Figure 40.7 illustrates the density variation in soil-plant fiber specimens relative to fiber content. For the three compaction pressures (2 MPa, 4 MPa, and 6 MPa), the dry density of the composites decreases with increasing fiber content. It varies from 1726 kg/m^3 for the reference sample (ST0-2MPa), to 1492 kg/m^3 for the composite containing 4.5% *WRP* fibers (ST- 4.5 *WRP*-2MPa). This is mainly due to the lightness of the

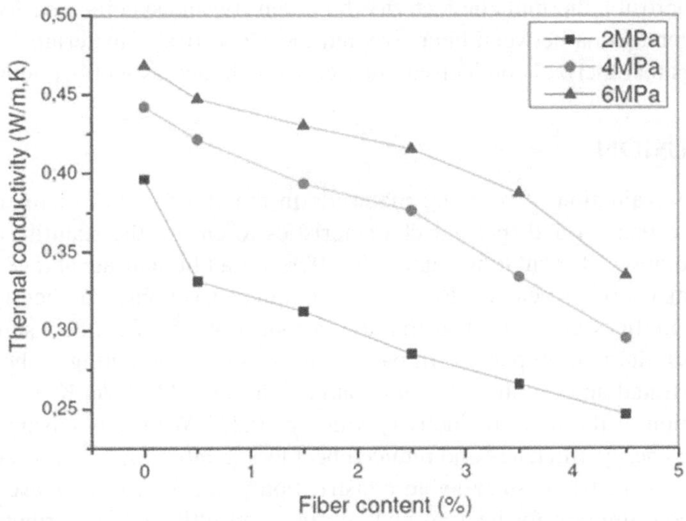

FIGURE 40.8 Changes in thermal conductivity as a function of fiber content.

incorporated fibers and their porous structure [24]. In addition, an increase in fiber content leads to an increase in the inclusion of air and voids in the composite [25, 26], making them considerably lighter. Similar results were found by [27] when they studied natural mortar reinforced with date palm fibers.

40.3.4 Thermal Conductivity

The thermal conductivities of sol-*WRP* blocks as a function of fiber content are shown in Figure 40.8.

Figure 40.8 shows a decrease in thermal conductivity with increasing *WRP* fiber content, resulting in a benefit in thermal insulation performance for the specimen containing 4.5% *WRP* fibers, corresponding to a thermal conductivity of 0.247 W/m.K. This value is lower than that of the control block, whose thermal conductivity is equal to 0.396 W/m.K. This decrease in thermal conductivity is explained firstly by the low thermal conductivity of the plant fibers compared with that of the matrix alone. The second reason is that the presence of the fibers in the matrix generates voids, leading to an increase in porosity and a decrease in density as air has much lower thermal conductivity than the solid material [24, 28]. Research has found that fiber size has a considerable impact on the porosity and density of materials (Table 40.3). Specifically, studies have demonstrated that different

TABLE 40.3
Comparison of Thermal Conductivity and Density of Matrix/*WRP* Composites with Some Materials Used for Thermal Insulation in Buildings

Materials	Density (kg/m³)	Thermal Conductivity (W/m.K)	References
Clay/*WRP*	1492	0.247	Current study
Clay/Hemp	347–523	0.089–0.120	[33]
Clay/Juncus acutus	1100	0.327	[34]
Lime/Alfa	1020	0.258	[35]
Clay/Posidonia-Oceanica	872	0.16	[36]
Clay/Date palm	1250	0.342	[27]
Clay/Straw	1393	0.294	[37]
Clay/Millet waste	1736	0.74	[38]

fiber sizes affect porosity, thermal conductivity, bulk density, and specific density [29]. These findings support the connection between fiber size and the properties of materials [30]. This is in line with other findings that thermal conductivity decreases with increasing fiber content [27, 31, 32].

40.4 CONCLUSION

In conclusion, the evaluation of building materials in the construction sector involves a careful examination of mechanical and thermal characteristics to ensure the stability of structures over time. The incorporation of insulating materials offers an additional advantage by reducing heat loss and optimizing energy efficiency. Based on a comprehensive physico-chemical analysis of the matrix (soil) used in this study, it is clear that the introduction of *WRP* fibers significantly impacts the resulting bricks' thermal properties. In particular, the brick containing a fiber fraction of 4.5% by weight demonstrated an optimum thermal conductivity of 0.223 W/m.K. In the context of sustainable construction, a thermal conductivity value of 0.223 W/m.K is advantageous because it signifies enhanced energy efficiency and reduced heat loss in buildings. By utilizing materials with such low thermal conductivity, sustainable construction practices can increase thermal comfort, decrease energy consumption for heating and cooling, and ultimately contribute to lower carbon emissions and energy savings. These results highlight the potential of fiber-reinforced compressed earth bricks to meet the thermal performance requirements of sustainable construction practices.

REFERENCES

1. Elmoudnia, H., Faria, P., Jalal, R., Waqif, M., Saâdi, L.: Effectiveness of alkaline and hydrothermal treatments on cellulosic fibers extracted from the *Moroccan Pennisetum Alopecuroides* plant: Chemical and morphological characterization. Carbohydrate Polymer Technologies and Applications. 5, 100276 (2023).
2. Giroudon, M., Laborel-Préneron, A., Aubert, J.-E., Magniont, C.: Comparison of barley and lavender straws as bioaggregates in earth bricks. Construction and Building Materials. 202, 254–265 (2019).
3. Caponetto, R., Cuomo, M., Detommaso, M., Giuffrida, G., Presti, A.L., Nocera, F.: Performance assessment of giant reed-based building components. Sustainability. 15, 2114 (2023).
4. Hussien, A., Maksoud, A., Abdeen, A., Nofal, E.: Exploring the use of wood pellets as a sustainable alternative for indoor insulation. Sustainability. 16, 25 (2023).
5. Boumaaza, M., Belaadi, A., Alshahrani, H., Khan, M.K., Jawaid, M.: Environmentally mortar development using Washingtonia/biochar waste hybrid: Mechanical and thermal properties. Biomass Conversion and Biorefinery. 14, 1–24 (2023)
6. Ajouguim, S., Abdelouahdi, K., Waqif, M., Stefanidou, M., Saâdi, L.: Modifications of alfa fibers by alkali and hydrothermal treatment. Cellulose. 26, 1503–1516 (2019).
7. Rajeshkumar, G.: Characterization of surface modified phoenix sp. fibers for composite reinforcement. Journal of Natural Fibers. 18, 2033–2044 (2021).
8. Shanmugasundaram, N., Rajendran, I., Ramkumar, T.: Characterization of untreated and alkali treated new cellulosic fiber from an Areca palm leaf stalk as potential reinforcement in polymer composites. Carbohydrate Polymers. 195, 566–575 (2018).
9. Manivel, S., Pannirselvam, N., Gopinath, R., Sathishkumar, T.: Physico-mechanical, chemical composition and thermal properties of cellulose fiber from hibiscus vitifolius plant stalk for polymer composites. Journal of Natural Fibers. 19, 6961–6976 (2022).
10. Jawaid, M., Khalil, H.A., Hassan, A., Dungani, R., Hadiyane, A.: Effect of jute fibre loading on tensile and dynamic mechanical properties of oil palm epoxy composites. Composites Part B: Engineering. 45, 619–624 (2013).
11. Lassoued, M., Mnasri, T., Hidouri, A., Younes, R.B.: Thermomechanical behavior of Tunisian palm fibers before and after alkalization. Construction and Building Materials. 170, 121–128 (2018).
12. Madhu, P., Sanjay, M.R., Jawaid, M., Siengchin, S., Khan, A., Pruncu, C.I.: A new study on effect of various chemical treatments on Agave Americana fiber for composite reinforcement: Physico-chemical, thermal, mechanical and morphological properties. Polymer Testing. 85, 106437 (2020).
13. Widodo, E., Widodo, T.D.: Comprehensive investigation of raw and NaOH alkalized sansevieria fiber for enhancing composite reinforcement. Case Studies in Chemical and Environmental Engineering. 9, 100546 (2024).

14. IMANOR: Reconnaissance et essais - Mesure de la capacit´e d'adsorption de bleu de m´ethyl`ene d'un sol ou d'un mat´eriau rocheux - D´etermination de la valeur de bleu de m´ethyl`ene d'un sol ou d'un mat´eriau rocheux par l'essai `a la tache (2018)
15. IMANOR: "Liquidity Limit Determination: Cone Penetrometer Method".
16. IMANOR: "Determination of the Atterberg Limits: Roller Plasticity limit/Liquidity Cup Limit".
17. IMANOR: Reconnaissance et essais - Détermination de la masse volumique des particules solides des sols - Méthode du pycnomètre à eau.
18. IMANOR: Reconnaissance et essais - Détermination de la masse volumique des sols fins en laboratoire - Méthodes de la trousse coupante, du moule et de l'immersion dans l'eau (2019).
19. IMANOR: reconnaissance et essais - Détermination des références de compactage d'un matériau - Essai Proctor Normal - Essai Proctor modifié (2019).
20. IMANOR: Sols, biodéchets traités et boues - Détermination du pH (2019).
21. AFNOR: reconnaissance et essais - Détermination de la teneur pondérale en matières organiques d'un matériau - Méthode par calcination (2019).
22. Mihiretie, B., Cederkrantz, D., Rosén, A., Otterberg, H., Sundin, M., Gustafsson, S., Karlsteen, M.: Finite element modeling of the hot disc method. International Journal of Heat and Mass Transfer. 115, 216–223 (2017).
23. Natarajan, T., Kumaravel, A., Palanivelu, R.: Extraction and characterization of natural cellulosic fiber from *Passiflora foetida* stem. International Journal of Polymer Analysis and Characterization. 21, 478–485 (2016).
24. Taallah, B., Guettala, A.: The mechanical and physical properties of compressed earth block stabilized with lime and filled with untreated and alkali-treated date palm fibers. Construction and Building Materials. 104, 52–62 (2016).
25. Shoukry, H., Kotkata, M., Abo-EL-Enein, S., Morsy, M., Shebl, S.: Enhanced physical, mechanical and microstructural properties of lightweight vermiculite cement composites modified with nano metakaolin. Construction and Building Materials. 112, 276–283 (2016).
26. Nitcheu, M., Njomo, D., Meukam, P., Talla, C.F.: Modeling of coupled heat and mass transfers in a stabilized earthen building envelope with thatched fibers. Fibers. 6, 75 (2018).
27. Khoudja, D., Taallah, B., Izemmouren, O., Aggoun, S., Herihiri, O., Guettala, A.: Mechanical and thermophysical properties of raw earth bricks incorporating date palm waste. Construction and Building Materials. 270, 121824 (2021).
28. Babé, C., Kidmo, D.K., Tom, A., Mvondo, R.R.N., Kola, B., Djongyang, N.: Effect of neem (*Azadirachta Indica*) fibers on mechanical, thermal and durability properties of adobe bricks. Energy Reports. 7, 686–698 (2021).
29. Rebolledo, P., Cloutier, A., Yemele, M.-C.: Effect of density and fiber size on porosity and thermal conductivity of fiberboard mats. Fibers. 6, 81 (2018).
30. Wijesinghe, K., Gunasekara, C., Law, D.W., Hidallana-Gamage, H., Wanasekara, N., Wang, L.: Thermal and acoustic performance in textile fibre-reinforced concrete: An analytical review. Construction and Building Materials. 412, 134879 (2024).
31. Charai, M., Salhi, M., Horma, O., Mezrhab, A., Karkri, M., Amraqui, S.: Thermal and mechanical characterization of adobes bio-sourced with *Pennisetum setaceum* fibers and an application for modern buildings. Construction and Building Materials. 326, 126809 (2022).
32. de Castrillo, M.C., Ioannou, I., Philokyprou, M.: Reproduction of traditional adobes using varying percentage contents of straw and sawdust. Construction and Building Materials. 294, 123516 (2021).
33. Mazhoud, B., Collet, F., Prétot, S., Lanos, C.: Effect of hemp content and clay stabilization on hygric and thermal properties of hemp-clay composites. Construction and Building Materials. 300, 123878 (2021).
34. Omrani, H., Hassini, L., Benazzouk, A., Beji, H., ELCafsi, A.: Elaboration and characterization of clay-sand composite based on *Juncus acutus* fibers. Construction and Building Materials. 238, 117712 (2020).
35. Garrouri, S., Lakhal, W., Benazzouk, A., Sediki, E.: Potential use of alfa fibers in construction material: Physico-mechanical and thermal characterisation of reinforced specimen. Construction and Building Materials. 342, 127787 (2022).
36. Braiek, A., Briki, C., Karkri, M., Settar, A., Jemni, A.: Thermo-physical and mechanical performances of a new lightweight construction material made with clay and Posidonia-Oceanica fibers. Case Studies in Construction Materials. 19, e02599 (2023).
37. Alioui, A., Kaitouni, S.I., Azalam, Y., Bendada, E.M., Mabrouki, M.: Effect of straw fibers addition on hygrothermal and mechanical properties of carbon-free adobe bricks: From material to building scale in a semi-arid climate. Building and Environment. 255, 111380 (2024).
38. Babé, C., Kidmo, D.K., Tom, A., Mvondo, R.R.N., Boum, R.B.E., Djongyang, N.: Thermomechanical characterization and durability of adobes reinforced with millet waste fibers (sorghum bicolor). Case Studies in Construction Materials. 13, e00422 (2020).

41 Electrochemical Corrosion Behavior of Titanium in Different Oral Environments

H. Hrir, O. Ait Layachi, A. Boudouma, A. Ghanam, A. Moufti, A. El Brouzi, M. El Marrakchi, and E. Khoumri

RESEARCH HIGHLIGHTS

A realistic approach employing varied oral simulations, diverse environments, integrated techniques (Office Chérifien des Phosphates (OCP), polarization curves, EIS, SEM), and advanced modeling unveils the dynamic electrochemical response of titanium alloy. The findings provide insights for potential implant enhancements, underscore the material's relevance, and contribute to biomechanical understanding.

41.1 INTRODUCTION

Titanium and its alloys are the materials of choice for biomedical applications, particularly dental implants, thanks to their exceptional resistance to corrosion. This resistance is attributed to the formation of a protective oxide film, mainly composed of TiO_2, which develops on contact with oxygen and gives titanium high performance in aggressive environments [1], making it a recommended choice for medical devices, orthopedic implants, and dental implants. In addition to their corrosion resistance, the properties of titanium that make it a preferred material for biomedical applications include its biocompatibility, high mechanical strength, low modulus of elasticity, and ability to promote osseointegration. Titanium's biocompatibility enables better integration with surrounding tissues, reducing the risk of rejection and inflammation. Its high mechanical strength ensures implant durability under repetitive loads, while its low modulus of elasticity is closer to that of human bone, minimizing stress on adjacent bone. In addition, titanium promotes osseointegration, improving the long-term stability of implants.

The fundamental aim of our research is to achieve a complete understanding of the electrochemical interaction of titanium alloy under conditions mimicking the oral environment. In the field of implantology, the longevity of biomedical implants is a major concern, with corrosion emerging as a key factor in the progressive degradation of materials, releasing metal ions into the body environment [2]. We set out to unveil these complex interactions by simulating the oral cavity as an electrochemical cell, using saliva as a weak electrolyte [3].

The careful selection of four distinct electrolyte media reinforces our realistic approach, reflecting diverse conditions such as lactic acidity, salivary physiology, bacterial presence, and soda alkalinity [4]. These strategic choices are guided by the aim of reproducing commonly encountered oral environments, incorporating elements such as lactate-containing milk, oral bacteria, and even the specific uses of soda in dental care. Koike and Fujii have carried out several studies on the behavior of titanium in the presence of organic acids (lactic, formic, and acetic acids). Beyond simulation, this study aims to closely examine titanium alloy corrosion under conditions similar to those frequently encountered in the human oral cavity [5–7]. The results obtained from electrochemical tests and scanning electron microscope (SEM) observations promise to shed new

DOI: 10.1201/9781003647263-41

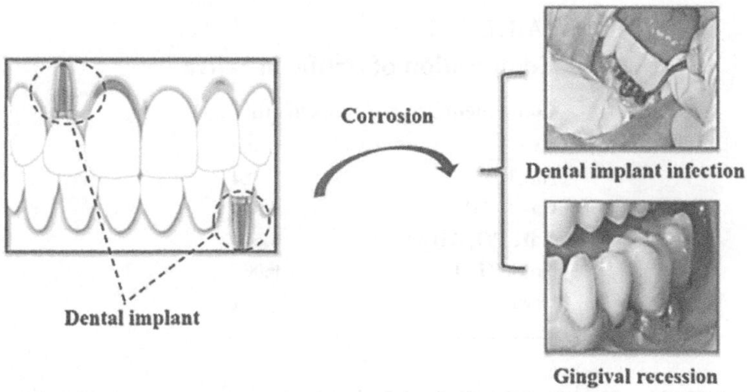

FIGURE 41.1 Effects of corrosion on dental implants [8].

light on titanium's alloy biocompatibility, highlighting the subtleties of protective layer formation under specific conditions, while also revealing the potential challenges posed by the presence of bacteria and alkaline environments. It is in this exploration of the electrochemical behavior of titanium alloys that the keys to optimizing the durability and efficacy of dental implants in a variety of clinical contexts lie.

This groundbreaking study is the first to examine titanium alloy in a variety of oral environments, including the presence of bacteria and alkaline conditions, in contrast to previous research which focused primarily on acidic or physiological environments. Our aim is also to develop solutions to improve the durability of dental implants in these corrosive environments, specifically considering the use of corrosion inhibitors as a potential approach (Figure 41.1) presents two distinct scenarios illustrating the consequences of titanium corrosion on dental implants. These case studies highlight potential complications such as infection and gingival recession, underlining the crucial importance of understanding and preventing dental implant corrosion [8].

41.2 EXPERIMENTAL PROTOCOL

41.2.1 Materials and Solutions

A α-Ti titanium alloy plate, fragmented into pieces, was cut into small thin plates measuring 5 mm in thickness and 40 mm in length. These fragments were meticulously polished using SiC paper up to 1500 grit, then cleaned by ultrasonication in a solution of ethanol and distilled water before being air-dried [9].

The tests were conducted in a solution of artificial saliva, starting with the addition of lactic acid to the artificial saliva to simulate an acidic environment with a pH of 3. For the physiological environment, artificial saliva was used as a neutral medium with a pH of 7. In the case of a simulated bacteriological environment, oral bacteria such as *S. mutans* were added to the artificial saliva. Finally, NaOH (1 M) was added to the saliva until the pH reached a value of 11 to mimic a basic environment. These choices of electrolytes were guided by previous studies [3, 10]. The simulation of the composition of the natural human saliva solution is summarized in Table 41.1.

41.2.1.1 Preparation of the Bacterial Strain

The oral environment is a highly complex polymicrobial system. However, we have chosen to study a single bacterial species involved in dental biofilm formation, *Streptococcus mutans*, a facultative, anaerobic, gram-positive, round, small-chain organized bacterium commonly found in dental plaque and human saliva. *S. mutans* is an acidogenic bacterium playing a major role in the development of

TABLE 41.1
Composition of Artificial Saliva

Component	Quantity (g/L)
KCl	0.4
NaCl	0.4
$CaCl_2.2H_2O$	0.906
$NaH_2PO_4.2H_2O$	0.690
$Na_2S.9H_2O$	0.005
Urea	1

dental caries in humans, metabolizing sucrose to lactic acid. *S. mutans* was isolated in our microbiology laboratory. Paraffin-stimulated saliva samples were collected from research collaborators treated for dental caries. Saliva was collected for 5 minutes, centrifuged at 5,000 rpm for 5 minutes and diluted 1/100 in sterile saline (NaCl 0.9%). Fifty-microliter aliquots of the dilution were spread on mitis-salivarius agar, supplemented with 0.2 units of bacitracin and sucrose (final concentration 20%). After 48 hours of anaerobic incubation at 37°C, colonies suspected of containing *S. mutans* were isolated (see Figure 41.2). The prepared cultures were then added to a cell containing 100 mL of sterile artificial saliva (SA) for subsequent experiments [11].

41.2.2 Electrochemical Testing

The samples designated for electrochemical tests were encapsulated in epoxy resin, leaving a working area of 1 cm^2. The tests were conducted within a three-electrode electrochemical cell, with α-Ti titanium alloy as the working electrode, a platinum rod as the counter electrode, and an Ag/AgCl reference electrode see (Figure 41.3). The solution used was a 100 mL solution of the medium studied, aerated by being exposed to air. Corrosion measurements were conducted using a PalmSens potentiostat.

After fixing the samples, an initial stabilization phase of 30 minutes was allocated, combined with heating the solution to 37°C to simulate oral conditions. Initially, the open circuit potential (OCP) evolution was monitored over a period of 3,600 seconds [3].

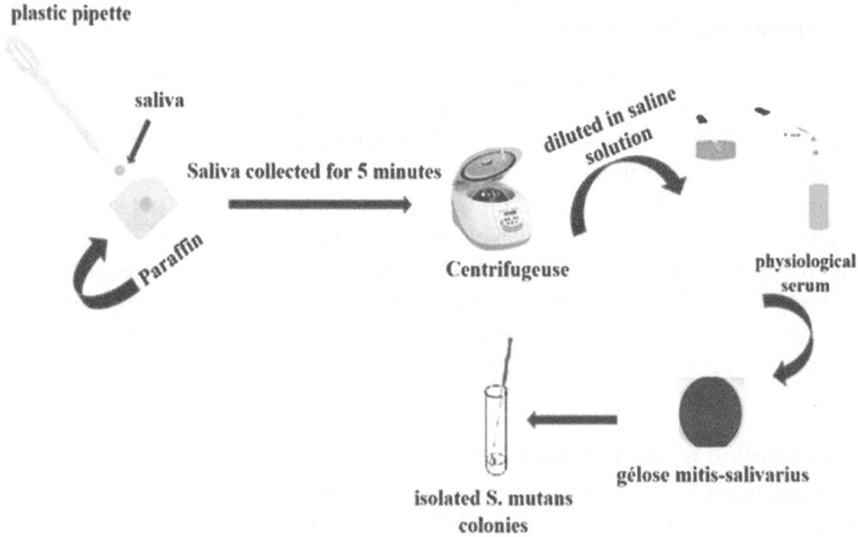

FIGURE 41.2 Illustrative diagram of streptococcus preparation.

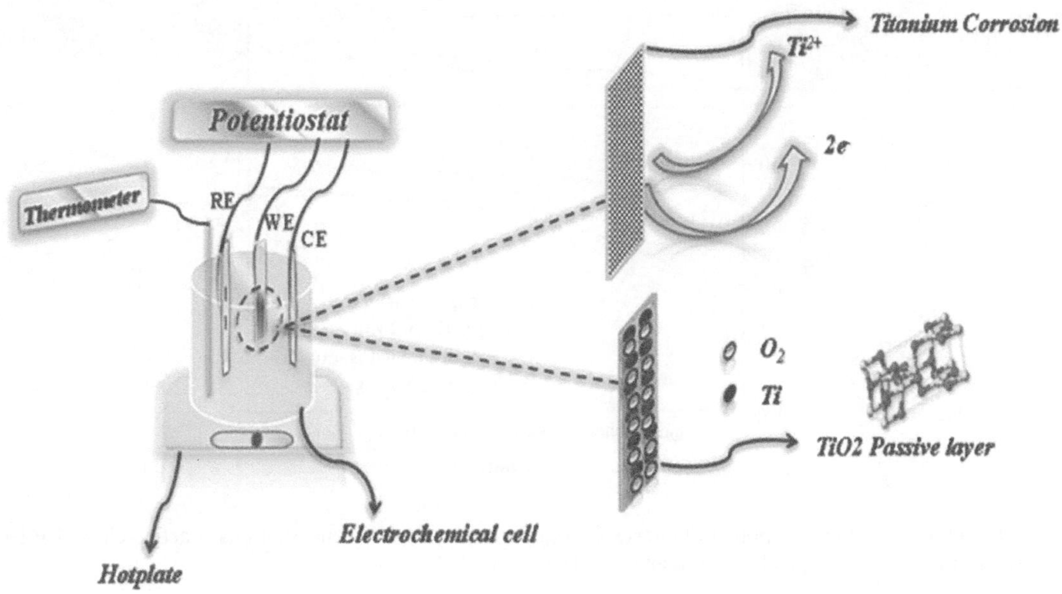

FIGURE 41.3 Electrochemical experimental setup.

The polarization curves were plotted in a range from –1 to 1 V (Ag/AgCl) with a scan rate of 1 mV/s [8], from these curves, Tafel lines are derived.

Electrochemical impedance spectroscopy (EIS) analysis was conducted following open circuit potential measurements. These EIS measurements covered a frequency range from 100,000 Hz to 0.01 Hz, with an alternating sine wave amplitude of 10 mV applied to the electrode when under its corrosion potential. The resulting data were used to plot the real (Z) and imaginary (Z") components of impedance, as depicted in the Nyquist plot [3, 11].

The impedance results were converted into equivalent electrical circuits using PS Trace 5.8 software, providing a comprehensive understanding of the impact of different environments on titanium alloy [3, 12].

41.3 IMMERSION TESTS AND SURFACE OBSERVATION

The surface morphologies of titanium alloy after immersion in different solutions for 7 days were characterized using field emission scanning electron microscopy (TESCAN VEGA3 SEM). This technique was employed to detect surface topographies of the samples operating at an acceleration voltage of 20 kV and under a pressure of 1.3×10^{-4} Pa.

41.4 RESULTS AND DISCUSSION

41.4.1 Open Circuit Potential

The OCP is a crucial parameter in corrosion studies. It represents the electrochemical potential of a material in contact with an electrolyte when no electric current flows through the interface.

Measuring the OCP allows for the assessment of a material's corrosion tendency and understanding of fundamental corrosion mechanisms [13, 14]. This is crucial for developing corrosion protection strategies and designing materials resistant to deterioration. The variation of the Titanium alloy electrode's OCP was monitored in the various electrolytes mentioned earlier, at a temperature of 37 degrees Celsius. The results of this observation are summarized in (Figure 41.4).

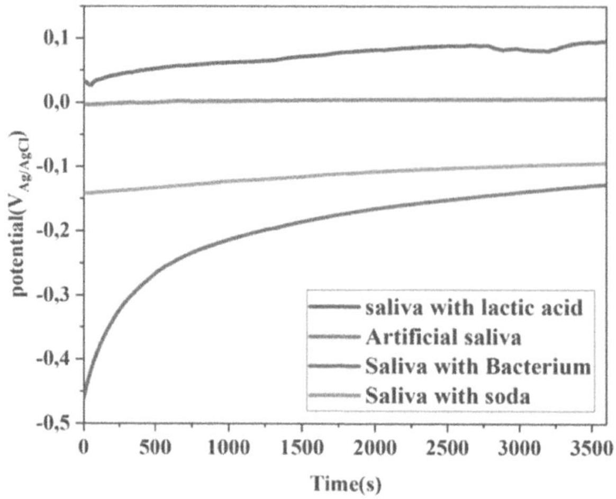

FIGURE 41.4 Open-circuit potential curves for titanium alloy in different solutions: Lactic acid (0.1 g/L), artificial saliva, bacteriological medium and NaOH(1M), at 37°C.

In the bacteriological environment, Titanium alloy exhibits a tendency toward corrosion, as suggested by the measured potential value of –0.15 V. In contrast, artificial saliva maintains titanium in a passive state, with a stable OCP around 0 V. The results obtained in both different environments are in line with the findings of Maruthamuthu et al. [11, 15, 16].

In the lactic acid solution, the OCP of titanium increases, suggesting the possible formation of a protective passive layer [17, 18]. Research by Schiff et al. [19, 20].

In 2002, these observations were confirmed, demonstrating that the passivation layer of titanium alloy remains functional in Fusayama artificial saliva acidified to a pH of 2.5.

In the soda solution, the measured corrosion potential is –0.11 V. Although slightly less negative than that of the bacteriological medium, it also indicates a certain tendency toward corrosion. The soda solution can cause chemical alteration of the α-Ti surface.

41.4.2 Potentiodynamic Polarization Curves

The Tafel method is a commonly used analysis technique to study the corrosion of metallic materials. It enables the determination of corrosion reaction kinetics by measuring the electrochemical response of the material under variable potential. This approach provides valuable insights into corrosion mechanisms, corrosion rate, and aids in designing corrosion-resistant materials [21].

We performed Tafel curves of a titanium alloy electrode in various environments (Figure 41.5). Analysis of the curves allowed us to determine parameters related to the corrosion phenomenon. The results obtained are summarized in Table 41.2.

According to the results table, the corrosion potential of titanium alloy in the acidic environment containing saliva and lactic acid with a pH of 3 is –0.012 V (Ag/AgCl). Compared to other environments, the corrosion rate value indicates that titanium alloy acquires high corrosion resistance with electrochemical stability. This result is consistent with previous studies [19]. It confirms titanium's ability to form a protective layer in lactate environments.

In the alkaline environment, the contradiction between a moderately negative corrosion potential and high polarization resistance on one hand, and high corrosion current on the other hand, can be attributed to the formation of an irregular protective layer on the surface of titanium alloy. This layer provides some protection against corrosion, explaining the high

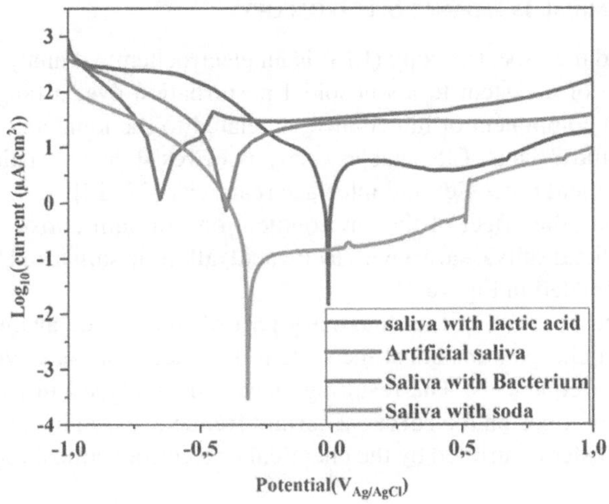

FIGURE 41.5 Potentiodynamic polarization curves for titanium alloy in different media solutions: Lactic acid (0.1 g/L), artificial saliva, bacteriological medium, NaOH(1M) at 37°C.

resistance. However, the non-uniformity of the protective layer exposes unprotected areas that may be subject to more aggressive localized corrosion, explaining the high corrosion current. This observation underscores the challenges to overcome in addressing corrosion in this particular environment.

In the case of saliva, the results obtained indicate a tendency for corrosion of titanium alloy. A slightly negative corrosion potential associated with high corrosion current and corrosion rate suggest significant corrosive activity in this environment. This is also explainable by the low polarization resistance indicating a lesser ability of the material to resist electrochemical corrosion. These observations converge to a conclusion that saliva creates corrosive conditions for titanium alloy. Our findings are consistent with the study by Abey, Biedunkiewicz et al. [3].

In the presence of bacteria, titanium alloy exhibits moderate resistance to corrosion. The measured corrosion potential is –0.65 V (Table 41.2), suggesting a tendency for corrosion with a polarization resistance of 5,269 Ω (Table 41.2). The results obtained in the presence of a bacterial medium indicate that titanium alloy may be vulnerable to corrosion attack [11].

Interpreting the findings is a complex matter, and interpretations will remain a topic of discussion among researchers. In some studies where titanium was not used alone but as an alloy with nickel, it was concluded that in a bacterial medium, there is a reduction in the corrosion current [15]. Other studies present consistency with our results [11, 22].

TABLE 41.2
Corrosion Parameters of Alloy Ti in Different Solutions: Lactic Acid, Artificial Saliva, Bacteriological Medium and NaOH, at 37°C

	E_{corr}(V)	I_{corr}(μA/cm²)	R_p (Ω)	V_{corr} (mm/year)
Lactic	–0,012	1,41	24,100	0,047
Saliva	–0,4	34,02	6,441	1,133
Bacteria	–0,65	9,046	5,269	0,338
Soda	–0,32	2,025	70,090	0,076

41.4.3 Electrochemical Impedance Spectroscopy

Electrochemical impedance spectroscopy (EIS) is an electrochemical analysis method that studies the electrical response of a system to a sinusoidal perturbation over a range of frequencies. The resistance, a measured component of impedance, is related to the ionic and electrical conductivity of the system under study. Thus, EIS enables analysis of resistance variations with frequency to understand electrochemical processes and interface reactivity [23, 24].

To further investigate the effect of the environment on titanium corrosion, EIS measurements were conducted in artificial saliva, saliva with lactic acid, alkaline saliva and bacterial medium. The obtained results are depicted in Figure 41.6.

The curves obtained by plotting the imaginary part of impedance against the real component suggest the presence of charge transfers at the system interface. For better visualization, the scales were adjusted at high frequencies. The resulting curves are included in Figure 41.6.a and b. We observe the formation of loops on the curves at certain frequencies, a phenomenon explained by the existence of charge transfer controlled by the chemical or electrochemical reaction at the electrode-electrolyte interface.

The presence of resistance at the solution and metal levels can influence the shape of these loops by altering system parameters such as time constant or frequency response. Analysis of these results leads to the representation of an equivalent circuit in Figure 41.6.c. The electrical parameters obtained from this circuit are summarized in Table 41.3.

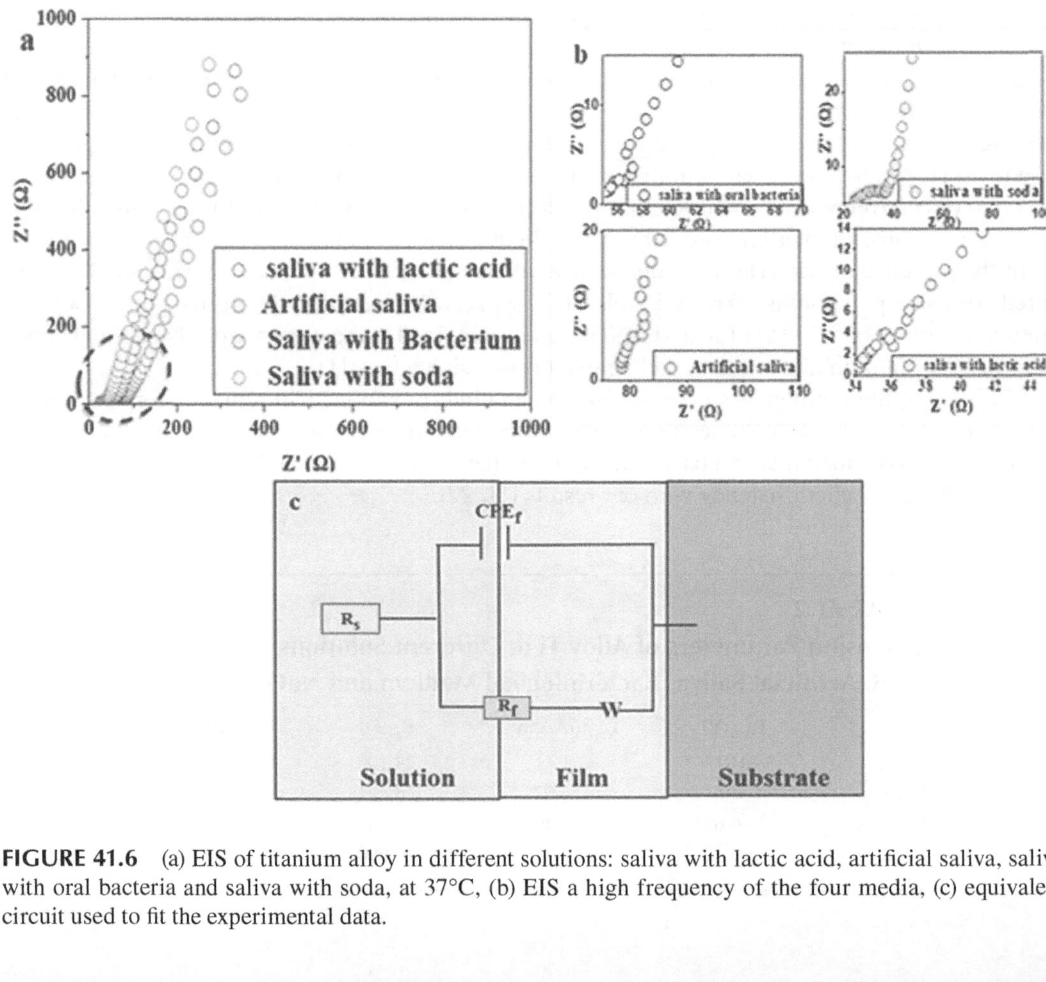

FIGURE 41.6 (a) EIS of titanium alloy in different solutions: saliva with lactic acid, artificial saliva, saliva with oral bacteria and saliva with soda, at 37°C, (b) EIS a high frequency of the four media, (c) equivalent circuit used to fit the experimental data.

TABLE 41.3
Adjusted Electrochemical Parameters for EIS of Titanium Alloy in Acidic, Physiological, Bacteriological and Basic Solutions

	Rs (Ω)	C (μF)	Rct (Ω)	W (Kσ)
Saliva with lactic acid	100	200	1,900	5
Artificial saliva	172	100	1,800	4.9
Saliva with oral bacteria	26	350	10	9
Saliva with soda	26,73	350,6	5	8,999

According to the values in Table 41.3, it is observed that the charge transfer resistance (Rct) values are low in the cases of the bacterial and alkaline media compared to those obtained for the salivary and lactate media. This observation suggests increased electrochemical reactivity, indicating oxidation of titanium alloy in both alkaline and bacterial media. Conversely, the increase in Rct in acidic (saliva with lactic acid) and physiological (artificial saliva) media indicates passivation of titanium alloy, confirming the formation of a protective layer. We observed an inconsistency between the results of the Tafel method and those of EIS for titanium alloy immersed in artificial saliva with soda. The Rp value obtained by Tafel is 70,090 Ω, indicating a low corrosion rate, while the Rct value obtained by EIS is 5 Ω, suggesting a very high corrosion rate. This discrepancy may be explained by the stability of the passivating layers and the presence of localized corrosion. It is possible that some areas of the titanium alloy surface were heavily corroded, leading to a low value of Rct in the EIS measurements, while other areas may have developed passivation, resulting in a high value of Rp in the Tafel measurements. This situation is characteristic of localized corrosion, where specific regions are subjected to intense corrosive attack while others remain protected by a passivating layer.

The parameters of the equivalent circuit, such as solution resistance (Rs), double layer capacitance (C) and Warburg factor (W), reinforce these observations consistent with previous results. In summary, the variations in parameters obtained by EIS highlight the different reactivity of Titanium alloy depending on the media, thus providing crucial information for evaluating its electrochemical behavior in various biomedical conditions.

41.5 OBSERVATION OF SURFACE MORPHOLOGY AFTER CORROSION

Scanning electron microscopy (SEM) is an imaging technique that employs an electron beam to observe samples at a microscopic scale. It provides high resolution and allows for the analysis of topography, chemical composition and material structure. SEM is widely utilized in various scientific and industrial fields for its effectiveness in providing detailed and accurate images of sample surfaces [20, 25].

To analyze the surfaces using SEM, we immersed the titanium alloy in four different environments (artificial saliva, lactate with saliva, bacteria with saliva, soda with saliva) for a duration of 7 days (168 hours). The analyses were conducted before and after immersion. The images obtained are summarized in Figure 41.7.

The analysis of the images reveals that titanium alloy exhibits a smooth and uniform surface before immersion (no corrosion) in image A. In the saliva with lactic acid solution (image B), no significant alteration is visible, suggesting the formation of a protective layer, which is in line with the results obtained in the OCP, Tafel and EIS studies. The artificial saliva solution (image C) shows slight modifications, indicating possible reactions. Immersion in a saliva solution with bacteria (image D) causes attack zones, illustrating a bacterial reaction. Finally, in the alkaline solution (image E), extensive corrosion is observed, highlighting the significant impact

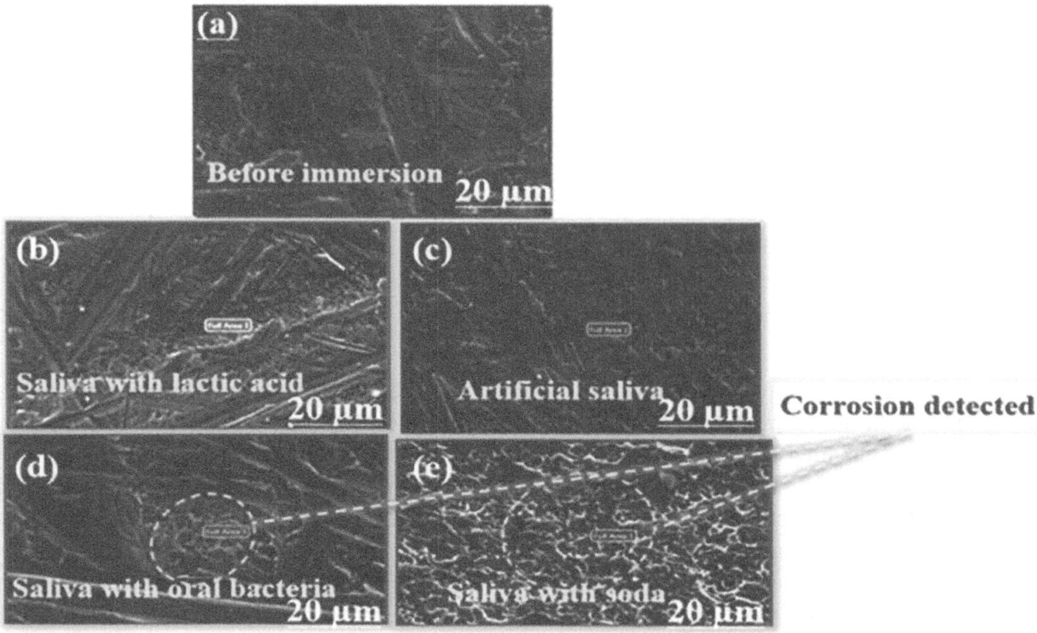

FIGURE 41.7 Surface morphology following corrosion of titanium alloy after immersion in solutions for 168 h: (a) Titanium alloy before immersion and (b) titanium alloy immersed in saliva with lactic acid, (c) titanium alloy immersed in saliva and (d) titanium alloy immersed in saliva with oral bacteria, (e) titanium alloy immersed in saliva with soda.

of the alkaline solution on titanium alloy. In this case, the surface texture shows localized corrosion predicted by the Tafel curves results.

41.6 PROPOSED MECHANISM FOR DISSOLUTION-PASSIVATION OF TITANIUM ALLOY IN DIFFERENT ACIDIC, PHYSIOLOGICAL, BACTERIOLOGICAL AND BASIC ENVIRONMENTS BASED ON POURBAIX DIAGRAM

The reaction mechanisms describe the chemical steps by which reactants transform into products during a chemical reaction [26, 27]. Their detailed understanding allows for the prediction and control of the products formed, which is essential for designing new reactions and optimizing experimental conditions.

The results obtained from EIS and SEM of titanium alloy behavior in four distinct environments, acidic (with artificial saliva and lactic acid), neutral (with artificial saliva), bacteriological (including artificial saliva and oral bacteria) and alkaline (with saliva and soda), have allowed us to propose a dissolution-passivation mechanism of this material in these environments Figure 41.8 illustrates this mechanism, referring to the Pourbaix diagram.

The Pourbaix diagram is a tool used in electrochemistry to predict the behavior of metals in aqueous environments based on pH and electrochemical potential. It identifies three main zones: passivation, immunity and corrosion. Passivation is characterized by the formation of a protective oxide layer and immunity indicates stability of the metal, while corrosion represents conditions where the metal is prone to active corrosion. These zones help understand and prevent metal corrosion in different applications (Figure 41.9) [28–31].

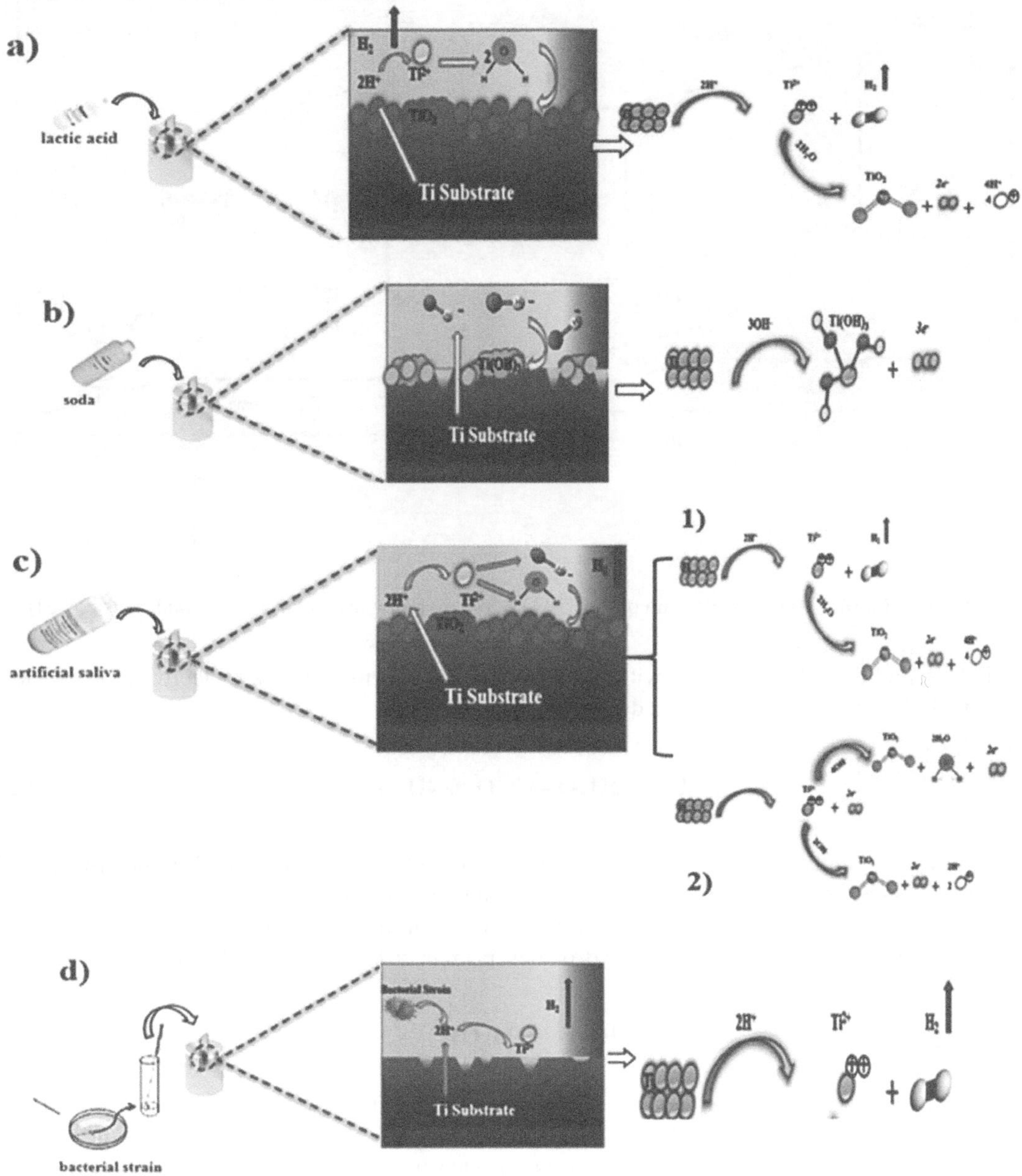

FIGURE 41.8 (a–d) Illustrative diagram of titanium alloy reaction mechanisms in various media: acidic, physiological, bacteriological, basic.

Figure 41.8a illustrates the reaction mechanism of titanium in an acidic environment. When titanium is immersed in an acidic electrolyte, it initially reacts with hydrogen ions to form titanium (II) ions, Ti^{2+}, as described by the equation:

$$Ti + 2H^{+} \rightarrow Ti^{2+} + H_2 \uparrow \quad (41.1)$$

At pH values below 6 and at potentials ranging from −1.8 V/SHE to −0.4 V/SHE according to the Pourbaix diagram (Figure 41.9a), reactions occur at the metal-electrolyte interface. The release

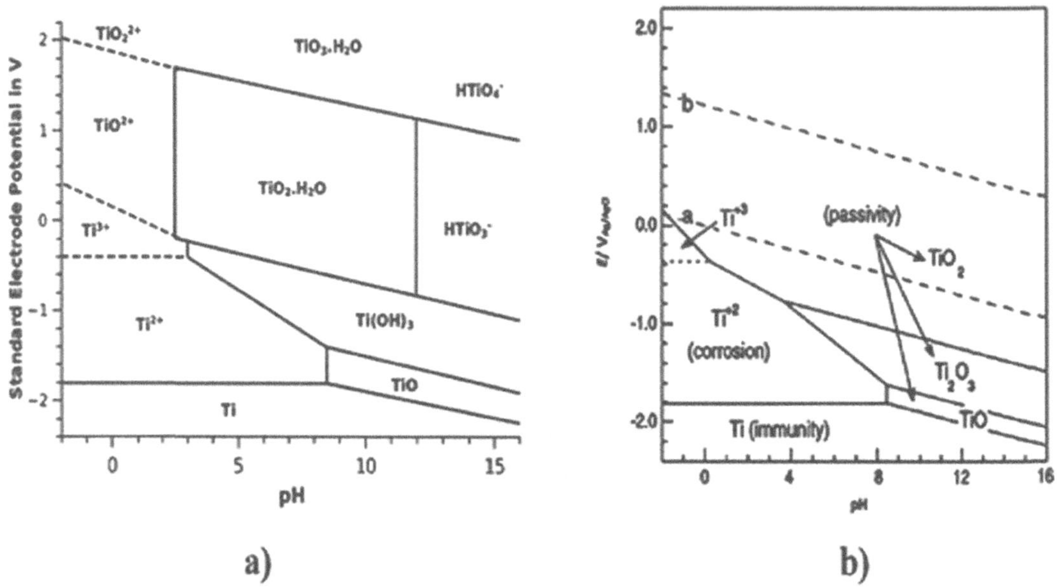

FIGURE 41.9 (a, b) Pourbaix diagram of the Ti-H_2O system: biological environment conditions [28–31].

of Ti^{2+} ions followed by a reaction with water results in the formation of a protective layer of TiO_2 (Figure 41.9b), inducing passivation, described by:

$$\mathbf{Ti^{2+} + 2H_2O \rightarrow TiO_2 + 4H^+ + 2e^-} \tag{41.2}$$

This layer is highly stable in an acidic environment. SEM observations confirm the presence of a protective layer, ensuring the resistance of titanium to corrosion.

Figure 41.8b depicts the reaction mechanism of titanium in a basic environment. It reacts with hydroxide ions (OH^-) to form titanium hydroxides, primarily $Ti(OH)_3$, according to the equation:

$$\mathbf{Ti + 3OH^- \rightarrow Ti(OH)_3 + 3e^-} \tag{41.3}$$

In the pH conditions used and for potentials ranging from –1.6 V/ESH to –0.6 V/ESH (Figure 41.9a), reactions occur at the metal-electrolyte interface. Although the formation of $Ti(OH)_3$ may initially constitute a protective layer, it is unstable (non-uniform) (Figure 41.8b), which can eventually expose titanium to continuous corrosion in a basic environment. In a basic environment with a pH of 10, the Pourbaix diagram suggests that the predominant form of titanium is as hydroxide ions, namely $Ti(OH)_3$.

Figure 41.8c presents two reaction mechanisms for a physiological (neutral) environment simulated by artificial saliva. Initially, in Figure 41.8c-**1**, a reaction with hydrogen ions leads to the formation of titanium II ions, Ti^{2+}, as follows:

$$\mathbf{Ti + 2H^+ \rightarrow Ti^{2+} + H_2 \uparrow} \tag{41.4}$$

At a pH of 7 and potentials between –0.4 V/Ag/AgCl and 0.8 V/Ag/AgCl, the formation of the TiO_2 layer according to the Pourbaix diagram (Figure 41.9.b), reactions occur at the metal-electrolyte interface. The release of Ti^{2+} ions followed by a reaction with water generates a protective layer of TiO_2, represented by:

$$Ti^{2+} + 2H_2O \rightarrow TiO_2 + 4H^+ + 2e^- \quad (41.5)$$

In Figure 41.8c-2, a second possible reaction involves Ti^{2+} ions and hydroxide ions OH^-, producing titanium dioxide TiO_2:

$$Ti^{2+} + 2OH^- \rightarrow TiO_2 + 2H^+ + 2e^- \quad (41.6)$$

This can also be written as:

$$Ti^{2+} + 4OH^- \rightarrow TiO_2 + 2H_2O + 2e \quad (41.7)$$

This layer proves to be highly stable in a neutral environment where the concentrations of H^+ and OH^- ions are of the same order of magnitude. SEM observations confirm the presence of a protective layer, ensuring titanium's resistance to corrosion in this environment.

Figure 41.7.d illustrates the reaction mechanism in a bacteriological environment, where oral bacterial strains produce acids. This acid production leads to a reaction between H^+ ions and Titanium, resulting in the formation of Ti^{2+} ions, which are responsible for the corrosion of the metal in the corrosion zone. As follows:

$$Ti + 2H^+ \rightarrow Ti^{2+} + H_2 \uparrow \quad (41.8)$$

At pH values below 6 and potentials ranging from –1.8 V/ESH to –0.4 V/ESH, according to the Pourbaix diagram (Figure 41.9.a), the formation of Ti^{2+} ions are observed. This prediction of Ti^{2+} ions formed in a bacteriological environment underscores the significance of these reactions in the severe corrosion of the studied material. Furthermore, SEM TESCAN VEGA3 [32] results show images of titanium corrosion in this environment.

41.7 CONCLUSION

In conclusion, this comprehensive study on titanium alloy corrosion in simulated oral cavity environments offers significant insights for optimizing titanium alloy dental implants. The results obtained through a series of electrochemical tests, polarization curve analyses, electrochemical impedance spectroscopy and SEM observations have provided a thorough understanding of titanium's alloy electrochemical behavior in representative conditions.

The observation of the formation of a protective passive layer in acidic (lactic acid with artificial saliva) and physiological (artificial saliva) environments suggests favorable biocompatibility, thus reinforcing the applicability of titanium alloy in the field of dental implants. However, the severe corrosion identified in bacteriological (oral bacteria with artificial saliva) and basic (sodium hydroxide with artificial saliva) environments highlights potential challenges related to bacterial presence and exposure to alkaline solutions.

Advanced modeling of electrochemical impedance data provides a unique approach to interpreting results, highlighting the differentiated impact of various environments on titanium alloy. Studying the reaction mechanisms of titanium alloy in different environments reveals a significant variation in its reactivity depending on the pH and chemical composition of the environment. These observations highlight the importance of understanding these mechanisms in detail to anticipate and control the reactivity of titanium alloy. These conclusions provide crucial insights for guiding the development of future dental implants, identifying optimal conditions and highlighting areas requiring particular attention. In summary, this research contributes not only to the advancement of knowledge in biomechanics but also to the design and improvement of medical devices, thereby offering promising prospects for more effective and safer use of titanium alloy in the field of dental implantology.

REFERENCES

1. Emeline, H.: Corrosion d'un type de minivis d'ancrage orthodontique en Ti6Al4V (No.02110192). [Thése de doctorat, Université de Lorraine], 2019.
2. Wang, L.: Enhanced corrosion resistance in an inflammatory environment and osteogenic properties of silicalite-1 coated titanium alloy implants. Journal of Colloids and Surfaces B: Biointerfaces, 2022. https://doi.org/10.1016/j.colsurfb.2022.112922
3. Biedunkiewicz, A., Dembowska, E.: Electrochemical behavior of titanium implants in artificial saliva after laser surface treatment. Journal of Archives of Metallurgy and Materials, 2020.
4. Koike, M., Fujii, H.: The corrosion resistance of pure titanium in organic acids. Journal of Biomaterials, 2001. https://doi.org/10.1016/S0142-9612(01)00040-0
5. Angela, S., Florian, J.: Cariogenic potential of oral nutritional supplements measured by intraoral plaque pH telemetry. Journal of Clinical Nutrition, 2020. https://doi.org/10.1016/j.clnu.2020.11.026
6. Dawood, I.: Saliva and Oral health. International Journal of Advanced Research in Biological Sciences, 2018. https://doi.org/10.22192/ijarbs.2018.05.07.001
7. John, F., McCabe., Angus, W.: Walls Applied Dental Materials (8eéd.). Blackwell Scientific, 1990.
8. Aherraki, F.: Combien coûte-t-il et comment peut-on se faire rembourser ? Clinique Dentaire la Colline. https://cliniquelacolline.com/combien-coute-un-implant-dentaire/. Accessed 16 Apr 2024.
9. Li, X., Wang, L., Fan, L.: Understanding the effect of fluoride on corrosion behavior of pure titanium in different acids. Corrosion Science, 2021. 192: p. 109812. https://doi.org/10.1016/j.corsci.2021.109812
10. Hrir, H., et al.: An in-depth review of corrosion pathways in titanium dental and orthopedic implants and their biological interactions. Biomedical Materials & Devices, 2025: pp. 446–466.
11. Bahije, L., Benyahia, H., El Hamzaoui, S.: Behavior of NiTi in the presence of oral bacteria: Corrosion by streptococcus mutans. International Orthodontics, 2011. 7(9): pp. 110–119. https://doi.org/10.1016/j.ortho.2010.12.001
12. Belrhazi, I., et al.: Catalytic transformation of plastic waste: Harnessing zeolite for enhanced energy product yield in pyrolysis. Energy Conversion and Management, 2024. 318: p. 118897.
13. Ait Layachi, O., et al.: Investigation of nucleation and growth mechanism of Cu2ZnSnS4 absorber layer electrodeposition on Indium Tin Oxide coated glass. Thin Solid Films, 2023. 782: p. 140019. https://doi.org/10.1016/j.tsf.2023.140019
14. Ait Layachi, O., et al.: Electrodeposition of Cu2NiSnS4 absorber layer on FTO substrate for solar cell applications. RSC Advances, 2024. 14(40): pp. 29439–29454.
15. Maruthamuthu, S., Rajasekar, A., Sathiyanarayanan, S.: Electrochemical behaviour of microbes on orthodontic wires, 2005. https://api.semanticscholar.org/Corpus ID:6204113
16. Benyahia, H., Ebntouhami, M., Forsal, I., Zaoui, F., Aalloula, E.: Corrosion resistance of NiTi in fluoride and acid environments. International Orthodontics, 2009. 7(4): pp. 322–334. https://doi.org/10.1016/S1761-7227(09)73506-5
17. Rochaix, C.: Electrochimie: thermodynamique, cinétique. Nathan (2021)
18. Ribereau-Gayon, S.: Etude de la corrosion du titane implantaire et des matériaux de la suprastructure prothétique: Analyse selon la méthode potentiocinétique [Th: Odonto, Bordeaux] (1995)
19. Schiff, N., Grosgogeat, B., Lissac, M., Dalard, F.: Influence of fluoride content and pH on the corrosion resistance of titanium and its alloys. Biomaterials, 2002. 23: pp. 1995–2002. https://doi.org/10.1016/S0142-9612(01)00328-3
20. Boudouma, A., et al.: Electrodeposition synthesis of Cu2ZnSnS4(CZTS) thin films as A promising material for photovoltaic cells: Fundamentals, methods, and future prospects - A comprehensive review. Materials Today Sustainability, 2024. 28: p. 101018.
21. Boudouma, A., Ait Layachi, O., Hrir, H., Khoumri, E.: A one-step electrodeposition method was used to produce monoclinic Cu2SnS3 thin films for the development of solar cells. Journal of Materials Science: Materials in Electronics, 2023. 34: p. 1903. https://doi.org/10.1007/s10854-023-11390-x
22. Rerhrhaye, W., Merzouk, N., Zaoui, F.: La corrosion biologique des alliagesorthodontiques. Disponiblesur, 2023. http://wjd.um5s.ac.ma/attachments/article/17/La%20Corrosion%20Biologique%20des%20alliages%20orthodontiques.pdf

23. Lazanas, A.C., Prodromidis, M.I.: Electrochemical impedance Spectroscopy–A tutorial. ACS Measurement Science Au, 2023. 3: pp. 162–193.
24. Hrir, H., et al: Electrochemical corrosion behavior of α-titanium alloys in simulated biological environments (comparative study). RSC Advances, 2024. 14(51): pp. 38110–38119.
25. Ali, A., Zhang, N., Santos, R.M.: Mineral characterization using scanning electron microscopy (SEM): A review of the fundamentals, advancements, and research directions. Applied Sciences, 2023. 13: p. 12600. https://doi.org/10.3390/app132312600)
26. Raphaël, C.: Microbial corrosion in titanium-based dental implants: How tiny bacteria can create a big problem? Journal of Bio-and Tribo-Corrosion, 2021. https://doi.org/10.1007/s40735-021-00590-9
27. Souza, J., Apaza-Bedoya, K.: A comprehensive review on the corrosion pathways of titanium dental implants and their biological adverse effects. Metals, 2020. 10(9): p. 1272. https://doi.org/10.3390/met10091272
28. Antoni, M.: Inorganic synthesis methods for nanostructured capacitors, 2015. https://www.researchgate.net/profile/Markus-Antoni/publication/309463827_Inorganic_Synthesis_Methods_for_Nanostructured_Capacitors/links/5811bb0a08ae009606be8d49/Inorganic-Synthesis-Methods-for-Nanostructured-Capacitors.pdf
29. Fattah-alhosseini, A., Naseri, M., Babaei, K.: A Review on the passive behavior of ultrafine-grained aluminum, copper, and titanium made using accumulative roll bonding process. Analytical and Bioanalytical Electrochemistry, 2019. 11: pp. 1353–1382.
30. Muñoz-Portero, M.J., García-Antón, J., Guiñón, J.L., Leiva-García, R.: Pourbaix diagrams for titanium in concentrated aqueous lithium bromide solutions at 25°C. Corrosion Science, 2011. 53: pp. 1440–1450. https://doi.org/10.1016/j.corsci.2011.01.013
31. Yang, X., Du, C., Wan, H.: Influence of sulfides on the passivation behavior of titanium alloy TA2 in simulated seawater environments. Applied Surface Science, 2018. 458: pp. 198–209. https://doi.org/10.1016/j.apsusc.2018.07.068
32. Zakir, O., Ait-Karra, A., Idouhli, R.: Effect of anodization time on the morphological, structural, electrochemical, and photocatalytic properties of anodic TiO2 NTs. Journal of Solid State Chemistry, 2023. 322: p. 123939. https://doi.org/10.1016/j.jssc.2023.123939

42 Treatment of Phosphoric Acid Production Effluent

K. Mouzdahir, B. Z. Elabidine, A. Zouhri, and K. Lalam

42.1 INTRODUCTION

Phosphoric acid, a vital component in various industrial processes and products, plays a pivotal role in global agriculture, food processing, and chemical manufacturing [1]. There are two types of production processes: the wet process and the thermal process, each with its unique requirements and advantages. In the wet process, phosphate rock is initially dissolved in sulfuric acid, to produce phosphoric acid. This method is widely employed due to its relatively low energy consumption and ability to produce high-purity phosphoric acid suitable for various applications, such as fertilizer production and food additives [2]. Alternatively, the thermal process involves the reaction of phosphate rock with high-temperature sulfuric acid to produce phosphoric acid and other by-products [3]. However, the wet process is the most commonly used.

As the demand for phosphoric acid continues to rise, optimizing its production processes becomes increasingly crucial. One of the key elements in the manufacture of phosphoric acid is sulfuric acid as a catalyst. Sulfuric acid serves as a key agent in the reaction sequence, facilitating the conversion of phosphate rock into phosphoric acid through a series of chemical reactions [4]. This catalytic action is essential for enhancing reaction kinetics and yield in the phosphoric acid production process. Additionally, sulfuric acid plays a crucial role in maintaining optimal pH levels and controlling impurities during various stages of production. However, the efficient utilization of sulfuric acid and optimization of process parameters pose significant challenges to manufacturers [5].

The phosphoric acid production as an important industry makes its valorization a crucial challenge for several reasons; on the one hand, to improve energy efficiency and valorize byproducts in order to minimize energy consumption and greenhouse gas emissions. On the other hand, to reduce the production of solid waste and highly acidic effluents in order to protect the environment. This is where modeling and simulation techniques emerge as indispensable tools in the industry [6]. By employing sophisticated computational models, manufacturers can simulate and analyze complex chemical reactions, reactor dynamics, and process conditions with high precision. Modeling and simulation enable predictive analysis, allowing for the optimization of reaction conditions, resource utilization, and overall process efficiency.

42.2 THERMODYNAMICS BACKGROUND

Thermodynamic models play a crucial role in predicting the behavior of chemical systems under various conditions and also in crafting chemical process models by forecasting the essential physical properties entailed in mass, heat, and momentum balance equations. These models are based on fundamental thermodynamic principles and equations of state, which describe the relationships between the properties of substances, such as temperature, pressure, volume, and composition [7]. Frequently, these models incorporate numerous unknown parameters, typically derived from experimental measurements.

In this project, the selection of the suitable thermodynamic for the phosphoric acid production model was meticulously curated to ensure optimal accuracy and efficiency. For the liquid phase, the ENRTL (extended non-random two-liquid) model was chosen, leveraging its robust capabilities in capturing non-ideal behavior and complex interactions within liquid solutions. Meanwhile, for the

 DOI: 10.1201/9781003647263-42

solid phase, the SOLIDS model was deemed most suitable, offering comprehensive representation and characterization of solid-state properties essential for the project's objectives [8].

42.2.1 Electrolyte-NRTL

The Electrolyte-NRTL model applied to characterize excess properties and vapor-liquid equilibrium in the system [9] has been utilized for the simulation of phosphoric acid process. The Electrolyte-NRTL model is an excess Gibbs energy model and operates on the basis of two core assumptions: firstly, in an electrolyte system, the local composition of cations around cations is assumed to be zero; secondly, the local net ionic charge of cations and anions around a central molecule is also considered to be zero [10]. The model's excess Gibbs free energy is expressed through Equation (42.1), which, upon the application of Equation (42.2), yields Equation (42.3):

$$\frac{G^{*ex}}{RT} = \frac{G^{*ex}_{PDH}}{RT} + \frac{G^{*ex}_{BORN}}{RT} + \frac{G^{*ex}_{NRTL}}{RT} \tag{42.1}$$

$$ln\gamma_i^* = ln\gamma_i^{*PDH} + ln\gamma_i^{*BORN} + ln\gamma_i^{*NRTL} \tag{42.2}$$

$$ln\gamma_i = \left[\frac{\partial\left(n_t G^{*ex}\right)/RT}{\partial n_i}\right]_{T,P,n_{j\neq i}} \tag{42.3}$$

where G^{*ex} is excess Gibbs energy of a mixed solvent electrolyte system, G^{*ex}_{PDH} is excess Gibbs energy calculated by the Pitzer–Debye– Huckel formula, G^{*ex}_{BORN} is excess Gibbs energy calculated by the Born expression, and G^{*ex}_{NRTL} is excess Gibbs energy calculated by a nonrandom two-liquid hypothesis developed by Renon and Prausnitz. The first two terms of Eq. (42.1) account for the long-range interactions and the third term accounts for the local interaction contribution.

The Pitzer–Debye–Huckel formula is given by:

$$\frac{G^{*ex}_{PDH}}{RT} = -\left(\sum_i x_i\right)\left(\frac{1000}{M_S}\right)^{1/2}\left(\frac{4A_\varphi I_x}{\rho}\right) ln\left(1+\rho I_X^{1/2}\right)$$

Where:

$$A_\varphi = \frac{1}{3}\left(\frac{2\pi N_A d_s}{1000}\right)^{1/2}\left(\frac{Q_e^2}{\varepsilon_s kT}\right)^{3/2}$$

$$I_x = \frac{1}{2}\sum_i x_i z_i^2$$

A_φ and I_x are Pitzer–Debye–Huckel parameter and ionic strength in mole fraction scale, respectively.

42.2.2 Solids

The SOLIDS model in Aspen Plus is grounded in a mathematical formalization of the physical phenomena and transfer mechanisms intrinsic to solid-liquid separation operations. These equations, rooted in the principles of fluid dynamics and particle kinetics, comprehensively encompass the intricate interactions between the solid and liquid phases as well as the involved physicochemical processes [8].

Within this formalization, particle kinetics are represented by expressions describing the dynamics of agitation, collision, and aggregate formation, under the influence of forces such as gravity, liquid viscosity, and particle-particle interactions. Forces of flocculation or aggregation are modeled, considering phenomena such as coagulation, coalescence, and fracturing, with equations integrating parameters such as particle concentration, viscosity, and surface forces.

Diffusion and advection phenomena are quantified by transport equations that describe the migration of solid particles through the liquid medium, influenced by concentration, pressure, and fluid velocity gradients. Furthermore, the properties of solids and liquids, crucial for system characterization, are determined by physical and chemical parameters such as particle size, shape, density, viscosity, thermal conductivity, and liquid solubility.

By integrating these diverse aspects within a coherent mathematical framework, the SOLIDS model facilitates a precise and detailed representation of solid-liquid separation processes, providing a robust analytical foundation for industrial operations simulation and optimization.

42.3 RESULTS AND DISCUSSION

The hemihydrate process for phosphoric acid production was modeled and simulated utilizing ASPEN PLUS software. Key process units integrated into the model encompassed chemical reaction, flash separation, and filtration. Furthermore, sensitivity analyses were conducted on select parameters to assess their impact on process performance and product quality. Sensitivity analysis was also performed to determine the optimum temperature for steam flow treatment.

42.3.1 Process Flowsheet

The process flowsheet, illustrated in Figure 42.2 through ASPEN PLUS simulation, delineates the operational sequence. The initial stages involve introducing phosphate rock, sulfuric acid, and water into the reaction system. There are two reactions taking place after the dissolution of rock. These are:

$$Ca_3\left(PO_4\right)_2 + 3H_2SO_4 + H_2O \rightarrow 3CaSO_4.\ H_2O + 2H_3PO_4$$

$$CaF_2 + H_2SO_4 + H_2O \rightarrow CaSO_4.\ H_2O + 2HF$$

Both reactions were assumed to attend 100% conversion.

The resulting mixture proceeds to a flash separation to separate the vapor phase into the solid-liquid phase. The solid-liquid phase enters filtration unit where calcium sulphate (gypsum) crystals are separated from phosphoric acid. Subsequently, the vapor phase enters a gas treatment by first cooler (to set it to an adequate temperature for sulfuric acid recuperation) to recover 12.22% sulfuric acid.

42.3.2 Model Simulation

Tables 42.A1 and 42.A2 provide an overview of the properties of all streams. The process shown in Figure 42.1 involves the addition of phosphate rock, sulfuric acid, and water to the reaction system within reactors R1 and R2. The resulting product from this reaction system is then directed to a flash separation unit FL1 to recover the steam generated during the reaction, which is inherently exothermic. Meanwhile, a filtration unit, F1, is used to separate the gypsum crystals from the phosphoric acid solution. The effluent steam is then transferred to the treatment unit (inset in red) for further processing.

It is evident from the data that the model generates 77,661 kg/h of phosphoric acid and 182,169 kg/hr of gypsum, this production generates 30,468 kg/hr of steam effluent that must be treated.

42.3.3 Sensitivity Analysis

Following the sensitivity analyses performed with the developed model to evaluate various operational parameters, including temperature, pressure, and flow rates, the focus of this study shifts

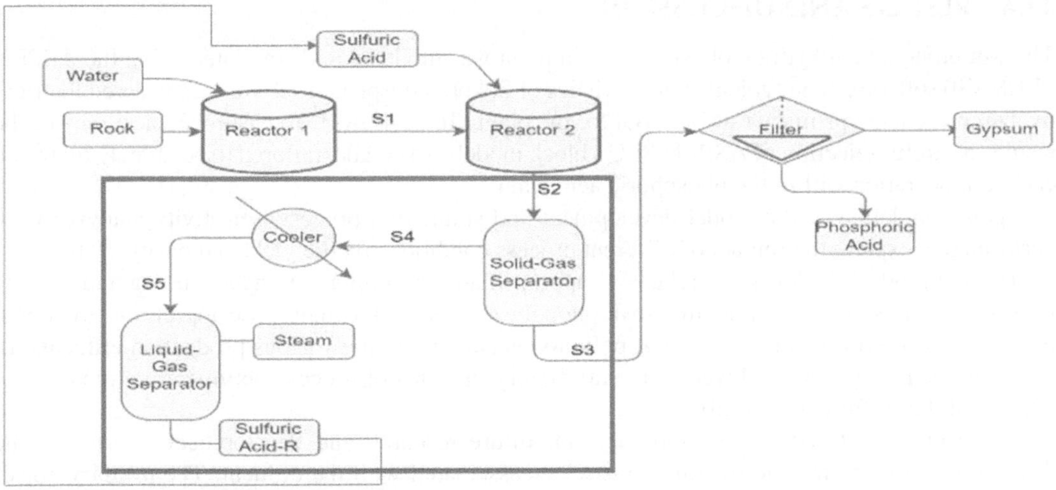

FIGURE 42.1 Process flowsheet of hemihydrates plant.

towards addressing the treatment of steam effluent. This effluent, a complex mixture of sulfuric acid, hydrofluoric acid, and water vapor, presents a significant challenge and opportunity for the industry.

The primary objective at this stage is to investigate the feasibility of recovering sulfuric acid in its liquid form from the vapor stream. By doing so, the industry stands to benefit from enhanced resource utilization and reduced waste generation.

Figure 42.2 presents the variation of sulfuric acid mass flow against the flash separator temperature in the treatment unit. The results obtained from the study are promising, indicating that sulfuric acid can indeed be reclaimed in liquid form at rates of up to 16,635 kg/hr, provided the temperature is maintained at point S equivalent to 150°C or below. This represents a considerable advancement, with a recovery rate of 12.22%, which can significantly bolster the industry's operational efficiency and economic viability.

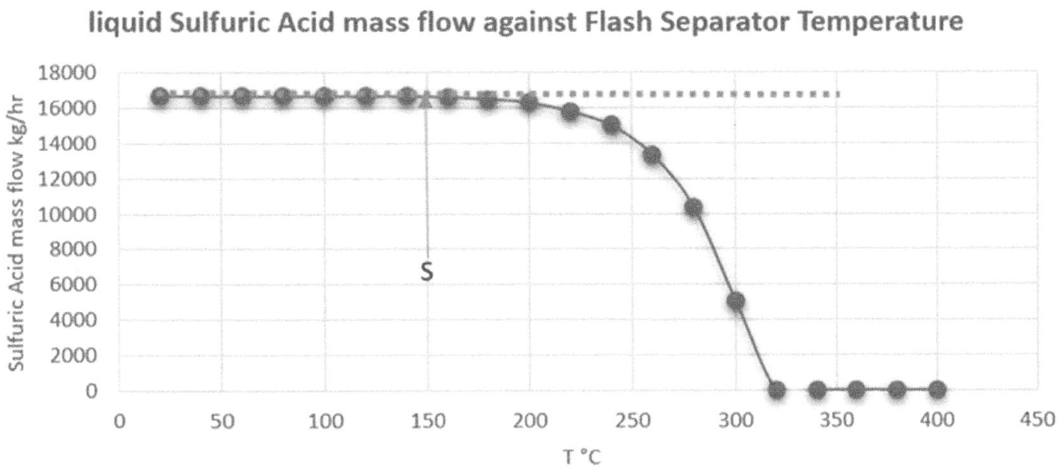

FIGURE 42.2 Determination of the optimum temperature (point S) for maximum recovery of liquid sulfuric acid (the dotted line shows the tangent of the curve).

42.4 RESULTS AND DISCUSSION

The wet process hemihydrate phosphoric acid plant was modeled and simulated using the ASPEN PLUS V10 software. The typical characteristics of Sokoto phosphate rock were considered to effectively represent the primary raw material for the plant (Table 42.A3). To ensure the accuracy of the results, a careful selection of ASPEN PLUS block models was made, tailored to accurately represent each unit operation within the phosphoric acid plant.

Upon completion of the model development and simulation process, sensitivity analyses were performed to explore the impact of different process conditions on the yield and purity of the phosphoric acid produced. This comprehensive approach aimed to improve understanding and provide the critical points to focus on in order to develop the process and then propose appropriate treatment processes to control the environmental pollution generated by the various production effluents. In our case, the results obtained were very satisfactory in terms of process optimization (recovery of 12.22% of the sulfuric acid used).

In the pursuit of continuous improvement, future research and development efforts will be directed toward refining the treatment processes associated with the effluent. The primary objective will be twofold: first, to improve the effectiveness of hydrofluoric acid and water vapor removal to ensure compliance with stringent environmental regulations and minimize adverse impacts. Second, there will be a concerted effort to explore innovative techniques for recovering any remaining valuable components in the effluent. Through this holistic approach, the industry reaffirms its unwavering commitment to sustainability and responsible resource management. Through continuous innovation and optimization, we strive not only to reduce our environmental footprint, but also to maximize the use of resources, thereby fostering a more sustainable and resilient industrial landscape.

APPENDIX

TABLE 42.A1
Summary of Stream Properties for Simulated Process

	Units	Water	Rock	S1	H_2SO_4	S2	S3
Phase		Liquid Phase			Liquid Phase		
Temperature	C	25	25	25	25	381.810952	381.810952
Pressure	bar	1	1	1	1	1	1
Mass Vapor Fraction		0	0	0.12426475	0	0.10495639	0
Mass Solid Fraction		0	0.06440001	0.87573525	0	0.62752716	0.70111349
Mass Flows	kg/hr	18143.7	136077.7	154221.4	136077.7	290299.1	259830.356
CA10(-01	kg/hr	0	126293.711	0	0	0	0
CALCI-01	kg/hr	0	0	116517.479	0	17529.6009	17529.6009
CALCI-02	kg/hr	0	0	9776.23215	0	1470.79605	1470.79605
SULFU-01	kg/hr	0	0	0	136077.7	31742.7732	15073.7834
WATER	kg/hr	18143.7	1020.58309	19164.2831	0	9582.14155	34.6738223
CALCI-03	kg/hr	0	0	0	0	154406.766	154406.766
ORTHO-01	kg/hr	0	0	0	0	62547.1501	62547.1501
HYDRO-01	kg/hr	0	0	0	0	4256.46633	4.1797255
HEMAT-01	kg/hr	0	2041.16618	2041.16618	0	2041.16618	2041.16618
ALUMI-01	kg/hr	0	2041.16618	2041.16618	0	2041.16618	2041.16618
SILIC-01	kg/hr	0	4681.07341	4681.07341	0	4681.07341	4681.07341

TABLE 42.A2
Summary of Stream Properties for Simulated Process

	Units	GYPSUM	H_3PO_4	S4	S5	STEAM	H_2SO_4-R
Phase				Vapor Phase		Vapor Phase	Liquid Phase
Temperature	C	406.810952	406.810952	381.810952	150	150.000052	150.000052
Pressure	bar	1	1	1	1	1	1
Mass Vapor Fraction		4.87E-10	0.00048706	1	0.43512352	1	0
Mass Solid Fraction		0.999999	2.35E-05	0	0	0	0
Mass Flows	kg/hr	182168.929	77661.4266	30468.7441	30468.7441	13257.6675	17211.0766
CA10-01	kg/hr	0	0	0	0	0	0
CALCI-01	kg/hr	17529.4256	0.17529601	0	0	0	0
CALCI-02	kg/hr	1470.78134	0.01470796	0	0	0	0
SULFU-01	kg/hr	0.035359	15073.748	16668.9898	16668.9898	33.7266563	16635.2632
WATER	kg/hr	8.1335E-05	34.673741	9547,46772	9547.46772	9010.65161	536.816109
CALCI-03	kg/hr	154405.222	1.54406766	0	0	0	0
ORTHO-01	kg/hr	0.14671861	62547.0034	1.99E-09	1.99E-09	1.45E-22	1.99E-09
HYDRO-01	kg/hr	9.80E-06	4.17971569	4252.2866	4252.2866	4213.28928	38.9973204
HEMAT-01	kg/hr	2041.14577	0.02041166	0	0	0	0
ALUMI-01	kg/hr	2041.14577	0.02041166	0	0	0	0
SILIC-01	kg/hr	4681.0266	0.04681073	0	0	0	0

TABLE 42.A3
Properties of Sokoto Phosphate Rock

Property	P_2O_5	CaO	F	SiO_2	Al_2O_3	Fe_2O_3	H_2O
weight %	36.25	52.3	3.84	3.44	1.5	1.5	0.78

REFERENCES

1. Z. Bakher, M. Kaddami. "Thermodynamic equilibrium study of phosphorus pentoxide-water binary system: The stability and solubility of 10H3PO4.H2O," Calphad, **63** (2018): page 148–155.
2. A. Bichri, M. A. Kamzon, S. Abderafi. "Modelling and optimization of P2O5 losses in phosphoric acid attack and filtration process," Materials Today: Proceedings, **51** (2022): page 1998–2004.
3. H. Maa, X. Fenga, B. Zeng. "Self-anticorrosion for the combustion tower of heat recovered thermal process phosphoric acid production," Process Safety and Environmental Protection, **118** (2018): page 330–347.
4. R. K. Sharma et al. "Revisiting the role of sulfur in crop production: A narrative review," Journal of Agriculture and Food Research, **15** (2024): page 101013.
5. C. Siminovicha, S. Joaoa. "Dynamic operator training simulators for sulphuric acid, phosphoric acid, and DAP production units," Procedia Engineering, **83** (2014): page 215–224.
6. A. S. Grema, Y. Y. Imam, H. I. Mohammed. "Modeling and simulation of hemihydrate phosphoric acid plant," Engineering, Technology and Environment, **14** (2018): page 169–171.
7. I. Bouchkira et al. "Thermodynamic modeling and parameter estimability analysis of a wet phosphoric acid process with impurities," Fluid Phase Equilibria, **524** (2023): page 113594.
8. Aspen Plus®. Aspen Plus® V11.0 Documentation (2019).
9. B. Messnaouia, A. Bouhaoussb, A. Derjaa. "A steady state model for describing the phosphate lattice loss," Procedia Engineering, **46** (2012): page 134–142.
10. B. K. Mondal et al. "Vapor–liquid equilibrium measurement and ENRTL modeling of CO2 absorption in aqueous hexamethylenediamine," Fluid Phase Equilibria, **402** (2015): page 102–111.

43 Integrating Advanced Biorefinery Approaches for Enhanced Bioethanol Production from *Opuntia ficus-indica* Wastes

A Circular Economy Perspective

I. Nori, K. Khallaki, and S. Touil

43.1 INTRODUCTION

This research evaluates the production of bioethanol from *Opuntia ficus-indica* (OFI), given its resilience and minimal maintenance needs, offering a sustainable solution for bioenergy in arid and semi-arid regions. By assessing various bioethanol production methodologies, the study investigates their scalability and economic feasibility, with a particular focus on the integration of circular economy principles. These principles guide the utilization of OFI waste, aiming to reduce waste and optimize resource utilization. The introduction of advanced biorefinery approaches could enhance the value derived from OFI, promote its broader use in biofuel production, and support economic growth in regions where conventional agriculture is limited by environmental conditions. The research includes a detailed analysis of these methodologies and their potential to transform OFI waste into a valuable resource, thereby contributing to a more sustainable bioenergy landscape.

Opuntia ficus-indica, often known as the prickly pear cactus, is a plant species that thrives in dry and semi-dry areas around the world. It belongs to the Cactaceae family [1, 2]. This family includes approximately 1,600 species across 174 genera, showcasing remarkable diversity [3]. The *Opuntia* species are most commonly found in the western Mediterranean basin, including southern Spain and Portugal, as well as in North African nations such as Tunisia, Algeria, and Morocco [4, 5]. This adaptability makes OFI an ideal biomass feedstock for sustainable energy solutions, particularly bioethanol production [6].

Bioethanol, a renewable energy source, has gained a major place in worldwide energy distribution [7]. Bioethanol can be used as a sustainable substitute for fossil fuels, providing many environmental advantages, such as decreased greenhouse gas emissions and a lower carbon footprint [8]. Studies have shown that OFI has the potential to produce bioethanol as a renewable energy source. Various parts of the cactus, such as its cladodes and fruits, contain different ethanol concentrations, making them suitable sources for bioethanol production [6].

The concept of a circular economy in bioethanol production advocates for sustainable biomass use, waste minimization, and resource recycling [9–11]. This approach not only reduces environmental impacts but also enhances resource efficiency in bioethanol production [12]. By converting agricultural residues and postharvest by-products into bioethanol, the circular economy approach reduces waste generation and promotes the efficient use of resources [13–17]. The biorefinery

 DOI: 10.1201/9781003647263-43

method, which combines different conversion methods on a single platform, is an important part of the circular economy in bioethanol production because it lowers the overall costs of the process and encourages a more environmentally friendly way of doing things [18].

Research has shown that OFI has high bioethanol production potential, making it an attractive option for sustainable bioethanol production [19]. Various parts of the cactus, such as its cladodes and fruits, yield bioethanol with different ethanol concentrations [5]. This feature makes it possible to use OFI waste in the production of bioethanol, which is an environmentally friendly and cost-effective solution. Integrating advanced biorefinery approaches could position this species as a cornerstone in biofuel development, particularly in regions where it is abundantly grown. OFI cultivation and utilization not only align with environmental sustainability goals, but they also bolster economic growth in areas where conventional agriculture faces challenges due to climatic conditions.

To produce bioethanol from biomass, several processes are involved, including pretreatment, hydrolysis, fermentation, and distillation [20–22]. These processes are essential for increasing the accessibility of biomass for the conversion of cellulose to glucose, making it more accessible to the fermentation process [23]. The pretreatment step increases biomass availability for hydrolysis by removing lignin and altering cellulose structures. Researchers have examined various pretreatment techniques, such as physical, chemical, and biological methods [24–26]. Some of the sugars that were released from biomass by pretreatment methods like diluted sulfuric, phosphoric, or organic acids can be broken down by known fermentation inhibitors [27, 28]. Hydrolysis breaks down cellulose into glucose molecules, which microbes can easily ferment [29–31]. Hydrothermal pretreatment with hot liquid water can also solubilize hemicellulose [32, 33]. Microorganisms in the fermentation process convert biomass carbohydrates into sugars, which are then transformed into ethanol through a series of chemical reactions [34, 35]. Microorganisms like *S. cerevisiae*, *Zymomonas mobilis*, *Klebsiella oxytoca*, *Escherichia coli*, *Thermoanaerobacter ethanolicus*, *Pichia stipitis*, *Candida shehatae*, and *Mucor indicus* have been studied for their ability to make bioethanol [36–38]. Distillation, the final step, typically separates ethanol from the fermented broth using a process called distillation [39, 40]. Azeotropic distillation using entrainers like benzene or cyclohexane can provide high ethanol purity, but it comes with high energy consumption costs [41, 42]. Researchers have studied pervaporation as an alternative method to reduce the total cost of the process by breaking down the azeotrope of ethanol and water [43]. Overall, the efficiency of the bioethanol production process depends on the optimal conditions and techniques used in each step, including pretreatment, hydrolysis, fermentation, and distillation [44].

43.2 PRELIMINARY ANALYSIS OF AVAILABLE LITERATURE ON THE PRODUCTION OF BIOETHANOL FROM OFI

A systematic search was performed in the Springer and Scopus® databases to gather published papers on the production of bioethanol from OFI. Customized search criteria included peer-reviewed articles, conference papers, and reviews published in English. Keywords such as "*Opuntia ficus-indica*," "bioethanol production," "biomass utilization," and "renewable energy" were employed. The selection criteria encompassed factors relevant to bioethanol production, the accuracy of the methodology, and the significance of the research in the field.

Research on bioethanol production from OFI has shown an improvement in methods and knowledge regarding the cost-effectiveness and scalability of this process over time.

The earliest recorded production of bioethanol from prickly pear occurred in Italy in the 19th century. Specifically, in 1865, a facility in Catania was producing 2,500 liters of alcohol daily from prickly pear fruit [45]. An initial study on the fermentation of various fruit juices from *Opuntia spp.* was conducted by Fowler and Gopalakrishnamurti [46]. A renewed interest in the 1980s spurred further research by Bustos [47], Retamal et al. [48], Lee et al. [49], Suescún [50], Arrizon et al. [51], and Castellar et al. [52]. Fowler and Gopalakrishnamurti also initiated the process of alcoholic fermentation using cladodes (*O. dillenii*) in 1923 [46]. These studies have significantly enhanced the understanding of the fermentation process and the potential of *Opuntia spp.* in bioethanol production, facilitating further research in this area.

The 1987 study by Retamal, Durán, and Fernández utilized both cellulose and acid hydrolysis methods, focusing on sugar yield and ethanol production. This dual approach potentially offers higher sugar yields but could increase operational complexity. The insights provided by this study into regional ethanol production are valuable, although their relevance may be constrained by their age [48].

In 1997, M. Hamdi conducted a comprehensive review of the potential applications of prickly pear cladodes and fruits in bioindustries. Although providing a broad perspective, the relevance of the findings may be limited by the study's age in the context of contemporary economic realities and technological advancements [53].

The 2009 study by Francisco Sánchez on bioethanol production from prickly pear cladodes focused on extracting fermentable sugars using mechanical and enzymatic methods, followed by fermentation with yeast or bacteria. Primarily concentrating on ethanol yield, the study underscored the potential of prickly pear cladodes for renewable energy, placing less emphasis on other bioactive compounds in the plant [54].

In 2014, Kuloyo et al. investigated the enzymatic hydrolysate of pretreated OFI cladodes using *Kluyveromyces marxianus* and *Saccharomyces cerevisiae*, which showed high ethanol productivity, particularly under low oxygen conditions. The need for precise control of aeration, however, adds complexity to the process [55].

A 2015 study by Gonçalves et al. explored bioethanol production through autohydrolysis pretreatment, achieving a high enzymatic hydrolysis conversion rate of 92.52%. Despite this efficiency, the high energy consumption of autohydrolysis may reduce its economic viability [56].

Nascimento Santos et al. (2016) assessed the potential of prickly pear biomass for both bioethanol and biogas production, noting the challenges in dry regions. The relatively low potential for ethanol production compared to traditional sources might limit its economic feasibility, although biogas production remains a viable alternative [57].

The 2016 study by Souza Filho et al. examined lignocellulosic biomasses with an emphasis on pretreatment methods. It was found that alkaline pretreatment yielded promising results for enzymatic saccharification, suggesting potential for scalability, although dependent on the cost of chemicals [58].

In 2018, Pérez-Cadena et al. conducted research to determine the impact of salt concentration on ethanol production through acid hydrolysis in OFI var. Atlixco cladodes. Although the method appeared cost-effective, the moderate yield indicated a need for further optimization [59].

Also in 2018, Alencar et al. enhanced the enzymatic hydrolysis of cactus pear, discovering that neither chemical pretreatment nor surfactants were necessary. This finding indicates a potentially cheaper and more scalable process [60].

The 2019 study by López-Domínguez et al. investigated the hydrolytic activities of *Acinetobacter pittii* and fermentation by *K. marxianus*. The use of wild microorganisms could potentially reduce costs, although controlling the process may pose challenges, and the relatively low alcohol yield (12.98 g/L) might affect its viability [61].

In 2022, Randremahitsimanana et al. explored the bioethanol potential of OFI from Madagascar, achieving various concentrations of bioethanol from cladodes and fruits. The need for rectification in commercialization indicates potential additional processing costs [6].

Also in 2022, a study conducted by Karalis et al. on enzymatic hydrolysis followed by alcoholic fermentation used termamyl and celluclast for hydrolysis, achieving an ethanol concentration of 21.54 g/L. Despite its effectiveness, the researchers identified scaling challenges due to the costs of enzymes and the necessity for precise fermentation control [62].

TABLE 43.1
Study Details and Methodology

Study	Study Title	Methodology
Retamal et al. [48]	Ethanol Production from Prickly Pear	Fermentation of cladodes and fruits using *Saccharomyces* strains with cellulose and acid hydrolysis
Hamdi [53]	Prickly Pear as Raw Material for Bioindustries	Review of potential uses of prickly pear cladodes and fruits in biotechnological processes
Kuloyo et al. [55]	Ethanol Production from Cladodes by *K. marxianus* and *S. cerevisiae*	Enzymatic hydrolysis of cladodes followed by fermentation using *Kluyveromyces marxianus* and *Saccharomyces cerevisiae*
Gonçalves et al. [56]	Bioethanol Production from Coconuts and Cactus	Autohydrolysis pretreatment on various biomasses including cactus, followed by enzymatic hydrolysis and fermentation
Santos et al. [57]	Potential for Biofuels from Prickly Pear Biomass	Chemical composition analysis and potential assessment for bioethanol and biogas production
de **Souza Filho et al. [58]**	SSF of Cactus Pear Biomass	Testing lignocellulosic biomasses of cactus pear for ethanol production using different pretreatments
Cadena et al. [59]	Ethanol Production from Acid Hydrolysis of Cladodes	Acid hydrolysis of *Opuntia ficus-indica* cladodes, followed by fermentation with wild microorganisms
Alencar et al. [60]	Enzymatic Hydrolysis of Cactus Pear for Bioethanol	Enzymatic hydrolysis of cactus pear varieties for bioethanol production
López-Domínguez et al. [61]	Hydrolysis by *Acinetobacter pittii* and Fermentation by *Kluyveromyces marxianus*	Use of *A. pittii* for hydrolysis and *K. marxianus* for fermentation under varying conditions
Randremahitsimanana et al. [6]	Bioethanol Potential of Cactus from Madagascar	Production of bioethanol from *Opuntia ficus-indica*, including grinding, fermentation, and distillation
Karalis et al. 62]	Enzymatic Hydrolysis and Fermentation of Prickly Pears	Advanced enzymatic hydrolysis followed by alcoholic fermentation using *Saccharomyces cerevisiae* E1A
Nori et al. [63]	Acid Hydrolysis of Cactus Pear Wastes	Optimizing conditions for acid hydrolysis of cactus pear wastes using sulfuric acid

The 2023 study by Nori et al. focused on optimizing acid hydrolysis conditions for cactus pear waste, emphasizing the sustainability benefits of utilizing waste materials. The scalability and economic feasibility of this approach depend on the availability of waste and the costs related to acid hydrolysis [63].

Table 43.1 provides a comprehensive overview of the diverse methodologies utilized in prior research, detailing innovative approaches to hydrolysis and fermentation processes.

Table 43.2 delves into the specific outcomes of these studies, with a particular focus on ethanol production metrics, significant advantages, and limitations encountered. The reported ethanol production values offer insights into the efficiency and scalability of the respective processes, which are critical for advancing commercial viability. The advantages and disadvantages listed for each study contribute to a more complete understanding of the current state of technology, including the high ethanol yields achievable with advanced enzymatic hydrolysis, the challenges associated with scaling up these processes, and the economic viability of producing bioethanol from cactus pear.

Figure 43.1 illustrates the advancements and improvements in bioethanol production from OFI. It demonstrates a trend of increasing ethanol yields over the years, reflecting the evolution of

TABLE 43.2
Ethanol Production, Pros, and Cons

Study	Ethanol Production (g/L)	Pros	Cons
Retamal et al. [48]	–	Pioneering study using *Opuntia* for ethanol	Limited by efficiency of hydrolysis methods and yeast strains
Hamdi [53]	0.86	Broadened understanding of prickly pear's potential	Lacked experimental data and specific methodologies
Kuloyo et al. [55]	20.6	High ethanol yield with *K. marxianus* and *S. cerevisiae*	Requires further bioprocess development for economic viability
Gonçalves et al. [56]	–	Efficient process for bioethanol production from green coconut shell	Limited data on ethanol yield from cactus
Santos et al. [57]	–	High methane production potential compared to traditional energy crops	Low ethanol but high methane production potential
de **Souza Filho et al. [58]**	–	Best yields with alkaline pretreatment	Specifics on ethanol yield not provided
Pérez-Cadena et al. [59]	–	Showed ethanol yield under oxygen-limited conditions	Lack of detailed methodology and scalability analysis
Alencar et al. [60]	–	Fermentation efficiency up to 82.8%	Requires detailed scalability and economic feasibility analysis
López-Domínguez et al. [61]	12.98	Maximum alcohol production of 12.98 g/L	Limited by temperature and pH conditions
Randremahitsimanana et al. [6]	–	Obtained bioethanol concentrations of 16° and 49°	Improvements needed for commercialization
Karalis et al. [62]	21.54	High ethanol yield using advanced enzymatic hydrolysis	Scalability and variability challenges
Nori et al. [63]	–	Optimal conditions for total reducing sugars identified	High water content in cactus pear affects hydrolysis

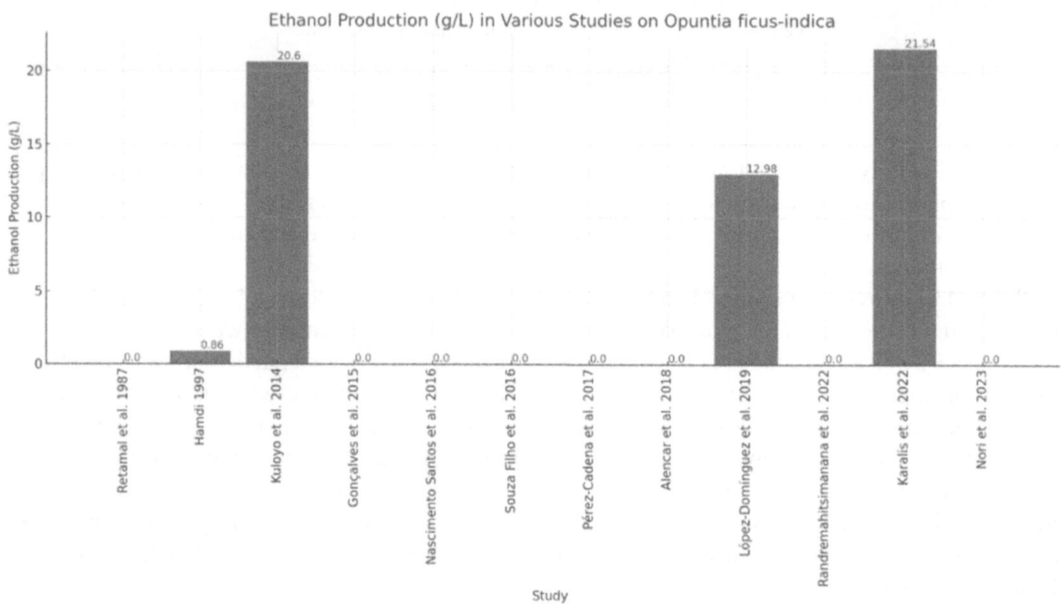

FIGURE 43.1 Ethanol production from *Opuntia ficus-indica* over the years.

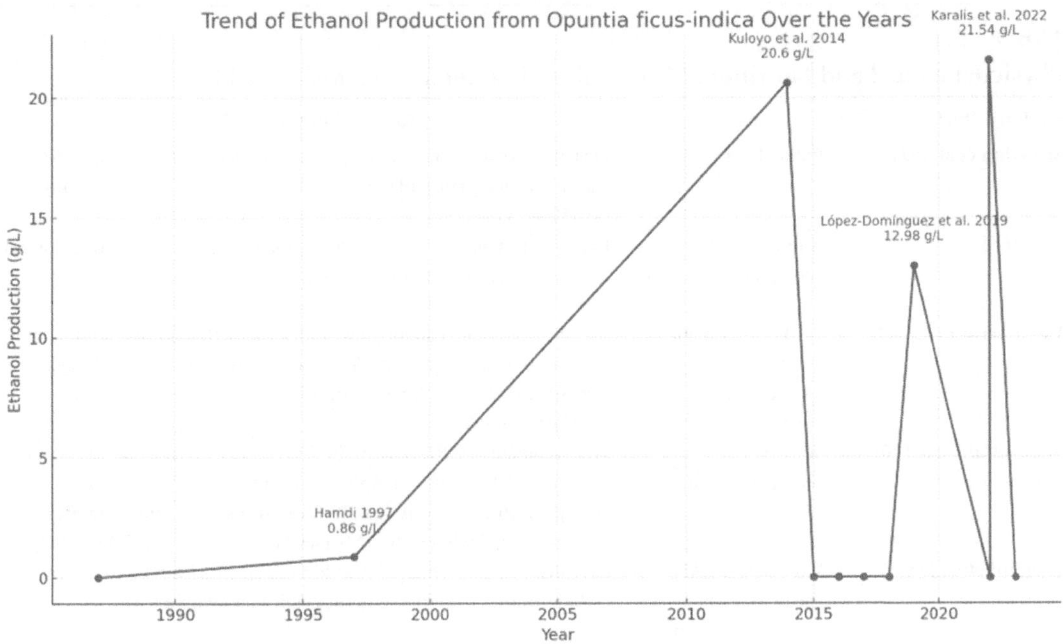

FIGURE 43.2 Trend of ethanol production from *Opuntia ficus-indica* over the years.

methodologies and technological advancements in this area of research. Figure 43.2 illustrates the ethanol production trend of OFI over the years from different studies.

43.3 COMPOSITION OF OFI

When assessing the valorization of OFI for bioethanol production, the initial step involves examining its composition. The physicochemical and nutritional composition of the prickly pear varies greatly due to the extensive range of OFI varieties and ecotypes, as well as its widespread distribution [64]. The fruit typically contains approximately 85% water, 15% sugars, 0.3% ash, and less than 1% protein, and is rich in minerals and various amino acids [65]. The composition of the plant material and its suitability for alcoholic fermentation have been studied under controlled conditions to evaluate the potential for ethanol production [66].

Table 43.3 presents a compilation of studies that have investigated the physical, chemical, and nutritional properties of OFI, also known as prickly pear. The table categorizes each study by author(s), year of publication, research focus, and key findings, providing a comprehensive overview of various aspects of prickly pear.

43.4 THE ECONOMIC FEASIBILITY OF OFI FOR BIOETHANOL PRODUCTION

Several studies have investigated the economic feasibility of using OFI for bioethanol production. Research focusing on different parts of OFI from Madagascar demonstrated that bioethanol can be obtained from its cladodes and fruits, yielding a 16° bioethanol from cladodes and a 49° bioethanol from fruits [6]. However, the relatively low concentration of fermentable sugars in the OFI biomass presents a technical constraint for commercial exploitation [66]. Another study conducted in the province of Almeria, Spain, estimated the total potential bioethanol production from OFI biomass at up to 502,927.8 tons per year, indicating significant potential for bioethanol production [72].

TABLE 43.3
Physicochemical and Nutritional Properties of *Opuntia ficus-indica* (OFI)

Study/Author	Focus	Key Findings and Values
Stintzing et al. [67]	Italian PP ecotypes properties	Density: 1.0543 g/cm^3 (red), 1.0479 g/cm^3 (orange-yellow), 1.0535 g/cm^3 (white-green); pH: 5.7–6.3; TSS: 12.3–13.7 °Brix; Acidity: 0.05%–0.11%. [58]
Kuti [68]	Antioxidant compounds in Texas *Opuntia*	Flavonoids: Quercetin 43.2 µg/g, Isorhamnetin 24.1 µg/g, Kaempferol 2.2 µg/g (green peel prickly pears).
El-Gharras et al. [69]	Prickly pear maturity stages and Moroccan PP maturation	pH, sugar, protein, calcium levels increase with maturity; acidity and moisture in peel and seeds decrease. Calcium: 111.11–263 mg/kg; Potassium: 963.9–2338 mg/kg; Sodium: 50.78 –236.78 mg/kg; Zinc: 0.62–0.29 mg/kg.
Salim et al. 2010 [70]	prickly pear components in Algeria	Water content: 94.40% (pulp), 90.33% (peel), 18.05% (seeds); Protein: 4.48% (seeds), 1.45% (pulp/peel); Lipids: 3.66% (seeds), 0.7% (pulp), 1.06% (peel); Ashes: 12.66% (seeds), 1% (pulp), 3.05% (peel); Carbohydrates: 61.15% (seeds), 4.11% (peel), 2.45% (pulp).
Khatabi et al. [71]	Moroccan prickly pear cultivars polyphenols and pigments	Polyphenols: 15.34 mg/kg (red skin juice), 17.81 mg/kg (red skin whole fruit), 15.03 mg/kg (yellow juice and whole fruit).
Martins et al. 2023 [64]	Physicochemical and nutritional profiles of prickly pear	Water content: 79.9–83.0% (fruits), 82.6–84.8% (mesocarps); Total carbs: 14–20%; Reducing sugars: 8–14%; Dietary fiber: 1.4% (fruits), 4.8% (mesocarps); Protein: ~0.5%; Fat: 0.12–0.15%; Ascorbic acid: 21–32 mg/100 g; Carotenoids: 18–43 mg/100 g; Polyphenols: 109–158 mg/100 g; Calories: 150.0–160.7 kcal/100 g; Potassium: 140–650 mg/100 g; Calcium: 18–286 mg/100 g; Magnesium: 13–96 mg/100 g; Phosphorus: 3.7–25 mg/100 g; Sodium: 1.9–4.3 mg/100 g.

The economic viability of bioethanol production from OFI biomass hinges on several factors, including the cost of biomass acquisition, conversion technology, market demand, the specific species, cultivation methods, and the use of chemical or organic fertilizers [73]. Economic and energetic analyses reveal that these elements significantly affect both the cost and the quantity of biomass produced, highlighting the complex interdependencies that influence the sustainability and profitability of cactus pear biomass production systems [74, 75]. Processing prickly pear biomass for bioethanol production is feasible at both large and small scales. However, while production costs decrease with the size of the plant, the required area for crop cultivation and the average distance for crop transport increase, leading to higher transport costs [62]. These dynamics should be considered in the economic feasibility analysis of bioethanol production from prickly pear biomass [76].

Scale efficiency in bioethanol production does not adhere to an absolute rule, as the distillation process itself is scalable [77]. Recent experiences have demonstrated that distilleries of various sizes can operate effectively and economically, ranging from "mini-distilleries" (20 million liters per year) to "micro-distilleries" (500,000 liters per year). Differentiations between mini-distilleries (medium size, typically 5,000–20,000 liters per day) and micro-distilleries (small size, typically 100–1,000 liters per day) are noted, with some definitions placing micro-distilleries at a capacity of 6,000–12,000 liters per day [78].

In 2023, worldwide production of fuel ethanol exceeded 111.7 billion liters, marking an increase of nearly 5.3 billion liters from the previous year and setting a new record high [79]. In response to rising production demands from renewable raw materials, researchers have intensively developed alternative bioethanol separation and purification processes [21, 80]. Optimization of the treatment train for producing bioethanol from source-separated food waste and its upscaling from lab to pilot scale have been conducted, demonstrating the potential for advanced bioethanol production [81, 82].

Prickly pear biomass cultivation is adaptable to various agricultural methods. The traditional or extensive cultivation approach, suitable for rain-fed areas with low-input management, typically yields between 2 and 10 tons of dry matter per hectare annually, with plant densities ranging from 2,500 to 5,000 plants per hectare. Conversely, the intensive cultivation system, ideal for irrigated or higher rainfall areas, supports denser plantings (10,000–50,000 plants per hectare) and can yield between 20 and 45 tons of dry matter per hectare annually with agricultural inputs [83]. Research in Brazil and studies on the use of treated domestic wastewater for irrigating 'Orelha-de-elefante Mexicana' prickly pear have demonstrated the plant's potential as a biomass feedstock in intensive systems, highlighting its viability across diverse climatic and water conditions [84].

43.5 ASSESSING THE ENVIRONMENTAL LOADS AND ECONOMIC PROFITABILITY OF PRODUCTION OF OFI

Research on OFI biomass has revealed its diverse economic and environmental possibilities. In Spain and Brazil, studies have investigated the economic viability of cultivating *Opuntia* for fresh food production, the profitability of cactus pear production, and the generation of biomass feedstock for bioethanol production [85]. Furthermore, OFI has been shown to have potential as a forage crop, capable of producing 20–30 times more forage than maize per unit of land, thus serving as a useful second-generation feedstock for bioenergy and biogas production [86]. Additionally, the environmental benefits of cactus cultivation have been documented, including ecosystem regulation through mitigation of wind and water erosion, reduced soil evaporation, lower water and maintenance requirements, and a reduced environmental footprint compared to other fruits. In addition, studies of the energy balance and economic indicators of spineless cactus biomass production in Brazil's semi-arid region have provided new insights into these topics [87]. These results help us learn more about the economic and environmental aspects of producing biomass from OFI. This is useful for figuring out how useful it could be for different purposes, such as making bioenergy and bioethanol [88, 89].

43.6 CHALLENGES AND FUTURE PERSPECTIVES

Addressing the challenges of producing bioethanol from OFI is crucial for its widespread adoption. These challenges include scalability, technological limitations in biomass processing and fermentation efficiency, and market acceptance. Achieving higher production levels without compromising efficiency or sustainability presents a significant scalability challenge. Technological advancements are required to enhance biomass processing and fermentation efficiency. Additionally, market acceptance poses a significant hurdle, as bioethanol from OFI is relatively new to the energy market and may face competition from established biofuels.

Future research and development directions could enhance the viability of OFI bioethanol. These directions include improving crop yields through agricultural research, developing more efficient processing technologies, and formulating strategies to boost market acceptance and integration into existing energy infrastructures. The potential integration of OFI bioethanol production within a broader biorefinery model, which would produce not only bioethanol but also other valuable by-products, is a promising future direction.

To enhance the economic viability and sustainability of bioethanol production from OFI, it is important to consider how OFI bioethanol production could be incorporated into a larger biorefinery model and to identify the additional research and development needed to overcome the aforementioned challenges.

43.7 CONCLUSION

The comprehensive study of OFI for bioethanol production underscores its significant potential within a biorefinery framework aligned with circular economy principles. The research demonstrates the capability of OFI as a sustainable and efficient biomass feedstock, presenting a promising alternative to conventional biofuel sources. The investigation revealed varying ethanol concentrations, with yields of up to 49° from fruits and 16° from cladodes, showcasing OFI's effectiveness as a versatile source of bioethanol. OFI cultivation requires minimal water and maintenance, reducing operational inputs by up to 40% compared to traditional bioenergy crops and significantly contributing to the production's sustainability. This study comprehensively evaluates the use of OFI as a sustainable feedstock for bioethanol production within a biorefinery framework emphasizing circular economy principles. Significant advances in bioethanol yields and process efficiency were demonstrated, positioning OFI as a promising alternative to traditional biofuel sources. The findings indicate that OFI can achieve ethanol concentrations of up to 49° from fruits and 16° from cladodes, significantly higher than many conventional bioenergy crops. The integration of advanced biorefinery techniques facilitated a reduction in operational inputs by up to 40% and waste generation by 30% compared to traditional methods, underscoring the technological advancements and efficiency gains achieved through novel processing techniques. The methodologies developed are scalable and demonstrate potential for commercial viability, particularly in arid and semi-arid regions where OFI thrives. Pilot studies suggest that scaling up these processes could effectively serve regional bioenergy needs, supporting economic growth and sustainability in underutilized agricultural zones. The role of OFI in enhancing the sustainability of bioenergy production was highlighted, with a potential increase in regional economic outputs by 25% in areas adopting these biorefinery techniques. The approach aligns with circular economy principles by optimizing resource use and minimizing waste, providing a blueprint for environmentally sustainable bioenergy production. Future research should explore optimized pretreatment processes to enhance yield efficiencies and reduce production costs. Addressing market acceptance and integration into existing energy infrastructures remains a priority to boost the commercial uptake of OFI-based bioethanol. Investigating the full life cycle impacts of OFI bioethanol production is crucial to understanding its long-term sustainability and environmental benefits. It is recommended that policymakers consider incentives for adopting sustainable bioenergy practices, including subsidies for bioethanol production from non-conventional sources like OFI. Regulatory frameworks can be designed to support the commercialization of innovative biorefinery technologies, promoting wider adoption and integration into national energy strategies. In conclusion, the use of OFI for bioethanol production offers a promising avenue for sustainable energy development, leveraging its high yield potential and alignment with circular economy principles. This study provides a foundation for future advancements in the field and a model for deploying renewable energy solutions in suitable geographies. The study facilitated a 30% reduction in waste generation by incorporating advanced biorefinery approaches, enhancing the overall resource efficiency of the bioethanol production process. The potential economic benefits of OFI bioethanol production were highlighted, with a projected increase in regional economic outputs by 25% in areas implementing these biorefinery techniques. The study also identified key challenges such as scalability, technological limitations in biomass processing, and market acceptance, indicating that addressing these issues is critical for the widespread adoption of OFI bioethanol. To make OFI bioethanol more commercially viable and long-lasting, future research should focus on improving pretreatment and fermentation processes to boost yields and lower costs, coming up with ways to make the market accept it more, and integrating OFI bioethanol into existing energy systems. Continued focus on the development of cultivation practices that further reduce water and land use is also recommended. This research serves as a foundational platform for stakeholders in academia, industry, and policymaking, guiding future innovations in sustainable bioenergy production. The integration of OFI within a broader biorefinery model not only supports environmental sustainability but also promotes economic growth, particularly in arid and semi-arid regions where traditional agriculture is challenging.

REFERENCES

1. Sipango, N., Ravhuhali, K.E., Sebola, N.A., Hawu, O., Mabelebele, M., Mokoboki, H.K., Moyo, B.: Prickly Pear (*Opuntia spp.*) as an Invasive Species and a Potential Fodder Resource for Ruminant Animals. Sustainability. 14(7), 3719 (2022). https://doi.org/10.3390/su14073719.
2. Mathur, M., Mathur, P.: Habitat Suitability of *Opuntia ficus-indica* (L.) MILL. (CACTACEAE): A Comparative Temporal Evaluation Using Diverse Bio-Climatic Earth System Models and Ensemble Machine Learning Approach. Environ. Monit. Assess. 196, 232 (2024). https://doi.org/10.1007/s10661-024-12406-7.
3. Aruwa, C.E., Amoo, S.O., Kudanga, T.: *Opuntia* (Cactaceae) Plant Compounds, Biological Activities and Prospects - A Comprehensive Review. Food Res. Int. 112, 328–344 (2018). https://doi.org/10.1016/j.foodres.2018.06.047.
4. Erre, P., Chessa, I., Nieddu, G., Jones, P.: Diversity and Spatial Distribution of *Opuntia spp.* in the Mediterranean Basin. J. Arid Environ. 73, 1058–1066 (2009). https://doi.org/10.1016/j.jaridenv.2009.05.010.
5. Yous, F.Z., Ferradous, A., Elgadi, S., Mostakim, L., Hafidi, M., Ouhammou, A., Alifriqui, M.: Effects of the Continentality and the Age of Prickly Pear (*Opuntia ficus-indica* (L.) Mill.) Plantations on Soil Properties Along a Coastal and Continental Transect in a Pre-Saharan Region of South-Central Morocco (Ait Baamrane). Appl. Ecol. Environ. Res. 21, 869–887 (2023).
6. Randremahitsimanana, F., Rabetokotany, N., Rasoanaivo, J., Rakotosaona, R.: Assessment of the Bioethanol Potential of Cactus ("*Opuntia ficus indica*") from Madagascar. Int. J. Prog. Sci. Tech. 34, 328 (2022). http://dx.doi.org/10.52155/ijpsat.v34.1.4550.
7. Kumar, A., Verma, A.: Recent Advances in Bioethanol: Current Scenario, Sources and Production Techniques. In: Microbiology-2.0 Update for a Sustainable Future (2024). https://doi.org/10.1007/978-981-99-9617-9_15.
8. Reshmy, R., Philip, E., Unni, R., Paul, S.A., Sindhu, R., Madhavan, A., Sirohi, R., Pandey, A., Binod, P.: Bioethanol—A Promising Alternative Fuel for Sustainable Future. In: Gupta, V.K., Tuohy, M., Ramteke, P., Nguyen, Q., Bhat, R. (eds.) Valorization of Biomass to Bioproducts. Elsevier, pp. 179–196 (2023). https://doi.org/10.1016/B978-0-12-822888-3.00006-2.
9. Devi, A., Bajar, S., Kour, H., et al.: Lignocellulosic Biomass Valorization for Bioethanol Production: A Circular Bioeconomy Approach. Bioenerg. Res. 15, 1820–1841 (2022). https://doi.org/10.1007/s12155-022-10401-9.
10. Scapini, T., Bonatto, C., Dalastra, C., Bazoti, S.F., Camargo, A.F., Alves Júnior, S.L., Venturin, B., Steinmetz, R.L.R., Kunz, A., Fongaro, G., Treichel, H.: Bioethanol and Biomethane Production from Watermelon Waste: A Circular Economy Strategy. Biomass Bioenergy. 170, 106719 (2023). https://doi.org/10.1016/j.biombioe.2023.106719.
11. Ezhumalai, G., Arun, M., Manavalan, A., et al.: A Holistic Approach to Circular Bioeconomy Through the Sustainable Utilization of Microalgal Biomass for Biofuel and Other Value-Added Products. Microb. Ecol. 87, 61 (2024). https://doi.org/10.1007/s00248-024-02376-1.
12. Althuri, A., Venkata Mohan, S.: Emerging Innovations for Sustainable Production of Bioethanol and Other Mercantile Products from Circular Economy Perspective. Bioresour. Technol. 363, 128013 (2022). https://doi.org/10.1016/j.biortech.2022.128013.
13. Rajkumar, R., Kurinjimalar, C.: Food Wastes/Residues: Valuable Source of Energy in Circular Economy. In: Sahay, S. (ed.) Handbook of Biofuels. Academic Press, pp. 147–163 (2022). https://doi.org/10.1016/B978-0-12-822810-4.00007-5.
14. Datta, B.: An Economic Analysis of Biofuels: Policies, Trade, and Employment Opportunities. In: Sahay, S. (ed.) Handbook of Biofuels. Academic Press, pp. 3–29 (2022). https://doi.org/10.1016/B978-0-12-822810-4.00001-4.
15. Hoang, A.T., Pandey, A., Huang, Z., Nižetić, S., Le, A.T., Nguyen, X.P.: Biofuels an Option for Agro-Waste Management. In: Hakeem, K.R., Bandh, S.A., Malla, F.A., Mehmood, M.A. (eds.) Environmental Sustainability of Biofuels. Elsevier, pp. 27–47 (2023). https://doi.org/10.1016/B978-0-323-91159-7.00011-4.
16. Capanoglu, E., Tomás-Barberán, F.A.: Introduction to Novel Approaches in the Valorization of Agricultural Wastes and Their Applications. J. Agric. Food Chem. 70(23), 6785–6786 (2022).
17. Yadav, R., Singh, S., Singh, A.N.: Cost-Benefit Analysis Act as a Tool for Evaluation of Agricultural Waste to the Economy: A Synthesis. In: Singh, P., Verma, P., Singh, R., Ahamad, A., Batalhão, A.C.S. (eds.) Waste Management and Resource Recycling in the Developing World. Elsevier, pp. 647–663 (2023). https://doi.org/10.1016/B978-0-323-90463-6.00009-9.

18. Sharma, S., Tsai, M.-L., Sharma, V., Sun, P.-P., Nargotra, P., Bajaj, B.K., Chen, C.-W., Dong, C.-D.: Environment Friendly Pretreatment Approaches for the Bioconversion of Lignocellulosic Biomass into Biofuels and Value-Added Products. Environments. 10(1), 6 (2023). https://doi.org/10.3390/environments10010006.
19. Yang, L., Lu, M., Carl, S., Mayer, J., Cushman, J., Tian, E., Lin, H.: Biomass characterization of *Agave* and *Opuntia* as Potential Biofuel Feedstocks. Biomass Bioenergy. 76, 10 (2015). https://doi.org/10.1016/j.biombioe.2015.03.004.
20. Tse, T.J., Wiens, D.J., Reaney, M.J.T.: Production of Bioethanol—A Review of Factors Affecting Ethanol Yield. Fermentation. 7(4), 268 (2021). https://doi.org/10.3390/fermentation7040268.
21. Bušić, A., Marđetko, N., Kundas, S., Morzak, G., Belskaya, H., Ivančić Šantek, M., Komes, D., Novak, S., Šantek, B.: Bioethanol Production from Renewable Raw Materials and Its Separation and Purification: A Review. Food Technol. Biotechnol. 56(3), 289–311 (2018). https://doi.org/10.17113/ftb.56.03.18.5546.
22. Ceaser, R., Montané, D., Constantí, M., et al.: Current Progress on Lignocellulosic Bioethanol Including a Technological and Economical Perspective. Environ. Dev. Sustain. (2024). https://doi.org/10.1007/s10668-024-04792-2.
23. Ko, C.-H., Yang, B.-Y., Lin, L.-D., Chang, F.-C., Chen, W.-H.: Impact of Pretreatment Methods on Production of Bioethanol and Nanocrystalline Cellulose. J. Cleaner Prod. 254, 119914 (2020). https://doi.org/10.1016/j.jclepro.2019.119914.
24. Baksi, S., Saha, D., Saha, S., Sarkar, U., Basu, D., Kuniyal, J.C.: Pre-treatment of Lignocellulosic Biomass: Review of Various Physico-Chemical and Biological Methods Influencing the Extent of Biomass Depolymerization. Int. J. Environ. Sci. Technol. 20(12), 13895–13922 (2023). https://doi.org/10.1007/s13762-023-04838-4.
25. Alawad, I., Ibrahim, H.: Pretreatment of Agricultural Lignocellulosic Biomass for Fermentable Sugar: Opportunities, Challenges, and Future Trends. Biomass Convers. Biorefin. 14, 1–29 (2022). https://doi.org/10.1007/s13399-022-02981-5.
26. Bansod, S., Makwana, K., Sarangi, P.K., Parikh, J.K.: Advanced Pretreatment Processes for Lignocellulosic Biomass to Biofuels Production: Path Towards Circular Bioeconomy. Sustain. Chem. Pharm. 39, 101514 (2024). https://doi.org/10.1016/j.scp.2024.101514.
27. Prasad, B.R., Padhi, R.K., Ghosh, G.: A Review on Key Pretreatment Approaches for Lignocellulosic Biomass to Produce Biofuel and Value-Added Products. Int. J. Environ. Sci. Technol. 20, 6929–6944 (2022). https://doi.org/10.1007/s13762-022-04252-2.
28. Falowo, O.A., Taiwo, A.E., Latinwo, L.M., Betiku, E.: Current Status of Substrate Hydrolysis to Fermentable Sugars. In: Betiku, E., Ishola, M.M. (eds) Bioethanol: A Green Energy Substitute for Fossil Fuels. Green Energy and Technology. Springer, Cham (2023). https://doi.org/10.1007/978-3-031-36542-3_4.
29. Thielemans, K., De Bondt, Y., Comer, L., Raes, J., Everaert, N., Sels, B.F., Courtin, C.M.: Decreasing the Crystallinity and Degree of Polymerization of Cellulose Increases Its Susceptibility to Enzymatic Hydrolysis and Fermentation by Colon Microbiota. Foods. 12(5), 1100 (2023). https://doi.org/10.3390/foods12051100.
30. Francis, F., Luthfi, A.A.I., Mohd Zaini, N.A.: The Influence of Enzymatic Hydrolysis Parameters of Oil Palm Empty Fruit Bunch on the Production of Bacterial Cellulose by *Komagataeibacter xylinus*. Biomass Convers. Biorefin. (2024). https://doi.org/10.1007/s13399-024-05444-1.
31. Samantaray, B., Mohapatra, S., Mishra, R., Behera, B.C., Thatoi, H.: Bioethanol Production from Agro-Wastes: A Comprehensive Review with a Focus on Pretreatment, Enzymatic Hydrolysis, and Fermentation. Int. J. Green Energy. 21, 1–27 (2023). https://doi.org/10.1080/15435075.2023.2253871.
32. Singh, A., Tsai, M.L., Chen, C.W., Singhania, R.R., Patel, A.K., Tambat, V., Dong, C.D.: Role of Hydrothermal Pretreatment Towards Sustainable Biorefinery. Bioresour. Technol. 367, 128271 (2023). https://doi.org/10.1016/j.biortech.2022.128271.
33. Toscan, A., Fontana, R.C., Camassola, M., Dillon, A.J.P.: Comparison of Liquid Hot Water and Saturated Steam Pretreatments to Evaluate the Enzymatic Hydrolysis Yield of Elephant Grass. Biomass Convers. Biorefin. 14, 1–14 (2022). https://doi.org/10.1007/s13399-022-02939-7.
34. Jayakumar, M., Gindaba, G.T., Gebeyehu, K.B., Periyasamy, S., Jabesa, A., Baskar, G., John, B.I., Pugazhendhi, A.: Bioethanol Production from Agricultural Residues as Lignocellulosic Biomass Feedstock's Waste Valorization Approach: A Comprehensive Review. Sci. Total Environ. 879, 163158 (2023). https://doi.org/10.1016/j.scitotenv.2023.163158.
35. Jahangeer, M., Rehman, M.U., Nelofer, R., Nadeem, M., Munir, B., Smułek, W., Qamar, S.A.: Biotransformation of Lignocellulosic Biomass to Value-Added Bioproducts: Insights into Bio-Saccharification Strategies and Potential Concerns. Top Catal. (2024). https://doi.org/10.1007/s11244-024-01941-9.

36. Adegboye, M.F., Ojuederie, O.B., Talia, P.M., et al.: Bioprospecting of Microbial Strains for Biofuel Production: Metabolic Engineering, Applications, and Challenges. Biotechnol Biofuels. 14, 5 (2021). https://doi.org/10.1186/s13068-020-01853-2.
37. Akinruli, F.T., Ibiyemi, M.F., Oluwasusi, V.O., Ajayi, F.: Isolation of Microorganisms from Cassava Peels for the Production of Bioethanol. GSC Biol. Pharm. Sci. (2022). https://doi.org/10.30574/gscbps.2022.19.1.0151.
38. Patil, N., Shendkar, T.G., Pardhi, A., Suthar, S.P., Patil, G.S., Pathak, P.D.: Bioethanol Production from Agricultural Biomass: Sources of Cellulose, Pretreatment Methods, and Future Prospects. In: Biorefinery: A Sustainable Approach for the Production of Biomaterials, Biochemicals and Biofuels. Springer Nature, Singapore, pp. 287–324 (2023).
39. Chen, H., Cai, D., Wen, J., Su, C., Wang, J.: Distillation Separation of Bioethanol: The Current Advances. In: Bioethanol Fuel Production Processes. II. CRC Press, pp. 393–408 (2023).
40. Wang, X., Cui, W., Guo, W., Sun, B., Huang, M., Li, J., Meng, N.: Separation Techniques for Manufacturing Fruit Spirits: From Traditional Distillation to Advanced Pervaporation Process. Compr. Rev. Food Sci. Food Saf. 23(1), 1–37 (2024).
41. Lasut, E.K., Hartanto, Y., Santoso, H.: Design and Optimization of Isopropyl Alcohol-Water Separation Using Heterogenous Azeotropic Distillation. CHEMICA J. Teknik Kimia (e-journal) (2022). https://doi.org/10.26555/chemica.v8i2.20849.
42. Yu, G., Dai, C., Lei, Z., Chen, B.: Chapter 3 - Azeotropic Distillation. In: Lei, Z., Dai, C., Chen, B., Ding, Z. (eds.) Special Distillation Processes (Second Edition). Elsevier, pp. 155–190 (2022). https://doi.org/10.1016/b978-0-12-820507-5.00006-3.
43. Zentou, H., Zainal Abidin, Z., Yunus, R., Awang Biak, D.R., Issa, M.A., Pudza, M.Y.: Modelling of Mass Transfer During Pervaporation of Ethanol/Water Mixture Using Polydimethylsiloxane Membrane. Chem. Eng. Res. Des. (2021). https://doi.org/10.1016/J.CHERD.2021.09.019.
44. Sevda, S., Garlapati, V.K., Datta, P., Chande, A., Pandey, L., Rathore, D., Singh, A., Sreekrishnan, T.: Bioethanol Production from Lignocellulosic/Algal Biomass: Potential Sustainable Approach. In: Algal Biofuel (2022). https://doi.org/10.1201/9781003363231-5.
45. Casas, A., Barbera, G.: Mesoamerican Domestication and Diffusion. In: Nobel, P. (ed.) Cacti: Biology and Uses. University of California Press, Berkeley, California, pp. 143–162 (2002). https://doi.org/10.1525/california/9780520231573.003.0009.
46. Fowler, G.J., Gopalakrishnamurti, B.: Contributions to the Prickly Pear Problem. Indian Inst. Sci. J. 6, 173–184 (1923).
47. Bustos, O.E.: Alcoholic Beverage from Chilean *Opuntia ficus indica*. In: Annual Meeting of the American Society of Enologists, Los Angeles, CA, (26 June 1981).
48. Retamal, N., Durán, J.M., Fernández, J.: Ethanol Production by Fermentation of Fruits and Cladodes of Prickly Pear Cactus (*Opuntia ficus-indica* (L.) Miller). J. Sci. Food Agric. 40, 213–218 (1987).
49. Lee, S.P., Lee, S.K., Ha, Y.D.: Alcohol Fermentation of *Opuntia ficus* Fruit Juice. J. Food Sci. Nutr. 5, 32–36 (2000).
50. Suescún, G.: Utilización Vinícola de *Opuntia dillenii*. Bistua. 3, 99–106 (2005).
51. Arrizon, J., Calderón, C., Sandoval, G.: Effect of Different Fermentation Conditions on the Kinetic Parameters and Production of Volatile Compounds During the Elaboration of a Prickly Pear Distilled Beverage. J. Ind. Microbiol. Biotechnol. 33(11), 921–928 (2006). https://doi.org/10.1007/s10295-006-0153-9.
52. Castellar, M.R., Obón, J.M., Alacid, M., Fernández-López, J.A.: Fermentation of *Opuntia stricta* (Haw.) Fruits for Betalains Concentration. J. Agric. Food Chem. 56(11), 4253–4257 (2008).
53. Hamdi, M.: Prickly Pear Cladodes and Fruits as a Potential Raw Material for the Bioindustries. Bioprocess Eng. 17, 387–391 (1997). https://doi.org/10.1007/s004490050402.
54. Sánchez, F., Curt, M., Fernández, J., Agüera, J., Uceda, M., Zaragoza, G.: Bioethanol Production from Prickly Pear (*Opuntia ficus-indica* (L) Mill.) Cladodes. (2009).
55. Kuloyo, O.O., du Preez, J.C., García-Aparicio, M.P., Kilian, S.G., Steyn, L., Görgens, J.: *Opuntia ficus-indica* Cladodes as Feedstock for Ethanol Production by *Kluyveromyces marxianus* and *Saccharomyces cerevisiae*. World J. Microbiol. Biotechnol. 30(12), 3173–3183 (2014). https://doi.org/10.1007/s11274-014-1745-6.
56. Gonçalves, F.A., Ruiz, H.A., dos Santos, E.S., Teixeira, J.A., de Macedo, G.R.: Bioethanol Production from Coconuts and Cactus Pretreated by Autohydrolysis. Ind. Crops Prod. 77, 1–12 (2015). https://doi.org/10.1016/j.indcrop.2015.06.041.

57. Santos, T.N., Dutra, E.D., do Prado, A.G., Leite, F.C.B., de Souza, R.F.R., dos Santos, D.C., de Abreu, C.A.M., Simões, D.A., Morais Jr, M.A., Menezes, R.S.C.: Potential for Biofuels from the Biomass of Prickly Pear Cladodes: Challenges for Bioethanol and Biogas Production in Dry Areas. Biomass Bioenergy. 85, 215–222 (2016). https://doi.org/10.1016/j.biombioe.2015.12.005.
58. de Souza Filho, P.F., Ribeiro, V.T., dos Santos, E.S., de Macedo, G.R.: Simultaneous Saccharification and Fermentation of Cactus Pear Biomass—Evaluation of Using Different Pretreatments. Ind. Crops Prod. 89, 425–433 (2016). https://doi.org/10.1016/j.indcrop.2016.05.028.
59. Cadena, R., Medina, S., Martinez, A., Lizardi-Jiménez, M., Espinosa-Solares, T., Tellez, A.: Effect of Concentration of Salts in Ethanol Production from Acid Hydrolysis of Cladodes of *Opuntia ficus indica* var. *Atlixco*. Revista Mexicana de Ingeniería Química. 17, 349–364 (2018). https://doi.org/10.24275/uam/izt/dcbi/revmexingquim/2018v17n1/PerezR.
60. Alencar, B., Damilano, E., Sampaio, E.V.D.S., Menezes, R., Morais, M.: Enzymatic Hydrolysis of Cactus Pear Varieties with High Solids Loading for Bioethanol Production. Bioresour. Technol. 250 (2018). https://doi.org/10.1016/j.biortech.2017.11.042.
61. López-Domínguez, C.M., Ramírez-Sucre, M.O., Rodríguez-Buenfil, I.M.: Enzymatic Hydrolysis of *Opuntia ficus-indica* Cladode by *Acinetobacter pittii* and Alcohol Fermentation by *Kluyveromyces marxianus*: pH, Temperature and Microorganism Effect. Biotechnol Rep (Amst). 24, e00384 (2019). https://doi.org/10.1016/j.btre.2019.e00384.
62. Karalis, I., Arapoglou, D., Markou, G., Eliopoulos, C., Papanikolaou, S.: Enzymatic Hydrolysis Followed by Alcoholic Fermentation of Prickly Pears (*Opuntia ficus-indica* L.) Cladodes and Fruits Using Saccharomyces cerevisiae E1A. (2022).
63. Nori, I., Touil, S., Khallaki, K.: Acid Hydrolysis of the Wastes of *Opuntia ficus-indica* (L.) Miller in Order to Produce Bioethanol. Mor. J. Chem. 11(1), 34–43 (2023). Available at: https://revues.imist.ma/index.php/morjchem.
64. Martins, M., Ribeiro, M.H., Almeida, C.M.M.: Physicochemical, Nutritional, and Medicinal Properties of *Opuntia ficus-indica* (L.) Mill. and Its Main Agro-Industrial Use: A Review. Plants. 12, 1512 (2023). https://doi.org/10.3390/plants12071512.
65. Cota-Sánchez, J.H.: Nutritional Composition of the Prickly Pear (*Opuntia ficus-indica*) Fruit. In: Nutritional Composition of Fruit Cultivars, 2015. https://doi.org/10.1016/B978-0-12-408117-8.00028-3.
66. Ferreira, R., Costa, A., Pinto, C., Silva, A., Saraiva, J., Cardoso, S.: Impact of Fermentation and Pasteurization on the Physico-Chemical and Phytochemical Composition of *Opuntia ficus-indica* Juices. Foods 12, 2096 (2023). https://doi.org/10.3390/foods12112096.
67. Stintzing, F.C., Schieber, A., Carle, R.: Evaluation of Colour Properties and Chemical Quality Parameters of Cactus Juices. Eur. Food Res. Technol. 216, 303–311 (2003).
68. Kuti, J.O.: Antioxidant Compounds from Four *Opuntia* Cactus Pear Fruit Varieties. Food Chem. 85(4), 527–533 (2004). https://doi.org/10.1016/S0308-8146(03)00184-5.
69. El-Gharras, H., Hasib, A., Jaouad, A., El-Bouadili, A.: Chemical and Physical Characterization of Three Cultivars of Moroccan Yellow Prickly Pears (*Opuntia ficus-indica*) at Three Stages of Maturity. CYTA J. Food. 5(2), 93–99 (2006). https://doi.org/10.1080/11358120609487677.
70. Salim, N., Chibani, A., Chadli, R., Bouznad, A.: Chemical Composition of *Opuntia ficus-indica* (L.) Fruit. Af J Biotech. 8 (2010).
71. Khatabi, O., Hanine, H., Elothmani, D., Hasib, A.: Extraction and Determination of Polyphenols and Betalain Pigments in the Moroccan Prickly Pear Fruits (*Opuntia ficus indica*). Arab. J. Chem. 9(Supplement 1), S278–S281 (2016).
72. Sanchez Lopez, J., Sánchez, F., Curt, M., Fernandez, J.: Assessment of the Bioethanol Potential of Prickly Pear (*Opuntia ficus-indica* (L.) Mill.) Biomass Obtained from Regular Crops in the Province of Almeria (SE Spain). Israel J. Plant Sci. 60, 301–318 (2013).
73. Rizwan, S., Cappai, M., Pia, G., Pilia, L.: An Updated Review: *Opuntia ficus indica* (OFI) Chemistry and Its Diverse Applications. Appl. Sci. (2023). https://doi.org/10.3390/app13137724.
74. da Silva, J.B.S., Santos Júnior, E.P., Pedrosa, J.G.T., Sales, A.T., Sampaio, E.V.S.B., Menezes, R.S.C., Dutra, E.D., Rotella Junior, P.: Energetic and Economic Analysis of Spineless Cactus Biomass Production in the Brazilian Semi-arid Region. Energies. (2022). https://doi.org/10.3390/en15145270.
75. da Silva, J.B., Santos Júnior, E.P., Siqueira e Silva, S.M., Maciel, V.G., Sales, A.T., Sampaio, E.V.S.B., Rotela Junior, P., Coelho Junior, L.M., Dubeux Jr, J.C.B., Menezes, R.S.C., Dutra, E.D.: Economic and Energetic Analysis of Cactus Pear Biomass Production Systems with Increasing Levels of Technological Intensity. Ind. Crops Prod. 208, 117883 (2024). https://doi.org/10.1016/j.indcrop.2023.117883.

76. Tellez, A., Texco, A., Cadena, A.: Optimization of the Acid Hydrolysis of Cladodes of *Opuntia ficus-indica* by Response Surface Methodology. Revista Mexicana de Ingeniería Química 17, 1095–1104 (2018). https://doi.org/10.24275/uam/izt/dcbi/revmexingquim/2018v17n3/Texco.
77. Hong, Q.: Advanced Bioethanol Production from Source-Separated Bio-waste in Pilot Scale. Sustainability. (2022). https://doi.org/10.3390/su141912127.
78. Imamoglu, E., Sukan, F.V.: Scale-up and Kinetic Modeling for Bioethanol Production. Bioresour. Technol. 144, 311–320 (2013). https://doi.org/10.1016/j.biortech.2013.06.118.
79. Statista: https://www.statista.com/markets/410/topic/443/petroleum-refinery/#overview (Accessed April 28, 2024)
80. Elsagan, Z.A., Ali, R.M., El-Naggar, M.A., El-Ashtoukhy, E.-S.Z., AbdElhafez, S.E.: New Perspectives for Maximizing Sustainable Bioethanol Production from Corn Stover. Renewable Energy. 209, 608–618 (2023). https://doi.org/10.1016/j.renene.2023.03.138.
81. Tsafara, P., Passadis, K., Christianides, D., Chatziangelakis, E., Bousoulas, I., Malamis, D., Mai, S., Barampouti, E.M., Moustakas, K.: Advanced Bioethanol Production from Source-Separated Bio-waste in Pilot Scale. Sustainability. 14(19), 12127 (2022). https://doi.org/10.3390/su141912127
82. Chatzimaliakas, F., Christianides, D., Malamis, D., Mai, S., Barampouti, E.M.: Piloting Bioethanol Production from Source-Separated Food Waste Boosts Technology Readiness. Sustainability. 15(23), 16349 (2023). https://doi.org/10.3390/su152316349.
83. Cortázar, V., Nobel, P.: Biomass and Fruit Production for the Prickly Pear Cactus, *Opuntia ficus-indica*. J. Am. Soc. Hort. Sci. 117, 558 (1992). https://doi.org/10.21273/JASHS.117.4.558.
84. Miguel, G., Fernández-Olmedilla, D., Sánchez-Godoy, F.: Drying of Prickly Pear (*Opuntia ficus-indica* (L.) Miller) and Its Potential as a Solid Biofuel. Agronomy (2022). https://doi.org/10.3390/agronomy12092231.
85. Junior, M., Neto, M., Medeiros, J., Sá, F., Lima, Y., Lenos, M., Queiroz, J., Batista, R.: Growth and Biomass Production of Prickly Pear in the Second Cycle Irrigated with Treated Domestic Sewage. Bioscience J. 36 (2019). https://doi.org/10.14393/BJ-v36n1a2020-42175.
86. Ghizlane, F., Mabrouki, J., Ghrissi, F., Azrour, M.: Proposal for a high-resolution particulate matter (PM10 and PM2. 5) capture system, comparable with hybrid system-based internet of things: case of quarries in the western Rif, Morocco. Pollution. 8(1), 169–180 (2022).
87. Dubeux, J.C.B., Ferreira dos Santos, M.V., de Andrade Lira, M., Cordeiro dos Santos, D., Farias, I., Lima, L.E., Ferreira, R.L.C.: Productivity of *Opuntia ficus-indica* (L.) Miller Under Different N and P Fertilization and Plant Population in North-East Brazil. J. Arid Environ. 67(3), 357–372 (2006). https://doi.org/10.1016/j.jaridenv.2006.02.015.
88. Jorge, A., Costa, A., Oliveira, M.: Adapting to Climate Change with *Opuntia*. Plants. 12, 2907 (2023). https://doi.org/10.3390/plants12162907.
89. Honorato-Salazar, J.A., Aburto, J., Amezcua-Allieri, M.A.: *Agave* and *Opuntia* Species as Sustainable Feedstocks for Bioenergy and Byproducts. Sustainability. 13(21), 12263 (2021). https://doi.org/10.3390/su132112263.

44 Transmissivity of the Glazing Surface of a Solar Still Single Slope Based on the Metrological Parameters of Agadir, Morocco

K. Zabour, M. Feddaoui, H. Meftah, and M. Hissouf

44.1 INTRODUCTION

Seas and oceans make up about 97% of the planet, while ice caps account for the remaining 1% and are believed to contain drinkable water. These days, the lack of a source of clean drinking water is a big concern, particularly for developing nations where industrial expansion and population growth are contributing factors. The global map shows that the world's water resources are dispersed unevenly. Furthermore, poverty and desertification in many nations would be significant factors, particularly in the upcoming decades. Then, reaching drinking water sources is getting more and more important.

The aim of innovative scientific desalination system research is to develop a mechanism with significant energy efficiency. Desalination techniques employing renewable energies, specifically, solar energy, are an efficient means of providing drinking water to suit human requirements. In arid to semi-arid areas, the application of solar desalination techniques could significantly improve the availability of drinking water while also having a positive economic and environmental impact.

Numerous studies in the subject have looked into the factors affecting the productivity of solar distillers, both theoretically and experimentally.

Rahbar and Esfahani [1] found that sun radiation and ambient temperature directly impact the efficiency of solar distillers. Aburideh et al. [2] conducted an experiment to investigate the impact of climatic conditions on a double-slope solar distiller in Bou Ismail-Algeria. The results show that the rise in production is significantly affected by the quantity of solar radiation received. El-Sebaii [3] examined the influence of wind speed on the performance of various water masses in basins and determined that performance enhances as wind speed increases. Nevertheless, research has demonstrated that when the wind speed escalated from 1 to 9 m/s, there was an associated decrease in the overall system efficiency by 13%.

Several researchers have studied the effect of changing ambient temperatures on the production of solar distillers. Al-Hinai et al. [4] pointed out that an increase in ambient air temperature of 10°C improves output by 8.2%.

Several experimental and numerical studies on different aspects of solar distiller design have been carried out. Rajaseenivasan and Murugavel [5] conducted both theoretical and experimental research on a double-slope distiller, as well as single-basin and double-basin distillers. The supplementary basin, sometimes known as the double basin, was incorporated into the design to enhance the efficiency of the single basin solar distiller. It was discovered that the inclusion of this addition resulted in a production gain of about 85% when compared to a single-basin counterpart operating under similar conditions. Naim and Abd El Kawi [6] succeeded in proving that the use of charcoal (grains) improves production by almost 15% compared to the wick distiller. Secondly, Nafey et al. [7] observed that utilizing black gravel

 DOI: 10.1201/9781003647263-44

as a storage medium assures rapid absorption and dissipation of solar energy, compared to black rubber. In addition, they determined that incorporating a layer of black gravel measuring 20–30 mm in thickness would result in a 19% increase in production when using 20 liters of salt water and a glass covering angle of 15 degrees. Nijmeh et al. [8] predicted that the single-slope distiller provides a 26% improvement with the use of potassium dichromate (combined with water to form an absorbent material).

The depth of water in the solar distillation tank is one of the parameters that strongly influences performance. Phadatare and Verma [9], Tiwari and Tiwari [10], and Tripathi and Tiwari [11] attempted to study the effects of water depth on heat and mass transfer coefficients for passive and active modes. The results showed a positive decrease in production when the water depth in the basin increased.

The efficiency of a solar distiller is highly dependent on the angle of inclination of the glass. This angle depends on the altitude and rotation of the lid, as well as on the latitude. It is expected that the covering, which has an inclination equal to the latitude angle, will receive normal solar radiation throughout the year [12]. Singh and Tiwari [13] conducted a computational study for latitudes (13–28°N), including the impact of solar radiation, wind speed, water depths, and cover tilt angle on production. They determined that a optimal angle for tilt the glass to achieve the highest annual production should be placed similar to the latitude of the location. Kumar et al. [14] conducted a study in India (latitude 28.36°N) and determined that a glass tilt angle of 15° resulted in the highest performance, based on their numerical analysis.

The glass surface is one of the most important elements of a solar distiller. It is designed to admit the maximum possible radiation and minimize heat loss.

Transparent glass is the most frequently used type of glass covering. The performance of the glazed surface depends on its transmission coefficient. The present work is a theoretical study applied to solar still in order to investigate the importance of glazing material, and to study the effect of glass transmissivity on solar still performance.

44.2 DESCRIPTION OF THE SYSTEM

Figure 44.1 shows a schematic diagram of a simple solar distiller. It consists mainly of a basin-type distiller. A waterproof box is the main component, with a transparent glass cover. The tank is typically constructed from metal and coated in black paint to maximize solar radiation absorption.

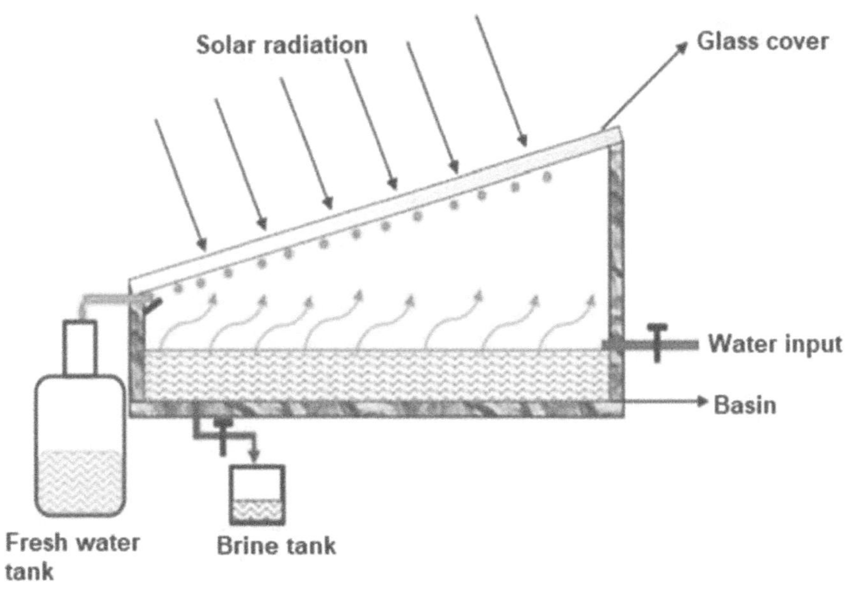

FIGURE 44.1 Schema of the solar still single slope.

Placing a layer of insulation under the absorber reduces heat loss to the bottom. The distiller is encased by a transparent cover that allows solar radiation to penetrate the absorber. The glass is inclined at a given angle to the horizon to facilitate the flow of condensate droplets. At the bottom of the glass, a plastic gutter collects the condensate and sends it to the collection bottle.

44.3 MATHEMATICAL MODEL

Modeling the distiller involves subdividing it into a set of control volumes, each representing a physical part of the system. Modeling requires an advanced understanding of the physical phenomena taking place in the system. This enables a judicious selection of the empirical correlations involved in estimating the heat and mass transfer coefficients.

The energy balance equations describing the various components of the solar distiller are obtained from the first principle of thermodynamics:

$$\rho V C p \frac{dT}{dt} = \Sigma Q_{input} - \Sigma Q_{output} \tag{44.1}$$

By analogy between thermal and electrical quantities, we can apply Ohm's and Kirchoff's laws. The energy balance reflecting the heat exchange between the three nodes of the distiller (glass, water, and absorber) can be expressed by the following analog electrical circuit (Figure 44.2): To simplify the calculations and reduce the number of variables, a set of simplifying assumptions has been taken into consideration:

- The regime is transitional;
- Condensation occurs only on the glass (no condensation on the side walls);
- The side walls are assumed to be adiabatic;

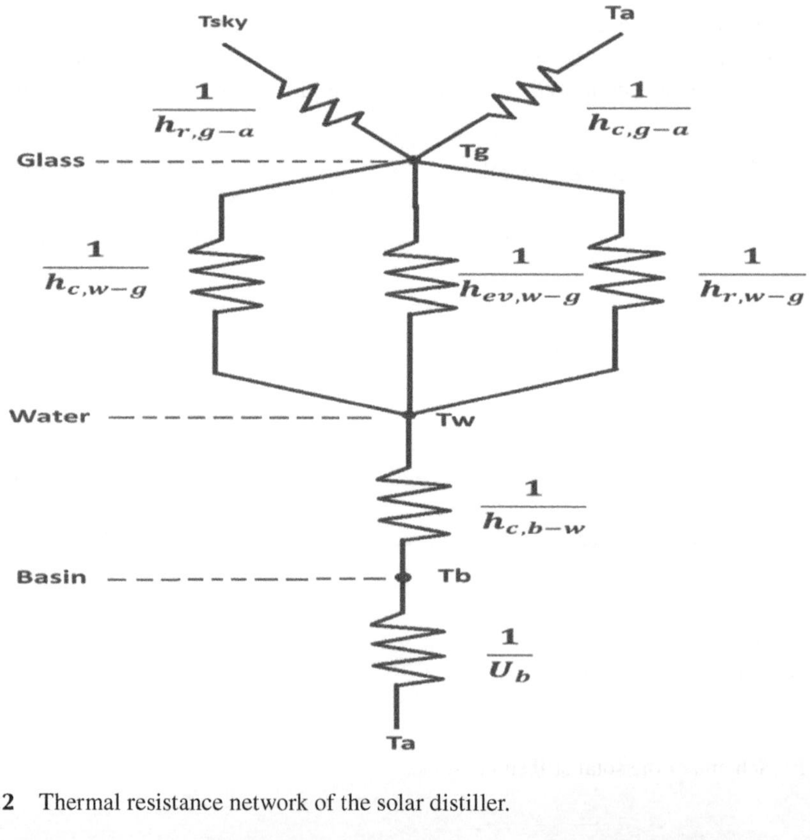

FIGURE 44.2 Thermal resistance network of the solar distiller.

- The temperature of each component is uniform;
- Condensation of water on the roof is homogeneous and continuous.

The heat balance equation for the single slope distiller is expressed for each component of the solar distiller, as shown below:

Energy balance for the glass cover [15]:

$$\frac{M_g cp_g dT_g}{dt} = Q_{c,w-g} + Q_{r,w-g} + Q_{ev,w-g} - Q_{c,g-a} - Q_{r,g\text{-}a} + P_g \tag{44.2}$$

Energy balance for the saline water [15]:

$$\frac{M_w cp_w dT_w}{dt} = Q_{c,b-w} - Q_{c,w-g} - Q_{r,w-g} - Q_{ev,w-g} - Q_{fw} + P_w \tag{44.3}$$

Energy balance for basin plate [15]:

$$\frac{M_b cp_b dT_b}{dt} = -Q_{loss} - Q_{c,b-w} + P_b \tag{44.4}$$

The hourly yield is given by the following equation [16]:

$$m_{ew} = \frac{h_{ev,w-g}(T_w - T_g) \times 3600}{h_{fg}} \tag{44.5}$$

where $Q_{ev,w-g}$ is the coefficient of evaporative heat transfer between water and glass and h_{fg} is the water latent heat.

The convective heat transfer between water and glass is given by:

$$Q_{c,w-g} = h_{c,w-g} A_w \left(T_w - T_g\right) \tag{44.6}$$

where the convective heat transfer coefficient between water and glass, $h_{c,w-g}$, is given by: [17]

$$h_{c,w-g} = 0.884 \times \left[T_w - T_g + \frac{\left(Pv_w - Pv_g\right) \times \left(T_w + 273.15\right)}{268900 - Pv_w}\right]^{\frac{1}{3}} \tag{44.7}$$

where Pv_g and Pv_w is the partial pressure of the glass and water, respectively.

The coefficient of radiation heat transfer from the basin to the glass cover is estimated [18]:

$$\mathrm{h}_{r,w-g} = \varepsilon_{eff} \times \sigma \times \frac{(T_w + 273.15)^4 - \left(T_g + 273.15\right)^4}{(T_w - T_g)} \tag{44.8}$$

where ε_{eff} is the water-glass effective emissivity and σ is the Stefan–Boltzmann constant.

The evaporative heat transfer between water and glass is given by:

$$Q_{ev,w-g} = h_{ev,w-g} A_w \left(T_w - T_g\right) \tag{44.9}$$

where the coefficient of evaporative heat transfer coefficient, $h_{ev,w-g}$, is calculated by [17]:

$$h_{ev,w-g} = 0.016237 \times h_{c,w-g} \times \frac{\left(Pv_w - Pv_g\right)}{T_w - T_g} \tag{44.10}$$

The convective heat transfer between the basin and water:

$$Q_{c,b-w} = h_{c,b-w} A_b \left(T_b - T_w\right) \tag{44.11}$$

where $h_{c,b-w}$ is the convective heat transfer coefficient, which is defined by the following equation [19]:

$$h_{c,b-w} = 0.54 \times \left(\frac{K_w}{D}\right) \times Ra^{0.25} \tag{44.12}$$

where K_w is the water thermal conductivity and D is the characteristic length.

The radiative heat transfer between glass and sky is given by:

$$Q_{r,g-a} = h_{r,g-a} A_g \left(T_g - T_a\right) \tag{44.13}$$

where the coefficient of radiative heat transfer between glass and sky, $h_{r,g-a}$, is given by [18]:

$$\mathrm{h}_{r,g-a} = \varepsilon_v \times \sigma \times \frac{(T_g + 273.15)^4 - (T_s + 273.15)^4}{(T_g - T_a)} \tag{44.14}$$

The convective heat transfer between the glass cover and the ambient can be characterized by:

$$Q_{c,g-a} = h_{c,g-a} A_g \left(T_g - T_a\right) \tag{44.15}$$

where $h_{c,g-a}$ is the coefficient of the convection heat transfer, dependent on the wind speed, as determined by the following equations:

$$h_{c,g-a} = \begin{cases} 5,7 + 3,8 \times V & \text{if } \mathrm{V} \le 5(\mathrm{m/s}) \\ 6.15 \times V^{0.8} & \text{if } \mathrm{V} > 5(\mathrm{m/s}) \end{cases} \tag{44.16}$$

The heat absorbed by the feed water is determined by [20]:

$$Q_{fw} = m_e Cp_w \left(T_w - T_a\right) \tag{44.17}$$

The heat losses by convection are given by [15]:

$$Q_{loss} = U_b A_b \left(T_b - T_a\right) \tag{44.18}$$

where $U_b = \frac{Ki}{Li}$ and Li and Ki are the thickness and thermal conductivity of the insulation, respectively.

44.4 COMPUTER PROCESSING

The numerical calculation code is developed using the MATLAB programming language. The transfer equation was solved using an iterative Runge-Kutta of order 4 method because the heat and mass transfer coefficients are a function of the temperature of the different components of the solar still, which are unknowns in the problem.

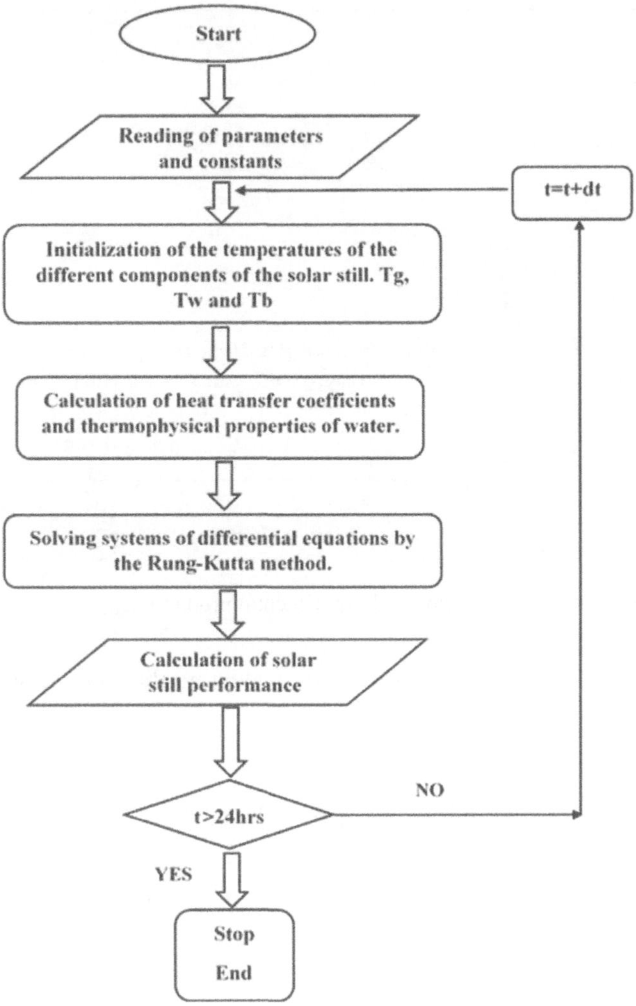

FIGURE 44.3 Flow chart for calculation.

Computations commenced at 5:00 a.m., assuming that the initial temperatures of different parts of the still were equivalent to the ambient temperature. By utilizing established initial values for the different temperatures, we computed distinct internal and exterior heat transfer coefficients. The program begins by calculating the basin, water, and glass cover temperature at a time interval of 1 second, using the provided heat transfer coefficients and meteorological parameters. The process was repeated for an additional time interval, Δt, and so forth. Figure 44.3 displays the simulation steps.

44.5 MODEL VALIDATION

The model is validated by comparing theoretical results obtained in the present work with the experimental data collected from Youssef's paper [21]. Figures 44.4 and 44.5 show theoretical and experimental comparisons of water temperature and hourly variation of freshwater productivity for a conventional solar still.

We carried out the validation using the same parameters and similar conditions as reported by the authors in the experimental study. Therefore, several factors, such as simplifying assumptions and uncertainties in the experimental data, can explain the difference between the two curves.

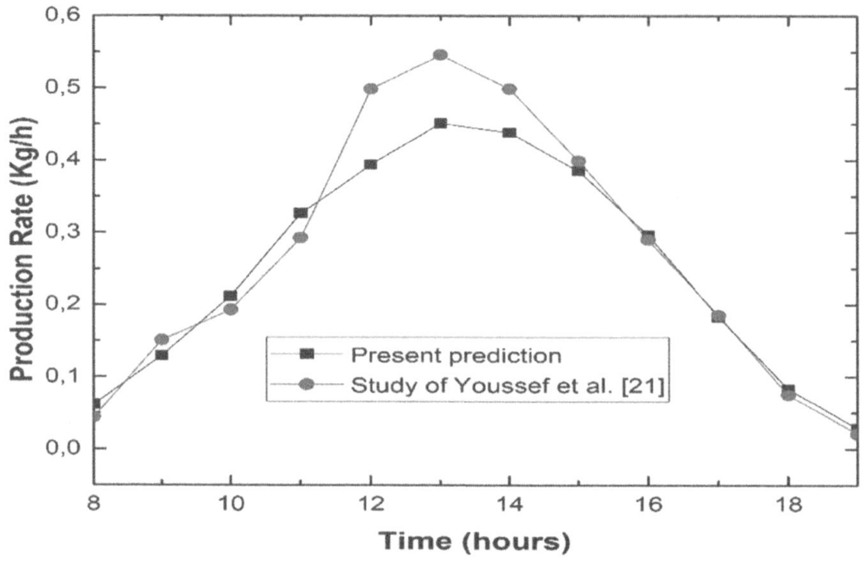

FIGURE 44.4 Comparisons of simulation and experimental productivity.

The relations shown below have been used for evaluating the root mean square of percentage deviation:

$$RMSE = \left[\sum_{i=1}^{N} (X_{fi} - X_{oi})^2 / N \right]^{1/2} \tag{44.19}$$

This comparison shows a good approach between calculated and experimental values. Indeed, RMSE does not exceed 10%.

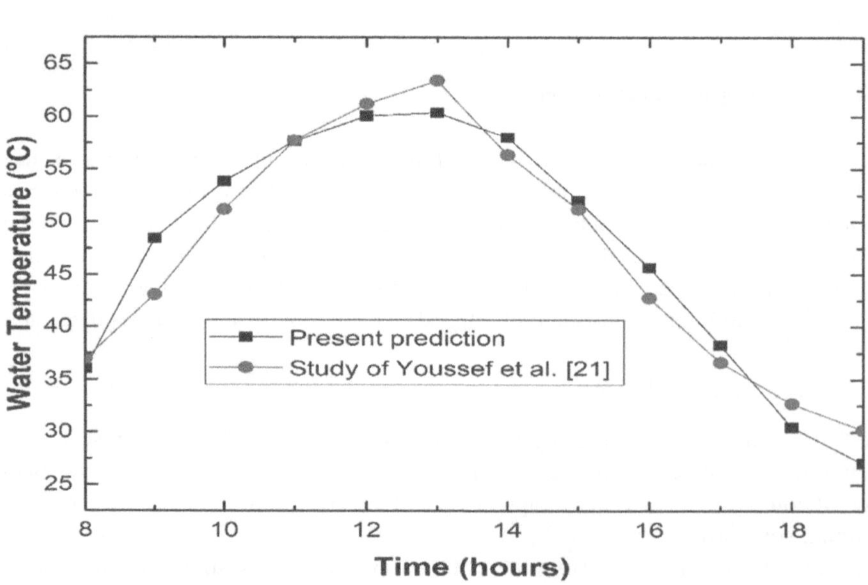

FIGURE 44.5 Comparisons of simulation and experimental water temperature.

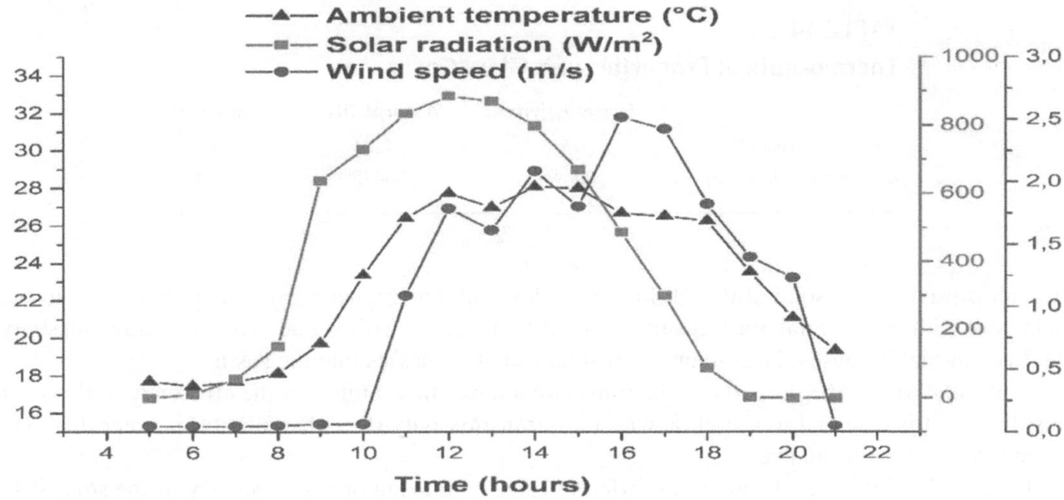

FIGURE 44.6 Weather conditions in Agadir for a typical day (21/08/2019).

44.6 RESULTS AND DISCUSSION

The simulation of the operation of the solar distiller is made using the weather data of the Agadir regions (Morocco) illustrated in Figure 44.6 and also using the operating parameters presented in Table 44.1.

The glass cover in a solar still must meet several specifications, including absorbing a minimum quantity of heat, reflecting a minimum quantity of energy from solar radiation, transmitting maximum energy, and having high thermal resistance to heat loss from the basin to the environment. Two types of glazing were chosen to study the effect of the optical properties of the glass on the performance of the solar distiller. The properties of these glazing types are illustrated in Table 44.2.

Figure 44.7 shows the hourly productivity of the solar still using two types of glazing. The performance of the glazed surface depends on its optical properties, such as transmissivity, reflexivity,

TABLE 44.1
Parameters of a Solar Still

Parameter	Value	Parameter	Value
$\rho_g\left(kg.m^{-3}\right)$	2,500	$Cp_g\left(J.kg^{-1}.K^{-1}\right)$	840
$\rho_b\left(kg.m^{-3}\right)$	7,800	$Cp_b\left(J.kg^{-1}.K^{-1}\right)$	460
Glass thickness (mm)	3	Basin thickness (mm)	2
Water thickness (cm)	1	Insolation thickness (cm)	5
$A_b\left(m^2\right)$	1	Θ (degree)	30
$K_b\left(W.m^{-1}.K^{-1}\right)$	73	$K_i\left(W.m^{-1}.K^{-1}\right)$	0.059
ε_g	0.88	ε_w	0.96
α_w	0.05	α_b	0.9
$\sigma\left(W.m^{-2}.K^{-4}\right)$	5,669.10-8	τ_w	0.95

TABLE 44.2
Thermo-optical Properties of a Glass Cover

	Transmissivity	Absorptivity	Emissivity
Low iron glass [22]	0.94	0.04	0.85
Commercial glass [23]	0.879	0.0475	0.88

and absorptivity. The solar still with low iron glass exhibits higher productivity compared to the solar still with commercial glass, resulting in a 9.3% increase. This is due to the high transmissivity of this material because a large quantity of solar radiation passes into the basin.

Transmissivity is the property of the transparent cover that influences the efficiency of the solar distiller, so it is essential to use glass with high transmissivity (low absorptivity) in solar distillers to achieve high performance.

Figure 44.8 illustrates the effect of different types of glazing on the efficiency of the solar still. The results obtained show that the highest efficiency is 59.64% for the solar still with the low iron glass cover, while the efficiency of the solar still with commercial glass is 54.69%. The glass transmits more than 90% of the solar radiation into the interior of the solar still. This results in an increase in the temperature of the system (Tg, Tw, and Tb). Table 44.3 displays the temperatures of different components of the system for two types of glazing.

Figure 44.9 demonstrates how increasing the transmissivity of glass affects the performance of solar stills. By increasing the transmissivity, daily production increases as more solar radiation passes through the basin and raises the temperature of the system. Cumulative production for transmissivities 0.75, 0.80, 0.85, 0.90, and 0.95 is 4.312 kg/h.m^2, 4.649 kg/h.m^2, 4.986 kg/h.m^2, 5.322 kg/h.m^2, and 5.672 kg/h.m^2, respectively. The increase in cumulative production for transmissivities of 0.8, 0.85, 0.9, and 0.95 is, respectively, 7.81%, 15.63%, 23.43%, and 31.53% compared to the 0.75 transmissivity. The rate of increase in average daily productivity is 7.11%.

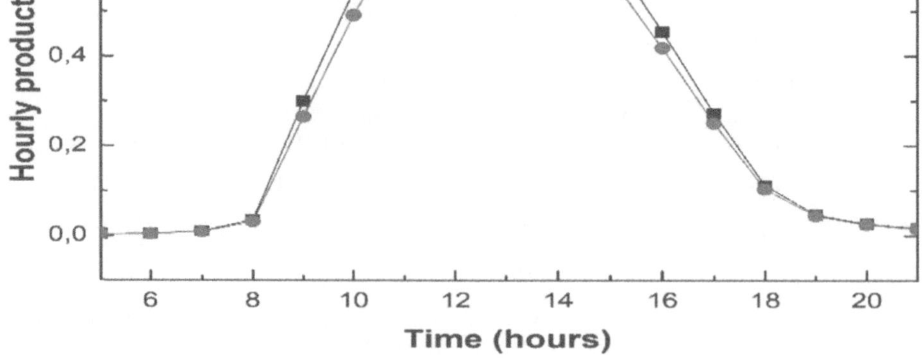

FIGURE 44.7 Hourly productivity of solar for two different glass covers.

TABLE 44.3
Maximum Temperature of Different Components of a Solar Still

Transmissivity	Tg_{max}	Tw_{max}	Tb_{max}
0.75	58.065	67.795	69.046
0.80	57.395	67.778	69.142
0.85	56.725	67.764	69.194
0.90	56.049	67.748	69.312
0.95	56.238	68.153	69.751

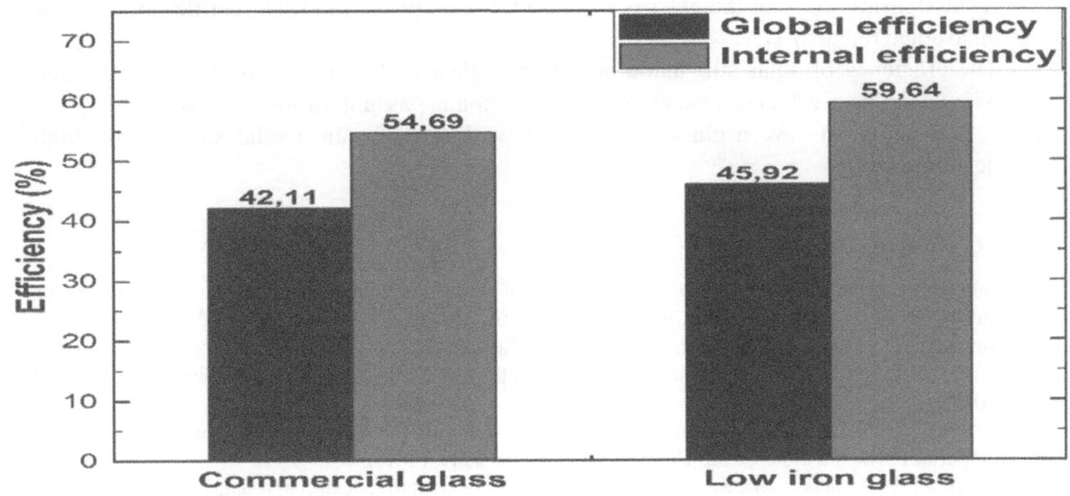

FIGURE 44.8 Efficiency of solar still for two different glass covers.

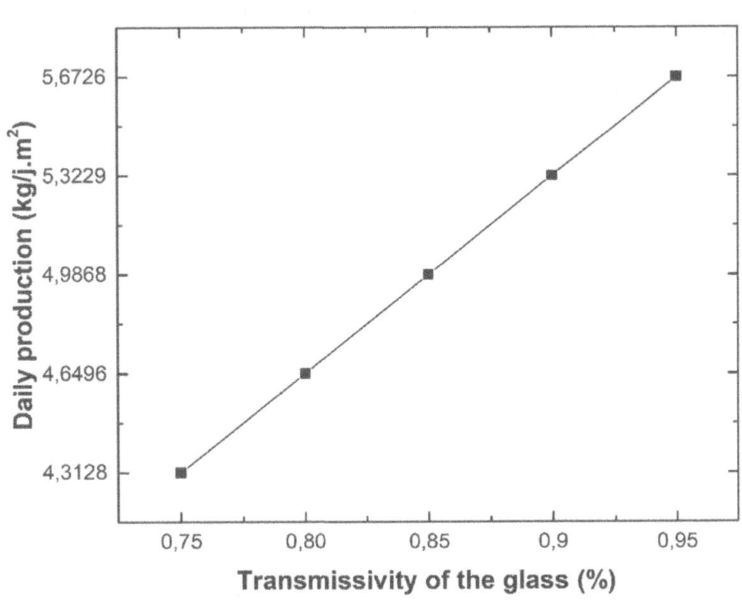

FIGURE 44.9 Daily productivity based on glass transmission.

It can be concluded that the cover acts as a radiation transmitter and a condenser. All the radiation used inside the solar distiller unit is transmitted through the cover, so its optical and thermal properties strongly affect the performance of the solar distiller.

44.7 CONCLUSION

The performance of the glazing surface depends on the magnitude of its transmissivity. For a given material, this optical property is a function of solar geometry that varies with geographic location. From the results obtained, the following conclusions were observed:

- The use of low iron glass improves the production of a solar still by 9.3% compared to commercial glass because of its highest transmission coefficient.
- If the transmissivity of the glass is increased, the output of the solar distiller increases at an average rate of 7.11%.
- The efficiency of solar still made of low iron glass and commercial glass is 59.64 and 54.69%, respectively, because more solar radiation is passing through the basin.
- It is essential to have a glass cover with low absorptivity in a solar still to have high performance.

REFERENCES

1. Rahbar, N., & Esfahani, J. A. Experimental study of a novel portable solar still by utilizing the heat pipe and thermoelectric module. Desalination, 284, 55–61. (2012).
2. Aburideh, H., Deliou, A., Abbad, B., Alaoui, F., Tassalit, D., & Tigrine, Z. An experimental study of a solar still: Application on the sea water desalination of Fouka. Procedia Engineering, 33, 475–484. (2012).
3. El-Sebaii, A. A. On effect of wind speed on passive solar still performance based on inner/outer surface temperatures of the glass cover. Energy, 36(8), 4943–4949. (2011).
4. Al-Hinai, H., Al-Nassri, M. S., & Jubran, B. A. Effect of climatic, design and operational parameters on the yield of a simple solar still. Energy Conversion and Management, 43(13), 1639–1650. (2002).
5. Rajaseenivasan, T., & Murugavel, K. K. Theoretical and experimental investigation on double basin double slope solar still. Desalination, 319, 25–32. (2013).
6. Naim, M. M., & Abd El Kawi, M. A. Non-conventional solar stills part 1. Non-conventional solar stills with charcoal particles as absorber medium. Desalination, 153(1–3), 55–64. (2003).
7. Nafey, A. S., Abdelkader, M., Abdelmotalip, A., & Mabrouk, A. A. Solar still productivity enhancement. Energy Conversion and Management, 42(11), 1401–1408. (2001).
8. Nijmeh, S., Odeh, S., & Akash, B. Experimental and theoretical study of a single-basin solar sill in Jordan. International Communications in Heat and Mass Transfer, 32(3–4), 565–572. (2005).
9. Phadatare, M. K., & Verma, S. K. Influence of water depth on internal heat and mass transfer in a plastic solar still. Desalination, 217(1–3), 267–275. (2007).
10. Tiwari, A. K., & Tiwari, G. N. Thermal modeling based on solar fraction and experimental study of the annual and seasonal performance of a single slope passive solar still: The effect of water depths. Desalination, 207(1–3), 184–204. (2007).
11. Tripathi, R., & Tiwari, G. N. Effect of water depth on internal heat and mass transfer for active solar distillation. Desalination, 173(2), 187–200. (2005).
12. Kabeel, A. E., & El-Agouz, S. A. Review of researches and developments on solar stills. Desalination, 276(1–3), 1–12. (2011).
13. Singh, H. N., & Tiwari, G. N. Monthly performance of passive and active solar stills for different Indian climatic conditions. Desalination, 168, 145–150. (2004).
14. Kumar, S., Tiwari, G. N., & Singh, H. N. Annual performance of an active solar distillation system. Desalination, 127(1), 79–88. (2000).
15. Kabeel, A. E., Omara, Z. M., & Essa, F. A. Numerical investigation of modified solar still using nanofluids and external condenser. Journal of the Taiwan Institute of Chemical Engineers, 75, 77–86. (2017).
16. Zarzoum, K., Zhani, K., & Bacha, H. B. Experimental validation of optimized solar still using solar energy. Fundamentals of Renewable Energy and Applications, 7, 1–4. (2017).

17. Altarawneh, I., Rawadieh, S., Batiha, M., Al-Makhadmeh, L., Alrowwad, S., & Tarawneh, M. Experimental and numerical performance analysis and optimization of single slope, double slope and pyramidal shaped solar stills. Desalination, 423, 124–134. (2017).
18. Bataineh, K. M., & Abbas, M. A. Performance analysis of solar still integrated with internal reflectors and fins. Solar Energy, 205, 22–36. (2020).
19. Sharshir, S. W., Peng, G., Elsheikh, A. H., Edreis, E. M., Eltawil, M. A., Abdelhamid, T., & Yang, N. Energy and exergy analysis of solar stills with micro/nano particles: A comparative study. Energy Conversion and Management, 177, 363–375. (2018).
20. Abdullah, A. S., Younes, M. M., Omara, Z. M., & Essa, F. A. New design of trays solar still with enhanced evaporation methods–comprehensive study. Solar Energy, 203, 164–174. (2020).
21. Yousef, M. S., & Hassan, H. An experimental work on the performance of single slope solar still incorporated with latent heat storage system in hot climate conditions. Journal of Cleaner Production, 209, 1396–1410. (2019).
22. Madhukeshwara, N., & Prakash, E. S. An investigation of heat transfer augmentation and friction characteristics in solar air heater duct with V-shaped wire as artificial roughness on absorber plate. International Journal of Energy and Environment (Print), 4, 519–538. (2013).
23. Mousa, H., & Gujarathi, A. M. Modeling and analysis the productivity of solar desalination units with phase change materials. Renewable Energy, 95, 225–232. (2016).

45 Critical Evaluation and Thermodynamic CALPHAD Reassessment of the Germanium-Antimony Binary System

M. Idbenali and A. Nait Alla

45.1 INTRODUCTION

Diamond (Ge)-rhombohedral (Sb)-based systems have garnered significant attention in recent decades due to their potential application as a novel type of memory reliant on crystalline-to-amorphous phase transitions, known as phase-change materials (PCMs). For an alloy to be considered a potential PCM, it must exhibit good thermal stability and rapid crystallization ability while avoiding phase segregation after the record/erase cycle [1]. Whatever the targeted technological application, basic knowledge such as the thermodynamic properties of the germanium-antimony is required for process development and optimization. The objectives of this study are twofold: first to assess an experimental phase diagram along with its corresponding thermodynamic data and second to compare the thermodynamic parameters for the solution phases. This comparison involves utilizing the excess energy with linear temperature-dependent interaction parameters and employing the CALPHAD (calculation of phase diagrams) method [2]. The calculations are conducted based on Gibbs free energy minimization using the thermo-calc software [3].

45.2 LITERATURE REVIEW

The phase equilibria in the Ge-Sb system have been experimentally explored by various research groups using thermal analysis methods [4–8]. All studies confirm that a simple eutectic reaction occurs in this system. The liquidus measured by Stohr and Klemm [5] is higher than that of Ruttewit and Masing [4] by up to 25°C. Later on, Zhurkin et al. [9] were determined liquidus using a crystal pulling. The first evaluation of the data available in the literature for this system (such as the phase diagram and thermodynamics) has been reported by Olesinski and Abbaschian [10], According to these authors, the solubility of Ge in Sb is 3.5 mol% and the solubility of Sb in Ge is negligible. They found the eutectic reaction to occur at 765 K and 85 mol% Sb. Optimization of thermodynamic parameters of the binary Ge-Sb system was carried out by [8, 11–13], but without considering the variation of enthalpy with temperature. The enthalpies of mixing of liquid at 1250 K and 1273 K were determined by Predel and Stein [14] and Alfer et al. [7], respectively, from direct calorimetry. In 1999, Kostov et al. [15] determined the activities of Ge and Sb in the liquid phase, at different temperatures (1073,1173, 1273 K), using Oelsen calorimetry activities and partial molar quantities for both components. The authors determined other thermodynamic

DOI: 10.1201/9781003647263-45

properties (partial enthalpies and chemical potential for both components). Their values differ very slightly from the results in [7, 14].

45.3 THERMODYNAMIC MODELS

45.3.1 Pure Elements

The Gibbs energy function element i (i=Ge, Sb) is taken from [16].

45.3.2 Solution Phases

The Gibbs energy of 1 mole of formula unit is expressed as the sum of three terms:

$$G^{\text{liq}} - H^{\text{SER}} = {}^{\text{ref}}G^{\text{liq}} + {}^{\text{id}}G^{\text{liq}} + {}^{\text{ex}}G^{\text{liq}} \tag{45.1}$$

$${}^{\text{ref}}G^{\Phi} = x_{Ge}\left[{}^{0}G^{\Phi}_{\text{Ge}} - H^{\text{SER}}_{Ge}\right] + x_{\text{Sb}}\left[{}^{0}G^{\Phi}_{Sb} - H^{\text{SER}}_{Sb}\right] \tag{45.2}$$

where $H_{\text{i}}^{\text{SER}}$ (298.15 K) is the molar enthalpy of the i element at 298.15 K in its standard element reference (SER) state.

$${}^{id}G^{\Phi} = RT\left[x_{Ge}Ln(x_{Ge}) + x_{Sb}Ln(x_{Sb})\right] \tag{45.3}$$

where R is the ideal gas constant; T is the temperature, in Kelvin; xGe and xSb are the mole fraction of Ge and Sb, respectively.

The ${}^{ex}G^{\Phi}$ energy part in Eq. (45.4) is given by the Redlich–Kister polynomial [17]:

$${}^{\text{ex}}G^{\Phi} = x_{\text{Ge}}x_{\text{Sb}}\sum_{\lambda=0}^{\lambda}{}^{\lambda}L^{\Phi}_{i,j}\left(x_{Ge} - x_{Sb}\right)^{\lambda} \tag{45.4}$$

x_i and x_j denote the mole fractions and i and j are the indices corresponding to the two species Ge and Sb.

The binary interaction parameters of the ${}^{\lambda}L^{\text{Liq}}_{\text{i,j}}$ type, assessed in the present work, were temperature-dependent, as follows:

$${}^{\lambda}L^{\text{Liq}}_{\text{i,j}} = a_{\lambda} + b_{\lambda}T + c_{\lambda}TLn\left(T\right) \tag{45.5}$$

45.4 RESULTS AND DISCUSSIONS

Thermodynamic assessment of the Ge-Sb system was made with the CALPHAD method and the thermo-calc software package [3], based on the data available in the literature cited in Section 45.2. The thermodynamic parameters used to construct the Gibbs energy functions for all the solution phases in the system are shown in Table 45.1. The calculated Ge-Sb phase diagram is depicted in Figure 45.1 and compared with the experimental measurements. This diagram is very similar to the experimental one established by [10].

The temperature and composition for the Liq <-> (Ge) + (Sb) eutectic reaction calculated at (865.5 K; 82.69 at.% Sb) also fit well with the experiment in ref [10] (see Table 45.2).

TABLE 45.1
The Optimized Thermodynamic Parameters for Solid Solution Phases of the Ge-Sb System

Phase	Parameters units in S.I. $^{\lambda}L^{Liq}_{i,j} = a_{\lambda} + b_{\lambda}T + c_{\lambda}T\ln(T)$		
	a_{λ}	**b_{λ}**	**C_{λ}**
Liquid $\lambda = 0$	14360.73	–74.09	8.91
$\lambda = 1$	–7033.96	3.65.01	–4.49
DIAMOND (**Ge,Sb**)	$G^{DIAM} - G^{RHOM} = G^{SERSb} + 11616.51$	7.4	0
	$\lambda = 0$ 33419.69	–0.26	
RHOMBOHEDRAL_A7	$G^{RHOM} - G^{DIAM} =$		0
(Sb,Ge)	$G^{SERGe} + 5705.63$	12.06	
	$\lambda = 0$ 3337.53	–14.59	

TABLE 45.2
Invariant Reactions of the Two-Phase Diagrams Binary Germanium-Antimony System

Invariant Reaction	**T/K**	**$X_{(Sb)}$**	**Ref**
Liq ⇔ (Ge)+(Sb)		X(Liq,Sb)	
	865	0.825	[10]
	865.5	0.826	This work
		X(DIAMOND, Sb)	
	865	0.001	[10]
	865.5	0.078	This work
		X(RHOMB, Sb)	
	865	0.938	[10]
	865.5	0.955	This work

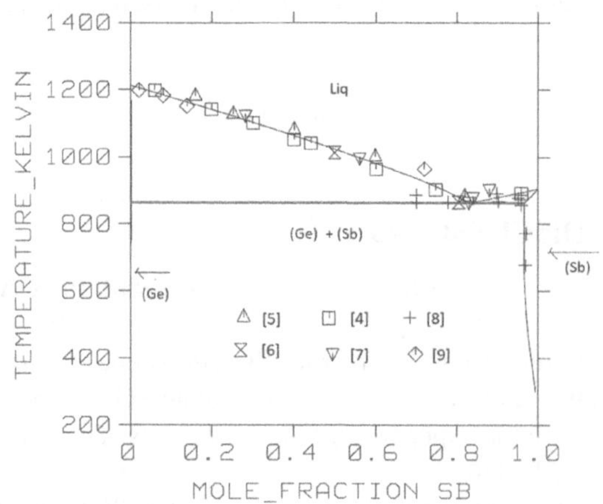

FIGURE 45.1 Calculated phase diagram of the Ge-Sb system compared with the experimental values in the literature.

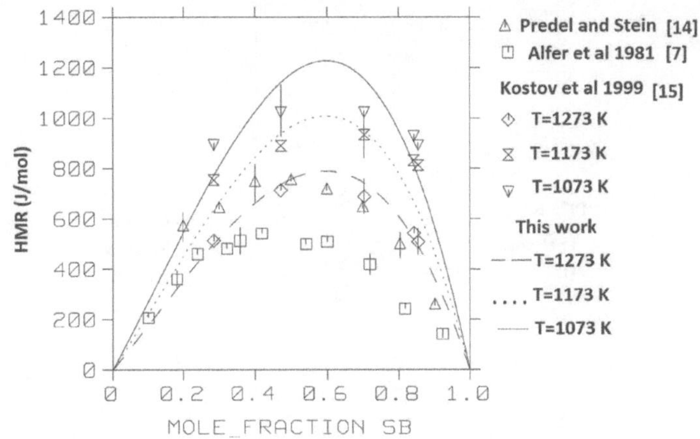

FIGURE 45.2 Calculated molar enthalpy of liquid compared with the experimental values in [7, 14, 15].

The calculated integral enthalpy of mixing of the liquid Ge-Sb alloys referring to liquid components at (T = 1073, 1173, and 1273 K) is plotted in Figure 45.2. Our values are in reasonable agreement with the experimental data in [14–15]. The measurements by [7] are less endothermic compared to those of other authors.

The Gibbs energies of mixing of the liquid phase calculated at 1073 K, 1173 K, and 1273 K were plotted with the appropriate experimental data in Figure 45.3. The calculated values are consistent with the experimental data from ref. [15].

In Figures 45.4 and 45.5(a, b, c), the calculated chemical potentials of the elements (Ge, Sb) are compared with the ones measured by Kostov et al. [15]. A satisfactory agreement is observed with the experimental data. Figures 45.6 and 45.7(a, b, c) show the activities of germanium and antimony in the liquid phase calculated at 1073, 1173, and 1273 K; the reference states were Sb (Liq) and Ge (Liq). The calculated values agree well with the experimental data in [15].

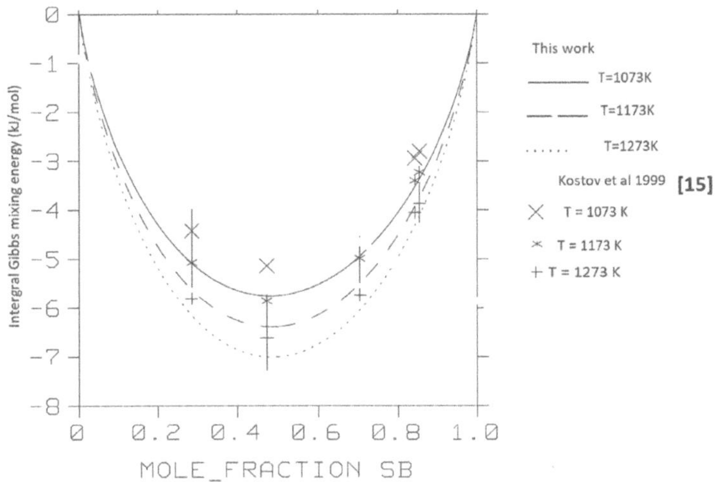

FIGURE 45.3 Calculated molar Gibbs energy of liquid compared with the experimental values in [15].

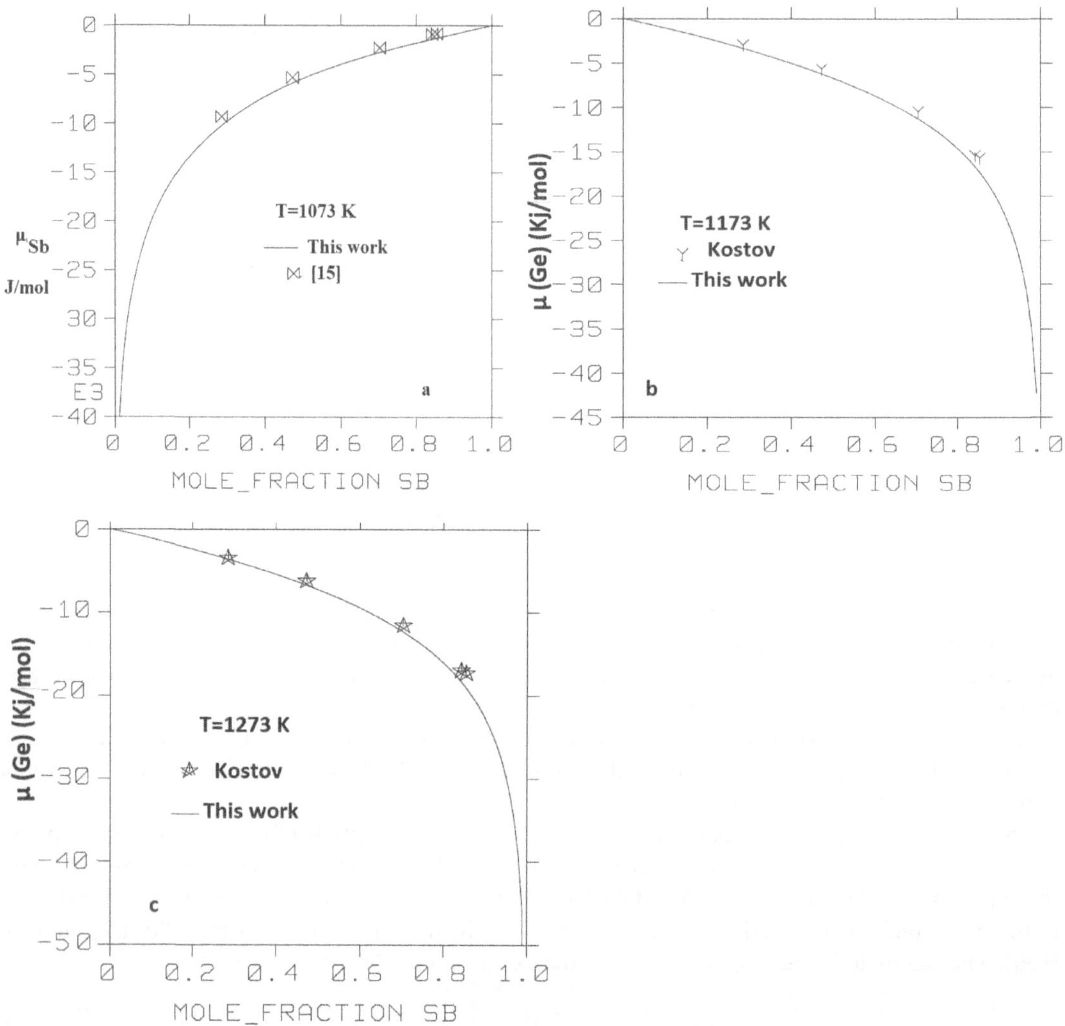

FIGURE 45.4 (a, b, c) The calculated chemical potential of germanium compared with the experimental values in [15].

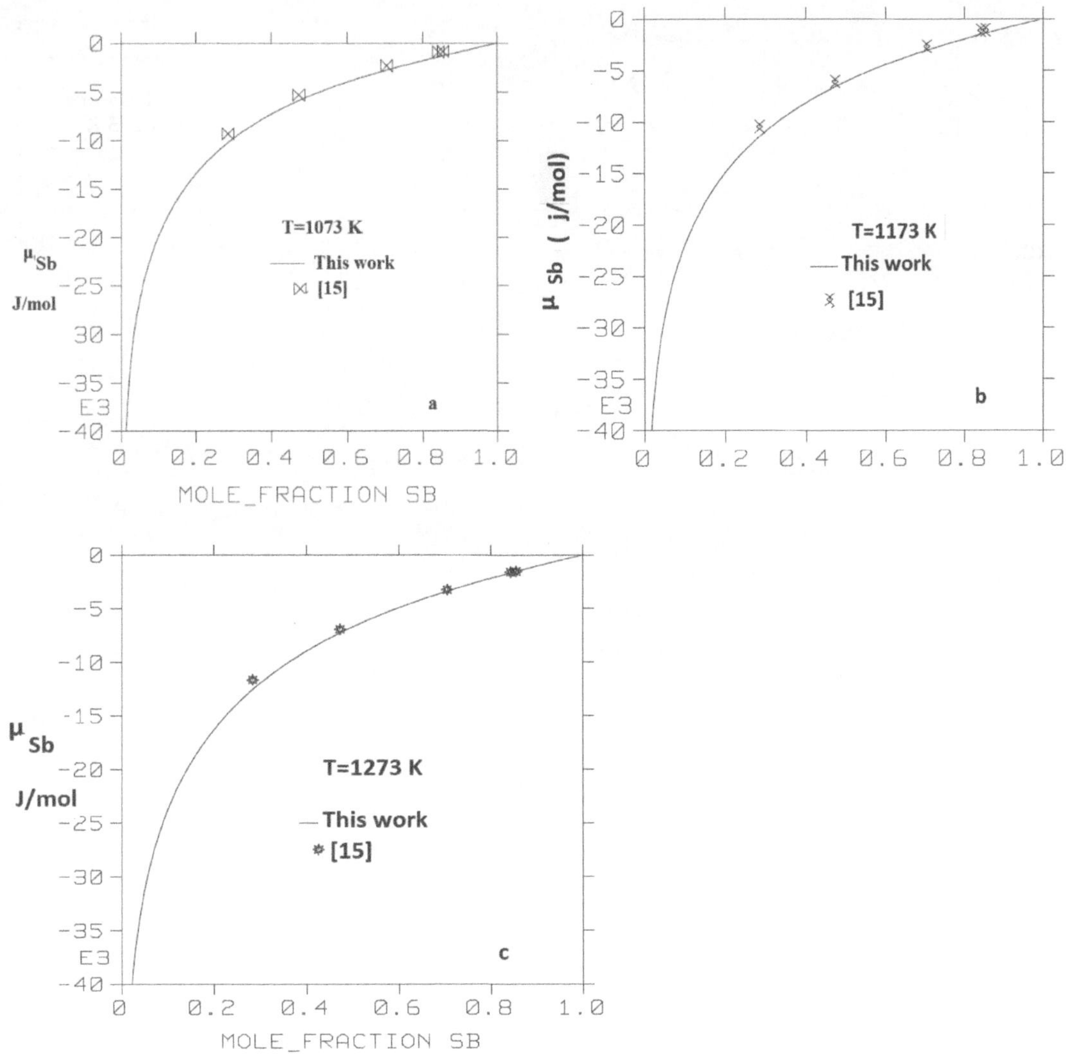

FIGURE 45.5 (a, b, c) The calculated chemical potential of antimony compared with the experimental values in [15].

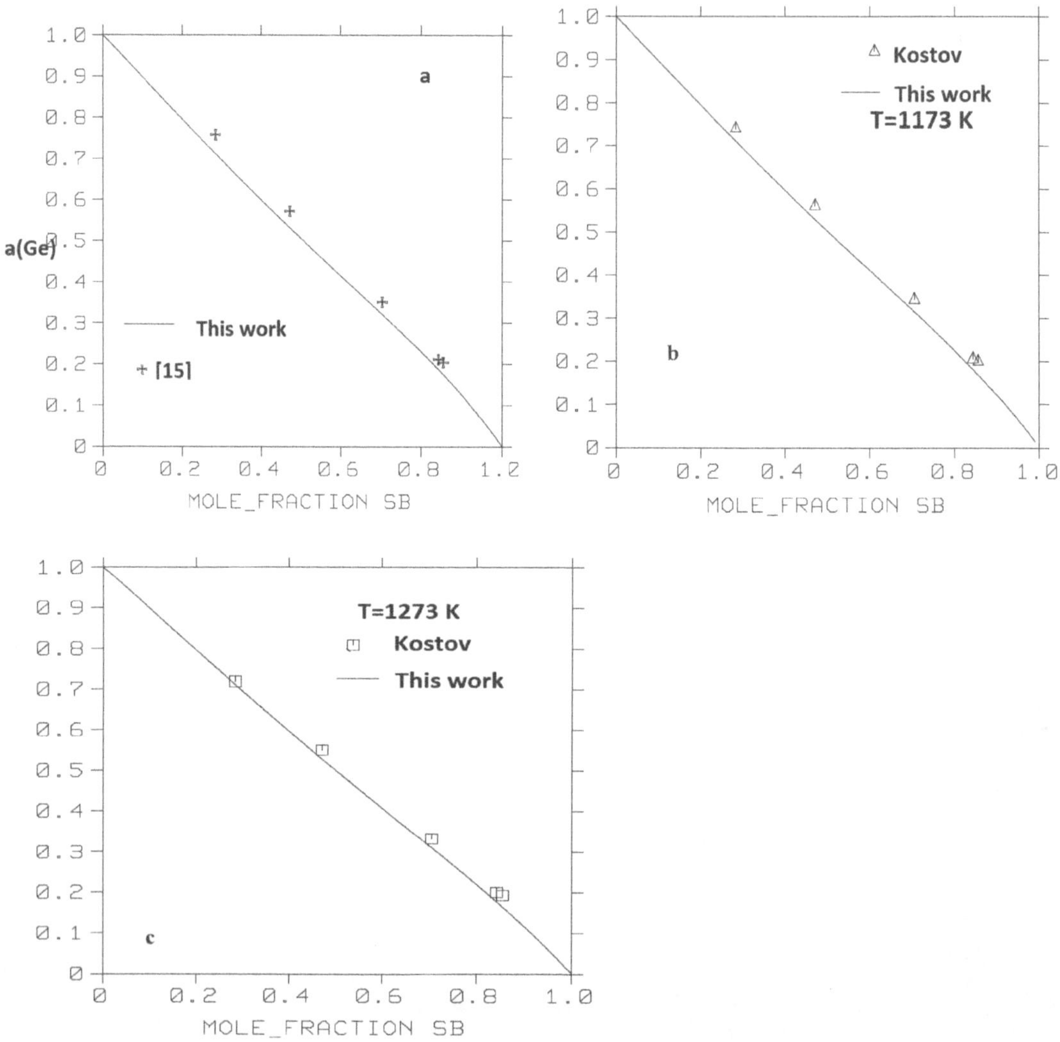

FIGURE 45.6 (a, b, c) The calculated activities of germanium compared with the experimental values in [15].

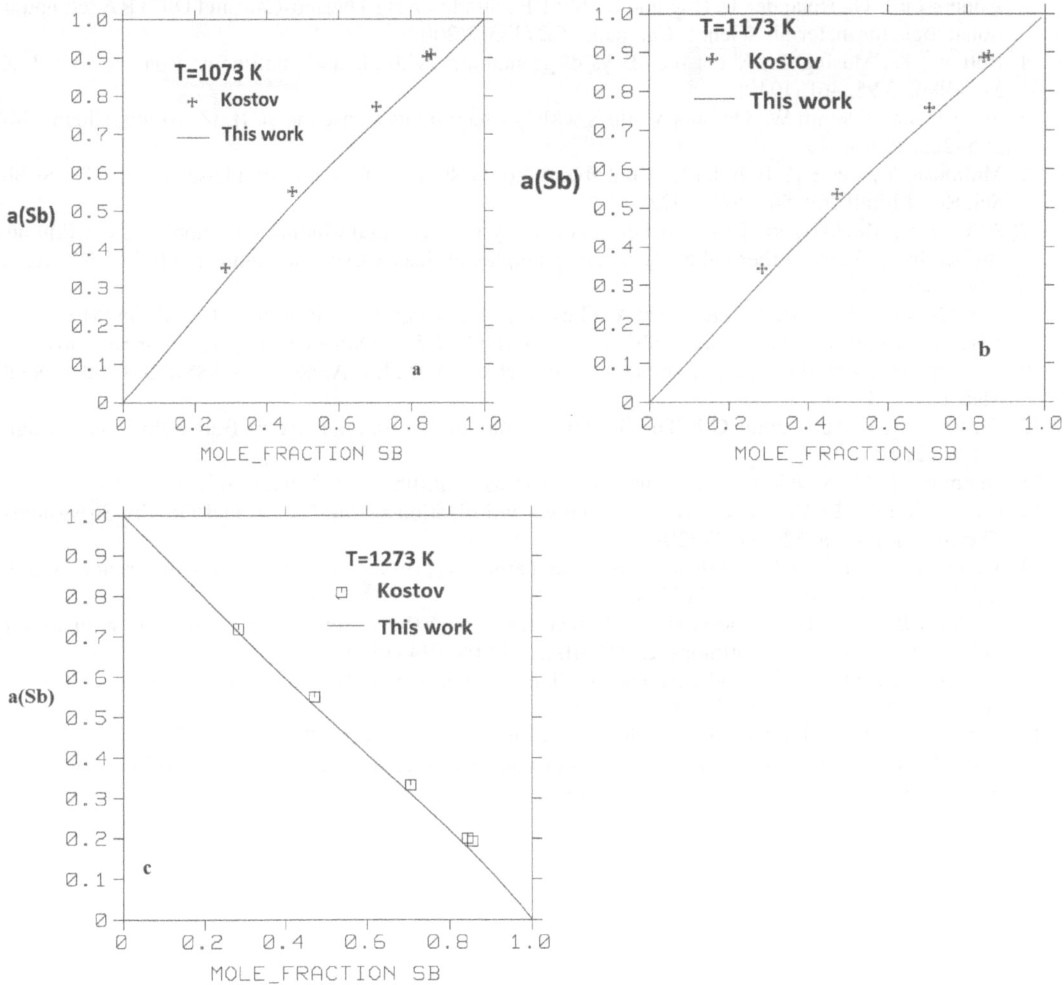

FIGURE 45.7 (a, b, c) The calculated activities of antimony compared with the experimental values in [15].

45.5 CONCLUSION

- Reassessment of the binary system Ge-Sb system is carried out. The calculated results are in good agreement with the most available experimental data in the literature.
- Thermodynamic optimizations have been performed on the binary system. The substitutional solution model was applied to describe the liquid phase and solid solution phase.
- This work enriches the thermodynamic databases of the germanium-antimony system.
- This study suggests another measurement on the enthalpies of the liquid phase because of the disagreement between in literature data.

REFERENCES

1. Premovic M., Minic D., Cosovic V., Manasijevic D., Zivkovic D., Experimental investigation and thermodynamic calculations of the Bi-Ge-Sb phase diagram, Metal. Mater. Trans. A, 45 4829–4841 (2014).
2. Lukas H.L., Fries S.G., Sundman B., Computational Thermodynamics: The Calphad Method, Cambridge University Press, 2007.

3. Andersson J.-O., Helander T., Höglund L., Shi P.F., Sundman B., Thermo-Calc and DICTRA, computational tools for materials science, Calphad, 26 273–312(2002).
4. Ruttewit K., Masing G., About the alloys of germanium with bismuth, antimony, iron and nickel, Z. Metallkd., 32 52–61 (1940).
5. Stohr H. and Klemm W., On binary alloys with germanium as a base metal. II., Z. Anorg. Chem., 244 205–223 (1940).
6. Malmejac Y., Desre P., Bonnier E., Contribution to the studies of the ternary phase diagram Ge-Si-Sb, Sei. Rev. Mdtall., 69, 565–577 (1972).
7. Alfer S.A., Mechkovskii L.A., Vecher A.A., Study of germanium-indium-antimony system liquidus surface by differential thermal analysis using simplex-lattice experiment planning, Zh. Fiz. Khim., 57 1292–1293 (1983).
8. Nasir N., Grytsiv A., Rogl P., Saccone A., Giester G., Phase equilibria in systems Ce–M–Sb (M= Si, Ge, Sn) and superstructure Ce12Ge9−xSb23+x (x = 3.8±0.1), J. Solid State Chem., 182 645–656 (2009).
9. Zhurkin B.G., Zemskov V.S., Petrov D.A., and Suchkov A.D., Izv. Akad. Nauk SSSR Otd. Tekh. Nauk Met. Topl., 5 86–90 (1959).
10. Olesinski R.W., Abbaschian G.J., The Ge–Sb (Germanium-Antimony) system, Bull. Alloy Phase Diagr., 7 219–222 (1986).
11. Okamoto H., Ge-Sb (Germanium-Antimony), J. Phase Equilib., 33 162–162 (2012).
12 Liu J., Guo C., Li C., Du Z., Thermodynamic optimization of the Ge–Sb and Ge–Sb–Sn systems, Thermochim. Acta, 520 38–47 (2011).
13. Wang J., Leinenbach C., Roth M., Thermodynamic description of the Au–Ge–Sb ternary system, J. Alloys Compd., 485 577–582 (2009).
14. Predel V.B., Stein D.W, Thermodynamic investigation of the systems germanium-zinc, germanium-indium, and germanium-antimony, Z. Metallkd., 61 909–914 (1970).
15. Kostov, A., Zivkovic, D., and Zivkovic, Z., Thermodynamic analysis of binary systems Ge–Ga and Ge–Sb, Thermochim. Acta, 338 35–43 (1999).
16. Dinsdale, A. T., SGTE data for pure elements, Calphad, 15 317–425 (1991).
17. Redlich, O., Kister, A. T., Thermodynamics of nonelectrolyte solutions-x-y-t relations in a binary system, Ind. Eng. Chem. Res., 40 341–345 (1948).

46 Evaluation of Reference Evapotranspiration Data in the CROPWAT Model with Local Climatic Data in Libya

E. Sagar, A. S. Sherif, M. Algohariani, A. Alraiani, J. Mabrouki, and Ahmed Ekhmaj

46.1 INTRODUCTION

There is a growing need to adopt sustainable agricultural strategies aimed at achieving food security and reducing pressure on natural resources. Water scarcity is one of the biggest challenges to agricultural development, especially in arid and semi-arid regions, where freshwater scarcity is a major barrier to achieving these goals (FAO, 2019a, 2019b). More than a third of the earth's surface (about 49 million square kilometers) falls within the classification of arid and semi-arid lands, and these lands receive rainfall rates of less than 400 mm per year, with 20% of the world's population living in these areas. In many cases, current management practices are unsustainable and lead to serious water-related problems such as depletion of aquifers, accumulation of pollutants to harmful levels, water disputes or uneconomical costs. The pressure on the land is such that between 50,000 and 70,000 square kilometers of useful productive land are lost each year due to desertification and climate.

In this context, evapotranspiration (ET) is one of the key factors to be considered in the management of water resources for irrigation areas, as it represents the largest consumption of water resources in irrigation, and is used as a biological index to understand the impact of climate change on agricultural production (Allen et al., 1998) and includes both direct evaporation from the soil surface and transpiration from plants, making it an important element in assessing water use efficiency and improving irrigation strategies. Libya ranks among the countries most exposed to water stress and desertification, ranking sixth globally in the level of water pressure, and most water extractions are unsustainable from groundwater sources (79%) and agricultural consumption (83.1%) of total freshwater resources. The evapotranspiration (ETo) measurement is becoming increasingly important as the effects of global warming worsen, as rising temperatures increase water loss rates from soil and plants, requiring the development of accurate irrigation techniques to reduce losses and enhance agricultural productivity (Zhang et al., 2020). In recent decades, studies have indicated a marked increase in global temperatures, leading to changes in ETo patterns because of changes in underlying climatic factors such as solar radiation, relative humidity, wind speed, and temperatures (IPCC, 2021). In arid and semi-arid regions, this change contributes to rainfall fluctuation, which affects cropping cycles and leads to changes in cultivation patterns. Due to the difficulty of obtaining direct field measurements of ET, experimental and statistical equations are used to estimate ETo based on climate data. Many equations have been developed for this purpose, but the validity of each depends on the climatic and environmental conditions from which they are derived. According to Burman et al. (1980a, 1980b), the definition of ETo dates back to Jensen et al. (1970), while Doorenbos and Pruitt (1975, 1977) adopted another definition that later became the basis for the FAO definition contained in FAO Paper 56, which describes ETo as the rate of ET from a standard

DOI: 10.1201/9781003647263-46

surface represented by an assumed herbaceous crop 12 cm high, with active and dense growth, an Albedo reflection coefficient of 0.23, and constant aerobic resistance at 70 seconds/m, without exposure to any water stresses (Allen et al., 1998), the ETo value provides an important criterion for comparing ET rates at different times of the year or between different regions and crops, helping to improve water use efficiency in agriculture and water resources management (Shuttleworth, 1993).

Studies show the effectiveness of using CLIMWAT with CROPWAT, as it is an effective tool for determining irrigation strategies according to local climatic conditions. Gaddikeri et al. (2024) also emphasized the need to adopt customized water management strategies to improve crop water use efficiency and reduce dependence on irrigation, especially in drought-prone areas. Accurate climate data are necessary to accurately estimate ETo, as different computational models rely on climate variables such as temperature, solar radiation, relative humidity, and wind speed. Kadkhodazadeh et al. (2022) noted that the use of reliable climate data enhances the accuracy of ETo estimates and contributes to more efficient water management, especially considering the effects of climate change. Also stressed that inaccurate data may lead to errors in the calculation of ETo, negatively affecting irrigation planning and water resources management. In addition, the analysis of future climate data helps to understand the impacts of climate change on ETo and take proactive measures. The study showed that the use of machine learning techniques depends mainly on the quality of the input data, which makes climate accuracy necessary to achieve the best results.

Assessing the quality of weather data and monitoring (correction) is essential to accurately calculate ETo rate and avoid biases in the data. The data collected may be affected by deviations caused by factors such as sensor failures, sensor obsolescence, software errors in data loggers, environmental influences surrounding the sensor, etc. (Allen, 2008). However, it seems that the process of data circulation and transfer plays an important role in the processing and correction processes.

The current research focuses on the analysis of differences between climate data provided by the Libyan National Center of Meteorology and the climate data available in the Agromet database of the Food and Agriculture Organization (FAO) of the United Nations, which is represented via the CLIMWAT program. Since this platform was originally designed to be a supporting database for the CROPWAT irrigation needs calculation program, it is necessary to understand the impact of these data on ETo estimates and the accuracy of agricultural water consumption calculations. This enables us to estimate the amount of water lost due to errors associated with estimating ETo in the CROPWAT, which may reach thousands of cubic meters during a single growing season. These errors may have a significant impact on water management efficiency and agricultural planning decisions in areas with water scarcity (Pereira et al., 2015).

46.2 MATERIALS AND METHODS

46.2.1 Study Area

Libya is located in the north of the continent of Africa, extending between latitudes 19° and 33° north, and longitudes 9° and 25° east, which gives it a strategic location overlooking the Mediterranean Sea from the north, while it is bordered by Egypt from the east; Tunisia and Algeria from the west; and Niger, Chad, and Sudan from the south (Figure 46.1). It has an area of about 1.76 million square kilometers.

The climate in Libya is characterized by a clear variation because of its wide geographical extension, as the dry desert climate prevails in most parts of the country, especially in the interior and desert areas that cover about 90% of the total area. The coastal areas have a semi-arid Mediterranean climate characterized by mild, rainy winters and hot, dry summers, with varying rainfall rates ranging from less than 100 mm in the western regions to more than 500 mm per year in the Jabal Akhdar region, where a semi-humid climate prevails due to the terrain and marine influences (El-Tantawi, 2005). ET rates in Libya are high due to high temperatures and low humidity, affecting the availability of surface and groundwater.

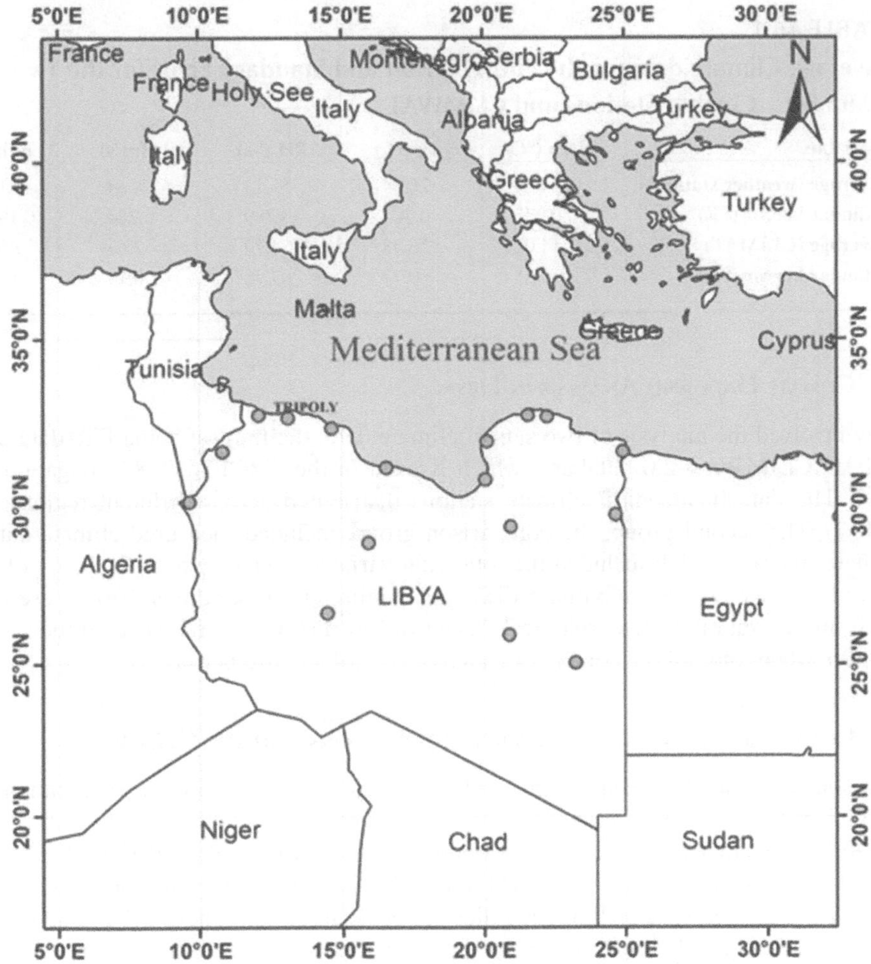

FIGURE 46.1 Map of Libya and the distribution of climate stations.

This study is based on climate data collected from official sources, including THE Libyan national meteorological center (LNMC), as well as data from the FAO database. Differences between local data and those contained in global databases will be analyzed, focusing on the main climate variables affecting ETo. This includes temperature, relative humidity, wind speed, and solar radiation, as these factors play a key role in determining water consumption rates in agricultural systems and the effectiveness of water resources management strategies. The effects of climate change on these variables and their impact on agricultural sustainability in arid and semi-arid areas in Libya will also be assessed.

The study area exhibits diverse natural features, including geological formations and climatic conditions that have influenced soil characteristics and distribution. Soils across much of the region tend to be weakly developed, varying in texture and water-holding capacity, reflecting the impact of environmental factors such as climate and parent material. Coastal and upland areas show some differences in structure and fertility, while soil. The interior is more affected by arid conditions, which can limit productivity in certain locations (Bockheim et al., 2014; FAO, 2006, 2016; Lal, 2006; Soil Survey Staff, 2014). These characteristics play a role in determining the accuracy of climatic and agricultural assessments, particularly when using models such as CROPWAT, where soil factors along with climate affect the estimation of crop water requirements.

TABLE 46.1
Average Climate data for the Entire Period and Standard Error for the Two Data Sets, Climate Stations and CLIMWAT

Variable	Tn (°C)	Tx (°C)	RH (%)	U (m/s)	n (h)
Average (weather stations)	14.90	26.84	55.13	4.44	9.12
Standard error (SE)	0.328	0.705	3.789	0.284	0.195
Average (CLIMAT)	14.02	26.98	54.79	2.63	7.99
Standard error (SE)	0.35	0.77	3.98	0.19	0.19

46.2.2 Climate Data and Analysis in Libya

This study involved the analysis of two sets of climate data, the first set being ETo data integrated into the FAO CLIMWAT 2.0 database, which is used in the CROPWAT 8.0 irrigation scheduling model. The data included 17 climate stations distributed across different regions of Libya (Figure 46.1). The second group, the comparison group, included measured climate data for the same climate stations, which included the following variables: wind speed (U), hours of sunshine (n), relative humidity (RH%), maximum (Tx), and minimum temperatures (Tn). These data were obtained from the National Meteorological Center of Libya (LMNC) and covered the period from 1991 to 2000, where monthly averages were analyzed for the entire period.

46.2.3 Comparison of Climate Data between Stations and the CLIMWAT

A comparison was made between the mean climate data derived from the actual stations and the CLIMWAT model, as shown in Table 46.1.

Table 46.1 shows that there are slight differences between the values derived from the actual stations and those extracted from the CLIMWAT. This is due to some variations in data sources, accuracy of measurements, as well as the influence of local climatic factors that may not be accurately represented in global climate models (Allen et al., 1998).

46.2.4 Mean Annual Precipitation Values from Meteorological Stations in Libya

Annual precipitation analysis is essential in assessing the water requirements of crops in arid and semi-arid regions such as Libya (Figure 46.2). Based on CLIMWAT and CROPWAT data, average

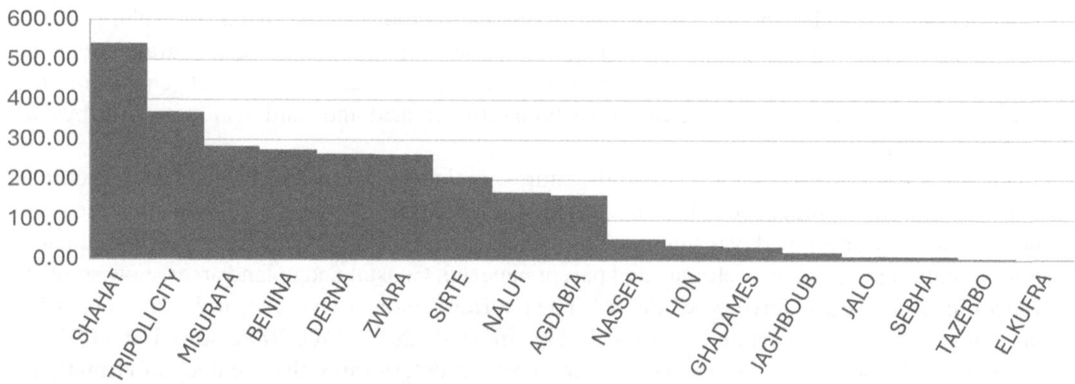

FIGURE 46.2 Average annual precipitation of climate stations (mm).

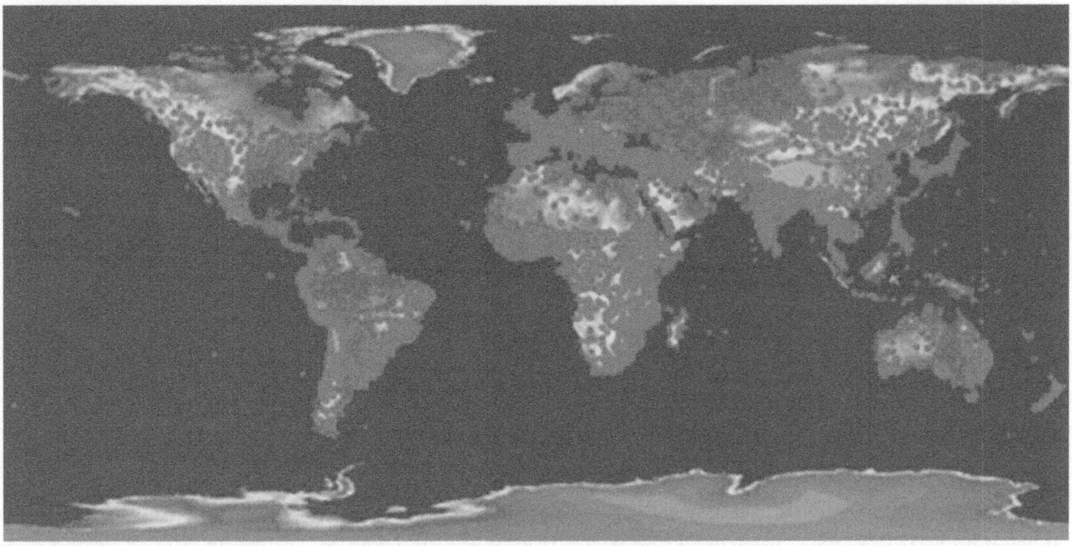

FIGURE 46.3 Distribution of climate stations around the world for climate data sets in the program.

annual precipitation shows great variation between different regions, with precipitation decreasing as we toward the south direction, underscoring the need for careful water resources management to ensure sustainable agriculture.

46.2.5 CROPWAT and CLIMWAT Software

CROPWAT is software developed by the FAO used to calculate aquatic crop requirements and schedule irrigation based on climate, soil and crop data. The program assists in estimating ETo using the FAO-Penman Monteith equation, requirements for aquatic crops according to their growth stages, and irrigation efficiency and scheduling according to water availability and type of system used analyze the impact of climate on agricultural production by comparing different scenarios.

$$ET_0 = \frac{0.408\Delta(R_n - G) + \gamma \frac{900}{T+273} u_2 (e_s - e_a)}{\Delta + \gamma(1 + 0.34u_2)} \tag{46.1}$$

Where: R_n: Net solar radiation (MJ/m2/s), G: soil heat flow (MJ/m^2/s), γ: is the psychrometric constant kPa/°C, Δ: slope of the vapor pressure curve (kPa/°C), U_2: wind speed m/day at an altitude of 2 m, e_s: saturated pressure of water vapor (kPa), e_a: actual pressure of water vapor (kPa), T: average temperature at an altitude of 2 m °C, (e_s-e_a): saturated vapor pressure deficit (kPa).

CLIMWAT provides weather data such as temperature, relative humidity, solar radiation, and wind speed for 5,000 weather stations worldwide (Figure 46.3). Data is imported directly into CROPWAT for use in calculating irrigation requirements and scheduling water for different crops.

46.3 RESULTS AND DISCUSSION

46.3.1 Statistical Analysis

The assumption of this research is that these differences are the result of the processing processes that were used on the same original (local) data set, and regardless of the usual data processing processes (filtering, cleaning, classification, completion), there is an approach specialized in correcting

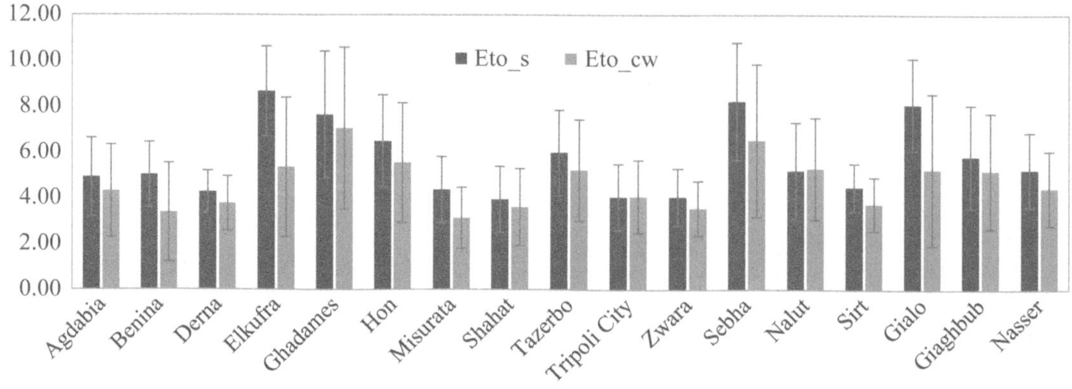

FIGURE 46.4 ETo data for climate stations and CLIMWAT.

the climate data necessary to calculate ETo from the Penman-Monthly equation, which is to correct these data as if they were measured under conditions bookmarked by definition. Based on the geographical locations of the climatic stations, they are divided into coastal with relatively high humidity and dry interior with high temperature. The differences between the minimum temperatures and dew point temperatures indicate the availability of moisture in the region, the greater these differences, the more drought conditions prevail over other climatic conditions, and the closer the value of these differences is to zero, this is an indication of the availability of moisture in the region and therefore does not need the corrections required to apply the equation.

Two ETo data sets were calculated using the CROPWAT software and the data showed clear differences between the two groups, especially in dry interior areas (Figure 46.4).

The existence of these differences is due to the effect of atmospheric humidity represented in the actual water vapor pressure (Figure 46.5), where the division of climate station data can be evaluated into two main groups based on the actual water vapor pressure (Figures 46.6 and 46.7), and the highest peak of the vapor pressure range was 2.32 at the Zwara station, and the lowest value was 0.60 at the Kufra station.

Climatic station data can be classified into two main groups based on actual water vapor pressure values:

- Highest value: 2.32 kPa at the Zwara station.
- Minimum value: 0.60 kPa at the Kufra station.

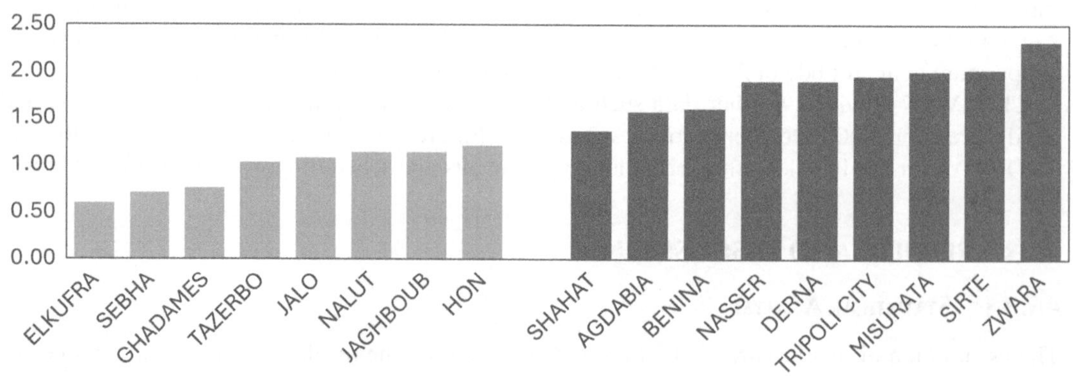

FIGURE 46.5 Actual water vapor pressure (KPa).

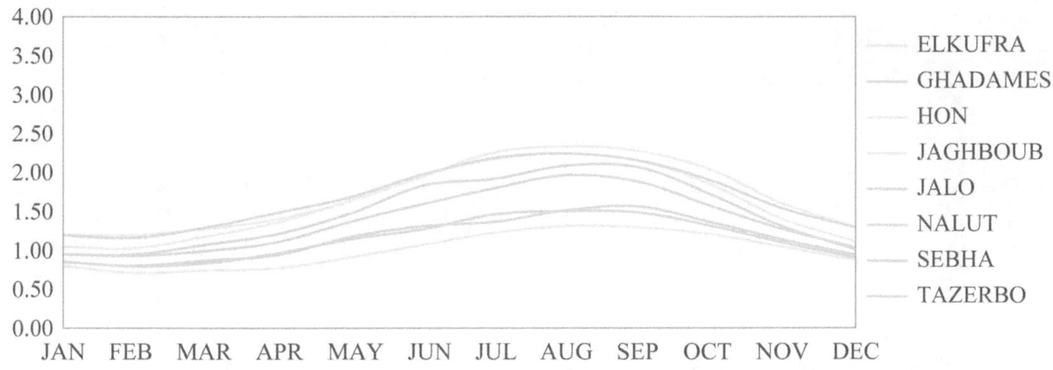

FIGURE 46.6 Actual water vapor pressure (kPa) for indoor climate stations.

46.3.2 Effect of Relative Humidity on the Calculation Accuracy

To determine the need to correct climate station data to accurately apply the Penman-Monteith model, the P/ETo ratio was used, which expresses the availability of soil moisture and its ability to support reference conditions (McMahon et al., 2013).

When the P/ETo ratio is ≥ 1, Tmin approaches Tdew, reflecting wet conditions that do not require additional corrections. When P/ETo is < 1, station drought results in significantly higher Tmin compared to Tdew, necessitating the correction of climate data before it can be used in ETo calculations.

46.3.3 Comparison between Arid and Irrigated Climatic Sites

The values of Tmin-Tdew were compared between two semi-arid sites in Idaho, USA, separated by about 200 km:

- Kimberley Station: A reference site located in the middle of a wide irrigated area.
- Boisi Station: An unreferenced location located at an airport, surrounded by a mix of irrigated and non-irrigated land.

The data showed that Tmin was very close to the Tdew at the Kimberley station, reflecting the correct reference conditions. At Boisei station, Tmin was 10°C higher than Tdew, suggesting the need to correct climate data before being used to calculate ETo using the FAO-PM equation (Allen et al., 1998). The results suggest that correcting climate data prior to use in ETo calculations is

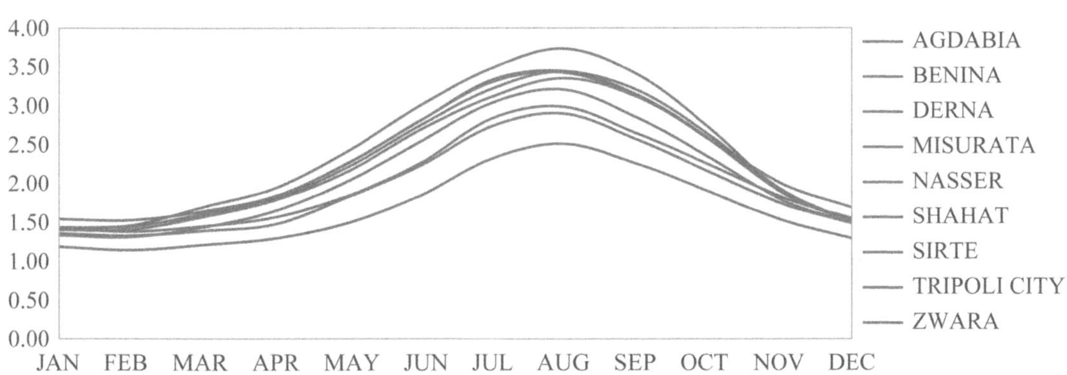

FIGURE 46.7 Actual water vapor pressure (kPa) for coastal climate stations.

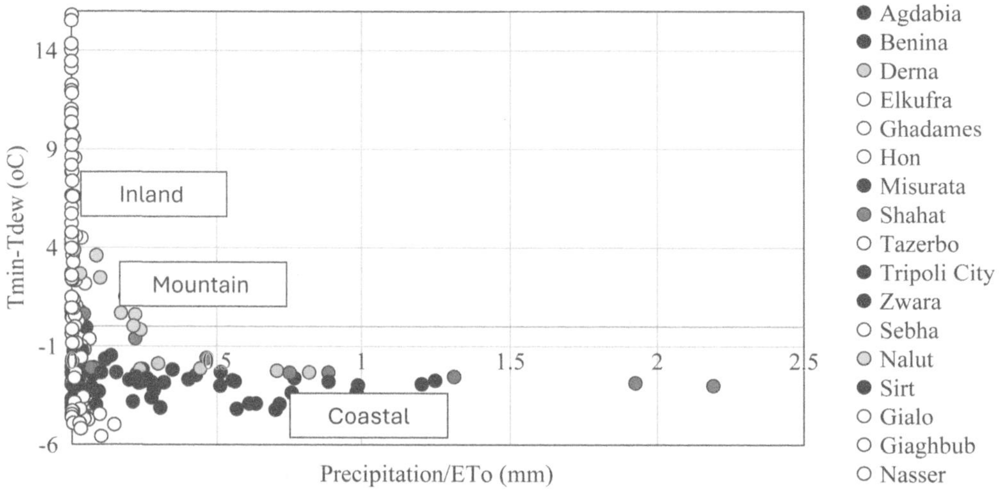

FIGURE 46.8 Tmin - Tdew and P/ETo climate station comparison.

necessary to ensure the accuracy of the results, especially in stations with dry conditions where T_{MIN} is high compared to T_{Dew} (Figure 46.9). The P/ETo assessment is also an effective tool in determining the need for climate corrections to ensure higher accuracy when using the Penman-Monteith equation to calculate ETo (Figures 46.8 and 46.9).

Based on the results of the statistical analysis of climate stations using the U-Theil's test, a significant decrease in the U-statistic value was recorded for most climate stations, reflecting an improvement in the level of compatibility of climate data extracted from two different sources. Values between 0 and 1 indicate suitability for testing, with values close to 0 expressing a significant similarity between the original and modified data, while values exceeding 1 indicate a gap between the two sets (Figure 46.10).

46.3.4 Analysis of U-Theil's Test Results for Climate Stations

The results of the analysis of climate data reveal the impact of different methods of data correction used in the CLIMWAT database, where U-statistic values reflect the extent to which these methods have succeeded in improving the accuracy of climate data. However, some stations have shown divergence from accepted values, suggesting that there may be additional factors affecting

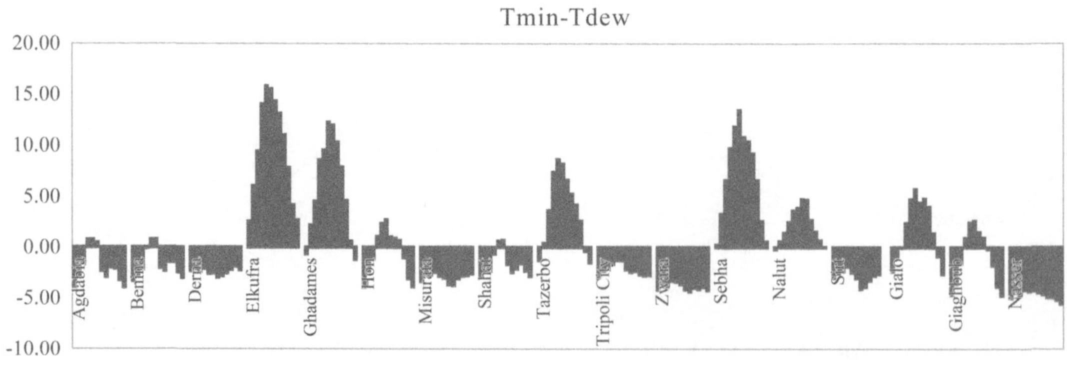

FIGURE 46.9 Tmin – Tdew.

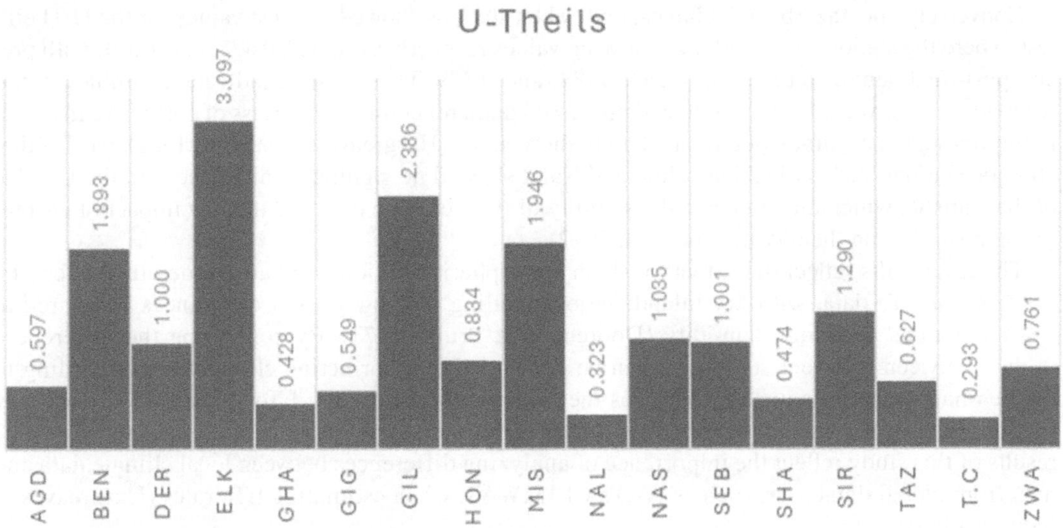

FIGURE 46.10 Results of statistical analysis of climate stations using U-Theil's test.

the accuracy of climate data, either through data correction methodologies or sources of technical errors in the original measurements (Allen et al., 1998) (Figure 46.11).

Out of eight climate stations, Kufra and Jalo stations recorded the highest U values, with pre-treatment of 3.097 and 2.386, respectively. After processing the data using FAO-approved methods, the values dropped to 2.203 and 2.252, respectively, indicating an improvement in the accuracy of the data after processing. However, these values remain relatively high compared to other stations, which may reflect two main sources of variation:

1. Use additional processing methods in the CLIMWAT database, which may include advanced correction techniques such as statistical estimation of missing data or climate modeling (Allen et al., 2006).
2. Possible measurement errors in the original data collected from climate stations, which may be caused by technical issues such as sensor accuracy or geographic location effects on data readings (Snyder et al., 2005).

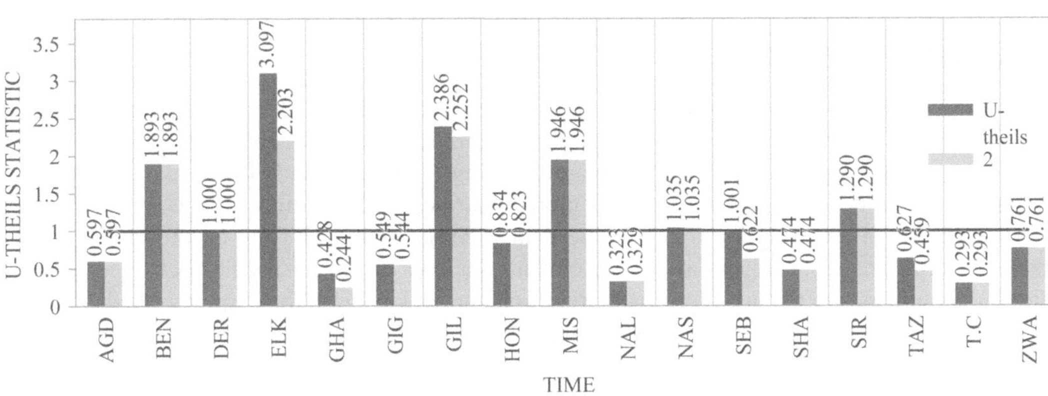

FIGURE 46.11 U-Theil's test results for climate stations.

Conversely, the Tazerbo, Ghadames, and Sabha stations showed the best values for the U-Theil's test, where the stations recorded the following values respectively: 0.627, 0.428, and 1.001, with pre- and post-treatment differences of 0.168, 0.184, and 0.379. These results indicate a significant correspondence between the measured and processed data, reflecting the success of corrective methods in improving the accuracy of climate data in these areas (Hargreaves & Samani, 1982), while other climatic stations such as Jaghbub, Hun, and Nalut showed no significant differences in the U value of the statistic, which may indicate the stability of their climate data and the low impact of corrective adjustments on them (Ghizlane et al. 2022).

The test results reflect the extent to which atmospheric humidity and air temperature affect the accuracy of ETo data, with dry inland stations tending to show greater differences compared to coastal stations with high humidity (Doorenbos & Pruitt, 1977). By comparing the differences in data between different stations, it can be concluded that correcting climate data according to the Penman-Monteith equation improves the accuracy of estimating ETo (Figure 46.12) but may require the use of additional methods to adjust for outliers in some critical climate settings. The results of this study reflect the importance of analyzing differences between local climate data and data from global databases, such as FAO's CLIMWAT, when estimating ETo rates (Hargreaves & Samani 1985).

The study proved that there are clear discrepancies between the values derived from the actual climatic stations in Libya and those in the CLIMWAT, indicating the possibility of errors in estimating the quantities of water required for irrigation when relying on non-local data.

46.4 DIFFERENCES BETWEEN CLIMATE DATA AND THEIR IMPACT ON ETo ESTIMATION

The results showed that the data of maximum and minimum temperature, relative humidity, wind speed, and the number of hours of sunshine differ between what was measured in Libyan climate stations and what is recorded in the CLIMWAT database (Hobbins et al. 2001). The differences were more pronounced in the variables of wind speed and number of hours of sunshine, as it was shown that CLIMWAT values tended to reduce wind speed and duration of sun brightness compared to local data, which may lead to a lower estimate of true ETo rates (Jenny 1941). The impact of climate variations on water resources management: Relying on inaccurate or unrepresentative data may lead to errors in calculating agricultural water consumption, which is critical in a water-scarce country such as Libya (LNMC). These errors affect agricultural planning decisions, as insufficient or excessive water may be allocated for irrigation, which can lead to either a lack of crop irrigation or significant water waste (Smith 1992).

46.5 RECOMMENDATIONS

To ensure greater accuracy in estimating irrigation needs, conduct further research on the relationship between climate change and ETo in arid and semi-arid areas. It is also advisable to develop advanced monitoring systems based on remote sensing techniques and climate data modeling to enhance the ability to predict water needs more accurately. The findings of this study highlight the importance of adopting accurate climate data when estimating ETo to ensure efficient water use in agriculture, especially in arid environments such as Libya. Disparities between local and global data underscore the need for further research and updates in ETo estimation models used in agricultural water management. By enhancing data accuracy and improving estimation methods, sustainable and effective water resources management can be achieved, contributing to food security in water-scarce areas.

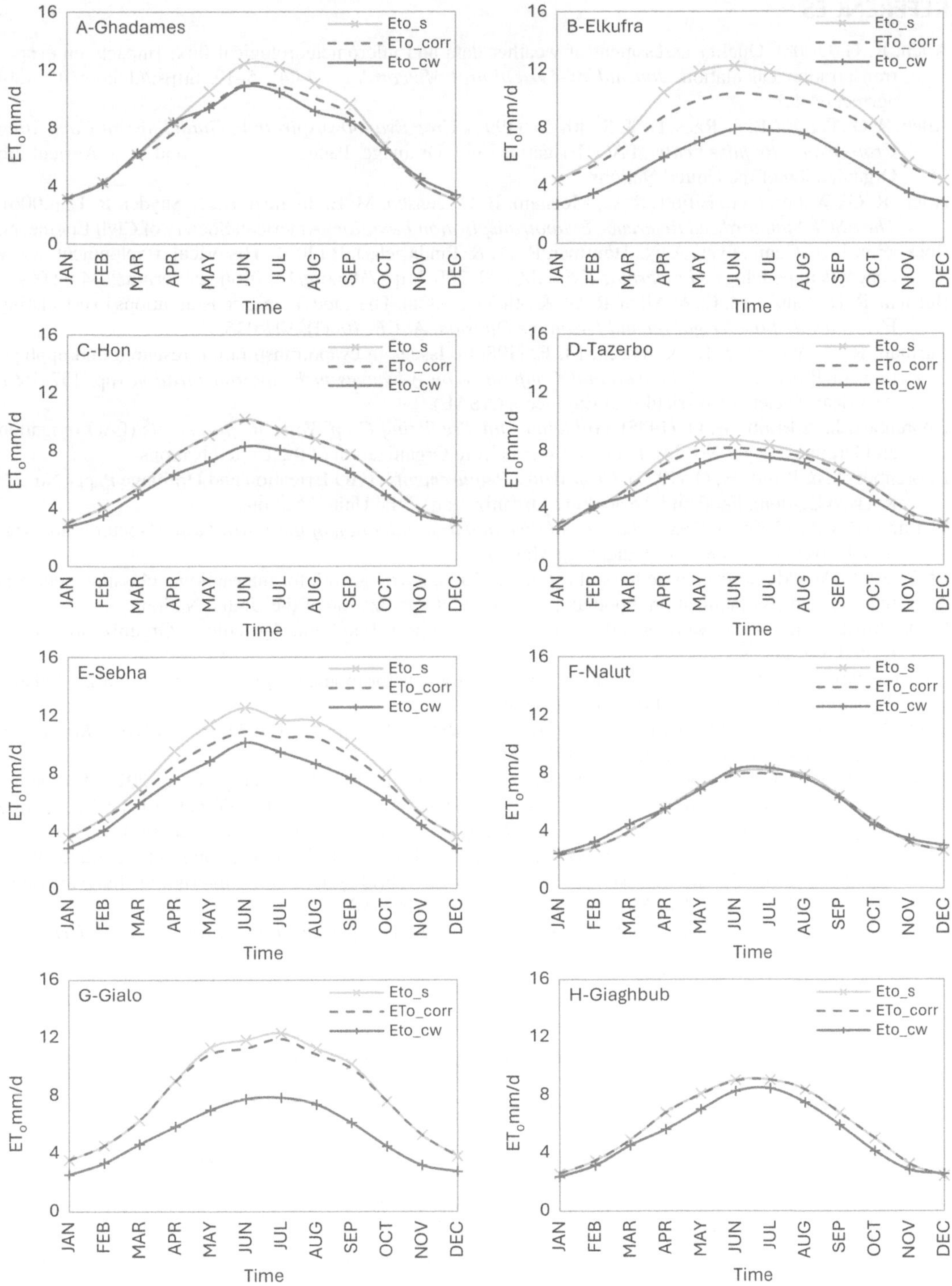

FIGURE 46.12 Corrected and original ETo data were compared with the ETo calculated by CROPWAT.

REFERENCES

Allen, R. G. (2008). Quality assessment of weather data and micrometeorological flux: Impacts on evapotranspiration calculation. *Journal of Agricultural Meteorology, 64*(4), 5–19. https://doi.org/10.2480/agrmet.64.4.5

Allen, R. G., Pereira, L. S., Raes, D., & Smith, M. (1998). *Crop Evapotranspiration: Guidelines for Computing Crop Water Requirements* (FAO Irrigation and Drainage Paper No. 56). Food and Agriculture Organization of the United Nations.

Allen, R. G., Walter, I. A., Elliott, R. L., Mecham, B. Q., Jensen, M. E., Itenfisu, D., & Snyder, R. L. (2006). *The ASCE Standardized Reference Evapotranspiration Equation*. American Society of Civil Engineers.

Bockheim, J. G., Gennadiyev, A. N., Hammer, R. D., & Tandarich, J. P. (2014). Historical development of key concepts in pedology. *Geoderma, 226–227*, 290–300. https://doi.org/10.1016/j.geoderma.2014.02.004

Burman, R. D., Jensen, M. E., & Allen, R. G., & others. (1980a). Thermodynamic considerations in estimating ET. *Journal of the Irrigation and Drainage Division, ASCE, 106*(1), 107–115.

Burman, R. D., Wright, J. L., & Jensen, M. E. (1980b). Issues in evapotranspiration research and application. In *Proceedings of the National Conference on Advances in Evapotranspiration* (pp. 157–166). American Society of Agricultural Engineers (ASAE).

Doorenbos, J., & Pruitt, W. O. (1975). *Guidelines for Predicting Crop Water Requirements* (FAO Irrigation and Drainage Paper No. 24). Food and Agriculture Organization of the United Nations.

Doorenbos, J., & Pruitt, W. O. (1977). *Crop Water Requirements* (FAO Irrigation and Drainage Paper No. 24, Revised Edition). Food and Agriculture Organization of the United Nations.

El-Tantawi, A. M. (2005). *Climate change in Libya and desertification of the Jifara Plain* (Doctoral dissertation, University of Johannes Gutenberg Mainz).

FAO. (2006). World reference base for soil resources 2006: A framework for international classification, correlation, and communication. Food and Agriculture Organization of the United Nations.

FAO. (2016). Status of the world's soil resources: Main report. Food and Agriculture Organization of the United Nations.

FAO. (2019a). The state of the world's land and water resources for food and agriculture (SOLAW). Food and Agriculture Organization of the United Nations.

FAO. (2019b). *The State of Food and Agriculture 2019: Moving Forward on Food Loss and Waste Reduction*. Food and Agriculture Organization of the United Nations.

Gaddikeri, V., Rajput, J., Dimple, Jatav, M. S., Kumari, A., Rana, L., Rai, A., & Gangwar, A. (2024). Estimating crop water requirement in Madhya Pradesh's agro-climatic regions: A CROPWAT and CLIMWAT software case study. *Environment Conservation Journal, 25*(1), 308–326. https://doi.org/10.36953/ECJ.26022353

Ghizlane, F., Mabrouki, J., Ghrissi, F., & Azrour, M. (2022). Proposal for a high-resolution particulate matter (PM10 and PM2. 5) capture system, comparable with hybrid system-based internet of things: Case of quarries in the Western Rif, Morocco. *Pollution*, 8(1), 169–180.

Hargreaves, G. H., & Samani, Z. A. (1982). Estimating potential evapotranspiration. *Journal of the Irrigation and Drainage Division, 108*(3), 223–230.

Hargreaves, G. H., & Samani, Z. A. (1985). Reference crop evapotranspiration from temperature. *Applied Engineering in Agriculture, 1*(2), 96–99.

Hobbins, M. T., Ramírez, J. A., & Brown, T. C. (2001). The complementary relationship in estimation of regional evapotranspiration: An enhanced advection-aridity model and its testing at the regional scale. *Water Resources Research, 37*(5), 1367–1383.

Intergovernmental Panel on Climate Change (IPCC). (2021). Climate change 2021: The physical science basis. Cambridge University Press.

Jenny, H. (1941). *Factors of Soil Formation: A System of Quantitative Pedology*. McGraw-Hill.

Jensen, M. E., Burman, R. D., & Allen, R. G. (1970). *Evapotranspiration and Irrigation Water Requirements* (ASCE Manuals and Reports on Engineering Practice No. 70). American Society of Civil Engineers.

Kadkhodazadeh, M., Valikhan Anaraki, M., Morshed-Bozorgdel, A., & Farzin, S. (2022). A new methodology for reference evapotranspiration prediction and uncertainty analysis under climate change conditions based on machine learning, multi-criteria decision making, and Monte Carlo methods. *Sustainability, 14*(5), 2601. https://doi.org/10.3390/su14052601

Lal, R. (2006). *Encyclopedia of Soil Science* (2nd ed.). CRC Press.

Libyan National Meteorological Center (LNMC). (n.d). Climatic data records for Libya.

McMahon, T. A., Peel, M. C., Lowe, L., Srikanthan, R., & McVicar, T. R. (2013). Estimating actual, potential, reference crop and pan evaporation using standard meteorological data: A pragmatic synthesis. *Hydrology and Earth System Sciences, 17*(4), 1331–1363.

Pereira, L. S., Cordery, I., & Iacovides, I. (2015). *Coping With Water Scarcity: Addressing the Challenges*. Springer.

Shuttleworth, W. J. (1993). Evaporation. In D. R. Maidment (Ed.), *Handbook of Hydrology* (pp. 4.1–4.53). McGraw-Hill.

Smith, M. (1992). *CROPWAT: A Computer Program for Irrigation Planning and Management* (FAO Irrigation and Drainage Paper No. 46). Food and Agriculture Organization of the United Nations.

Snyder, R. L., Spano, D., Duce, P., & Baldocchi, D. D. (2005). Surface renewal estimates of evapotranspiration: Current status and future challenges. *Agricultural and Forest Meteorology, 132*(1–2), 1–13.

Soil Survey Staff. (2014). *Keys to Soil Taxonomy* (12th ed.). United States Department of Agriculture, Natural Resources Conservation Service.

Zhang, B., Li, Y., Sun, H., Wang, Y., & Liu, H. (2020). Impact of climate change on crop evapotranspiration and irrigation water requirements in arid and semi-arid regions. *Agricultural Water Management, 241*, 106394. https://doi.org/10.1016/j.agwat.2020.106394

47 Comprehensive Vision and Diverse Tracking Methods for Humans and Objects

Expansive Perception and Multifaceted Strategies

A. Ariss, I. Ennejjai, J. Mabrouki, and S. Ziti

47.1 INTRODUCTION

Monitoring systems are essential for examining dynamic human activities and object behaviors through their data. Many research teams have introduced a range of tracking strategies to mitigate the challenges of monitoring a large number of individuals. This study assesses the results of various tracking techniques, acknowledging that no single method is universally superior. The widespread data loss stemming from inadequate monitoring and management of data generated by humans or objects highlights the critical importance of effective tracking. A multifaceted approach to tracking humans or objects is essential for managing irregular movements, changes in appearance, and detecting abnormal behaviors. This approach is maintained until enough information is collected to make informed decisions [1]. Implementing a tracking system facilitates comprehensive learning from activities and optimizes data utilization [2]. These systems can be crucial in crime prevention, epidemic management, and various other applications.

Achieving accurate tracking in both confined and expansive spaces demands robust methods that can effectively interpret and analyze object or human activities to support informed decision making. While specific extraction and processing models are essential to eliminate data ambiguity and ensure precise tracking, these models must also be general and straightforward to maintain overall robustness.

Numerous studies have explored the realm of multiple tracking of individuals and objects. Remarkable contributions involve the utilization of graph theory [3,4], technological tools, Internet of Things (IoT) [5], big data [6], and artificial intelligence [7], either independently or in synergy, to achieve improved results.

Our research falls in line with these various categories and methodologies. This chapter offers an analytical examination of tracking systems, offering a thorough comprehension of the techniques employed within this domain. It serves as a supplement to investigating alternative paths for future research on this topic. A fundamental aspect of our research involves term extraction to generate word clouds from strongly labeled domains within tracking papers. We outline a term extraction approach based on word clouds, which are visual tools for representing textual data. These clouds effectively highlight key textual elements, showcasing the frequency and significance of specific words. Figure 47.1 illustrates the most prevalent terms in published papers addressing resolved tracking issues (e.g., tracking, system, person, people, object, technology, graph, big data, IoT, artificial intelligence, etc.).

Our experience in extracting terms from various paper abstracts demonstrates a noticeable enhancement in word cloud quality with each additional abstract. Furthermore, unlike other

 DOI: 10.1201/9781003647263-47

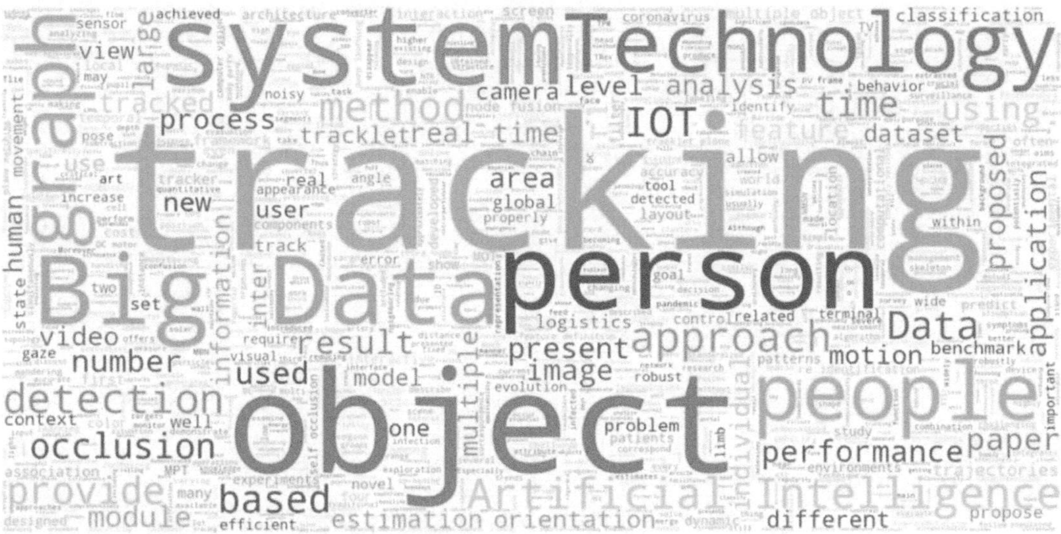

FIGURE 47.1 The extraction of words from articles about tracking thematics to create word clouds.

cloud generation methods sensitive to data scarcity, our results emphasize the robustness of this methodology.

The remainder of this chapter is structured as follows. The tracking is explained in the second section. A brief summary of some relevant work is included in the third section. Taking into account that the fourth section describes the analysis and discussion, the fifth and final section summarizes our findings and details our ongoing research.

47.2 DEFINITION OF TRACKING

The expansive nature of the tracking domain is marked by its multitude of available definitions. Consequently, researchers from diverse fields tailor their descriptions of this theme to align with their respective areas of expertise and professional domains. Termed variably as a tracking system, this technology serves the purpose of detecting and monitoring the movement of individuals or objects, providing subsequent location data for further analysis, as per commonly accepted definitions. It is worth noting that a tracking system, often synonymous with surveillance systems, serves as a valuable resource for gathering information about mobile entities, be it people or other subjects of interest.

Our research group advocates for the strategic utilization of tracking systems to focus on specific targets, whether they be objects or individuals. This targeted tracking is achieved through a meticulous analysis of the target's activities. It is pertinent to highlight that, as of now, no author has devised a tracking system endowed with the desired flexibility.

Having elucidated various initiatives undertaken in this field, we underscore that tracking systems can function across three dimensions. Illustrated in Figure 47.2, these dimensions, at a given time (T), may manifest 1 to N different implementations. These dimensions are interconnected by bidirectional, nonexclusive relationships. Moreover, the method allows for flexibility at the starting point, enabling tracking initiation from any dimension and imparting a backward-compatible force to the system based on the input.

Several implementations are made accessible at the conclusion of the data qualification process; their purpose is to construct an integrated system whose implementations can coexist with one another.

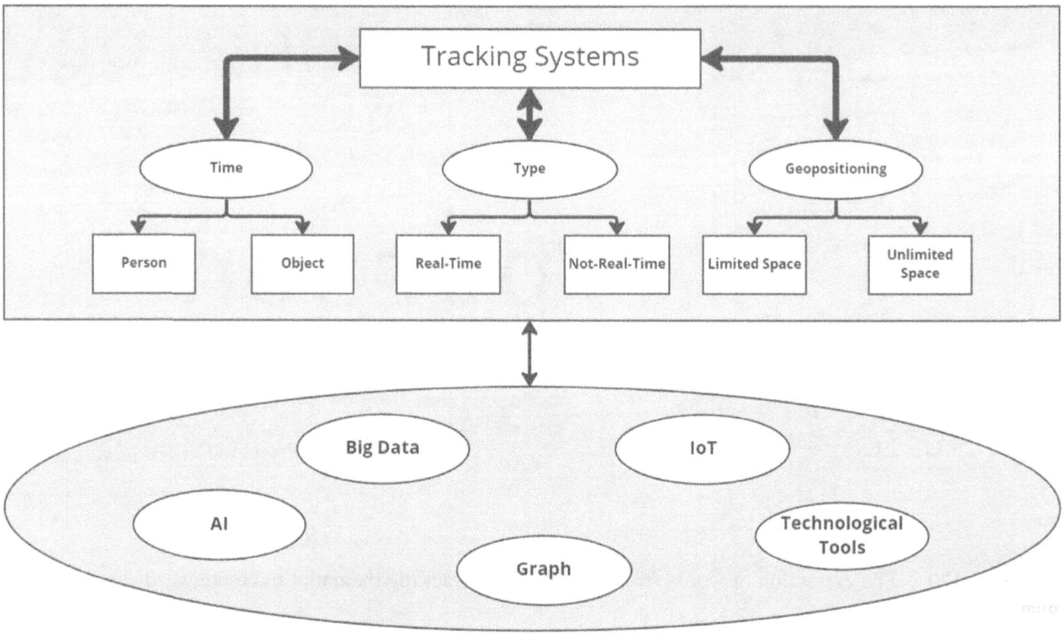

FIGURE 47.2 Tracking systems architecture.

47.3 RELATED WORK

Research in the field of tracking [8] is considered not only highly significant but also inherently complex. This complexity arises from the fact that each proposed solution approaches the issue in a unique manner, and notably, there is a lack of standardized support for tracking systems.

Wu et al. made a significant contribution to multiperson online tracking [9] by introducing a method based on deep learning [10,11]. Their approach centers on acquiring nuanced representations of instances through a multibranch neural network (MBN) [12]. The aim is to leverage learning and intricate instance representations to monitor numerous individuals online. This approach represents a shift in paradigm from detection to association, addressing detection challenges by learning representations tailored to individual instances. This departure from conventional methods, which often rely on generic human detectors at the category level, marks a notable advancement.

An impressive result of this method is the development of a straightforward yet efficient solver for constructing an association matrix using the outputs of the MBN. This solver facilitates joint state inference of targets, aiming to assign a distinct tracker to each moving individual, thereby minimizing ambiguities in intricate scenes.

Addressing the broader challenge of tracking individuals, especially those with Alzheimer's disease prone to wandering, Janusz Wojtusiak and Reyhaneh Mogharab Nia propose a predictive model for determining the person's location during routine activities and in the event of potential loss [13]. The model utilizes unsupervised and supervised learning, alongside other procedures [14,15], whereby detected clusters are passed on to supervised learning for identifying usual locations. While bearing similarities to prior research, this approach distinguishes itself in various aspects, particularly in its methodology based on spatial clustering duration. Algorithms are applied to construct models predicting typical locations based on date, time, and geolocation data features, resulting in an improvement with an enhanced AUC of 78% for predicting location attributes extracted from day, hour, location, and duration.

Djamal Merad et al. propose a solution for tracking multiple objects, with a focus on handling identity changes due to occlusions and interactions between tracked objects [16]. The inspiration for this method is derived from studying customer navigation in busy retail environments. The approach involves reidentifying individuals through non-overlapping cameras, particularly addressing challenges posed by prolonged severe occlusions. The tracking strategy consists of two stages: firstly, retrieving trajectory segments (tracklets) using a particle filter, and secondly, merging these tracklets using a reidentification technique to obtain global trajectories for tracked individuals.

In order to attain a thorough trajectory for individuals tracked by a single-camera system, the authors introduce a tracking system comprising two modules: a tracking module and an association module. At the core of this system lies the matching module, which enables the identification of correspondences between tracked individuals. A dedicated tracking engine, utilizing particle filters tailored for each person, extracts trajectory segments. These correspondences are subsequently leveraged to merge trajectory segments pertaining to the same individual, thereby establishing the comprehensive trajectory. The outcomes of this study are encapsulated within these two modules, demonstrating the effectiveness of the tracking engines and the association module in obtaining a global trajectory.

In response to the global coronavirus pandemic [2, 17,18], Benreguia et al. [19] present a notable solution aimed at tracking asymptomatic patients. Acknowledging these individuals as significant contributors to the rapid spread of coronaviruses, the authors introduce an architecture tailored to record the continuous trajectory of people using IoT devices. This architecture prioritizes tracking highly contagious viruses rather than individuals themselves. The proposed model rapidly identifies potentially infected individuals and efficiently manages pandemics, especially for those who are confined to their homes or vehicles.

To ensure dependable real-time tracking of individuals across diverse scenarios, Ruiz-del-Solar et al. [20] present a computer system based on two primary methodologies: feature-based and image-based approaches. The feature-based method involves utilizing low-level investigations, analysis of active form entities, and modeling. Conversely, the image-based approach employs techniques from sub-based linear space, neural networks, and statistical approaches. The integration of these approaches has exhibited superior performance in comparison to feature-based methods.

Thome et al. [21] propose a robust methodology for tracking and analyzing human movements, distinct from existing approaches based on motion, models, appearance, or features. Their hybrid technique, independent of viewpoint and human position, focuses on localizing and identifying visible body parts in images. This inspired the development of an algorithm that identifies image segments corresponding to body parts, generates a graph encoding the silhouette's shape, and compares it to a 3D model of the human skeleton, allowing tracking of each pair of graph nodes over time.

In the realm of real-time COVID-19 tracking, Menni et al. [22] introduce a predictive model that integrates symptoms to forecast probable infections. Significantly, they acknowledge odor and taste loss as potential indicators of COVID-19, alongside more conventional symptoms such as high temperature and persistent cough. Their secondary contribution involves identifying a cluster of symptoms, including anosmia, fatigue, chronic cough, and loss of appetite, which, when considered collectively, can aid in diagnosing COVID-19 in individuals.

47.4 EXPLORATION AND DISCOURSE

In this section, we will delve into a comprehensive analysis and engage in detailed discussions concerning the various approaches explored earlier. We will categorize and scrutinize the works of Wu et al. [23], Wojtusiak and Nia [24], Cihan Küçükkeçeci and Adnan Yazici (2016) [25], M'hand et al. [26], Benreguia et al. [27], Ruiz-del-Solar et al. [20], Nicolas Thome et al. [21], and Menni et al. [22]. Let us begin by examining each work's methodology, contributions, and potential implications.

Starting with Wu et al.'s work [9], they proposed a novel method for multiperson online tracking based on in-depth learning, emphasizing the importance of learning instance-sensitive representations. Moving to Wojtusiak and Nia's contribution [13], their model addresses the prediction of individual locations during routine and potential loss scenarios, utilizing unsupervised and supervised learning.

Merad et al.'s approach [16, 26,27] tackles the challenge of identity changes in tracking multiple objects, specifically addressing occlusions and interactions. The methodology involves a two-step process of retrieving trajectory segments and merging them using a reidentification technique.

In the logistics domain, M'hand et al. [28] tailored a tracking architecture for Roll-On/Roll-Off terminals, introducing automated solutions for supervision, control, and tracking of rolling goods. Benreguia et al.'s work [19] is pivotal in addressing the global coronavirus pandemic by proposing an architecture to track asymptomatic patients using IoT devices.

Ruiz-del-Solar et al. [20] present a computer system for reliable real-time tracking, combining feature-based and image-based methodologies. Thome et al.'s solution [21] focuses on categorizing human body components for robust tracking using a hybrid technique. Finally, Christina Meny et al.'s work [22] contributes to real-time tracking of COVID-19 instances, offering a predictive model based on a combination of symptoms.

This analysis provides insights into the diverse approaches and contributions within the field, highlighting the advancements and challenges addressed by each work.

We arrange our agreed-upon criteria in accordance with the problems addressed and the solutions suggested, so that we can talk about the problems addressed, consider the suggestions for solutions, and evaluate the working methods.

Following the examination, we can divide the aforementioned work into two major areas. The first relates to methods of tracking people, while the second group focuses on tracking objects. As shown in Figure 47.2, these methods can be used in real time or virtually utilizing the IoT, graph theory, big data, or artificial intelligence, in a place with a finite surface area or one with an infinite number of dimensions.

We can infer from Table 47.1 that, starting with the tracking model we chose, we can create a tracking system by combining all the work's advantages [9, 13, 16, 19–22, 28]. Some methods for analyzing the work make use of artificial intelligence models [9, 13, 22], where the authors launch a subnet dynamically for each instance of a person in order to predict their next location and use the intersection on the union and the detection confidence in the calculation of association costs. This method is still subject to improvement, and the basic subnet of our MBN network will be strengthened in order to improve the robustness and discrimination of its extracted feat. By adapting the feature extraction process to different object sizes, designing a subinstance network model that is more efficient, and finally studying more efficient terms for composing the association matrix's elements, the approach can better manage small objects and will make use of new data association algorithms. The chapter's authors also plan to develop their work to include complete category detection and create a unified framework [26,27, 29,30]. These techniques typically address the issue of scale variation by using multicolumn architectures to improve functionality learning, where the input is processed with convolutive kernels of different sizes in each column to extract features on different scales. These techniques have the powerful ability to learn representations of deep convolutional neural network (CNN) models. Nonetheless, they experience a few issues. First, the restricted number of columns cannot account for the large-scale fluctuation, and the rapid growth in columns will lead to a tremendous overload of parameters that is readily unstable at various scales. Second, when they learn increasingly abstract properties, these networks typically provide a low-resolution map of crowd density. Hence, when used in challenging crowd settings with a large-scale variation or heavy congestion, they continue to have the issue of substantial precision loss [31–33].

In order to forecast potential routes for people with dementia (often known as Alzheimer's disease, or AD), the authors [13] offer a study on the efficacy of machine learning approaches applied in GPS data. To learn about people's routines, they combined supervised and unsupervised models [20]. They employ an unsupervised model in their method to ascertain the location where a person spends most of their time. Then, based on the day of the week and the time of day, they use a supervised model to forecast where people are. With a precision of 0.662 and an accuracy of

TABLE 47.1
Comparison of Studied Approach

Studied Approach	Problem Treated	Proposed Solution	Evaluation Methodology
Wu et al. [9]	Multiperson Tracking Problem (MPT).	An MPT approach based on deep learning that learns the representations sensitive to the instances of the people followed and deduces robustly online the states of these people.	Case study and practical case
Wojtusiak and Nia [13]	Problem of exploring the possibility of using GPS devices to track people with Alzheimer's disease (AD) and find patterns of movement that may possibly lead to the prediction of wandering.	Machine learning methods applied to real-time GPS tracker data to create individualized models that describe movement patterns and predict typical locations of individuals with dementia and to detect movements that do not follow these patterns and may correspond to wandering.	Case study and practical case
Merad et al. [16]	Multiple object tracking issue is the identity change that occurs between tracked targets due to occlusions and interaction between tracked objects.	Use of a reidentification strategy in the context of occlusion management for a single-camera tracking system.	Case study and practical case
M'hand et al. [28]	Problem of tracking and tracing the goods, to ensure the control and management of the various logistic operations.	A new architecture based on automatic identification technologies such as QR code, barcode, magnetic card ID to support real-time tracking of vehicles in a RORO terminal (Roll-On/Roll-Off) and provide their stay time as well as processing time in each service.	Case study
Benreguia et al. [19]	Addressing the global coronavirus pandemic.	An IoT investigation system specially designed solution that allows determining areas of high contamination and also provides a mechanism for rapid detection of undocumented patients that are known to be the main and most important factor in propagation rapid coronavirus, as well as all people in contact with recently tested positive patients, and determine the exact trajectories of infected people with the exact coordinates marked over time.	Case study and practical case
Ruiz-del-Solar et al. [20]	Make a robust tracking of people in real environments and in real time.	A computer system for the real-time tracking of several people in natural environments. The system incorporates state-of-the-art methodologies for the analysis of movements and colors, as well as for face detection, which is complemented by a heuristic-based face tracking module developed by the authors.	Case study and practical case
Thome et al. [21]	Track the parts of the human body and robust motion interpretation.	One method that uses graph matching is to correctly label parts of the human body in video sequences.	Case study and practical case
Menni et al. [22]	Identification of the main symptoms and problems caused by this coronavirus.	Survey of 2,618,862 participants reported potential COVID-19 symptoms on smartphone app.	Practical case

0.631, the authors were able to achieve an average receiver operating characteristic (ROC) (area under the curve, AUC) area of 0.778. They did not address the detection of anomalous behavior and merely took into account temporal-spatial information [34–37]. According to how frequently a person wears the laptop so that GPS data can be acquired, the scientists propose that prediction models can be created for each individual. The biggest flaw in this research is how quickly the GPS depletes the IoT device's battery, which means that the data is not just confined to AD patients and that no information is available about the carriers of the device.

However, [19] a lot of work has started to be done in the fields of health, tracking, preventive, and various other applications, employing an IoT system to handle COVID-19. The system performs well when real-world indicators are used; it does, in fact, track the movements of infected individuals a few days before the onset of their symptoms, identify those who had close contact with infected patients, and pinpoint dark areas that may be the primary sites of viral spread. The proposed system can be set up to identify all uninfected individuals who have gone to locations that have already been frequented by infected individuals in order to assess whether or not citizens respect social distance and to verify that those suspected of being infected adhere to the ordered internment. This system can also be used to lessen the economic harm brought on by the suspension of all activities. Any prospective outbreak with a higher potential strength and rate of spread can be dealt with using the suggested approach. Eventually, the system can have extra characteristics like deep learning or machine learning capabilities [38]. A minor flaw in this strategy is that it is impossible to track the path of infected individuals who have not crossed paths with them along the way, as well as if the individual disregards the warnings and safety precautions.

Some of the strategies considered to address this issue include technology instruments as the primary actors [16, 28] conducted a study of consumer behavior based on the volume of computer vision technology and the presence of shoppers. The tracking of individuals in the video recordings has been done using a variety of algorithms; in this case, they have categorized the appearances of pedestrians in front and back postures as well as segmented the detected humans into different components, such as the head and body. Unfortunately, when the target is completely obscured or far away, part-based tracking algorithms find it difficult to overcome the difficulties. By incorporating new behavioral analysis techniques based on artificial vision that are under development, we can enhance this study. Using eye tracking and gesture recognition as examples, one of these options. Together with emphasizing how this association module can be utilized in a tracking system to stop identity alterations brought on by extreme occlusions and interactions amongst customers being tracked in busy stores. In RORO (Roll-On/Roll-Off) terminals, [28] presented a real-time tracking and surveillance architecture for logistics and transport. The system's goals include managing fluid logistics operations and improving cost-related factors like traffic flow and recording time. Identification, monitoring, and tracking of rolling freight is the system's specialized mission. The ability to capture data in a matter of seconds; the reduction of recording time and the increase in traffic flow; the use of the necessary technological tools to track a variety of data, such as barcodes or QR codes; as well as the adoption of identity scanning cards, which are affordable and do not need a person present to perform the scanning operation, are some of the strengths of this solution proposed in [28]. This work requires the actualization of the projected through an assessment using data from, for example, a port and a real-world case study. There are specialized models to follow people who are motivated by the theory of graphs, among other ways to make use of graphs. A silhouette model is used in [21] to determine the points of a human skeleton. For limb labeling, a topology of the human body's structure is used rather than estimating the 3D position of the skeleton points. The suggested approach might take into account several perspectives of a person, including front, back, and profile. Also, it offers the proper limb labeling for hypothetical human positions. To address the circumstance when the arms are fused with the torso, however, they merely employ the topology of the associated graph to label various body sections.

By examining the suggestions made in each article, we can see that no method can be applied to both items and people. For example, in [9, 13, 19–21] and [22] is considered as the sub-level of

TABLE 47.2
Detailed Comparison of the Approach Studied through a Classification of Approaches

	Type		Geopositioning		Time		Implementation					Evaluation Methodology	
Studied Approach	Person	Object	Limited Space	Unlimited Space	Real-Time	Non-Real-Time	IoT	Graph	Technological tools	Big Data	AI	Case study	Practical case
Wu et al. [9]	✓	✓	✓			✓			✓		✓	✓	✓
Wojtusiak and Nia. [13]	✓			✓	✓				✓		✓	✓	✓
Merad et al. [16]	✓	✓	✓			✓			✓			✓	✓
M'hand et al. [28]		✓	✓		✓				✓	✓		✓	
Benreguia et al. [19]	✓			✓	✓		✓		✓	✓		✓	✓
Ruiz-del-Solar et al. [20]	✓		✓		✓		✓		✓		✓	✓	✓
Thome et al. [21]	✓		✓			✓		✓	✓			✓	✓
Menni et al. [22]	✓			✓		✓					✓		✓

tracking for human beings, but in [16] and [28] is considered the sub-level of tracking for objects. Our team agrees with the researchers who claim that we must find a method that integrates every technological component employed in the papers [9, 13, 16, 19–22, 28] and should have a high-performance, all-encompassing strategy and adaptive solution with Figure 47.1's granularities.

Table 47.1 provides specific information about the approaches suggested in the articles, the issue raised, and the suggested fix, while also evaluating the types of methodologies used in each approach [9, 13, 16, 19–22, 28].

In contrast, Table 47.2 provides a classification of approaches [9, 13, 16, 19–22, 28] according to Figure 47.2. Approaches are based on a case study, but they can also use a practical case as shown [9, 13, 16, 19–21]. An approach [28] is based only on a case study, but [15] is based just on a practical case.

Since the topic of tracking has been forcefully pushed recently, this study will make it easier to choose the model to use for tracking. In addition, despite all the ways stated in this article, many of them have not yet generated results at the tracking level. All the tracking methods that have been suggested so far are also specific and not all-encompassing. We might, however, employ additional strategies not covered in this chapter.

We highlight the primary granularities of a tracking strategy based on the papers researched to wrap up our study and discussion of the various approaches analyzed:

- Tracking for people or objects.
- Tracking in limited or unlimited space.
- Tracking in real or unreal time.
- Tracking using technology tools, IoT, graphs, big data, and artificial intelligence.

47.5 CONCLUSION AND FUTURE WORK

This chapter covers tracking as a whole, examining and comparing various methods while shedding more light on the most recent. Few strategies have been put forth thus far, and most of them are incomplete.

Our research team wants to develop a tracking system that continuously monitors the target. This system's goal is to continuously monitor the target. This is because certain topics, like terrorism, diseases, and criminology-related kidnapping, are sensitive and accurate.

In our upcoming work, we will supplement the strategy under discussion by putting forth fresh, all-encompassing ideas, covering all likely scenarios, and emphasizing the various components that have not yet been developed.

REFERENCES

1. Han, Mei, Sethi, Amit, Hua, Wei, and Gong, Yihong, A detection-based multiple object tracking method, 2004 International Conference on Image Processing. ICIP'04., 5, 3065–3068, 2004, IEEE.
2. McKenna, Stephen J, Jabri, Sumer, Duric, Zoran, Rosenfeld, Azriel, and Wechsler, Harry, Tracking groups of people, Computer Vision and Image Understanding, 80, 1, 42–56, 2000, Elsevier.
3. Bollobás, Béla, Modern graph theory, Springer Science & Business Media, 184, 2013.
4. Bondy, JA and Murty, USR, Théorie des graphes, Springer, 2008.
5. Buyya, Rajkumar and Dastjerdi, Amir Vahid, Internet of Things: Principles and paradigms, 2016, Elsevier. https://books.google.co.ma/books?hl=fr&lr=&id=_k11CwAAQBAJ&oi=fnd&pg=PP1&dq=Buyya,+Rajkumar+and+Dastjerdi,+Amir+Vahid,+Internet+of+Things:+Principles+and+paradigms,+2016,+Elsevier.&ots=MeXsjSqBnm&sig=O95sfmHamCOopQ5l73KnF6nE5aA&redir_esc=y#v=onepage&q=Buyya%2C%20Rajkumar%20and%20Dastjerdi%2C%20Amir%20Vahid%2C%20Internet%20of%20Things%3A%20Principles%20and%20paradigms%2C%202016%2C%20Elsevier.&f=false
6. Furht, Borko and Villanustre, Flavio, Big data technologies and applications, Springer, 2016.
7. Russell, Stuart J, and Norvig, Peter, Artificial intelligence: A modern approach, Prentice Hall Upper, Saddle River, NJ, USA, 2002.
8. Leal-Taixé, Laura, Milan, Anton, Schindler, Konrad, Cremers, Daniel, Reid, Ian, and Roth, Stefan, Tracking the trackers: An analysis of the state of the art in multiple object tracking, 2017, arXiv preprint arXiv:1704.02781.
9. Wu, Hefeng, Hu, Yafei, Wang, Keze, Li, Hanhui, Nie, Lin, and Cheng, Hui, Instance-aware representation learning and association for online multi-person tracking, Pattern Recognition, 94, 25–34, 2019, Elsevier.
10. He, Feng, Deng, Yu, and Li, Weina, Coronavirus disease 2019: What we know? Journal of Medical Virology, 92, 7, 719–725, 2020, Wiley Online Library.
11. Headey, Derek D and Ruel, Marie T. 2020. The COVID-19 nutrition crisis: What to expect and how to protect. In COVID-19 and Global Food Security, eds. Johan Swinnen and John McDermott. Part Two: Diets and nutrition, Chapter 8, 38–41. Washington, DC: International Food Policy Research Institute (IFPRI).
12. Mabrouki, J, Azrour, M, Farhaoui, Y, and El Hajjaji, S, Intelligent system for monitoring and detecting water quality, in Big data and networks technologies, Springer International Publishing, 3, 172–182, 2020.
13. Wojtusiak, Janusz and Nia, Reyhaneh Mogharab, Location prediction using GPS trackers: Can machine learning help locate the missing people with dementia? Internet of Things, 13, 100035, 2019, Elsevier.
14. Berry, Michael W, Mohamed, Azlinah, and Yap, Bee Wah, Supervised and unsupervised learning for data science, Springer, 2019.
15. Liu, Qiankun, Chen, Dongdong, Chu, Qi, Yuan, Lu, Liu, Bin, Zhang, Lei, Yu, Nenghai, Online multi-object tracking with unsupervised re-identification learning and occlusion estimation, Neurocomputing, 483, 333–347, 2022, Elsevier.
16. Merad, Djamal, Aziz, Kheir-Eddine, Iguernaissi, Rabah, Fertil, Bernard, and Drap, Pierre, Tracking multiple persons under partial and global occlusions: Application to customers' behavior analysis, Pattern Recognition Letters, 81, 11–20, 2016, Elsevier.
17. Lee, Hyeon Chang, Luong, Duc Thien, Cho, Chul Woo, Lee, Eui Chul, and Park, Kang Ryoung, Gaze tracking system at a distance for controlling IPTV, IEEE Transactions on Consumer Electronics, 56, 4, 2577–2583, 2010, IEEE.
18. Benchrifa, M, Elouardi, M, Fattah, G, Mabrouki, J, and Tadili, R, Identification, simulation and modeling of the main power losses of a photovoltaic installation and use of the internet of things to minimize system losses, in Advanced technology for smart environment and energy, Springer International Publishing, Cham, 49–60, 2023.
19. Benreguia, Badreddine, Moumen, Hamouma, and Merzoug, Mohammed Amine, Tracking COVID-19 by tracking infectious trajectories, IEEE Access, 8, 145242–145255, 2020, IEEE.
20. Ruiz-del-Solar, Javier, Shats, Alon, and Verschae, Rodrigo, Real-time tracking of multiple persons, 12th International Conference on Image Analysis and Processing, 2003, 109–114, IEEE.

21. Thome, Nicolas, Merad, Djamel, and Miguet, Serge, Human body part labeling and tracking using graph matching theory, 2006 IEEE International Conference on Video and Signal Based Surveillance, 38–38, 2006, IEEE.
22. Menni, Cristina, Valdes, Ana M, Freidin, Maxim B, Sudre, Carole H, Nguyen, Long H, Drew, David A, Ganesh, Sajaysurya, Varsavsky, Thomas, Cardoso, M Jorge, Moustafa, Julia S El-Sayed, Visconti, Alessia, Hysi, Pirro, Bowyer, C. E. Ruth, Mangino, Massimo, Falchi, Mario, Wolf, Jonathan, Ourselin, Sebastien, Chan, Andrew T., Steves, Claire J. and Spector, Tim D., Real-time tracking of self-reported symptoms to predict potential COVID-19, Nature Medicine, 26, 7, 1037–1040, 2020, Nature Publishing Group.
23. Huo, Feifei and Hendriks, Emile A, Multiple people tracking and pose estimation with occlusion estimation, Computer Vision and Image Understanding, 116, 5, 634–647, 2012, Elsevier.
24. Wengefeld, Tim, Lewandowski, Benjamin, Seichter, Daniel, Pfennig, Lennard, Müller, Steffen, and Gross, Horst-Michael, Real-time person orientation estimation and tracking using colored point clouds, Robotics and Autonomous Systems, 135, 103665, 2021, Elsevier.
25. Küçükkeçeci, Cihan and Yazici, Adnan, Multilevel object tracking in wireless multimedia sensor networks for surveillance applications using graph-based big data, IEEE Access, 7, 67818–67832, 2019, IEEE.
26. Lukasczyk, Jonas, Weber, Gunther, Maciejewski, Ross, Garth, Christoph, and Leitte, Heike, Nested tracking graphs, Computer Graphics Forum, 36, 3, 12–22, 2017, Wiley Online Library.
27. Xiao, Chenhui, Li, Zhenzhou, Lu, Jianfeng, Wang, Jinyan, Zheng, Haoteng, Bi, Zuyue, Chen, Mengyang, Mao, Rui, and Lu, Minhua, A new deep learning method for displacement tracking from ultrasound RF signals of vascular walls, Computerized Medical Imaging and Graphics, 87, 101819, 2021, Elsevier.
28. M'hand, Mouna Amrou, Boulmakoul, Azedine, Badir, Hassan, and Lbath, Ahmed, A scalable real-time tracking and monitoring architecture for logistics and transport in RoRo terminals, Procedia Computer Science, 151, 218–225, 2019, Elsevier.
29. Widanagamaachchi, Wathsala, Christensen, Cameron, Pascucci, Valerio, and Bremer, Peer-Timo, Interactive exploration of large-scale time-varying data using dynamic tracking graphs, IEEE Symposium on Large Data Analysis and Visualization (LDAV), 9–17, 2012, IEEE.
30. Meijering, Erik, Dzyubachyk, Oleh, and Smal, Ihor, Methods for cell and particle tracking, Methods in Enzymology, 504, 183–200, 2012, Elsevier.
31. Rachiq, T, Abrouki, Y, Mabrouki, J, Samghouli, N, Fersi, C, Rahal, S, and El Hajjaji, S, Evaluation of the efficiency of different materials to remove specific pollutants from landfill leachate, Desalination and Water Treatment, 238, 240–250, 2021.
32. Yilmaz, Alper, Javed, Omar, and Shah, Mubarak, Object tracking: A survey, ACM Computing Surveys (CSUR), 38, 4, 13–es, 2006, ACM New York, NY, USA. https://dl.acm.org/doi/abs/10.1145/1177352.1177355
33. Walter, Tristan and Couzin, Iain D, TRex, a fast multi-animal tracking system with markerless identification, and 2D estimation of posture and visual fields, Elife, 10, e64000, 2021, eLife Sciences Publications Limited.
34. Tryggvason, Grétar, Bunner, Bernard, Esmaeeli, Asghar, Juric, Damir, Al-Rawahi, N, Tauber, W, Han, J, Nas, S, and Jan, Y-J, A front-tracking method for the computations of multiphase flow, Journal of Computational Physics, 169, 2, 708–759, 2001, Elsevier.
35. Peng, Jinlong, Wang, Tao, Lin, Weiyao, Wang, Jian, See, John, Wen, Shilei, and Ding, Erui, TPM: Multiple object tracking with tracklet-plane matching, Pattern Recognition, 107, 107480, 2020, Elsevier.
36. Ghizlane, F, Mabrouki, J, Ghrissi, F, and Azrour, M, Proposal for a high-resolution particulate matter (PM10 and PM2. 5) capture system, comparable with hybrid system-based internet of things: Case of quarries in the western Rif, Morocco, Pollution, 8, 1, 169–180, 2022.
37. Scopus preview - Scopus - Welcome to Scopus, https://www.scopus.com/, 2021- 06-27.
38. Goodfellow, Ian, Bengio, Yoshua, Courville, Aaron, Bengio, Yoshua, Deep learning, MIT Press Cambridge, 1, 2, 2016.

Index

Note: Bold page numbers refer to tables and Italic page numbers refer to figures.

D

E

F

G

H

I

O

P

Q

R

S

T

V

W

Y

Z

For Product Safety Concerns and Information please contact our
EU representative GPSR@taylorandfrancis.com Taylor & Francis
Verlag GmbH, Kaufingerstraße 24, 80331 München, Germany